# PHYSICS OF ELECTRONIC AND ATOMIC COLLISIONS

# Physics of Electronic and Atomic Collisions

Invited Papers of the XII International Conference on
the Physics of Electronic and Atomic Collisions
Gatlinburg, Tennessee, 15-21 July, 1981

*Edited by*

Sheldon DATZ

*Oak Ridge National Laboratory*

1982

NORTH-HOLLAND PUBLISHING COMPANY

AMSTERDAM · NEW YORK · OXFORD

*ISBN: 0 444 86323 0*

*Publishers*
NORTH-HOLLAND PUBLISHING COMPANY
AMSTERDAM · NEW YORK · OXFORD

*Sole distributors for the U.S.A. and Canada*
ELSEVIER SCIENCE PUBLISHING COMPANY, INC.
52 VANDERBILT AVENUE
NEW YORK, N.Y. 10017

Library of Congress Cataloging in Publication Data

International Conference on the Physics of
    Electronic and Atomic Collisions. (12th :
    1981 : Gatlinburg, Tenn.)
    Physics of electronic and atomic collisions.

    Sponsored by the International Union of Pure
and Applied Physics, and others.
    Includes index.
    1. Collisions (Nuclear physics)--Congresses.
I. Datz, Sheldon.  II. International Union of Pure
and Applied Physics.  III. Title.
QC794.6.C6I57  1981      539.7'54      81-22445
ISBN 0-444-86323-0 (Elsevier North-Holland).
                                AACR2

PRINTED IN THE NETHERLANDS

# PREFACE

The Twelfth International Conference on the Physics of Electronic and Atomic
Collisions, held at Gatlinburg, Tennessee, from July 15 to July 21, 1981,
demonstrated the continuing growth in magnitude, scope, and quality of research
being carried out in this area of atomic and molecular science. Almost 800
attendees from 32 nations actively participated during the five days of the
conference.

Six hundred and thirty contributed papers were presented during four three-hour-
long poster sessions, in which a bustling market-place atmosphere prevailed and
much information was exchanged. Extended abstracts of these papers (1183 pp)
have been published and are avilable through North-Holland Press.

The invited papers of the conference, presented in this volume, indicate both
the breadth of the subject, with collision partners ranging from photons,
electrons, positrons, and muons to atoms, heavy ions, and molecules at energies
ranging from thermal to many millions of electron volts, and the unity of
scientific approach to the manifold phenomena which occur in these collisions.

Following a plenary invited lecture by Professor A. Dalgarno and a plenary
symposium dedicated to Professor U. Fano, the remaining invited papers consisted
of review talks, progress reports, and symposia, which were presented in two
parallel sessions. Although he was unable to attend the conference, the paper
by Devdariani on collisional electron detachment has been included for the sake
of completeness. Only four of the invitees failed to meet the deadline
necessarily applied for the timely production of this volume. Since the papers
presented here were invited they have not been reviewed; in a sense they are
reviewed for scientific quality by the process of selection by the program
committee. The use of photo-ready manuscripts reduces costs and expedites
rapid publication but oft-times makes extensive editing difficult because of the
turnaround time for papers from around the world. What is gained is immediacy
and spontaneity. However, even within this framework we anticipate that this
volume, along with its predecessors, will be a valuable resource for many years
to come.

The XII ICPEAC is indebted to the following sponsors:  the International Union
of Pure and Applied Physics, the Department of Energy, the National Science
Foundation, the Office of Naval Research, and the National Aeronautics and
Space Administration.  Generous contributions have also been made by the
following private organizations:  Dow Chemical Corporation, Hewlett-Packard
Corporation, Nuclear Data Corporation, Texas Instruments Corporation,
International Business Machines Corporation, EG & G ORTEC, and Union Carbide
Corporation-Nuclear Division.

Sheldon Datz
Oak Ridge National Laboratory
(Editor)

TABLE OF CONTENTS

# INVITED LECTURE

## REVIEWS (R) AND PROGRESS REPORTS (P)

### ELECTRON-ATOM/MOLECULE COLLISIONS

### ION-ATOM COLLISIONS

## COLLISIONS INVOLVING EXCITED STATES
### (Dedicated to V. Čermák)

## COLLISIONAL ELECTRON DETACHMENT

## EXOTIC SPECIES

## ATOM/ION-MOLECULE COLLISIONS

## COLLISIONS INVOLVING PHOTONS

# SYMPOSIA

## FANO SYMPOSIUM

## NEW THEORETICAL METHODS

## NEW EXPERIMENTAL METHODS

## ELECTRON-MOLECULE COLLISIONS

## COLLISIONS OF MULTICHARGED IONS WITH ELECTRONS

## COLLISIONS OF MULTICHARGED IONS WITH ATOMS
### (Dedicated to J. R. Macdonald)

## REACTIVE SCATTERING

## ATOMIC COLLISION PROCESSES IN FUSION

## ATOMIC COLLISIONS IN SOLIDS

# INTERNATIONAL CONFERENCE ON THE PHYSICS OF ELECTRONIC AND ATOMIC COLLISIONS ORGANIZATION 1979-1981

## EXECUTIVE COMMITTEE

*Chairman*
Ronald F. Stebbings--USA

*Vice Chairman*
Frank H. Read--United Kingdom

*Secretary*
John S. Risley--USA

*Treasurer*
Guy Watel--France

*Members*
Vadim V. Afrosimov--USSR
Sheldon Datz--USA
Ingolf V. Hertel--Germany
Jaap Kistemaker--Netherlands
Eugen Merzbacher--USA
Kazuo Takayanagi--Japan
J. Peter Toennies--Germany

## GENERAL COMMITTEE

AUSTRALIA
Erich Weigold

BELGIUM
F. Brouillard

CANADA
Chris E. Brion
William J. McConkey

DENMARK
Knud Taulbjerg

FRANCE
Thomas R. Govers
Jean Durup
Richard I. Hall
Vo Ky Lan
Guy Watel

GERMANY
John S. Briggs
Ingolf V. Hertel
Hartmut Hotop
Frans Linder
Werner Mehlhorn
Paul H. Mokler
J. Peter Toennies

INDIA
S. P. Khare

ITALY
Franco A. Gianturco

JAPAN
N. Oda
K. Takayanagi
H. Tawara

NETHERLANDS
Henk G. M. Heideman
Jaap Kistemaker
J. Los

UNITED KINGDOM
Brian H. Bransden
H. B. Gilbody
W. Roy Newell
Frank H. Read

USA
Benjamin Bederson
David Crandall
Sheldon Datz
Alan Gallagher
David E. Golden
Eugen Merzbacher
James R. Peterson
David Pritchard
John S. Risley
Arnold Russek
Ronald F. Stebbings
Howard S. Taylor

USSR
Vadim V. Afrosimov
Robert J. Damburg
Yurii N. Demkov
E. E. Nikitin
E. A. Yukov

YUGOSLAVIA
Leposava Vuskovic

## LOCAL COMMITTEE

*Chairman:*   Sheldon Datz
*Secretary:*   Robert N. Compton
*Treasurer:*   Philip D. Miller
*Members:*   David H. Crandall
           W. Ray Garrett
           Herbert F. Krause
           Fred W. Meyer
           Charles D. Moak
           David J. Pegg

*Address:*   Oak Ridge National Laboratory
          P. O. Box X
          Oak Ridge, Tennessee 37830, USA

## PROGRAM COMMITTEE

Executive Committee, Local Committee,
and
B. Bederson
T. A. Carlson
J. Eichler
I. A. Sellin
N. Stolterfoht

## SYMPOSIUM ORGANIZERS

| | |
|---|---|
| *Fano Symposium* | J. Macek |
| *Theoretical Methods* | E. Merzbacher |
| *Experimental Methods* | Y. Kaneko |
| *Electron-Molecule Scattering* | S. Trajmar |
| *Multicharged Ions and Electrons* | G. Dunn |
| *Multicharged Ions and Atoms* | P. Hvelplund |
| *Reactive Scattering* | Y. T. Lee |
| *Atomic Collision Processes in Fusion* | D. E. Post |
| *Atomic Collisions in Solids* | N. Tolk |

PHYSICS OF ELECTRONIC AND ATOMIC COLLISIONS
S. Datz (editor)
© North-Holland Publishing Company, 1982

APPLICATIONS OF ATOMIC COLLISIONS PHYSICS
TO ASTROPHYSICS

A. Dalgarno

Harvard-Smithsonian Center for Astrophysics
Cambridge, Massachusetts
U.S.A.

A review is presented of the applications of atomic collision
physics to the interpretation of astrophysical phenomena.  The
cosmological significance of accurate measures of the deuterium
and helium abundances is pointed out and the uncertainties which
arise from the electronic, atomic and molecular processes entering into the analysis of the observational data are noted.
Brief mention is made of some observational consequences of
massive neutrinos.  Limits on the variability of the fundamental
constants obtained from atomic emission and absorption line data
are summarized.

INTRODUCTION

Most of the knowledge we have about the Universe resides in the form of photons.
To interpret the message they bring in their journey to us we must reconstruct
the events in which they participated.  By reconstructing their histories we hope
to gain insight into the nature of the astronomical entities at the earlier times
when the photons were originally created and we hope to learn about the physical
environments of the intervening intergalactic and interstellar material through
which the photons passed.  The processes which produce the photons and the processes which modify them belong usually to the domain of electron, atomic and
molecular physics and the subject material of the International Conference on the
Physics of Electronic and Atomic Collisions is an essential and substantial
component of astronomical research.

The applications of atomic collisions physics to astronomy cover a large canvas
and I cannot hope to do more than illustrate a few of its features.  As a guide to
my remarks, I will attempt to use a cosmological thread, though with digressions.
I will, at least, begin at the beginning.

1.  THE EARLY UNIVERSE

According to the standard cosmology of today[1,2], the universe expanded from an
initial singularity.  As it did so, the densities of matter and of radiation
decreased.  Particle Physics determined the earliest phase of the evolution as
the universe cooled in about one hundred seconds from $10^{12}$K to $10^{9}$K.  Between
$5 \times 10^{9}$K and $5 \times 10^{8}$K, Nuclear Physics became important as nucleosynthesis occurred
leading to the production of $^{4}$He and small amounts of deuterium, $^{3}$He and $^{7}$Li.
As the expansion continued, Thompson scattering of photons with electrons maintained the matter and the radiation in thermal equilibrium with the temperature
and density falling until at some time before a million years had elapsed the
temperature reached values of about 4000K.  Atomic Physics took control of the
evolution at this point as recombination of photons and electrons occurred
rapidly by the radiative process

$$H^+ + e \rightarrow H + h\nu \tag{1}$$

and matter changed from a fully-ionized state to a largely neutral condition. Thermal contact was lost and matter and radiation evolved independently. At the recombination epoch, the universe was composed mostly of hydrogen and helium atoms with some electrons, a nearly equal number of protons and many photons. Out of this material was formed the first generation of stars, in which heavy elements were manufactured. The process of electronic, atomic and molecular physics played a critical role in creating the conditions necessary for the formation, growth and ultimate gravitational collapse of pre-galactic clouds. Following the dawn of atomic physics, signalled by the radiative recombination reaction (1), there occurred in quick succession, a rich assembly of processes:

radiative attachment

$$H + e \rightarrow H^- + h\nu \ , \tag{2}$$

associative detachment

$$H + H^- \rightarrow H_2 + e \ , \tag{3}$$

radiative association

$$H + H^+ \rightarrow H_2^+ + h\nu \ , \tag{4}$$

ion-atom interchange

$$H_2^+ + H \rightarrow H^+ + H_2 \tag{5}$$

and charge transfer

$$H_2^+ + H \rightarrow H_2 + H^+ \ . \tag{6}$$

The sequences, which lead to the formation of molecular hydrogen, are interrupted by photoionization

$$H + h\nu \rightarrow H^+ + e \ , \tag{7}$$

photodetachment

$$H^- + h\nu \rightarrow H + e \ , \tag{8}$$

mutual neutralization

$$H^+ + H^- \rightarrow H + H \ , \tag{9}$$

and dissociative recombination

$$H_2^+ + e \rightarrow H + H \ . \tag{10}$$

The molecule may be destroyed by photoionization

$$H_2 + h\nu \rightarrow H_2^+ + e \tag{11}$$
$$\rightarrow H^+ + H + e \ , \tag{12}$$

by photodissociation

$$H_2 + h\nu \rightarrow H + H \ , \tag{13}$$

by electron impact to excited electronic states which lead to dissociation

$$H_2 + e \rightarrow H + H \ , \tag{14}$$

by direct collision-induced dissociation

$$H + H_2 \rightarrow H + H + H \tag{15}$$

and by chemical reactions

$$H_2^+ + H_2 \rightarrow H_3^+ + H \ . \tag{16}$$

These processes modify the behavior of the fractional ionization as a function of redshift and change its residual value.

The presence of helium added to the diversity of processes with Penning ionization, associative ionization and radiative charge transfer:

$$He^* + H \rightarrow He + H^+ + e \tag{17}$$

$$He^* + H \rightarrow HeH^+ + e \tag{18}$$

$$He^+ + H \rightarrow He + H^+ + h\nu \tag{19}$$

In an uncertain scenario, embedded perturbations and anisotropies produced density enhancements and initial condensations into clouds in the cooling gas. In the pre-galactic clouds, the presence of hydrogen molecules diminished the equilibrium value of the kinetic temperature. Because gravitational collapse does not occur unless the gravitational energy exceeds the thermal energy, the formation of $H_2$ substantially reduces the minimum cloud mass required for collapse. Thus in the absence of $H_2$, cooling occurs by electron impact excitation of H atoms,

$$e + H \rightarrow e + H' . \tag{20}$$

The excited atoms emit photons which escape from the system. Of greatest importance is the excitation of the 2p level of hydrogen followed by emission of Ly$\alpha$ at 1216 Å,

$$H(2p) \rightarrow H(1s) + Ly\alpha . \tag{21}$$

As the collapse proceeds and the density of H increases, photon trapping occurs as the resonance Ly$\alpha$ photons are reabsorbed. Cooling still proceeds by excitations of the 2s level followed by two-photon decays

$$H(2s) \rightarrow H(1s) + h\nu_1 + h\nu_2 \tag{22}$$

The continuous emission is not absorbed by the gas. The 2s level is also populated by

$$e + H(2p) \rightarrow e + H(2s) \tag{23}$$

and more rapidly by

$$H^+ + H(2p) \rightarrow H^+ + H(2s) . \tag{24}$$

The threshold for the excitation of the n = 2 levels is 10.2 eV and a hydrogen gas composed of atoms cools very slowly below 8,000 K. Molecular hydrogen has rotational and vibrational levels which may be excited by electron, photon and neutral atom impacts and cooling is rapid down to about 300 K. Because $H_2$ is a homonuclear molecule, the probability that an infrared photon emitted by the excited rotation-vibration levels is reabsorbed in the gas is negligible.

Various studies of the evolution of pre-galactic gas clouds have been presented, most recently by Hutchins[3], Shchekinov and Edel'man[4] and Khersonskii and Varshalovich[5] (though none employs an entirely accurate description of the atomic and molecular collision processes).

2.1 The background radiation temperature

The radiation created at the beginning retained its blackbody spectrum but underwent enormous redshifts. It now fills the space around us and its intensity is characterized today by a temperature of 2.8 K. Although unrecognized at the time, the temperature of the blackbody background radiation was first obtained from measurements of the absorption of starlight by the interstellar molecule CN. Absorption lines at 3874.0 and 3874.6 Å originate in the J = 0 and J = 1 rotational levels of the ground electronic and vibrational state of CN. From the strengths of the absorptions, column densities, N(J), of CN in the two rotational levels can be derived and an excitation temperature $T_{ex}$ may be defined by the relationship

$$\frac{N(J = 1)}{N(J = 0)} = 3 \exp(-\Delta E/kT_{ex}) \tag{25}$$

where $\Delta E$ is the energy separation between the rotational levels. The J = 1 level is populated by impact excitations from the J = 0 level and depopulated by

spontaneous radiative transitions.  Typical interstellar conditions produce a
population ratio of not more than 0.03 whereas the observed ratio is 0.45.  The
additional source of J = 1 molecules, it is now widely believed, is absorption
of the universal background radiation, the relic of the original fireball.  The
CN excitation temperature of 2.83 $\pm$ 0.15 K[7] is the temperature at 2.6mm of the
blackbody radiation.


3.  THE ABUNDANCE OF DEUTERIUM

Depending upon the original matter density the universe is open and will continue
to expand forever or it is closed and will eventually come to a halt and then
contract back to a final catastrophic singularity.  The mass observed in galaxies
corresponds to a density well below that required to close the universe but there
may exist much larger amounts of invisible material and there is dynamical evidence that the universe is near to being closed.  One measure of the matter
density is the abundance ratio [D]/[H] of deuterium nuclei to hydrogen nuclei,
provided that deuterium is made only in the primordial universe and not by
stellar or galactic processes.

3.1  Observations of atomic deuterium

The ratio of atomic deuterium to atomic hydrogen in the interstellar gas can be
obtained from measurements of the strengths of ultraviolet absorption lines in
the Lyman series seen toward hot stars.  There appear to be variations by a
factor of about five in the derived D/H ratios.

The variations may reflect departures from spatial uniformity in the relative
abundances arising from radiation pressure[8] or they may reflect differences in
the fractions of deuterium and hydrogen which are in the form of neutral atoms.

The ions $H^+$ and $D^+$ are both removed by radiative recombination

$$H^+ + e \rightarrow H' + h\nu \tag{26}$$

$$D^+ + e \rightarrow D' + h\nu \tag{27}$$

but there is a difference in the effective rates.  Recombinations into the
ground states produce photons with enough energy to ionize the atoms.  The
effective recombination coefficient for $H^+$ is given by

$$\alpha_{eff}(H) = \alpha - \alpha_{1s} \tag{28}$$

where $\alpha$ is the total recombination coefficient and $\alpha_{1s}$ is the value for capture
into the ground state.  At a temperature of 4000 K, $\alpha$ = 5.35 x $10^{-13}$ $cm^3$ $s^{-1}$ and
$\alpha_{1s}$ = 2.50 x $10^{-13}$ $cm^3$ $s^{-1}$.  Because the photon which is produced by recombination into D(1s) will preferentially ionize the more abundant H atoms, the
effective recombination coefficient for $D^+$ is simply $\alpha$.

The ions $H^+$ and $D^+$ also react at different rates with atomic oxygen in the charge
transfer processes

$$H^+ + O \rightarrow H + O^+ \tag{29}$$

$$D^+ + O \rightarrow D + O^+ \tag{30}$$

For $O(^3P_2)$ atoms, the energy defect of reaction (29) is equivalent to 228 K and
of reaction (30) to 184 K.  The rate coefficients have been calculated by
Chambaud et al.[9] and Roueff[10].  Significant differences appear at temperatures
below 50 K.

Any departures from uniformity in the fractional ionization of D and H tend to be
removed by the charge transfer reaction

$$H^+ + D \rightarrow H + D^+ \tag{31}$$

and its reverse. Reaction (31) is endothermic by an energy equivalent to 44 K. Rate coefficients which take account of the energy differences in the forward and back reactions have been computed by Watson, Christensen and Deissler[11] from the theoretical cross sections of Hunter and Kuriyan[12].

Substantial departures from uniformity will occur in low temperature ionized gas. However, in cool regions the steady-state fractions of D and of H in ionized form are very small. If the variations in D/H are to be attributed to atomic reactions, it seems necessary to invoke time-dependent cooling recombining plasmas. Bruston et al.[8] argue that the observational data can be interpreted to give [D]/[H] = 2.25 x $10^{-5}$. We may obtain further information from observations of HD.

## 3.2 Observations of HD

In cool diffuse clouds in which some of the hydrogen has been converted to molecular form, the analysis is more complicated. Molecular hydrogen is formed, not by the reactions appropriate to warm ionized gases, but by association on the surfaces of grains. In contrast, HD is produced by the gas phase reaction[13,14]

$$D^+ + H_2 \rightarrow HD + H^+ \qquad (32)$$

The $D^+$ ions come from cosmic ray ionization of H and $H_2$ to give $H^+$ ions, followed by reaction (31).

The amount of D which is converted to HD is proportional to the cosmic ray ionizing flux $\zeta$ and the abundance ratio of HD to $H_2$ depends upon $\zeta$ and on [D]/[H]. Cosmic ray ionization followed by reaction (29) leads to the formation of the interstellar molecule OH[13,15] so that the abundance ratio of OH to $H_2$ provides a measure of $\zeta$. Then in conjunction with observations of HD, [D]/[H] can be derived. The observational data are limited to clouds lying in front of three stars. According to Hartquist, Black and Dalgarno[16], for the three stars, [D]/[H] is consistent within the observational uncertainties with a value of 2 x $10^{-5}$. The derived values are to some extent model-dependent and they ignore any production of OH from shocked regions[17] which may lie along the line of sight. They are in harmony with the abundance ratio of 2.25 x $10^{-5}$ inferred by Bruston et al.[8]

## 3.3 Observations of deuterated molecules

If deuterium is of cosmological origin, because it is consumed by stellar processing we should expect to find a gradient in the abundance of deuterium with distance from the galactic center where stellar activity is greatest. The galaxy can be sampled over extensive regions by radiofrequency observations of interstellar molecules. Table 1 is a list of interstellar molecules detected and identified so far. The chemistry of diffuse clouds implies that in the dense molecular clouds where molecules are shielded from ultraviolet radiation, almost all the hydrogen is in the form of $H_2$, and almost all the deuterium is in the form of HD. Although neither $H_2$ nor HD has a radiofrequency spectrum, several of the molecules in Table 1 have been observed with deuterium substituted for one of the hydrogen atoms, most recently $DC_5N$ in the interstellar cloud TMC 1.[18,19]

The interpretation of the emission and absorption data involves a complicated array of collision processes and radiative transfer processes. The collision processes mostly involve rotational excitation by impact with ortho and para-hydrogen molecules. Theoretical estimates are available for several systems[20-26] but the important differences between ortho and para $H_2$ collisions have not been fully explored.

Table 2[27] is a list of the derived ratios of the abundances of $DCO^+$ and $HCO^+$ and of DCN and HCN in various locations. There is apparent a general tendency for the ratios to decrease inward to the galactic center, which suggests that

Table 1

Interstellar Molecules

| | | | |
|---|---|---|---|
| $H_2$ | hydrogen | CH | methylidyne |
| $CH^+$ | methylidyne ion | OH | hydroxyl |
| $C_2$ | carbon | CN | cyanogen |
| CO | carbon monoxide | NO | nitric oxide |
| CS | carbon monosulphide | SiO | silicon monoxide |
| SO | sulphur monoxide | NS | nitrogen sulfide |
| SiS | silicon sulphide | $CO^+$ | carbon monoxide ion |
| $H_2O$ | water | $C_2H$ | ethynyl |
| HCN | hydrogen cyanide | HNC | hydrogen isocyanide |
| HCO | formyl | $HCO^+$ | formyl ion |
| $N_2H^+$ | Protonated nitrogen | $H_2S$ | hydrogen sulphide |
| HNO | nitroxyl | OCS | carbonyl sulphide |
| $SO_2$ | sulphur dioxide | $O_3$ | ozone |
| $HCS^+$ | thioformyl ion | | |
| $NH_3$ | ammonia | $H_2CO$ | formaldehyde |
| HNCO | isocyanic acid | $H_2CS$ | thioformaldehyde |
| $C_3N$ | cyanoethynyl | HNCS | isothiocyanic acid |
| HOCN | cyanic acid     or | $HOCO^+$ | protonated carbon dioxide |
| $CH_2NH$ | methanimine | $CH_2CO$ | ketene |
| $NH_2CN$ | cyanamide | HCOOH | formic acid |
| $C_4H$ | butadinyl | $HC_3N$ | cyanoacetylene |
| $CH_3OH$ | methyl alcohol | $CH_3CH$ | methyl cyanide |
| $NH_2CHO$ | formamide | $CH_3SH$ | methyl mercaptan |
| $CH_3NH_2$ | methylamine | $CH_3C_2H$ | methyl acetylene |
| $CH_3CHO$ | acetaldehyde | $CH_2CHCN$ | vinyl cyanide |
| $HC_5N$ | cyanodiacetylene | $HCOOCH_3$ | methyl formate |
| $CH_3CH_2OH$ | ethyl alcohol | $(CH_3)_2O$ | dimethyl ether |
| $HC_7N$ | cyanohexatriyne | $CH_3CH_2CN$ | ethyl cyanide |
| $HC_9N$ | cyano-octatetra-yne | | |

deuterium is not created in galactic events. The ratio derived from ultraviolet absorption studies, $2 \times 10^{-5}$, if it is the primordial value, implies that the universe is open. The presence of deuterium at the galactic center where astration occurs efficiently indicates that an infall into the galaxy of new unprocessed material has occurred[28].

The deuterated molecules are detected only because their concentrations have been enhanced by fractionation processes. According to ion-molecule reaction theories of the formation of interstellar molecules[29,30,31], $HCO^+$ and $DCO^+$ are produced in the reactions

$$H_3^+ + CO \rightarrow HCO^+ + H_2 \qquad\qquad (33)$$

$$H_2D^+ + CO \rightarrow DCO^+ + H_2 . \qquad\qquad (34)$$

Table 2

Relative Abundances of Deuterated Molecules

| Source | Distance (kiloparsec) | 1000 DCN/HCN | 1000 $DCO^+/HCO^+$ |
|---|---|---|---|
| Sqr A | 0.0 | 1.1 ± 0.4 | |
| Sqr B | 0.1 | 0.8 ± 0.5 | 0.4 ± 0.3 |
| W33 | 5.7 | 2.3 ± 0.4 | |
| W51 | 7.6 | 1.8 ± 0.7 | |
| M17 | 8.0 | 1.8 ± 0.6 | 0.7 ± 0.3 |
| DR21(OH) | 9.9 | 2.9 ± 1.2 | |
| DR21 | 9.9 | 1.2 ± 0.3 | 5.3 ± 0.5 |
| Orion A | 10.9 | 2.9 ± 0.6 | 1.7 ± 0.6 |
| | | 6.8 ± 0.6 | |
| NGC 2264 | 11.1 | | 10.2 ± 0.7 |
| W3(OH) | 12.2 | 4.3 ± 0.9 | |
| NGC 7538 | 12.7 | 4.6 ± 1.0 | |

Chemical fractionation occurs through the reaction

$$H_3^+ + HD \rightarrow H_2D^+ + H_2. \tag{35}$$

The forward and back reaction rate coefficients are related by

$$\frac{k_\rightarrow}{k_\leftarrow} = \exp(-T^*/T) . \tag{36}$$

Quantal calculations by Carney[32] of the $H_3^+$ potential energy surface give $T^* = 550$ K so that a substantial enhancement of $H_2D^+$ is possible in cold clouds.

In a cold gas, the reverse of (35) proceeds slowly and the $HD^+$ ions are removed by dissociative recombination

$$H_2D^+ + e \rightarrow H + D + H \tag{37}$$

and by reactions with neutral species X such as CO, O and $N_2$,

$$H_2D^+ + X \rightarrow H_2 + DX^+ \tag{38}$$
$$\rightarrow HD + HX^+ . \tag{39}$$

On equating the formation and destruction rates of $H_2D^+$ we obtain for the number density ratio

$$\frac{n(H_2D^+)}{n(H_3^+)} = \{ \frac{\exp(T^*/T)}{1 + \frac{\alpha}{k_\leftarrow}\frac{n_e}{n(H_2)} + \frac{\Sigma k_x n(X)}{k_\leftarrow n(H_2)}} \} \frac{n(HD)}{n(H_2)} \tag{40}$$

where $\alpha$ is the rate coefficient of (37) and $k_x$ is the sum of the rate coefficients of (38) and (39)[33]. Thus variations in the ratio of $DCO^+$ to $HCO^+$ which directly reflect the ratio of $H_2D^+$ to $H_3^+$ could be attributed alternatively to a gradient in the factor inside the brackets and not in the [D]/[H] ratio. A similar formula applies to other simple molecular species including the DCN to HCN ratio. Because the greater stellar activity at the galactic center presumably enhances the cosmic ray ionization frequency, a gradient in the fractional ionization increasing toward the center is not implausible. To infer [D]/[H] from molecular observations, an independent estimate of $n_e/n(H_2)$ is needed or a molecule must be identified whose fractionation is enhanced above that of (40) by some additional reaction. Radiative association with HD is probably more rapid than with $H_2$ and may lead to fractionation in some of the heavier molecular species.

If we accept that [D]/[H] is $2 \times 10^{-5}$, expression (40) gives useful limits to the fractional ionization in dense interstellar clouds. Because of the coupling of the cloud to the surrounding interstellar medium, the electron density is a critical parameter in the evolution of a collapsing cloud. In some clouds, the measured ratio $n(DCO^+)/n(HCO^+)$ is as large as 0.1[34,35] so that $n_e/n(H_2)$ is less than $10^{-7}$.

In molecular clouds, molecular ions are removed rapidly by dissociative recombination until the electron density is so reduced that charge transfer reactions to metal atoms such as

$$HCO^+ + Mg \rightarrow HCO + Mg^+ \tag{41}$$

become more rapid.[36] The metal ions are removed only slowly by radiative recombination and a fractional ionization less than $10^{-7}$ appears to demand a large metal depletion[37,38] though recombination onto grains[36,39,40] may lessen the requirement.

## 3.4  Observations of shocked $H_2$

Shocks are ubiquitous phenomena in the interstellar medium. In a shock front, the  directed energy of the shock is converted to random thermal energy and the gas is heated. Emissions from molecular hydrogen in the v = 1 excited vibrational level have been detected at wavelengths between 2 and 3μ from a wide range of astrophysical objects including molecular clouds, the T Tauri star, planetary nebulae, supernova remnants and a Seyfert galaxy. In the Orion molecular cloud emissions from ·the v = 2 vibrational level have been seen in the region of the Kleinman-Low nebula. Recently[41] emissions in the 0-0 band from rotational levels up to J = 17 have been detected.

The distribution of rotation-vibration level populations indicate kinetic temperatures T of about 2000 K. At 2000 K, the interchange of H and D in the neutral reaction

$$D + H_2 \rightleftharpoons H + HD \tag{42}$$

occurs rapidly and the abundances of HD and $H_2$ will be in chemical equilibrium. A measurement of the HD abundance should provide an unambiguous value of the [D]/[H] ratio.

Because of the long radiative lifetimes, the populations of the rotation-vibration levels of $H_2$ are in thermal equilibrium and the emission rate is proportional to the number density $n(H_2)$. The deuterated molecule HD has a small electric dipole moment and the radiative lifetimes are much shorter than for $H_2$. Thermal equilibrium is probably not achieved and the photon emission rate from level vJ equals the population rate $n(H)n_{vJ}q_{vJ}$ cm$^3$ s$^{-1}$ where $q_{vJ}$ is the rate coefficient at 2000 K for the excitation process

$$H + HD(0,0) \rightarrow H + HD(v,J) \ . \tag{43}$$

An accurate estimate of $q_{vJ}$ would be valuable in assessing the feasibility of detecting HD in heated gases.

Vibrationally excited molecular hydrogen has also been detected by fluorescence scattering[42,43] of atomic emission lines. The process was identified in sunspot umbrae[42,43]. There are near coincidences of Ly α and lines of Si IV and C II with absorption lines of $H_2$ originating in the excited vibrational levels v" = 2, 3 and 5. The absorptions populate the excited vibrational levels of the $B^1\Sigma_u^+$ and $C^1\Pi_u$ states of $H_2$ which radiate. The same process in which Ly α is the pump has been observed in T Tauri[44] where infrared emission from v " = 1 has also been detected.  Similar fluorescence mechanisms have been observed for carbon monoxide in the solar spectrum[45] and in the spectrum of the red giant Arcturus[46].

Lyman alpha photons probably constitute a large part of the precursor radiation produced by a fast interstellar shock wave[47].  The fluorescence produced by Ly α absorption has an emission component into the vibrational continuum of the ground electronic state and Ly α pumping may be a rapid destruction mechanism of vibrationally excited $H_2$ molecules.

A large variety of electronic, atomic and molecular excitation, ionization and dissociation processes occur in shocked gases[48] and for many of them reliable quantitative data are not available.  Often the internal energy modes are not in thermal equilibrium and laboratory data are not directly applicable.  One example is the dissociation of molecular hydrogen at low densities where a process of radiative stabilization occurs[49,50].

Far infra-red and sub-millimetre observations of heated regions[51,52] are a rapidly developing source of data on shock-heated and X-ray heated regions and collisional excitation of high-lying rotational and vibrational levels is an important mechanism.


4.  THE ABUNDANCE OF HELIUM

Because the fractional abundance of deuterium is small, the amount now present is sensitive to weak creation and destruction mechanisms which may not have been recognized.  The cosmological abundance of helium is predicted to be large, of the order Y = 0.25 by weight.  It is enriched by chemical evolution in galaxies and a correction will be necessary.  The initial fraction Y is sensitive to the expansion rate of the universe but not to the matter density.  A recent discussion has been presented by Olive et al.[52]

Because their resonance transitions lie in the photoionization continuum of hydrogen, neither neutral nor ionized helium is detectable in absorption and the abundance of helium has to be derived from studies of its emission lines.  The interpretation of emissions from stellar atmospheres is perhaps too complex to yield abundances with sufficient precision and helium (and heavy element) abundances are more reliably obtained from studies of gaseous nebulae, both in our galaxy and in other galaxies.

4.1  Gaseous nebulae

Gaseous nebulae are low density regions of interstellar gas ionized by ultra-violet radiation from stars.  In them, excited states of atoms and ions are populated by recombination processes and by electron impacts.  From the intensities of the resulting emission lines, the individual element abundances can be derived.

In this way, it has been shown that the helium abundance is well-correlated with the heavy element abundance.  Lequeux et al.[54] have obtained data on the helium to hydrogen and the oxygen to hydrogen ratios in irregular and blue compact galaxies which they have extrapolated back to zero oxygen content to obtain a primordial [He]/[H] ratio by weight of Y = 0.228 ± 0.014.  This ratio is barely compatible with the standard cosmology and only if the universe is open.

The error in the determination of Y seems remarkably small.  To obtain the density of the emitting species from the measured line emission intensities, the electron density and the electron temperature which determine the population rates must be known.  They are usually obtained from the intensities of lines emitted by metastable levels of species such as O III and N II, analyses which make use of radiative transition probabilities and electron impact excitation rate coefficients.

Because any particular element is often not detected in more than one or two of its ionization stages, corrections must be applied for the element content of the unobserved stages. It is customary to employ semi-empirical correction factors based upon the similarities between the ionization potentials of different ionic systems[55,56,57]. An example is the relationship

$$\frac{N(O)}{N(H)} = \frac{N(He^{++} + He^{+})}{N(H^{+})} \frac{N(O^{+} + O^{++})}{N(H^{+})} \tag{44}$$

The ionization correction factors are broadly consistent with theoretical models of homogeneous gaseous nebulae ionized by central stars in which the electrons are produced by photoionization by stellar ultraviolet radiation and removed by radiative and dielectronic recombination.

Uncertainties in the atomic physics of gaseous nebulae are a source of error in the derivation of the element abundances. Substantial progress has been made in recent years in the experimental and theoretical determination of collision cross sections for electron impact excitation of neutral and ionic systems which is reviewed elsewhere in this volume. For the lighter elements, the more serious errors in the model nebulae probably lie in the treatment of dielectronic recombination, charge transfer and ionization processes.

### 4.1 Dielectronic recombination

Burgess[58] drew attention to the importance of dielectronic recombination as an electron removal process in the solar corona and worked out an approximate formula for the rate coefficients based upon the recognition that at high temperatures dielectronic recombination proceeds mostly by captures into doubly-excited autoionizing states with high principal quantum numbers[59]. Various versions of the formula have been developed and in astrophysics widespread use has been made of a table of rate coefficients compiled by Aldrovandi and Péquignot[60].

More accurate values for high temperature plasmas have been calculated for a few isoelectronic sequences[61,62,63,64,65]. Dielectronic recombination is accompanied by a complex cascading process and the emitted photons provide a powerful diagnostic probe of hot plasmas[66,67].

Dielectronic recombination at lower temperatures has been considered recently by Storey[68] following the suggestion of Harrington et al.[69] that the process might be important in planetary nebulae both as a recombination mechanism and as a source of excited levels of C III. The dielectronic recombination coefficients calculated by Storey[68] for C II, C III, N III and N IV are much larger at nebular temperatures than the values estimated by Aldrovandi and Pequignot[60]. Using the results of Storey[68], Harrington, Lutz and Seaton[70] have confirmed that dielectronic recombination into low-lying autoionizing states is important in establishing the distribution of ionization stages of carbon in planetary nebulae and as a mechanism for producing the C III emission line at 2297 Å.

Dielectronic recombinations into the $3s3p3d$ $^{2}F^{o}$ levels of Si II have been suggested as the source of a broad ultraviolet emission in the supergiant P Cygni by Underhill[71] who argues that the ionization limit is depressed so that the $^{2}F^{o}_{5/2}$ level, which lies below the ionizing limit in laboratory conditions, lies above it in the atmospheres of B-type stars.

4.12  Charge transfer

Because the photoionization cross section of atomic hydrogen decreases rapidly
with photon frequency, high frequency ionizing sources incident upon a gas of
cosmic composition produces plasmas in which hydrogen is only partly ionized.
In such a gas charge transfer of multiply-charged ions

$$X^{m+} + H \rightarrow X^{(m-1)+} + H^{+} \tag{45}$$

may dominate the ionization structure of heavy elements.

Bates and Moiseiwitsch[72] showed that some reactions of the kind (45) proceed
rapidly at thermal energies.  Steigman[73] pointed out that because of charge
transfer, multiply-charged ions which are created in the interstellar gas[74,75,76]
are rapidly converted to lower ionization stages and hence do not provide a
useful signature of interstellar high energy ionizing sources.  The importance
of charge transfer in planetary nebulae was established empirically by Péquignot,
Aldrovandi and Stasinski[77] and the importance of the reverse processes as
ionization sources in high temperature plasmas theoretically by Baliunas and
Butler[78].

Extensive theoretical calculations have been carried out in recent years of the
rate coefficients of thermal charge transfer processes[79-92] which have established that almost all systems more than three-times ionized undergo rapid charge
transfer but that doubly-ionized systems behave in specific ways.  Thus the
reaction of Ne III with H is very slow, indeed proceeds radiatively,

$$Ne^{++} + H \rightarrow Ne^{+} + H^{+} + h\nu \tag{46}$$

with a rate coefficient at $10^{4}$ K of about $10^{-15}$ cm$^3$ s$^{-1}$ whereas O III reacts
quite rapidly

$$O^{++} + H \rightarrow O^{+} + H^{+} \tag{47}$$

with a rate coefficient at $10^{4}$ K of about $8 \times 10^{-10}$ cm$^3$ s$^{-1}$.

The ionization correction factor used to correct for the unseen Ne II and Ne I[93]
relates the neon and oxygen ionization distributions.  It tends to overestimate
the neon abundance and the difference in charge transfer rate coefficients for
O III and Ne III is the probable explanation of the apparent overabundance of
neon observed for planetary nebulae[94].

In thermal charge transfer reactions of multiply-charged ions with neutral atoms,
the product ion is often produced preferentially in an excited state which decays
radiatively[95].  An example is the reaction[91]

$$O \; IV + H \rightarrow O \; III + H^{+} \tag{48}$$

which populates several levels of O III, and in particular, the 2p3p$^1$P level
which decays by emitting a photon at 5592 Å.  An emission line at 5592 Å is seen
in many planetary nebulae[96].  Emissions produced by charge transfer may have
particular utility as diagnostic probes of quasars and X-ray photoionized
nebulae[97,98].

4.13  Ionization processes

Direct photoionization is the main source of ionization in nebulae.  It is
usually incorporated into models of nebulae, using cross sections based upon
some form of hydrogenic approximation.  More accurate data on some highly ionized
systems have been calculated[99].  The effects of autoionization, the reverse of
dielectronic recombination, and of double photoionization, have not been fully
included and could be significant in producing highly-ionized systems.

## 4.2  Radio recombination lines

In our galaxy a more direct measure of the [He]/[H] ratio may be obtained from
the study of radio recombination lines emitted by highly excited Rydberg atoms.
Although the interstellar medium is a hostile environment for compound systems,
there is space enough for atoms in high Rydberg states to survive until they
radiate.  For hydrogen, emission has been detected from Rydberg levels up to
n = 390 though the record may belong to carbon.  A line at 26.131 MHz observed
in absorption towards the young supernova remnant Cassiopeia A by Konovalenko
and Sodin[100] and attributed by them to a hyperfine transition of [14]N may
instead be the C631α line[101].

Rydberg atoms have contributed substantially to astrophysics because they emit and
absorb radiation in a great variety of physical conditions.  Recombination lines
have been observed from hydrogen, helium, carbon and heavier elements in galactic
H II regions, planetary nebulae, C II regions, supernova remnants and external
galaxies.  An essential part of the value of the recombination line studies is
the existence of a comprehensive picture of the physical processes which control
the populations of high n-levels.

Observations of helium and hydrogen recombination lines from H II regions in the
galaxy have established that a gradient exists in the ratio of $He^+$ to $H^+$ as a
function of distance from the galactic center[102].  The conversion of the $He^+/H^+$
ratio to a [He]/[H] ratio is not straightforward because the fractions of
hydrogen and helium which are present in ionized form depend upon the central
ionizing source and the amount of dust.  It is believed that because of
increasing metallicity in the stars, the supply of energetic photons diminishes
toward the galactic center and so does the fraction of $He^+$.  Panagia[103] has
made the appropriate (large) corrections and estimates that Y decreases from
0.34 at the galactic center fo a limiting value of 0.24 at large distances, which
is in harmony with the primordial ratio derived by Lequeux et al.[54] and others[53].

## 4.3  Molecules containing helium

No molecule containing helium has been detected in astronomical objects.
Dabrowski and Herzberg[104] have drawn attention to the molecule $HeH^+$ and have
suggested it might be formed in gaseous nebulae.  The suggestion has been further
considered by Black[105], Flower and Roueff[106] and Roberge and Dalgarno[107].  The
analysis by Roberge and Dalgarno concluded that $HeH^+$ may be detectable in
emission in the pure rotational transition at 149.1μ and the 1-0 vibrational
transition at 3.364μ in X-ray sources powered by accretion of a dense cloud onto
a compact object.  The sources and sinks of $HeH^+$ are numerous and its successful
detection would not provide a useful estimate of the helium abundance itself.

## 5.  MASSIVE NEUTRINOS

The cosmological conclusions drawn from the deuterium and helium abundances will
be modified if massive neutrinos exist, a possibility which has received much
attention recently.  The massive neutrinos may supply the missing mass.  There
is some experimental support for the existence of neutrinos with a rest-mass of
$34.3 \pm 4$ eV/$c^2$[108] obtained from measurements of the β-decay of tritium.  A
massive neutrino is predicted to decay into a light neutrino and a photon with
about half the energy.  Neutrinos in the halo of the galaxy will produce a
Doppler-broadened emission line near 730 Å if the 34 eV rest mass is correct.
If neutrinos have decayed over cosmological times, a broad background emission
will appear.  The observed ultraviolet emission at high galactic latitudes can
in fact be used to place a lowerlimit on the lifetime of the massive
neutrino[109-111].  More speculatively, Melott and Sciama[112] have argued that
the existence of high velocity clouds of neutral hydrogen at high latitudes
places a limit on the intensity of photoionizing radiation produced by decaying

neutrinos which corresponds to a neutrino lifetime in excess of $10^{24}$ s, orders of magnitude longer than the age of the universe which is less than $10^{18}$ s.

If the neutrinos are sufficiently massive, the neutrino decay photons will produce highly-ionized systems. Savage and de Boer[113,114] detected C IV and Si IV in absorption in the spectra of several stars belonging to the Magellanic Clouds and concluded that the ions are located in a galactic corona several kiloparsecs above the galactic plane. Additional evidence was obtained by Ulrich et al.[115] from ultraviolet observations of the quasar 3C 273. Such a corona had been predicted by Spitzer[116]. The temperature is about $3 \times 10^5$ K and the ionization distribution is determined by electron impact ionization and radiative and dielectronic recombination. Detailed calculations of the ionization distributions of hot gases have been carried out on numerous occasions[117]. Autoionization, the reverse of dielectronic recombination, may make a major contribution to the ionization balance, though for highly-ionized systems it is limited by the efficiency of stabilizing radiative transitions[118]. For some systems, charge transfer ionization will modify the calculated distributions[78]. Jakobsen and Paresce[119] have calculated emission line intensities of the hot coronal gas which are consistent with rocket observations of the spectrum of the diffuse far-ultraviolet background, suggesting the presence near the north galactic pole of a weak emission line component[120].

Sciama and Mellott[121] argued that the high latitude C IV and Si IV found by Savage and de Boer[113,114] is produced through photoionization by neutrino decay photons. Charge transfer processes will tend to remove highly stripped systems created by ionization of a cool neutral gas and the possible detection of emission lines tends to confirm the existence of a hot coronal gas in which electron impact ionization is the source of the highly ionized material. If a reliable determination of the neutrino mass is made, it may be possible to identify some very specific fluorescence process or absorption edge for the detection of the decay photon.

With the advent of X-ray observations with high spectral resolution for objects other than the Sun, the study of atomic processes in hot astrophysical plasmas requires more attention. Winkler et al.[122] have obtained a high resolution spectrum of the supernova remnant Puppis A and Canizares and Winkler[123] have concluded that oxygen and neon are enriched in the remnant. A theoretical model of the emitting plasma is an essential part of the analysis of the spectral data.

The X-rays emitted by hot plasmas may be absorbed by cooler gas and the response of the intervening gas is potentially a powerful indication of the nature of the X-ray emitter and of the absorbing gas. A detailed accounting is needed of the processes that occur. They have been discussed recently by Shapiro and Bahcall[114] for an atomic gas. For a molecular gas, specific molecules may be created in response to the X-rays. Black[105] has considered molecular formation in ionized plasmas and Roberge and Dalgarno[107] have suggested that $HeH^+$ may be formed and emit detectable radiation.

6. FUNDAMENTAL CONSTANTS

I have discussed so far aspects of atomic physics as they relate to cosmological questions. Cosmology in turn has information to give on atomic physics questions about the constancy in time and space of the fundamental constants[125].

Absorption lines are detected in the spectra of distant objects with large redshifts. In a few, both components of a fine-structure doublet have been observed, such as Mg II for which the absorptions

$$Mg^+(^2S_{1/2}) + h\nu \rightarrow Mg^+(^2P_{1/2}) \tag{49}$$
$$Mg^+(^2S_{1/2}) + h\nu \rightarrow Mg^+(^2P_{3/2}) \tag{50}$$

occur.  From the measured line frequencies, redshifts z can be determined according to

$$\frac{\nu(\text{observed})}{\nu(\text{laboratory})} = \frac{1}{(1+z)} \quad . \tag{51}$$

Any difference between the two values of z may be attributed to a variation of the fine structure constant $\alpha$ which controls the fine-structure splitting.  No difference has been detected and it has been shown that $\Delta\alpha/\alpha < 0.05$ for a quasar at z = 1.95[126], $\Delta\alpha/\alpha < 0.02$ for five radio galaxies at z < 0.2[127] and $\Delta\alpha/\alpha < 0.03$ for a BL Lac-type object at z = 0.52[128].  The most severe of these limits corresponds to a mean rate of change of

$$\left|\frac{d}{dt}\ \ell n\alpha\right| \leq 2x10^{-12}\ yr^{-1} \quad .$$

Similar comparisons of resonance absorption lines of hydrogen and heavy elements observed in quasar spectra lead to a limit

$$\left|\frac{d}{dt}\ \ell n\,(m/M)\right| \leq 5x10^{-11}\ yr^{-1}$$

for the ratio of the electron and proton mass.[129]

The strictest limits on variability are obtained from comparisons with hydrogen 21 cm absorption[128,130].  The hyperfine splitting of hydrogen depends upon the product $\alpha^2 g_p m/M$ where $g_p$ is the nuclear g factor for the proton.  The derived limit, based upon data for the quasar 3C 286[130,131] is

$$\left|\frac{d}{dt}\ \ell n\,(\alpha^2 g_p m/M)\right| \leq 7x10^{-15}\ yr^{-1} \quad .$$

The observations demonstrate also that $\alpha^2 g_p m/M$ is spatially uniform to a few parts in $10^4$ through the observable universe[125].

In no sense either in content or in references have I given more than a cursory account of recent applications of atomic collisions physics in astrophysics.  I have ignored much of great importance and interest.  For the immediate future, the increase in sophistication of astronomical observational techniques in conjunction with the extension of the wavelength range of high resolution spectroscopy into the sub-millimetre and X-ray regions of the spectrum and the increase in precision of numerical techniques in astronomical theories create new demands for more and for better data on atomic and molecular properties.  In my view there is a greater need for a deeper understanding of atomic and molecular processes so that those which are relevant to the interpretation of astronomical phenomena may be identified, subjected to experimental and theoretical study and incorporated into the astronomical interpretations.

Acknowledgements.  I am indebted to the Division of Astronomical Sciences of the National Science Foundation (Grant AST-79-06373) and to the Division of Chemical Sciences of the Department of Energy (DE-AC02-76ER-02887) for support of research into the applications of atomic physics.

REFERENCES:

[1] Weinberg, S., Gravitation and Cosmology (Wiley, New York, 1972).
[2] Harrison, E.R., Ann. Rev. Astron. Ap. 11 (1973) 155.
[3] Hutchins, J.B., Ap. J. 205 (1976) 103.
[4] Shchekinov, Yu. A. and Edel'man, M.A., Soviet Astron. Letters 4 (1978) 234.
[5] Khersonskii, V.R. and Varshalovich, D.A., Soviet Astron. 22 (1978) 279.
[6] McKellar, A., Publ. Dominion Astrophys. Obs. 7 (1941) 251.
[7] Bortolot, V.J., Clauser, J.R. and Thaddeus, P., Phys. Rev. Letters 22 (1969) 309.
[8] Bruston, P., Audouze, J., Vidal-Madjar and Laurent, C., Ap. J. 243 (1981) 161.
[9] Chambaud, G., Launay, J.M., Levy, B., Millie, P., Roueff, E. and Tran Minh, F., J. Phys. B. 13 (1980) 4205.
[10] Roueff, E., Astron. Ap. 99 (1981) 394.
[11] Watson, W.D., Christensen , R.B. and Deissler, R.J., Astron. Ap. 69 (1978) 159.
[12] Hunter, G. and Kuriyan, M., Proc. Roy. Soc. A 358 (1977) 321.
[13] Black, J.H. and Dalgarno, A., Ap. J. Letters 184 (1973) L101.
[14] O'Donnell, E.J. and Watson, W.D., Ap. J. 191 (1974) 89.
[15] Watson, W.D., Ap. J. Letters 181 (1973) L129.
[16] Hartquist, T.W., Black, J.H. and Dalgarno, A., M.N.R.A.S. 185 (1978) 643.
[17] Elitzur, M. and Watson, W.D., Ap. J. 236 (1980) 172.
[18] McLeod, J.M., Avery, L.W. and Broten, N.W., Ap. J. Letters (1981) in press.
[19] Langer, W.D., Schloerb, F.P., Snell, R.L. and Young, J.S., Ap. J. Letters (1981) in press.
[20] Green, S. and Thaddeus, P., Ap. J. 191 (1974) 653.
[21] Green, S. and Thaddeus, P., Ap. J. 205 (1976) 766.
[22] Green, S., Ap. J. 201 (1975) 366.
[23] Green, S., Ap. J. Suppl. 42 (1980) 103.
[24] Green, S. and Chapman, S., Ap. J. Suppl. 37 (1978) 167.
[25] Green, S., Garrison, G.J., Lester, N.A. and Miller, W.A., Ap. J. Suppl. 37 (1978) 321.
[26] Mehrotra, S.C., Astrophys. Spa. Sci 71 (1980) 507.
[27] Penzias, A., Science 208 (1980) 663.
[28] Audouze, J., Lequeux, J., Reeves, H. and Vigroux, L., Ap. J. Letters 208 (1976) L151.
[29] Herbst, E. and Klemperer, W.B., Ap. J. 185 (1973) 505.
[30] Watson, W.D., Rev. Mod. Phys. 48 (1976) 413.
[31] Dalgarno, A. and Black, J.H., Reports Progr. Phys. 39 (1976) 573.
[32] Carney,E.D., Chemical Physics 54 (1980) 103.
[33] Watson, W.D. in: Audouze, J. (ed.), CNO Isotopes in Astrophysics (Reidel, Dordrecht, 1977).
[34] Guelin, M., Langer, W.D., Snell, R.L. and Wootten, A.H., Ap. J. Letters 217 (1977) L165.
[35] Watson, W.D., Snyder, L.E. and Hollis, J.M., Ap. J. Letters 222 (1978) L145.
[36] Oppenheimer, M. and Dalgarno, A., Ap. J. 192 (1974) 29.
[37] Mitchell, G.F., Astron. J. 83 (1978) 1612.
[38] de Jong, T., Dalgarno, A. and Boland, W., Astron. Ap. 91 (1980) 68.
[39] Elmegreen, B.G., Ap. J. 232 (1979) 729.
[40] Umebayashi, T. and Nakano, T., Publ. Astron. Soc. Japan 32 (1980) 405.
[41] Knacke, R.F. and Young, E.T., Ap. J. Letters (1981) in press.
[42] Jordan, C., Brueckner, G.E., Bartoe, J-D. F., Sandlin, G.D. and Van Hoosier, M.E., Nature 270 (1977) 326.
[43] Jordan, C., Brueckner, G.E., Bartoe, J-D. F., Sandlin, G.D. and Van Hoosier, M.E., Ap. J. 226 (1978) 687.
[44] Brown, A., Jordan, C., Millar, T.J., Gondhalekar, P. and Wilson, R., Nature 290 (1981) 34.
[45] Bartoe, J-D. F., Brueckner, G.E., Sandlin, G.D., Van Hoosier, M.E. and Jordan, C., Ap. J. Letters 223 (1978) L51.
[46] Ayres, T.R., Moos, H.W. and Linsky, J.L., Ap. J. Letters (1981) in press.
[47] Hollenbach, D.J. and McKee, C.F., Ap. J. Suppl. 41 (1979) 555.

[48] McKee, C.F. and Hollenbach, D.W., Ann. Rev. Astron. Ap. 18 (1980), 219.
[49] Dalgarno, A. and Roberge, W., Ap. J. Letters 233 (1979) L25.
[50] Roberge, W. and Dalgarno, A., Ap. J. (1981) in press.
[51] Watson, D.M., Storey, J.W.V., Townes, C.H., Haller, E.E. and Hansen, W.L.,
     Ap. J. Letters 239 (1980) L129.
[52] Storey, J.W.V., Watson, D.M. and Townes, C.H., Ap. J. Letters 244 (1981) L27.
[53] Olive, K.A., Schramm, D.N., Steigman, G., Turner, M.S. and Yang, J.,
     Ap. J. 246 (1981) 557.
[54] Lequeux, J., Peimbert, M., Rayo, J.F., Serrano, A. and Torres-Peimbert, S.,
     Astron. Ap. 80 (1979) 155.
[55] Seaton, M.J., M.N.R.A.S. 139 (1968) 129.
[56] Peimbert, M., Rodriguez, L.F. and Torres-Peimbert, S., Rev. Mexicana Astro.
     Astrof. 1 (1974) 129.
[57] Peimbert, M. and Torres-Peimbert, S., Ap. J. 203 (1976) 581.
[58] Burgess, A., Ap. J. 139 (1964) 776.
[59] Burgess, A., Ap. J. 141 (1965) 1588.
[60] Aldrovandi, S.M.V. and Péquignot, D., Astron. Ap. 25 (1973) 137.
[61] Roszman, L.J., Phys. Rev. 20 (1979) 673.
[62] Gau, J.N. and Hahn, Y., J. Quant. Spectr. Rad. Trans. 23 (1980) 121.
[63] Gau, J.N., Hahn, Y. and Retter, J.A., J. Quant. Spectr. Rad. Trans. 23 (1980)
     131 and 147.
[64] Hahn, Y., Gau, J.N., Luddy, R. and Retter J.A., J. Quant. Spectr. Rad. Trans.
     23 (1980) 65.
[65] Jacobs, V.L., Davis, J., Rogerson, J.E., Blaha, M., Cain, J., and Davis, M.,
     Ap. J. 239 (1980) 1119.
[66] Bely-Dubau, F., Dubau, J., Faucher, P. and Gabriel, A.H., M.N.R.A.S. (1981)
     in press.
[67] Dubau, J., Gabriel, A.H., Loulerque, M., Steenman-Clark, L. and Volonte, S.,
     M.N.R.A.S. 195 (1981) 705.
[68] Storey, P.J., M.N.R.A.S. 195 (1981) 27P.
[69] Harrington, J.P., Lutz, J.H., Seaton, M.J. and Stickland, D.J., M.N.R.A.S.
     191 (1980) 13.
[70] Harrington, J.P., Lutz, J.H. and Seaton, M.J., M.N.R.A.S. 195 (1981) 21P.
[71] Underhill, A.B., Astron. Ap. 97 (1981) L9.
[72] Bates, D.R. and Moiseiwitsch, B.L., Proc. Phys. Soc. A67 (1954) 805.
[73] Steigman, G., Ap. J. 199 (1975) 642.
[74] Weisheit, J.C. and Dalgarno, A., Astrophys. Lett. 12 (1972) 103.
[75] Weisheit, J.C., Ap. J. 185 (1973) 877.
[76] Steigman, G., Rees, M.J. and Kozlovsky, B.Z., Astron. Ap. 30 (1974) 87.
[77] Péquignot, D., Aldrovandi, S.M.V. and Stasinska, G., Astron. Ap. 63 (1978)
     313.
[78] Baliunas, S. and Butler, S.E., Ap. J. Letters 235 (1980) L45.
[79] Blint, R.J., Watson, W.D. and Christensen, R.B., Ap. J. 205 (1976) 634.
[80] Butler, S.E., Guberman, S.L. and Dalgarno, A., Phys. Rev. A 16 (1977) 500.
[81] Christensen, R.B., Watson, W.D. and Blint, R.J., Ap. J. 213 (1977) 712.
[82] Butler, S.E., Bender, C.F. and Dalgarno, A., Ap. J. Letters 230 (1979) L59.
[83] McCarroll, R. and Valiron, P., Astron. Ap. 44 (1975) 465.
[84] McCarroll, R. and Valiron, P., Astron. Ap. 53 (1976) 83.
[85] McCarroll, R. and Valiron, P., J. de Phys. 39 (1978) C1, 52.
[86] McCarroll, R. and Valiron, P., Astron. Ap. 78 (1979) 177.
[87] Watson, W.D. and Christensen, R.B., Ap. J. 231 (1979) 627.
[88] Butler, S.E. and Dalgarno, A., Ap. J. 241 (1980) 838.
[89] Butler, S.E., Heil, T.G. and Dalgarno, A., Ap. J. 241 (1980) 442.
[90] Heil, T.G., Butler, S.E. and Dalgarno, A., Phys. Rev. A 23 (1981) 1100.
[91] Dalgarno, A., Heil, T.G. and Butler, S.E., Ap. J. 245 (1981) 793.
[92] Christensen, R.B. and Watson, W.D., Phys. Rev. A (1981) in press.
[93] Peimbert, M. and Costero, R., Bol. Obs. Tonantzintla y Tacubaya 5 (1969) 3.
[94] Hawley, S.A. and Miller, J.S., Ap. J. 212 (1977) 194.
[95] Dalgarno, A. and Butler, S.E., Comments At. Mol. Phys. 7 (1978) 129.
[96] Kaler, J.B., Ap. J. Suppl. 31 (1976) 517.
[97] Hatchett, S., Buff, J. and McCray, R.A., Ap. J. 206 (1976) 847.

PHYSICS OF ELECTRONIC AND ATOMIC COLLISIONS
S. Datz (editor)
© North-Holland Publishing Company, 1982

# THEORY OF ELECTRON-ATOM SCATTERING AT INTERMEDIATE ENERGIES

M. R. C. McDowell

Mathematics Department, Royal Holloway College,
(University of London), Egham Hill, Egham,
Surrey TW20 OEX, England

Recent advances in theoretical studies of electron
scattering by hydrogen and helium are reviewed.
Particular attention is paid to perturbative methods,
especially exact second Born calculations. Optical
Potential methods are discussed. Non-exchange close
coupling calculations are treated in some detail. A
recent second Born calculation for ionization is also
discussed. Where possible comparison is made with
experiment.

INTRODUCTION

I shall concern myself with electron scattering by neutral atoms at
impact energies above the ionization threshold. The discussion will
center on advances since the Kyoto meeting, and necessarily must be
highly selective. It will be concerned primarily with calculations
published in that period, and not with new, but untested, theoretical
techniques. The main topics will be

1. Perturbative methods for

   (a) elastic scattering by hydrogen;

   (b) inelastic scattering by hydrogen.

2. Variable charge distorted wave models.

3. Optical potential studies of elastic scattering by hydrogen.

4. Non-exchange close coupling for

   (a) elastic and inelastic scattering by hydrogen;

   (b) coherence parameters for H (2p).

5. R-matrix and close coupling calculations for elastic and
   inelastic scattering by helium.

6. Second Born approximation for the ionization of helium.

These are not the only fields in which there have been advances
(notably, I will not deal with recent relativistic R-matrix results)
but reflect my personal tastes and interests.

1.  PERTURBATIVE METHODS FOR ELECTRON SCATTERING BY ATOMIC HYDROGEN

A very significant advance in the last two years has been the <u>exact</u>
evaluation of the second Born approximation for the 1s - 1s and
1s - 2s transitions by Walters and Ermolaev [1,2,3]. Results had
previously been obtained by several authors for the elastic amplitude
at $\theta = 0°$, but closure, or other simplifying approximations, had
been used at other angles. In their first paper Ermolaev and Walters
remark that the breakdown of perturbative methods at low energies
(< 50 eV) may be due to a combination of the use of closure in the
second order term, the neglect of third and higher orders in the
direct scattering and neglect of second and higher order exchange
effects. In my view the third cause will not be important above
about 30 eV in hydrogen, as I hope to demonstrate.

(a)  <u>Elastic scattering</u>

The second Born direct elastic amplitude may be written

$$f^{B2}(\theta,k_o^2) = -\frac{1}{8\pi^4} \sum_n \lim_{\eta \to o^+} \int \underline{dk} \; \frac{T_{1s,n}(\underline{k}_f,\underline{k}) \, T_{n,1s}(\underline{k},\underline{k}_o)}{k_n^2 - k^2 + i\eta} \qquad 1.1$$

the sum over n including the continuum, where

$$T_{i,j}(\underline{k}',\underline{k}) = < \underline{k}',\psi_i |V| \underline{k},\psi_j > \qquad\qquad 1.2$$

and $k_n^2 = k_o^2 + 2(\varepsilon_{1s} - \varepsilon_n)$ .                1.3

In a closure approximation $k_n^2$ is replaced by

$$\bar{k}^2 = k_o^2 + 2(\varepsilon_{1s} - \bar{\varepsilon}) \qquad\qquad 1.4$$

and in the most sophisticated closure models [4] different values
of $\bar{\varepsilon},(\varepsilon_i,\varepsilon_r)$ are chosen for the real and imaginary parts of the
amplitude, the initial and final states being treated "exactly"
(e.g. in one or two state close coupling). It is usual to choose
$\varepsilon_r$ to reproduce the exact static dipole polarizability of the
ground state, $\varepsilon_i$ to give the first Born total cross section [5].
At 30 eV this gives

$$\varepsilon_r = -0.0556 \text{ a.u.}, \qquad \varepsilon_i = 0.1133 \text{ a.u.} \qquad 1.5$$

Ermolaev and Walters evaluate (1.1) at 30 eV by numerical methods,
without other approximation, and find that the closure approximation (1.5) gives a very fair approximation to $f^{B2}(\theta,k^2)$, the
largest error being ≈ 30% for Imf at 50°. However, the
amplitude $f^{B1} + f^{B2}$ gives a poor approximation to the measured
differential cross section [6] at 30 eV (though a similar calculation [4] is accurate at 50 eV). Thus the discrepancy does not
primarily arise from the use of closure.

Walters [2] has again used numerical techniques (not yet described in the literature) to evaluate the second Born exchange
amplitude $g^{B2}(\theta,k^2)$ for elastic scattering, at the same energy
(30 eV), and also a closure approximation to it in which
$\varepsilon_r = \varepsilon_i = \varepsilon(n=3)$. The closure approximation is poor for the
imaginary part but this choice of $\bar{\varepsilon}$ may not be optimum. The
calculation implies that Glauber exchange ($\equiv \bar{\varepsilon} = \varepsilon(n=1)$ <u>must</u>
be poor. He combines $g^{B1} + g^{B2}$ with Kingston and Walters [4]
DWBA closure calculations of $f(\theta,k^2)$, which as we have seen

above is a good approximation to $f^{B1} + f^{B2}$ to obtain the elastic differential cross section (Figure 1) and finds the discrepancy at small angles is much reduced.

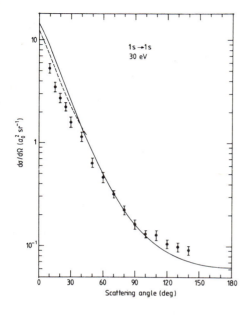

Figure 1.

Elastic differential cross sections for $e^-$-H (1s) at 30 eV in $a_0^2$ - str.$^{-1}$. The solid curve is the DWSBA model [4], the dashed curve is an estimate [2] of the modification due to second order exchange. The experimental results (I) are those of Williams [6]. (Reproduced by permission of the Institute of Physics.)

Rather surprisingly, exchange is unimportant at large angles. The remaining, and significant, discrepancy would appear to be due to neglect of third and higher order effects, though it is difficult to be sure until $g^{B2}$ is calculated with distorted waves. I would tentatively conclude that for this process second order exchange is clearly important below 30 eV, and it is almost certain that so are higher order direct terms. Experiments with spin polarized electrons on spin polarized atoms are urgently required.

(b) <u>Inelastic scattering</u>

In a subsequent paper [3] Ermolaev and Walters have carefully examined the exact second Born approximation for electron impact excitation of the $1s \rightarrow 2s$ transition at 54.4 eV, comparing their results with a closure approximation ($\varepsilon_r = \varepsilon_i = 0.0404$) and the DWSBA calculation [4] ($\bar{\varepsilon} = -0.0125$). Their closure approximation is extremely accurate for Im $f^{B2}_{1s-2s}$ at all angles but poor for the smaller Re $f^{B2}_{1s-2s}$ for $\theta > 10°$. The DWSBA model [4] of Kingston and Walters chooses $\bar{\varepsilon}$ to reproduce the transition dipole polarizability, $\alpha_{1s,2s}$, exactly, and treats the 1s,2s,2p states by close coupling. Again it fails badly for Re $f^{B2}_{1s,2s}$ (which has a deep minimum near 40°) but the DWSBA results are a good approximation to Im $f^{B2}_{1s,2s}$ except at small angles. They conclude that allowing for the use of distorted waves in the DWSBA, the closure approximation is satisfactory to

within 20% for  $d\sigma/d\Omega_{1s-2s}$  at this energy.  Given therefore that
the closure approximation is a reasonable one (for the second
order term) for elastic scattering above 30 eV and for  1s - 2s
excitation at and above 54.4 eV, we note that the resultant
differential cross section is much higher than the 3CC result in
the forward direction, and very much lower for  $\theta > 30°$.  The
authors attribute this to omission of states higher than  n = 2
in the 3CC model.

This is confirmed in Figure 2 where the DWSBA [4] differential
cross section is compared with the 13CC (non-exchange) results of
Morgan [7].  There is now very close agreement in the forward
direction, though the 13CC NE results are 40% higher at 160°.
Most noticeably the local minimum in the DWSBA results near 50°

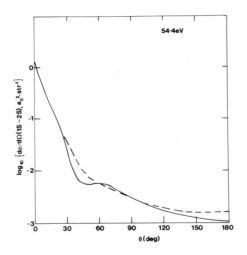

Figure 2.   Differential cross section for  $e^- + H$ (1s) $\rightarrow e^- + H$ (2s)
            at 54.4 eV.  The solid curve is the DWSBA [3] and the
            dashed curve the 13CCNE [7].

is no longer present.  Very similar differences have been found
earlier by Willis et al. [8] for positron excitation of the
$1^1S \rightarrow 2^1S$  transition of helium at 100 eV, in comparing 3CCNE with
the EBS results, though the local minimum in the EBS values has
vanished by 200 eV.  These results strongly suggest that third
and higher order effects must be important for  S - S  inelastic
direct amplitudes below about five times the excitation threshold.

## 2.   VARIABLE CHARGE DISTORTED WAVE MODELS

At higher energies recent work by Schaub-Shaver and Stauffer [9] and
Sharma and Mathur [10] using the variable-charge Coulomb-projected
Born approximation has confirmed that accurate results for  S - S
transitions in atomic hydrogen can be obtained in a distorted wave
formalism above 200 eV.  They take

$$H_O = H - V$$

$$= H_{atom}(1) - \tfrac{1}{2} \nabla_2^2 - \frac{\zeta(r_2)}{r_2} \qquad 2.1$$

with

$$\zeta(r_2) = Z - \int_0^{2\pi} \int_0^{\pi} \int_0^{r_2} |\phi(r_1)|^2 \, r_1^2 \, dr_1 d\cos\theta \, d\phi \qquad 2.2$$

so that

$$V(r_1, r_2) = -\{1 - \exp(-2Zr_2)[2(Zr_2)^2 + 2Zr_2 + 1]\} r_2^{-1} + r_{12}^{-1} \qquad 2.3$$

and evaluate the T-matrix with distorted waves satisfying

$$\left[ \tfrac{1}{2} \left( \frac{d^2}{dr^2} - \frac{\ell(\ell+1)}{r_2^2} \right) + \frac{\zeta(r_2)}{r_2} + \tfrac{1}{2} k_f^2 \right] U_\ell(k_f, r_2) = 0. \qquad 2.4$$

For elastic scattering Sharma and Mathur's results [10] are in better agreement with experiment at 200 eV than either the EBS [11] or DWSBA [4] values. (Figure 3). However, the method, in its present stage of development, is less intellectually satisfying for inelastic transitions as it is not time reversal invariant.

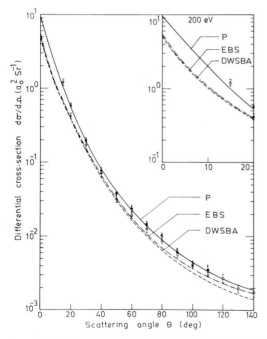

Figure 3.

Differential cross section for elastic scattering of $e^-$ by H (1s) at 200 eV. P: Variable charge CPB approximation [10], EBS: Eikonal Born-Series [11], DWSBA [4]. Experimental data of Williams. (From reference [10].)

3.  OPTICAL POTENTIAL STUDIES OF ELASTIC SCATTERING BY HYDROGEN

A wide range of optical potential models have been investigated in the last two years. One important result [12] is that McCarthy and his colleagues have shown that the long-range polarization interaction $V(r) = -\alpha/r^4$ which is derived in an adiabatic model (e.g. from

Re $f_{1s-1s}^{B2}$) is both sinusoidal and energy dependent in a non-adiabatic treatment. They find, after localization, that for any state $j = (nm)$,

$$V_j^{pol}, \text{local} (r) = -\frac{1}{r^4}\left[ \frac{-(3/4\pi)g_j^2(o)}{\varepsilon_o \quad \varepsilon_j} + \frac{(k-w^{\frac{1}{2}})}{k} \times \right.$$

$$\left\{ \frac{(3/4\pi)g_j(o)g_i((k-w^{\frac{1}{2}})^2)\cos(k-w^{\frac{1}{2}})r}{(k+w^{\frac{1}{2}})^2} \right.$$

$$\left. \frac{-(3/4\pi)g_j(o)g_j((k+w^{\frac{1}{2}})^2)\cos(k+w^{\frac{1}{2}})r}{(k+w^{\frac{1}{2}})^2} \right\}$$

$$\left. \frac{+(3/2\pi)g_j(o)g_j((k+w^{\frac{1}{2}})^2)\cos(k+w^{\frac{1}{2}})r}{(k+w^{\frac{1}{2}})^2} \right] + O(r^{-5}) \qquad 3.1$$

where $g_j(q^2)$ is the momentum wave function of state $j$, and $W = k^2 - 2^j(\varepsilon_o - \varepsilon_j)$. The simpler form is obtained if.f. $k_i, k_f \sim k$ and $K \simeq 0$.

Scott and Bransden [13] have used the second order optical potential method without closure to investigate elastic scattering by electrons of atomic hydrogen from 20 to 100 eV. They incorporate sixteen pseudostates (5s, 4p, 4d and 3f orbitals) as well as the exact 1s state and solved

$$(\nabla^2 + k_o^2) F_o^{\pm} (\underline{r}) = V_{oo}(\underline{r}) F_o^{\pm}(\underline{r}) + \int W_{oo}(\underline{r},\underline{r}'(F_o(\underline{r}')\underline{dr}'$$

$$+ \int K_{oo}(\underline{r},\underline{r}') F_o^{\pm}(\underline{r}')\underline{dr}'$$

$$+ \int L_{oo}^{\pm}(\underline{r},\underline{r}') F_o^{\pm}(\underline{r}')\underline{dr}' \qquad 3.2$$

where the first two terms on the right hand side are the first order direct and exchange potentials, the second order direct potential is

$$K_{oo}(\underline{r},\underline{r}') = \sum_{m=1}^{\infty} V_{om}(\underline{r})G_o(\underline{r},\underline{r}',k_m^2) V_{mo}(\underline{r}'), \qquad 3.3$$

where $G_o$ is the free particle Green's function. The second order exchange operator was omitted. Their model is an approximation, including only first order exchange, to a seventeen-state close coupling calculation. The effect of the pseudostates in $K_{oo}$ is to approximate the continuum integral by an integration rule. The model gives good overall agreement with the experimental differential cross sections of Williams [6] (except at 100 eV where the data are suspect): however it tends to overestimate the differential cross sections at small angles, between 20 and 30 eV. Table 1 compares the total inelastic cross section $\sigma_{inel} = \sigma_{tot} - \sigma_{el}$ obtained in their model with the estimates of this by de Heer et al. [14]. The optical potential model tends to overestimate by about 20%, apparently due to the imaginary part of $K_{oo}$ being too large, or in other words to an overestimate of the inelasticity due to ionization.

Byron and Joachain [15] have carried out a similar calculation including the leading third order direct term, but with localization of exchange, at impact energies from 50 to 500 eV. The second order

Table 1.  Values of  $\sigma_{inel}$  from (a) de Heer et al. [14], (b) Scott and Bransden [13] in units of  $\pi a_o^2$ .

| $E_i$ (eV) | 20 | 30 | 50 | 100 |
|---|---|---|---|---|
| (a) | 1.48 | 1.98 | 2.09 | 1.76 |
| (b) | 2.23 | 2.65 | 2.56 | 1.87 |

Table 2.  Total elastic cross section for electron scattering by atomic hydrogen (in units of  $\pi a_o^2$ ).  (a) Scott and Bransden [13], (b) Byron and Joachain [15], experiment, (c) van Wingerden et al. [16].

| $E_i$ (eV) | (a) | (b) | (c) |
|---|---|---|---|
| 50 | 1.22 | 1.32 | - |
| 100 | 4.6,-1 | 4.9,-1 | 5.8,-1 |
| 200 | - | 2.0,-1 | 2.5,-1 |

optical potential is obtained by using the Fourier transform of the second Born amplitude, in which the 1s, 2s and 2p state are treated exactly and closure applied to the rest.  They choose  $\bar\varepsilon_i$  to give Im $f^{B2}_{1s-1s}$ (o) exactly, at 400 eV, thus  $\bar\varepsilon_j = 0.465$ , but found surprisingly, that there was no such  $\bar\varepsilon_r$ .  However, including a  $\overline{3p}$  pseudostate with threshold at -0.10 ryd, they found  $\bar\varepsilon_r = 1.61$ . Subtracting the 1s - state, the second order localized optical potential is given by

$$V^{(2)} (r) = - \frac{1}{\pi r} \int_0^\infty K \sin Kr \ f^{B2}_{opt} (K) \ dK \ . \qquad 3.4$$

To obtain the leading third order term they write the eikonal amplitude

$$f_E = \frac{k}{2\pi_i} \int e^{i\underline{K}\cdot\underline{b}} \left\{ e^{\frac{i}{k}[\lambda\chi_1 + i\lambda^2\chi_2 + \lambda^3\chi_3 + \cdots]} -1\right\} d^2b \qquad 3.5$$

where  $\lambda$  is a coupling constant and expand in powers of  $\lambda$ : remember that (3.5) is in principle exact.  Similarly the Glauber many-body amplitude is written as

$$f_G = \frac{1}{2\pi} \int e^{i\underline{K}\cdot\underline{b}} \Big[ \lambda <0|\chi_G|0> + \frac{i\lambda^2}{2k} <0|\chi_G^2|0>$$

$$- \frac{1}{6k^2} \lambda^3 <0|\chi_G^3|0> + \cdots\Big] d^2b \qquad 3.6$$

and since  $\chi_G$  is known, a Glauber approximation to  $\chi_3$  is obtained, say  $\chi_3^G$ , giving on inversion

$$V^{(3)} \simeq \frac{1}{\pi} \int_r^\infty (b^2 - r^2)^{\frac{1}{2}} \frac{d}{db} \chi_3^G (k,b) db \ . \qquad 3.7$$

Finally, exchange is approximated by the Furness-McCarthy potential.

Their total elastic cross sections are summarized in Table 2. Clearly they differ little from Scott and Bransden's values [13]. At 100 eV the third order term increases the forward differential cross section by about 10%, in good agreement with other earlier

estimates of the effect. However, in this model the average disagreement in differential cross section with experiment is of order
25% from 50 to 200 eV. This is contrary to expectation as it is a
"high-energy" model and should be better at 200 eV than at 50 eV.
The model is in fact inferior to the EBS treatment over this energy
range.

These two treatments suggest to me that optical potential methods
will in general be inferior to many-state close-coupling treatments
for some time to come: it is simply too difficult to construct an
accurate optical potential; while at the same time no progress has
been made in tackling the many-channel problem.

It is worth remarking that Joachain and Winters [16] have also
tackled elastic scattering from H (2s), by a similar method, but
omitting the third order terms. They worked at 200 and 400 eV only,
and found that polarization effects were negligible compared with
absorptive effects. Their results are in close agreement with those
obtained using the static potential (or the FBA)

$$V^{(1)} = < 2s \vert V \vert 2s >$$

3.8

alone, except for very small angles where that underestimates, and at
large angles where it overestimates by up to 16%. The earlier EBS
and simplified second Born results are much larger, especially at
large angles. This suggests that for H (2s) perturbative methods
are very slowly convergent for large momentum transfer.

4.  NON-EXCHANGE CLOSE COUPLING FOR ELECTRON-HYDROGEN SCATTERING

Morgan [7] and Edmunds et al. [17] have been investigating the use of
non-exchange close coupling at intermediate energies. By comparison
with the well known 3CC data, Morgan found that exchange was a small
effect at impact energies of 54.4 eV and above. This conclusion
agrees with that of Walters [2] in the second Born, and may well be
valid down to 30 eV.

It is of interest to compare Morgan's results for the orientation and
alignment parameters with the recent measurements of Slevin et al.
[18]. They give

$$\lambda = \frac{\sigma_O(\theta)}{\sigma(\theta)}$$

4.1

and

$$R = \frac{Re < a_O a_1 >}{\sigma(\theta)}$$

4.2

at energies down to 35 eV, at angles out to 120°. The comparison for
$\lambda$ is shown in Figure 4 which also shows 3CC and 6CC non-exchange
calculations. These differ little, but are in clear disagreement
with experiment for $\theta > 60°$. However, the 12CC results are in quite
remarkable agreement with experiment. This difference can be
ascribed to the inclusion of two s-type, three p-type and one d-type
pseudostate (all with thresholds below 35 eV).

A similar effect is found at 54.4 eV, where comparison with 9CC and
11CC results indicates that the 13CC results (cf. section 1 above)
have converged. However, lest it be thought that the problem is now
solved, a similar comparison of theory and experiment for R(θ) is
made in Figure 5. There is no overall agreement with experiment,

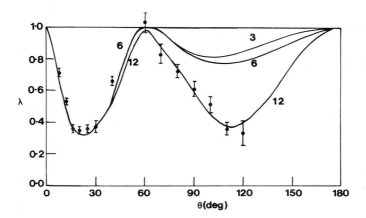

Figure 4.  Calculated non-exchange close coupling values of  $\lambda(\theta)$
for  H (2p)  at 35 eV [7].  The number above each
curve is the number of states included.  The
experimental data are those of Slevin et al. [18].
(By permission of Dr. L. A. Morgan.)

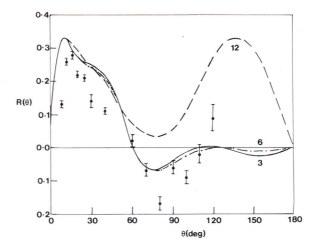

Figure 5.  Calculated non-exchange close coupling values of  $R(\theta)$
for  H (2p)  at 35 eV [7].  The number above each
curve is the number of states included.  The
experimental data are those of Slevin et al. [18].
(By permission of Dr. L. A. Morgan.)

and no evidence of convergence, and this lack of convergence to
experiment is confirmed by calculations at 54.4 eV (not shown).
Furthermore at 54.4 eV, $R$ (nCCNE,$\theta$) (n $\leq$ 13), remains positive at
all angles, whereas the experimental data are negative for $\theta$ > 75°.
The discrepancy is not due to neglect of exchange. Clearly $R$ is
the more sensitive parameter. Morgan suggests that treating the
experimental data, i.e. the number of coincidences, as if exchange
were absent, one can deduce the modulus of the relative phase, $|\chi|$,
as for He ($2^1P$). Her 12CCNE results at 35 eV have the same general
behaviour as the data, suggesting that the poor results for $R$
arise from it being a very sensitive function of $\chi$, i.e. small
errors in $\chi$ lead to large errors in $R$.

## 5. R-MATRIX AND CLOSE COUPLING FOR ELASTIC AND INELASTIC SCATTERING BY HELIUM

Bhadra et al. [19] have published a five-state ($1^1S,2^1S,2^1P,2^3S,2^3P$)
close-coupling study of electron scattering by helium at energies
from 30 to 100 eV, while Fon et al. [20] have carried out a similar
five-state R-matrix calculation, which differs not only in the
approximate wave functions used, but includes some allowance for
short range correlations. We shall refer to these as the 5CCE and
5CCRE models, respectively. Willis et al. [8] found that a similar
calculation using localized exchange of the Furness-McCarthy type,
gave excellent agreement with the total elastic cross section
measurements [22],[23], from 50 to 200 eV. Willis and McDowell [21]
remarked that neither the 5CCE [19] nor 5CCRE [20] calculations were
in any sort of agreement with experiment for the $1^1S \rightarrow 2^1S,2^1P$
transitions below 200 eV. They then carried out a similar five-
state calculation using localized exchange (5CCEEP), and found good
agreement with the 5CCRE values. The 5CCE [19] calculation gave a
poor $1^1S \rightarrow 2^1P$ oscillator strength, which accounts for its much
higher cross section, whereas the 5CCRE value of 0.2794 and the
5CCEEP value of 0.2875 are in reasonable accord with the "exact"
value of 0.2759.

Willis and McDowell then showed that the inclusion of $3^1S$ and $3^1P$
states had little effect, but that the inclusion of a $k^1P$ pseudostate, with threshold 3.5 eV about the ionization threshold, had a
very significant effect, in bringing the calculated values into
agreement with experiment. (Figures 6,7). It is of interest to
note that the earlier distorted wave polarized orbital calculation
(DWPO II) of Scott and McDowell is in agreement with experiment in
both cases, suggesting that much cancellation occurs in few-state
close coupling or R-matrix calculations of electron scattering by
neutral atoms at intermediate energies. Willis and McDowell also
used a five-state basis ($1^1S,2^1S,2^1P,3^1S,3^1P$) to carry out a non-
exchange close coupling calculation for the benchmark case of
$1^1S \rightarrow 3^1S$. Their results were, as expected, in much poorer agreement with the experimental values of Van Zyl et al. [31] than were
the DWPO I results [24]. Further work is in progress with additional
$3^1D,4^1S,k^1P$ states added to the basis.

## 6. TRIPLE DIFFERENTIAL CROSS SECTIONS FOR IONIZATION

There has been a long-standing failure to explain the ratio $\mathbb{R}$ of
forward ("binary") to backward ("recoil") peaks in the famous
experiments by Erhardt and his colleagues on electron impact ionization of helium [26]. A wide variety of calculations in the first
Born approximation (which differ only in the details of the wave

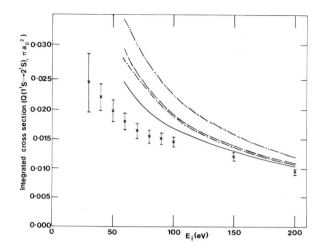

Figure 6.   Inelastic cross section for the $1^1S \rightarrow 2^1S$ transition in helium.   The —··— curve is the 5CCEEP and the solid curve the 7CCEEP [21] including $\overline{4^1S}$ and $\overline{k^1P}$ pseudostates.   The experimental data are those of de Heer and Jansen [28].

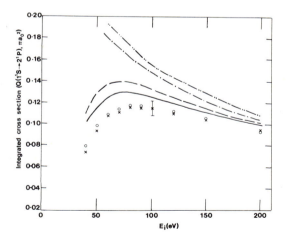

Figure 7.   Inelastic cross section for the $1^1S \rightarrow 2^1P$ transition in helium.   The —··— curve is the 5CCEEP and the solid curve the 7CCEEP [21] including $\overline{4^1S}$ and $\overline{k^1P}$ pseudostates.   The experimental data points are due to Westerveld et al. [29] (✕) and Donaldson et al. [30](O).

functions used for the ground state, and for those for the incident, scattered, and ejected electrons) are in general agreement with each other, and in disagreement with experiment.  Recent unpublished measurements by the Kaiserslautern group show that $\mathbb{R}_{exp}/\mathbb{R}_{FBA} \simeq 2$ and does not tend to unity with increasing impact energy.  Byron, Joachain and Piraux  27  have now solved the problem in principle, by showing in a calculation for atomic hydrogen, that a double scattering is involved.  If the first Born amplitude for the process is written

$$f^{B1} = - \frac{2}{K^2} M(\underline{K}, \underline{k}_{ej})$$

where  $\underline{K}$  is the momentum transfer of the incident electron (the binary peak being nearly symmetric about $\hat{\underline{K}}$), then the second Born approximation may be written, after closure, as

$$f^{B2} = \frac{2}{\pi^2} \int \frac{d\underline{q}}{q^2 - p^2 - i\varepsilon} \frac{1}{K_i^2 K_f^2} [M(\underline{K},\underline{k}_{ej}) - M(-\underline{K}_i,\underline{k}_{ej}) - M(\underline{K}_f,\underline{k}_{ej})]$$

with  $p^2 = k_0^2 - 2\bar{\varepsilon}$.  This was evaluated numerically for  $\bar{\varepsilon} = 0.5, 0.75,$ 1.0.  The binary peak moves to larger angles with increasing  $\bar{\varepsilon}$ (Figure 8) and the recoil peak is enhanced by about 50% relative to the FBA.  Calculations for helium are in progress.

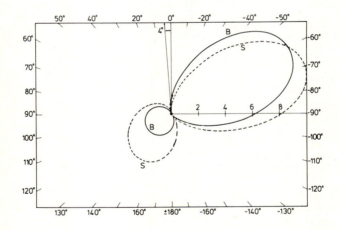

Figure 8.  Triple differential ionization cross section for  $e^-$,H at 250 eV, the slower ejected electron having 4 eV, the faster being scattered through $4^o$.  Solid curve FBA, dashed curve SBA with  $\bar{\varepsilon} = 0.5$.  (Adapted from Byron et al. [27].)

ACKNOWLEDGEMENTS

I am indebted to Dr. L. A. Morgan for valuable comments and for allowing me to use her unpublished results, and to many colleagues for preprints of recent work.

REFERENCES

[1]   Ermolaev, A. M. and Walters, H. R. J., Exact second Born amplitude for elastic electron-hydrogen scattering, J. Phys. B. 12 (1979) L779-84.

[2]   Walters, H. R. J., Exact second Born exchange amplitude for elastic electron-hydrogen scattering, J. Phys. B. 13 (1980) L749-55.

[3]   Ermolaev, A. M. and Walters, H. R. J., Exact second Born amplitude for the electron impact excitation of the 1s → 2s transition in atomic hydrogen, J. Phys. B. 13 (1980) L473-78.

[4]   Kingston, A. E. and Walters, H. R. J., Electron scattering by atomic hydrogen: the distorted wave second Born approximation, J. Phys. B. 13 (1980) 4675-82.

[5]   Woollings, M. J., Collisions of fast electrons with helium II: Identities satisfied by the second Born scattering amplitude, J. Phys. B. 5 (1972) L164-66.

[6]   Williams, J. F., Electron scattering from atomic hydrogen III: Absolute differential cross sections for elastic scattering of electrons from 20 to 680 eV, J. Phys. B. 8 (1975) 2191-99.

[7]   Morgan, L. A., Thirteen state close coupling non-exchange calculations of e⁻-H scattering. (Private communication 1981).

[8]   Willis, S. L., McDowell, M. R. C., Hata, J., Joachain, C. J. and Byron, F. W. Jr., Positron collisions with helium at intermediate energies, J. Phys. B. 14 (1981) in press.

[9]   Schaub-Shaver, J. and Stauffer, A. D., The variable-charge Coulomb-projected Born approximation, J. Phys. B. 13 (1980) 1457-69.

[10]  Sharma, R. K. and Mathur, K. C., Elastic scattering of electrons by hydrogen at intermediate energies, submitted to Phys. Rev. A.

[11]  Byron, F. W. Jr. and Joachain, C. J., Elastic scattering of electrons and positrons by atomic hydrogen and helium at intermediate energies, J. Phys. B. 10 (1977) 207-26.

[12]  McCarthy, I. E., Saha, B. C. and Stelbovics, A. T., The polarization potential for intermediate-energy electron-atom scattering, FIAS-12-73 December 1980, to be submitted to J. Phys. B.

[13]  Scott, T. and Bransden, B. H., A second order optical potential for the elastic scattering of electrons by atomic hydrogen, J. Phys. B. 14 (1981) in press.

[14]  de Heer, F. J., McDowell, M. R. C. and Wagenaar, R. W., Numerical study of the dispersion relation for e⁻-H scattering. J. Phys. B. 10 (1977) 1945-53.

[15]   Byron, F. W. Jr. and Joachain, C. J., A third-order optical
       potential theory for elastic scattering of electrons and
       positrons by atomic hydrogen, J. Phys. B. $\underline{14}$ (1981) in press.

[16]   Joachain, C. J. and Winters, K. H., An optical model approach
       to the elastic scattering of electrons by  H (2s), J. Phys. B.
       $\underline{13}$ (1980) 1451-56.

[17]   Edmunds, P., McDowell, M. R. C., Noble, C. and Morgan, L. A.,
       Non-exchange close coupling calculations for  $n - n^1$
       transitions in hydrogen  (n,n' $\leq$ 4), in preparation.

[18]   Slevin, J., Eminyan, M., Woolsey, J. M., Vassilev, G. and
       Porter, H. Q., Electron-photon angular correlation measure-
       ments for excitation of the  2p state of hydrogen at 35 eV,
       J. Phys. B. $\underline{14}$ (1981) in press.

[19]   Bhadra, K., Callaway, J. and Henry, R. J. W., Electron-impact
       excitation of the  n = 2  levels of helium at intermediate
       energies, Phys. Rev. A $\underline{19}$ (1979) 1841-57.

[20]   Fon, W. C., Berrington, K. A. and Kingston, A. E., The
       $1^1S \rightarrow 2^1S$  and  $1^1S \rightarrow 2^1P$  excitation of helium by electron
       impact, J. Phys. B. $\underline{13}$ (1980) 2309-25.

[21]   Willis, S. L. and McDowell, M. R. C., Electron impact
       excitation of helium at intermediate energies, J. Phys. B. $\underline{14}$
       (1981) in press.

[22]   Jansen, R. H. J., de Heer, F. J., Luyken, H. J.,
       van Wingerden, B. and Blaauw, H. J., Absolute differential
       cross sections for elastic scattering of electrons by helium,
       neon, argon and molecular nitrogen, J. Phys. B. $\underline{9}$ (1976)
       185-21.

[23]   Register, D. F., Trajmar, S. and Srivastava, S. K., Absolute
       elastic differential cross sections for  He : a proposed
       calibration standard from 5 to 200 eV, Phys. Rev. A $\underline{21}$ (1980)
       1134-51.

[24]   Scott, T. and McDowell, M. R. C., Electron impact excitation
       of the  $n^1S$  and  $n^3S$  states of  He  at intermediate energies,
       J. Phys. B. $\underline{8}$ (1975) 1851-65.

[25]   Scott, T. and McDowell, M. R. C., Electron impact excitation of
       the  $n^1P$  (n = 2 - 5)  and  $2^3P$  states of helium at intermediate
       energies, J. Phys. B. $\underline{9}$ (1976) 2235-54.

[26]   Erhardt, H., Hesselbucher, H., Jung, K. and Willman, K.,
       Differential cross sections in electron impact ionization, in
       Case Studies in Atomic Collision Physics 2 (Ed. McDaniel, E. W.
       and McDowell, M. R. C., North-Holland, Amsterdam, 1972) Ch.3,
       161-210.

[27]   Byron, F. W. Jr., Joachain, C. J. and Piraux, B., Triple
       differential cross sections for the ionization of atomic
       hydrogen by fast electrons:  a second Born treatment,
       J. Phys. B. $\underline{13}$ (1980) L673-76.

[28]   de Heer, F. J. and Jansen, R. H. J., Total cross sections for
       scattering by  He,  J. Phys. B. $\underline{10}$ (1977) 3741-58.

[29] Westerveld, W. B., Heideman, H. G. M. and van Eck, J., Electron impact excitation of $1^1S \rightarrow 2^1P$ and $1^1S \rightarrow 3^1P$ of helium: excitation cross sections and polarization fractions obtained from XUV radiation, J. Phys. B. 12 (1979) 115-35.

[30] Donaldson, F. G., Hender, M. A. and McConkey, J. W., Vacuum ultraviolet measurements of the electron impact excitation of helium, J. Phys. B. 5 (1972) 1192-210.

[31] Van Zyl, B., Dunn, G. H., Chamberlain, G. and Heddle, D. W. O., Benchmark cross sections for electron-impact excitation of $n^1S$ levels of He, Phys. Rev. A 22 (1980) 1916-29.

PHYSICS OF ELECTRONIC AND ATOMIC COLLISIONS
S. Datz (editor)

# THEORY OF LOW-ENERGY ELECTRON-ATOM SCATTERING

## R. K. Nesbet[*]

IBM Research Laboratory
San Jose, California 95193
U.S.A.

Methodology and applications of the quantitative theory of low-energy electron-atom scattering are reviewed. Calculations of absolute cross sections have provided fundamental data for rate processes and for calibration of experiments. Calculations of resonance and threshold structures have become essential to the interpretation of experimental data.

## INTRODUCTION

The quantitative theory of electron-atom scattering has achieved a new level of maturity during the past decade. In this period, developments of methodology have made it possible to incorporate most of the relevant effects of electronic interaction into viable computational procedures. Although formal difficulties remain in the intermediate energy range (above the ionization threshold of the target atom) it appears that reliable quantitative results will be obtained in the bound excitation region, by evolutionary refinement of present methods.

The fundamental concepts of atomic physics have been understood since the discovery of quantum theory. As in other branches of physics, successive improvements in experimental technique have revealed a great complexity of detailed phenomena. While these phenomena are consequences of the fundamental concepts, the chain of logical connection can no longer be followed trivially. Sophisticated quantitative theory is required in order to draw valid qualitative conclusions from the fundamental theoretical postulates.

Unless this logical connection is established, it can be difficult to distinguish valid experimental data from subtle artifacts, or to study elementary processes that occur in complex environments. Examples to be described here indicate that the methodology of electron-atom scattering theory has reached the point where it can begin to fill this essential role as an adjunct to current experimental research.

In addition to providing a necessary confirmation of experimental results, quantitative theory in some cases can go beyond current experimental capabilities. Published results of theoretical calculations for low energy electon scattering include predictions of structural cross section features that have not yet been studied experimentally or that are beyond the current limits of energy or angular resolution. Recent calculations have established a quantitative standard absolute differential cross section that can be used for experimental calibration.

Selected examples of these results will be discussed here. More details can be found in recent publications [1,2,3]. Theoretical results for electron-impact excitation of positive ions, which will not be discussed in detail here, have recently been reviewed by Henry [4]. Theory and experiment for electron-atom scattering resonances have been reviewed by Golden [5].

## QUANTITATIVE METHODS

Because of the general absence of restrictive selection rules at low collision energies, and because low-energy cross sections are large, electron-impact excitation of atoms is an important elementary process in many applications. Multichannel theory is required because several states of the target atom must be taken into account. For atoms beyond hydrogen, accurate N-electron variational target wave functions are required.

For a neutral target atom in a spherically symmetric state, low energy electron scattering is dominated by the electric dipole polarization potential. This is a dynamical correlation effect, not described in the $N+1$-electron static exchange (Hartree-Fock) approximation. One of the principal advances of quantitative theory in the past decade has been the implementation of methods for incorporating the polarization potential into multichannel scattering formalism.

The polarization potential results from mutual polarization of the target atom and the external electron. For a single scattering channel, the implied effective potential can be modeled by the polarized orbital method [6,7]. For multichannel scattering, Green's function theory [8] provides useful approximations to the nonlocal matrix potential that describes target polarization response to an external electron. In the close-coupling formalism, the polarization response can be treated by including closed scattering channels that correspond to polarization pseudostates, computed variationally to give accurate values of target state polarizabilities [9,10,11].

In order to provide an accurate treatment of the very strong polarizabilities of alkali atoms, Mittleman [12] proposed to treat correlation between the series and external electron explicitly by solving a continuum Bethe-Goldstone equation. A hierarchy of n-particle continuum Bethe-Goldstone equations can be defined in analogy to methods that have been successful in bound-state calculations [13]. To implement this structure computationally, the $N+1$-electron Hamiltonian matrix is partitioned between bound and continuum variational basis functions. Exact expressions for the coefficients of the bound components are substituted into the equations for the continuum components. The resulting terms in these equations define a generalized matrix optical potential [3]. When certain approximations are made, these terms reduce to the effective potentials of the polarized orbital or Green's function methods.

When the radial coordinate r of the external electron becomes large, the matrix optical potential is equivalent to a multichannel local potential function whose components vary as inverse powers of r. These potentials define a system of ordinary differential equations that can readily be solved by direct numerical integration. The full integrodifferential close-coupling equations are required only for smaller values of r.

In the R-matrix method, as implemented for electron-atom scattering [14], the $N+1$-electron wave function is expanded in a set of basis functions only inside some finite channel radius $r_0$. The R-matrix, which connects orbital wave function values and their radial derivatives, is computed at $r_0$ and matched to external wave functions obtained by integrating the asymptotic close-coupling equations. This approach makes it possible to use bound state methods when nonlocal exchange and optical potentials are important, while taking advantage of the simplified form of the asymptotic equations.

In the matrix variational method [3,15], bound state computational techniques are used over the full range of r. The asymptotic wave functions are expanded in a basis set of continuum orbital functions. In the method of numerical asymptotic functions [16,17], the continuum basis functions are taken to the numerical solutions of the asymptotic equations. This combines computationally favorable aspects of the R-matrix and matrix variational methods. It has recently been shown that these two methods become almost computationally identical when a suitable common choice of bound basis functions is made [18].

Improved direct methods of solution have also been developed for the integrodifferential close-coupling equations, converted to equivalent integral or algebraic equations before solving. Applications of such methods, especially to electron-ion scattering, have been reviewed by Henry [4].

## ABSOLUTE CROSS SECTIONS

Accurate phase shifts for e⁻-H elastic scattering, which is a two-electron problem, can be computed for the leading partial waves ($\ell \leq 3$) by variational methods. For low energy scattering, higher order phase shifts are given to sufficient accuracy by the partial wave Born formula, which depends only on the atomic polarizability and on $\ell$. The original s-wave calculations by Schwartz [19], who used relative coordinates in a variational trial function, were followed by other calculations of higher-order phase shifts [20,21,22,23,24]. At least two independent variational calculations have been carried out for each phase shift with $\ell \leq 2$.

Absolute measurements of the e⁻-H differential elastic cross section for incident energies 0.5 to 8.7 eV were made by Williams [25]. The experimental error was given as ±6%. The data agreed within error limits with theoretical models, variational, polarized orbital [7], or close-coupling with polarization pseudostates [10], that take into account the full polarizability of ground-state hydrogen.

In an interesting recent development, optical potential methods designed for the intermediate energy range have been extended to lower energies. Calculations by Scott and Brandsden [26], using a second-order optical potential built from intermediate pseudostates, agree roughly within experimental error with the e⁻-H differential elastic cross section measured by Williams [25].

Because helium is chemically inert and because the e⁻-He cross section is free of resonance and excitation structure up to 19.3 eV, this cross section can provide a reference standard for other measurements. Recent progress in both experiment and theory have combined to realize such a standard.

Until recently, there has been a large unexplained discrepancy between different experimental measurements. This situation was clarified by measurements of the relative differential cross section, to 5% accuracy [27]. Absolute values of the total and momentum transfer cross section were deduced by a phase shift analysis. From this work it appeared that beam scattering data agreed with earlier swarm data [28] within estimated error limits. Recent total cross section measurements [29,30,31,32] confirm this result and are consistent with the narrow error limits assigned to the swarm data [28], which have been extended up to 12 eV [33].

Although the He ground state wave function cannot be expressed in closed form, variational approximations can be highly accurate, yet simple enough to be used in scattering calculations. The first electron scattering calculation to include both target atom electronic correlation and quadrupole polarizability was reported by O'Malley et. al. [34], who used the R-matrix method. This work was verified by variational calculations designed to achieve 1% accuracy for the differential cross section below 19 eV [35].

Figure 1 shows a summary of computed and observed values of the e⁻-He total elastic and momentum transfer cross sections ($\sigma_T$ and $\sigma_M$, respectively) below 19 eV. The calculations do not include the narrow resonance at 19.37 eV. The full curves are obtained by a spline fit to estimates based on calculated points [35]. The points labeled OBB are the calculated values and best estimates, taking residual errors into account, of O'Malley *et al.* [34]. SN refers to an earlier variational calculation [33] which did not include target atom correlation or the quadrupole polarization potential. Beam experimental data points of Kauppila *et al.* [29] and of Andrick and Bitsch [27] are shown, as well as swarm data of Crompton *et al.* [28,33]. The agreement

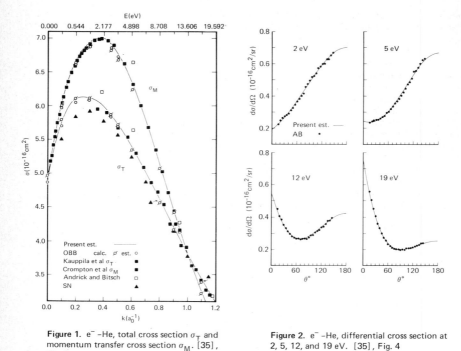

**Figure 1.** $e^-$ –He, total cross section $\sigma_T$ and momentum transfer cross section $\sigma_M$. [35], Fig. 3

**Figure 2.** $e^-$ –He, differential cross section at 2, 5, 12, and 19 eV. [35], Fig. 4

between experiment and theory is within the combined error limits, and is compatible with the 1% limit assigned to the variational calculations.

A comparison with directly measured angular distribution data [27] is shown in Figure 2. The data points shown are scaled by the ratio between computed $\sigma_T$ and the values used by Andrick and Bitsch to normalize their data. The full curves are theoretical estimates extrapolated from the variational calculations [35]. The agreement is excellent for all angles and energies. The comparison, shown in Figure 1, with $\sigma_T$ deduced by Andrick and Bitsch from their angular distribution data is much less satisfactory. This can be attributed to inaccuracy inherent in phase shift analysis of the experimental data, as discussed by Steph *et al.* [36].

Calculations of comparable accuracy have not yet been carried out for other rare gas target atoms. For $e^-$-Ne scattering at low energies, drift-velocity data have been fitted by use of an effective range formula for $\eta_0$, the s-wave phase shift [37]. This gives phase shifts up to 2 eV and a well-determined value of the scattering length, $0.215 \pm 0.005$ $a_0$. Values of $\eta_0$ computed by Thompson [38] straddle the fitted values. More accurate calculations are needed to give definitive results.

For heavier rare gas target atoms, the observed Ramsauer minima are of interest. In the case of argon, there has been a large discrepancy between different experimental measurements near the Ramsauer minimum. The momentum transfer cross section has recently been redetermined from mobility and drift velocity data [39]. The implied cross section appears to be more accurate than previous results, because it reproduces transport coefficient values much more accurately [39]. Calculations by Thompson [38] and by Garbaty and La Bahn [40], in the adiabatic exchange approximation, do not include enough computed points to determine the shape of Ramsauer minimum. Lack of uniqueness in fixing this shape limits the accuracy of the analysis of swarm data.

Theoretical calculations of e⁻-H inelastic scattering have been reviewed by Callaway [2]. Variational calculations, in the algebraic close-coupling formalism, treated 1s, 2s, and 2p states as open channels, and included a 1s→p polarization function [41,42,43]. Variational calculations for low partial wave states (total L≤3) were augmented by DWPO (distorted wave polarized orbital) calculations for high L-values. Figure 3 shows some results of these calculations, for the 1s→2s excitation cross section. Resonances calculated below the n=3 threshold are indicated. Experimental data points [44] are shown, with typical error bars. Agreement between theory and experiment is quantitative except for details of the resonance structure.

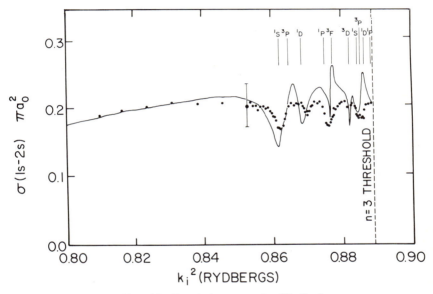

**Figure 3.** e⁻ –H, total 1s→2s excitation cross section. [2], Fig. 4

Theoretical calculations of electron scattering by alkali metal atoms are reviewed in reference [3]. The near degeneracy of valence states ns and np produces very large ground state polarizabilities. This causes the polarized orbital method (based on perturbation treatment of polarization) to be inadequate unless reformulated for strong interactions, but simplifies close-coupling calculations, because np acts as a polarization pseudostate. Very similar results are obtained with two-state close-coupling and with the matrix variational method, applied to the two-electron continuum Bethe-Goldstone equation for valence and external electrons. For Na and K, theoretical and experimental cross sections, including spin-exchange, are in good agreement, although the experimental data do not extend to low enough energies to probe the low-lying $^3P^o$ resonance peak predicted by theory.

For e⁻-Li scattering, definitive experimental data have not been available prior to recent measurements of the total cross section between 2 and 10 eV [45], using the atomic recoil technique. These data are in good agreement (10% or better) with the sum of the close-coupling excitation cross section [46] and the elastic cross section computed by a modified polarized orbital method [47]. In this modified method, the polarized orbital function is constrained to give both the correct target electron affinity and correct polarizability.

In another recent experiment, the relative differential cross section for 2s→2p excitation of Li has been measured in the range 10-200 eV, and then normalized to an absolute scale by use of the optical oscillator strength [48]. This requires extrapolation of the generalized oscillator strength to zero momentum transfer (forward scattering). The lower end of the energy range studied overlaps the range of total cross section measurements in reference [45]. It was found that the generalized oscillator strength, when fitted as a function of even powers of momentum transfer, was compatible with the theoretical cross section computed in an approximation expected to be valid at low scattering angles. This was not true for the corresponding fit in odd powers of momentum transfer.

Recent algebraic close-coupling calculations, by Wakid and Callaway [49], of the 1s→2s excitation cross section of $He^+$ agree with the earlier quantitative calculations [17,50] and confirm a substantial disagreement with experimental data [51] in the energy range 40-60 eV.

Hata et al. [52] computed excitation cross sections between n=1, 2, and 3 levels of hydrogen, using the multichannel numerical asymptotic function method of Morgan [17]. For n=2→n=3 transitions, a considerable discrepancy was found in comparison with earlier six-state close-coupling calculations [53], which had omitted certain exchange terms. The new results agree well with six-state calculations that include these terms [52]. In consequence of this work, a semiempirical formula of Johnson [54] for the 2→3 excitation cross section must be reduced by approximately 50%. This formula has been used to derive rate coefficients important for plasma diagnostics.

## RESONANCE AND THRESHOLD EFFECTS

Theoretical studies of resonances in $e^-$-H scattering have been reviewed recently, in references [2] and [3]. The degeneracy of nl states of hydrogen produces an effective dipole potential. Such a potential would support an infinite series of resonances at each successive threshold. Because these resonances become progressively narrower as they approach the threshold, only a few have been observed and characterized experimentally. Some idea of computed and observed resonance structure, at the n=3 threshold, is indicated in Figure 3.

The study of resonances and threshold effects in $e^-$-He scattering has been particularly rewarding. Accurate theoretical and experimental work is feasible. The observed structures are sufficiently complex that some points of controversy over interpretation still remain, even in the n=2 excitation region.

The $^2$S $He^-$ resonance at 19.37 eV, below the first excitation threshold, has been the subject of numerous theoretical and experimental studies. This was the first narrow resonance to be observed in electron-atom scattering [55]. The resonance profile has recently been measured by Kennerly et al. [56]. This is the first work in which the experimental energy resolution, 5 meV, is less than the resonance width. From differential cross section measurements at 90° and 135°, the background s and p phase shifts were found to be 1.813±0.017 and 0.309±0.013 radians, respectively, and the resonance width to be 11.0±0.5 meV. Comparing with earlier results, summarized in reference [56], these background phase shifts agree within combined error limits with recent measurements by Williams [31] and with accurate variational calculations [35]. The resonance width is significantly larger than several previous experimental values, which appeared to indicate a true value near 9 meV. The larger width is in good agreement with recent theoretical calculations [57,58,59]. Calculations comparable to those of references [34] and [35], including target atom electronic correlation as well as dipole and quadrupole polarization response, have not yet been carried out.

In the n=2 excitation region, e⁻-He scattering exhibits several structural features associated with resonances, threshold effects, and a virtual state. Theory and experiment in this region are reviewed in reference [3]. These energy-dependent cross section features can be understood in terms of the sums of eigenphases for the three lowest total symmetry states $^2$S, $^2$P⁰, and $^2$D. Earlier variational calculations by Oberoi and Nesbet [60], in the frozen-core Bethe-Goldstone approximation, are currently being repeated with larger variational basis sets in an improved variational formalism. Preliminary results for the eigenphase sums are shown in Figure 4, which agrees in its principal qualitative features with Figure 1 of reference [60]. The eigenphase sums are plotted against electron momentum k in the $2^3$S channel.

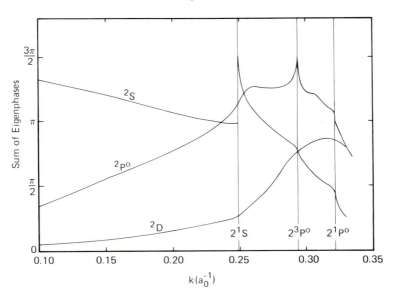

**Figure 4.** e⁻–He, sums of eigenphases for $^2$S, $^2$P⁰, and $^2$D scattering states.

The $^2$S eigenphase sum descends from the $2^3$S threshold, following its rapid rise through $\pi$ radians at the narrow $^2$S resonance just below this threshold. When applied to this two-channel example, formal analysis of multichannel threshold structures shows that the resonance below threshold should be associated with a threshold cusp structure, appearing as a descending rounded step in the elastic cross section, and with a relatively broad threshold excitation peak [61]. The rounded step has been observed directly [62]. As shown in R-matrix calculations by Sinfailam [63], the threshold excitation peak observed in $2^3$S excitation at 90° is produced by interference between $^2$S and $^2$D components of the scattering wave function.

At the $2^1$S threshold, the $^2$S eigenphase sum rises rapidly through nearly $\pi/2$ radians. This behavior is characteristic of a virtual state. Three-channel threshold analysis [61] associates this virtual state with a narrow threshold peak in the $2^1$S excitation channel, and with cusp and rounded step structures in the background channels. More detailed calculations of this threshold excitation peak [64,65] indicate that its width is less than 1 meV. For comparison with experiment, the ratio R($\Delta$E) of the integrals of the $2^1$S and $2^3$S excitation cross sections up to $\Delta$E was computed as a function of the energy $\Delta$E above either threshold [65]. This ratio can be compared with the relative signal measured in trapped-electron experiments. Recent measurements of this ratio by Spence [66], for $\Delta E$ between 3 and 40 meV, show quantitative agreement with the theoretical curve.

The $^2P^o$ eigenphase sum, as shown in Figure 4, rises through $\pi$ radians in a broad resonance below the $2^1S$ threshold, then exhibits a strong Wigner cusp structure at the $2^3P^o$ threshold and a rounded step at the $2^1P^o$ threshold. The $^2S$ eigenphase sum is strongly perturbed at both $2^3P^o$ and $2^1P^o$ thresholds, but these are not true cusp structures. The $^2D$ eigenphase sum exhibits a broad resonance with overlaps the $2^3P^o$ threshold.

In general, these structural features correspond to transient excited valence states of He$^-$ [3]. Configurations $(1s2s^2)^2S$ and $(1s2p^2)^2S$ mix strongly to produce the narrow resonance at 19.37 eV. The upper state of this pair has not been identified, but may be associated with the virtual state at the $2^1S$ threshold. Of the two states described by $(1s2s2p)^2P^o$, one corresponds to the broad $^2P^o$ resonance betwen the $2^3S$ and $2^1S$ thresholds, and the other has not been identified. A second $^2P^o$ resonance in this energy range would not be compatible with the eigenphase sum shown in Figure 4, which rises through only approximately $\pi$ radians except for the prominent cusp at the $2^3P^o$ threshold. The broad $^2D$ resonance corresponds to the state $(1s2p^2)^2D$ of He$^-$.

Because of interference between different partial-wave components, the differential cross sections that result from the energy dependent features shown in Figure 4 can be very complex. Differential cross sections were computed in a five-state representation, using the R-matrix method [67]. Comparison with absolute differential cross section measurements [68] shows detailed agreement in energy location and shape for the principal structural features. This agreement appears to be quantitative except at $90°$ and above the $2^3P^o$ threshold, where the computed $2^3S$ and $2^1S$ excitation cross sections are larger than those observed. The most prominent features are the broad $^2P^o$ and $^2D$ resonances.

Phillips and Wong [69] have recently reported new absolute measurements of the e$^-$-He n=2 excitation cross sections. Some discrepancies remain between these results and two other absolute measurements [68,70]. In particular, the maximum of the $2^3S$ excitation peak at $90°$ varies by a factor of two between different experiments. Definitive results are needed so that this cross section can be used as a reference standard for other threshold excitation structures. New variational calculations, comparable to references [34] and [35], are needed to resolve this issue.

Phillips and Wong find a threshold peak or abrupt rise of the $2^3P^o$ excitation cross section at threshold. The observed structure varies with scattering angle. In contrast, the $2^1P^o$ excitation cross section shows a gradual onset at all angles. The $2^3S$ excitation cross section shows strong angular dependence between the $2^1S$ and $2^3P^o$ thresholds, at 20.61 and 20.96 eV, respectively. Phillips and Wong suggest that this might arise from a second $^2P^o$ resonance near 20.8 eV, which would interfere with the $^2D$ resonance. From Figure 4, the rise of the $^2P^o$ eigenphase sum is not compatible with a second resonance in this energy range. A more plausible explanation of the experimental data might be that the strong $2^3P^o$ cusp structure interferes both with the $^2D$ resonance and with the relatively rapidly varying $^2S$ component of the scattering wave function. The cusp structure predicted by the theoretical calculations shows up prominently in the observed $2^3S$ excitation cross section, especially at $55°$, where the direct $^2D$ outgoing wave does not contribute.

No structure is observed in e$^-$-He scattering between the $2^1P^o$ threshold and 22.4 eV, just below the first n=3 state, $3^3S$. Figure 5 shows structural features observed with energy resolution 15 meV in the metastable excitation cross section $(2^3S+2^1S)$ by Brunt et al. [71], in the n=3 excitation region. Matrix variational calculations that included n=3 basis orbitals were carried out [72], extending the calculations of reference [60] into this region. Total symmetry states $^2S$, $^2P^o$, and $^2D$ were included. The computed total cross section for excitation of the $2^3S$ and $2^1S$

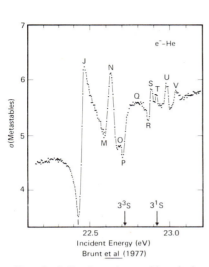

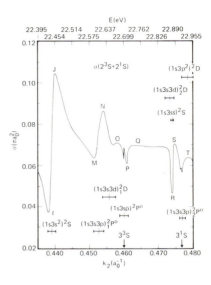

**Figure 5.** e⁻–He, observed metastable excitation cross section. [71]

**Figure 6.** e⁻–He, computed $2^3S+2^1S$ excitation cross section. [72], Fig. 1

states is shown in Figure 6, plotted against $k_2$, electronic momentum in the $2^3S$ channel. Since $^2F^o$ and higher scattering states are omitted, a background rising with energy but not expected to have sharp structural features should be added to the theoretical cross section.

Comparison of Figures 5 and 6 indicates close correspondence between theory and experiment for all structural features that are experimentally resolved. The theoretical calculations also show fine details at the $3^3S$ and $3^1S$ thresholds that are not apparent in the experimental data. The $3^3S$ threshold structure shows inverted cusps in both the $2^3S$ and $2^1S$ excitation cross sections. This is accompanied by a rapid initial rise of the $^2S$ eigenphase sum, characteristic of a virtual state, and by a narrow threshold excitation peak. There is an excitation peak also at the $3^1S$ threshold, associated not with a virtual state but with a narrow $^2S$ resonance just below threshold.

A systematic search for resonances was carried out and dominant He⁻ configurations were identified in each case. The results are shown in Figure 6. The computed resonance positions and widths are shown graphically by small vertical and horizontal bars, respectively. The general agreement between theory and experiment is very good. In the particular case of the resonance labeled NO, it has been shown by Andrick [73], from measurements of the $2^3S$ differential excitation cross section, that the width is 43-50 meV, in agreement with the theoretical width, calculated to be 44.7 meV, rather than the width, 20 meV, deduced by Brunt *et al.* from their data. Because this $^2D$ resonance overlaps the adjacent $^2P^o$ resonance, it is inherently difficult to determine the width from a total cross section measurement.

Two classes of resonances are found. For the first, exemplified by $(1s3s^2)^2S$, two outer electrons, both in the n=3 shell, are temporarily bound in the Coulomb field of the He⁺ core. These are excited valence shell (VS) resonances, analogous in structure to the n=2 resonance states. A second class, nonvalence (NV) resonances such as $(1s3s\bar{p})^2P^o$ and $(1s3s\bar{s})^2S$, lie close to excitation thresholds and are not associated with n=3 valence-shell configurations of He⁻. These states are presumably formed by attachment of an electron in the strong polarization potential of the parent state.

## OPEN-SHELL TARGET ATOMS

Although low energy electron scattering from C, N, and O atoms is important in atmospheric physics and in astrophysics, very little reliable laboratory data is available. Earlier experimental data was reviewed by Bederson and Kieffer [74] and theoretical results are reviewed in reference [3].

In the case of $e^-$-O scattering, there are no resonances below the $n=3$ excitation region, because the ground configuration $1s^2 2s^2 2p^5$ of $O^-$ produces only a single state, the $^2P^o$ bound state. However, the absolute value of the low-energy cross section is of considerable interest. For theoretical calculations, it is essential to represent target state polarizability and short-range correlation effects accurately. Several calculations have been carried out, using methods that describe these effects.

Thomas and Nesbet [75] reported variational calculations in the valence shell Bethe-Goldstone approximation. The computed total cross section was found to be in close agreement with earlier polarized orbital results [76]. Tambe and Henry [77] reported close-coupling calculations that included polarization pseudostates but omitted explicit short-range correlation terms. Calculations of similar structure were carried out by Le Dourneuf [78] using the R-matrix method. Differences between these computed results are most noticeable in the energy range 2-6 eV, where the Bethe-Goldstone cross section is consistently lower than the pseudostate calculations. The computed differential cross sections at 5 eV differ significantly in the forward direction. Available experimental data do not resolve these discrepancies.

The ground electronic configurations of carbon and nitrogen negative ions each contain several states. Quantitative calculations indicate that all of these valence states exist either as true bound states, as metastables, or as electron scattering resonances. A strong $^2P^o$ $e^-$-C scattering resonance is predicted in the energy range 0.4 to 0.6 eV [78,79], but there are no experimental results for comparison.

In the case of $N^-$, theoretical calculations [78,80,81] indicate that the lowest state, $^3P$, should be a narrow resonance, approximately 0.10 eV above threshold. The excited $^1D$ and $^1S$ states of the ground configuration should be metastable. Experiments on dissociative attachment of $N_2$ and NO [82,83] show that as $N^-$ separates from the molecule it goes to an autodetaching state, whose energy and width agree with the predicted low-lying $^3P$ resonance. Long-lived $N^-$ has been observed in a direct experiment [84]. The detected species is identified only by mass and charge. It may correspond to one or both of the expected metastable excited states.

Shape resonances are observed in low energy electron scattering from Mg, Zn, Cd, and Hg [85]. These can be interpreted as $(ns^2 np)^2P^o$ valence states of the negative ions. Theoretical calculations, including a simplified polarization potential and using Dirac wave functions for relativistic effects have been carried out for Hg [86] and for Be, Mg, Zn, and Cd [87]. These calculations verify identification of the resonances, but the results are not yet quantitative.

## REFERENCES

*Work supported in part by the U.S. Office of Naval Research.

[1]     Burke, P. G. and Williams, J. F., Phys. Reports 34 (1977) 325.

[2]     Callaway, J., Phys. Reports 45 (1978) 89.

[3]     Nesbet, R. K., Variational methods in electron-atom scattering theory (Plenum, New York, 1980).

[4]   Henry, R. J. W., Phys. Reports 68 (1981) 1.

[5]   Golden, D. E., Adv. At. Mol. Phys. 14 (1978) 1.

[6]   Temkin, A., Phys. Rev. 107 (1957) 1004.

[7]   Drachman, R. J. and Temkin, A., in: McDaniel, E. W. and McDowell, M. R. C. (eds.), Case studies in atomic collision physics, vol. 2 (North-Holland, Amsterdam, 1972) 399.

[8]   Csanak, Gy., Taylor, H. S., and Yaris, R., Adv. At. Mol. Phys. 7 (1971) 287.

[9]   Damburg, R. J. and Karule, E., Proc. Phys. Soc. (London) 90 (1967) 637.

[10]  Burke, P. G., Gallaher, D. F., and Geltman, S., J. Phys. B 2 (1969) 1142.

[11]  Vo Ky Lan, Le Dourneuf, M., and Burke, P. G., J. Phys. B 9 (1976) 1065.

[12]  Mittleman, M. H., Phys. Rev. 147 (1966) 69.

[13]  Nesbet, R. K., Phys. Rev. 156 (1967) 99.

[14]  Burke, P. G. and Robb, W. D., Adv. At. Mol. Phys. 11 (1975) 143.

[15]  Lyons, J. D., Nesbet, R. K., Rankin, C. C., and Yates, A. C., J. Comput. Phys., 13 (1973) 229.

[16]  Oberoi, R. S. and Nesbet, R. K., J. Comput. Phys. 12 (1973) 526.

[17]  Morgan, L. A., J. Phys. B 13 (1980) 3703.

[18]  Nesbet, R. K., J. Phys. B 14 (1981) Lxxx; Phys. Rev. A 24 (1981) xxx.

[19]  Schwartz, C., Phys. Rev. 124 (1961) 1468.

[20]  Armstead, R. L., Phys. Rev. 171 (1968) 91.

[21]  Gailitis, M., in Abstracts, IV ICPEAC (Science Bookcrafters, Hastings-on-Hudson, New York, 1965) 10.

[22]  Shimamura, I., J. Phys. Soc. Japan 30 (1971) 1702.

[23]  Register, D. and Poe, R. T., Phys. Lett. 51A (1975) 431.

[24]  Callaway, J., Phys. Lett. 65A (1978) 199.

[25]  Williams, J. F., J. Phys. B 8 (1975) 1683.

[26]  Scott, T. and Brandsden, B. H., to be published (1981).

[27]  Andrick, D. and Bitsch, A., J. Phys. B 8 (1975) 393; Adv. At. Mol. Phys. 9 (1973) 207.

[28]  Crompton, R. W., Elford, M. T., and Jory, R. L., Aust. J. Phys. 20 (1967) 369; Crompton, R. W., Elford, M. T., and Robertson, A. G., ibid. 23 (1970) 667.

[29]  Kauppila, W. E., Stein, T. S., Jesion, G., Dababneh, M. S., and Pol, V., Rev. Sci. Instrum. 48 (1977) 822;
Stein, T. S., Kauppila, W. E., Pol, V., Smart, J. H., and Jesion, G., Phys. Rev. A 17 (1978) 1600.

[30]  Kennerly, R. E. and Bonham, R. A., Phys. Rev. A 17 (1978) 1844.

[31]  Williams, J. F., J. Phys. B 12 (1979) 265.

[32]  Register, D. F., Trajmar, S., and Srivastava, S. K., Phys. Rev. A 21 (1980) 1134.

[33]  Milloy, H. B. and Crompton, R. W., Phys. Rev. A 15 (1977) 1847.

[34]  O'Malley, T. F., Burke, P. G., and Berrington, K. A., J. Phys. B 12 (1979) 953.

[35]  Nesbet, R. K., Phys. Rev. A 20 (1979) 58.

[36]  Steph, N. C., McDonald, L., and Golden, D. E., J. Phys. B 12 (1979) 1507.

[37]  O'Malley, T. F. and Crompton, R. W., J. Phys. B 13 (1980) 3451.

[38]  Thompson, D. G., Proc. Roy. Soc. (London) A 294 (1966) 160; J. Phys. B 4 (1971) 468.

[39]  Milloy, H. B., Crompton, R. W., Rees, J. A., and Robertson, A. G., Aust. J. Phys. 30 (1977) 61.

[40]  Garbaty, E. A. and La Bahn, R. W., Phys. Rev. A 4 (1971) 1425.

[41]  Callaway, J. and Wooten, J. W., Phys. Rev. A 9 (1974) 1924.

[42]  Callaway, J., McDowell, M. R. C., and Morgan, L. A., J. Phys. B 8 (1975) 2181; ibid. B 9 (1976) 2043.

[43]  Morgan, L. A., McDowell, M. R. C., and Callaway, J., J. Phys. B 10 (1977) 3297.

[44]  Williams, J. F., J. Phys. B 9 (1976) 1519.

[45]  Jaduszliwer, B., Tino, A., Bederson, B., and Miller, T. M., Phys. Rev. A 24 (1981) in the press.

[46]  Burke, P. G. and Taylor, A. J., J. Phys. B 2 (1969) 869;
Norcross, D. W., J. Phys. B 4 (1973) 1458.

[47]  Bhatia, A. K., Temkin, A., Silver, A., and Sullivan, E. C., Phys. Rev. A 18 (1978) 1935.

[48]  Vuskovic, L., Trajmar, S., and Register, D. F., J. Phys. B 14 (1981) in the press.

[49]  Wakid, S. A. and Callaway, J., J. Phys. B 13 (1980) L605.

[50]  Henry, R. J. W. and Matese, J. J., Phys. Rev. A 14 (1976) 1368.

[51]  Dolder, K. T. and Peart, B., J. Phys. B 6 (1973) 2415.

[52]  Hata, J., Morgan, L. A., and McDowell, M. R. C., J. Phys. B 13 (1980) L347; 4453.

[53]  Burke, P. G., Ormonde, S., and Whitaker, W., Proc. Phys. Soc. (London) 92 (1967) 319.

[54]  Johnson, L. C., Astrophys. J. 174 (1972) 227.

[55]  Schulz, G. J., Phys. Rev. Lett. 10 (1963) 104.

[56]  Kennerly, R. E., Van Brunt, R. J., and Gallagher, A. C., Phys. Rev. A 23 (1981) 2430.

[57]  Hazi, A. U., J. Phys. B 11 (1978) L259.

[58]  Junker, B. R. and Huang, C. L., Phys. Rev. A 18 (1978) 313;
Junker, B. R., Phys. Rev. A 18 (1978) 2437.

[59]  Foster, G., Hummer, D. G., and Norcross, D. W., Bull. Am. Phys. Soc. 24 (1979) 1183.

[60]  Oberoi, R. S. and Nesbet, R. K., Phys. Rev. A 8 (1973) 2969.

[61]  Nesbet, R. K., J. Phys. B 13 (1980) L193.

[62]  Cvejanovic, S., Comer, J., and Read, F. H., J. Phys. B 7 (1974) 468.

[63]  Sinfailam, A. L., J. Phys. B 9 (1976) L101.

[64]  Berrington, K. A., Burke, P. G., and Sinfailam, A. L., J. Phys. B 8 (1975) 1459.

[65]  Nesbet, R. K., Phys. Rev. A 12 (1975) 444.

[66]  Spence, D., J. Phys. B 13 (1980) L73.

[67]  Fon, W. C., Berrington, K. A., Burke, P. G., and Kingston, A. E., J. Phys. B 11 (1978) 325.

[68]  Pichou, F., Huetz, A., Joyez, G., Landau, M., and Mazeau, J., J. Phys. B 9 (1976) 933.

[69]  Phillips, J. M. and Wong, S. F., Phys. Rev. 23 (1981) 3324.

[70]  Seng, G. H., Ph.D. thesis, University of Kaiserslautern (1975).

[71]  Brunt, J. N. H., King, G. C., and Read, F. H., J. Phys. B 10 (1977) 433.

[72]  Nesbet, R. K., J. Phys. B 11 (1978) L21.

[73]  Andrick, D., J. Phys. B 12 (1979) L175.

[74]  Bederson, B. and Kieffer, L. J., Rev. Mod. Phys. 43 (1971) 601.

[75]  Thomas, L. D. and Nesbet, R. K., Phys. Rev. A 11 (1975) 170.

[76]  Henry, R. J. W., Phys. Rev. 162 (1967) 56.

[77]  Tambe, B. R. and Henry, R. J. W., Phys. Rev. A 13 (1976) 224.

[78]  Le Dourneuf, M., Ph.D. thesis, University of Paris VI (CNRS Report A012658, 1976).

[79]  Thomas, L. D. and Nesbet, R. K., Phys. Rev. A 12 (1975) 2378.

[80]  Thomas, L. D. and Nesbet, R. K., Phys. Rev. A 12 (1975) 2369.

[81]  Le Dourneuf, M. and Vo Ky Lan, J. Phys. B 10 (1977) L97.

[82]  Spence, D., Huebner, R. H., and Burrow, P. D., Bull. Am. Phys. Soc. 23 (1978) 143.

[83]  Mazeau, J., Gresteau, F., Hall, R. I., and Huetz, A., J. Phys. B 11 (1978) L557.

[84]  Hiraoka, H., Nesbet, R. K., and Welsh, L. W., Jr., Phys. Rev. Lett. 39 (1977) 130.

[85]  Burrow, P. D., Michejda, J. A., and Comer, J., J. Phys. B 9 (1976) 3225.

[86]  Sinfailam, L. T., J. Phys. B 13 (1980) L427.

[87]  Sinfailam, L. T., J. Phys. B 14 (1981) in the press.

PHYSICS OF ELECTRONIC AND ATOMIC COLLISIONS
S. Datz (editor)
© North-Holland Publishing Company, 1982

ELECTRON SCATTERING FROM ATOMIC OXYGEN

W.R. Newell

Department of Physics and Astronomy
University College London
Gower Street
London WC1E  6 BT.

INTRODUCTION :

   This paper presents a review of the measurements of  elastic
and (total and differential) electron scattering cross sections
for atomic oxygen.  Measurement and calculation of atomic oxygen
cross sections are important since oxygen is an major contributor
to the chemistry of the Earths upper atomosphere and is responsible
for observed emissions (eg 130.4nm and 135.6nm) in Planetary
Atmospheres where electron impact is probably an important
mechanisum in producing the excited atomic states.  Processes
involving O are also important in laboratory plasmas.  In addition
the $2p^4$ outer electron configuration poses a non-trivial theoretical problem in the construction of different scattering models
and the various approximations used in calculating atomic properties.  Although this article in primarly concerned with experimental effort and progress in e - O scattering the comparative
status of current calculations of e - O collision processes are
included where appropriate with the discussion of the experimental
results.

   Several recent reports on different aspects of electron atom
scattering have been published;  Callaway (1980) on theoretical
proceedures with the emphasis on the intermediate energy range,
Nesbet (1977) on low energy electron scattering and a comprehensive review by Bransden and McDowell (1978).  References to
earlier reviews are given in these works.  No detailed discussion
of reactive scattering, nor heavy particle excitation involving
atomic oxygen will be given.  However comparisions with electron
scattering processes will be made where appropriate.  Dissociative
processes and photo-excitation of O will also be excluded from
detailed discussion (see Dyke 1979  for  a recent review).
Studies of branching ratios and decay schemes in atomic oxygen
discharges will also not be considered;  some new emission studies
of this type have recently been reported by Christenson and
Cunningham (1978).

TOTAL CROSS SECTIONS :

   The first total elastic cross section for electron scattering
from O was made by Neynaber et al (1961) using a modulated
molecular beam apparatus.  The cross section was determined by
monitoring the reduction in the primary electron beam current
over the energy range 2.3eV to 11.6eV.  The experiment measured
the ratio of the molecular to the atomic cross section with the
dissociation fraction of the target gas being determined by a
mass spectrometer.  The data was made absolute by normalization

to the total molecular cross section of Brucke (1927) and a least
squares fit to this data yielded a cross section value of
$6.2 \pm 0.5 \Pi a_0^2$ over the entire energy range.

Total cross section measurements have also been made by
Sunshine et al (1967) using the atomic beam recoil method at 22
electron energies in the range 0.5 to 11.3eV and for higher
values of $n^2 eV$ $(n > 4)$ up to 100eV. In the experiment a modulated
electron beam crosses a partially dissociated (25%) $O_2$ beam and
the deflection of the atomic oxygen flux is monitered using an
ionizer and mass spectrometer located after the interaction region.
An absolute measurement of e + O total scattering cross section
is obtained by virtue of the experimental arrangement which allows
an absolute determination of the total cross section for e + $O_2$
to be made. Although the total elastic molecular data is 10%
to 20% higher than that of Brucke the shapes of the cross section
curves are in good agreement and hence by normalization the
e + O total cross section was determined. A variation in the total
cross section from 5.3 $\AA^2$ at 0.5eV to 8.3 $\AA^2$ at 11.3eV was recorded
with a fall off to approximately 6 $\AA^2$ at 100eV. Within the quoted
experimental errors there is reasonable agreement between the data
of Sunshine et al and Neynaber et al at energies < 6.2eV while the
data of Sunshine et al is consistently higher than that of Neynaber
et al for the energies 6.2eV to 11.3eV. A single data point of
Lin and Kivel (1959) obtained in a shock tube experiment is a
factor of $\sim$ 3 below that of Sunshine et al. The experimental data
is of insufficient quantity to be of use in distinguishing between
the different calculations (eg Thomas and Nesbet 1975, Henry 1967,
Rountree et al 1974 Saraph 1973, and Nesbet 1977).

EXCITATION CROSS SECTIONS:

Absolute cross sections for the excitation of OI 3s ($^3S^0$)
and OI 3s ($^5S^0$) states by electron impact on atomic oxygen have
been measured by Stone and Zipf (1971, 1974) for incident
electron energies from threshold to 300eV. Atomic oxygen was
produced in a microwave discharge and the emitted radiation was
observed at right angles to the electron beam with a 1-m normal
incidence vacuum monochromator. A solar blind photomultiplier
tube was operated in a pulse counting mode. An absolute determination of the OI ($^3P$) number density was made by performing an
additional absorption experiment on the resonance transition
OI ($^3P \rightarrow {}^3S$) at 130.4mm and using the known optical oscillator
strength measurements of Lawrence and Savage (1966), Lawrence (1970)
and Lin et al 1970 to normalize the result. Data obtained in
this experiment was accumulated as a difference signal by slowly
pulsing the discharge on and off with a period of 30s. Stone
and Zipf took particular care in considering the probability of
emission at the observed wavelengths (130.4nm and 135.6nm)
arising from excited species produced by the discharge. With the
discharge off the experiment was repeated to measure the
contribution of dissociative electron impact excitation of
oxygen molecules to the emission of the 130.2nm radiation. It
was not possible to discriminate against the contribution of
130.2nm emission from dissociative electron impact exciation
of a'$_{\Delta g}$ oxygen molecules or from cascade processes from excited
oxygen atoms. Hence they repeated the experiment (1974) and
removed the oxygen atoms using a sleeve of aluminium foil in the gas
flow system (this removed only the oxygen atoms, and not the
a'$_{\Delta g}$ molecules) and thus determined the contributions to the

130.2nm signal from the excited molecules. The absolute cross sections for the excitation of the OI ($^3$S$^O$) stater is given in Figure (1) and includes contributions from cascade transitions.

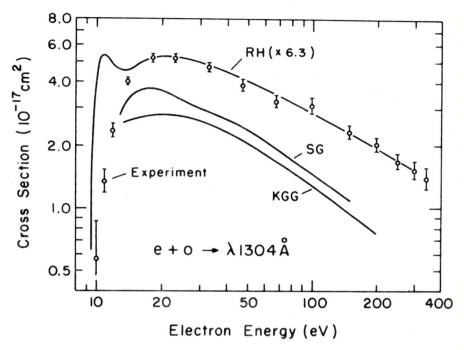

Rountree and Henry (RH), (1972) Swada and Ganas (SG), (1973) Kazaks et al (KGG), (1972)

Figure 1

The data, for the $^3$p → $^3$s excitation compares well in shape with the theoretical results of the total excitation cross sections for the $^3$p → $^3$s transition of Rountree and Henry (1972), the sum of the $^3$p → ($^3$S + $^3$P) of Kazaks et al (1972) and the curve of Sawada and Ganas (1973) which includes cascade contribution from the 3p ($^3$p$^O$) and 4p ($^3$p$^O$) states. However the absolute magnitudes of these theoretical results are consistently lower than the experimental measurements for all impact energies. The data of Rountree and Henry (non-exchange Born close-coupling calculation) are a factor of 6.3 smaller whilst the data of Kazaks et al and Sawada and Ganas are both smaller by approximately a factor of 1.75. Kazaks et al used the independent particle model of Green et al (1969) to compute approximate potentials and wave-functions for the oxygen atom which were in turn used to calculate the total cross sections using the first Born approximation. Sawada and Ganas extended the range of the Kazaks et al Born calculations to energies below 150eV by using the distorted

wave Born approximation. They also used extra electron configurations, of the excited states of atomic oxygen, in order to produce
better wave-functions for the target. This produced the only a
marginal improvement over the calculations of Kazaks et al,
when compared to the experiment data of Stone and Zipf (1974).
However, in general it is difficult to ascertain which (experiment
or theory) is nearest the truth since the experimental data
contains cascade contributions and all theoretical results are
based on a frozen-core approximation in which the active electron
is built on a $^4S^o$ configuration. If coupling to the other
configurations $^2P^o$ and $^2D^o$ is included it would probably raise the
values of the calculated cross section. Since the life time of
the OI ($^5S^o$) state is long i.e. $\sim$190µs (see for example Nicholaides
(1972) then the excited atoms will drift away from the centre of
the interaction region before they decay. Hence allowing for the
thermal drift of OI ($^5S^o$) atoms and assuming that none rebound from
the walls of the collision chamber then it is possible by comparing
the ratio of the 130.2nm and 135.6nm signals, measured under
identical experimental condition, to obtain the absolute cross
section for the OI ($^3P$ - $^5S$) excitation. Again the experimental
values exceed the theoretical cross section of Sawada and Ganas
(1973) by a factor of $\sim$2.


IONIZATION :

     Measurements of electron impact ionization of atomic oxygen
were made two decades ago by Fite and Brackmann (1959), Rothe et al
(1962) and Boksenberg (1961). Fite and Brackmann and Rothe et al
used a modulated crossed beam experimental arrangement and all
three groups produced atomic oxygen using an r.f. discharge with
typically 20% to 30% dissociation. The experimental proceedure
in each case was to measure the ratio of $O^+$ produced from O,
using a mass spectrometer to analyse the ionization products,
to the total molecular ionization cross section. This ratio
was then multiplied by an absolute molecular ionization cross
section (Tate and Smith 1932) to yield the atomic oxygen ionization
cross section. In order for these experiments to yield reliable
data there must be 100% ion collection efficiency and although
there is a certain doubt about this there is reasonable agreement
between the data of Fite and Brackmann and Rothe et al which is
lower than the data of Boksenberg (by a factor of 1.7). Keiffer
and Dunn (1966) have commented that this higher value of Boksenberg
is probably due to poor ion collection efficiency producing a
low value for the measured total ionization cross section of $O_2$.
Beam composition as a possible source of error was investigated
by Fite and Brackmann who estimated from appearance potentials
that 3% of the partially dissociated beam was comprised of excited
species.

     The first calculation of the atomic oxygen ionization cross
section was made by Seaton (1959) who used a scaling proceedure
and neon ionization data. Subsequently the Born approximation
was used in association with effective potential techniques by
McGuire (1971), Omidvar et al (1972) and Kazaks et al (1972).

     Brook et al 1978 preformed a detailed measurement of the
ionization cross section of O from threshold to 1keV. In this
experiment the O beam was prepared using a charge exchange method
with $O^+$ ions at 2keV and 4keV. Short lived excited species
decayed in the drift time, 3 - 5µs, between the charge exchange

cell and the interaction region and excited states with principle
quantum number n > 15 were field ionized. Metastable states and
states with 8 < n <12 remained in the beam and their contribution
to the $O^+$ signal was made by measuring the $O^+$ signal below 12eV.
This experiment yielded an absolute cross section without recourse
to secondary normalization and the data is presented in Figure (2)
with some calculations of the ionization cross section.

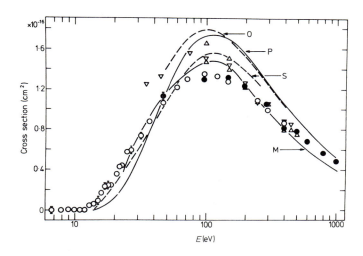

V Fite and Brackman (1959), Δ Rothe et al (1962), P Peach
(1970, 1971), O Omidvar et al (1972), M McGuire (1972),
S Seaton (1959) O and ● Brook et al 1978.

Figure 2

The cross section for the double ionization of O has been
reported by Ziegler et al 1981 using an experimental arrangement
similary to that of Brook et al. The $O^+$ cross section measurement,
was normalized to the $O^+$ cross section of Brook et al. In
addition there are the recent extensive calculations of Burnett
and Rountree (1979) who employed a six term configuration interaction wavefunction for the initial state and a close coupling
wavefunction for the final state. The interaction between the
product ion and the ejected electron was treated in detail in
this work by including the three states $^4S^o$ $^2D^o$ $^2P^o$ of the residual
ion in the close coupling expansion. There is good agreement
between between the experimental work of Fite and Brackmann and
Rothe et al and the calculation of Burnett and Rountree and
McGuire for electron energies greater than 100eV. Paradoxically
the data of Boksenberg, when rescaled, gives excellent agreement
with these calculations at all energies. To date there have been
no measurements of differential ionization cross sections of
atomic oxygen.

RESONANCE SCATTERING :

Feshback resonances have been observed by Spence and Chupka
(1974) and by Spence (1975) (using improved data collection
facilities) in atomic oxygen at energies in the range 8eV to 16eV.
The apparatus used was a transmission electron spectrometer of
the type described by Sanche and Burrow (1972) in which a trochoided
monochromator selected a well resolved electron beam ($\Delta E \sim$ 75 meV)
before it was accelerated into the collision region.  Electrons
which suffer a change in momentum are prevented from leaving the
collision region by a retarding potential.  In this experiment
the primary electron beam was modulated with a 20meV amplitude
sine  wave and the derivative of the transmitted current plotted.
In this work atomic oxygen was produced by a microwave discharge
but no estimation of the number density of O nor the dissociation
fraction was given.  It is also assumed that ions and electrons
produced in the discharge are trapped in the collision region
and do not contribute to the detected resonance signal.

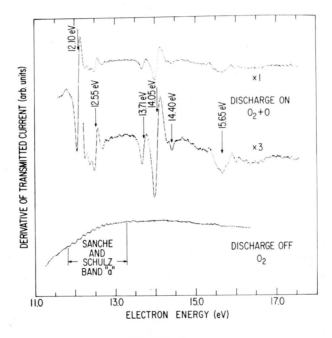

Figure 3

Figure (3) shows part of the resonant structure observed
by Spence in a partially dissociated $O_2$ beam.  The energy scale
calibration, accurate to ± 50meV, was accomplished using the
observed molecular   resonances and normalizing  the energy scale
to the molecular resonance work of Sanche and Schulz (1972).
Atomic oxygen resonances (marked by arrows) all exhibit widths
which are characteristic of the instrument, except the feature
at 15.65eV which appears to have a natural width greater than the
instrumental resolution.

All resonances observed in atomic oxygen are listed, with their assignments, in Table (1) and all the features except that

| Resonance configuration | Spence | Experimental Spence and Chupka | Edwards and Cunningham | Theory Matese |
|---|---|---|---|---|
| $(^4S)3s^2;\,^4s$ | 8.78 | 8.80 | | 8.68 |
| $(^4S)3s3p;\,^2P$ | | | 9.50 | 9.50 |
| $(2s2p^6)\,^2S$ | | 10.10 | 10.11 | |
| $(^4S)3p^2;\,^6P$ | 10.73 | | | 10.63 |
| $(^4S)3p^2;\,^2P$ | 10.90 | 10.90 | 10.87 | 10.88 |
| $(^2D)3s^2;\,^2D$ | 12.10 | 12.11 | 12.12 | 12.05 |
| $(^2D)3s3p$ | 12.55 | | | |
| $(^2D)3p^2;\,^4F$ | 14.05 | 14.06 | | 14.09 |
| $(^2P)3s^2;\,^2P$ | 13.71 | 13.71 | 13.71 | 13.65 |
| $(^2P)3s3p$ | 14.40 | | | |
| $(^2P)3p^2;\,^4D$ | 15.65 | | | 15.75 |

Table 1

at 10.10eV are associated with a doubly excited state of $O^-$. The four doublet resonances at 9.50, 10.87, 12.12, and 13.71eV have been observed by Edwards and Cunningham (1973) in an ion molecule experiment and the observed positions are in good agreement with the theoretical of work Matese (1974). However the feature at 9.50eV is not observed in the electron impact excitation. Its absence is attributed to obscuring effects of molecular spectra. The quartet resonance states are readily excited by electron impact but not by ion molecule collisions where the collision dynamics are such that charge exchange or spin dependent interactions are unlikely. The resonance feature at 10.11eV, initially reported by Edwards and Cunningham was also detected by Spence and Chupka at 10.11eV. This was attributed to electron scattering from the $O(^1D)$ state producing the $O^-$ $2s2p^6(^2S)$ single electron excitation resonance. That this resonance feature was not observed from the ground state $O\ (^3P)$ at $(10.10 + 1.97)$eV was attributed by Spence and Chupka to the rigidity of L-S coupling.

This feature is however absent in the later work of Spence where the microwave discharge source was 300mm further from the interaction region which could account for the removal of the $O(^1D)$ atoms from the beam as a result of wall collision along the quartz tube. The resonance feature at 15.65eV has a width 3 times that of the other nine resonances which complete the triad of resonance configurations $3p^2$ $3s3p$ $3p^2$ associated with each of the $^4S$, $^2D$ and $^2P$ positive ion grandparent states. The shape and nature of this feature was carefully checked to ensure that it was not an experimental artifact by repeating the scans at different magnetic field strengths.

Excellent agreement exists between the results of different experiments and the theoretically predicted resonance positions any discrepancy being greatest where the electron affinity is largest and this generally occurs when the orbital angular momentum of both electrons is zero. All of the observed resonances, except $^2D^O$ (3s3p) and $^2P^O$ (3s3p) were predicted using the method of configuration interaction with a frozen $O^+$ core ($^4S^O, ^2D^O$. $^2P^O$) and two electrons in excited orbitals. In addition resonances in which the excited electron has $n = 4$ have been predicted but not yet observed (Matese 1974).

DIFFERENTIAL SCATTERING :

(a)   Elastic Scattering

The first experiment to measure electron scattering from atomic oxygen was performed by Dehmel et al (1974) who measured the ratio of forward scattering ($20^O$ to $88^O$) to back scattering ($92^O$ to $160^O$) of electrons from atomic oxygen for impact energies of 3eV to 20eV using a cross beam experiment. An r.f. discharge source produced the O beam of which 47% was atomic oxygen as determined by mass analysis after the interaction region. No account was taken of the effects of $O_2$ ('$\Delta g$) molecules in the gas beam. A strong forward scattering of electrons at low energies was observed but this experiment suffered from poor signal to noise ratios and from an inherent difficulty in accounting accurately for the reflections of electrons scattered into the two Faraday detectors employed. However, the matrix variational calculations of Thomas and Nesbet (1975), integrated in the above angular ranges for forward and backward scattering, are in good agreement with the experimental measurements.

The first differential cross section measurements of electron scattering from atomic oxygen were also made by Dehmel et al (1976) at impact energies of 5eV and 15eV and scattering angles from $15^O$ to $150^O$. In this experiment the atomic oxygen was produced using a microwave source and the primary electron beam ($\Delta E = 120$meV) was monochromatized using a 127 degree cylindrical electrostatic analyser. The scattered electrons were detected using a moveable channel electron multiplier. There was no discrimination between elastically and inelastically scattered electrons in this experiment, nor was the effect of excited molecules taken into account. However the calculations of Thomas and Nesbet (1975) at 4.9eV and 11eV agree quite well with this experimental data at 5eV and 15eV respectively as do the polarised pseudostate R-matrix calculations of Tambe and Henry (1976). However, both sets of theoretical results fail to reproduce the experimentally observed dips in the differential cross sections at small angles, see Figure (4), nor produce the proper small angle slope. To date this is the only published data on elastic scattering of electron from atomic oxygen.

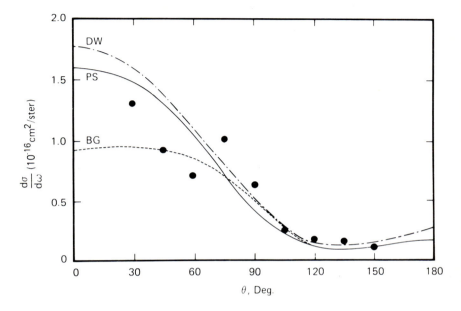

Distorted Wave (DW) Blaha and Davis (1975), Polarized Pseudostate )(PS) Tambe and Henry (1976), Bethe Goldstone (BG) approx, Thomas and Nesbet (1975), Experiment ● Dehmel et al (1976).


Figure 4


(b)   Inelastic Scattering

Inelastic electron scattering measurements from atomic oxygen have been reported by Khakoo et al (1979, 1980) who used a double hemispherical electron spectrometer ($\Delta E \sim$ 90meV) scattering apparatus and a microwave discharge source to produce ground state atomic oxygen. To date only the $^3P$ - $^3S$ transition at 9.51eV has been reported for scattering angles of $0^\circ$ to $20^\circ$ and for incident electron energies of 100eV to 500eV.

The apparatus consists of an energy-loss electron spectrometer and an oxygen source comprising a pyrex glass gas flow tube and a microwave cavity. The electron beam enters the tube through an aperature and scatters from the atomic oxygen, after which it leaves through a slot in the tube and is detected in the analyser section of the electron spectrometer. An estimated 10% dissociation fraction of $O_2$ into O was obtained using a 2.5GHz 60W microwave discharge source which provided a $10^{13}$ $cm^{-3}$ number density of oxygen atoms in the interaction region.

One difficulty encountered in producing beams of atmoic oxygen is the high reactivity which causes it to recombine on contact with the walls of the containing vessel or other atomic and molecular species in the gas discharge.  A fast-flowing gas discharge will reduce the recombination of oxygen atoms since they make fewer wall and gas collisions between the discharge source and the interaction region as the mass of gas is pumped along the tube.  Flowing gas discharge sources produced with microwaves have been used by White and Ross (1976) for the production of atomic nitrogen and hydrogen and by Stone and Zipf (1971) for the production of atomic oxygen.  The discharge source was stable, after an initial stabilisation time of 20s, for periods in excess of four hours.  In addition to atomic oxygen the source produced charged particles ($\sim 2 \times 10^{-9}$ A), (which were removed by applying 75V across pusher plates in the source tube) and the excited molecular state of $O_2$ (a'$\Delta$g) which was detected by superelastic scattering.  By placing a cylindrical lining of aluminium foil (40mm long) in the source tube it is possible to remove completely the $O(^3P - ^3S)$ feature in the observed

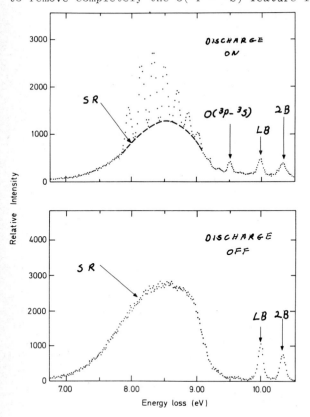

spectra because of recombination of oxygen atoms on the aluminium surface (Stone and Zipf 1974) and yet retain the molecular metastable state at the interaction region. No other atomic or molecular metastable states were detected.

An energy-loss spectrum is shown in Figure 5.  The

Figure 5

$O(^3P - {}^3S)$ transition at 9.51 eV is clearly resolved from the Schumann-Runge, longest band and second band features of molecular oxygen. In addition to the 9.51eV line other structure superimposed on the Schumann-Runge continuum is also detected when the discharge is on. These vibrational features probably arise from transitions from the '$\Delta g$ state to higher levels of the molecule. The measured relative line intensities (RLI) of $O(^3P - {}^3S)$ were normalised to the RLI of the longest band in $O_2$ measured at the same time under identical experimental conditions. Since the differential cross sections for the longest band are known (Newell et al 1980,) the RLI of $O(^3P - {}^3S)$ are thus automatically corrected for the unknown variation of the effective scattering volume.

In order to place these relative data on an absolute scale, the RLI were converted to relative generalised oscillator strengths (GOS) using the formula

$$f_{O_n}^{REL} (E, \Theta) = \tfrac{1}{2}\Delta E K^2 (k_o/k_n) (RLI)_{On}$$

where $\Delta E$ is the excitation energy of the nth state $k_o$ is the incident electron momentum $k_n$ is the scattered electron momentum and $K = k_o - k_n$ is the momentum transfer. The $f_{On}^{REL}$ were then subjected to a Lassettre-Vriens fit (Lassettre (1965) and Vriens (1967) at each energy and extrapolated to $K^2 = 0$ in accordance with the theorem of limiting oscillator strengths (Lassettre and Skerbele 1974) and normalised to an optical oscillator strength (OOS) measurement (see Newell 1979). This procedure gives the GOS curves, shown in Figure (6) for the $O(^3P - {}^3S)$ transitions.

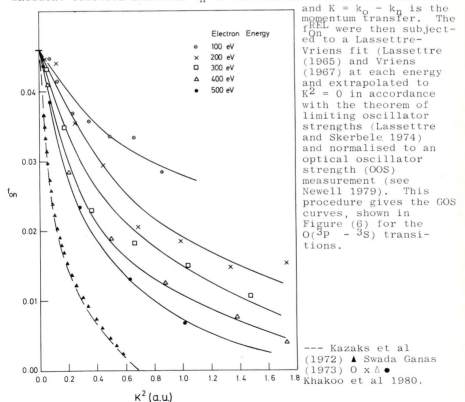

Electron Energy

| | |
|---|---|
| ⊙ | 100 eV |
| × | 200 eV |
| □ | 300 eV |
| △ | 400 eV |
| ● | 500 eV |

--- Kazaks et al (1972) ▲ Swada Ganas (1973) O x △ ● Khakoo et al 1980.

The OOS value used in this work was determined from the lifetime
measurement of Lawrence (1970; $f_o = 0.046$).

The GOS data obtained by Khakoo et al 1980 at different
incident electron energies do not lie on a universal GOS curve,
which indicates the non-applicability of the Born approximation
over this total energy and angular range.   Only for $K^2$ values
less than about 0.1au will the GOS data be independent of the
incident electron energy.   Khakoo et al (1979) have also reported
GOS values fo zero-angle scattering with $K^2$ values less than
0.02au.   Nevertheless, within the experimental errors of the
present experiment, the GOS curves at 500eV and 400eV incident
electron energies coincide for $K^2 < 0.5$au.

There is poor agreement between the theoretical GOS curves
of Kazaks et al (1972) (Born approximation) and Sawada and Ganas
(1973) (distorted-wave approximation) and the experiment.   The
only other available calculations of electron impact excitation
of the $^3P$ - $^3S$ transition in atomic oxygen corresponding to the
energies employed in the present experiment are the total
excitation cross section calculations of Rountree and Henry
(1972), McGuire (1976), Smith (1976), and Rountree (1977)
and on comparing these calculations with the experimental measurements of Stone and Zipf (1974), considerable disagreement between
the magnitudes of the theoretical and experimental results is
apparent.   However, these theoretical data, when converted to
total cross sections, are lower than the experimental data of
Stone and Zipf (1974) by a factor of approximately 1.75.   Hence
multiplying the curves of Sawada and Ganas and Kazaks et al by
1.75 gives 'corrected' theoretical GOS curves which are in
reasonable agreement with the GOS data determined in the present
work at 400 and 500eV.   Since the GOS is related to the differential
cross section it is possible to extract the absolute differential
cross section from the absolute GOS curves.   These differential
cross sections are plotted in Figure 7.   There are no calculations
of the differential cross sections for electron impact
excitation of the $O(^3P$ - $^3S)$ transition known to the authors
with which to make comparisons.

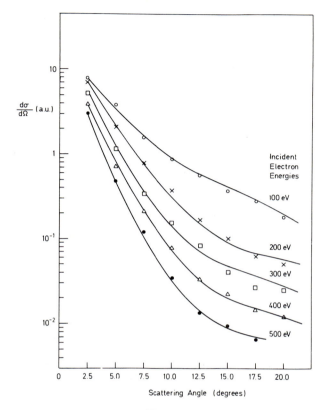

Figure 7

REFERENCES :

Blaha M. and Davis J. 1975: Phys. Rev. A12 2319.

Boksenberg A., 1961: Thesis, University of London.

Brooke, Harrison M.F.A., Smith A.C.H., 1978 : J. Phys. B: Atom. Molec. Phys. 11 3115.

Bransden B.H., and McDowell MR.C., 1978: Physics Report 46 No.7 p251.

Bruche E., 1928: Ann. Phys. 13 1065.

Burnett T, and Rountree S.P., 1979: Phys. Rev. A. 20 1468.

Callaway J. 1980: Advances in Physics No. 5 29 p771.

Christenson A.B., and Cunningham A.J. 1978, Journal Geophysical Res. 83 4393.

Dehmel R.C., Fineman M.A., Miller D.R., 1974: Phys. Rev. A9 1564.

Dehmel R.C., Fineman M.A., Miller D.R., 1976: Phys. Rev. A 13 115.

Dyke J.M., Jonathan N., and Morris A., 1979:  Electron Spectroscopy 3 189 ed. Brundle C.R., and Baker A.D.

Edwards A.K., and Cunningham D.C., 1973: Phys. Rev. 113 815.

Fite W.L. and Brackmann R.T., 1959: Phys. Rev. 113 815.

Green A.E.S., Sellin, D.L., Zachor A.S., 1969: Phys. Rev. A4
182.

Henry R.J.W., 1967: Phys. Rev. 162  56.

Kazaks P.A., Ganas P.S., Green A.E.S., 1972: Phys. Rev. A6 2169.

Khakoo M.A., Newell W.R., Smith A.C.H., 1979: J. Phys. B. 12 L425..

Khakoo M.A., Newell W.R., and Smith A.C.H., 1980: J. Phys. B :
Atom. Molec. Phys. 13, 4263.

Kieffer L.J. and Dunn G.H., 1966: Rev. Mod. Phys. 38 1.

Lassettre E.N., 1965: J. Chem. Phys. 43 12 4479.

Lassettre  E.M. and Skerbele  A., 1974: Methods of Experimental
Physics (ed. Williams, D. New York and London: Academic Press)
3B 868.

Lawrence G.M., 1970: Phys. Rev. A22 397.

Lawrence G.M. and Savage B.D., 1966: Phys. Rev. 171 67.

Lin S.C. and Kivel B., 1959: Phys. Rev. 114 1026.

Lin C.L., Parkes D.A., Kaufman F., 1970: J. Chem. Phys. 53  3898.

Matese J.J. 1974: Phys. Rev. A10 454.

McGuire E.J. 1971: Phys. Rev. A3 267.

Nesbet R.K. 1977: Advances in atomic and Molecular Physics
Vol. 13 p.315 ed. Bates D.R. and Bederson B. (Academic Press).

Newell  W.R., 1979: Progress in Atomic Spectroscopy B (ed. Hanle
W., and Kleinpoppen H., (New York and London : Plenum Press))
p. 1075.

Newell W.R., Khakoo M.A., Smith A.C.H., 1980: J. Phys. B: Atom.
Molec. Phys. 13 4877.

Neynaber R.H., Marino L.L., Rothe E.W., Trujillo S.M., 1961:
Phys. Rev. 123 148.

Nicholaides C., Sinanoglu O., Westhaus P., 1971: Phys. Rev. A4 1400.

Omidvar K., Kyle H.L., Sullivan E.C. 1972: Phys. Rev. A5 1174

Peach G. 1970: J. Phys B: Atom. Molec. Phys. 3 328.
            1971: '    '   '    '       '       '   4 1670

Rothe E.W., Marino L.L., Neynaber R.H. Trujillo S.M. 1962:
Phys. Rev. 125 582.

Rountree  S.P., Smith E.R., Henry R.J.W., 1974:  J. Phys. B: Atom. Molec. Phys. 7 L167.

Rountree S.P. and Henry R.J.W., 1972: Phys. Rev. A6 2106.

Rountree S.P., 1977: J. Phys. B. Atom. Molec. Phys. 10 2719.

Sanche L. and Burrow P.D. 1972: Phys. Rev. Lett 29 1639.

Sanche L. and Schulz G.J., 1972: Phys. Rev. A6 69.

Saraph H.E. 1973: J. Phys. B6 L243.

Sawada T., and Ganas P.S. 1973: Phys. Rev. A7 617.

Seaton M.J. 1959 : Phys. Rev. 113 814.

Smith E.R. 1976: Phys. Rev. A13 65.

Spence D. and Chupka W.A. 1974: Phys. Rev. A10 71.

Spence D. 1975 : Phys. Rev. A12 721.

Stone E.J. and Zipf E.C., 1971: Phys. Rev. A4 610.

Stone E.J. and Zipf E.C., 1974: J. Chem. Phys. 60 4237.

Sunshine G., Aubrey B.B., Bederson B., 1967: Phys. Rev. 154 1.

Tambe B.R. and Henry R.J.W., 1976: Phys. Rev. A14 512.

Tate J.T. and Smith P.T. 1932: Phys. Rev. 39 270.

Thomas L.D. and Nesbet R.K., 1975: Phys. Rev. A11 170.

Vriens L., 1967 : Phys. Rev. 160 100.

White M.D. and Ross K.J. 1976: J. Phys. B: Atom. Molec. Phys. 9 2147.

Ziegler D.L., Newman J.H., Smith K.A., Stebbings R.F., 1981: Abstracts XII ICPEAC Gatlinburg P267.

**PHYSICS OF ELECTRONIC AND ATOMIC COLLISIONS**
S. Datz (editor)
© North-Holland Publishing Company, 1982

ELECTRON INTERACTIONS WITH POLAR MOLECULES

W. R. Garrett

Chemical Physics Section, Health and Safety Research Division
Oak Ridge National Laboratory
Oak Ridge, Tennessee 37830
U.S.A.

A description is given of a number of the features of
discrete and continuous spectra of electrons interacting
with polar molecules. Attention is focused on the extent to
which theoretical predictions concerning cross sections,
resonances, and bound states are strongly influenced by
the various approximations that are so ubiquitous in the
treatment of such problems. Similarly, threshold
scattering and photodetachment processes are examined for
the case of weakly bound dipole states whose higher members
overlap the continuum.

## I. INTRODUCTION

Problems associated with the interaction of low-energy electrons with polar
molecules have received a great deal of theoretical and experimental attention
during the past few years. Several features of quantum mechanical desciptions
of such problems follow directly from the long-range nature of the electron-
polar molecule interaction potential. In the context of usual molecular theory
the nature of this interaction term poses new issues in theoretically describing
electron scattering, electron binding, and photodetachment processes involving
polar molecules. Since the general category of strongly polar systems automatically implies rather complex members (certainly as compared to $H_2$) detailed
quantum treatments represent fairly ambitious challenges to current techniques.
However, much of the physics in discrete and low-energy continuum problems can
be drawn from solutions to simple model systems, e.g., charged particles interacting with a "pure dipole" potential or a simple dipole rotor. Moreover,
approximations can often be made in treating scattering by a specific polar
molecule such that a given portion of the problem may be conveniently described
by a result from a simple dipolar system. Thus, it is both instructive and
productive to examine some of the features of solutions to model problems, then
to test the predictions on real systems.

In this progress report no attempt will be made to review the subject matter
covered by the title. However, a general perspective will be established on the
features of problems involving electron interactions with polar molecules followed
by some concrete examples involving Li halide and Li hydride molecules. In Part
II characteristics of the bound state spectra of polar negative ions are discussed
noting the overlapping of bound states with adjacent continuua. In Part III
characteristics of scattering cross sections in various approximations are discussed, and near threshold resonances for strongly polar systems are demonstrated
by calculations for LiH. In Part IV the more general problems of threshold
energy dependences of cross sections for multichannel electron scattering by
polar target systems are discussed.

## II.   BOUND STATES AND OVERLAPPING RESONANCES IN THE CONTINUUM

As is well known in quantum chemistry, where the Born-Oppenheimer (BO) approximation (fixed nuclei) is ubiquitous in molecular structure calculations, the spectrum of electronically excited states for negative ions of polar molecules having a permanent dipole moment $\mu$ greater than $0.639$ $ea_0$, will include an infinite number of levels [1,2] (e = electronic charge, $a_0$ = Bohr radius).  If, for simplicity, we restrict the discussion to linear molecules the dipole bound state theorem allows the following assertion:  Within the BO approximations a breakdown of the total wave function for an N-electron polar negative ion into the form

$$\Upsilon(\vec{r}_1\vec{r}_2 \cdots \vec{r}_N, R_{z_1}R_{z_2}) = \chi(R_{z_1}R_{z_2})\phi(\rho\vec{r}_1\vec{r}_2 \cdots \vec{r}_N) , \qquad (1)$$

where $r_i$ are electronic and $R_{z_1}$ nuclear coordinates, produces a potential energy surface for the negative ion (eigenvalue for electronic motion in the presence of nuclei at separation $\rho$) which lies at least slightly below the corresponding surface for the neutral molecule over that region of nuclear separation, $\rho$, in which $\mu(\rho) > 0.639$ $ea_0$ [2] [where $\mu(\rho)$ is the dipole moment of the neutral as a function of $\rho$].  This point is illustrated in Fig. 1, where $\mu(\rho)$ and the surface $E(\rho)$ for a hypothetical case are shown schematically.

One of the interesting features of electron interactions with polar molecules is the important role played by non-Born-Oppenheimer terms in the Hamiltonian for any member of this class of problems.  If, in contrast to the fixed nucleus approximation, one includes rotational degrees of freedom in the Hamiltonian, then discrete and continuum quantum solutions for the entire class of problems are changed in character [3,4].  Thus, we include such terms in the Hamilitonian and write the total wave function in the form

$$\Psi^J_{\eta\nu}(\vec{r}_1\vec{r}_2 \cdots \vec{r}_N\vec{R}_{z_1}\vec{R}_{z_2}) = \sum_{j\ell} C_{j\ell} \, \chi^j_\nu(R_1R_2\hat{S}) \, \psi^\ell_\nu(\vec{r}_1\vec{r}_2 \cdots \vec{r}_N\vec{\rho}) \qquad (2)$$

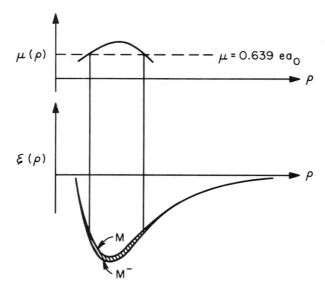

Fig. 1.  Upper curve:  dipole moment as a function of nuclear coordinate $\rho$, illustrating region where $\mu > \mu_c$. Lower curve:  schematic potential energy surfaces for a neutral molecule M and its negative ions $M^-$.

where the system is in vibrational state $\nu$, electronic state of "principal" quantum number $\eta$ and total angular momentum $\vec{J} = \vec{j} + \vec{\ell}$ (i.e., sum of total orbital angular momentum $\vec{\ell}$ plus rotational angular momentum, $\vec{j}$, of the nuclei). If these are coupled such that $J^2$ and $J_z$ are good quantum numbers, then the predictions on the bound state spectrum of the N-electron negative ion are quite different (as are cross sections and resonance structures to be discussed below). In order to be more specific about the issues of present interest we choose a rotational close coupling description to treat problems involving excited states of polar negative ions and electron scattering by these species. Thus, within an independent particle framework we can consider the interaction of the excited valence (or scattered) electron to be represented by an interaction potential:

$$V(\vec{r},\vec{s}) = V_\mu(\vec{r},\vec{s}) + V_\alpha(\vec{r},\vec{s}) + V_{Q_{zz}}(\vec{r},\vec{s}) + V_{SR}(\vec{r},s) \tag{3}$$

where we have depicted the long-range terms respectively as:

$$V_\mu(\vec{r},\vec{s}) = - \sum_{\lambda=1,3\,\ldots} \frac{-\mu e r_<^\lambda}{s r_>^{\lambda+1}} P_\lambda(\cos\theta) \quad, \tag{4}$$

where $\lambda = 1$ is the dipole term and $r_<(r_>)$ is the lesser (greater) of the charge separation, $s$, and electron coordinate, $r$.

$$V_\alpha(\vec{r},\vec{s}) = - \frac{\alpha_0 e^2}{2r^4} f(r) - \frac{\alpha_2 e^2}{2r^4} f(r) P_2(\cos\theta) \quad, \tag{5}$$

is the induced dipole potential where $\alpha_0$ and $\alpha_2$ are the spherical and quadrupole polarizabilities of the linear molecule;

$$V_{Q_{zz}}(\vec{r},\vec{s}) = \frac{-e Q_{zz}}{r^3} f(r) P_2(\cos\theta) \tag{6}$$

is the potential due the permanent quadrupole moment, $Q_{zz}$, and the factor $f(r)$ is a cutoff function designed to remove the singularities at the origin in (5) and (6). In the present study

$$f(r) = [1-e^{(r/r_0)^6}] \quad.$$

The above forms ensure proper asymptotic behavior for $V(\vec{r},\vec{s})$ with a parameterized form [due to $f(r)$] at small distances (origin chosen at the center of mass of the molecule). The final term, $V_{SR}(\vec{r},\vec{s})$ is a "short range," nonlocal potential which can in principle compensate for everything not contained in Eqs. (4-6), including all exchange and electrostatic contributions plus compensation for the cutoff function $f(r)$.

Note that the form chosen in Eq. (2) for the total wave function implicitly assumed separation of vibrational and rotational degrees of freedom. In the present context our interest is directed, as usual, toward the effect of angular momentum coupling on electronic energy levels. The difference here is the greatly exaggerated form which this effect exhibits for the case of dipole bound states. Thus, if one examines the average asymptotic potential experienced by an excited (or continuum) electron, an integration over the total charge distribution, but including only vibrational degrees of freedom of the nuclei, will yield a function dominated by the average dipole moment term. However, an integration over rotational coordinates would produce a zero average dipole field since the dipole potential averages to zero when taken over all directions in space. Thus, we will take equilibrium positions for nuclei, along with the average dipole moment that results from integration over electronic and nuclear vibrational coordinates and write the Schrödinger equation for present purposes as

$$[H_{rot}\, Y_j^{m_j} - \frac{\hbar^2}{2m} \nabla_r^2 + V(\vec{r},\vec{s})]\ \Psi(\vec{r},\vec{s}) = E\Psi(\vec{r},\vec{s})\ . \tag{7}$$

The term $H_{rot}$ is the rotational Hamiltonian of the molecule which has eigenfunctions satisfying

$$H_{rot}\, Y_j^{m_j}\, (\hat{s}) = j(j+1)\hbar^2/2I\ Y_j^{m_j}\, (\hat{s})\ , \tag{8}$$

where I is the moment of inertia. In a single center representation, $\Psi$ is expressed as

$$\Psi^J\, (\vec{r},\vec{s}) = \sum_{j,\ell} r^{-1} U_{j\ell}^J(r)\ Y_{j\ell}^{JM}(\hat{r},\hat{s}) \tag{9}$$

where $Y_{j\ell}^{JM}\, (\hat{r},\hat{s})$ is a coupled spherical harmonic which is an eigenfunction of $J^2$ and $J_z$ [3]. Note that an independent particle picture has been adopted with no antisymmetrization between core and valence (or scattered) electrons. Since exchange terms are "short range" in character we choose instead to represent the effect of Pauli exclusion through a pseudopotential technique. Thus, $V_{SR}$ is formulated accordingly [5].

By expanding $V(\vec{r},\vec{s})$ in a multipole expansion

$$V(\vec{r},\vec{s}) = \sum_{\lambda=0}^{\infty} v_\lambda(r,s)\, P_\lambda(\cos\theta)\ , \tag{10}$$

(where $\theta$ is the angle between electron coordinate $\vec{r}$ and internuclear direction $\hat{s}$), one gets the familiar coupled radial equations whose solutions yield the components $U_{j\ell}^J(r)$ of the single center wave function [9]. Thus,

$$\left[\frac{d^2}{dr^2} + E_{j'} - \ell'(\ell'+1)/r^2\right] U_{j'\ell'}^J =$$
$$\sum_{j''\ell''}\ \sum_\lambda v_\lambda(r,s) f_\lambda(j'\ell',j''\ell'';J) U_{j''\ell''}^J(r) \tag{11}$$

The factors $f_\lambda$ are constants that result from integrations over angle variables [3]. Note that the potential terms (4-6) are already in the appropriate multipole form. Solutions to Eq. (11) are obtained with expontentially decaying boundary conditions for $E_j = E_0 - j(j+1)/2I < 0$ or continuum type (R-matrix) boundary conditions for $E_j > 0$ [3,4].

Considerable insight into questions concerning excited state spectra of polar ions can be gained from consideration of electron binding by a simple dipole rotor. This primitive system is represented by deleting all terms other than $V_\mu$ [i.e., Eq. (4)] in $V(\vec{r},\vec{s})$. One can then determine the minimum dipole moment required to sustain electronically and rotationally excited states of the electron-dipole rotor systems. Results of such a study are displayed in Fig. 2. We define a critical moment $\mu_n^J(I,R)$ as that which produces zero total energy for the system having total angular momentum J and total number of nodes n in the reduced radial eigenfunction components [6] (n=2 is first excited state, etc.). These critical moments are functions of the moment of inertia and charge separation of the dipole rotor. Note that rather large dipole moments are required for binding in an electronically excited state.

In order to examine the characteristics of actual excited state spectra of strongly polar molecular ions, we apply present methodology to Li hydride and Li halide anions. Thus, we ensure proper form for the long-range interaction terms in the interaction potential through the use of Eq. (3), with experimental or theoretical values for $\mu$, $\alpha_0$, $\alpha_2$, and $Q_{zz}$ in Eqs. (4-6). Then a pseudopotential choice is made for the remaining short-range contribution to the total

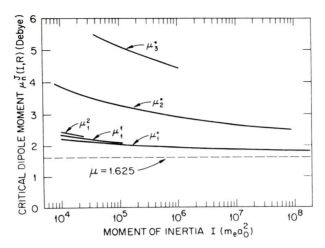

Fig. 2. Critical dipole moments for a simple dipole rotor of charge separation 4 $a_0$ as a function of moment of inertia I.

interaction potential, represented by $V_{SR}$. In this approach, the combined effect of Pauli exclusion, electrostatic interactions, and corrections for inadequate representations at small distances of the long-range terms in Eqs. (4-6) are all handled by a parameterized function which is adjusted to yield one or more experimental quantities [5]. In the present instance we have utilized a pseudopotential of the form

$$V_{SR}(r) = V_0 e^{-(r/r_c)^6} . \tag{12}$$

Here $V_0$ is a "strength" parameter and $r_c$ a "range" parameter for the short-range interaction. Justification for use of a pseudopotential method in studies of excited state spectra and of low-energy electron scattering phenomena rests on the dominance of long-range forces in determining the relevant physics and on the insensitivity of the physical quantities of interest to details in choices of $V_{SR}$ [5]. In the present study, combinations of $V_0$ and $r_c$ were chosen ($r_c \cong$ 2 $a_0$) to yield the known electron affinity of LiCl [7] as the ground state of LiCl$^-$ [5]. The same choice of $V_{SR}$ was then used for other LiX$^-$ ions, but with dipole moments appropriate to the individual species. This procedure yielded an electron affinity for LiH which agreed well with that from an elaborate configuration interaction calculation [8]. Several different choices for parameters in $V_{SR}$ (and also a different choice in the functional form of $V_{SR}$) all yielded excited state spectra which differed insignificantly ($\sim$5%) from each other.

Energy levels for negative ions of LiH, LiF, LiCl, and LiI are tabulated in Table I and represented schematically (not to scale) in Fig. 3. Dipole moments and moments of inertia of these molecules are listed in the table. Experimental values are not available for the polarizabilities and quadrupole moments of these strongly polar species. Thus, polarizability values $\alpha_0 = 15.3$ $a_0^3$ and $\alpha_2 = 1.1$ $a_0^3$ were used with $Q_{zz} = 3.28$ $ea_0^2$ [5].

Starting with the most strongly polar member of the set, LiI, note that two electronically excited states are bound. [$E_1$ = ground state with radial quantum number n = 1 corresponding to one node in the reduced radial functions (the pseudopotential technique yields a ground state eigenfunction nodeless except at the origin); n = 2 (two nodes) is the first excited state, etc., each orthogonal to the ground electronic state. Rotational levels are labeled with J = 0,1, ... etc.]. Only the J = 0 level of the second excited state was calculated, though one or two additional rotational levels of this state are undoubtedly bound in LiI$^-$.

Table I.    *Negative Ion States (Li Hydride and Li Halides)*

| | LiH⁻ | LiF⁻ | LiCl⁻ | LiI⁻ |
|---|---|---|---|---|
| $\mu$ (Debye) | 5.9 | 6.3 | 7.1 | 7.4 |
| $I(m_e a_0^2)$ | $2.6 \times 10^4$ | $8.2 \times 10^4$ | $1.5 \times 10^5$ | $2.4 \times 10^5$ |
| $E_1$ j=0 (eV) | -0.32 | -0.38 | -0.61 | -0.68 |
| $E_2$   j=0 | -0.00262 | -0.00458 | -0.102 | -0.0139 |
|      j=1 | -0.00196 | -0.00447 | | -0.0138 |
|      j=2 | -0.00089 | -0.00397 | | -0.0137 |
|      j=3 | 0 | -0.00347 | | -0.0135 |
| $E_3$   j=0 | 0 | $-1.19 \times 10^{-5}$ | $-1.33 \times 10^{-4}$ | $-2.72 \times 10^{-4}$ |
|      j=1 | 0 | 0 | | |

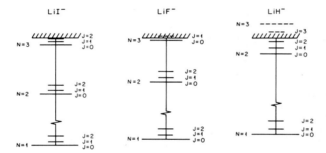

Fig. 3.  Schematic representation of energy levels for negative ions of LiI, LiF, and LiH at their equilibrium nuclear positions. Here N=1 is the electronic ground state, N=2 the first excited state, etc.

Progressing across the table to LiF⁻ note that there are still two bound excited states but the second (N = 3) state is, in this instance, only stable in the rotational ground state (J = 0). Thus, J = 1,2 ... of the N = 3 electronic state are in the continuum slightly above E = 0.

Finally, note that LiH⁻ is bound only in the ground state and in one excited state. Furthermore, this lowest excited state is so near the continuum that only the three lowest rotational levels are bound. Thus J = 3 and greater levels of the $E_2$ state cross over into the continuum, and the second excited state, $E_3$, has crossed over completely. Any or all such states offer possibilities for the appearance of scattering resonances in the near threshold continuum, or in near threshold photodetachment of these molecular anions, as will be discussed below.

The results in Table I were obtained with pseudopotential parameters, $V_0$ = 2.0 $R_y$, $r_0$ = 2.2 $a_0$. The range of $V_{SR}$, chosen to yield the proper electron affinity for LiCl, was $r_c$ = 2.828 $a_0$. Note that $R_0$ and $r_c$ both effect the range of attractive potentials and are thus not independent in the present procedure. Results were quite insensitive to all the details of the short-range interactions. A change in the "strength" $V_0$ by a factor of 2 (with corresponding change in the range of $V_{SR}$) left excited state energies unaffected within 2 or 3%.

III.  SCATTERING

A.  General Features. - Continuing in the same theme as above we can note some very general features of electron scattering cross sections for polar molecules. The properties follow directly from the long range nature of the interaction (or in certain instances from the singularity at the origin in the case of a point dipole scatterer) and from the level of approximation in which the quantum scattering problem is treated.  In fact, one can summarize as in Table II, several of the general features of scattering problems involving real molecules or simple dipole scatterers.

*Table II.  Characteristics of Theoretical Scattering Cross Sections from Polar Systems*

| Fixed Scatterer | Freely Rotating Scatterer |
|---|---|
| Born Approximation | |
| Momentum transfer cross section, $\bar{\sigma}_m$, (averaged over all orientations) is well defined for all $\mu$, with any charge separation. | Momentum transfer and total cross sections are well behaved for all $\mu$, at any charge separation (includes point dipole rotor).  The rotational excitation and deexcitation cross sections diverge in the limit as $I \to \infty$ [4]. |
| Total cross section, $\sigma$, is infinite for all $\mu > 0$. | |
| Exact Treatment | |
| $\bar{\sigma}_m$ is defined for all dipolar systems of finite charge separation.  For *point* dipoles, $\bar{\sigma}_m$ diverges for $\mu \geqslant \mu_c$ [9]. Total cross section is infinite for all $\mu > 0$ independent of charge separations [4]. | Momentum transfer, total elastic and inelastic cross sections are all well behaved for all $\mu$ except for the pathological *point dipole rotor* where all cross sections become undefined for $\mu > \mu_c$ [10]. In the limit as $I \to \infty$ cross sections for rotational excitation and deexcitation diverge [4]. |

Other general features of momentum transfer cross sections from polar target molecules have been elucidated by Fabrikant [11,12] through an extension of effective range theory to include long range $r^{-2}$ potentials.  This analysis was carried through for a fixed dipole, and thus is applicable where collision times are short as compared to rotational frequencies, but where an expansion in terms of $kr_0$ is valid ($r_0 \sim$ molecular size) i.e., $(kr_0)^2 \ll 1$.  In this energy region, $(\mu^{1/2} j/I) \ll k^2 \ll r_0^{-2}$, (i.e., $\sim 0.005$ to $\sim 0.5$ eV) analytic formulas for $\sigma_m(k)$ yield weak oscillations as a function of k with a functional dependence $\sin^{-1}(2 \mu \ln k + \alpha)$.  The cross sections contain the parameter $\alpha$ and other constants that depend on $\mu$ and on short-range forces.  The characteristics of these oscillations in $\sigma_m$ are determined by the magnitude of $\mu$ (the phases are sensitive to short range contributions to the interaction potential) and thus are a general feature of the scattering problem, over the range of validity of the effective range expansions.

Other important and only partially understood characteristics of the cross sections, namely threshold behavior, will be discussed below.

B.  Cross Section Calculations. - As was mentioned in the introduction, detailed treatment of electron scattering by the simplest polar molecules represents a problem of considerable complexity.  Consequently, theoretical progress in dealing with such systems has been centered primarily on investigations of a number of approximate methods for treating exchange and for solving the large sets of equations which result from angular momentum coupling. This work has been summarized in recent reviews by Collins and Norcross [13], Itikawa [14], and Lane [15]. In this report we will only note some of the general features of problems associated with calculations of scattering cross sections for specific polar systems without attempting to review all salient points which have been made in the large body of studies on this subject.

Thus, we observe a characteristic of frame transformation techniques [16], whereby the coupled equations that represent the scattering problem are solved in a body fixed (BF) coordinate system at points "inside" the target molecule, then transformed to a space fixed (SF) or laboratory frame at some point "outside" the molecular charge distribution. The point is that for polar systems such techniques are not only convenient but, because of a large number of coupled channels in the lab system, these techniques, or hybrid mixtures of such methods [17,18] are almost obligatory [13,15]. However, we also note that detailed treatment of exchange effects is quite troublesome for any "multiple region" treatment of a scattering problem. A number of different approximate methods have been investigated for their efficacy in describing scattering by polar systems including local exchange potentials, orthogonalization procedures and iterative techniques [15,19]. A sizeable bag of tricks has also been accumulated for use in solving the coupled radial equations that pervade this field [20].

Two noteworthy features of these detailed calculational efforts, which are essentially independent of the particular method employed in the studies, are: (1) The appearance of Wigner cusps in an open channel at the threshold for the opening of a new channel to which the first is coupled. Such expected cusps were obtained by Gianturco and Rahman [21] in the rotationally elastic channel of very low energy electron scattering by HF at the threshold for opening of a rotationally inelastic channel and by Rudge [22] in the $\nu = 1$ excitation cross section of HF at the threshold for $\nu = 2$ excitation. (2) The appearance in a study by Collins, Norcross, and Schmid [23] of the predicted sin (a $\ln k + \alpha$) oscillatory behavior [11,12] in momentum transfer cross sections for KOH and CsOH. The latter study was based on a very simple cut-off dipole molecular model, but the behavior, which derives from the $r^{-2}$ interaction, should nevertheless be valid insofar as this general feature [11,12] would be manifested.

On the basis of the formalism introduced in Section II above for investigating the bound state spectrum of strongly polar negative ions, further progress can also be made in describing low energy scattering cross sections from the same molecules. Thus, the pseudopotentials that were utilized to describe the excited state spectra of the LiX$^-$ molecules have been utilized in the rotational close coupling scattering equations [Eq. (11) with R-matrix boundary conditions] to yield elastic and rotational excitation cross sections. Of particular interest are possible appearances of resonances associated with bound states of the negative ion which cross over into the continuum, and the characteristics of cross sections in the range of validity of the effective range expansion technique [11,12] which predicts oscillatory behavior in $\sigma_m$.

We examine the latter question first. In Figs. 4 and 5 are shown cross sections $\sigma^J(j,j')$ for scattering by LiH initially in the ground rotational state. We depict separately the $J = 0$ and $J = 1$ components of the elastic, $\sigma^J(0,0)$ and rotational excitation, $\sigma^J(0,1)$, cross sections

$$[\sigma(j,j') = \sum_{J=0}^{\infty} \sigma^J(j,j')].$$

Note, aside from the sharp resonance features, that sin (a $\ln k + \alpha$) oscillatory behavior is exhibited by the $J = 0$ component of the elastic and inelastic cross sections. This behavior is not so evident in the $J = 1$ component, and is completely absent in higher J contributions. However, the diffusion cross section, $\sigma_m$, is dominated by its lowest angular momentum components, thus the ERT predictions based on a stationary target approximation, are born out in the full rotational coupling treatment. Similar results are also found in preliminary calculations of LiI cross sections.

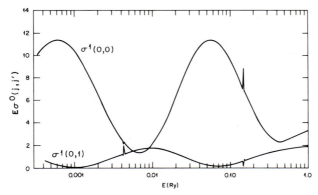

Fig. 4. The J = 0 component of the elastic, $\sigma(0,0)$, and rotational excitation $\sigma(0,1)$ cross sections for LiH initially in the ground rotational state.

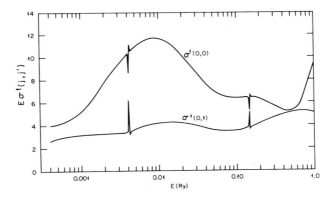

Fig. 5. The J = 1 components of LiH cross sections in Fig. 4.

Note that the oscillatory behavior in the lowest angular momentum contributions to $\sigma(j,j')$ will be obiterated in total scattering cross sections, due to the very large contribution to $\sigma(j,j')$ from higher partial waves. Thus this feature of the cross sections will be evident in momentum transfer or differential cross section measurements or in photodetachment from negative ions where the combination of partial wave contributions and high energy resolution would offer exceptional visibility to cross sectional features.

C. Resonances. - Independent of any oscillatory behavior of the momentum transfer cross sections (which has not been experimentally observed), large resonance cross sections have been measured at the threshold for vibrational excitation for a number of polar (and some nonpolar) molecules [24]. Resonant threshold scattering from polar molecules has been investigated in a fixed nucleus approximation [25,26], and in model potential studies [22,27], but a quantitative *ab initio* description of this phenomenon has yet to be realized. We defer to the following section our consideration of threshold behavior and consider other possible resonant behavior in the low energy scattering region.

With reference to the excited state spectrum of LiH⁻ in Fig. 2 we note that all excited electronic states above the lowest two cross over into the continuum, though an infinite number of such levels would move down below $\varepsilon = 0$ if the moment of inertia were allowed to increase to an infinite value. Since this behavior will be characteristic of all strongly polar molecules, a question of intense interest concerns the possible appearance of low energy resonances associated with this spectral behavior (higher values of "n" in the above notation). In Figs. 4 and 5 we note that in addition to the broad oscillatory characteristics of $\sigma^J$ (j,j') there are also at least two very sharp resonances in the LiH cross sections [these resonances also appeared in sample calculations of the J = 2 contribution to $\sigma$(j,j')]. We can infer that these features are associated with bound states in the continuum of LiH⁻. The lower energy resonance is very narrow (width $\Gamma \sim 0.00002$ Ry), and the higher energy resonance, though 100 times broader, is still quite narrow. It is quite possible that other undiscovered features exist, yet were not uncovered in the tedious point by point search procedure used here. Also, other features could exist nearer E = 0 where the presence of closed rotation channels prevented cross section determinations with the code used in these calculations. Thus, we cannot say that the features are associated specifically with n = 3 and 4 in the spectral series, the two lowest members of which are depicted in the figures.

In the LiH results reported here, vibrational excitation channels have not been included in the analysis. Thus one might expect to find resonances similar to the above features in vibrational excitation cross sections for this species. Indeed, it was indicated in an earlier study of elastic rotational resonances [23] that low energy scattering by polar molecules provides multiple opportunities for the occurrence of resonance scattering processes. The present results reinforce this contention.

IV.   THRESHOLD BEHAVIOR

Of course, it is well known that the case for a dipole potential is not included in the classic analysis of scattering threshold laws by Wigner [28]. The problem has yet to be solved in a form of sufficient generality for application to threshold electron scattering by polar molecules or threshold photodetachment by polar negative ions. However, a body of information has been accumulated on solutions to coupled equations that are approximations to the set (11) containing $r^{-2}$ coupling.

Thus, we first note that a truncation of the J=0 set to only two terms (j,$\ell$ = 0,1) *plus* a neglect of the energy difference between levels $E_j$ and $E_{j'}$ (degenerate approximation) yields equations appropriate to the two state atomic hydrogen excitation problem (neglecting the Lamb shift splitting). The threshold law for the degenerate two-state hydrogen-like system was solved by Gailitis and Damburg [29], and by O'Malley [30] who showed that these equations can be diagonalized and solved analytically. The threshold law is found to depend on the strength of the coupling coefficients in the radial equations. The threshold cross section has the form $\sigma \propto$ constant $k^{2\lambda+1}$ where $\lambda(\lambda+1)$ is the characteristic value of the matrix $\ell(\ell+1) + \alpha$ of the truncated Eq. (11) and $\alpha$ is the coupling constant [$\alpha = f_\lambda(j\ell_j,j'\ell';0)$ $\mu e^2$]. This form holds when $\lambda(\lambda+1) > -1/4$. For stronger coupling (larger $\mu$) such that $\lambda(\lambda+1) < -1/4$, the cross section is non-zero at threshold and oscillates with a frequency proportional to *constant* times $\ln k$. Thus, in a truncated two-state, degenerate approximation, threshold scattering by a polar molecule behaves in the same fashion. (The critical dipole moment in this approximation is $\mu_c = 0.65$ $ea_0$ rather than the proper 0.639 $ea_0$ which results with the inclusion of other terms in the coupled equations. Thus, the threshold law switches form at this value of $\mu$ in the approximation of a representation including only j,$\ell$ = 0,1 terms.)

Next, we note that in the limit I → ∞, all the channel energies $E_j = E_0 - j(j+1)/2I$ in Eq. (11) become degenerate and the set can be transformed to the same scattering problem in the fixed nucleus approximation (where the quantum numbers j no longer appear) [3]. Starting with the approximation of a nonrotating dipole scatterer, Fabrikant [25] and later Domke and Cederbaum [26] treated threshold behavior of vibrationally elastic and inelastic scattering. In the fixed dipole treatment the asymptotic equation can be separated in spherical polar coordinates [9] with the appearance of a Bessel type radial equation and an unorthodox angular eigenvalue problem. The threshold behavior of cross sections for the fixed polar scatterer again separates into two regimes. For $\mu < \mu_c = 0.639$ $ea_0$ the scattering amplitude goes as $k^\lambda$ where $\lambda$ is the order of the Bessel function, which is real. For $\mu > \mu_c$ the order of the Bessel function becomes complex, $\lambda = -1/2 + in$, and the scattering amplitude is non-zero at threshold and oscillatory. (The total cross sections are infinite for the fixed dipole scatterer.)

At this point we note the obvious fact that it is almost certainly necessary to include the rotational energy spacings of a target molecule in formulating the threshold scattering problem involving such systems. Unlike the hydrogen atom where the Lamb shift spacing between s and p levels is extremely small as compared to electronic excitation thresholds, the threshold scattering problem in the case of a polar target molecule involves questions concerning the rotational thresholds themselves. Thus the spacings, rather than being negligible, form a subset of the thresholds of interest. Added to this are the facts, discussed above, of the strong influence of rotational degrees of freedom on this class of problems.

To examine the true threshold behavior of scattering by polar molecules we consider the asymptotic form of coupled [Eqs. (11)] for scattering by molecules in the ground state. For large r and arbitrary total angular momentum J, the members of the set, in increasing order of rotational energy, become:

$$\left[\frac{d^2}{dr^2} + E - J(J+1)/r^2\right] U_{0J}^J \cong$$

$$\frac{\mu}{r^2} [f_1(0,J,1,J-1;J)U_{1,J-1} + f_1(0,J,1,J+1;J)U_{1,J+1}]$$

$$\left[\frac{d^2}{dr^2} + E_0 - \frac{2}{2I} - (J-1)J/r^2\right] U_{1,J-1}^J \cong$$

$$\frac{\mu}{r^2} [f_1(1,J-1,0,J;J)U_{0J} + f_1(2,J,1,J-1;J)U_{2,J} + f_1(2,J-2,1,J-1;J)U_{2,J-2}]$$

$$\left[\frac{d^2}{dr^2} + E_0 - \frac{2}{2I} - (J+1)(J+2)/r^2\right] U_{1,J+1}^J \cong$$

$$\frac{\mu}{r^2} [f_1(1,J+1,0,J;J)U_{0J} + f_1(2,J,1,J+1;J)U_{2,J} + f_1(2,J+2,1,J+1;J)U_{2,J+2}]$$

The set can be stopped at level j where E, the incident energy, is smaller than $j(j+1)/2I$, since the energy $E_j = E-j(j+2)/2I$ is then negative for this and all successively higher $j\ell$ components of the coupled set. These closed channels decay exponentially and are consequently negligible at large r as compared to open channel components. Thus, we immediately conclude that in the totally elastic energy regime, below the first rotational excitation threshold (or $0 \leq E < 1/I$), *no* dipole coupling exists in the asymptotic region. This follows since all $U_{j\ell}$ except $U_{0J}$ decay exponentially as $r \to \infty$, leaving only the equation

$$\left[\frac{d^2}{dr^2} + E - J(J+1)/r^2\right] U_{0J} = 0$$

Thus, below the first inelastic threshold the elastic cross section behaves identically to the case of atomic threshold scattering. The analysis of O'Malley et al. [29,30] thus applies with the dominant long range forces being due to

quadrupole and induced dipole moments. The dipole moment is completely ineffec-
tive in its influence on the cross section in this energy region. Physically one
might say that the collision times are sufficiently long to allow the dipole
contributions to average to zero over the rotational period of the target molecule.

Now consider the energy regime from the first rotational excitation threshold to
the second, $\Delta j=2$, threshold. In this instance dipole coupling in the asymptotic
region exists in the coupled equations for $U_{0j}$ and $U_{1,J\pm1}$, but these are not
coupled to higher j components $(U_{2,\ell}, U_{3,\ell} \ldots)$ all of which decay exponentially.
Thus, the asymptotic form of the exact solution to the polar molecule scattering
problem reduces in this energy region to the two state nondegenerate threshold
excitation investigated by Damburg [31] and by Gailitis [32] in the context of
threshold excitation of atomic levels. (Actually, it is only the coupled set for
J=0 that reduces asymptotically to two equations in this energy regime, but the
threshold behavior is dominated by this lowest angular momentum contribution.)
On the basis of an intuitively constructed equation, Damburg showed that the
threshold law for exciting what corresponds here to $j=1\to2$ rotational transitions
behaves according to the normal Wigner law, but with a second order correction
term in $k_1$ which goes as $\ln k_2$. Later Gailitis showed, for the case J=0, that a
direct solution of the coupled equations for two open channels coupled by an $r^{-2}$
interaction led to Damburg's same result [32]. (It was unnoticed in Gailitis'
paper that his solutions became invalid, due to collapse into the origin for
values of the coupling term, $\alpha$, greater than 3/4. This corresponds to $\mu = 0.65$
$ea_0$, the critical moment for binding in s + p representation, and to the collapse
into the origin for a point dipole rotor in this limited representation [10].)
Thus, we conclude that over $E_1 \leqslant E \leqslant E_2$, where $E_1$ is the first rotational excitation threshold and $E_2$ the second, the cross section for rotational excitation of
a polar molecule goes as

$$\sigma^\circ(0,1) = constant \times k^3 \left[ 1 + \frac{4\beta^2}{15} k_{01}^2 \ln k_{01} + \sigma(k^2) \right] .$$

Note that the dipole moment only appears in the term $\beta^2 = \mu^2 f_1^2(0,0,1,1,0)/(k_{00}^2 - k_{01}^2)$. (No consideration has been given to contribution from induced dipole terms
which will further modify the Damburg correction term.) Finally, we note the
results of a model study by Faisal in which two and three channel threshold
excitation was studied through numerical calculations [33]. These studies also
confirmed the applicability of the Wigner type law for dipole coupled systems
over a range of values of the coupling coefficients.

In summary, we can conclude a few things about threshold behavior of the cross
sections for electron scattering (or photodetachment) from polar molecules and we
can infer on indirect, but unproven, evidence a few other items. (1) In the
totally elastic regime, below $E_{threshold}$ for rotational excitation, the cross
sectional behavior is similar to the atomic problem. It is governed by induced
dipole forces, essentially independent of the permanent dipole moment. (2) In
the region where only two channels are open, the cross section for the second
channel is that given by Damburg and by Gailitis. (3) Numerical studies indicate
that the behavior remains conventional (i.e., Wigner law) for three open channels.
(4) The oscillatory threshold behavior predicted by fixed nuclei treatments is
missing at threshold, but oscillations do appear further above threshold where
the effective range expansion of Fabrikant is applicable. Further studies of
these phenomena, both theoretical and experimental, are very appealing.

## V.    ACKNOWLEDGMENT

Research is sponsored by the Office of Health and Environmental Research, U.S.
Department of Energy under contract W-7405-eng-26 with the Union Carbide
Corporation.

## REFERENCES

[1] Crawford, O. H., Mol. Phys. 20 (1971) 585.
[2] Garrett, W. R., Chem. Phys. Lett. 62 (1979) 325.
[3] Garrett, W. R., Phys. Rev. A 3 (1971) 961.
[4] Garrett, W. R., Phys. Rev. A 4 (1971) 2229.
[5] Garrett, W. R., J. Chem. Phys. 69 (1978) 2621; 71 (1979) 651.
[6] Garrett, W. R., J. Chem. Phys. 73 (1980) 5721.
[7] Carlsten, J. L., Peterson, J. R., and Lineberger, W. C., Chem. Phys. Lett. 37 (1976) 5.
[8] Liu, B., O-Ohata, K., and Kirby, K., J. Chem. Phys. 67 (1977) 1850.
[9] Mittleman, M. H., and Von Holdt, R. E., Phys. Rev. 140 (1965) A726.
[10] Garrett, W. R., Phys. Rev. A 23 (1981) 1737.
[11] Fabrikant, I. I., Zh. Eksp. Teor. Fiz. 71 (1976) 148; JETP 44 (1976) 77.
[12] Fabrikant, I. I., J. Phys. B: Atom. Molec. Phys. 10 (1977) 1761.
[13] Collins, L. A., and Norcross, D., Phys. Rev. A 70 (1978) 4.
[14] Itikawa, Y., Phys. Rept. 46 (1978) 117.
[15] Lane, N., Rev. Mod. Phys. 52 (1980) 29.
[16] Chang, E. S., and Fano, U., Phys. Rev. A 6 (1970) 173.
[17] Clark, C. W., and Siegel, J., J. Phys. B: Atom. Molec. Phys. 13 (1980) L31.
[18] Siegel, J., Dehmer, J. L., and Dill, D., Phys. Rev. A 23 (1981) 632.
[19] Morrison, M. A., and Collins, L. A. Phys. Rev. A 23 (1981) 127.
[20] Morrison, M. A., in Electron-Molecule and Photon-Molecule Collisions, Rescigno, T., McKoy, V., and Schneider, V. (Eds.) Plenum Press, New York, (1978) 15
[21] Gianturco, F. A., and Rahman, N. K., J. Phys. B: Atom. Molec. Phys. 10 (1977) L219.
[22] Rudge, M.R.H., J. Phys. B: Atom. Molec. Phys. 13 (1980) 1264.
[23] Garrett, W. R., Phys. Rev. A 11 (1975) 509.
[24] Rohr, K., and Linder, F., J. Phys. B: Atom. Molec. Phys. 9 (1976) 2521.
[25] Fabrikant, I. I., Sov. Phys. JETP 46 (1977) 693.
[26] Domke, W., and Cederbaum, L. S., J. Phys. B: Atom. Molec. Phys. 14 (1980) 149.
[27] Dube, L., and Herzenberg, A., Phys. Rev. Lett. 38 (1977) 820.
[28] Wigner, E., Phys. Rev. 73 (1948) 1002.
[29] O'Malley, T. F., Rosenberg, L., and Spruch, L., Phys. Rev. 125 (1962) 1300.
[30] O'Malley, T. F., Phys. Rev. A (1965) 1668.
[31] Damburg, R. J., J. Phys. B: Atom. Molec. Phys. 1 (1968) 1001.
[32] Gailitis, M., Theor. and Math. Phys. 3 (1970) 572 [Translated from Teor i Mat. Fiz. 3 (1970) 364].
[34] Faisal, F.H.M., J. Phys. B: Atom. Molec. Phys. 1 (1968) 181; ibid, 1 (1908) 187.

PHYSICS OF ELECTRONIC AND ATOMIC COLLISIONS
S. Datz (editor)
© North-Holland Publishing Company, 1982

ELECTRONIC AND ATOMIC IMPACTS
ON LARGE CLUSTERS

Jürgen Gspann

Institut für Kernverfahrenstechnik der Universität
und des Kernforschungszentrums Karlsruhe,
Postfach 3640, 7500 Karlsruhe
Federal Republic of Germany

Describing first the generation and properties of molecular
beams of large Van der Waals clusters such as speed distribution, cluster size distribution, and internal temperature
of the clusters, the review then features the results of electronic impacts on large clusters: metastable electronic cluster excitations, ejection of positive cluster ions of less
than 100 atoms from much larger parent clusters, and ionization of the large clusters. Atomic impacts at thermal energies
are treated with respect to the scattering cross section of
the clusters, their drag coefficient in free molecular flow,
and the peculiarities of impacts on helium clusters of either
isotope.

INTRODUCTION

Impact phenomena involving the clusters of molecules or atoms in condensed molecular beams /1/ most often occur when the clusters are to be ionized by electron impact for application or diagnostic purposes, or when they interact with carrier,
background, or crossjet gas. Accordingly, the energies of impact to be considered
here are about 20 to 200 eV in the case of electronic impacts, and thermal energies
in the case of atomic or molecular impacts, respectively. Applications of cluster
beams aim at so widespread goals as, e. g., fuel injection into nuclear fusion
devices /2/, epitaxial film growth /3/, or internal targets in storage rings for
particle physics /4/. Usually, one is interested in beams of high mass flow density
which, as a rule, contain rather large clusters. In the following, impacts on Van
der Waals clusters of more than 1000 atoms or molecules are considered.

GENERATION AND PROPERTIES OF MOLECULAR BEAMS OF LARGE CLUSTERS

An experimental set-up for the generation and investigation of beams of large Van
der Waals clusters is shown schematically in Figure 1. A partly condensing nozzle
flow of expanding precooled gas is repeatedly skimmed in differential pumping
stages to yield a sharply bounded beam of clusters in high vacuum. Typical dimensions of the nozzle are 0.1 mm throat diameter, 25 mm length of the diverging part,
and 10⁰ full angle of divergence.

Nozzle and feed gas are often cooled by the liquid phase of the respective cluster
substance under atmospheric pressure. The nozzle inlet pressures have then to be
held below 1 bar. Higher nozzle temperatures are used as well, requiring higher
inlet pressures for condensation to occur in the expanding flow /5/.

Velocity distribution

As indicated in Figure 1, the cluster beam can be investigated by means of a time-

*J. Gspann*

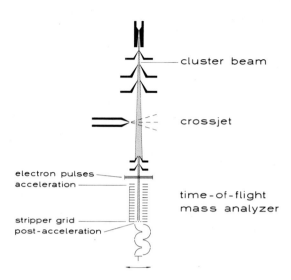

Figure 1
Experimental set-up for cluster beam generation
and investigation

of-flight spectrometer which is especially designed for the mass analysis of large
clusters.  An electron beam pulse of typically 2 µs duration intersects the cluster
beam, ionizing some of the clusters whose times of arrival at the end of the flight
path serve to determine the cluster velocity distribution.  Cluster beams generally have very narrow time-of-flight distributions, the full width at half maximum
intensity being about 3 to 5 percent of the most probable time of flight in our
experiments. (Example cases can be seen in Figures 4 and 9).  In terms of the speed
ratio, i. e. the ratio of the average speed to the most probable thermal speed in
the moving frame, this corresponds to values between 55 and 33.

## Size distribution

The mass-to-charge ratio of the cluster ions is derived from the shift of the arrival times which results from accelerating the cluster ions in front of the last
part of their flight path /6/.  For the case of a $N_2$ cluster beam, Figure 2 shows
the measured distribution of cluster sizes,  together with a fitted lognormal distribution.  In order to obtain the size distribution of the original neutral clusters, the cluster beam is deflected by a carbondioxide crossjet, as indicated in
Figure 1.  The cluster sizes are separated since smaller clusters are more strongly deflected than bigger ones /7/.  Combining measurements of the spatial distribution of the cluster size N with those of the molecular intensity, as obtained by
a stagnation pressure gauge movable in and out of the plane of the beam and jet
axes, the size distribution f(N) can be derived without having to assume a particular size dependence of the ionization cross section.  The result illustrated in
Figure 2 is thought to be rather typical, with the full width at half maximum of
the distribution being about as large as the most probable size.  This may seem a
rather broad distribution but if, as is often the case, only the order of magnitude of the cluster sizes matters, i. e. on a logarithmic scale, the size distribution in a cluster beam is indeed quite narrow.

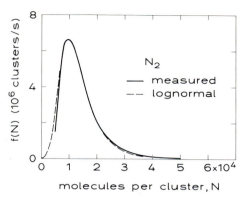

Figure 2
Measured distribution of $N_2$ cluster sizes
and deviations of fitted lognormal distribution

As shown in Figure 2, a lognormal distribution (or more correctly: its probability density function) may be scaled to reproduce the experimental curve rather well:

$$f(N) = A(\sqrt{2\pi}\ \sigma\ N)^{-1}\ \exp\ \{-(\ln N - \mu)^2/2\sigma^2\} \tag{1}$$

with $A = 8.610 \times 10^6$ clusters/sec; $\mu = 9.399$; $\sigma = 0.479$. Sufficiently deviating portions are indicated by dashed lines in Figure 2. The "gap" in the experimental curve at the small cluster sizes is expected as nucleation stops before cluster growth and late-coming nuclei re-evaporate.

Finally, it has to be noted, however, that we observe very often cluster size distributions more complicated than that shown in Figure 2, predominantly with two size fractions which may also differ somewhat in their average speed /6/. The rather complicated supersonic expansion with heat addition due to condensation which takes place in the slender nozzle cone is thought to be the reason for these phenomena.

## Cluster temperatures

The internal temperatures of the clusters can be derived from electron diffraction measurements in which a beam of electrons of about 50 keV energy intersects the cluster beam, generating Debye Scherrer diffraction patterns /8/. The diameter of the diffraction rings gives the lattice parameter which allows to determine the lattice temperature by comparison with bulk lattice data if possible effects of the finite cluster size are accounted for by extrapolating to infinitely large sizes.

This method of evaluation is only possible for clusters with more than about 1000 atoms or molecules since smaller clusters solidify in non-crystalline icosahedral structures. The observed temperatures of the large clusters are practically independent of the conditions of generation, obviously representing a kind of final temperature characteristic for the cluster substance.

We can try to estimate the cluster temperature in the following way: The average lifetime $\tau$ of an atom before it evaporates from a surface of temperature T is given by /9/

$$\tau = \tau_0 \exp (u_0/kT) \tag{2}$$

where $\tau$ is the period of vibration of the surface atom, and $u_0$ its heat of sublimation. A cluster becomes unable to cool down by evaporation during its time of
flight from the source region to the electron beam intersection region, which is
typically of the order of $10^{-3}$ s, if the residence time $\tau$ of its surface atoms is
much longer, say $10^{-2}$ s. With a typical period of vibration of $10^{-12}$ s one then
obtains

$$T = (k \ln 10^{10})^{-1} u_0. \tag{3}$$

Using bulk data for the heat of sublimation /10/, Figure 3 illustrates that this
relation fits quite nicely the observed temperatures of clusters of the noble
gases except helium which have recently been reported by the Orsay group /8/.

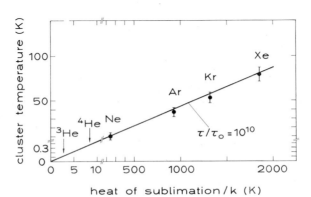

Figure 3
Internal temperatures of rare gas clusters

These authors have related their data to the well depth of the respective interatomic potentials not giving, however, an estimate of the constant of proportion-
ality. Equation (3) is thought to provide also reasonable estimates for the temperatures of clusters of the helium isotopes.


ELECTRONIC IMPACTS

The impact of electrons of sufficient energy primarily leads to the ionization of
the clusters. Two other effects more recently observed with large helium clusters
will be described first, however: the excitation of metastable electronic cluster
states /11/ and the ejection of "minicluster ions" with less than 100 atoms from
parent clusters of some million atoms /12/. Figure 4 gives examples of time-of-
flight distributions showing these effects while Figure 5 illustrates the supposed
underlying processes.

Let us consider Figure 1 again in order to understand how these effects show up:
The end of the cluster flight path is given by either the wires of a copper mesh
of about 45 % transparency or, for those clusters passing throught the mesh holes,
the first dynode of an electron multiplier. Positively charged clusters are accelerated between the mesh and the multiplier by up to 4.5 kV. This is by far too
small an acceleration for clusters of millions of atoms to become fast enough to

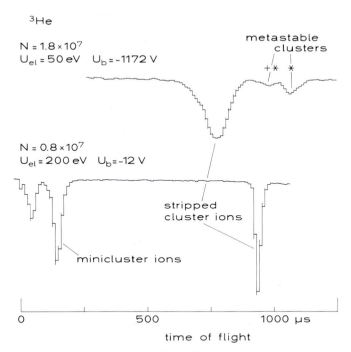

Figure 4
Time-of-flight distribution of $^3$He clusters of N atoms
average size indicating metastable species (upper trace)
and minicluster ions of 85 atoms (lower trace). $U_{el}$ is
the energy of the impinging electrons, $U_b$ the ion acceleration voltage. Electron current: 100 µA

liberate electrons at the first dynode. However, large ionized clusters hitting
the mesh wires are stripped of most of their mass there, the very small fragment
ions then being able to release secondary electrons at the first multiplier dynode
after a practically negligible additional flight time.

It turns out, however, that signals are generated also at the times when the large
clusters passing through the mesh holes reach the first dynode. Figure 4 shows in
its upper part an exemplary time-of-flight distribution obtained with $^3$He clusters
of $1.8 \times 10^7$ atoms, on the average, which are hit by electrons of 50 eV. The last
peak appears at the time when clusters of the original speed have travelled the
distance from the electron beam to the first dynode, irrespective of any electric
field. It is therefore ascribed to neutral clusters, while the middle peak is
caused by ionized clusters of the original size. The first large peak is due to
mass stripped cluster ions. As will be discussed later, the last two peaks become
possible by excitation energy transfer to the multiplier dynode from metastable
cluster excitations.

The lower trace of Figure 4 is obtained with $^3$He clusters of nearly the same size
but only 12 V instead of about 1.2 kV accelerating voltage $U_b$ in front of the
13,6 cm field-free flight distance ending at the copper mesh. While the large
cluster ions are practically unaffected by this low potential, another peak of
much smaller cluster ions is clearly resolved. More detailed investigation shows

that they contain 85 ± 5 atoms, on the average, and have passed the mesh holes.
We call them "minicluster ions" in order to distinguish them from the background
atomic ions forming the first peak of the lower trace of Figure 4 as well as from
the much larger cluster ions corresponding in size to the original neutral clusters.  Crossjet cluster beam deflection shows that a neutral equivalent to the
minicluster ions is not yet present in the original cluster beam.  With $^4$He cluster beams, the minicluster ions contain 68 ± 5 atoms, on the average.

An explanatory model for the observed phenomena is outlined in Figure 5. An electron
of, e. g., 50 eV energy has a mean free path for elastic scattering in liquid $^4$He
of the order of 1/10 of the diameter of a liquid helium droplet containing $10^6$
atoms, which amounts to about 45 nm.  The impinging electron therefore travels
along a random path within the cluster, eventually exciting or ionizing an atom.

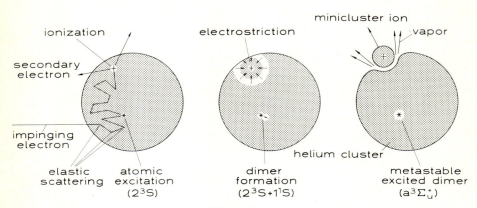

Figure 5
Metastable electronic excitation and minicluster ion ejection as
result of electron impact on large helium clusters

## Metastable electronic cluster excitations

The respective atomic cross sections used to estimate the excitation probabilities
are shown in Figure 6 /11/. They have been compiled from the work of various
authors /13-19/ and are simplified by interpolating more detailed structures
exemplified by the crosses in Figure 6.  The largest atomic excitation cross section is that leading to the triplet state $2^3$S. It is about two orders of magnitude smaller than the elastic scattering cross section, and about one order of
magnitude below the atomic ionization cross section.

The observed energy dependences of the helium cluster excitations can indeed be
explained by using only these three atomic cross sections /11/. Figure 7 shows,
by solid lines, the results of corresponding calculations taking into account the
energy lost to secondary electrons which had previously /11/ been disregarded.
In this latter case, indicated by broken lines, a probability scale unit 15 %
smaller than shown applies.  The fit to the experimental data (circles) is obtained
by choosing that total length L of the random electron path within the cluster
which provides at 70 eV the observed ratio of the probabilities for ionized and
neutral metastable cluster excitation.  The energy dependence of L is then assumed
to be proportional to that of the elastic scattering mean free path and is shown
in the upper part of Figure 7.  It might be mentioned, however, that a reduction
of the ionization cross section by a constant factor of order one would also allow
to fit the measured excitation probability ratio if a correspondingly lengthened

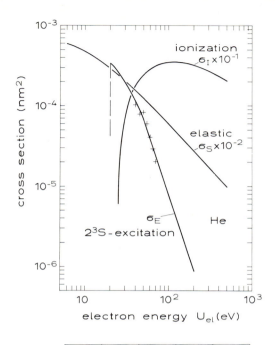

Figure 6
He atom cross sections in electron collisions. Crosses from Ref. 18

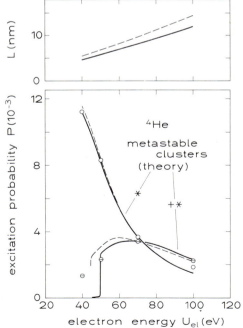

Figure 7
Calculated excitation probabilities for neutral and ionized $^4$He clusters considering (solid) or disregarding (dashed) the energy lost to secondary electrons. L is the assumed length of the random electron path in the cluster

path L is assumed.  A more exact absolute determination of the excitation probability which is now missing due to a lack of knowledge of the electron multiplier
efficiency would allow to fix L, and in turn the effective ionization cross section.

The lifetime of the metastable atomic $2^3S$ state in liquid helium is known to be
only 15 μs /20/, however, which is much shorter than the experimental flight time
of the metastable clusters of about 1 ms.  Such long lifetimes are only known for
the metastable molecular triplet state $a^3\Sigma^+_u$ /21/ into which the $2^3S$ state atoms
are thought to decay by reaction with a ground state atom /20/.  The present results offer strong support for this conjecture as they show longlived metastable
states to be formed according to the atomic excitation probability.  The metastable
excited dimer is thought to sit in a little bubble within the liquid helium which
gives way to the large electron orbit.


Minicluster ion ejection

The other experimental feature depicted in Figure 5, the minicluster ion ejection,
is believed to proceed as follows /12/: Positive charge carries in liquid helium
are known to be rather massive entities bound by electrostrictive polarization of
the surroundings of the charge /22/.  Immediately after the ionization, this elec-
trostriction must lead to a temporarily depleted region farther outwards which is
heated, and therefore pressurized, by the released polarization energy.  If this
process occurs sufficiently near to a free surface, the electrostrictively bound
minicluster ion may be expelled.

An estimate shows that the released energy is about an order of magnitude larger
than the energy required for the ejection from the parent cluster.  Moreover, the
sizes of the observed minicluster ions can be explained amazingly well along these
lines /12/:

Let us assume that the expelled minicluster ion is at first in pressure and temperature equilibrium with its carrying vapor plume.  The left hand side of the
classical formula for the vapor pressure of a singly charged droplet of radius R
of a dielectric medium of polarizability $\alpha$, /23/

$$kT \ln(p/p_{sat}) = (2\sigma/n_1 R)-(\alpha e^2/2R^4),  \tag{4}$$

then vanishes, determining an equilibrium droplet radius.  The number of atoms per
droplet, $(4\pi/3)n_1 R^3$, has the equilibrium value

$$N_{eq} = \pi\alpha e^2 n_1{}^2/3\sigma  \tag{5}$$

which depends on the temperature T only through the atom number density in the liquid, $n_1$, and the surface tension $\sigma$.  Figure 8 shows the corresponding variation
as calculated using bulk values for $n_1$ and $\sigma$.  The vertical arrows indicate the
temperatures leading to saturation pressures which correspond to the respective
experimental vacuum chamber pressures.

Evidently, at sufficiently low temperatures the equilibrium size approaches a constant final value which already agrees fairly well with the respective observed
size.  In the $^3He$ case, the calculated final size is 85.4 atoms per minicluster
ion, in perfect agreement with the experimental value of 85 ± 5 atoms.  In the $^4He$
case, the calculated final equilibrium size of 61.7 atoms is somewhat low compared
to the experimental value of 68 ± 5 atoms per minicluster ion.

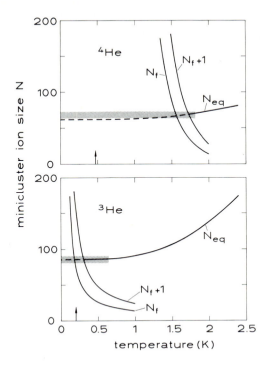

Figure 8
Equilibrium sizes $N_{eq}$ and
adiabatic final and penultimate sizes $N_f$ and $N_{f+1}$
of helium minicluster ions.
Hatched area: experimental
mean sizes

Considering, however, that the very low temperatures are in reality reached by
free evaporation of the minicluster ion after its carrying plume has faded away,
this latter discrepancy becomes also explicable. The minicluster ion evaporates
adiabatically as long as its heat content permits. Another final size $N_f$ is
therefore given by the number of atoms whose combined heat content at the temperature T just equals the latent heat h which would be needed to vaporize one further atom

$$N_f = h(T)/\int_0^T c_s(T)dT \qquad (6)$$

where $c_s$ is the specific heat per atom. Thus, miniclusters of $N_f$, or fewer, atoms
are absolutely stable at the temperature T if they are adiabatically isolated.
The temperature jump $\Delta T$ connected with the evaporation of the last possible atom
is implicitly given by the relation

$$\int_T^{T+\Delta T} c_s dT = h(T+\Delta T)/(N_f+1) \qquad (7)$$

Both the final and the penultimate sizes, $N_f$ and $N_f+1$, are indicated in Figure 8.
Now, if a $^4$He minicluster ion follows the equilibrium curve due to the interaction
with its carrying plume until only one last atom remains to be evaporated adiabatically, this last jump, along a nearly horizontal path in Figure 8, leads to a final
size of 68 atoms. Again, a perfect agreement with the measured size of 68 ± 5

atoms per minicluster ion is observed.  Obviously, the lower specific heat of liquid $^4$He prevents only the $^4$He minicluster ions from reaching the limiting equilibrium size.

## Ionization

According to the described model of the minicluster ion ejection, an ionization of
the large parent cluster can only result from atomic or molecular ionization processes taking place sufficiently deep inside the large cluster, so that the near
surrounding of the charge is isotropic.  Of course, the secondary electron has to
escape from the cluster in order to produce a net positive charge.  In that respect, the minicluster ion ejection model seems to conflict another model of the
cluster ionization assuming a rather restricted depth of secondary electron escape
of only a few atomic layers /24/.  This model was conceived to explain the observed
deficiency of the absolute ionization cross section of clustered compared to free
molecules.  If, however, the escape depths of secondary electrons and of minicluster ions were comparable, large clusters should nearly always remain uncharged, in
contrast to the experimental findings.  There is, however, a number of arguments
against a thin secondary electron escape layer:

- As soon as the secondary electrons become subexcitation electrons /25/, i. e.,
their energies are lower than that required for electronic excitation of the medium, they can travel long distances without losing  sufficient energy to get ther-
lalized.

- Model calculations /26/ apparently confirming the concept of a thin secondary
electron escape layer are found to overestimate the secondary electron energy loss
below the excitation threshold by up to two orders of magnitude /27/.

- There are experimental data with size separated nitrogen clusters showing a linear size dependence of the ionization cross section of clusters with up to $4 \times 10^4$
molecules /7/.

- Moreover, experiments with hydrogen clusters show a 40 % deficiency of the effective ionization cross section per molecule already for clusters of less than
thousand molecules which should be completely comprised within the escape layer
/28/.  Figure 9 presents these data together with the values expected for an escape depth D of 5.5 molecular diameters.

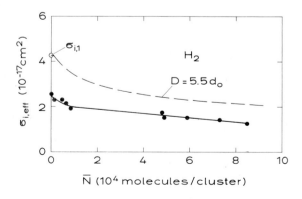

Figure 9
Measured molecular
ionization cross sec-
tions of hydrogen
clusters (solid curve)
and expected values if
the secondary electron
escape depth is 5.5
molecular diameters
(dashed)

Thus, a full understanding of the observed effective ionization cross section of clustered atoms or molecules seems to be lacking still. While with respect to charging a given cluster the secondary electron trapping plays the same role as ejection of both the secondary electron and a minicluster ion, it is not easy to imagine how in the integral determinations of the ion currents /24,28/ many of the minicluster ions could have escaped being measured, in which case they would have given rise to an apparent deficiency of the ionization probability.

Possibly, a certain fraction of the secondary electrons is trapped nearly regardless of their origin. The escape depths of the minicluster ions and of the escaping secondary electrons should at least be very disparate, the latter being many atomic or molecular layers thick, presumably.

In any case, the ovservation of the ejected minicluster ions has added a new degree of freedom for interpreting experimental results. Let us consider, for example, the generation of negatively charged clusters. As evidenced by Figure 10, inverting the polarity of the accelerating voltage reveals negatively charged clusters coexisting with positive cluster ions of the same mass /6/. (The negative post-

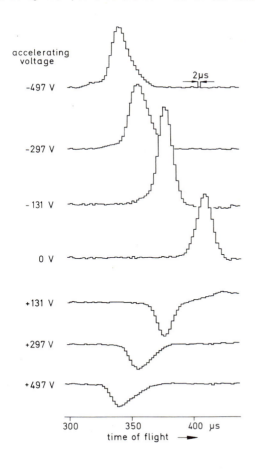

Figure 10
Negatively and positively charged hydrogen clusters of $6.5 \times 10^4$ molecules are shown to coexist by changing the polarity of the accelerating voltage $U_b$. Electron energy $U_{el}$: 44 eV

acceleration voltage used with the electron multiplier prevents the observation of
negatively charged clusters when measuring with the multiplier). These negative
cluster ions may result from primary electron capture as discussed previously /6/
but also from trapping a secondary electron with the primary electron and a minicluster ion escaping.

Furthermore, the hitherto disregarded  release of energy due to the medium polarization should markedly influence the results of multiple cluster ionization. Heavy
electron bombardment of a cluster should lead, in the limit, to a bursting swarm
of minicluster ions in a cloud of vapor.


## ATOMIC IMPACTS

With respect to atomic collisions at thermal energies the larger clusters occupy
an interesting intermediate position between molecules and the small particles
treated, e. g., in aerosol physics: the condensed matter extension is comparable
to the spatial range of the intermolecular forces. Correspondingly, a description
of the gas-cluster interaction has to embody gas kinetic aspects which are not
taken care of in conventional rarefied gas dynamics.

Figure 11 illustrates some idealized cases of interaction /29/. The hatched area
visualizes approximately the range of the intermolecular forces. In the upper
part of the figure trajectories of elastic collisions with a rigid target are
shown. Atoms or molecules with impact parameters b larger than that called
$b_{orbiting}$ are affected only by the attractive part of the potential while those
with smaller impact parameters probe also the repulsive part. The dashed trajectories would be followed in corresponding collisions with a hard smooth sphere.

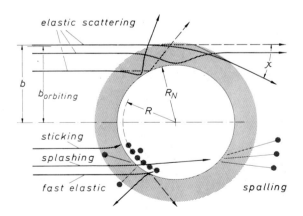

Figure 11
Interaction mechanisms between a cluster and impinging molecules or atoms

In the lower part of Figure 11 the lowermost  trajectory represents a collision
which is sufficiently energetic to probe the molecular roughness of the cluster
surface. If the impact energy is large compared to the binding energy of the cluster constituents, a splashing of cluster material and a more or less deep penetration of the projectile may result, eventually leading also to spallation at the
cluster backside. Softer impacts may result in completely inelastic sticking of

the projectile. For the molecule-cluster interaction potential, disregard of the atomistic cluster structure and integration over a sphere of homogeneously distributed Lennard-Jones 12-6 potentials gives the following expression: /7/

$$V(r) = \frac{N}{(r^2-R^2)^3} 4\varepsilon\sigma^6 \left\{ \frac{\sigma^6(r^6+(21/5)r^4R^2+3r^2R^4+(1/3)R^6)}{(r^2-R^2)^6} -1 \right\}$$  (8)

for $r > R$

The radius R is chosen to be

$$R = R_N - R_1$$

with $R_N = (3mN/4\pi\rho)^{1/3} = N^{1/3}R_1$

This choice ensures the correct limits: for N = 1 the Lennard-Jones potential,

$$V(r) = 4\varepsilon \left\{ (\sigma/r)^{12} - (\sigma/r)^6 \right\}$$  (9)

and for N → ∞ the corresponding plane surface potential /30/

$$V(z) = \frac{4\pi\varepsilon\sigma^3}{3} \frac{\rho}{m} \left\{ \frac{1}{15}(\frac{\sigma}{z})^9 - \frac{1}{2}(\frac{\sigma}{z})^3 \right\}$$  (10)

with $z = r-R$

Figure 12 shows examples for these potentials /29/. The solid curves indicate the

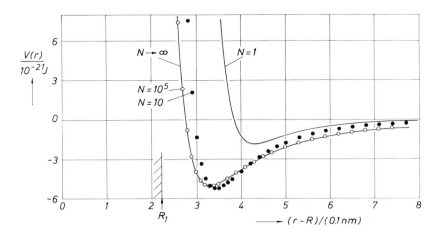

Figure 12
Intermolecular potentials of clusters of N molecules as a function of the distance r-R from the centers of the outermost cluster molecules located at the radial position R

limiting cases while the dots are calculated for clusters of 10 and $10^5$ molecules
per cluster, respectively. The numerical values are taken to represent the interaction of carbondioxide with nitrogen clusters: $R_1 = 0.222$ nm, $\varepsilon = 1.82 \times 10^{-21}$ J,
$\sigma = 0.384$ nm. Potentials as given by equation (8) have been used in evaluating
total scattering cross sections as well as calculating cluster drag coefficients.

## Total scattering cross section of clusters

The total scattering cross section of clusters, $\tilde{\sigma}$, depends on the weak elastic interaction with colliding atoms or molecules at large distances. Inserting the potential $V(r)$, equation (8), into the classical deflection function $\chi(b)$, one obtains, approximately /31/

$$\tilde{\sigma} = \pi R^2 + N^{1/3} \tilde{\sigma}_1 \tag{11}$$

where $\sigma_1$, denotes the corresponding cross section of the uncondensed atoms or molecules of the cluster. Evidently, the long-ranging intermolecular forces add a
contribution proportional to the cluster circumference, $N^{1/3}\sigma_1$, to the geometrical
part, $\pi R^2$, of the cluster scattering cross section.

Due to the mutual shielding of the cluster constituents, their effective scattering cross section, $\tilde{\sigma}/N$, is much smaller than that of the free constituents, $\sigma_1$.
Figure 13 shows the measured effective total scattering cross section of nitrogen
molecules for potassium atoms as a function of the nozzle inlet pressure $p_0$ and

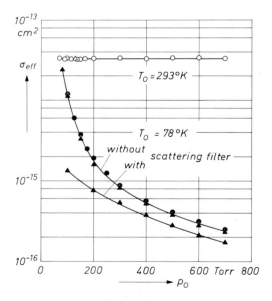

Figure 13
Effective total scattering cross section per $N_2$
molecule from K beam attenuation as a function
of the nozzle inlet pressure $p_0$ for uncondensed
(nozzle temperature $T_0 = 293$ K) and condensed
($T_0 = 78$ K) molecular beams of nitrogen

the temperature $T_0$. With the nozzle at room temperature, the effective scattering cross section does not depend on the inlet pressure and agrees fairly well with literature data for the $N_2$ molecule if the angular resolution of the apparatus ($0.75^\circ$) is taken into account. On the other hand, at 78 K nozzle temperature, the effective scattering cross section per clustered nitrogen molecule decreases from 5 to 0.5 percent of the free molecule value if the nozzle inlet pressure is increased from 100 to 700 Torr. Using equation (11), an average cluster size N increasing from 160 to 6500 has been derived from these data and found to be in fairly good agreement with later mass spectrometric determinations of the approximate mean size of singly ionized clusters of the same beams /32/. (The "scattering filter" mentioned in Figure 13 serves to distinguish clusters from residual uncondensed molecules by taking advantage of the higher persistence of the clusters when the beam passes a gas-filled scattering chamber).

## Drag coefficients of clusters

The drag coefficient $C_D$ of a body immersed in a flow of mass density $\rho_f$ and average velocity of the relative motion u is defined by the relation for the drag force D:

$$D = C_D \, \rho_f (u^2/2)F \tag{12}$$

where F is the body cross section perpendicular to the flow. Since the momentum flow density of the flow is $\rho_f u^2$, a drag coefficient $C_D = 2$ means that all the intercepted momentum flow is transferred to the body.

Using the potential V(r), equation (8), one can calculate the drag exerted on a cluster in free molecular flow, i. e., when the cluster diameter is small compared to the mean free path in the flow, as is usually the case. Figure 14 shows calculated drag coefficients of nitrogen clusters in xenon or carbondioxide flows as a function of the cluster radius $R_N$ and the relative velocity u, as well as measured data derived from crossjet cluster beam deflections /29/.

Obviously, the long-ranging intermolecular forces enhance the drag on the smaller clusters appreciably, attracting momentum flow which would not be intercepted geometrically. The probably more realistic results in the given case are denoted as "sticking" in Figure 14. They are obtained assuming sticking with complete momentum transfer to occur in all collisions with impact parameters smaller or equal than that leading to the orbiting trajectory, while elastic interaction is assumed otherwise. The half-filled dots representing carbondioxide crossjet experiments indicate even higher momentum transfer, however, possibly resulting from backward ejection, or splashing, of cluster molecules, cf. Figure 11.

The calculations discussed so far refer to impinging flows of infinitely high speed ratio, viz. uniform molecular velocity. Corresponding drag enhancements due to the longranging intermolecular forces have been shown to apply also at finite or vanishing speed ratio, however /33/.

## Peculiarities of helium clusters

In scattering as well as in crossjet deflection experiments helium clusters show peculiar velocity and size dependences which distinguish them from other Van der Waals clusters /34-36/. Here I want to discuss as an example the size dependence of the drag coefficients. As shown in Figure 15, the drag coefficients of $^3$He as well as $^4$He clusters are found to be smaller than 2, at relative velocities of the impinging free molecular flow of about 370 m/s. According to the foregoing discussion, that means that the geometrically intercepted momentum flow is not fully transferred to the helium clusters. A simple explanation would be that part of the impinging molecules or atoms pass through the whole cluster without being com-

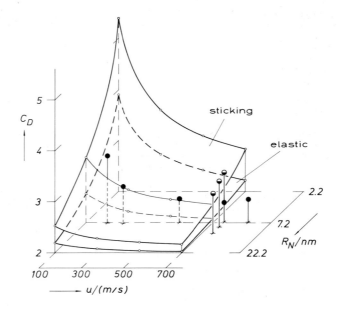

Figure 14
Measured and calculated drag coefficients of nitrogen
clusters as function of the relative speed u and the
cluster radius $R_N$ (= 2.22 $N^{1/3}$Å).  Black dots: Xe crossjet experiments; half-filled dots: $CO_2$ crossjet experiments

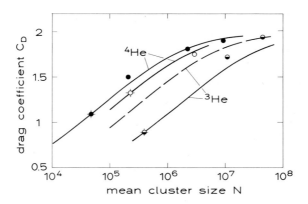

Figure 15
Drag coefficients of helium clusters as a function
of the cluster size.  Black and half-filled dots:
$CO_2$ crossjet; empty circles: Xe crossjet

pletely decelerated.  As helium remains liquid under its own vapor pressure even at zero temperature, there is at least some plausibility for this explanation. The solid curves of Figure 15 represent corresponding calculations fitted at the crossed experimental points by assuming a suitable constant drag coefficient for the immersed molecule or atom.  The dashed curve results from using the $^4$He fit at the full dots to predict the $^3$He values shown as half-filled dots.  Obviously, the overall trend of the experimental data is indeed reproduced.

On the other hand, estimating the penetration of a Xe atom of 400 m/s velocity into normalfluid $^4$He of 36 μP viscosity leads to a final depth of 50 Å, which is only about a tenth of the diameter of a $^4$He cluster of $10^6$ atoms.  For these estimate the whole set of semi-empirical sphere drag relations /37/ has been used, considering Mach and Reynolds number corrections because of the initially supersonic atomic speed as well as Knudsen number corrections since the projectile dimensions are comparable to the interatomic distances in liquid helium.

The Xe atom deceleration is much slower, however, and the penetration accordingly deeper, if the viscosity is assumed to vanish.  Moreover, the fit to the experimental points (empty circles) in Figure 15 corresponds to a deceleration, as shown by the dashed curve in Figure 16, which is rather similar to the vanishing viscosity case.  Hence, the experimental data seem to indicate that the helium clusters consist of an inviscid compressible continuum fluid.  This statement is preliminary, however, derived from work still in progress.

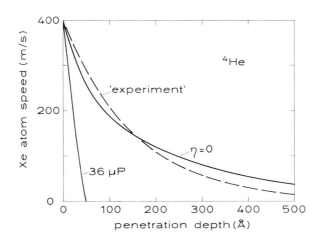

Figure 16
Calculated deceleration of a Xe atom in liquid $^4$He of indicated viscosity.  Dashed curve corresponds to solid curve through empty circles in Figure 15

CONCLUDING REMARK

In conclusion, I would like to note that in the described studies of impacts on large clusters one takes advantage of all the features of molecular beam work, using isolated species, mass spectrometers, time-of-flight analyzers, crossed beams etc., but the information obtained pertains primarily to the properties of condensed matter and its surface.

REFERENCES

/1/ E.W. Becker, K. Bier and W. Henkes, Z. Phys. 146 (1956) 333
/2/ E.W. Becker, H. Falter, O.F. Hagena, P.R.W. Henkes, R. Klingelhöfer,
    H.O. Moser, W. Obert and I. Poth, Nuclear Fusion 17 (1977) 617
/3/ T. Takagi, I. Yamada and H. Takaoka, Surface Sci. 106 (1981) 544
/4/ J. Gspann and H. Poth, KfK-Report 3198 (1981)
/5/ O.F. Hagena, in: P.P. Wegener (ed.), Molecular Beams and Low Density Gas
    Dynamics (Marcel Dekker, New York, 1974) 93
/6/ J. Gspann and K. Körting, J. Chem. Phys. 59 (1973) 4726
/7/ J. Gspann and H. Vollmar, in: K. Karamcheti (ed.), Rarefied Gas Dynamics
    (Academic, New York, 1974) 261
/8/ J. Farges, M.F. de Feraudy, B. Raoult and G. Torchet, Surface Sci. 106 (1981)
    95
/9/ J. Frenkel, Kinetic Theory of Liquids (Oxford University, London, 1946) Chap.I
/10/ R.K. Crawford, in: M.L. Klein, J.A. Venables (eds.), Rare Gas Solids (Academ-
     ic, New York, 1977) Chap. 11
/11/ J. Gspann and H. Vollmar, J. Chem. Phys. 73 (1980) 1657
/12/ J. Gspann and H. Vollmar, J. Low Temp. Phys. 45 (1981) 343
/13/ P.T. Smith, Phys. Rev. 36 (1930) 1293
/14/ R.W. LaBahn and J. Callaway, Phys. Rev. 147 (1966) 28; also L.G. Christophorou,
     Atomic and Molecular Radiation Physics (Wiley, London, 1971) p. 282
/15/ R.H.J. Jansen, unpublished thesis, Amsterdam 1975, as cited by B.H. Bransden
     and M.R.C. McDowell, Phys. Rep. 46 (1978) 249
/16/ H.H. Bongersmaa, F.W.E. Knoop and C. Backx, Chem. Phys. Lett. 13 (1972) 16
/17/ S. Trajmar, Phys. Rev. A8 (1973) 191
/18/ G.B. Crooks, R.D. DuBois, D.E. Golden and M.E. Rudd, Phys. Rev. Lett. 29
     (1972) 327
/19/ L. Vriens, J.A. Simpson and S.R. Mielczarek, Phys. Rev. 165 (1968) 7
/20/ J.W. Keto, M. Stockton and W.A. Fitzsimmons, Phys. Rev. Lett. 28 (1972) 792
/21/ R. Mehrotra, E.K. Mann and A.J. Dahm, J. Low Temp. Phys. 36 (1979) 47
/22/ K.R. Atkins, Phys. Rev. 116 (1959) 1339
/23/ M. Volmer, Kinetik der Phasenbildung (Steinkopff, Dresden, 1939) Chap. 4B;
     See also Ref. 9, Chap. VII
/24/ H. Falter, O.F. Hagena, W. Henkes and H.v. Wedel, Int. J. Mass Spectrom. Ion
     Phys. 4 (1970) 145
/25/ R.L. Platzman, Radiation Research 2 (1955) 1
/26/ F. Bottiglioni, J. Coutant and M. Fois, Phys. Rev. A6 (1972) 1830
/27/ J. Gspann, Europ. Conf. on Atomic Physics, Heidelberg, 1981, Europhysics
     Conf. Abstracts, Vol. 5A, 423
/28/ W. Henkes and F. Mikosch, Int. J. Mass Spectrom. Ion Phys. 13 (1974) 151
/29/ J. Gspann and H. Vollmar, in: R. Campargue (ed.), Rarefied Gas Dynamics
     (CEA, Paris, 1979) 1193
/30/ T.L. Hill, J. Chem. Phys. 16 (1948) 181
/31/ H. Burghoff and J. Gspann, Z. Naturforschung 22a (1967) 684
/32/ J. Gspann, Entropie No. 42 (1971) 129
/33/ J. Gspann, Progress in Astronautics and Aeronautics 74 (1981) 959
/34/ E.W. Becker, J. Gspann and G. Krieg, Proc. 14th Int. Conf. on Low Temp. Phys.
     Otaniemi 1975 (North Holland, Amsterdam, 1975) Vol. 4, 426
/35/ J. Gspann and H. Vollmar, J. Physique 39 (1978) C6-330
/36/ J. Gspann, Physica 108B (1981) 1309
/37/ C.B. Henderson, AIAA-J. 14 (1976) 707

PHYSICS OF ELECTRONIC AND ATOMIC COLLISIONS
S. Datz (editor)
© North-Holland Publishing Company, 1982

Alignment and Orientation in Ion-Atom/Molecule Collisions[*]

D.H. Jaecks, A. Goldberger, M. Natarajan, D. Montgomery, D. Mueller

Behlen Laboratory of Physics
University of Nebraska
Lincoln, NE 68588-0111

Photon correlation experiments provide new and detailed information
about inelastic process in ion-atom/molecule collisions.  Parameters
used to describe the emitted radiation can be written in terms of
source parameters which in turn reflect the collision dynamics.
Some of these parameters are:  the Stokes vectors, $P_1$, $P_2$, $P_3$ the
degree of polarization $P^2 = P_1^2 + P_2^2 + P_3^2$, the relative phase of
orthogonal radiation components, and the overall symmetry of the
radiation angular distribution.  The use of these parameters in
gaining new insights into inelastic processes will be illustrated.

INTRODUCTION

The electron charge distribution of an excited atom or molecule reflects the
collision process by which it was formed; the results of the study of the sizes
and shapes of these electron charge distributions have led to new insights into
excitation mechanisms and continue  to provide new tests for various theoretical
collision models.

The atomic shape is, of course, manifested by different relative populations of
magnetic substates.  And the radiation from the decay of such states has a specific character being, in general, polarized and anisotropic.

The idea of studying the shape, or the relative populations of magnetic substates
by means of polarization analysis of the radiation is not new, for it was already
implicit in the early work of J.A. Smit in 1935.[1]  From simple symmetry considerations it is clear that the polarization and angular distributions of 2p - 1s
radiation from different magnetic substates directly reflect the shape of the
excited state.  This is illustrated in Fig. 1, where we consider a $2p_0$ - 1s transition in a hydrogen - like atom.  Here the radiation is preferentially emitted
perpendicular to the aligned $2p_0$ orbital with a polarization parallel to this
alignment.  It is clear that either polarization or angular distribution measurements will determine the direction of alignment and shape of the electron cloud.

We will see that in examples to be discussed here, this direction of alignment
provides critical tests of theoretical collision models.

Added information can be obtained if both the $2p_0$ and $2p_{\pm 1}$ states are coherently
excited.[2]  In this case, the emitted radiation carries with it the quantum mechanical difference between the excitation amplitudes, [2] and appears as an interference between the two degenerate states.  For 2p - 1s transitions this results
in a "tilting" or rotation of the dipole intensity or polarization pattern as
illustrated in Fig. 2.

We will limit our discussion to p - s transitions because they are the only types
that have been to date utilized in the study of ion-atom collisions.  This

limitation is due in part to the ease of analyzing p - s transition, but also because of the large number of important collision systems that lead to p - state excitation.

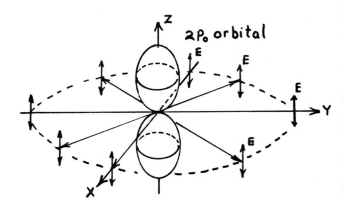

Fig. 1

Radiation from $2p_o$ emitted preferentially perpendicular to Z - axis with polarization parallel to Z.

Thus it is clear that measures of the angular distribution or polarization are specific signatures of the radiating atomic source. For coherent excitation and radiation, some of these important signatures are: the excitation intensities of each magnetic substate, $\sigma_o$ and $\sigma_1$, and the phase difference between the excitation amplitudes. In addition, the overall symmetry of the distribution is important; different theoretical collision models predict very specific symmetry axes.

The molecular orbital picture, extended by Fano and Lichten to diabatic states of many electron systems, [3] has provided a framework for interpreting a wide range of collision experiments. We shall see that the results of the measurements of the polarization or angular distribution of the dipole radiation often give unambiguous answers to important questions concerning the applicability of this model.

RADIATION MEASUREMENTS AND SOURCE PARAMETERS

Much has been written in the last decade on a variety of parameters to describe atomic radiation from atomic collisions.[2] The discussion that follows is the one the authors finds most useful and natural in thinking about ion atom/molecule collisions. There are three steps linking the measured radiation to the collision process: there is the measurement of the radiation from the collisionally produced source, there is the interpretation of the radiation properties in terms of source parameters, and then the interpretation of the source parameters in terms of a collision model.

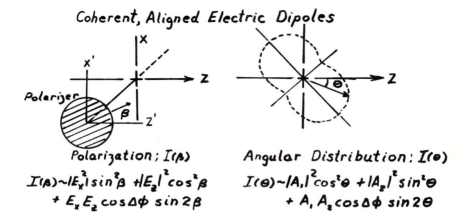

Fig. 2

The radiation is completely defined by a set of normalized Stokes's parameters $P_1$ $P_2$, and $P_3$.[4]

$$P_1 = \frac{I(0°,0°)-I(90°,0°)}{I(0°,0°)+I(90°,0°)}; \quad P_2 = \frac{I(45°,0°)-I(135°,0°)}{I(45°,0°)+I(135°,0°)}, \quad \text{and} \quad P_3 = \frac{I(45°,90°)-I(135°,90°)}{I(45°,90°)+I(135°,90°)}$$

Here $I(\beta,\alpha)$ is the intensity of the radiation polarized along a direction $\beta$, and $\alpha$ is an added phase shift before the intensity is measured. In ion beam experiments, it is generally most convenient to measure $I(\beta)$ in a direction perpendicular to the scattering plane and $\beta$ from the beam axis. Since we are primarily interested in interpreting the collision excitation in terms of the molecular orbital picture of the collision, it is most convenient to rewrite the Stokes parameters in terms of amplitudes and phase differences of magnetic substates, since the atomic amplitudes are the infinite separation limit of molecular orbital state amplitudes. The essential features of the Stokes parameters written in terms of source parameters can be illustrated by considering 2p - 1s transitions with no fine or hyperfine interaction.

If we view the radiation perpendicular to the x - z collision plane as in Fig. 2 we note a one - to - one correspondence between $\sigma_0 \sim |E_z|^2$ and $2\sigma_1 \sim |E_x|^2$, therefore we can write

$$I(\beta) = \sigma_0 + (2\sigma_1 - \sigma_0) \sin^2 \beta + (2\sigma_0\sigma_1)^{\frac{1}{2}}\cos\Delta\phi \, \sin2\beta$$

The Stokes parameters, written in terms of the source parameters become

$$P_1 = \frac{\sigma_0 - 2\sigma_1}{\sigma_0 + 2\sigma_1} \quad , \quad P_2 = \frac{2(2\sigma_1\sigma_0)^{\frac{1}{2}}\cos\Delta\phi}{(\sigma_0 + 2\sigma_1)} \quad , \quad P_3 = \frac{2(2\sigma_1\sigma_0)^{\frac{1}{2}}\sin\Delta\phi}{(\sigma_0 + 2\sigma_1)}$$

When a single final state is observed with a single phase $\Delta\phi$, the above satisfy the relationship $P_1^2 + P_2^2 + P_3^2 \equiv P^2 = 1$. The above expressions must be modified to include fine and hyperfine depolarization,[6] and the possibility of not observing pure states. The coherency matrix, P, is the ratio of the intensity of polarized light to the total intensity.[5]

If in a collision, more than one final state of the total system of target plus projectile is accessable, P is generally less than 1.

However, the above coherency condition can also, under certain conditions, be satisfied more generally if one observes radiation from more than one final state. In the collision of two particles, $M^+ + N$, resulting in the excitation of both N and $M^+$, one can think of the observed radiation from $N^*$ as coming from a statistical distribution of pure states, designated by $|n\rangle$ [5]

$$[M^+ + N^*]_1 \quad : \quad |1\rangle \quad : \quad I_1\,(\beta)$$

$$M^+ + N \rightarrow$$

$$[M^+ + N^*]_n \quad : \quad |n\rangle \quad : \quad I_n\,(\beta)$$

where n represents the collection of relevant quantum variables and the final momentum of the scattered particle. Writing the intensity $I\,(\beta) = \Sigma\, I_n\,(\beta)$ one can show the coherency matrix for this statistical distribution is

$$P^2 = 1 - \frac{4 \sum_{j<i}[2\sigma_1^i\,\sigma_0^j + 2\sigma_1^j\,\sigma_0^i - 2\,[2\sigma_1^i\,\sigma_0^j]^{\frac{1}{2}}\,[2\sigma_1^j\,\sigma_0^i]^{\frac{1}{2}}\cos(\Delta\phi_i - \Delta\phi_j)]}{[\,\sum_i 2\sigma_1^i + \sigma_0^i\,]^2}$$

The above expression assumes reflection symmetry in the collision plane. One can readily extend this to the case of no reflection symmetry if needed. Thus $P^2 \le 1$ since the sum is a positive quantity. $P^2 = 1$ if $\Delta\phi_j = \Delta\phi_i$ and $\sigma_0^i/\sigma_1^i = \sigma_0^j/\sigma_1^j$. That is, we speak of having a pure radiating, coherent state when the phase differences in each final state $|n\rangle$ are the same and when the overall intensities from each pure state of the statistical distributions differ from each other by a multiplicative constant. A statement that $|P| = 1$, for the case when many states of $M^+$ are accessable in a collision, would imply that the excitation of $N^*$ is independent of the variables of $M^+$.


EXPERIMENTAL RESULTS

A variety of ion/atom - atom/molecule collisions have been studied by photon co-incidence measurements over approximately the past decade. One underlying connection between all of the ion/atom - atom experiments has been the interpretation of the data within the framework of the molecular orbital picture of the Fano - Lichten model. There is no longer any doubt concerning the general efficacy of this picture, however the extent or range of its applicability has not been established. For example, experimental evidence to be discussed below, indicates the independent electron picture of the Fano - Lichten model does not work in cases of two electron excitation.


$He^+ + He$ COLLISIONS

The results of polarized photon measurements for this collision pair has been discussed many times before.[7] However, since the general results relate to subsequent measurements on other systems to be discussed, we review them here.

The linear polarization, $I(\beta)$, for radiation measured in a direction perpendicular to the scattering plane, for the collision process $He^+ + He \rightarrow He(3^3P) + He^+$ can be written

$$I(\beta) = C[(28\sigma_0 + 26\sigma_1) + 15\,[2\sigma_0\sigma_1]^{\frac{1}{2}}\cos\Delta\phi\,\sin 2\beta + (30\sigma_1 - 15\sigma_0)\sin^2\beta]$$

where $\Delta\phi$ is the phase difference between the magnetic substate amplitudes $a_o$ and $a_1$ where $|a_o|^2 \sim \sigma_o$ and $|a_1|^2 \sim \sigma_1$. From measurements of $I(\beta)$ for the $3^3P \to 2^3S$ transition at a variety of scattering angles of $He(2^3S)$, $\Delta\phi$ was found to be equal to $\pm 90°$.[7]

This rather general result can be understood in terms of the $He^+$ - He correlation diagram. The principle mechanism for the He (1s 3p) population is a $\Sigma_g$ - $\Sigma_g$ transition occurring at the crossing, shown in Fig. 3, which occurs as the two nuclei approach and recede from each other at a distance of about 4 a.u. Subsequent coupling between the two degenerate $\Sigma$ - $\pi$ states that lead to He(1s 3p) allows the electron distribution to rotate in the molecular frame as the collision proceeds. This $\Sigma$ - $\pi$ coupling is a necessary requirement for the electron distribution to be "frozen" in the fixed lab frame. The freezing of the electron distributions at two internuclear axis orientations leads to a 90° phase difference in the final state amplitudes of $\sigma_o$ and $\sigma_1$.[7] This model does not require that transitions occur only at the $\Sigma$ - $\Sigma$ crossings. One can show that as long as the collision occurs symmetrically about the crossing, and if back coupling can be neglected, a phase difference of 90° degrees results.

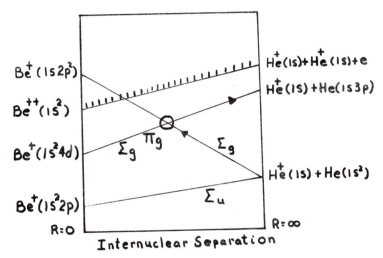

Fig. 3

## DOUBLE EXCITATION IN He - He COLLISIONS

Recently Fayeton, Houver, Brenot, and Barat [8] reported measurements of the 2'P - 1'S radiation angular distributions in 500 - 1000 eV He - He collisions. The results of these measurements illustrate well the specific nature of the dipole radiation and how it can be used to answer important questions about the dynamics of the collision. Using a unique multiarray detector system [9] the authors were able to measure the angular distribution of the 2'P - 1'S radiation for the single and double excitation reactions.

$$He + He \to He(2'P) + He$$

$$He + He \to He(2'P) + He(2'P)$$

at various scattering angles. The internal energy changes were large enough to allow a separation of the two reactions by a time of flight technique.

Within the framework of the independent particle model, the excitation mechanisms for single and double excitation should be the same; namely a $\sigma_u \rightarrow \pi_u$ rotational coupling of the single particle orbitals, between the initial $^1\Sigma_g \sigma_g^2 \sigma_u^2$, and $^1\Pi_g \sigma_g^2 \sigma_u \pi_u$ and $^1\Delta_g \sigma_g^2 \pi_u^2$ configurations as shown in the He - He correlation diagram in Fig. 4.

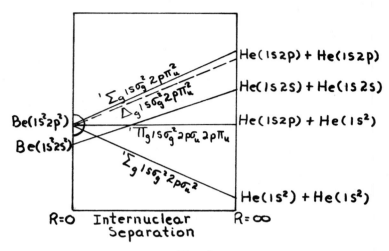

Fig. 4

The results of Fayeton, et al, for single electron excitation in 500 eV collisions are shown in Fig. 5, and indeed show the pattern of 2P - 1S radiation having an angular distribution minimum perpendicular to the final direction of the inter- nuclear axis. The symmetry of the distribution is characteristic of the 2P($\pm$1) state decay, indicating the excitation mechanism is a rotational coupling between the single electron, $\sigma$ - $\pi$ orbitals, near the united atom limit.

The angular distributions for 1000 eV collisions measured at 4.2°, 5.2°, and 7° scattering for double excitation show clear departures from a minimum at 90° to the internuclear axis, indicating a clear departure from the distribution predic- ted by the independent particle model and appreciable coupling to molecular states leading the $2P_0$ state of Helium. Within the context of the Fano - Lichten model, the two electron excitation mechanism should be the same as the one elec- tron excitation and should result in similar photon distributions. Clearly this is not the case; the conclusion that one can draw from these data is that a simple application of the independent particle model to the case of two electron excita- tion is not appropriate. The experimental work of Fayeton, et al illustrates well the power of the photon correlation experiments; a simple deviation of the photon angular distribution, from a symmetry axis, concisely answered the question the authors put to the independent particle model.


# $H^+$ + He COLLISIONS

Proton-Helium collisions leading to excited hydrogen or helium are interesting, as a class, in that the intitial mechanism for charge transfer is not a curve cros- sing but rather a $\Sigma$ - $\Sigma$ radial coupling of some 10 eV. From Fig. 6 we can see that these transitions start to occur at a separation of about 4 a.u. Subsequent $\Sigma$ - $\Pi$ rotational coupling near the united atom limit and $\Sigma$ - $\Sigma$ coupling at larger internuclear separation should then populate the H(2p) state. The first question

we address ourselves to is the relative importance of these two secondary couplings.

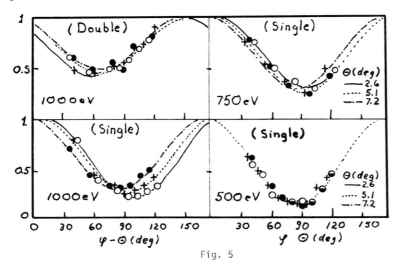

Fig. 5

To examine this question in detail, we have made initial measurements of the linear polarization of $L_\alpha$ radiation for the reaction $H^+ + He \rightarrow H(2p) + He^+$ for scattering angles near $1°$ at 8 keV. The polarizer was a LiF crystal employed at the
Brewster reflection angle.[10]

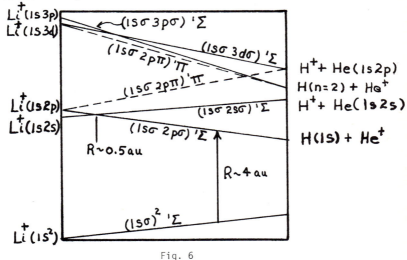

Fig. 6

If the main secondary process is a rotational coupling between the $(1s\sigma \ 2p\sigma)'\Sigma$
and $(1s\sigma \ 2p\pi)'\pi$ states near the united atom limit we would expect an unambiguous
polarization signature of a $2p_{\pm1}$ state, namely an intensity pattern with a maximum perpendicular to the final internuclear axis (in this case the $\pi$ orbital

would follow the rotating internuclear axis because of the lack of $\Sigma$ - $\Pi$ degeneracies).

Typical polarization measurements are shown in Fig. 7. Other scattering angles show approximately the same "tilt" angle of 54°. Also shown is the expected polarization intensity for the case of $\Sigma$ - $\Pi$ transitions dominating the secondary mechanism. It is clear from the data that $\Sigma$ - $\Sigma$ transitions play as an important role as the $\Sigma$ - $\pi$ transitions for this energy and range of scattering angles.

Fig. 8 shows the measured probabilities of $\sigma_0$ and $\sigma_1$ for impact parameters ranging from about 0.2 au - 0.45 a.u. These were obtained by a least squares analysis of the expression relating the polarization intensity $I(\beta)$ to $\sigma_0$, $\sigma_1$ and $\Delta\phi$

$$I(\beta) = C[(5\sigma_0 + 4\sigma_0) + (6\sigma_1 - 3\sigma_0)\ \sin^2\beta + 3\ (2\sigma_1\sigma_0)^{\frac{1}{2}}\cos\Delta\phi\ \sin2\beta]$$

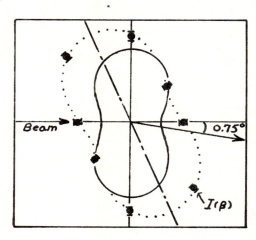

Fig. 7

The relative values of $\sigma_1$ and $\sigma_0$ as a function of impact - parameter indicate that secondary $\Sigma$ - $\pi$ and $\Sigma$ - $\Sigma$ couplings are of the same order of magnitude. The relative phase of the scattering amplitudes, $a_0$ and $a_1$, vary rapidly with impact parameter, about a value of 45° as shown in Fig. 9. The origin of the rapid variations of $\sigma_0$, $\sigma_1$, and $\Delta\phi$ are not yet understood.

It should be pointed out that $\sigma_1$ shows much smaller variations than $\sigma_0$, and increases slowly with increasing impact parameter. This slow increase in $\sigma_1$ is characteristic of the calculated probabilities by Taulberg, Briggs, and Vaaben [11] for rotational coupling at this velocity.

The calculated probability for rotational coupling by Taulberg, et al has a characteristic shape that is velocity dependent. (Fig. 10) Our measurements and the theoretical tabulated values [11] are consistent with $\sigma_1$ being on the increasing portion of this characteristic curve.

The increasing values of $P(\sigma_0 + 2\sigma_1)$ with b, shown in Fig. 8 are also consistent with the early measurements of McKnight and Jaecks [12] taken at 6.25 keV, although the range of common impact parameters is limited. Both show a 2 - 3 times increase in $P(\sigma_0 + 2\sigma_1)$ over this range. The average constant value of $P(\sigma_0)$ in this impact parameter range is also consistent with the earlier $P(2s)$ measurements of Crandell and Jaecks. [13]

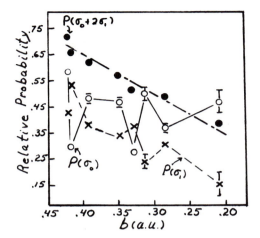

Fig. 8

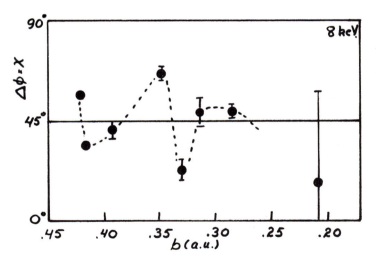

Fig. 9

Taking all of the evidence at hand we can conclude the secondary $\Sigma$ - $\Sigma$ coupling is independent of impact parameter and that the $\Sigma$ - $\Pi$ coupling shows the expected increase in probability as calculated by theory. The smooth behavior of $P(\sigma_0 + 2\sigma_1)$ suggests that oscillations in $P(\sigma_0)$ and $P(\sigma_1)$ results from some mechanism that removes amplitude from one channel and adds it to the other. Such coupling seems to alter the phase about a value of 45°.

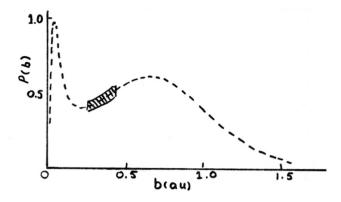

Fig. 10.

He$^+$ - H$_2$/D$_2$

The charge transfer process, He$^+$ + H$_2$/D$_2$ → He($3^3$P) + H$_2^+$/D$_2^+$ has been studied by
measuring P$_1$, P$_2$, and P$_3$ for the $3^3$P → $2^3$S radiation emitted perpendicular to the
scattering plane. The apparatus is essentially the same as that described by
Jaecks.et al[4] where the initial linear polarization measurements for these reactions were reported.

The emphasis of this present study is to determine what differences might exist
in the collision process for H$_2$ and D$_2$ targets. The measured values of P$_1$, P$_2$,
and P$_3$, from $1^0$ - $7^0$ (c.m.) scattering, for H$_2$ were found to be completely different than for D$_2$. From the measured values of P$_1$ we can plot an effective
$\lambda = \Sigma\sigma_0^1/\Sigma(\sigma_0^1 + 2\sigma_1^1)$, for H$_2$ and D$_2$, as shown in Fig. 11, where the index is a sum
over all the states of the system at a specific scattering angle.

This data clearly shows that the electron cloud shape around the excited He is
different for the two targets, as a function of scattering angle. For small
angles ($1^0$) the electron distribution is given mostly by an m$_\ell$ = 1 state, or electron distribution with a maximum perpendicular to the molecule - atom axis
(approximately the beam axis) whereas at larger angles, the distribution is principally along the beam axis.

Figures 12 and 13 shows the significant differences in the behavior of P$_2$, and
P$_3$ for H$_2$ and D$_2$. Here the ratio of these two parameters are plotted as a function of scattering angle. For fully coherent radiation and reflection symmetry
P$_3$/P$_2$ = tan$\Delta\phi$, where $\Delta\phi$ is the relative phase angle between the $\sigma_0$ and $\sigma_1$ scattering amplitudes. It is clear the results of the two isotope systems are dif-
ferent. If one neglected any differences in the rotational and vibrational
structure one would expect the same results for H$_2$ and D$_2$. It is clear that if
one is to understand the excitation mechanisms in ion-molecule collisions one
must take into account more than the electronic structure.

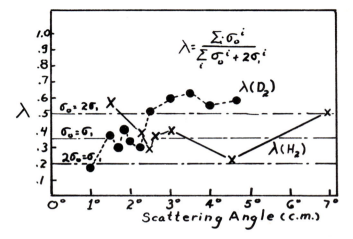

Fig. 11.

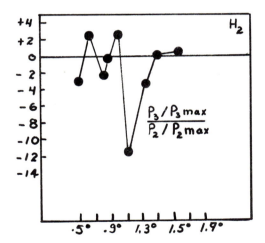

Fig. 12.

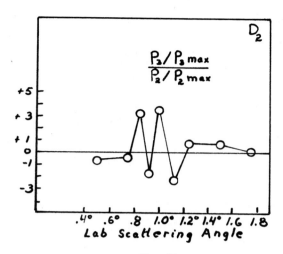

Fig. 13.

REFERENCES

*The authors acknowledge the support of the National Science Foundation.

1.  J.A. Smit  Physica 2, 104 (1935).

2.  J. Macek and D.H. Jaecks Phys. Rev A 4, 2288, (1971).

3.  U. Fano, W. Lichten, Phys. Rev. Lett 23, 157 (1969).

4.  "Coherence and Correlation in Atomic Collisions," Edited by H. Kleinpoppen and J.F. Williams Plenun Press, New York 1980.

5.  "Principles of Optics", M. Born and E. Wolf.  Pergamon Press, 1970.

6.  U. Fano and J.H. Macek Rev. Mod. Phys. 45, 553 (1973).

7.  F.J. Eriksen, D.H. Jaecks, W. deRijk and J. Macek Phys. Rev. A 14, 119 (1976).

8.  To be published, J. Phys. B.

9.  J.C. Brenot, J.A. Fayeton, and J.C. Houver, Rev. Sci Inst. 51, 1623 (1980).

10. W. Kaupilla, P.J.O Teubner, W.L. Fite, and R.J. Girnius, Phys. Rev. A 2, 1759, (1970).

11. K. Taubjerg, J.S. Briggs and J. Vaaben J. Phys. B 9 1351, (1976).

12. R.H. McKnight and D.H. Jaecks Phys. Rev. A 4, 2281 (1971).

13. D.H. Crandall and D.H. Jaecks Phys. Rev. A 4, 2271, (1971).

PHYSICS OF ELECTRONIC AND ATOMIC COLLISIONS
S. Datz (editor)
© North-Holland Publishing Company, 1982

PROTON HYDROGEN COLLISIONS
ENERGY-LOSS SPECTROMETRY

John T. Park

Department of Physics
University of Missouri-Rolla
Rolla, Missouri 65401
U.S.A.

Differential cross sections for elastic scattering, charge
transfer, and excitation of atomic hydrogen by intermediate
energy protons have been measured by energy-loss spectrometry.
The resulting differential cross sections show general
agreement with the available theoretical treatments; however,
differences of more than a factor of two are common. At the
largest scattering angles studied, the theories tend to
diverge from one another and from the experimental results.

INTRODUCTION

Experimental tests of the foundations of quantum scattering theory are only
possible in a few areas of quantum physics. Ion-atom collisions provide an
opportunity to construct with known forces a quantum mechanical treatment of the
scattering process in the nonrelativistic limit. Comparisons of experimental and
theoretical results in ion-atom collisions illuminate the weaknesses in the
quantum mechanical description of the process without introducing the complication
produced by imprecise knowledge of the fundamental interaction forces. In this
search for understanding, the proton-atomic hydrogen collision has been called the
"Rosetta Stone of Ion-Atomic Collisions". This basic collision system holds the
key for understanding three-body collision physics. Although the two-body atomic
physics problem is understood in some detail, the three-body problem has not yet
yielded its secrets. The three-body scattering problem--nucleus-nucleus-electron--
is, therefore, a very unique one in the field of physics. The proton-atomic
hydrogen collision system is the simplest example of the three-body problem.

Prior to 1975, the only available data on intermediate energy proton-atomic
hydrogen collisions was to be found in the pioneering work of Fite[1], who used
crossed beam techniques. Later, similar crossed beam measurements were made by
groups directed by Stebbings[2,3], and Gilbody[4,5]. For their work[1-7], they depended
on the "window" in the absorption spectrum of molecular oxygen, which provided a
Lyman-alpha detector. Total cross sections for Lyman-alpha emission from atomic
hydrogen excited to the 2p state by 5 to 30 keV protons were the result. These
measurements could not easily be extended to higher impact energies, because the
detected signal of a crossed beam experiment decreases as the square of the impact
velocity.

The University of Missouri-Rolla ion energy-loss spectrometer[8-17] (Fig. 1) was
developed for the ultimate purpose of attacking the proton-atomic hydrogen
collision problem. This specially designed spectrometer currently has an energy
resolution of 0.7 eV and an angular resolution of $1.2 \times 10^{-4}$ radians throughout
its 15 to 200 keV energy range. It provides energy-loss spectra for ions
scattered through a definite angle by target atoms. Information from these
spectra, when combined with a knowledge of the target density and geometry, yields
cross sections that are differential in both energy loss and scattering angle. If

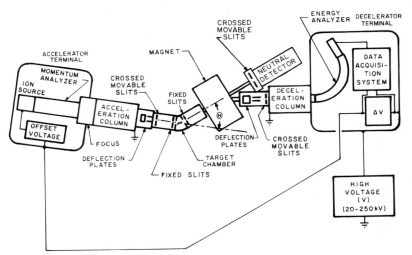

Fig. 1.   Schematic diagram of the UMR differential energy-loss spectrometer

the pertinent structure can be resolved, differential cross sections can be obtained for all ion-atom collision processes that do not involve a change in charge of the incident ion.  In addition, some differential charge transfer cross sections can be measured.

The energy-loss spectrometry technique eliminates any ambiguity in the identity of the particle responsible for the collision event because the projectile itself is studied.  The results are not affected by postcollision events involving the target.  The cross sections are absolute in the sense that the detector's efficiency does not enter into their determination.  The results are not dependent on the detection of a particular secondary particle.

EXPERIMENTAL METHOD

Because the energy-loss spectrometer and the general method employed in ion energy-loss spectrometry have been discussed in detail elsewhere[8-17], only a brief outline of the technique is provided here.

Ions are produced in a low voltage discharge source.  The energy distribution produced by this source is estimated to have a 0.2 eV full width at half maximum. The mass-selected ions are accelerated and steered through the entrance collimator. Additional collimation slits define the angular extent of the beam entering the collision chamber, which is constructed of coaxial tungsten tubing[18].  The high temperature necessary to dissociate molecular hydrogen is provided by joule heating.

Ions exiting from the collision chamber pass through the exit collimator, which consists of a fixed exit slit and a pair of movable collimating slits.

The transmitted ion beam is magnetically analyzed to remove any products of charge-changing collisions.  This removes any ambiguity as to the detected ion. Following the magnetic analysis, the ions enter the deceleration column and are decelerated by a well defined potential.  The decelerated ions are energy analyzed by an electrostatic analyzer.

Spectra differential in energy loss are obtained by increasing the potential difference between the accelerator and decelerator terminals. Whenever the increased potential energy compensates for a discrete energy loss of the projectile-target system, a peak is detected in the energy-loss spectrum. See Fig. 2. The energy-loss scale can be determined to an accuracy of ±0.03 eV[10]. Angular distributions of the scattered ion current corresponding to a particular scattering process can be measured by setting the internal energy loss at the calculated value while pivoting about the scattering center provided that the process is resolved in its energy-loss spectrum. The relative angular position of the accelerator is known to within 3.3 x 10$^{-6}$ radians[13].

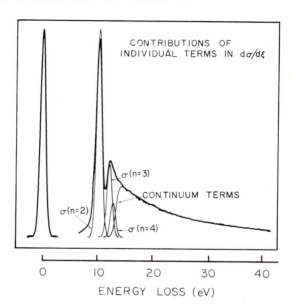

Fig. 2. Energy-loss spectrum for 50 keV protons incident on atomic hydrogen

The raw data obtained with the technique described above were analyzed by using the method described in detail in Ref. 13. This program of data analysis was used to extract the "real" differential cross section from the apparent differential cross section by determining the effects of geometrical factors and the angular distribution of the incident beam. Because the description of the method is quite involved, it is not repeated here.

A significant problem with the proton-atomic hydrogen collision work is the difficulty of determining the target density. In the investigations of the UMR laboratory, a Born approximation calculation[19] for total excitation of atomic hydrogen to the n = 2 state by 200 keV protons is used to determine the target density. The calculated target density is then used to determine the cross sections for other processes. There are significant differences in the total excitation cross section for 200 keV proton excitation of atomic hydrogen between the various theoretical calculations, and the comparisons between a particular theory and experiment are frequently improved significantly by renormalizing to the theory to which the data is being compared.

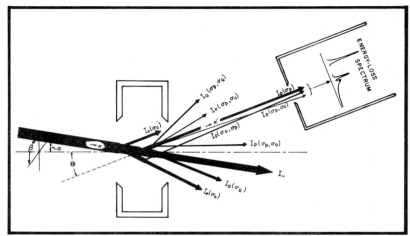

Fig. 3.    Ion beam scattering schematic. $I_u$ represents the unscattered beam,
           $I_e(\sigma_e)$ represents ions that have been elastically scattered with a
           cross section $\sigma_e$. The term $\sigma_c$ represents the cross section for charge
           transfer, and $I_n$ is the flux of atoms produced by charge transfer.
           $I_p(\sigma_p)$ and $I_Q(\sigma_Q)$ represent ions that have undergone an energy loss
           corresponding to excitation of the target to the states p and Q,
           respectively, with corresponding cross sections $\sigma_p$ and $\sigma_Q$. Multiple
           collisions are indicated by the fact that they involve more than one
           cross section. The term $\hat{K}$ represents a part of the incident ion beam
           entering the chamber at angles $\alpha$ and $\beta$ relative to the accelerator
           axis, and $\hat{K}'$ indicates the direction of the detected ions that have
           been scattered into the detector window at the angle $\theta$ from the
           accelerator axis.

## ELASTIC SCATTERING OF PROTONS FROM ATOMIC HYDROGEN

The problem of elastic scattering of protons from an atomic hydrogen target
provides an indication of the strengths of the energy-loss technique. The process
of interest is $H^+ + H(1s) \rightarrow H^+(\theta) + H(1s)$. Much of the reported work on "elastic"
scattering really provides a sum of elastic and inelastic cross sections. The
cross sections established by energy-loss spectrometry have been determined from
the "true" elastic scattering peak, which is completely resolved from any
inelastic or charge transfer processes. The elastically scattered proton beam
has undergone a kinematic energy loss, which for the equal mass projectile and
target reduces to $\Delta\xi = 4E \sin^2\theta$. At small angles, however, this small kinematic
energy loss is not sufficient to permit the separation of the transmitted
unscattered protons from those protons scattered through very small angles.

The energy-loss technique makes it possible to label each proton by both angle and
energy loss. See Fig. 3. A rate equation can then be written for each beam
component identified by its angle and energy. Analysis of the scattering process[20]
provides a rate equation for the detected ion current, I, that has not suffered
any inelastic energy loss. This equation, which includes second order processes,
is given by

$$\frac{dI(\theta)}{d(n\ell)} = \int d\Omega \;\; \frac{dI(\hat{k})}{d\Omega} \int_{\Delta\Omega} d\Omega' \;\; \frac{d\sigma_e(\hat{k}\cdot\hat{k}')}{d\Omega'} - I(\theta)\sigma_T. \qquad (1)$$

In Eq. 1, $\hat{k}$ is a unit vector in the direction of the incident ion, $\hat{k}'$ is a unit vector directed toward the detector of the scattered ion, $d\sigma_e/d\Omega$ is the differential elastic scattering cross section, $\Delta\Omega$ is the solid angle subtended by the detector, and $\sigma_T$ is the total cross section for all processes, elastic and inelastic. The angle, $\theta$, is the "measurement" angle corresponding to the angle between the accelerator axis and the direction to the center of the detector as measured from the center of the collision region, and $n\ell$ is the product of the target density, $n$, and the collision length, $\ell$. This deceptively simple equation is the fortuitous result of the cancellation of the several terms that appear in its derivation.

When this set of equations is solved simultaneously with data taken at various pressures and angles, the elastic cross section can be extracted for all measurement angles. Unfortunately, at small angles, the solution to the equation involves a small difference between two large numbers, and the cross section is lost in the statistical noise. Fig. 4 shows the preliminary data for the differential elastic scattering cross section for $H^+ + H(1s) \rightarrow H^+(\theta) + H(1s)$. These data span the angular range from 0.55 to 4.7 mrad and fall three orders of magnitude within this angular range. The data point at 0.55 mrad is located at the smallest scattering angle, where statistically significant results can be obtained.

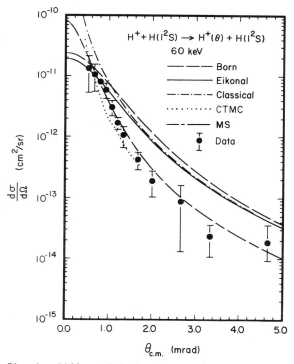

Fig. 4. Differential elastic scattering cross section for 60 keV protons by atomic hydrogen.

Theoretical calculations based on classical, Born and eikonal techniques are shown in Fig. 4. The agreement between experiment and theory is not good. It is to be noted that if the experimental values of the elastic, inelastic, and charge transfer scattering are added together, the agreement with the classical calculation is much better.

The preliminary results of Olson[21], who used the Classical Trajectory Monte Carlo (CTMC) method, are also shown. The agreement in magnitude between experiment and theory is quite good; however, the curve shapes appear to be slightly different. The CTMC calculation unambiguously provides the cross section for the "true" elastic scattering process, $H^+ + H(1s) \rightarrow H^+(\theta) + H(1s)$. The CTMC calculation includes the elastic, in-elastic charge transfer channels so that it could be expected to give good results for elastic scattering and the agreement with the experimental results reinforces this prediction.

There do not appear to be any published sophisticated quantum mechanical calculations available for the "true" elastic scattering process, $H^+ + H(1s) \rightarrow H^+(\theta) + H(1s)$. R. Shakeshaft has provided elastic amplitude for his multistate, MS, calculation[23]. An elastic differential scattering cross section was then obtained from the MS amplitude using a Bessell function transform, Fraunhoffer integral, technique[24]. The resulting MS differential cross section is also shown in Fig. 4. The agreement with the data is very good. The curve shape is almost exactly the same as the data. The agreement in magnitude is excellent. It is likely that most of the required information is available from some of the other multistate calculations so that the differential scattering cross section could be obtained by using a Bessel function transform technique[24].

EXCITATION OF ATOMIC HYDROGEN BY INCIDENT PROTONS

Analysis of the scattering processes, which result in the excitation of a target atom, provides a rate equation for differential target excitation of the state "p"[20], which is given by

$$\frac{dI_p(\theta)}{d(n\ell)} = \int_{\Delta\Omega} d\Omega \int d\Omega' \frac{dI(\hat{k})}{d\Omega} \frac{d\sigma_p(\hat{k}'\cdot\hat{k})}{d\Omega'}$$
$$+ \int_{\Delta\Omega} d\Omega \int d\Omega' \frac{dI_p(\hat{k})}{d\Omega} \frac{d\sigma_e(\hat{k}'\cdot\hat{k})}{d\Omega'} - \sigma_T I_p(\theta). \quad (2)$$

In Eq. 2, $I_p$ is the detected "current" of ions that has excited the target to the state p, and $d\sigma_p/d\Omega$ is the excitation cross section. The remaining terms are as defined for Eq. 1. The equation is valid to second order. Data acquisition is performed under single collision conditions, which are defined for this experiment by the requirement that

$$\int_{\Delta\Omega} d\Omega \int d\Omega' \frac{dI_p(\hat{k})}{d\Omega} \frac{d\sigma_e(\hat{k}'\cdot\hat{k})}{d\Omega'} - \sigma_T I_p(\theta)$$

is negligible relative to

$$\int_{\Delta\Omega} d\Omega \int d\Omega' \frac{dI(\hat{k})}{d\Omega} \frac{d\sigma_p(\hat{k}'\cdot\hat{k})}{d\Omega'} \quad .$$

The dominant feature of the proton-atomic hydrogen energy-loss spectrum is the peak corresponding to the excitation of the n = 2 state. With the current resolution, the $2^1S$ and $2^1P$ states cannot be separated. Differential cross section data for excitation of atomic hydrogen to the n = 2 level by proton impact are shown by the solid octagons in Fig. 5.

The experiments[16,22] in which differential energy-loss spectrometry is used to measure differential cross sections for the excitation of atomic hydrogen by intermediate energy proton impact provide the only available data. The only other experiment in which angular differential cross sections were measured for this particular collision system was published by Houver et al.[25]. They measured the incident proton energies between 250 and 2000 eV.

Shakeshaft[26] and Bransden and Noble[27] obtained differential cross sections from their coupled-state impact parameter calculations. They accomplished this by multiplying the transition amplitudes at a given impact parameter by a suitable Bessell function and integrating over all the impact parameters. Thus, many

Fig. 5.  Differential cross
sections for
excitation of the
n = 2 state of
atomic hydrogen
by 25, 50, and
100 keV protons.

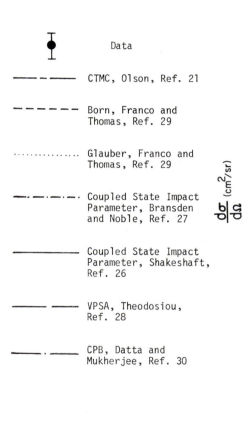

●       Data

————— · —————   CTMC, Olson, Ref. 21

— — — — —   Born, Franco and
Thomas, Ref. 29

·················   Glauber, Franco and
Thomas, Ref. 29

—— · —— · —— ·   Coupled State Impact
Parameter, Bransden
and Noble, Ref. 27

—————————   Coupled State Impact
Parameter, Shakeshaft,
Ref. 26

———— ——   VPSA, Theodosiou,
Ref. 28

———— · ——   CPB, Datta and
Mukherjee, Ref. 30

$\dfrac{d\sigma}{d\Omega}$ (cm²/sr)

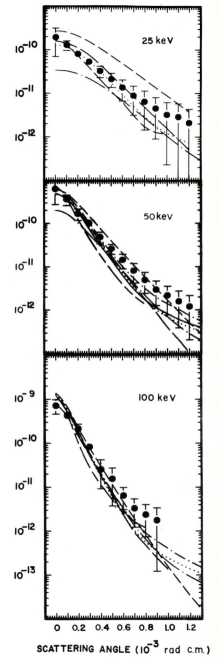

impact parameters contribute to a given scattering angle. Shakeshaft[26] used a scaled hydrogenic basis set with 35 basis functions centered about each proton. He chose the scale factors so that the resulting basis energy eigenvalues would almost coincide with the energies of the ls, 2s, 2p, 3s, and 3d states and overlap the low-energy part of the continuum spectrum of the hydrogen atom. Bransden and Noble[27] used two sets of single-center expansions, which included eigenfunctions to represent the discrete target states and pseudostates chosen to represent the continuum. Their basis set contained the exact ls, 2s, and 2p hydrogenic wave functions. The theoretical results of Shakeshaft for the differential cross section for direct excitation of the n = 2 level is in reasonably good agreement with this experiment over the entire energy range measured from 15 to 145 keV. The agreement is quite good near 50 keV. The theoretical results of Bransden and Noble are also reasonably good over the energy range calculated, i.e., 50 to 150 keV.

The Vainshtein-Presnyakov-Sobel'man approximation (VPSA) has recently been applied by Theodosiou[28] to the calculation of differential cross sections for direct excitation of atomic hydrogen by proton impact. Theodosiou's calculation includes the contributions of the projectile-target core interaction to the transition matrix element. The projectile-target core interaction makes a significant contribution to the derived differential cross sections. The agreement between the VPSA and both the total and differential experimental results is good. The VPSA theory diverges from experiment at the largest angles studied.

Olson[21] has very recently provided unpublished results for the cross section for excitation of atomic hydrogen by proton impact that he obtained by using the CTMC technique. The CTMC results are slightly lower than the data, but the agreement in curve shape is reasonably good.

Theoretical results of the first Born and Glauber approximation calculations of Franco and Thomas[29] are shown in Fig. 5. The first Born approximation calculations give differential cross sections that are in fairly good agreement with the experimental cross section curve over all scattering angles measured for incident proton energies greater than 50 keV. For incident proton energies greater than 70 keV, the Born approximation calculations yield results that are more sharply peaked with respect to angle than the experimental results. At the larger scattering angles, the Born approximation results cross the experimental results so that the former are below the latter, and they continue to decrease quite rapidly in magnitude as the scattering angle continues to increase.

The Glauber approximation calculations[29] yield differential cross sections that are also in reasonably good agreement with the experimental differential cross sections over the entire range of incident proton energies reported here. The Glauber approximation does remarkably well considering its computational simplicity. It is interesting to note that the Glauber approximation results for the differential cross section approach the experimental results from below as do the Glauber approximation results for the total n=2 excitation cross section.

Datta and Mukherjee[30] have used the Coulomb-projected-Born (CPB) method to determine differential cross sections for excitation of atomic hydrogen to the 2s and 2p states. At 100 keV the CPB method provides good results except at the large scattering angles. At 50 keV the CPB results provide satisfactory agreement with the experiment, however, at 25 keV the CPB cross-sections underestimate the observed values except at the largest angles.

The coupled-state calculations[26,27], the VPSA calculation, the CPB results[30], and the CTMC calculation[21] as well as the simpler Glauber approximation calculation[29], yield a curve shape for the differential cross section that is in reasonable agreement with the experimental results for the angles measured. All of the theoretical treatments differ from the data and from each other in curve shape. At the larger scattering angles, the cross sections obtained from the theoretical treatments tend to fall off more rapidly than the experimental data. More data are obviously needed at the larger scattering angles in order to reduce the size of the error bars and to determine the differential cross section curve shape better.

It should be emphasized that the differential cross sections reported here were normalized to a first Born approximation total cross section for excitation of atomic hydrogen to its n = 2 level at 200 keV. Renormalization might provide slightly better agreement between the experimental results and a particular theoretical result.

There are many more total calculations than differential cross sections for proton impact. Experimental results from the application of energy-loss spectrometry techniques have been obtained for the excitation of atomic hydrogen to its n = 2, 3, and 4 states by proton impact[14]. These results were arrived at by increasing the angular acceptance of the spectrometer deceleration-analyzer system to capture essentially all of the scattered protons. The total cross sections for excitation of atomic hydrogen are discussed in Ref. 14. References 23, 31 to 36 contain discussions of the theoretical efforts that have been put forth since the publications of Ref. 14.

## ANGULAR DIFFERENTIAL CHARGE TRANSFER FOR PROTON-ATOMIC HYDROGEN COLLISIONS

Angular differential charge transfer does not really require the full capabilities of the differential energy-loss spectrometer. The ability to make charge transfer measurements, however, requires only the addition of a fast-atom detector with defining slits to establish an appropriate solid angle mounted on the zero degree port of the analyzing magnet. See Fig. 1. The angular resolution of the apparatus in the laboratory system is 120 μrad.

Because the detection efficiency of the scattered neutral particle detector for hydrogen atoms has not been measured, the present results have been normalized to the value of the total cross section for electron capture into all bound states reported in Ref. 37.

In Fig. 6, the experimentally determined differential cross sections for electron capture in collisions of protons with hydrogen atoms are presented for incident proton velocities of 1, 1.55, and 2.24 a.u. (25, 60, and 125 keV). The experimental results for capture into all bound states of hydrogen are compared with the theoretical results for capture into the ground state calculated in the two-state, two-center, atomic expansion method (TSAE)[38-40], the continuum distorted wave approximation (CDW)[41], the multistate two-center, and the coupled-state approximation (MS)[26,42-43]. All three calculations were formulated in the impact parameter approximation by using the eikonal approximation to obtain the reported differential cross sections. Olson[21] has provided a Classical Trajectory Monte Carlo calculation for the capture into <u>all</u> bound states of hydrogen.

The shapes of the differential cross sections predicted by Shakeshaft's MS calculation[26] shown in Fig. 6 are in reasonably good agreement with the shapes of the experimentally determined differential cross sections. However, it does not appear that the addition of capture into excited states to the theory would fully account for the differences between theory and experiment. Both the total cross section for capture into H(2s)[4,7,44-46] and the angular distribution of the probability for electron capture into H(2s) have been measured[47]. Bayfield[47] observed an off zero maximum in the angular distribution of the probability centered near 3.5 mrads in the center of mass system. Although such an angular distribution would contribute to the observed differences between the present experimental results and the MS results, the magnitude of the probability is too small (of the order of 0.05) to account for the observed differences between the MS theory and the experiment. Adjusting the theory with a multiplicative factor of 1.2 (to account for a $1/n^3$ excited state population) does not bring the theoretical and experimental results into confluence.

Fig. 6.  Differential Charge Transfer
         for 25, 60, and 125 keV
         protons incident on atomic
         hydrogen.  The data are for
         charge transfer into all
         states.  The multistate
         calculation (MS) Shakeshaft[26],
         the Continuum Distorted Wave
         Calculation (CDW) of Belkic,
         Gayet and Salin[41], and the
         Two-State Two-Center Atomic
         Expansion (TSAG) of Lin[38-40]
         are for capture to ground
         state.  Olson's[21] Classical
         Trajectory Monte Carlo (CTMC)
         calculation is for capture
         to all states.

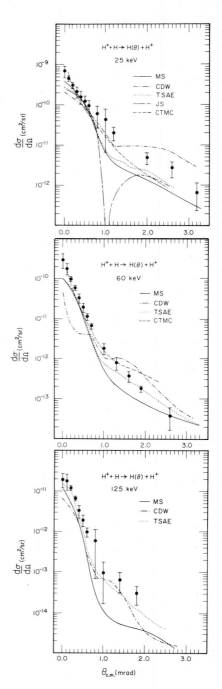

The results of the TSAE calculation of Lin[48] are in good agreement with the
results of both the present work and the MS calculation. This agreement shown in
Fig. 6 and the agreement obtained between TSAE results and experimental results
for collisions of protons with helium atoms[17] demonstrates the utility of the
relatively simple TSAE calculation in the range of intermediate incident velocities in the proton-hydrogen atom collision system.

The results of the CDW calculation[41,49-51] compare less favorably than the MS and
the TSAE results with the present experimentally determined differential cross
sections. For angles greater than approximately 1.0 mrad the CDW results overestimate the experimental results for all but the highest energy. For angles less
than approximately 0.6 mrad the CDW results underestimate the experimental results
more than the results of either the TSAE or the MS calculation. The worst overall
agreement between the results of the CDW calculation and the present experiment
is obtained at 60 keV as shown in Fig. 6. The structure in the differential cross
sections predicted by the CDW approximation is not observed in the MS or TSAE
approximation results nor is there any evidence for such structure in the experimentally determined differential cross sections.

An explanation for the observed "discrepancy" between the CDW and the experimentally determined differential cross sections cannot be offered at the present
time. However, it should be pointed out that when the total cross sections
obtained in the CDW approximation for capture into the $n = 1$, $n = 2$, and $n = 3$
states are added together the resulting total cross section is in good agreement
with the experimentally accepted result for capture into all states in the intermediate velocity range. (See Ref. 41, Fig. 7).

The CTMC calculation of Olson[21] is for capture into all bound states. The CTMC
results, therefore, provide an unambiguous comparison with the experiment. The
agreement is good. The CTMC theory does not provide a good fit at angles near
zero. The curvature of the CTMC theory is always concave and tends to underestimate the cross section at larger scattering angles. However, the magnitudes
of the CTMC calculation provide the best overall agreement with experiment.

The results of a Jackson-Schiff (JS) calculation[52] of the differential cross
section for electron capture are shown in Fig. 6. A computer simulation of the
present experiment demonstrated that although the Jackson-Schiff minimum could not
be resolved, the presence of the minimum shown in Fig. 6 would be detected by a
significant decrease of two orders of magnitude in the differential cross section
in an interval form 0.9 to 1.8 mrads. Although the total cross section predicted
by the JS model is in reasonably good agreement with the experimentally accepted
result[37], the nonphysical Jackson-Schiff minimum, a mathematical artifact of the
calculated angular distribution, which has not been observed in the present
experimental results, indicates the weakness of this approximation.

With the exception of Olson's CTMC calculation the available theoretical calculations are for capture into the ground state. Because the magnitude of the
differential cross sections for capture into excited states is unknown, it is not
possible to judge the effect of the failure to include them, however, the good
agreement with Olson's CTMC calculation indicates that including all states will
improve the agreement between experiment and theory.

DISCUSSION

Differential cross sections for elastic scattering, excitation, and charge transfer are now available for proton-atomic hydrogen collisions in the intermediate
energy range. The comparisons of theory and experiment are less satisfactory for
the differential measurements than for the total measurements. The total cross
sections involve the integral of $\sin \theta \, d\sigma/d\Omega$. Because $d\sigma/d\Omega$ falls very rapidly
with increasing angle, $\sin \theta \, d\sigma/d\Omega$ is only significant over a very limited
angular range. In general, experiment and theory are in better agreement at

angles near the maximum in $\sin \theta \, d\sigma/d\Omega$ than at larger scattering angles. Even the Jackson-Schiff calculation provides reasonable total cross sections for charge transfer although the differential cross section displays dramatic structure which is not experimentally observed.

The reported differential cross sections vary in magnitude by as much as three orders of magnitude. The disagreement between theory and experiment is often masked by the necessity to display cross sections that fall by several orders of magnitude in a single small graph. Differences between theory and experiment greater than a factor of two go almost unnoticed. There is a need for a great deal of additional work on these systems. In examining the comparisons with experiment and theory, the largest differences are at the largest and smallest angles. These are of course the regions where the experiments are the most uncertain. It is also the region in which theories disagree most dramatically.

## ACKNOWLEDGMENTS

The support of this work by the National Science Foundation is gratefully acknowledged.

I wish to thank R. E. Olson and R. Shakeshaft for permitting the use of their results prior to publication.

The assistance of J. L. Peacher, T. J. Kvale, E. Redd and D. M. Blankenship in preparing the data for publication and L. Caudle and J. Koenig for typing and editing the manuscript is greatly appreciated.

## REFERENCES

1. Fite, W. L. and Brackman, R. T., Phys. Rev. 112, 1151 (1958).
2. Stebbings, R. F., Young, R. A., Oxley, C. L., and Ehrhardt, H., Phys. Rev. 138, A 1312 (1965).
3. Young, R. A., Stebbings, R. F., and McGowan, J. W., Phys. Rev. 171, 85 (1968).
4. Morgan, T. J., Geddes, J., and Gilbody, H. B., J. Phys. B6, 2118 (1973).
5. Morgan, T. J., Geddes, J., and Gilbody, H. B., J. Phys. B7, 142 (1974).
6. Kondow, T., Girnius, R. J., Chong, Y. P., and Fite, W. L., Phys. Rev. A10, 1167 (1974).
7. Chong, Y. P. and Fite, W. L., Phys. Rev. A16, 933 (1977).
8. Park, J. T. and Schowengerdt, F. D., Rev. Sci. Instrum. 40, 753 (1969).
9. Schowengerdt, F. D. and Park, J. T., Phys. Rev. A1, 848 (1970).
10. York Jr., G. W., Park, J. T., and Miskinis, J. J., Crandall, D. H., and Pol, V., Rev. Sci. Instrum. 43, 230 (1972).
11. Pol, V., Kauppila, W., and Park, J. T., Phys. Rev. A8, 2990 (1973).
12. Park, J. T., Pol, V., Lawler, J., George, J., Aldag, J., Parker, J., and Peacher, J. L., Phys. Rev. A11, 857 (1975).
13. Park, J. T., George, J. M., Peacher, J. L., and Aldag, J. E., Phys. Rev. A18, 48 (1978).
14. Park, J. T., Aldag, J. E., George, J. M., and Peacher, J. L., Phys. Rev. A14, 608 (1976).
15. Park, J. T., Aldag, J. E., and George, J. M., Phys. Rev. Lett. 34, 1253 (1975).
16. Park, J. T., Aldag, J. E., Peacher, J. L., and George, J. M., Phys. Rev. Lett. 40, 1646 (1978).
17. Martin, P. J., Arnett, K., Blankenship, D. M., Kvale, T. J., Peacher, J. L., Redd, E., Sutcliffe, V. C., and Park, J. T., Lin, C. D., and McGuire, J. H., Phys. Rev. A23, 2858 (1981).
18. Lockwood, G. J., Helbig, H. F., and Everhart, E., J. Chem. Phys. 41, 3820 (1964).
19. Bates, D. R., and Griffing, G., Proc. Phys. Soc., London 66, 64 (1953).
20. Park, J. T., IEEE Trans. Nucl. Sci. NS-26, 1011 (1979).
21. Olson, R. E., private communication.

22. Park, J. T., Aldag, J. E., Peacher, J. L., and George, J. M., Phys. Rev. A21, 751 (1980).
23. Shakeshaft, R., private communication.
24. Wilets, L. and Wallace, S. J., Phys. Rev. 169, 84 (1968).
25. Houver, J. C., Fayeton, J., and Barat, M., J. Phys. B7, 1358 (1974).
26. Shakeshaft, R., Phys. Rev. A18, 1930 (1978). Note that the scale on Fig. 2 is mislabeled. The scattering angle should be in units of $10^{-4}$ rad.
27. Bransden, B. H. and Noble, C. J., Phys. Lett. A70, 404 (1979).
28. Theodosiou, C. E., Phys. Rev. A22, 2556 (1980).
29. **Franco, V. and Thomas, B., Phys. Rev. A4, 945 (1971), private communication.**
30. Datta, S. and Mukherjee, S. C., Phys. Rev. A23, 1780 (1981).
31. Fitchard, E., Ford, A. L., and Reading, J. F., Phys. Rev. A16, 1325 (1977).
32. Bransden, B. H. and Dewangan, D. P., J. Phys. B12, 1377 (1979).
33. Shakeshaft, R., Phys. Rev. A14, 5 (1976).
34. Shakeshaft, R., Phys. Rev. A18, 5 (1978).
35. Lodge, J. G., Percival, I. C., and Richards, D., J. Phys. B9, 239 (1976).
36. Storm, D. and Rapp, D., Phys. Rev. A14, 1 (1976).
37. Barnett, C. F., Ray, J. A., Ricci, E., Wilker, M. I., McDaniel, E. W., Thomas, E. W., and Gilbody, H. B., "Atomic Data for Controlled Fusion Research", ORNL-5206 (1977).
38. Lin, C. D., Soong, S. C., and Tunnell, L. N., Phys. Rev. A17, 1646 (1978).
39. Lin, C. D. and Soong, S. C., Phys. Rev. A18, 499 (1978).
40. Martin, P. J., Blankenship, D. M., Kvale, T. J., Redd, E., Peacher, J. L., and Park, J. T., Phys. Rev. 23, 3357 (1981).
41. Belkic, Dz., Gayet, R., and Salin, A., Physics Reports 56, 279 (1979).
42. Shakeshaft, R., Phys. Rev. A14, 1626 (1976).
43. Shakeshaft, R., Phys. Rev. A18, 307 (1978).
44. Bayfield, J. E., Phys. Rev. 185, 105 (1969).
45. Hill, J., Geddes, J., and Gilbody, H. B., J. Phys. B12, L341 (1979).
46. Morgan, T. J., Stone, J., and Mayor, R., Phys. Rev. A22, 1460 (1980).
47. Bayfield, J. E., Phys. Rev. Lett. 25, 1 (1970).
48. Lin, C. D., private communication.
49. Cheshire, I. M., Proc. Phys. Soc. 84, 89 (1964).
50. Gayet, R., J. Phys. B5, 483 (1972).
51. Belkic, Dz. and Gayet, R., J. Phys. B10, 1911 (1977).
52. Jackson, J. D. and Schiff, H., Phys. Rev. 89, 359 (1953).

PHYSICS OF ELECTRONIC AND ATOMIC COLLISIONS
S. Datz (editor)
© North-Holland Publishing Company, 1982

ATOMIC REARRANGEMENT COLLISIONS AT
ASYMPTOTICALLY HIGH IMPACT VELOCITIES

Robin Shakeshaft [†]

Physics Department
New York University
New York, New York 10003
U.S.A.

The basic theory underlying atomic rearrangement collisions
at asymptotically high velocities is described. In the
asymptotic regime there are two natural expansion parameters.
Several examples are given where these parameters are small for
impact energies of a few hundred eV/amu.

INTRODUCTION

This paper is a review, though of rather limited scope, of some theoretical aspects of the generic rearrangement collision

$$P + (A + T) \rightarrow (P + A) + T \quad , \tag{1.1}$$

in the regime where P is incident with an asymptotically high speed v relative to
$(A + T)$. It will be assumed that relativistic corrections, of order $(v/c)^2$, are
negligible. The active particle A, the target "nucleus" T to which A is initially
bound, and the projectile P will be treated as particles without structure, and T
will be treated as infinitely massive and always at rest. Specific examples of
reaction (1.1) for which these restrictions more or less apply are the simple
charge transfer reactions

$$e^+ + H \rightarrow Ps + H^+ \quad , \tag{1.2}$$

$$He^{++} + H \rightarrow He^+ + H^+ \quad , \tag{1.3}$$

(where A is an electron, T is a proton, and P is either a positron or an alpha
particle) and the more complicated atom transfer reaction

$$H^+ + CH_4 \rightarrow H_2^+ + CH_3 \quad , \tag{1.4}$$

(where A is an entire hydrogen atom, T is the composite "nucleus" $CH_3$, and P is a
proton). Nuclear reactions, eg. $^3He + {}^2H \rightarrow {}^4He + {}^1H$ , are not covered since, in
the asymptotic regime, the deBroglie wavelength of the projectile is comparable
to the nucleon size, and nucleon structure becomes important.

The meaning of the term "asymptotically high speed" rests on the assumption that
the scattering amplitude has a diagrammatic expansion with an expansion parameter
which decreases as the speed v increases and which is sufficiently small when v
is "asymptotically high" for the expansion to converge. It is worth noting that
the Faddeev-Watson expansion of the scattering amplitude for reaction (1.1) is
known (Hagedorn 1979) to converge at sufficiently high v for a wide class of interactions -- but a proof which encompasses Coulomb interactions has not yet been
given. Now a peculiarity of diagrammatic expansions of the amplitude for rearrangement scattering is that the simplest diagram is not the dominant one in
the limit that the expansion parameter approaches zero. This peculiarity will be
discussed qualitatively (with no pretense of rigor) in the following three sections. For detailed mathematical expressions of some of the following statements
the reader is referred to the classic paper by Dettmann (1971). In sections V
and VI the formal discussion of sections II - IV will be illustrated with the

examples of electron capture from a hydrogenlike ion by a bare ion, and atom capture from a molecule.

For a broader review of electron capture processes at high impact velocities see Basu et al. (1978) and Belkić et al. (1979). Since these reviews there have been extensive applications of the impulse approximation (Kocbach 1980, Amundsen and Jakubassa 1980) and the eikonal approximation (Eichler and Chan 1979), and there are also now several large, general computer programs (eg. Ford et al. 1981). A detailed comparison of experimental data with theoretical results has been made by Horsdal-Pederson in a companion paper in this volume.

## II.  DIGRESSION ON DIRECT EXCITATION

To appreciate the peculiarity just mentioned, it is helpful to digress for a moment and comment briefly on the <u>direct</u> inelastic process

$$P + (A + T) \rightarrow P + (A + T)^* \tag{2.1}$$

where (A + T) is simply excited by P from one bound state to another.  If $\Delta E$ is the difference in the binding energies of (A + T) in its <u>initial</u> and <u>final</u> states, i and f, the characteristic time required for (A + T) to make a <u>complete</u> transition from i to f, or vice versa, is $\hbar / \Delta E$.  In the calculation <u>of the integrated</u> cross section for reaction (2.1), two expansion parameters naturally arise.  One parameter, $\lambda_s$, is the collision <u>strength</u>, which is independent of the speed v. If, for example, P were a bare ion impinging on a hydrogen atom, $\lambda_s$ would be the atomic number of P.  The other parameter, $\lambda_t$, is the ratio of two <u>t</u>imes, the characteristic value, $\tau_{coll}$, of the effective <u>collision</u> duration, to $\hbar / \Delta E$.  The simplest diagram, labelled (D1) (labels D and R are used to distinguish between diagrams for <u>d</u>irect and <u>r</u>earrangement processes), is:

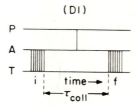

A vertical line represents a single two-body interaction.  A block of closely spaced vertical lines represents schematically a bound state; when two particles are bound together they interact an infinite number of times during an orbital period.  The equally simple diagram

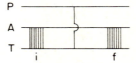

(where the indented vertical line represents a single P-T interaction) makes no contribution since T, being infinitely massive, is at the center of mass of the target and therefore the P-T interaction cannot directly affect the internal state of the target.

Diagram (D1) describes <u>impulsive</u> excitation -- the perturbation of the target and projectile during the time $\tau_{coll}$ is neglected. Now a small value of either $\lambda_s$ or $\lambda_t$ implies that the collision is impulsive since the interaction of the projectile with the target is too weak (if $\lambda_s$ is small) or does not act long enough (if $\lambda_t$ is small) to cause an appreciable perturbation. It is therfore reasonable to expect that in the limit that either of these two parameters approaches zero (with the other one held fixed) the simplest diagram, (D1), is the dominant one.

III.   DIAGRAMS FOR REARRANGEMENT PROCESSES

Returning to reaction (1.1), the two expansion parameters which naturally arise in the calculation of the integrated cross section are $\lambda_i \equiv \tau_{coll} / \tau_i$ and $\lambda_f \equiv \tau_{coll} / \tau_f$, where $\tau_i$ and $\tau_f$, respectively, are the characteristic internal orbital periods of (A + T) in state i and (P + A) in state f. If $a_i$ and $a_f$ are the characteristic radii of (A + T) and (P + A), respectively, the characteristic effective collision duration is

$$\tau_{coll} = a/v \quad , \quad a = \min(a_i, a_f) \quad , \tag{3.1}$$

and if $v_i$ and $v_f$ are the characteristic internal orbital speeds of (A + T) and (P + A), respectively,

$$\tau_i = a_i/v_i \quad , \quad \tau_f = a_f/v_f \quad . \tag{3.2}$$

If $\lambda_i$ and $\lambda_f$ are <u>both</u> small, A may be considered a free particle during the time $\tau_{coll}$, and the rearrangement process is an impulsive one.

To simplify the discussion, assume for the moment that P and T do not interact. Then the simplest diagrams are:

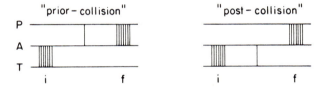

Since the vertical line can be incorporated into a block, each of these diagrams is physically equivalent to the single diagram, labelled (R1):

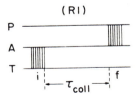

As drawn, diagram (R1) can give no contribution, for it represents the capture of A without a collision actually occurring. This diagram can give a significant contribution only if the two blocks overlap for a significant fraction of the initial and final orbital periods:

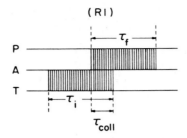

In this form, (R1) represents a full <u>three-body</u> collision. With $\tau_i$ and $\tau_f$ fixed, maximum overlap occurs when $\lambda_i = \lambda_f = 1$. In the limit that either $\lambda_i$ or $\lambda_f$ approaches zero, the overlap approaches zero, and there is no reason to expect that (R1) is the leading diagram in an expansion whose parameters are $\lambda_i$ and $\lambda_f$.

The next simplest diagram, labelled (R2), is:

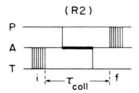

This diagram represents the following process, originally discussed within the framework of classical mechanics by Thomas (1927): first P <u>impulsively</u> knocks A out of the bound system (A + T); then A propagates as a free particle (represented by the thick horizontal line); finally, T <u>impulsively</u> knocks A into the bound system (P + A). Since small values of both $\lambda_i$ and $\lambda_f$ imply that the capture process is an impulsive one, it is reasonable to expect that (R2) is the leading diagram in the limit that <u>both</u> $\lambda_i$ and $\lambda_f$ approach zero; detailed calculations indicate that this is correct. (It is interesting to recall that for direct inelastic processes the expansion converges to the leading diagram in the limit that <u>either</u> of the two expansion parameters approaches zero.) During the interval between the two collisions, A propagates as a free particle but it need not propagate on the energy shell; from the uncertainty principle, the deviation, $\Delta\varepsilon$, off the energy shell has a characteristic value $\hbar / \tau_{coll} = \hbar v / a$. Far-off-shell contributions from (R2), that is, contributions with $\Delta\varepsilon$ much greater than $\hbar / \tau_{coll}$, arise when A propagates in the vicinity of both P and T. Since this requires block overlap, the far-off-shell contribution from (R2) is expected to be, and generally is, of the same order of magnitude as the contribution from (R1). The next simplest diagrams, labelled (R3a) and (R3b), are:

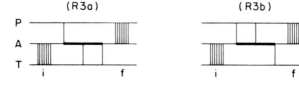

Diagram (R3a) differs from (R2) only in that after A is impulsively knocked out of the bound system (A + T), it undergoes <u>two</u> impulsive collisions with T before binding with P. The two free-particle propagations occur near to the energy shell (that is, $\Delta\varepsilon \lesssim \hbar / \tau_{coll}$) and the additional impulsive collision implies that the contribution from (R3a) is a factor of order $\lambda_i$ smaller than that from (R2). Similarly, the contribution from (R3b) is a factor of order $\lambda_f$ smaller than that from (R2). For capture from and to <u>isotropic</u> bound states, the contributions from (R1) and (R3a) + (R3b), and the <u>far-off-shell</u> contribution from (R2), are all of the same order. More complicated diagrams, involving a greater number of impulsive two-body collisions, will be of higher order in $\lambda_i$ or $\lambda_f$.

Suppose now that one of the expansion parameters, say $\lambda_f$, is not small. Then a diagrammatic expansion must be found which has only one expansion parameter, $\lambda_i$. This can be achieved, of course, by replacing the Greens function for free-particle propagation by the Greens function for propagation in the strong potential (in the present case, the A-P potential). Diagram (R2) is replaced by the following diagram, labelled (R2P):

( R2P )

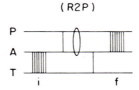

Here the loop represents propagation in the strong potential. The fact that there is now only one expansion parameter is a hint that the sum of diagrams (R1) and (R2P) might be viewed as a direct excitation process. To see that this picture is <u>roughly</u> correct, note first that the exact equation in diagrammatic form for the state vector of three particles, only two of which interact (namely, A and P), is, if A and P form a scattering state i' (represented by a block of lines)

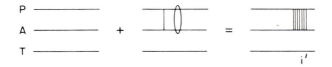

This equation is correct only on the energy shell. However, the equation will now be used with the block i inserted in the left-hand-side, that is, A will be described as initially bound to T. While this introduces a deviation from the energy shell which is small (roughly the binding energy of A and T) the resulting error is not small, as first noted by Macek and Taulbjerg (1981). The error would be small if A could be treated as a free particle during the collision time $\tau_{coll}$. However, while A is effectively free of T during this time (since $\lambda_i$ is small) it is not free of P (since $\lambda_f$ is large). The resulting error is therefore large, and increases with increasing $\lambda_f$. Bearing this in mind, the above equation may be used to roughly combine (R1) in its "post-collision form" with (R2P) to yield a diagram labelled (RIMP):

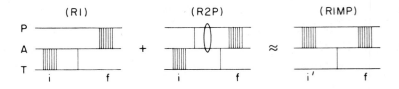

Diagram (RIMP) represents the direct impulsive (de)excitation of (P + A) from a
continuum state i' to a bound state f, and this diagram corresponds to the so-
called impulse approximation (Briggs 1977). While the relative error of (RIMP)
is not small, the physical picture which (RIMP) represents leads one to expect
that the correct combination of (R1) and (R2P) is the leading contribution in the
limit that $\lambda_i$ approaches zero.

Up to now the P-T interaction has been neglected; it will now be included. The
simplest diagram, labelled (R1'), that involves the P-T interaction is:

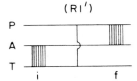

Since, with T infinitely massive, the P-T interaction is one between P and the
center of mass of (A + T), it cannot directly affect the internal state of (A + T).
This interaction can indirectly affect the internal state of (A + T) since the
deflection of P by T modifies the effect of the primary A-P interaction. Hence
(R1') can make a nonvanishing contribution only if the two blocks overlap -- and
then (R1') represents a full three-body collision. If the P-T, A-T, and A-P
interactions are of the same kind, (R1) and (R1') yield contributions of the same
order of magnitude. The next simplest diagram, labelled (R2'), that involves the
P-T interaction only once is:

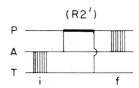

This diagram is similar in nature to (R2), and if all three interactions are of
the same kind, and if the masses of P and A are comparable, (R2) and (R2') yield
contributions of the same order of magnitude. (For the reaction $e^+ + H \rightarrow Ps + H^+$,
the near-on-shell amplitudes corresponding to (R2) and (R2') have the same absolute magnitude, and if their relative sign -- which depends on the orbital angular
momentum in the initial and final states -- is negative, they cancel (Shakeshaft
and Wadehra 1980).) Suppose, however, that the mass of P is much larger than the
mass of A. In the limit that the mass of P becomes infinite, the deflection of P

during the collision becomes negligible, and the integrated cross section must become independent of the sum of <u>all</u> diagrams involving the P-T interaction (Mott 1931; see also the footnote due to Wick on p. 359 of Jackson and Schiff 1953). This does not at all imply that in this limit the sum of such diagrams is zero, but it does turn out that when all three interactions are of the same kind, (R1') and (R2') cancel through leading order in $\lambda_i$ and $\lambda_f$.

To summarize: when $\lambda_i$ and $\lambda_f$ are both small, the dominant diagram(s) does not require block overlap. Those diagrams which require block overlap to yield non-vanishing contributions represent three-body collisions, and they are small corrections associated with large deviations from the energy shell. Appreciable block overlap occurs only between those components of the initial and final orbital distributions that have effective orbital periods comparable to $\tau_{coll}$ and therefore much smaller than the <u>characteristic</u> periods $\tau_i$ and $\tau_f$. In other words, appreciable overlap occurs only between the high velocity components of the initial and final bound states. Now if the orbital angular momentum quantum number of the initial and/or final bound state is large, the relative weight of the high velocity components of that state is extremely small. This is because two particles, bound together, can acquire a high orbital velocity only when they are close together (provided that their interaction is singular, not necessarily unbounded, at the origin); but a large centrifugal barrier will prevent the two particles from moving together. Therefore, diagrams which require block overlap, such as (R1), are extremely small corrections when the internal orbital angular momentum of the initial and/or final bound state is large.

Radiative capture may also be significant in the asymptotic regime (Raisbeck and Yiou 1971, Schnopper et al. 1972, Briggs and Dettmann 1974). For more details on radiative capture, and a somewhat different discussion of some of the points raised above, see Shakeshaft and Spruch (1979).

IV. THE THOMAS MODEL

Thomas (1927) treated the capture of an electron from an atom by an impinging ion within a classical model. Later, Bates et al. (1964) generalized his model to treat atom capture from molecules, for a restricted set of mass ratios. Recently Shakeshaft and Spruch (1980) built the relevant quantum-mechanical features into this model, with the ratio of the masses of P and A arbitrary; the quantum version of the model corresponds to diagrams (R2) and (R2').

Thomas noted that if P is incident with a high velocity $\vec{v}$, so that $\lambda_i$ and $\lambda_f$ are both small, capture can effectively occur only if A and P emerge from the collision with almost the same velocity. He pointed out that this could be achieved via the double-scattering event illustrated in Fig. 1. The projectile P, which is incident with an impact parameter b relative to T, first impulsively knocks A towards T; P is deflected through an angle $\beta$, known (Dettmann 1971) as the "critical angle", and emerges with a speed $u_1$, while A (whose initial orbital speed is negligible compared to v) emerges with a speed $v_1$ and in a direction making an angle $\alpha_1$ with $\hat{v}$ (where $\vec{v} \equiv v\hat{v}$). Then A impulsively scatters from T through an angle $\alpha_2$; this second collision is elastic since T is infinitely massive and so the speed of A remains $v_1$. Capture may occur if $v_1 \approx u_1$ and $\alpha_2 \approx \alpha_1 + \beta$. Let $m_P$ and $m_A$ denote the masses of P and A, respectively, and define $\xi \equiv m_P/(m_P + m_A)$. If the two collisions are treated as separate binary collisions, it follows from the equations of energy and momentum conservation and the requirement $v_1 = u_1$ that (Shakeshaft and Spruch 1980)

$$u_1 = v_1 = \xi^{1/2} v \quad , \tag{4.1}$$

$$\beta = \cos^{-1} [ (3\xi - 1)/(2\xi^{3/2}) ] \quad , \tag{4.2}$$

$$\alpha_1 = \cos^{-1} (\tfrac{1}{2} \xi^{-1/2}) \quad . \tag{4.3}$$

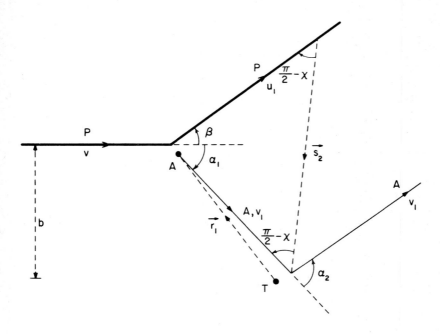

Figure 1

Schematic diagram depicting the motions of A and P (with
T at rest). The paths of A and P are indicated by thin
and thick lines, respectively. Each collision is a close
encounter. The position vector of A relative to T just
before the first collision is $\vec{r}_1$ and the position vector
of A relative to P just after the second collision is $\vec{s}_2$.

Note that $\beta$ is real only if $\frac{1}{4} \leq \xi \leq 1$. In a spherical coordinate system with
polar axis parallel to $\hat{v}$, the initial position vector, $\vec{r}_1$, of A relative to T is

$$\vec{r}_1 = (r, \pi - \alpha_1, \phi) \quad , \tag{4.4a}$$

$$r = b/\sin \alpha_1 \quad , \tag{4.4b}$$

and the position vector, $\vec{s}_2$, of A relative to P immediately after the second collision is

$$\vec{s}_2 = (s, \tfrac{1}{2}\pi - \chi + \alpha_1, \phi) \quad , \tag{4.5a}$$

$$s = r/\xi^{\frac{1}{2}} \quad , \tag{4.5b}$$

$$\chi = \sin^{-1} (\tfrac{1}{2} \xi^{-\frac{1}{2}}) \quad . \tag{4.5c}$$

The momentum transfers in the first and second collisions are $\xi^{\frac{1}{2}} m_A v$ and $m_A v$, respectively.

Thomas (1927), and Bates et al. (1964), assumed that: (i) A is initially at a
fixed distance from A, that is, r of Eqs. (4.4) is fixed; (ii) the differential
cross sections for the first and second collisions may be determined classically;
and (iii) the solid angle into which A is scattered after the second collision may

be determined from the condition that the final relative speed of A and P be less than the classical escape speed of A in the field of P at the distance s from P. The quantum modifications to the model amount to: (i) describing the spatial distribution of A with the initial and final bound-state wavefunctions, $\psi_i(\vec{r})$ and $\psi_f(\vec{r})$, respectively; (ii) using the quantum differential cross sections, and (iii) determining the solid angle into which A is scattered after the second collision by using the uncertainty principle (with some hindsight) and noting that the final relative speed of A and P should not greatly exceed the characteristic orbital speed of (P + A). The quantum version of the model yields a cross section for the double-scattering capture of A which, when integrated over impact parameter b (or, equivalently, over r), is (Shakeshaft and Spruch 1980):

$$\sigma_{ds} = \frac{2\pi}{\xi^{3/2}} \left( \frac{2\pi h}{\xi \, m_A v} \right)^3 \frac{d\sigma_1}{d\Omega_1{}'} \frac{d\sigma_2}{d\Omega_2{}'} \int_0^\infty dr |\psi_i(\vec{r}_1) \, \psi_f(\vec{s}_2)|^2 \; , \qquad (4.6)$$

where $d\sigma_1/d\Omega_1{}'$ and $d\sigma_2/d\Omega_2{}'$ are the two-body center-of-mass differential cross sections for P and A to exchange momentum $\xi^{\frac{1}{2}} m_A v$ and for A and T to exchange momentum $m_A v$, respectively. This cross section is, to leading order in $\lambda_i$ and $\lambda_f$, identical to the near-on-shell contribution from diagram (R2).

The two collisions may be regarded as separate binary collisions provided that the deBroglie wavelength, $\hbar/(\xi^{\frac{1}{2}} m_A v)$, of A as it propagates between collisions is much smaller than the distance r between collisions. Since $\frac{1}{4} \le \xi \le 1$, this condition amounts to

$$b \gg \hbar/(m_A v) \; . \qquad (4.7)$$

The contribution to the integral of Eq. (4.6) from the range of impact parameters $b \lesssim \hbar/(m_A v)$ is small, and is a correction of order $\lambda_i$ or $\lambda_f$ or smaller. However, it should be noted that when $b \lesssim \hbar/(m_A v)$ the capture event is a full three-body collision.

The analogue of diagram (R2') is the double-scattering event in which P, rather than A, scatters from T. In this case $\beta$ and $\alpha$ are simply interchanged in Eqs. (4.2) and (4.3), $d\sigma_2/d\Omega_2{}'$ in Eq. (4.6) is the differential cross section for P and T to exchange momentum $m_P v$ (not $m_A v$), and $m_A$ is replaced by $m_P$ in Eq. (4.7). Note that if $m_P$ is greater than $m_A$, the momentum transfer in the second collision is greater by a factor of $m_P/m_A$ in this second case than it is in the first case, and the total capture cross section is correspondingly smaller.

The formulation of the quantum version of the Thomas model when either $\lambda_i$ or $\lambda_f$ is not small would be useful, but has not yet been given.

V. ELECTRON CAPTURE FROM A HYDROGENLIKE ION BY A BARE ION

In this section, T is an infinitely heavy bare ion of charge $Z_T e$, P is a bare ion of charge $Z_P e$ and mass M, and A is an electron, to be denoted as e, of charge $-e$ and mass m. Since m/M is small, the integrated cross section is almost independent of the sum of diagrams involving the P-T interaction, and these diagrams will be ignored in this section. Let n, $\ell$, and $\mu$ denote the principal and orbital angular momentum quantum numbers of the initial bound state and let n', $\ell$', and $\mu$' denote the corresponding quantum numbers of the final bound state. The magnetic quantum number $\mu$ will be averaged over, and $\mu$' will be summed over. Note that

$$a_i = n^2 a_0/Z_T \quad , \qquad a_f = n'^2 a_0/Z_P \quad , \qquad (5.1)$$

$$v_i = Z_T e^2/(\hbar n) \quad , \qquad v_f = Z_P e^2/(\hbar n') \quad , \qquad (5.2)$$

where $a_0 \equiv \hbar^2/me^2$ is the Bohr radius. In the context of electron capture, diagram (R1) is usually called the Brinkman-Kramers amplitude. The sums (R1) + (R2) and (R1) + (R2) + (R3a) + (R3b) will be called the second and third Born amplitudes, respectively. The sum of (R1) and either (R2P) or (R2T), where (R2T) is similar to (R2P) but with the A-T potential the strong one, will be called[1] the strong-potential second Born amplitude. The cross section for capture from state $n\ell$ to $n'\ell'$ will be written as $\sigma(n\ell \to n'\ell')$. A subscript on $\sigma$ will indicate a particular contribution to $\sigma$; the subscripts BK, B2, B3, and spB2 indicate contributions from the Brinkman-Kramers amplitude, the second Born amplitude, the third Born amplitude, and the strong-potential second Born amplitude, respectively.

In the asymptotic limit where both $\lambda_i$ and $\lambda_f$ approach zero, $\sigma_{B3} \sim \sigma_{spB2} \sim \sigma_{B2} \sim \sigma_{ds}$, where $\sigma_{ds}$ is the double-scattering contribution given by Eq. (4.6), and

$$\sigma_{BK}(n\ell \to n'\ell') \sim \text{const.}(e^2/\hbar v)^{12} \, (v_i/v)^{2\ell} \, (v_f/v)^{2\ell'} \quad , \tag{5.3}$$

$$\sigma_{ds}(n\ell \to n'\ell') \sim \text{const.}(e^2/\hbar v)^{11} \quad . \tag{5.4}$$

The factors of $(v_i/v)^{2\ell}$ and $(v_f/v)^{2\ell'}$ in $\sigma_{BK}$ can be understood by recalling that only high orbital velocity components contribute to $\sigma_{BK}$, but the centrifugal barrier prevents e from gaining an orbital velocity much greater than the characteristic orbital velocity. Note that $\sigma_{ds}$ behaves as $v^{-11}$ with increasing v for all $\ell$ and $\ell'$. Hence for $\ell$ and/or $\ell'$ sufficiently large, $\sigma_{BK}$ is negligible compared to $\sigma_{ds}$ over a wide range of speeds where relativistic corrections are small. However, when $\ell = \ell' = 0$, the correction to $\sigma_{ds}$ is large at all nonrelativistic speeds. Briggs and Dubé (1980), extending an earlier result of Drisko (1955), have shown that, through the $v^{-12}$ term,

$$\sigma_{B2}(no \to n'o) \sim \sigma_{ds}(no \to n'o) + 0.295 \, \sigma_{BK}(no \to n'o) \quad , \tag{5.5}$$

where n and n' are arbitrary. The second term on the right of Eq. (5.5) contains both the contribution from (R1) and the far-off-shell contribution from (R2). The asymptotic expansion of $\sigma_{B3}(no \to n'o)$, through the $v^{-12}$ term, differs from the right-hand-side of Eq. (5.5) by the inclusion of the term

$$(Z_p + Z_T) \, \pi \, (e^2/\hbar v) \, \sigma_{ds}(no \to n'o) \quad ,$$

which is the near-on-shell contribution of (R3a) and (R3b); this result has been proved (Dettmann 1971, Shakeshaft 1978) only for n = n' = 1 but it is probably correct for arbitrary values of these quantum numbers. The exact value of $\sigma_{B2}(10 \to 10)$ was recently calculated by several groups (Miraglia et al. 1981, Simony and McGuire 1981, Wadehra et al. 1981). For the reaction $H^+ + H(1s) \to H(1s) + H^+$, at a lab energy of 10 MeV, Simony and McGuire find that, in units of $\pi a_0^2$, $\sigma_{BK} = 1.2 \times 10^{-11}$, $\sigma_{B2} = 8.0 \times 10^{-12}$, and the r.h.s. of Eq. (5.5) is $4.6 \times 10^{-12}$. At this energy, $\lambda_i = \lambda_f = 0.05$.

In their investigation of $\sigma_{B2}$, Briggs and Dubé (1980) found that when $\ell$ and/or $\ell' \geq 1$ the first correction to $\sigma_{ds}$ is a $v^{-13}$ term; this term is part of the near-on-shell contribution from (R2), and is, in fact, just the first correction to the leading part of the near-on-shell contribution.

Several examples will now be given for which one or both of $\lambda_i$ and $\lambda_f$ can be extremely small.

1.  Asymmetric ($Z_T \gg Z_p$) and nonclassical

Assume that n = n' = 1, $\ell = \ell' = 0$, and that $Z_T \gg Z_P$. Then $a_i < a_f$ and the expansion parameters are

$$\lambda_i = Z_T(e^2/\hbar v) \quad , \quad \lambda_f = (Z_p/Z_T)(Z_p e^2/\hbar v) \tag{5.6}$$

For $Z_T \geq 10$, $\lambda_i$ cannot be considered small until v is so large that relativistic corrections are important. On the other hand, since $\lambda_f$ decreases as $Z_T$ increases, $\lambda_f$ may be small if $Z_T$ is large even when $Z_p e^2/\hbar v$ is of order unity. The parameter $\lambda_i$ can be eliminated by incorporating the strong potential (the A-T potential) into the Greens function, and $\sigma_{spB2}$ is presumably an accurate estimate of the exact nonradiative capture cross section when $\lambda_f$ is very small. Recently Macek and Taulbjerg (1981), following earlier work of Macek and Shakeshaft (1980), reduced $\sigma_{spB2}$ to a one-dimensional integral and they explicitly demonstrated that the off-energy-shell effect that is neglected in deriving the impulse approximation (Briggs 1977) is an important effect when $\lambda_i$ is not small. In fact, they showed that if corrections of order $(Z_pe^2/\hbar v)^2$ are neglected, $\sigma_{spB2}$ is smaller than the cross section obtained from the impulse approximation by a factor

$$2 \ [(1 + \lambda_i^2)(1 + e^{-2\pi\lambda_i})]^{-1} \ .$$

In their reduction of $\sigma_{spB2}$ to a one-dimensional integral, Macek and Taulbjerg used a peaking approximation which is accurate only when the inequalities $Z_p/Z_T \ll 1$ and $(Z_pe^2/\hbar v) \ll 1$ are separately satisfied. Therefore their estimate of $\sigma_{spB2}$ is inaccurate when $(Z_pe^2/\hbar v)$ is not small even though $\lambda_f$ may be small. Results of their calculation of $\sigma_{spB2}$ for electron capture from the ground-state of a hydrogenlike ion with $Z_T = 18$ to the ground-state of a hydrogen ($Z_p = 1$) are shown in Fig. 2. Also shown are the results obtained by Kocbach (1980) using the impulse approximation in conjunction with a peaking approximation somewhat analogous to that used by Macek and Taulbjerg. For further comparison, the experimental results of McDonald et al. (1974) for K-shell electron capture from argon by protons are shown. Note that $\sigma_{spB2}$ and the experimental data peak at roughly the same energy, whereas the cross section obtained using the impulse approximation peaks at a significantly lower energy.

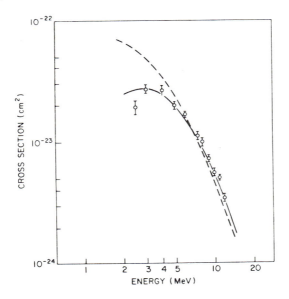

Figure 2

Cross sections for capture of argon K-shell electrons by protons. Experimental data are from McDonald et al., the dashed curve is the impulse approximation result of Kocbach, and the solid curve is the strong-potential second Born result of Macek and Taulbjerg.

## 2.   Symmetric and classical

Assume now that $Z_P \approx Z_T$ and that the initial state is a high Rydberg state with
$n \gg 1$ and $\ell \approx n - 1$. Then the initial state is classically describable. Assume
that the final bound states are summed over. Since capture occurs primarily to
states with $n' \approx n$ and $\ell' \approx \ell$, those states which provide the major contribution
to the sum over final bound states are also classically describable. In this case
the expansion parameters are, with Z denoting the average of $Z_P$ and $Z_T$,

$$\lambda_f \approx \lambda_i \approx (Z/n)(e^2/\hbar v) \tag{5.7}$$

Therefore, with bd denoting the sum over final <u>bound</u> states, $\sigma_{ds}(n\ell \rightarrow bd)$ is an
accurate estimate of the exact nonradiative capture cross section for projectile
ions incident with energies in the eV range when n is sufficiently large, and
$\ell \approx n - 1$. Further, as shown by Spruch (1978), Thomas' classical estimate of
$\sigma_{ds}(n\ell \rightarrow bd)$ becomes <u>exact</u>. Spruch has derived the following result:

$$\sigma_{ds}(n\ell \rightarrow bd) = \frac{2^{13/2} Z_P^{7/2} Z_T^{11/2}}{3n^7} \left(\frac{e^2}{\hbar v}\right)^{11} (\pi a_0^2) \quad , \tag{5.8}$$

valid when $n \gg 1$ and $\ell \approx n - 1$. This result may be compared with:

$$\sigma_{BK}(n, \ell = n-1 \rightarrow bd) = \frac{\beta Z_P^5}{n^{5/2}} \left(\frac{4Z_T e^2}{n\hbar v}\right)^{2n+3} \left(\frac{e^2}{\hbar v}\right)^7 (\pi a_0^2) \quad , \tag{5.9}$$

where $\beta$ is weakly dependent on n when n is large and is of the order of $10^2$.
Evidently $\sigma_{BK}(n\ell \rightarrow bd)$ is utterly negligible compared to $\sigma_{ds}(n\ell \rightarrow bd)$ when $n \gg 1$
and $\ell \approx n - 1$.

## 3.   Capture to the continuum

In "capture to the continuum" the electron emerges with a velocity $\vec{v}_e$ almost equal
to the velocity $\vec{v}$ of the (essentially undeflected) projectile ion. More precisely,
the internal energy of (P + e) after the collision is nonnegative but very small
compared to the ground-state binding energy of (P + e). Thus e emerges from the
collision almost bound to P but it eventually escapes. This process was first
identified experimentally by Rudd et al. (1966) and early theoretical work was
done by Salin (1969), Macek (1970), and Bonsen and Banks (1971). Recently, attention has focussed on the singly differential cross section, $d\sigma/dv_e$, for the electron to emerge into a narrow forward cone with a speed $v_e$ close to v. This cross
section can be expressed as

$$\frac{d\sigma}{dv_e} = \int d\Omega_e \frac{f(\vec{v}_e - \vec{v})}{|\vec{v}_e - \vec{v}|} \tag{5.10}$$

where the integration is over the solid angle of the cone and where $f(\vec{v}_e - \vec{v})$ is a
function, implicitly dependent on v, that is finite at $\vec{v}_e = \vec{v}$. Owing to the de-
nominator $|\vec{v}_e - \vec{v}|$ of the integrand, $d\sigma/dv_e$ exhibits a cusp centered at $v_e = v$
(Dettmann et al. 1974), as indicated in Fig. 3.

Since $a_f = \infty$ the collision time is $\tau_{coll} = a_i/v$ and

$$\lambda_i = (Z_T/n)(e^2/\hbar v) \quad . \tag{5.11}$$

The orbital time $\tau_f$ is undefined and the natural generalization of $\lambda_f$ is $\lambda_f \equiv v_f/v$
where now $v_f$ is the characteristic value of $|v_e - v|$, that is, the characteristic
width of the cusp. The characteristic distance of e relative to P just after the
collision is $a_i$, and therefore $v_f$ is defined by

$$\frac{1}{2} mv_f^2 - \frac{Z_P e^2}{a_i} = 0 \quad , \tag{5.12}$$

which is the requirement that the characteristic internal energy of (P + e) be zero. Therefore $v_f = (2Z_pZ_T)^{1/2}$ $(e^2/n\hbar)$ and, dropping the 2,

$$\lambda_f = (Z_pZ_T/n^2)^{1/2} \ (e^2/\hbar v) \quad . \tag{5.13}$$

Since $\lambda_f$ increases only as $Z_p^{1/2}$ with increasing $Z_p$, it can be small even when $Z_p$ is reasonably large.

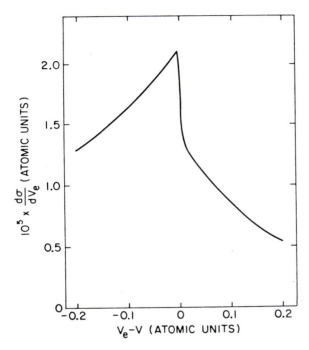

Figure 3
Estimate (within the second Born peaking approximation) of $d\sigma/dv_e$ for $C^{6+}$ incident at 2 MeV/amu on H. The cone semiangle is 1.85°.

Recent experimental (Breinig et al. 1980, and references therein) and theoretical (Shakeshaft and Spruch 1978, Chan and Eichler 1979, Miraglia and Ponce 1980, Garibotti and Miraglia 1981) work indicates that the cusp of $d\sigma/dv_e$ is highly asymmetric about $v_e = v$. If $f(\vec{v}_e - \vec{v})$ were an isotropic function of $\vec{v}_e - \vec{v}$, that is, if $f(\vec{v}_e - \vec{v}) = f(|\vec{v}_e - \vec{v}|)$, the cusp would be almost symmetric since the integral of $f(|\vec{v}_e - \vec{v}|)/|\vec{v}_e - \vec{v}|$ is, up to a correction of order $\delta v_e/v$ where $\delta v_e = v_e - v$, symmetric about $v_e - v$. Now the major part of the contribution from (R1) is to the $\ell' = 0$ component of the continuum since the contribution to a component with $\ell' \neq 0$ is smaller by a factor of $(v_f/v)^{2\ell'}$ -- see Eq. (5.3). Therefore the contribution from (R1) to $f(\vec{v}_e - \vec{v})$ is essentially isotropic. It follows that the <u>asymmetric</u> contribution to the cusp resulting from (R1) and, for similar reasons, from the far-off-shell part of (R2), behaves with increasing v as

$$(e^2/\hbar v)^{12} \ (v_i/v)^{2\ell} \ (\delta v_e/v)$$

On the other hand, the near-on-shell contribution from (R2) is not restricted to the $\ell' = 0$ component since it behaves with v as $v^{-11}$, with a correction that behaves as $v^{-13}$, for all $\ell'$ (and $\ell$). Hence the near-on-shell part of (R2) yields a highly nonisotropic contribution to $f(\vec{v}_e - \vec{v})$ and a highly asymmetric contribution to the cusp which behaves with increasing v as

$$(e^2/\hbar v)^{11} + const.(e^2/\hbar v)^{13} \quad .$$

Note especially that near the tip of the cusp (where $\delta v_e \approx 0$) the asymmetry arising from (R2) dominates over that from (R1). This is important since the sharp drop in $d\sigma/dv_e$ as $v_e - v$ increases from zero (see Fig. 3) has been observed (Breinig et al.) and is probably a signature of (R2).

VI.  ATOM TRANSFER

As a final example, consider the atom transfer reaction $H^+ + CH_4 \rightarrow H_2^+ + CH_3$. Here P is a proton of mass M, A is a "structureless" hydrogen atom with almost the same mass M, and T is the "structureless" particle $CH_3$ whose mass, which is almost 15M, may be considered infinite. The "orbital" periods are now periods of vibration, and they are very large:

$$\tau_i \sim \tau_f \sim (M/m)^{\frac{1}{2}} (\hbar/E_e) \quad , \tag{6.1}$$

where $E_e \sim \hbar^2/(ma^2)$ is a typical ground-state electronic energy (a few eV), and a is a typical length (about 1 Å), of a diatomic molecule, and m is the electron mass. It follows that

$$\lambda_i \sim \lambda_f \sim (m/M)^{\frac{1}{2}} (\hbar/mva) \quad . \tag{6.2}$$

Therefore, the asymptotic regime begins at impact energies of about 100 eV.

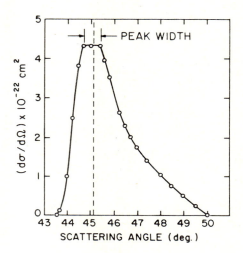

Figure 4

Cook et al. (1975) have measured the cross section for $H_2^+$ formation, via this reaction, for incident proton energies in the range 70 to 150 eV. Their data for the angular distribution of the emergent $H_2^+$, at an incident proton energy of 85 eV, are shown in Fig. 4. From Eq. (4.2) the critical angle is $45°$ since $\xi = \frac{1}{2}$. The critical peak at $45°$ shown in Fig. 4 confirms the validity of the basic double-scattering mechanism proposed by Thomas (1927) and Bates et al. (1964). Quantum effects are, of course, important in the estimation of the size of this peak. The

value of the integrated cross section calculated from the classical model is about
a factor of 30 larger than the measured value (Cook et al. 1964) whereas the value
calculated from the quantum version of the model (Shakeshaft and Spruch 1980) is
about a factor of 6 larger than the measured value. (To obtain further improvement,
proper account would have to be taken of the structure of the particles and of the
interference between diagrams (R2) and (R2').)

ACKNOWLEDGMENTS

It is a pleasure to thank L. Spruch for many helpful suggestions and comments.
This work was supported by the National Science Foundation, Grant Nos. PHY79-09954
and PHY79-10413, the Office of Naval Research, Contract No. N00014-76-C-0317, and
the Texas A & M Center for Energy and Mineral Resources.

FOOTNOTES

$^+$ Present address:  Physics Department, University of Southern California,
                      Los Angeles, California  90007

1.This terminology was suggested by J. Macek.

REFERENCES

[ 1]    Amundsen, P.A. and Jakubassa, D., J. Phys. B13 (1980) L467-72.
[ 2]    Basu, D., Mukherjee, S.C., and Sural, D.P., Physics Reports 42 (1978) 145-234.
[ 3]    Bates, D.R., Cook, C.J., and Smith, F.J., Proc. Phys. Soc. London 83 (1964)
        49-57.
[ 4]    Belkić, Dz., Gayet, R., and Salin, A., Physics Reports 56 (1979) 279-369.
[ 5]    Bonsen, T.F.M. and Banks, D., J. Phys. B4 (1971) 706-714.
[ 6]    Breinig, M., Elston, S., Sellin, I., Liljeby, L., Thoe, R., Vane, C.R.,
        Gould, H., and Marrus, R., Phys. Rev. Lett. 45 (1980) 1689-92.
[ 7]    Briggs, J.S., J. Phys. B10 (1977) 3075-89.
[ 8]    Briggs, J.S. and Dettmann, K., Phys. Rev. Lett. 33 (1974) 1123-5.
[ 9]    Briggs, J.S. and Dubé, L., J. Phys. B13 (1980) 771-84.
[10]    Chan, F.T. and Eichler, J., Phys. Rev. A20 (1979) 367-8.
[11]    Cook, C.J., Smyth, N.R.A., and Heinz, O., J. Chem. Phys. 63 (1975) 1218-23.
[12]    Dettmann, K., Springer Tracts Mod. Phys. 58 (1971) 119-206.
[13]    Dettmann, K., Harrison, K.G., and Lucas, M.W., J. Phys. B7 (1974) 269-87.
[14]    Drisko, R.M., Thesis, Carnegie Institute of Technology (1955).
[15]    Eichler, J. and Chan, F.T., Phys. Rev. A20 (1979) 104-12.
[16]    Ford, A.L., Reading, J.F., and Becker, R.L., Phys. Rev. A23 (1981) 510-18.
[17]    Garibotti, C.R. and Miraglia, J.E., J. Phys. B14 (1981) 863-8.
[18]    Hagedorn, G.A., Commun. Math. Phys. 66 (1979) 77-94.
[19]    Jackson, J.D. and Schiff, H., Phys. Rev. 89 (1953) 359-65.
[20]    Kocback, L., J. Phys. B13 (1980) L665-71.
[21]    MacDonald, J.R., Cocke, C.L., and Edison, W.W., Phys. Rev. Lett. 32 (1974)
        648-51.
[22]    Macek, J., Phys. Rev. A1 (1970) 235-41.
[23]    Macek, J. and Shakeshaft, R., Phys. Rev. A22 (1980) 1441-6.
[24]    Macek, J. and Taulbjerg, K., Phys. Rev. Lett. 46 (1981) 170-4.
[25]    Miraglia, J.E. and Ponce, V.H., J. Phys. B13 (1980) 1195-207.
[26]    Miraglia, J.E., Piacentini, R.D., Rivarola, R.D., and Salin, A., J. Phys.
        B14 (1981) L197-202.
[27]    Mott, N.F., Proc. Cambridge Phil. Soc. 27 (1931) 553-60.
[28]    Raisbeck, G. and Yiou, F., Phys. Rev. A4 (1971) 1858-68.
[29]    Rudd, M.E., Sautter, C.A., and Bailey, C.L., Phys. Rev. 151 (1966) 20-7.
[30]    Salin, A., J. Phys. B2 (1969) 631-6.
[31]    Schnopper, H.W., Betz, H.D., Delvaille, J.P., Kalata, K., Sohval, A.R.,
        Jones, K.W., and Wegner, H.E., Phys. Rev. Lett. 29 (1972) 898-901.
[32]    Shakeshaft, R., Phys. Rev. A17 (1978) 1011-17.
[33]    Shakeshaft, R. and Spruch, L., Phys. Rev. Lett. 41 (1978) 1037-40.
[34]    Shakeshaft, R. and Spruch, L., Rev. Mod. Phys. 51 (1979) 369-405.

[ 35]    Shakeshaft, R. and Spruch, L., Phys. Rev. $\underline{A21}$ (1980) 1161–72.
[ 36]    Shakeshaft, R. and Wadehra, J.M., Phys. Rev. $\underline{A22}$ (1980) 968–78.
[ 37]    Simony, P.R. and McGuire, J.H., submitted for publication (1981).
[ 38]    Spruch, L., Phys. Rev. $\underline{A18}$ (1978) 2016–21.
[ 39]    Thomas, L.H., Proc. Roy. Soc. $\underline{114}$ (1927) 561–76.
[ 40]    Wadehra, J., Shakeshaft, R., and Macek, J., submitted for publication (1981).

PHYSICS OF ELECTRONIC AND ATOMIC COLLISIONS
S. Datz (editor)
© North-Holland Publishing Company, 1982

ELECTRON CAPTURE FROM K SHELLS BY LIGHT IONS

Erik Horsdal-Pedersen

Institute of Physics, University of Aarhus
DK-8000 Aarhus C, Denmark

# 1. INTRODUCTION

Electron capture or charge transfer, taking place in ion-atom collisions when electrons are exchanged between the collision partners, is
a fundamental process in atomic scattering. It has been studied for
more than 50 years, but in spite of this, a general understanding of
the phenomenon is still lacking.

Aspects of the theory for electron capture relevant to the high-energy
region, in which orbital velocities are small compared to the collision velocity, were discussed on classical grounds already in 1927
by Thomas, who introduced a double-scattering mechanism, in which the
active electron is first accelerated to move at about the speed of
the projectile in a close collision with this particle, and after
that deflected to move almost parallel to the projectile in a second
collision with the target nucleus. The classical theory of Thomas and
its relation to quantum theory has been reviewed by Shakeshaft and
Spruch (1979). Strict   high-velocity quantum approximations, which include  the double-scattering mechanism, were reviewed by Belkić et al.(1979).

At intermediate velocities, the classical picture   of two isolated
and localized scatterings no longer applies. Instead, nonclassical
models have been suggested, in which the active electron is scattered
into a bound state of the projectile, either directly or through intermediate states interacting weakly with the target ion. This mecha-
ism can in lowest approximation be treated by first-order perturbation
theory. It was considered already in 1928 by Oppenheimer and in 1930
by Brinkman and Kramers. The approximation, which is referred to as
the OBK approximation, is not satisfactory because of difficulties
originating from the nonorthogonality of the initial and final states.
Various attempts have been made to solve this problem by adding certain functions of the internuclear distance to the perturbation potential, but such procedures are not satisfactory (Horsdal-Pedersen 1981).
Dewangan (1977) and Chan and Eichler (1979) recently extended the OBK
theory by introducing an eikonal phase factor in the OBK-transition
matrix element. The connection between this eikonal approximation and
the Born expansion with free motion between interactions was discussed
by Eichler and Narumi (1980), who showed that the prior form of the
eikonal approximation includes the interaction between the projectile
and the active electron to first order and weak interactions between
the electron and the target nucleus to all orders. The eikonal phase
factor lowers the OBK cross section at intermediate energies to bring
theory into better overall agreement with experimental data. At high
energies, the eikonal theory reduces to the OBK theory, which in this
limit is believed not to provide the correct asymptotic behaviour (by
neglecting close double collisions reflecting the classical double-
scattering mechanism of Thomas).

For <u>asymmetric</u> collisions (i.e., $Z_1 \ll Z_2$, where $Z_1$ and $Z_2$ are the
atomic numbers of projectile and target, respectively), it is possible to take advantage of the different strengths of the atomic potentials in developing theories for electron capture, which should be
valid at both intermediate and high velocities. These are the Impulse
Approximation (Briggs 1977, Jakubassa-Amundsen and  Amundsen 1980,
Kocbach 1980) and the 2nd Born Approximation with the Coulomb Green's
function (Macek and Shakeshaft 1980). A discussion of  the relation
between the two theories has been given by Macek and Taulbjerg (1981).
Both theories represent expansions to 1st order in the weak potential
while treating the strong one to all orders. This means that the double-
scattering mechanism of Thomas is included but also that problem.
due to nonorthogonality are avoided since the electron is first excited into (off-the-energy-shell) continuum states of the <u>target</u> ion
and subsequently projected onto bound states of the projectile. The
local perturbation by the projectile on the intermediate states of
importance for the capture process representing higher-order perturbations in $Z_1$ was not taken into account in the above theories.  A
symmetric form of the impulse approximation, which includes approximately this distortion, was given was Briggs (1980), but numerical
calculations have not yet been made.

Another line of progress was pursued by Shakeshaft (1978) and in particular by Ford et  al. (1981), who aim at very precise <u>numerical</u> solutions of the entire scattering problem, including excitation to
bound and continuum states as well as charge transfer. The calculations are performed using the method of coupled states, with a large
set of eigenstates centered on the projectile or the target. Apart
from yielding inelastic scattering amplitudes, the theory provides
illustrative mappings of how the scattering wave function evolves
in time (Shakeshaft 1978) and standards for judging more approximate
theories, but it also allows estimates to be made of the effect of
multielectron processes in collisions involving multielectron atoms
(Ford et al. 1981).

The merits of the above theories have been tested mainly by comparison with experimental total cross sections because only very few differential measurements have been published. However, to better understand the nature of the electron-capture process, it is important
that theoretical models be developed, which account for not only the
energy dependence and absolute magnitude of total cross sections but
also the impact-parameter and angular dependence of the process. Systematic experimental studies of these dependences are therefore
needed.

The present contribution concentrates on experimental electron capture from inner (or occasionally from outer) shells of atoms by fully
stripped, light ions. Capture from an inner shell fulfils the requirement of asymmetry set forth in the Impulse Approximation and the 2nd
Born Approximation mentioned above. Further advantages are the simplicity of wave functions for inner-shell electrons and the strong
electrostatic repulsion between projectile and target, which, even at
large collision velocities and impact parameters, results in measure-
able deflections, thus facilitating the differential measurements.
Disadvantages are related to the many-electron nature of the targets.
The relaxation of the passive electrons (Carlson 1973) may influence
the capture process to some degree, depending on the collision velocity, and multielectron processes can result in increased or reduced experimental apparent cross sections, depending on circumstances
(Ford et al. 1981).

The primary experimental data usually are cross sections differential in scattering angle. To display variations due to the capture process itself, it is useful to divide these cross sections by appropriate elastic differential cross sections to obtain angular-dependent probabilities. Further, in the limit of small deflections, the angular dependence is mainly an effect of the corresponding change of impact parameters. The classical impact parameter is therefore a more illustrative parameter than is the scattering angle. For these reasons, we have often preferred to present experimental data in the form of impact-parameter-dependent probabilities.

The experimental data and their interpretation are discussed in Sec. 2. Comparisons with theoretical results are discussed in Sec. 3. Finally, in Sec. 4, we indicate new experiments, which may help advancing the understanding of the electron-capture process.

## 2. ELECTRON CAPTURE FROM INNER SHELLS

Most of the experimental data discussed in the following section refer to capture from inner shells of atoms by fully stripped, light ions. Figure 1 illustrates such a collision. A fast proton hits an atom (neon) and captures a K-shell electron. The resultant K-shell hole subsequently decays by the emission of a KLL-Auger electron or a K-x ray, which provides a possibility for identifying the K-shell-capture events. This is done by detecting (scattered) hydrogen atoms in coincidence with KLL-Auger electrons or K-x rays. The latter possibility was originally used by Macdonald et al. (1974), who investigated total cross sections for capture of K-shell electrons from argon by protons. Corresponding differential cross sections were later reported by the same group (Cocke et al. 1976). Total cross sections for capture from the K shells of C, N, O, and Ne by $H^+$, $He^{++}$, and $Li^{+++}$ measured by the KLL-Auger branch were recently published by Cocke et al. (1977) and by Rødbro et al. (1979).

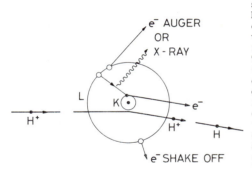

CAPTURE OF K-SHELL ELECTRONS

$e^-$ AUGER
OR
X-RAY

L

K

$e^-$

$H^+$

$H^+$

H

$e^-$ SHAKE OFF

PHYSICAL EFFECTS:
 X-RAY EMISSION
 OR
 AUGER ELECTRON EMISSION
EXPERIMENTAL METHODS:
 X-RAY
 OR          } IN COINC. WITH
 AUGER ELECTRON } SCAT. H ATOM

Fig. 1

Since the first measurements by Macdonald et al. (1974), it was realized that there are important questions as to the interpretation of the experimental data, which include the totality of processes in which a target K-shell vacancy is formed and an electron is transferred to a bound state of the projectile. The captured electron can be the K-shell electron, but it could also be an electron from a shell further out if the projectile simultaneously excited a K-shell electron. However, this is true only if the x ray or the Auger spectrometer does not resolve diagram lines from satellites. In case a diagram line is indeed resolved, the experiment ensures that the outer shells are left

fully relaxed. This leaves out most multiple processes but not all.
The registration of a true K-shell-capture event may be blocked if
an outer-shell electron is simultaneously deexcited to fill the K-
shell vacancy. Such a process will affect K-shell-capture probabilities measured as the ratio between the number of scattered particles
in coincidence and the total number of scattered particles. If, however, (differential) cross sections are extracted by normalizing the
rate of coincidences to the rate of detected Auger electrons or x
rays, the process is in practice of no consequence. Processes, in
which an outer-shell electron is captured and a K-shell electron is
excited to fill the outer-shell vacancy formed, are not discriminated
for, and they add to the measured cross sections or probabilities. On
the assumption that the cross section for the particular excitation
process referred to above is much smaller than the total K-shell vacancy-production cross section, one can show that this multielectron
process is negligible at the (high) collision velocities studied so
far. Finally, there is good reason to believe that capture from inner K shells by fully stripped, light ions is dominated by transitions to the 1s state of the projectile (Jakubassa-Amundsen and
Amundsen 1980). The comparison to be made between experimental and
theoretical data is based on this belief.

The impact-parameter-dependent K-shell-capture probabilities for $H^+$,
$He^{++}$, and $Li^{+++}$ in carbon, neon and argon presented in the following
paragraph are all measured by the KLL-Auger branch with a spectrometer resolution sufficient to resolve the major KLL-diagram line
(Rødbro et al. 1979). The rate of real coincidences for a given projectile scattering angle was normalized to the rate of KLL-Auger
electrons. In this way, a K-shell-Capture cross section, differential
in scattering angle and Auger-electron energy $d^2\sigma_{KC}/d\Omega dE$, is measured
relative to the total K-shell Vacancy-production cross section, differential in Auger-electron energy, $d\sigma_{KV}/dE$. We now assume that the
outer-shell excitation is independent of the K-shell process (capture or ionization) and the impact parameter, when this is of the
order of the K-shell radius. We also assume that the K-shell process
depends only weakly on outer-shell excitation. Under these plausible
assumptions, the measured ratio

$$\frac{d^2\sigma_{KC}}{d\Omega dE} \bigg/ \frac{d\sigma_{KV}}{dE} \quad \text{reduces to} \quad \frac{d\sigma_{KC}}{d\Omega} \bigg/ \sigma_{KV} \quad ,$$

where $\sigma_{KV}$ is the total vacancy-production cross section, which is
taken from the literature (Rødbro et al. 1979). The derived differential cross section therefore measures K-shell capture "no matter
what else happens". The main absolute uncertainty is due to $\sigma_{KV}$ and
is estimated to be 10-15%, based on quoted uncertainties and the
scattering of independent absolute measurements. The statistical errors are mainly due to the coincidence counting statistics, which in
most cases is better than 10% standard deviation.


## 3. COMPARISON WITH THEORY

We first consider the capture of K-shell electrons by protons at
large and intermediate velocities and then the $Z_1$ and $Z_2$ scalings at
a constant, scaled velocity.

3.1  K-Shell Capture by $H^+$ at Different Velocities. Bratton et al.
(1977) have measured differential capture cross sections for helium
in fast collisions with $H^+$. The experimental data as well as theoretical estimates are shown in Fig. 2. All theories use an impact-

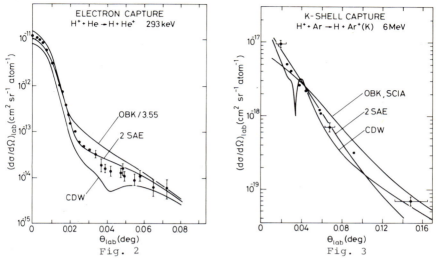

Fig. 2          Fig. 3

parameter description, in which the heavy nuclei follow classical trajectories, whereas the motion of the electrons is described in different quantum approximations. Differential cross sections are found by using the eikonal transformation, including the phase from the scattering potential. Rogers and McGuire (1977) described the electronic motion in the Brinkman-Kramers (OBK) approximation, which over-estimates the total cross section by a factor of 3.55 at this energy but reproduces the angular variation of the differential cross section relatively well. Lin and Soong (1978) used a 2-State, Atomic-Expansion model (2SAE) and obtained good agreement in absolute magnitude as well as dependence on scattering angle. Finally, Rivarola et al. (1980) used the Continuum-Distorted Wave approximation (CDW), which also fits the experimental data reasonably well.

Experimental data by Cocke et al. (1976) for capture from the K shell of argon in fast collisions with protons are shown in Fig. 3. The OBK results of Belkić and Salin (1976) reproduce the trend of the data except at the smallest scattering angles. The good agreement in absolute magnitude is fortuitous. The 2SAE results of Lin and Soong (1978) fit the experimetnal data quite well. This is also the case for the CDW results of Belkić and Salin (1978) obtained by using a hydrogenic, initial wave function but <u>neglecting</u> the outer screening. The origin of the sharp structure of the theoretical data is not clear. The results of the SemiClassical Impulse Approximation (SCIA) by Jakubassa-Amundsen and Amundsen (1980), also evaluated by using a hydrogenic description but <u>including</u> the outer screening, are finally seen to be in good agreement also with the experimental data.

On the basis of these comparisons, it appears that the 2SAE, the CDW, and the SCIA all describe the capture process quite reliably. As one goes towards smaller velocities, the CDW no longer applies, but the SCIA and the 2SAE still do.      Experimental data for 500-keV H$^+$ on Ne(K), which represents an intermediate velocity, are shown in Fig. 4 and compared with estimates based on the 2SAE. The strong disagreement exposes a serious deficiency of the 2SAE (Horsdal-Pedersen 1981).

In Fig. 5 is shown impact-parameter-dependent probabilities for capture of neon K-shell

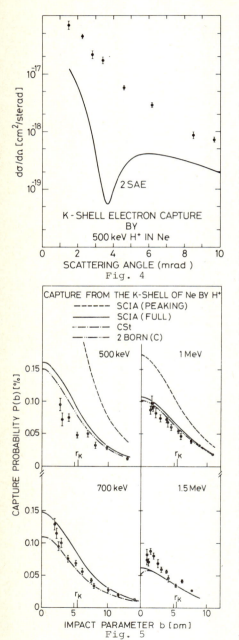

Fig. 4

Fig. 5

electrons by protons at various energies (Horsdal-Pedersen, Folkmann, and Pedersen 1979, unpublished). The experimental data are compared with theoretical results of the SCIA, the 2nd Born approximation with the Coulomb Green's function (2Born(C)), and Coupled-States calculations (CSt) with a large set of basis functions. The full SCIA (Jakubassa-Amundsen and Amundsen 1980) was evaluated using hydrogenic wave functions but with no further approximations. The remaining, relatively small, deviations from the experimental data, in particular with respect to energy variation, may to some degree be corrected by using more accurate initial wave functions. However, it has been argued by Macek and Taulbjerg (1981) that the SCIA is inherently uncertain by neglecting terms of the order of the ratio between the initial K-shell velocity and the collision velocity $Z_2/v$, which for the cases shown in Fig. 5 is rather large, ranging from 1.3 to 2.2. This apparent defect of the SCIA is remedied by the 2Born(C), which is related to the SCIA simply by a factor which depends on the parameter $Z_2/v$ only (Macek and Taulbjerg 1981). When this relation is used for the evaluation of the 2Born(C), which represents a systematic expansion of the scattering amplitude in powers of $Z_1/Z_2$, a simplified rather than the full SCIA should be applied. The appropriate form is obtained by a peaking approximation introduced by Briggs (1977) and further discussed by Kocbach (1980) (approximation (i)). Numerical results are plotted in Fig. 5. When evaluating the 2Born(C) as indicated above, the remaining error is of the order of $Z_1/v$, which for the example shown in Fig. 5 amounts to about 20%. With this and the absolute uncertainty of the experimental data (10-15%) in mind, we find reasonable agreement. The similarity between the results of the full SCIA and the 2Born(C) is probably fortuituous. It remains to be seen which approximation gives the most accurate description of a more extended set of experimental data. The CSt calculations (Ford et al. 1980, private communication) and the full SCIA agree at 1.5 MeV, but there is a difference at 700 keV, which, however, is no larger than the remaining uncertainty of the SCIA. In comparing the approximations, it should be borne in mind that the SCIA and the 2Born(C) are evaluated in a single-electron model with hydrogenic wave functions and experimental binding energies, whereas the CSt calculations use accurate wave functions and allow for distortions as well as multielectron processes. The CSt results and the experimental data agree within the absolute uncertainty of the data points.

3.2 K-Shell Capture by Light Ions at Constant Scaled Velocity. The Brinkman-Kramers theory predicts simple scaling laws at high collision velocities. These are approximately obeyed by total cross sections for capture from K shells by fully stripped, light ions (Rødbro et al. 1979). The same scaling laws are implied by the SCIA (Jakubassa-Amundsen and Amundsen 1980). For the impact-parameter-dependent capture probability $P(b)$, they are

$$(Z_2/Z_1)^5 \, P(b,v,Z_1,Z_2) \;\simeq\; \hat{P}(bZ_2,v/Z_2) \quad .$$

Thus, at fixed, scaled impact parameter $bZ_2 = b/r_K$ and velocity $v/Z_2 = v/v_K$, the probability increases in proportion to $(Z_1/Z_2)^5$. This scaling has been tested at a relatively low, scaled velocity. The insert of Fig. 6 gives a survey of the cases studied with values of

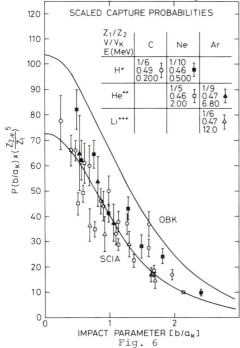

Fig. 6

the $Z_1/Z_2$ ratio, the scaled velocity $v/(Z_2-0.3)$ (Slater screening), and the kinetic energy of the beams. The data fall into two groups, one with $Z_1/Z_2 \simeq 1/10$ and another with $Z_1/Z_2 \simeq 1/6$. $H^+$ on argon is not measured yet but should be included. The experimental data are shown in Fig. 6. The theoretical curves represent the Brinkman-Kramers approximation (OBK) and the Semiclassical Impulse Approximation (SCIA) for $He^{++}$ on neon. In spite of the small, scaled collision velocity, the simple scalings describe the experimental data well. They scatter around a common curve by less than ±25%. The SCIA for the other collision partner in Fig. 6 tends to lie higher than the curve shown by as much as 70%, indicating the theoretical uncertainty of the scaling at the low collision velocity studied.

The comparisons in this section between theoretical and experimental data leads to the following provisional conclusions as regards capture of K-shell electrons. The OBK tends to overestimate probabilities at intermediate velocities but reproduces approximately the impact-parameter dependence, the 2SAE agrees poorly in absolute magnitude and impact-parameter dependence at the same velocities, the full SCIA, the 2Born(C), and the CSt all describe capture in asymmetric collisions quite well, and the CDW provides a satisfactory description at large velocities both for asymmetric and nearly symmetric systems.

## 4. FURTHER EXPERIMENTS

The experimental data presented above touch on the impact-parameter and velocity dependence of K-shell electron capture by fully stripped, light ions at intermediate and high velocities. The characteristics of capture probabilities for initial states other than 1s states and the dependence on the scattering angle, as distinct from the impact parameter, have not yet been studied experimentally. In this final section, we present preliminary data pointing in these directions and indicate new experiments. In relation to the widespread use of (higher-order) perturbation theory to describe electron capture at high velocities, it is of special interest to search for particular effects of multiple-scattering terms.

4.1  Capture from the L Subshells. According to existing theories, the probability for electron capture at intermediate velocities depends sensitively on the momentum distribution of the initial bound state of the captured electron. A natural outgrowth of the measurements of capture from inner K shells would therefore be to study capture from inner L shells and, in particular, the $L_{2,3}$ subshells of electrons occupying single-electron orbitals of 2p character (Jakubassa-Amundsen 1981, private communication). The $L_{2,3}$ subshells of argon may be studied by using an extension of the Auger-electron, scattered-projectile coincidence technique developed for the K-shell process. Since the 2p hole formed is most certainly not isotropic, it will be necessary to measure the angular distribution of the coincident Auger emission at each impact parameter. Such a measurement may be carried to the point of determining the absolute values of the amplitudes for the magnetic substates $2p_0$ and $2p_{\pm 1}$ and the phase angle between them.

A less time-consuming but probably also less accurate method consists of detecting the charge-state distribution of the target ions left behind after the capture process. The method has been used previously for total cross sections by Horsdal-Pedersen and Larsen (1979). The interpretation of the charge-state distributions in terms of inner-shell processes is based on the different average charge state $\bar{q}$ resulting from capture of electrons in different (sub)shells. For example, the capture of an M-shell electron of argon results in $\bar{q} \simeq 1.2$, whereas capture of an $L_1$ or $L_{2,3}$ electron results in $\bar{q} \simeq 2.4$ and $\bar{q} \simeq 3.4$, respectively. Hence the method to some extent distinguishes between capture of 2s and 2p electrons, but it does not resolve the magnetic substates of the 2p state. In that sense, it is complementary to the former method mentioned above. In Fig. 7 (following page), we show preliminary charge-state distributions for 1-MeV $H^+$ on argon. One corresponds to a small impact parameter (of $\simeq 7.5$ pm or half the "radius of maximum charge density" for the L shell), while the other is the averaged charge-state distribution for all impact parameters (total cross section). The close collision results in a higher mean charge state or relatively more capture from the L shell as compared to the M shell (capture from the K shell is negligible).

4.2  Angular-Dependent Capture Probabilities. The individual K-shell electron-capture processes discussed in the preceding sections of this report depend on collision velocity and impact parameter. The angular dependence of the electron-capture process may be studied by measuring the neutral fraction of protons scattered through large angles, corresponding to virtually zero impact-parameter collisions. Preliminary data from such measurements are shown in Fig. 8 (Horsdal-Pedersen, Loftager, and Rasmussen 1981, unpublished). The capture probability  for helium (capture from a K shell) is independent of scattering

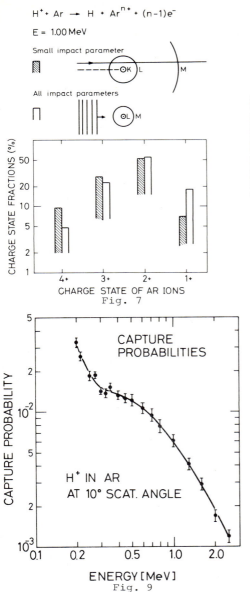

$$H^+ + Ar \rightarrow H + Ar^{n+} + (n-1)e^-$$

E = 1.00 MeV

Small impact parameter

All impact parameters

Fig. 7

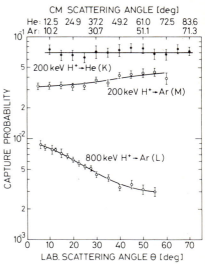

ANGULAR DEPENDENT PROBABILITIES
FOR CAPTURE BY H⁺ IN He AND Ar

Fig. 8

CAPTURE
PROBABILITIES

H⁺ IN AR
AT 10° SCAT. ANGLE

Fig. 9

angle. Figure 9 shows the energy dependence of capture probabilities for $H^+$ on argon at $\theta$ = 10°. The pronounced "bump" around 500 keV is believed to signify the transition from M-shell capture at energies less than about 200 keV to L-shell capture at energies above 800 keV. Figure 8 shows angular-dependent probabilities for argon at 200 and 800 keV. The (M-shell) probabilities at 200 keV increase slowly with increasing scattering angle, whereas the (L-shell) probabilities at 800 keV decrease sharply. Using the SCIA, Amundsen and Jakubassa-Amundsen (1981, private communication) have recently shown that the angular dependence of the capture probability for an initial orbital with angular momentum $\ell_i$ and a final orbital with angular momentum $\ell_f$ is described by Legendre polynomials in $\cos\theta$ of orders $\ell$, satisfying the relation $|\ell_i - \ell_f| \leq \ell \leq \ell_i + \ell_f$.

4.3 <u>Effects of Multiple Scattering</u>. We have repeatedly referred to the classical double-scattering mechanism of Thomas. However, effects of double scattering or even still higher terms of perturbation theory

in capture to <u>bound</u> states have not yet been demonstrated experimen-
ally. A number of experiments aiming at     providing clear evidence
for such terms (or the opposite) has been suggested by Shakeshaft
and Spruch (1978). One such possibility consists of detecting effects
of projectile → electron, electron → electron double scattering. This
mechanism was recently discussed in detail by Briggs and Taulbjerg
(1979) for $H^+$ on helium $(1s^2)$. Classically, and for stationary target
electrons, an electron is ejected perpendicular to the beam direction
with speed equal to the speed of the beam whenever the above mechanism operates. In quantum theory, this $\delta$ function in direction and
speed turns into a ridge with a peak at the position of the classical
$\delta$ function. This peak might be observable. Alternatively, one could
measure the charge-state distribution of target helium ions after
the capture process as a function of collision velocity and look for
an increase of doubly charged ions at high velocities where the
double scattering is considered to be of importance (Briggs 1980,
private communication). Published data (Horsdal-Pedersen and Larsen
1979), which have been extended in energy to 1 MeV, exhibit a constant level of about 3% doubly charged ions, which is accounted for
by the "shake-off" mechanism. The measurements probably have to be
extended towards 10 MeV $(v/v_K \simeq 10)$ for the effect to be seen if it
is there at all. Effects due to terms beyond first order of perturbation theory <u>have</u> probably been seen experimentally in the important case of capture to <u>continuum</u> states of the projectile. We refer
to the literature for discussions of this point (Shakeshaft and
Spruch 1978, Macek et al. 1981).

## 5. SUMMARY

The first experiment on electron capture by protons from an inner K
shell was done by Macdonald and coworkers (1974) at Kansas State University. Further studies of total and differential cross sections
have been carried out since then with fully stripped light ions on
different target gases such that systematic data are now available,
showing the dependence of the K-shell-capture process on collision
velocity, impact parameter, and projectile as well as target atomic numbers.

This experimental development has been parallelled by theoretical
progress along two lines. For asymmetric collisions, first-order perturbation theory in the <u>weak</u> projectile potential has been developed
(the Impulse Approximation and the 2nd Born Approximation with the
Coulomb Green's function of the strong target field). The other approach uses a coupled-states formulation with classical nuclear motion. A large set of bound and unbound pseudostates centered on the
target     needs to be included to give an adequate description. When
this is done, the two theoretical formulations imply the same physical interpretation. The capture process is seen mainly as ionization
to a set of target continuum states centered at zero momentum relative to the projectile. The width in momentum space is essentially
given by the momentum distribution of the final bound state.

When applied to the capture of inner K-shell electrons, both formulations give results in agreement with available experimental data, but
it still remains to be seen how successful the theories are in predicting the many details of capture from inner L subshells and in
predicting the behaviour at large scattering angles. The significance
of the classical Thomas scattering at high collision velocities is
also an unsettled experimental question.

ACKNOWLEDGEMENTS

The author is grateful to P.A. Amundsen, J.S. Briggs, A.L. Ford, D.H. Jakubassa-Amundsen, and K. Taulbjerg for discussions and for communicating unpublished results.

REFERENCES

Belkić Dz. and Salin A. (1976) J.Phys.B $\underline{9}$, L397
Belkić Dz. and Salin A. (1978) J.Phys.B $\underline{11}$, 3905
Belkić Dz. , Gayet R., and Salin A. (1979) Phys.Reports $\underline{56}$, 279
Bratton T.R., Cocke C.L., and Macdonald J.R. (1977) J.Phys.B $\underline{10}$, L517
Briggs J.S. (1977) J.Phys.B $\underline{10}$, 3075
Briggs J.S. and Taulbjerg K. (1979) J.Phys.B $\underline{12}$, 2565
Briggs J.S. (1980) J.Phys.B $\underline{13}$, L717
Brinkman H.C. and Kramers H.A. (1930) Proc.Acad.Sci.Amsterdam $\underline{33}$, 973
Carlson T.A. (1973) Inv.Lectures and Prog.Reports VIII ICPEAC (Inst. Phys., Belgrade) p. 205
Chan F.T. and Eichler J. (1979) Phys.Rev.A $\underline{20}$, 1841
Cocke C.L., Macdonald J.R., Curnutte B., Varghese S.L., and Randall R. (1976) Phys.Rev.Lett. $\underline{36}$, 782
Cocke C.L., Gardner R.K., Curnutte B., Bratton T., and Saylor T.K. (1977) Phys.Rev.A $\underline{16}$, 2248
Dewangan D.P. (1977) J.Phys.B $\underline{10}$, 1083
Eichler J. and Narumi H. (1980) Z.Phys.A $\underline{295}$, 209
Ford A.L., Reading J.F., and Becker R.L. (1981) Phys.Rev.A $\underline{23}$, 510
Horsdal-Pedersen E. and Larsen L. (1979) J.Phys.B $\underline{12}$, 4085
Horsdal-Pedersen E. (1981) J.Phys.B $\underline{14}$ L249
Jakubassa-Amundsen D.H. and Amundsen P.A. (1980) Z.Phys.A $\underline{297}$, 203
Kocbach L. (1980) J.Phys.B $\underline{13}$, L665
Lin C.D. and Soong S.C. (1978) Phys.Rev.A $\underline{18}$, 499
Macdonald J.R., Cocke C.L., and Eidson W.W. (1974) Phys.Rev.Lett. $\underline{32}$, 648
Macek J.H. and Shakeshaft R. (1980) Phys.Rev.A $\underline{22}$, 1441
Macek Joseph, Potter J.E., Duncan M.M., Menendez M.G., Lucas M.W., and Steckelmacher W. (1981) Phys.Rev.Lett. $\underline{46}$, 1571
Macek J.H. and Taulbjerg K. (1981) Phys.Rev.Lett. $\underline{46}$, 170
Oppenheimer J.R. (1928) Phys.Rev. $\underline{31}$, 349
Rivarola R.D., Piacentini R.D., Salin A., and Belkić Dz. (1980) J. Phys.B $\underline{13}$, 2601
Rogers S.R. and McGuire J.H. (1977) J.Phys.B $\underline{10}$, L497
Rødbro M., Horsdal-Pedersen E., Cocke C.L., and Macdonald J.R. (1979) Phys.Rev.A $\underline{19}$, 1936
Shakeshaft R. (1978) Phys.Rev.A $\underline{18}$, 1930
Shakeshaft R. and Spruch L. (1978) J.Phys.B $\underline{11}$, L457
Shakeshaft R. and Spruch L. (1978) Phys.Rev.Lett. $\underline{41}$, 1037
Shakeshaft R. and Spruch L. (1979) Rev.Mod.Phys. $\underline{51}$, 369
Thomas L.H. (1927) Proc.Roy.Soc.(London) $\underline{114}$, 561

PHYSICS OF ELECTRONIC AND ATOMIC COLLISIONS
S. Datz (editor)
© North-Holland Publishing Company, 1982

CHARACTERISTIC X RAY PRODUCTION
IN HIGH-ENERGY HEAVY-ION COLLISIONS

R. Schuch

Physikalisches Institut
Universität Heidelberg
6900 Heidelberg, Fed. Rep. of Germany

Mechanisms of inner shell (particularly K-shell) vacancy
production in ion atom collisions with projectile velocities
smaller than inner shell electron velocities are discussed
for two regimes of collision systems ($Z_T - Z_p \gg 1$ and $Z_p/Z_T \simeq 1$;
$Z_p$ and $Z_T$ are projectile and target nuclear charge). In the
regime ($Z_T - Z_p \gg 1$) K vacancy probabilities are compared
with first order perturbation theory including the "binding
effect". In the regime $Z_p/Z_T \simeq 1$ the impact parameter and
projectile charge state dependent cross sections are discussed
within the molecular orbital model.

## 1. INTRODUCTION

Characteristic x rays are emitted with a high yield from energetic heavy ion atom
collisions indicating a large probability for producing inner shell vacancies in
those collisions. In this review I want to concentrate on our present understanding of the collision mechanism which produces the states decaying by characteristic
x-ray emission. The characteristic x rays are used as a probe to investigate the
mechanism of inner shell vacancy production in energetic heavy ion collisions.

One kind of information from characteristic x rays is obtained by determining
their energy with high resolution. Because the energy and fine structure of the
x-ray lines depend on the population of inner atomic shells, they therefore give
detailed information about the states of the ions after the collisions when decaying by characteristic x radiation (see e.g. P. Richard, ref.1, p.99). Another kind
of information from characteristic x rays is obtained by converting the x ray
intensity with a fluorescence yield into an inner shell vacancy production cross
section. The experimental procedure for determining inner shell vacancy cross
sections is presented in part 2. The dependence of this cross section on the different collision parameters like projectile-velocity $v_p$ , -charge state q, projectile
and target nuclear charge $Z_p$ and $Z_T$ and impact parameter b, will be considered
here for determining the mechanisms of producing inner shell vacancies in energetic
ion atom collisions.

When a projectile hits a target atom in a close collision the electron bound in
an inner atomic shell is perturbed by the time changing Coulomb field $V(Z_p, Z_T,$
$R(t))$ and can be removed from this shell by an energy transfer $\Delta E$. $R(t)$ is the
time dependent internuclear distance. According to the different final states of
the electron one distinguishes ionization where the electron is brought into the
continuum (unbound state), excitation if the electron ends up in a higher bound
state of the same atom and charge transfer if the electron is transferred by the
Coulomb interaction from an inner shell state of one collision partner to a state
in the other collision partner. This last process is important if the projectile
is in a high charge state and has therefore empty states.

In order to remove the electron from the inner shell the energy transfer $\Delta E$ from the time changing Coulomb field ($V(Z_p, Z_T, R(t))$) to the electron, must be large enough. Therefore the basic frequency of $V(Z_p, Z_T, R(t))$ which is $v_p/b$ has to be larger than the frequency change of the electron, which is $\Delta E/\hbar$, when it is excited or ionized. This gives an upper limit for the impact parameter: $b < \hbar v_p/\Delta E =$ :$b_{ad}$. $b_{ad}$ is called the adiabatic impact parameter, because for $b > b_{ad}$ the electron adjusts adiabatically to the time changing field V and is not likely removed from its initial state. This very simply derived condition is basic for vacancy production in atomic collisions[2]. It gives an estimate in which impact parameter region the inner shell vacancies are produced.

The probability for producing an inner shell vacancy is determined by the height of the perturbation of the electron in comparison with the energy transfer $\Delta E$, necessary to ionize or excite the electron. The inner shell vacancy production process is theoretically treated so far mainly for the following extreme cases (see e.g. Madison a. Merzbacher, ref.1, p.59): The atomic approach, where the initial electronic wavefunction is assumed to be a one center atomic wavefunction and the electron (velocity $v_e$) is only slightly perturbed by $V(Z_p, Z_T, R(t))$, so that first order perturbation theory is applicable. For inner shell electrons this condition should be fulfilled if $Z_p \ll Z_T$ or $v_p \gg v_e$.

In the molecular approach the electronic states are calculated in a two center potential of projectile and target nucleus (quasimolecular states) and excitation may occur by electron promotion via couplings between quasimolecular levels. The molecular approach should be applicable if $Z_p \simeq Z_T$ and $v_p \ll v_e$.

This distinction in the theoretical treatment of the inner shell vacancy producing process is more a matter of convenience than a matter of principle. Although we use a similar classification into asymmetric collision systems (part 3) and near symmetric collision systems (part 4), one should, however, recognize theoretical schemes and experimental results which attempt to bridge the gap between the atomic- and molecular-approach.

## 2. EXPERIMENTAL

A principle set-up for the measurement of total- and impact parameter dependent differential- x ray emission cross sections is shown in Fig. 1. The ion beam is collimated by two sets of collimators and scraper slits to reduce slit scattering. It then hits on a solid- or gas target in a scattering chamber.

The characteristic x rays are detected in most experiments by a Si(Li) detector (2 - 20 keV) or a Ge(I) (5 - 100 keV) which have a reasonable resolution of a few percent; for a much better resolution crystal spectrometers are used. At lower x-ray energies proportional counters and channeltron electron multipliers with filters or gratings for good resolution are used (see e.g. ref.1, part II).

A particle monitor dectector is very often used for determining the product of target thickness and beam intensity which is necessary for the measurement of absolute cross sections. A main source for uncertainties in these experimental results, especially with gas targets, are the x ray- and particle-detector solid angles. A method which avoids the knowledge of these solid angles is described in ref.3.

The impact parameter dependent differential cross sections are determined with a coincidence requirement for characteristic x rays and scattered particles. A detailed description of the coincidence techniques is given by C.L. Cocke in ref.4. As particle detectors one uses surface barrier detectors with ring apertures[5], radial position sensitive parallel plate avalanche detectors[6] and two dimensional position sensitive multi-wire chambers[7]. The scattering angle $\vartheta$ can be converted into impact parameter b using a Coulomb potential at not too small angles

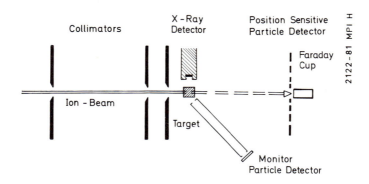

Figure 1
Schematic experimental set-up for measuring total- and impact parameter dependent
x ray emission cross section

$\vartheta(R_0(\vartheta) < r_K$, $R_0$ distance of closest approach, $r_K$ K shell radius); for smaller
$\vartheta$ a screened potential has to be used to convert $\vartheta$ into b. The probability
for producing an inner shell vacancy at impact parameter b is then determined
in the following way: The number of true coincident x-ray events at impact parameter b corrected with fluorescence yield, x-ray detector solid angle and -effi-
ciency gives the number of inner shell vacancies created during the measuring
time in collisions with impact parameter b. By dividing this number by the total
number of particles detected at the same scattering angle one gets the inner
shell vacancy production probability.

## 3. ASYMMETRIC COLLISION SYSTEMS

Many experiments have been performed[1,4] to test the validity of first order
perturbation theory, especially for K shell excitation and ionization. Here
the atomic wavefunction is most reliably known, furthermore the condition for
first order perturbation theory, that the perturbation of the electronic state
by the projectile is small and for treating the electrons as independent particles
e.g. taking into account the electron-electron interaction by a screening constant
(effective charge), can be best fulfilled. The models most commonly used[1] are
the semiclassical approximation (SCA), plane wave born approximation (PWBA) and
binary encounter approximation (BEA). The total cross sections can partly be very
well described by these models (see e.g. ref.1). A more sensitive test, however,
is the impact parameter dependent K vacancy probability $P_K(b)$. This allows to
investigate in detail e.g. trajectory effects in the ionization process at small
impact parameters[8,9]. A large interest in these effects has arisen recently by
measuring K x-ray emission probabilities at large scattering angles to obtain informations about lifetimes on simultaneous occuring nuclear reactions[10-12]. Another
very important effect in the ionization probability is the deviation from the $Z^2$-
scaling of first order perturbation theory by the "binding effect"[13-14].This is
discussed here in more detail as it introduces schemes which may help to bridge the
gap between atomic- and molecular approaches.

The first $P_K(b)$ measurements which investigated the "binding effect" were performed by Andersen et al[14] and Schmidt-Böcking et al[15]. As an example the K vacancy production probability of Cu by impact of 2 MeV/u p,Be,C,O measured by Andersen et al[14] is shown as function of impact parameter in fig. 2. First we concentrate on the data measured with protons. $P_K(b)$ decreases strongly with increasing b in the region of b investigated here, which is in the order of the adiabatic impact parameter. A comparison is made here with a prediction of RHFS-SCA calculation of Trautmann et al[16]. The Cu K wavefunctions used are Relativistic Hartree Fock Slater (RHFS). The agreement with the data of p on Cu is quite reasonable by considering that calculation gives absolute probabilities. Within first order perturbation approach the matrix element is proportional to $Z_p$, therefore the ionization probability proportional $Z_p^2$. The question to the experiment with different projectiles was, how does the Cu K vacancy production probabilities fit this scaling law with increasing $Z_p$. The normalized Cu-K vacancy probabilities $P(b)/Z_p^2$ measured with Be, C, and O-beams of the same velocities should fall on a common curve, if first order perturbation approach with the Cu K - atomic wavefunction is appropriate. With increasing $Z_p$ one observes, however, a systematically decreasing $P(b)/Z_p^2$. This indicates that the Cu-K electron feels the additional charge of the projectile, and has an increased binding energy. At impact parameters $b < r_K$ where the projectile enters the K-shell, and projectile velocities $v_p < v_e$, the electron can adjust to the additional charge of the projectile, and the description of the initial state by the target K-shell state is not longer appropriate. The K-shell state of the atom with $Z_p + Z_T$ nuclear charge called united atom K-shell, could be more appropriate as an unperturbed electronic initial state. For a test of this Ansatz target (which is here the heavier collision partner) K vacancy probabilities are measured with systems of the same sum of projectile and target nuclear charge[17]. The Ni-Zr data were taken at GSI[18] at little higher velocity of 1.4 MeV/u and scaled down to 1.26 MeV/u. These $P_K(b)$ data of the same united atom charge normalized by $Z_p^2$ are presented in Fig. 3 and one finds that these normalized data points are falling

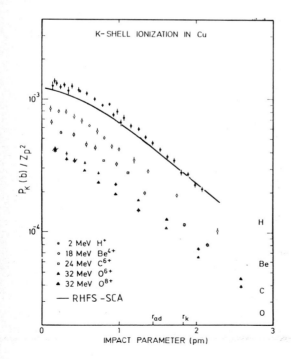

Figure 2
Normalized Cu K ionization probability $P_K(b)/Z_p^2$ induced by impact of 2 MeV/u p, Be, C, and O (Ref. 14)

nearly together on a common curve. Considering that the original target x-ray
probabilities differed by orders of magnitude the still remaining deviations are
small. The adiabatic impact parameter $b_{ad} = \hbar v/E_B$, $E_B$ is the united atom K binding
energy, is 150 fm; above this impact parameter all $P_K(b)$ are decreasing rapidly.
Also a comparison with the RHFS SCA calulation of Trautmann et al[16], here with
K wavefunction and binding energy of the "united atom" (u.a.) with the nuclear
charge $Z = Z_p + Z_T$ , shows reasonable agreement.

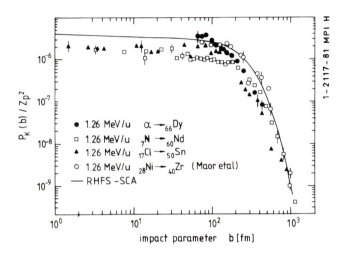

Figure 3
Target K-vacancy production probabilities $P_K(b)/Z_p^2$ as function of impact parameter
b (ref. 17)

In the calculation of Trautmann et al the u.a. wavefunction is centered at the
target nucleus. The calculations are performed with a Coulomb trajectory for the
collision system Cl-Sn, including the finite mass of the electron (recoil effect).
Due to the trajectory effects no common curve is expected at small b. For the other
collision systems the calculated probabilities can change at small b up to 50%.
Furthermore, it seems to be more reasonable that the u.a. wavefunction is not centered at the target nucleus but at the center of charge, at a distance $R \cdot Z_p/Z$ from
the target nucleus along the internuclear axis, as it was proposed by Briggs[19].
The expected reduction of the ionization probability due to this transformation
is partly included in the calculation of Trautmann et al[16] by the recoil effect.

At large impact parameters or small projectile energies so that $R_0 > r_K$ the approximation of the initial wavefunction by a u.a. wavefunction can not longer be appropriate. In the limit of very large internuclear distances R the electronic state
has to approach the K-shell state of the heavier collision partner. For a smooth
transition between both regimes Andersen et al[14] approximated the two center electronic state by an R dependent atomic 1s wavefunction, centered on the target nucleus. The charge the K electron feels is determined by minimizing the electron
binding energy at every internuclear distance. Anholt[20] extended this model by
allowing to vary also the center of the atomic wavefunction as function of R from
the center of charge when R=0 (Briggs model) to the target nucleus when $R \to \infty$
(Andersen model). An interesting scheme to calculate inner shell vacancy production
probabilities particularly in a regime where binding effects are important is

given by a variational principle[21]. This method is not restricted to low velocities.
But still all these models can be considered as atomic approaches in which the
binding effect is incorporated. In order to be able to apply one of the other
models the collision velocity must be small ($v_p \ll v_e$) so that the electron can
adjust to the additional nuclear charge.

In a molecular approach the initial and final electronic wavefunctions are described with respect to a two center Coulomb potential called molecular orbitals
(MO). In the molecular approach the binding effect is incorporated  most realistic,
if $v_p \ll v_e$. As an example those MO are shown for the collision system[22] Br on Zr
in fig. 4a. In asymmetric collision systems ($|Z_p - Z_T| > 10$) the K shell vacancy
in the heavier collision partner is made  within the MO model in the $1s\sigma$ MO during
the collision. The collision system  Br + Zr with a u.a. of Z = 75 is somewhat
heavier as the u.a. of the systems presented in fig. 3. However, one can see that
even in this heavier system the binding energy of the $1s\sigma$ state is already very
close to the one of the u.a. system for R < 800 fm, this is consistent with the
results presented in fig.3 where the $P_K(b)/Z_p{}^2$ fall upon another at b < 1000 fm.
This result can lead to the conclusion that the Briggs u.a. model together with
a good first order perturbation calculation is quite useful in determining $1s\sigma$-
excitation probabilities at low collision velocities ($v_p < v_e$).

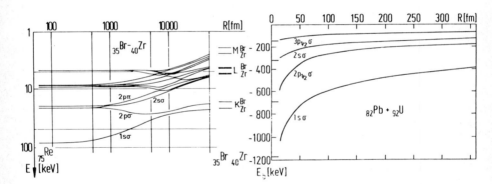

Figure 4a
Molecular orbital diagramm for the
system Br + Zr (ref.22)

Figure 4b
Molecular orbital diagramm for the
system Pb + U (ref.22)

By turning this method around the K-vacancy production probability can tell something about the formation of a united atom system. This becomes of special interest
in the region of very heavy collision partners where $\alpha \cdot Z > 1$ and the formation
of electronic states in superheavy atoms can be investigated. The main difference
to the intermediate heavy systems is that in very heavy collision systems the
electrons behave extremely relativistic. This has the consequence that the electron binding energy is still changing at small internuclear distances of the collision partners and must therefore be considered in a two center potential. In
a very heavy system like Pb + U (fig. 4b) the calculated[22] $1s\sigma$-binding energy
still changes at R < 50 fm very dramatically, so that the use of a u.a. binding
energy cannot be appropriate here.

One can still get a universal representation of the $1s\sigma$ excitation probabilities
$P_{1s\sigma}(b)$ by using a scaling law which has also been derived in first order perturbation theory[23], however, for systems $\alpha \cdot Z > 1$. Within this scaling law the $P_{1s\sigma}$

are normalized by a function $D(Z)$[23] and the impact parameter scale is replaced by a universal scale $R_0 \cdot E_{1s\sigma}/\hbar v$. This is the ratio of distance of closest approach and adiabatic distance, where $E_{1s\sigma}(R_0)$ are calculated $1s\sigma$ binding energies[24]. Figure 5 shows a semilogarithmic plot of normalized $P_{1s\sigma}$ measured[25] with four different collision systems. All data, except those for Pb + Cm at $R_0 < 50$ fm[25] fall within some scatter on a straight line as it is expected by the first order perturbation theory. The $P_{1s\sigma}(b)$ measured at two different velocities but same $R_0$

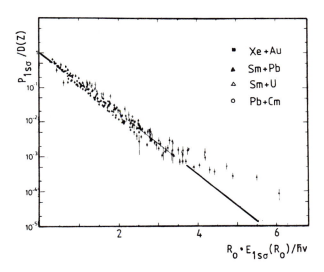

Figure 5
$1s\sigma$ excitation probabilities[25] normalized with $D(Z)$[23] as function of the ratio of minimum distance of approach and adiabatic distance. Full line represents first order perturbation theory.

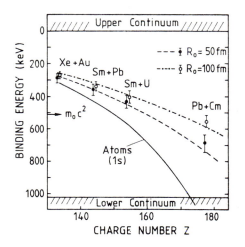

Figure 6
$1s\sigma$ binding energies as function of $Z_P + Z_T$ extracted from $P_{1s\sigma}(b)$[25]. The dashed lines are calculated $1s\sigma$ binding energies, the full line is the united atom K shell binding energy[26].

can be used[25] to extract $1s\sigma$ binding energies as function of $R_0$. Fig. 6 shows the
so determined $1s\sigma$ binding energies for two values of $R_0$ as a function of Z. This
representation suggests that indeed some spectroscopic information about the most
strongly bound electronic state in superheavy quasimolecules can be obtained. Even
by taking into account a large uncertainty in this method, one can still extract
very large $1s\sigma$ binding energies which exceed even the rest mass energy of the electron.

## 4. NEAR SYMMETRIC COLLISIONS

Up to now we only considered K vacancy production in the heavier collision partner
in asymmetric collision systems where the perturbation approach was appropriate.
With the experimental results[27] shown in fig.7 an overview on the K vacancy cross
sections as function of target nuclear charge is given, from very asymmetric systems where the projectile, in these experiments 47 MeV I, is the heavier collision
partner up to above the region of symmetric collision systems. The target K cross
section decreases by many order of magnitude with increasing $Z_T$. The projectile
is at small $Z_T$ the heavier collision partner so that the projectile K vacancy cross
section increases first proportional to $Z_T^2$ according to perturbation theory and
decreases at larger $Z_T$ because of the strongly increasing $1s\sigma$ electron binding
energy. This collision regime has been considered in part 3. Approaching symmetric
collision systems the projectile K cross section increases very strongly indicating
another mechanism for K shell vacancy production.

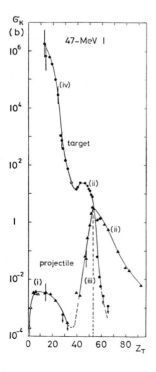

Figure 7
K-vacancy cross section of the projectile
(47 MeV I) and the target as function of
target nuclear charge $Z_T$ (ref.27)

Such an indication for a different excitation mechanism of inner shell vacancies in near symmetric collision systems has been first observed by Specht[28], Afrosimov et al[29] and by Everhardt and Kessel[30]. A current way of treating this rather complex process of inner shell vacancy production in slow ($v_p \ll v_e$) near symmetric ion atom collision is based on the molecular orbital (MO) model of Fano and Lichten[31]. A MO correlation diagram like the one shown in fig. 4a may visualize the possibility to obtain large inner shell cross section when two electron shells with equal binding energies interpenetrate. The process of K vacancy production is described within this model in the following way: It can be assumed that in the weakly bound $2p\pi$ MO vacancies are produced in an early stage of the collision, or these vacancies are brought into the collision in the projectile L shell. At small internuclear distances R the $2p\pi$ state comes energetically very close to the $2p\sigma$ state which corresponds to the K shell state of the lighter collision partner. The letters $\sigma$ and $\pi$ mark the projection quantum number of the electronic angular momentum on the internuclear axis which is the quantisation axis. By the rapid rotation of the internuclear axis in the collision at $R \simeq R_0$ a vacancy in $2p\pi_x$ can be transfered to $2p\sigma$, called $2p\pi$-$2p\sigma$ rotational coupling[32]. If the collision system is near symmetric so that the energy difference between the two K-shells is small, the vacancy in $2p\sigma$ can be furtheron transferred by radial coupling[32] into $1s\sigma$. This gives a large K vacancy cross section also in the heavier collision partner and explains the rise of the projectile cross section (fig.7) at symmetric collision systems. The predicted $2p\sigma$ - $1s\sigma$ coupling probability, also called K shell vacancy sharing[33], was found in very good agreement with experimental cross sections[3,33]. Even for the predicted impact parameter dependence[34] agreement with the experiment[35] was reached. (A special case of this process where a $2p\sigma$ vacancy on incoming part of the collision is present will be considered later). The sum of projectile- and target K vacancy cross section should therefore be compared with the calculated $2p\sigma$ excitation cross section because $1s\sigma$-excitation and -ionization is negligible (part 3).

According to the model of $2p\pi$-$2p\sigma$ rotational coupling, this cross section should increase as soon as there are L vacancies brought by a highly charged projectile into the collision.

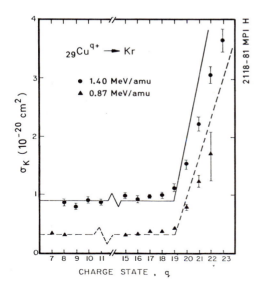

Figure 8
K vacancy production cross section for Cu on Kr as function of projectile charge state[37]. Solid and dashed lines are are predictions of $2p\pi$-$2p\sigma$ rotational coupling

Experimental results as presented in Fig. 8 have been obtained in many laboratories (see e.g. J.R. Macdonald in ref.4, p.279, and rfs. 36, 37) and as in all experiments of this type one observes an increase in the K vacancy cross section when the projectile bears L-vacancies. This result is consistent with the prediction of the 2pπ-2pσ rotational coupling model, but it alone does not show the existence of the 2pπ-2pσ rotational coupling process. There are experimental data[36],[37] which show a similar dependence on projectile charge states also in very asymmetric collision systems where the 2pπ-2pσ rotational coupling process should not operate and only direct excitation or ionization of the projectile K shell can occur.

A very sensitive and unambiguous test for the importance of the 2pπ-2pσ rotational coupling mechanism in near symmetric collision systems is the impact parameter dependent K vacancy probability because the theory predicts[32] a characteristic dependence of the 2pπ-2pσ rotational coupling probability on impact parameter $P_{rot}(b)$. A peak at small b, corresponding to a center of mass scattering angle of 90°, caused by the transfer of a vacancy in $2p\pi_x$ into the 2pσ state by a rapid 90° rotation of the internuclear axis.

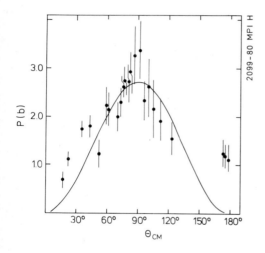

Figure 9
K vacancy production probability for 30 MeV Ca+Ti as
function of center of mass
scattering angle $\theta_{cm}$ in relative units. The solid line
represents the rotational coupling prediction

This maximum in the K vacancy production probability at very small b has indeed been found in two experiments: For 60 MeV I on Ag[38] and for the system 30 MeV Ca on Ti[39]. The result of the latter experiment is shown in fig.9. This evidence for the existence of the kinematic maximum is an important hint for the importance of the 2pπ-2pσ rotational coupling process for K vacancy production in such slow near symmetric heavy ion atom collisions.

Another maximum in $P_{rot}(b)$ called the adiabatic peak is at large b, at the adiaba-tic impact parameter. Most $P_K(b)$ data in near symmetric collision systems were taken in the b-region of the adiabatic peak. To compare all these data with each other and the theoretical prediction we want to use a universal scale for the impact parameter. Because in the region of the adiabatic peak the scattering angle is small ($\vartheta \leq 5°$) the trajectory can be approximated by a straight line. Mainly with this approximation a universal solution of the two state coupled equations for 2pπ-2pσ rotational coupling and a reduced impact parameter scale b' was ob-tained[40]:

$$b' = \left[ \frac{\alpha(Z_p \, , \, Z_T)}{\hbar \, v_p} \right]^{1/3} b$$

All available $P_K(b')$ data measured with gas targets of collision systems from
O - O and Ne - Ne at 400 MeV and 360 keV[41], S - Ar at 32 MeV[42], Cl - Ar at 15,
30 MeV[43] up to Br - Kr at 108, 156 MeV[44] are shown in fig. 10 and fall with a
reasonable scatter on a common curve. The universal solution of the two state
coupled equation is represented by the solid line, the agreement is reasonable.
The small deviations might be partly due to the approximations introduced to get
the universal solution and partly (at small b) due to contributions of ionization
and direct excitation of the K shell[42]. These b- and q- dependences lead to the
conclusion that in these types of collision systems K vacancies are predominantly
produced by 2pπ-2pσ rotational coupling as soon as there are L vacancies present
in the collision. It should be noted that "these types of collision systems" were
near symmetric ($Z_p \simeq Z_T$), slow with respect

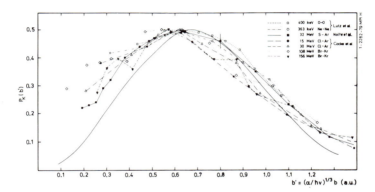

Figure 10
K vacancy production probabilities P(b') normalized in height to one $2p\pi_x$ vacancy,
as function of reduced impact parameter b'. Solid line represents the adiabatic
peak of 2pπ-2pσ rotational coupling.

to K electron velocity, and all targets were very thin gas targets. This last
specification is in fact very important since all experimental K vacancy probabilities measured up to now with solid targets show a very significant deviation from
the predicted 2pπ-2pσ rotational coupling probability in the region of the adiabatic
maximum, whereas they agree in the region of the kinematic maximum (fig.9, ref.38).
This can be seen by an example in fig. 11 where absolute K-vacancy production
probabilities measured in gas and solid targets are compared as function of impact
parameter. The 156 MeV Br beam from the MP accellerator of the Max Planck Institut
für Kernphysik Heidelberg was poststripped and charge state selected for the experiments with a Kr gas target. The Br projectile of charge state 25+ (fig. 11,
triangles) has no L vacancy and Br of charge state 28+ (fig.11, closed points)
bears three L-vacancies. $P_K(b)$ increases in the maximum by nearly a factor of three
going from 25+ to 28+, which is consistent with the results of total cross sections
discussed before (fig.8 and refs. 4 p.279,37). Also the shape (particularly the
position of the adiabatic maximum) agrees well with the prediction of 2pπ-2pσ rotational coupling from Taulbjerg et al[40] (fig. 11 solid line normalized to give the
same total cross section). With the solid targets Se and Rb one also observes a
maximum in $P_K(b)$ but at about a factor two smaller b. Such a shift of the maximum
has been found in all solid target experiments up to now[39,44-47]. Also surprising
is that $P_K(b)$ of Se, Rb solid targets is in the maximum much higher than $P_K(b)$

even compared to 28+ Br on the Kr gas target. The mean charge state of 156 MeV Br
after a foil is, however, only 24+ and the fraction of the beam bearing L vacancies
after a foil is very low[48]. In a solid L vacancies could be produced by a multi-
-collision process prior to the K vacancy producing collision. But in those collisions the L shell must be nearly emptied to reach these high measured probabilities of the solid target. It is very unrealistic to assume an empty L shell of
such a projectile in a solid, also it would not explain[49] the position of the
maximum at lower b. So the mechanism of K-vacancy production in near symmetric
collisions in the high density of a solid target, where many outer shell affecting
collisions occur is not understood at present.

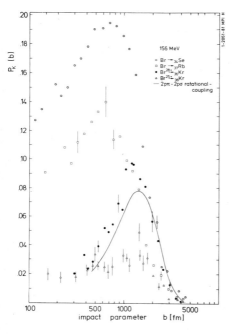

Figure 11
Comparison of K vacancy production
probabilities measured with 156
MeV Br projectiles on Kr gas target
and Se-, Rb-solid targets

In the following part some examples for the importance of inner shell charge
transfer for characteristic x-ray production in near symmetric heavy ion atom collisions are given. An extensive treatment of the charge transfer process between
inner shells is given by Lin and Richard ref.50 and Macdonald in ref.4. Therefore
only special aspects of this process will be discussed here. When the projectile
is very highly charged the transfer of a target inner shell electron to a bound
state in the projectile can become the dominant process for target characteristic
x ray production. This is demonstrated in fig. 12 by the dependence of the cross
section for target (Ar) K ray emission on the projectile (S) charge state. The
S-beam up to the very high charge state of 16+ (bare S nuclei) was obtained from
the four stage accel-decel facility at Brookhaven National Laboratory[51].

With increasing S projectile charge state up to 13+ (increasing number of L vacancies) one observes a "weak" ($\approx$ factor 3) increase of $\sigma_x$ , which may partly be due
to an increasing number of $2p\pi$ vacancies and rotational coupling into $2p\sigma$ and
partly to an increasing Ar K fluorescence yield (see discussion above and in ref.
42). Charge state 14+ is not "clean" because of metastable $1s^1 2s^1$. At q = 15+
one K vacancy in S and at 16+ two K vacancies in S are brought into the collision,
and one observes an increase by more than two orders of magnitude in the Ar K x-ray
cross section due to K-shell to K-shell charge transfer. This tremendous large

K x-ray cross section reached by K -shell to K -shell transfer could be of interest for x-ray laser developments. A very detailed information about the K-shell to K-shell transfer process can be obtained from the impact parameter dependence. In such an experiment the K shell vacancy sharing ratio for 32 MeV $S^{15+}$ on Ar was measured[52]. A very pronounced interference structure of the transfer probability as function of impact parameter was observed due to the coherence of the transition amplitude on incoming and outgoing part of the collision. This interference structure is, however, not in quantitative agreement with the prediction of the two state coupled equations for $2p\sigma$-$1p\sigma$ vacancy transfer[34].

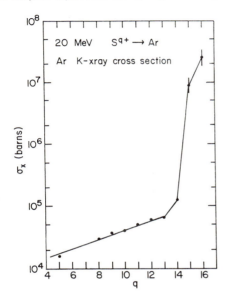

Figure 12
Ar K x ray cross section as function of S projectile charge state. Line is drawn to guide the eye.

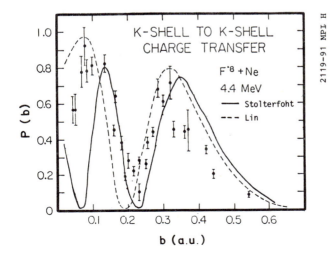

Figure 13
Ne K vacancy production probability as function of impact parameter[53]. Dashed line shows the prediction of ref.54 and solid line of ref.55

Another experiment[53] of this type was performed with a 4.4 Mev $F^{8+}$ (one K vacancy) beam on a Ne. Ne K vacancy production was observed by detecting Ne KLL-Auger electrons, which indicated the transfer of the K vacancy from $F^{8+}$ to Ne.

The observed very large K-shell K-shell charge transfer cross sections (Fig.12, $S^{15+}$ and $S^{16+}$) are due to the fact that the transfer probability is of the order of 10-20% up to very large impact parameters of 0.5 a.u. Also an oscillatory behaviour like in $S^{15+}$ on Ar (ref.52) is observed in the Ne K vacancy production probability. The agreement with the predictions of two theoretical treatments of the K-shell to K-shell vacancy transfer process, one by Lin[54], dashed line, and one by Stolterfoht[55], full line, shows that one is already close in understanding this important process of K-shell to K-shell charge transfer.

## 5. CONCLUSION

This paper attempts to give a review on the actual understanding of the main mechanisms by which inner shell vacancies in energetic ion atom collisions are produced. Because of the continuous expansion of this field the paper can, however, not be complete in touching all the different problems which are under investigation in this field.

A high degree of quantitative agreement between theory and experiment is obtained in so called "asymmetric" collision systems (atomic regime) where the ionization process to the continuum gives the predominant contribution to the inner shell excitation. Even in collision systems with $Z_p/Z_T \lesssim 1$, $v_p < v_e$ a reasonable description of the K-vacancy production in the heavier collision partner (also called $1s\sigma$ excitation) is given by first order perturbation theory when the "binding effect" is included.

In slow near symmetric collision systems ($Z_p/Z_T \simeq 1$, $Z \lesssim 100$) K-vacancy production is found to be well described by $2p\pi-2p\sigma$ rotational coupling of the molecular orbital model. Severe deviations from a prediction of this model is only found is solid targets at large impact parameters, which is probably due to a density effect.

Very large K x ray cross sections are reached by K-shell to K-shell charge transfer when the projectile brings K-shell vacancies into the collision. Even a distinct interference structure in the impact parameter dependence of this process is explained by two state coupling models.

## ACKNOWLEDGEMENT

The author wish to thank for the contribution of his co-workers, particularly of H. Ingwersen and H. Schmidt-Böcking to this work.

References:

1) Atomic Inner Shell Processes, ed. by B. Crasemann,
Academic Press, New York 1975

2) Bohr, N., Det Kgl. Danske Vid. Sel. Math. Fys. Medd. XVIII, No.18, NO.8 (1948)
see also: Bang, J., and Hansteen, J.M.: Det Kgl. Danske Vid. Sel. Math.Fys.
Medd. XVIII, No. 31, NO.13 (1959)

3) Schuch, R., Gaukler, G., Schmidt-Böcking, H.,
Z. Physik A 290, 19 (1979)

4) "Methods of Experimental Physics"; Vol. 17 Atomic Physics,
Accelerators; ed. P. Richard, Academic Press, New York 1980

5) Stein, N.J., Lutz, H.O., Mokler, P.H., Armbruster, P.,
Phys. Rev. A5, 2126 (1972)

6) Gaukler, G., Schmidt-Böcking, H., Schuch, R., Schulé, R., Specht, H.J.,
Tserruya, I.; NIM 141, 115 (1977)

7) Stähler, J., Hemmer, G., Presser, G.; NIM 164, 306 (1979)

8) Chemin, J.S., Andriamonje, S., Roturier, J., Saboya, B., Gayet, R.,
Salin, A., Phys. Lett. 67A, 116 (1976)

9) Schmidt-Böcking, H., Stiebing, K.E., Schadt, W., Löchter, N., Gruber, G.,
Kelbsch, S., Bethge, K., Schuch, R., Tserruya, I.,
Proceedings of Workshop on Coulomb Ionization, Linz 1981, appears in NIM.

10) Blair, J.S., Dyer, P., Shower, K.A., Trainar, T.A.,
Phys. Rev. Lett. 41, 1712 (1978)

11) Chemin, J.F., Andriamonje, S., Roturier, J., Saboya, B., Thibaud, J.P.,
Joly, S., Plattard, S., Uzureen, J., Laurant, H., Maison, J., Shapira, J.P.,
Nucl. Phys. A33, 407 (1979)

12) Röhl, S., Hoppenau, S., Dost, M.,
Phys. Rev. Lett. 43, 1300 (1979)

13) Brandt, W., Laubert, R., Sellin, I., Phys. Rev. 151, 56 (1966)
and Basbas, G., Brandt, W., Laubert, R., Phys. Rev. A7, 983 (1973)

14) Andersen, J.U., Laegsgaard, E., Lund, M., Moak, C.D.,
NIM 132, 507 (1976)

15) Schmidt-Böcking, H., Schulé, R., Stiebing, K.E., Bethge, K., Tserruya, I.,
Zekl, H., J. Phys. B 10, 2663 (1977)

16) Trautmann, D., Rösel, F., Baur, G., private communication
see also: Rösel, F., Trautmann, D., Baur, G.,
Proceedings of Workshop on Coulomb-Ionization, Linz 1981, appears in NIM

17) Gaukler, G., PhD - thesis, Heidelberg 1981 (unpublished)

18) Maor, D., Liesen, D., Mokler, P.H., Schmidt-Böcking, H., Schuch, R.,
(to be published)

19) Briggs, J.S., J. Phys. B 8, L485 (1975)

20) Anholt, R., Z. Phys. A 295, 201 (1980)

21) Kleber, M., Unterseer, K.,
Z. Phys. A 292, 311 (1979), see also invited talk in these proceedings

22) Müller, B., Greiner, W.,
Z. Naturforsch. 31a, 1 (1976)
and Soff. G., Betz, W., Kirsch, J., Oberacker, V., Reinhardt, J.,
Wietschorke, K., Müller, B., Greiner, W.,
GSI-Report, M-6-78

23) Müller, B., Soff, G., Greiner, W., Ceausescu, V.,
    Z. Phys. A 285, 27 (1978)
    Bosch, F., Liesen, D., Armbruster, P., Maor, D., Mokler, P.H.,
    Schmidt-Böcking, H., Schuch, R.,
    Z. Phys. A296, 11 (1980)
    Bang, J., Hansteen, J.M.,
    Phys. Scripta 22, 609 (1981)

24) Soff, G., Reinhardt, J., Betz, W., Rafelski, J.,
    Phys. Scripta 17, 417 (1978)

25) Liesen, D., Armbruster, P., Bosch, F., Hagmann, S., Mokler, P.H.,
    Schmidt-Böcking, H., Schuch, R., Wilhelmy, J.B., Wollersheim, H.J.,
    Phys. Rev. Lett. 44, 983 (1980)
    Armbruster, P., Bosch, F., Liesen, D., Maor, D., Mokler, P.H.,
    Warcak, A., Schmidt-Böcking, H., Schuch, R.,
    Jahresbericht GSI 1980

26) Fricke, B., Soff, G.,
    At. a. Nucl. Data Tables 19, 83 (1977)

27) Meyerhof, W.E., Anholt, R., Saylor, T.K., Lazarus, S.M., Little, A.,
    Phys. Rev. A 14, 1653 (1976)

28) Specht, H.J., Z. Phys. 185, 301 (1965)

29) Afrosimov, V.V., Gordeev, Yu.S., Panov, M.N., Fredorenko, N.V.,
    Soviet Phys. - Tech. Phys. 9, 1248 (1965)

30) Everhardt, E., Kessel, Q.C.,
    Phys. Rev. Lett. 14, 247 (1965)

31) Fano, U., Lichten, W.,
    Phys. Rev. Lett. 14, 627 (1965)

32) see e.g. Briggs, J.S., Rep. Prog. Phys. 39, 217 (1976)

33) Meyerhof, W.E., Phys. Rev. Lett. 31, 1341 (1973)

34) Briggs, J.S., Techn. Rep. Harwell No. 594 (1074)

35) Schuch, R., Schmidt-Böcking, H., Schulé, R., Tserruya, I.,
    Phys. Rev. Lett. 39, 79 (1977)

36) Tserruya, I., Johnson, B.M., Jones, K.W.,
    Phys. Rev. Lett. 45, 894 (1980)

37) Lennard, W.N., Mitchell, I.V., Ball, G.C., Mokler, P.H.,
    Phys. Rev. A 23, 2260 (1981)

38) Anholt, R., Stoller, Ch., Meyerhof, W.E.,
    J. Phys. B 13, 3807 (1980)

39) Pflanz, E., Schuch, R., Neumann, J., Nolte, G., Schmidt-Böcking, H.,
    Contributed Paper to X 80, Stirling 1980

40) Taulbjerg, K., Briggs, J.S., Vaaben, J.,
    J. Phys. B. 9, 1351 (1976)

41) Lutz, H.O., Luz, N., Sackmann, S., Jitschin, W., Hippler, R.
    in Physics of Atoms and Molecules: Coherence and Correlations in Atomic
    Collisions, ed. by Kleinpoppen and Williams, Plenum Press N.Y. 1980

42) Nolte, G., Volpp, J., Schuch, R., Specht, H.J., Lichtenberg, W.,
    Schmidt-Böcking, H., J. Phys. B 13, 4599 (1980)

43) Cocke, C.L., Randall, R.R., Varghese, S.L., Curnutte, B.,
    Phys. Rev. A 14, 2026 (1976)

44) Müller, K., Pflanz, E., Schmidt-Böcking, H., Schuch, R., Specht, H.J.,
Book of Abstracts, European Conference on Atomic Physics, Heidelberg 1981

45) Annette, C.H., Curnutte, B., Cocke, C.L.,
Phys. Rev. A 19, 1038 (1979)

46) Schuch, R., Nolte, G., Schmidt-Böcking, H.,
Phys. Rev. A 22, 1447 (1980)

47) Presser, G., Stähler, J., contributed paper to this conference

48) Betz, H.D., Rev. Mod. Phys. 44, 465 (1972)

49) Anholt, R., preprint and contributed paper to this conference

50) Lin, C.D., and Richard, P. in Advances of Atomic and Molecular
Physics 17, ed. by D.R. Bates and B. Bederson, Academic Press N.Y. 1981

51) Barrette, J., Johnson, B.M., Jones, K.W., Schuch, R., Tserruya, I.,
Kruse, T.H., contributed paper to this conference

52) Schuch, R., Nolte, G., Schmidt-Böcking, H., Lichtenberg, W.,
Phys. Rev. Lett. 43, 1104 (1979)

53) Hagmann, S., Cocke, C.L., Macdonald, J.R., Richard, P., Schmidt-Böcking, H.,
Schuch, R., to be published

54) Lin, C.D., private communication

55) Stolterfoht, N., J. Phys. B 13, L 651 (1980)

PHYSICS OF ELECTRONIC AND ATOMIC COLLISIONS
S. Datz (editor)
© North-Holland Publishing Company, 1982

PRODUCTION OF CONTINUOUS ELECTRON SPECTRA
IN HEAVY ION - ATOM COLLISIONS

Yu.S. Gordeev, A.N. Zinoviev

A.F. Ioffe Physico-Technical Institute
Leningrad, USSR

Energy spectra of electrons emitted in slow
(v < 1 a.u.) collisions of heavy ions and atoms
contain a continuous part. Two physical reasons
of these continuous spectra are discussed:
Auger transitions in quasimolecule and direct
electron transitions from quasimolecular levels
into continuum due to nonadiabatic coupling.

Production of continuous electron spectra in slow atomic collisions
was observed in [1-3]. Rudd et al. [2] showed that for proton-atom
collisions the continuous spectra could be described by impact
ionization model, but for heavy atom collisions this model did not
work. Statistical models of ionization also could not explain these
results [4].

Detailed studies of continuous electron spectra were performed in
our works since 1976 [5-7]. We studied the cases when vacancies
appeared on deep molecular orbitals (MO) and Auger transitions in
quasimolecules at small internuclear distances were possible. In
this case continuous electron spectra are due to the dependence of
electron energy upon the internuclear distance. In a simple quasi-
static model [6] for a definite trajectory we obtain:

$$\frac{dG}{dE_e} = \int_0^{\tilde{\infty}} 2\pi b f(E_e) db \quad , \quad f(E_e) = \frac{W_A(R)}{V_R} \left| \frac{dR}{dE_e} \right| \tag{1}$$

where $W_A(R)$ is Auger transition probability, $E_e(R)$ is the dependence of the emitted electron energy on the internuclear distance.
$V_R$ is radial velocity of the colliding particles and b is the impact parameter.

The cross section (1) has a peculiarity in the point $V_R = 0$. This
peculiarity disappears in a more accurate quantum mechanical description. Such a description of Auger transitions in quasimolecules
was developed by Devdariani et al. [8]. For the case of the para-
bolic time dependence of the term $E(t) = -\alpha t^2 + E_o + i\frac{\Gamma}{2}$ ($\Gamma$ is the
term width) the electron energy spectra was described in [8] by
Airy function Ai:

$$f(E_e) = G \Gamma Ai \left[ \alpha^{-1/3} \left( E_o + i\frac{\Gamma}{2} - E_e \right) \right] \tag{2}$$

169

where factor G is found from the normalization of the spectra on the initial population of the term.

For a definite trajectory an electron spectrum consists of two parts: an allowed one at $E_e < E_o$ and a forbidden one at $E_e > E_o$. Due to interferention of the amplitudes for the Auger transitions in symmetric points of the trajectory the cross section oscillates at $E_e < E_o$ and decreases exponentially at $E_e > E_o$. A similar approach was developed independently by Gerber and Niehaus [9].

In the electron spectra integrated over impact parameters oscillations are averaged and contributions from the allowed and forbidden regions overlap. Nevertheless one can divide the continuous spectra into two parts: $E_e < E_o$ and $E_e > E_o$. In the case of a demoting orbital the largest value of $E_o$ corresponds to the trajectory with zero impact parameter. These considerations were used in [5-7,10] to obtain $E_o(R_o)$ where $R_o$ is the smallest internuclear distance which can be achieved at the studied collision energy. Figure 1 shows the molecular orbitals for $Kr^+$ - Kr case obtained from the experiment [10] and calculated by Eichler [11] and by Nikulin and Guschina [12]. The calculated $4p\pi$ orbitals are in reasonable agreement with the experiment.

Probabilities of Auger transitions in quasimolecules $W_A(R)$ can be obtained from absolute measurements of the electron emission cross sections. The probabilities $W_A(R)$ appeared to be very large, $\sim 10^{16}$ sec$^{-1}$ [5-7], which was one or two orders of magnitude larger than $W_A$ for Auger transitions in isolated atoms. This increase of $W_A$ can result from the increase of the number of electrons participating in transitions and from large overlap of wave functions. This effect is similar to what we see in the case of Coster-Kronig transitions.

New possibilities in studying $E_o(R)$ and $W_A(R)$ appeared due to the experiments in which the trajectory of collision was fixed [13,14]. Fig. 2 illustrates the idea of these experiments. Studying the emitted electrons in coincidence with the ions scattered at definite angle $\vartheta$ one could directly obtain the electron spectra corresponding to a definite trajectory. Fig. 3 shows some results of the

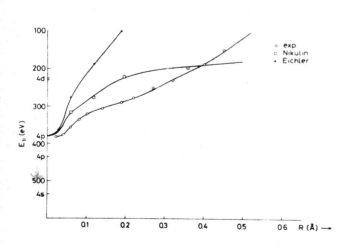

Fig. 1. Molecular orbital of Kr-Kr system:
○ - experiment [10]
● - calculation of $4p\pi$-orbital [11]
□ - calculation of $4p\pi$-orbital [12]

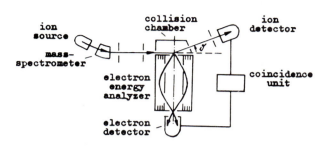

Fig. 2. Schematic diagram of the installation [14] for studying electron spectra by electron-scattered ion coincidence method.

experiments [13,14]. One can see that the first maximum in Airy function shifts to large electron energies with the impact parameter decrease. Describing the electron spectra by Airy function the authors of [13,14] found $E_o$ at different impact parameters b and obtained the molecular orbital more accurately than in earlier

experiments. Fig. 4 shows $E_e(R)$ obtained from coincidence measurements and from the measurements of the cross sections integrated over b. These results are in satisfactory agreement. The values of $W_A$ measured in coincidence experiments also agree well with the earlier data.

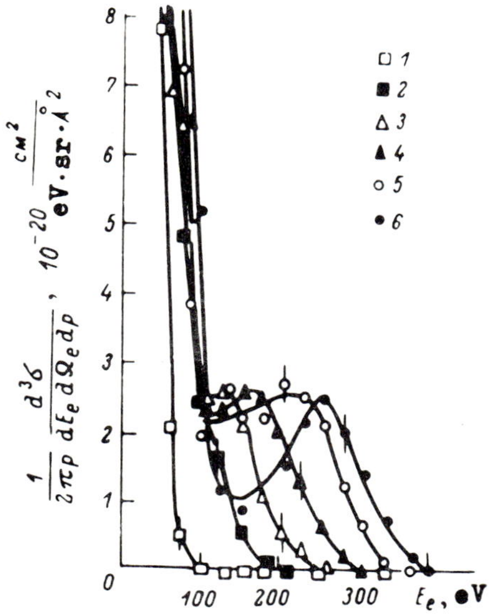

The considered data indicate that in the cases like Kr-Kr and Ar-Kr the continuous electron spectra result from Auger transitions on demoting MO in quasimolecules.

A system like Kr-Kr is too complicated for an analysis. The situation is much simpler for systems like Ne-Ne where there is also a demoting orbital $2p\pi$. The experimental studies of $Ne^+$, $Ne^{2+}$ - Ne and $O^+$ - $O_2$ collisions [15,16] showed that in these cases the electron spectra for definite trajectories exhibit some specific peculiarities (fig.5). An analysis of these spectra also enabled us to perform spectroscopy of quasimolecular levels (fig. 6). The experimental MO agree well with the calculation. The experimental uncertainty in fig. 6 results mainly from the uncertainty in the energy of the upper level from which the transition occurs. The values of $W_A$ for these cases are lower than for Kr-Kr case.

Fig. 3. Electron energy spectra [14] for $Kr^+$-Kr collisions, 50 keV, at different impact parameters b. b(Å): 1 - 0.52, 2 - 0.45, 3 - 0.4, 4 - 0.34, 5 - 0.3, 6 - 0.26

For Ne-Ne case $W_A = 10^{15}sec^{-1}$ and for $O^+-O_2$ $W_A = 10^{14}sec^{-1}$ which agrees with the transition probabilities for 2p vacancy in the united atom ($W_A = 3.3 \cdot 10^{14}sec^{-1}$ for Ca and $W_A = 10^{14}sec^{-1}$ for S [17]).

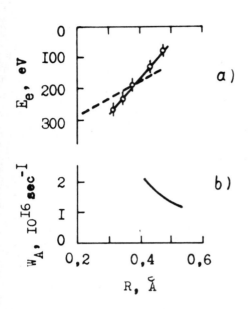

Fig. 4. Auger electron energy and transition probability as function of internuclear distance for Kr-Kr system.
a) experiment [14], b) experiment [7].

It confirms the conclusion that the large value of $W_A$ for Kr-Kr system results from participation of many electrons in the transitions.

The spectra in fig. 5 in addition to the continua due to the transitions on 2p$\pi$ orbitals show exponential tails at large $E_e$. These exponential continua were also observed in integrated cross sections. Systematic studies of these continua were started by Woerlee et al. [18]. Fig. 7 shows some of their results. The authors of [18] showed that the continua could be described by a simple expression

$$\frac{d\sigma}{dE_e} \sim exp\left[-\frac{a\,(E_e-E_o)}{v}\right] \quad (3)$$

These exponential continua were assumed [18] to result from nonadiabatic transitions into continuum due to radial or rotational coupling. Theoretical description of these transitions was developed in [19-21]. We shall call these transitions the dynamical ionization.

Like in the case of Auger transitions in quasimolecules (formula (2)), the probability of the transition into continuum is described by Airy function [19]. If we consider the emission of electrons with

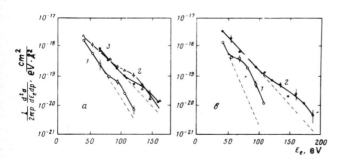

Fig. 5. Electron energy spectra [15] for the collisions: a) $O^+-O_2$; 1 - 25 keV, b = 0.18 Å
2 - 50 keV, b = 0.16 Å
3 - 50 keV, b = 0.09 Å
b) Ne$^+$-Ne; 1 - 15 keV, b = 0.19 Å
2 - 50 keV, b = 0.09 Å

large energy so that $\left|E - E_e\right| \gg \frac{v}{a}$ where $a$ is a characteristic length for the studied term, then the electron spectra can be described by a simple expression:

$$f(E_e) = \left|V_p\right|^2 \left|\frac{dt(E_e)}{dE_e}\right| exp\left[-2\,\text{Jm}\int_{E_0(b)}^{E_e} t(E)dE\right] \qquad (4)$$

where $t(E)$ is the inverse function to the time dependence of the term energy $E(t)$, $E_0(b)$ is the term energy in the turning point and $t(E_e)$ is the analytical continuation of the function $t(E)$ into the region $E > 0$.

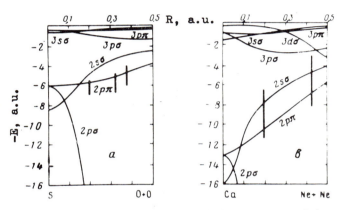

Fig. 6. MO diagrams for O-O (a) and Ne-Ne (b).
 Vertical lines: experimental E(R) [15].

It was shown in our paper [22] that in the case of close collisions electron spectra can be described with the accuracy of factor 2 using the results of [19]. According to [19] for the short-range potential $\left|V_p\right|^2 = 4E_e$. In the case of Auger transitions $\left|V_p\right|^2 \sim \Gamma$. Therefore if transitions occur from an autoionizing term, then at large electron energies $E_e - E_0 \gg \Gamma$ the dynamical ionization dominates.

Integrating (4) over impact parameters we obtain [22]:

$$\frac{d\sigma}{dE_e} = \int_0^\infty 2\pi b\,f(E_e)db = \frac{4\pi E_e \frac{dR}{dE_e}\,exp\left[-\frac{2}{v}\int_{E_0}^{E_e} R(E)dE\right]}{\int_{E_0}^{E_e} dE\,\text{Jm}\,\frac{1}{R_p(E)}} \qquad (5)$$

where $E_0 = E_0$ (b = 0), $R_p(E) = vt(E)$, $v$ is the collision velocity. Here the straight line trajectory approximation was used.

Using (5) we can obtain parameters of the term from experimental electron spectra. Indeed [22]:

$$\mathrm{Im}\, R(E_e) = \frac{1}{2}\left|\frac{1}{v_1} - \frac{1}{v_2}\right|^{-1} \frac{\partial}{\partial E_e}\, \ell n\, \frac{\dfrac{d\sigma}{dE_e}\Big|_{v_1}}{\dfrac{d\sigma}{dE_e}\Big|_{v_2}} \tag{6}$$

and using the dispersion relation from Im $R(E_e)$ we obtain Re $R(E_e)$.

Figure 8 shows the term obtained from experimental data [18] and the electron spectra calculated [22] using this term. The calculated spectra agree well with the experiment at all collision energies. It shows that ionization in Ne$^+$ - Ne collisions is due mainly to dynamical ionization of the term which is formed at R $\to \infty$ from the outer shells and is promoted into continuum at R $\approx$ 0.7 a.u. The comparison of this term with the MO diagram for Ne-Ne system shows that this term includes 4f$\sigma$ orbital.

In [23] electron spectra for Ne$^+$ - Ne collisions were studied at lower collision energies. The spectra were explained as results of Auger transitions on the demoting 3p$\pi$ and 3s$\sigma$ orbitals. MO obtained from the experiment using zeros of Airy function indicate autoionizing state and agree with 3p$\pi$ and 3s$\sigma$ orbitals. As was shown above, indeed in the region close to $E_o$ contribution of autoionization can be large, but at $E_e \gg E_o$ dynamical ionization dominates.

There is an interesting aspect in the Ne$^+$ - Ne case. The contributions of transitions from different MO into the integrated cross section are proportional to $R_c^2 V_p^2$, and because of a large $R_c$ transitions from the 4f$\sigma$ orbital dominate. But if trajectories are fixed the contributions of different transitions are proportional to $V_p^2$ and can be comparable. This conclusion is confirmed by the results of coincidence measurements [24]. The dependence of the electron emission cross section on the collision velocity and on $E_e$ can be described taking into account transitions from 4f$\sigma$ and 3d$\sigma$ orbitals.

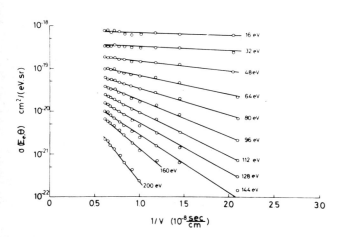

Fig. 7. Differential cross sections of electron emission for Ne$^+$ - Ne collisions as functions of inverse collision velocity [18]. The parameter - electron energy $E_e$.

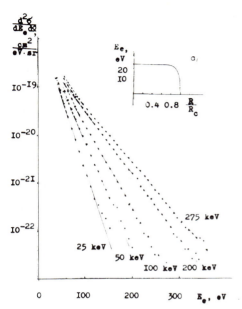

Fig. 8. Electron energy spectra for Ne$^+$-Ne collisions at different collision energies. Points: experiment [18]; lines: calculations [22] using the term (fig. 8a) obtained from the experiment.

But at small b transitions from deeper MO can be seen (fig. 9), probably from the 2p$\sigma$ orbital.

Therefore the main results of the analysis of continuous electron spectra for the cases like Ne-Ne are the following:

a) at $E_e > 100$ eV dynamical ionization of the 4f$\sigma$ orbital dominates;

b) at $E_e < 100$ eV at fixed trajectories the contributions of transitions from several MO (4f$\sigma$, 3d$\sigma$ etc.) are comparable;

c) at $E_e < 100$ eV the contribution of autoionizing transitions is also considerable, in particular, at low collision energies.

There are cases when the process of dynamical ionization is the only process of ionization. It occurs when the initial state is not an autoionizing state. This situation was analyzed in [22, 25].

Usually in the limit of the united atom a term depends parabolically on the internuclear distance:

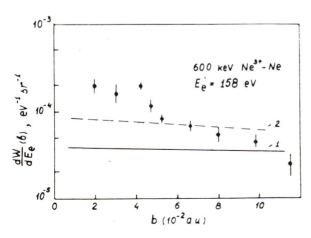

Fig. 9. Probability of electron emission as a function of impact parameter. Points: experiment [24]. Lines: our calculation.
1 - ionization of 4f$\sigma$ orbital.
2 - ionization of 4f$\sigma$ and 3d$\sigma$ orbitals.

$E(R) = E_o(1 - \frac{R^2}{R_o^2})$. In this case the total cross section for ionization of the inner shell due to the transitions into continuum is [22,25]:

$$\sigma = \int \frac{d\sigma}{dE_e} dE_e = \sigma_o \, f(\frac{v}{v_o}) \tag{7}$$

where $\sigma_o = \pi R_o^2$, $v_o = E_o R_o$, $f(\frac{v}{v_o}) = \int_0^\infty \frac{x \, dx}{x+1} \, exp\left[-\frac{4}{3} \frac{v_o}{v}(x+1)\right]$.

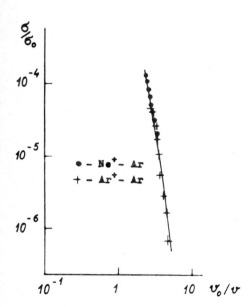

In [25] the formula (7) was used to describe cross sections of ionization of any quasimolecular level due to the coupling with the continuum. The parameters $E_o$ and $R_o$ can be found for any system and for any level: $E_o$ - from electron energies in the united atom and $R_o$ - by scaling MO diagram of $H_2^+$ -like system.

Fig. 10 shows that formula (7) describes very well experimental cross sections [26] of k-ionization of Ne in $Ne^+$ - Ar collisions. In both cases electron transitions from the 2pб orbital into continuum take place.

The authors are thankful to A.P.Shergin and to S.Yu.Ovchinnikov for useful discussions.

Fig. 10. Dependence of reduced ionization cross section on the reduced collision velocity for 2pб orbital [25].
Points: experiment [26].

REFERENCES

1. Rudd, M.E.  and Jorgensen, T., Phys. Rev. 131 (1963) 666.
2. Rudd, M.E., Jorgensen, T. and Volz, D.J., Phys. Rev. 151 (1966) 28.
3. Ogurtsov, G.N., Flaks, I.P., Avakyan, S.V. and Fedorenko, N.V., ZhETF Pisma 8 (1968) 541.
4. Cacak, R.K. and Jorgensen, T., Phys. Rev. A2 (1970) 1322.
5. Afrosimov, V.V., Gordeev, Yu.S., Zinoviev, A.N., Rasulov, D.H. and Shergin, A.P., Abstr. II Int. Conf. Inner Shell Ioniz. Phen. (Post-Deadline Papers) (Freiburg, 1976).
6. Afrosimov, V.V., Gordeev, Yu.S., Zinoviev, A.N., Rasulov, D.H. and Shergin, A.P., ZhETF Pisma 24 (1976) 33.
7. Afrosimov, V.V., Gordeev, Yu.S., Zinoviev, A.N., Korotkov, A.A. and Shergin, A.P., Abstr. X ICPEAC p. 924 (Paris, 1977).

8. Devdariani, A.Z., Ostrovsky, V.N. and Sebyakin, Yu.N., ZhETF 73 (1977) 412.
9. Gerber, G. and Niehaus, A., J. Phys. B 9 (1976) 123.
10. Gordeev, Yu.S., Woerlee, P.H., de Waard, H. and Saris, F.W., J. Phys. B 14 (1981) 513.
11. Eichler, J., Wille, N., Fastrup, B. and Taulbjerg, K., Phys. Rev. A 14 (1976) 706.
12. Nikulin, V.K. and Guschina, N.A., J. Phys. B 11 (1978) 353.
13. Afrosimov, V.V., Meskhi, G.G., Tsarev, N.N. and Shergin, A.P., Abstr. VI ICAP p. 398 (Riga, 1978).
14. Afrosimov, V.V., Meskhi, G.G., Tsarev, N.N. and Shergin, A.P., ZhTF Pisma 5 (1979) 897.
15. Afrosimov, V.V., Meskhi, G.G., Tsarev, N.N. and Shergin, A.P., ZhETF Pisma 31 (1950) 729.
16. Meskhi, G.G., Tsarev, N.N. and Shergin, A.P., Abstr. I ECAP p. 772 (Heidelberg, 1981).
17. McGuire, E.J., Phys. Rev. A 3 (1971) 587.
18. Woerlee, P.H., Gordeev, Yu.S., de Waard, H. and Saris, F.W., J. Phys. B 14 (1981) 527.
19. Solovyev, E.A., ZhETF 70 (1976) 872.
20. Chaplik, A.V., ZhETF 45 (1963) 1518; 47 (1964) 126.
21. Watanabe, T., Woerlee, P.H. and Gordeev, Yu.S., Abstracts XI ICPEAC p. 652 (Kyoto, 1979).
22. Zinoviev, A.N., Ovchinnikov, S.Yu. and Gordeev, Yu.S., Abstr. XII ICPEAC p. 900 (Gatlinburg, 1981).
23. Meskhi, G.G., Tsarev, N.N. and Shergin, A.P., ZhTF Pisma 7 (1981) 207.
24. Zinoviev, A.N., Liesen, D. and Saris, F.W., Abstr. I ECAP p. 774 (Heidelberg, 1981).
25. Zinoviev, A.N., Ovchinnikov, S.Yu. and Gordeev, Yu.S., ZhTF Pisma 7 (1981) 139.
26. Woerlee, P.H., Fortner, R.I. and Saris, F.W., J. Phys. B 14 (1981).

PHYSICS OF ELECTRONIC AND ATOMIC COLLISIONS
S. Datz (editor)
© North-Holland Publishing Company, 1982

# SPECTROSCOPY OF HIGH ENERGY δ-RAYS EMITTED
## IN COLLISIONS OF VERY HEAVY IONS

C. Kozhuharov*

Gesellschaft für Schwerionenforschung mbH
P.O. Box 11 05 41
6100 Darmstadt, W. Germany

Investigations of the momentum distributions of strongly
bound electrons in heavy quasi-atoms by means of high-energy
δ-ray spectroscopy are discussed, with special emphasis on
collision systems with combined charge :  $(Z_p + Z_t) \cdot \alpha \geq 1$.

INTRODUCTION

The motivation of our investigations is to establish and further develop δ-ray
spectroscopy as one of the tools to study the properties of strongly bound quasi-
atomic states, formed transiently during heavy-ion collisions. We concentrate
furthermore on very heavy collision systems, for which the combined charge of projectile and target nucleus $Z_u = Z_p + Z_t$ are much higher than the charge of the
heaviest elements known at present. Following theoretical predictions we expect
a strong increase of the binding energies with the nuclear charge : For $Z \gtrsim 150$
the 1s-binding energy exceeds the rest mass of the electron itself and for $Z \gtrsim 173$
the innermost 1s-state becomes a resonance embedded in the negative energy continuum of the Dirac equation /1/.

Let us consider a Pb-Pb encounter at relative velocity $v_{po}/c = 0.10$ - much smaller
than the relativistic velocities of K-electrons in the Pb-atom. The bombarding
energy is 4.7 MeV/u ($E_{lab} \gtrsim 1$ GeV) and leads to a distance of closest approach for
a head-on collision of 20 fm, which is also much smaller than the radius of the
Pb-K-atomic shell (∿ 650 fm). During the collision time of $\tau_{coll} \lesssim 10^{-20}$ sec the
electrons are exposed to a two-centre Coulomb field with a time dependent two-
centre separation R(t), which is even smaller than the K-shell radius of the quasi-
atom with $Z_u = 164$. This strong field determines in this case the energy and the
wave function of the bound electron. The time variation of the field may induce
transitions from inner bound states to the positive energy continuum. The ionized
electrons are ejected with continuous momentum distributions, the high energy tail
of which - above 100 keV up to 2 MeV - is the object of our investigations. The
reason for our interest in the high momentum component of the spectrum is easy to
be understood qualitatively : For pure kinematical reasons high momentum can be
transferred in a heavy ion collision only to electrons with very high initial
momentum, i.e. only if the electron "stems" from those portions of the initial
orbit, which are very close to the charge centre. It is also evident that δ-rays
with extremely high energies (while of low absolute intensities) originate mostly
from the sharper localized innermost shells. But even more important, a rapid increase of the high momentum components can be expected for superheavy quasi-atoms
considered, since the extremely strong Coulomb field leads to enormous relativistic
contraction of the electronic wave functions. Hence, the high energy tail of the
δ-ray spectrum can be attributed to the high energy component of the momentum distribution of strongly bound quasiatomic electrons and, thus, provide information
about their wave function and energy.

It should be pointed out, however, that the observed high energy part of the δ-ray spectrum represents a superposition of contributions from several inner shells. Indeed the K-shell is more sharply localized than the L-shells - i.e. the value of the high energy component is higher in the region of initial momenta considered. However, the ionization probability depends also on the energy transfer needed ($\Delta E = E_{kin} + |BE|$) to induce transition from a state with binding energy BE to continuum state with $E_{kin}$. Thus, the stronger binding can cancel to some extent the effects of sharper localization up to δ-ray kinetic energies $E_{kin} \gg |BE|$, at which the binding can be neglected. A preferential observation of the contributions from a particular quasi-atomic shell to the total δ-ray spectrum is possible in some cases, if only δ-rays, associated with a characteristical X-ray emission of one of the collision partners are being detected. The vacancy left in a shell after δ-electron emission decays in a time of $\tau_{hole} \approx 10^{-17}$ sec, which is much longer than the collision time ($\tau_{coll} \lesssim 10^{-20}$ sec) - i.e. the decay occurs after the nuclei are well separated and the quasi-molecular picture does not hold. Thus, the hole decays predominantly via X-ray emission from a separated atom. It was shown /2/ that a vacancy produced in the 1sσ-state of the quasi-atom becomes a K-hole in the heavier atom, whereas a $2p_{1/2}$-vacancy occurs as a K-hole in the lighter collision partner. In symmetrical or nearly symmetrical collisions ($Z_p \approx Z_t$) vacancies produced in higher shells may be transferred to lower, stronger bound states in the outgoing phase of the collision, where at large two-centre distances the energy splitting between those levels becomes small /3/.

Thus, for asymmetric collisions a measurement of high energy δ-rays associated with the characteristic K-X-rays of the heavier collision partner leads to spectroscopy of the innermost 1sσ-quasi-atomic shell /4/. (For very low δ-ray energies and heavy collision systems, where the multiplicity of δ-ray emission per collision is higher than unity, it is possible, however, to detect δ-electron from a higher shell emitted in a collision, in which a K-hole has been also produced.)

TEST OF THE BASIC CONCEPT

"LIGHT" COLLISION SYSTEMS WITH $Z_u < 107$**

In order to prove the significance of the proposed method /7/, first experiments have been carried out with "lighter" $^{16}$O- or $^{32}$S-projectiles and $^{197}$Au- or $^{108}$Pb-targets, respectively /4,5/. The charge numbers of the quasi-atoms studied remain thus, in the region of known elements, where the theoretical description should be easier. Double-differential cross sections ($d^2\sigma/d\Omega dk_f$ or $d^2\sigma/d\Omega dE$) have been measured both for single electron spectra as well as for δ-rays in coincidence with the K-X-rays from gold. The energy spectra were calculated within the framework of the Born approximation, using wave functions in the united atom limit and is given by :

$$\frac{d^2\sigma}{d\Omega\,dk_f} = \left(\frac{2Z_p e^2}{\hbar v_\infty \zeta}\right)^2 \; k_f^2 \int_{qm}^{\infty} \frac{q\,dq}{q^4} \; |F(k_i)|^2 \tag{1}$$

with

$$\zeta = \frac{Z_t}{Z_p + Z_t}$$ reflecting the correlation of the quasi-atomic wave function to the centre of charge.

$\hbar\vec{k}_i$ and $\hbar\vec{k}_f$ - the initial and final momentum of the electron ejected, related to the momentum $\hbar q$ transferred in the collision by :

$$\vec{k}_f = \vec{k}_i + \vec{q} \tag{2}$$

In the integral (from $q_m = q_{min}/\zeta$ to $\infty$) the factor $1/q^4$ reflects the probability for momentum transfer q in the collision, whereas the form factor given by :

$$\left|F(k_i)\right|^2 = \int_0^\infty d\phi_q \left| <f(\vec{k}_f)|e^{\,i\vec{q}\cdot\vec{r}}|i> \right|^2 \tag{3}$$

(with $\phi_q$ - the azimuthal angle of q) contains the information about the momentum $k_i$ of the bound electron in the united atom. The minimum momentum transfer $q_{min}$ to an electron with a binding energy $|BE|$ ejected to a continuum state ($E_{kin}$) can be easily obtained from the energy and momentum conservation laws :

$$\hbar q_{min} = \frac{|BE| + E_{kin}}{v_\infty} \tag{4}$$

valid for small changes of the initial momentum $\mu v_\infty$ of the ions due to the ionization (this condition is practically always fulfilled).

Figure 1 shows the $\delta$-ray spectrum from $^{32}S$-$^{208}Pb$ collisions at 120 MeV bombarding energy, corresponding to a relative velocity of $v_\infty/c = 0.09$ and to a collision diameter $2a = 18.2$ fm /5/. Double differential cross section ($d^2\sigma/d\Omega dE$) has been measured with good statistics over 5 orders of magnitude and up to very high kinetic energies. The data are in a very good agreement with calculations using Born approximation with *relativistic* wave functions of the *united atom* ($Z_u = 98$). It should be noted that no fitting parameters have been used. Also shown are calculations in the target atom limit (dashed line) as well as results obtained using non-relativistic wave functions (dashed-dotted line) - both underestimate the data significantly. In all three cases the cross section calculated represents a sum taken over all electrons of the K- and L-shell.

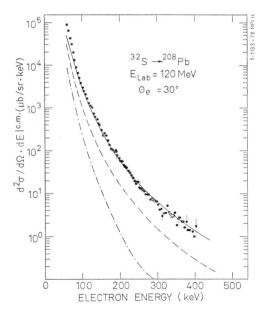

Figure 1

$\delta$-ray spectrum from $^{32}S \rightarrow {}^{208}Pb$ collisions at 120 MeV bombarding energy /5/;
(——) *united atom* limit ($Z_u$=98);
*relativistic* Born calculation;
(---) *target atom* limit ($Z_u$=82);
*relativistic* Born calculation;
(-,-) *united atom* limit ($Z_u$=98),
*non-relativistic* Born calculation.

The data clearly show that the Cf quasi-atom is formed transiently during the collision, where - at larger $Z_u$ than for the target atom - the relativistic shrinking of the wave function overcompensates the effects of stronger electronical binding. The very good agreement with the calculated cross sections demonstrates the direct relation between the spectral distribution of high energy $\delta$-rays and the form factor of the inner shell electrons of the quasi-atom.

Since the first measurements were carried out, considerable results have been achieved in systematic studies of high-energy $\delta$-ray emission, covering various target projectile combinations in the region $66 \leq Z_u \leq 145$ and relative velocities $0.07 \leq (v_\infty/c) \leq 0.10$ /8/. For purposes of systematic comparison of the data from different systems a so-called spectral function S $(q_{min}, Z_u)$ has been introduced :

$$S(q_{min}, Z_u) = \int_{qm}^{\infty} \frac{qdq}{q^4} \; |F(k_i)|^2 \tag{5}$$

From Equation (1) one can see that $S(q_{min}, Z_u)$ can be derived from the measured cross sections by a reduction of the pure kinematic effects, such as projectile velocity ($v_\infty$) or transformation into the centre of charge system ($\zeta$). Keeping in mind that the dominant contribution to the cross section comes from $q \approx q_{min}$ one can replace in Equation (2) $q$ by $q_{min}$. Furthermore, from the fact that the relative ion velocity $v_\infty$ is much smaller than this of the relativistic $\delta$-rays and from Equation (4) one can obtain $q_{min} \gg k_f$, which gives $k_i = |\vec{k}_f - \vec{q}| \approx q_{min}$. Thus, using the spectral function $S(Z_u, q_{min})$ one can describe the experimental data as a function of the united atom charge $Z_u$ and the minimum momentum transfer only. At given $q_{min}$ one observes a strong increase of the spectral function by increasing $Z_u$, thus, convincingly demonstrating the quasi-atomic origin of the $\delta$-rays. The slope of the spectral function is well reproduced up to $Z_u < 137$ by a power law in $1/q_{min}$

$$S(q_{min}, Z_u) \propto 1/q_{min}^{6+2\gamma} \quad \text{with } \gamma = \sqrt{1-(Z_u \cdot \alpha)^2} \tag{6}$$

extracted by expending the form factor in power series in $1/k_i$, using hydrogen-like wave functions for a point-like charge and integrating the first term in Equation (5) ($k_i \approx q_{min}$ is assumed). The dominant role of the relativistic effects is evident, since with the much steeper fall off: $S \propto 1/q_{min}^{10}$, calculated in the non-relativistic limit, no similarity with the data can be achieved.


EXPERIMENTAL ARRANGEMENTS

The data presented in Figure 1 also demonstrate the necessity of a flexible experimental set-up, capable of operating in a large range of electron energies, with sufficient background suppression and reasonable detecting efficiency for measurements of small cross sections and for coincidence measurements. Two experimental set-ups possessing somewhat complementary features were utilized :  An iron-free "orange"-type    $\beta$-spectrometer /9/ and an achromatic electron channel /5/.

Figure 2a shows a schematic view of the "orange"-type $\beta$-spectrometer, which uses a toroidal magnetic field produced by 60 current coils to momentum analyze and focus electrons emitted from target onto a small, cone-shaped plastic scintillator counter The momentum direction of the emitted $\delta$-rays is defined by the entrance slits. Two  configurations have been used - one shown in Figure 2a, where all electrons emitted between 30° and 50° ($\theta_\delta$) relative to the beam direction and within a momentum band of $(\Delta p/p) = 0.018$ were focussed and the second configuration - with $50° \leq \theta_\delta \leq 70°$ and  $(\Delta p/p) = 0.014$. The transmission efficiency in both cases run to 0.08 of $4\pi$ (1 srad). The large transmission renders possible the detection of small $\delta$-ray intensities and the performance of coincident measurements with X-rays and for scattered particles. Characteristical K-X-rays were detected by two $\phi$ 3" x 1 cm NaI scintillator counters, mounted 4.5 cm away from the target. All particles, scattered between 9.5° and 27° (arrangement shown in Figure 2a) or between 16° and 48° (second configuration, not shown here) relative to the beam axis were detected by an annular parallel-plate avalanche counter. Its active area was subddivided into 8 concentric rings with individual read-out, each capable of operating up to counting rates of $10^6$ $sec^{-1}$. Kinematical particle coincidences between particular rings were also possible, allowing an unambiguous definition of scattered particle trajectories for asymmetric collision systems. The magnetic field of the spectrometer was swept repeatedly up and down focussing δ-rays with energies typically between 150 keV and 2.4 MeV. The measuring time after each field adjustment was normalized to the number of elastically scattered particles, detected by a Si-surface barrier monitor counter (BM). The multi-parameter data were recorded event-by-event, together with the normalization spectra and the corresponding instantaneous value of the spectrometer magnetic field. Self-supported ∿ 1 mg/cm² metalic foils of $^{238}$U, $^{232}$Th, $^{208}$Pb, $^{197}$Au, and $^{120}$Sn were bombarded with $^{208}$Pb-ions, accelerated up to 4.7 MeV/u ($v_\infty/c$ = 0.10) at GSI Darmstadt.

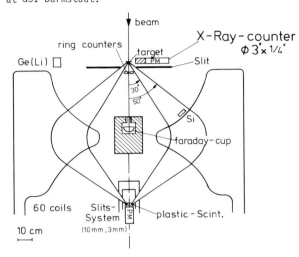

Figure 2a
Experimental set-up in the ion-free "orange"-type β-spectrometer. Momentum reso-lution (Δp/p) = 0.08. Trans-mission T = 1 srad. Coinci-dent measurements with scattered particles and/or X-rays can be performed.

The experiments with the achromatic magnetic electron channel, shown in Figure 2b were performed both at the Heidelberg MP Tandem van de Graaff accelerator : C, S, and Ni-beam (and in combination with the postaccelerator - Ni, Br, and I-ions, respectively) as well as at UNILAC accelerator: (Pb-ions). Zr, Pr, Pb, and U foils of thicknesses 0.5 ÷ 1mg/cm² were used as targets. The channel separates a momentum

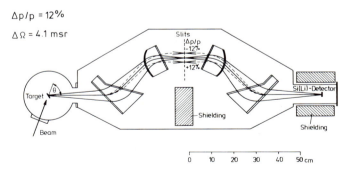

Figure 2b
Achromatic magnetic electron channel (the parallel-plate avalanche particle counter and the NaI-counters for X-rays are not shown in the Figure). The variable electron detection angle θ allows measurements of δ-angular distribution.

byte of ($\Delta p/p$) = 0.24 and focusses it with a transmission of 4 msrad onto a cooled Si(Li) counter, which energy resolution ran to 2.5 keV at 1 MeV. Similar as in the "orange" spectrometer, X-ray counters (two $\phi$ 2" x 1 cm NaI scintillator counters) and particle-counters (annular parallel-plate avalanche counter with a $\theta$-sensitive anode and a $\phi$-sensitive kathode) were mounted in the scattering chamber (not shown in the Figure). Unlike the "orange" spectrometer the relative small solid angle allows better definition of the $\delta$-ray emission angle relative to the beam axis or (combined with the $\theta$- and $\phi$-information from the particle counter) relative to the trajectory of the scattered particles. The channel can be rotated in the plane around the target, thus allowing angular distribution measurements.

With the experimental set-upts presented both double-differential cross sections for positron emission ($d^2\sigma/d\Omega dE$) as well as energy differential yields (emission probabilities per collision - $\Delta P/\Delta E\Delta\Omega$) were investigated as a function of the $\delta$-ray kinetic energy $E_{kin}$, relative ion velocity $v_\infty$, united atom charge $Z_u$, and of the projectile $\theta_p$ and target $\theta_t$ scattering angles. The emission probability for a $\delta$-ray with energy E per collision leading to particle scattering angle $\theta$ is defined as :

$$\frac{\Delta P(\theta)}{\Delta\Omega\Delta E} = \frac{N_\delta^{coin}(\theta)}{N_{part}(\theta)} \cdot \frac{1}{\varepsilon_\delta} \qquad (7)$$

with :  $N_\delta^{coinc}(\theta)$  -  number of $\delta$-rays with kinetic energy E, detected in coincidence with particles scattered through angle $\theta$

$N_{part}(\theta)$  -  number of particles, scattered through angle $\theta$ and detected

$\varepsilon_\delta$  -  overall electron detection efficiency

K-X-rays, associated with $\delta$-electrons (both single as well as in coincidence with scattered particles) were recorded, as already mentioned, simultaneously.

For asymmetric collision systems the Rutherford trajectory of the projectile and target nuclei can be determined unambiguously - i.e. impact parameter (b) and distance of closest approach ($R_{min}$) can be obtained easily by :

$$b = a \; ctg \frac{\theta_{CM}}{2} \qquad R_{min} = a(1+csc\frac{\theta_{CM}}{2}) \qquad (8)$$

with :  $2a = \frac{Z_p Z_t e^2}{E_{CM}}$  - collision diameter.

From the investigation of $\delta$-ray production probabilities as a function of the distance of closest approach $R_{min}$ or the impact parameter b, one expects not only more detailed information about the mechanism of Coulomb ionization. The ion trajectory must be taken into consideration for very heavy collision systems, since the binding energies of the innermost shells are expected to vary strongly with the internuclear separation, especially at very small distances /10/. On the other hand the impact parameter dependence of the $\delta$-ray emission probabilities represents a very sensitive test for scaling laws, as proposed for ionization probabilities determined via measurements of X-ray yields /11,12/. Unlike X-ray measurements the energy transfer $\Delta E = E_{min} + |BE|$ is determined for fixed $\delta$-ray energy, provided that the binding energy is known. And vice versa - if the validity of the scaling law is established - unknown binding energies of electrons in superheavy colliding systems can be experimentally determined to some extent.

$\delta$-RAY SPECTROSCOPY IN THE HIGH $Z_u$ REGION

Extending the experiments outlined before to  much heavier systems, we ask first of all whether the observed $\delta$-electrons still originate from the quasi-atom, as

already demonstrated for lower united charge $Z_u$. Let us, therefore, look first at the $Z_u$-dependence of the $\delta$-ray emission. The principal features reflected in the data of Figure 3 are :
( i) the very high kinetic energy of the $\delta$-ray observed : Up to 750 keV for the Pb $\to$ Sn collision system with $Z_u$ = 132 and over 1.7 MeV for the Pb $\to$ Pb system ($Z_u$ = 164). These are energies 8.5 or even 20 times higher than the binding energy of K-electrons in Pb-atom and correspond to extremely high momentum transfer;
( ii) the large difference in the intensities of the Pb-Sn and Pb-Pb spectra of more than one order of magnitude;
(iii) the much steeper fall-off of the Pb-Sn spectrum, reflecting the relative lack of high momentum component of the bound state wave functions compared to the Pb-Pb distribution.

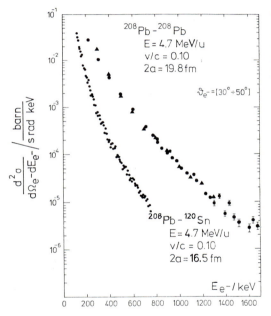

Figure 3
$\delta$-ray spectra from Pb+Pb ($Z_u$=164) and Pb+Sn ($Z_u$=132) collision systems measured at the same relative velocity ($v_\infty/c$) = 0.10. Also indicated are the distances of closest approach for a head-on collision, 2a (data from /17/).

All of these imply that the electrons have been emitted from bound quasi-atomic states, where, due to the $Z_u$-dependent relativistic contraction of the wave function, large momentum transfers to strongly localized bound electrons become possible. This is also corroborated by the data in Figure 4, in which the double-differential cross section ($d^2\sigma/d\Omega dE$) at given kinetic energy $E^\delta_{kin}$ = 470 MeV is displayed versus the united charge $Z_u$ for the systems C+Pb; Ni+Pb; Pb+Sn; Pb+Pr; Pb+Au; Pb+Pb, and Pb+Th measured also at the same relative ion velocity. (The data for C+Pb; Ni+Pb, and Pb+Pr are from /13/, the point for Ni+Pb has been extrapolated from ($v_\infty/c$) = 0.091 to 0.10 using Equation (1).) The increase of the cross section over four order of magnitude, whereas $Z_u$ only doubles, demonstrates again the characteristic behaviour of bound state wave functions in superheavy quasi-atoms. For lower $Z_u$ the ionized electrons originate predominantly from the L-shell /13/ with binding energies much smaller than the kinetic energy observed - i.e. at given $\delta$-ray energy the energy transfer required ($\Delta E = |BE| + E_{kin}$) is determined by $E_{kin}$. This does not apply to the heavier systems, where the saturation of the $Z_u$-dpendence observed is partially due to the stronger binding.

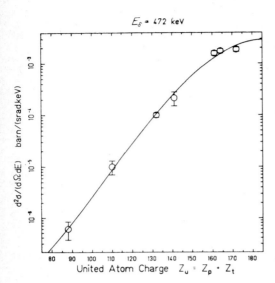

Figure 4
Double-differential cross section
for emission of δ-rays of 472 keV
kinetic energy for the collision
systems : C+Pb; Ni+Pb; Pb+Sn;
Pb+Pr; Pb+Au; Pb+Pb, and Pb+Th as
a function of the united charge Zu
for the relative ion velocity of
$v_\infty/c$ = 0.10 (Ni+Pb extrapolated
from v/c = 0.091). (The line is
to guide the eye.)
(Data from /13,17/).

## ASYMMETRIC COLLISION SYSTEMS WITH $Z_u \cdot \alpha \approx 1$

For two systems with similar united charge $Z_u$ δ-rays associated with K-X-rays of
the heavier collision partner have been measured. In Fig. 5a the results for
$^{208}$Pb → $^{120}$Sn collisions ($Z_u$=132) are shown together with the single spectrum ($\Sigma$).
The measurements were performed with the "orange"-type β spectrometer at GSI
Darmstadt. The relative ion velocity ran to $(v_\infty/c)$ = 0.10. The corresponding
spectra for the system $^{127}$J → $^{208}$Pb are presented in Figure 5b. The data were
taken at MPI Heidelber at slightly different ion velocity $(v_\infty/c)$ = 0.09 with the
achromatic electron channel. The solid lines in Figures 5a and 5b represent results of coupled channel calculations performed by Soff et al. /14/ for δ-ray
emission in coincidence with an 1sσ-vacancy creation. Also included are multistep
processes - i.e. the ejected electron can be re-scattered during the collision to
higher energy. These calculations reproduce very well both the absolute δ-ray
yields as well as their energy distribution. It should be emphasized again, that
the good agreement between the theoretical calculations (which adopt the quasi-
atom-picture) with measurements performed with two different experimental arrangements and with inverse projectile-target combinations strongly corroborate the
expectations to exploit in this way the characteristic behaviour of the innermost
1sσ bound state in the Pb-Cm quasi-atom ($Z_u$=178).

The contributions of the K-shell to the total cross section ($\Sigma$) are very small at
low kinetic energies (less than 1 % at 40 keV) and do not become significant
until the energy reaches very high values. They exhibit a slope not as steep as
for the single spectrum but similar to that of the Pb-Pb distribution in Figure 3.
As already mentioned, the lower intensity can be explained by the larger momentum
transfer required to emit a K-electron with respect to the much weaker bound
L-electrons. On the other hand we also know that higher momenta can be transferred
at rather smaller impact parameters b, due to the higher Fourier frequencies
available in the Coulomb collision. Figure 6a shows δ-ray emission probabilities
per collision as a function of the distance of closest approach $R_{min}$ for two fixed
δ-ray kinetic energies 540 and 720 keV, at which the K-shell contributions are more
significant.

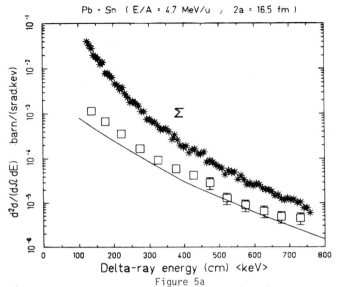

### Figure 5a

δ-ray spectrum from Pb+Sn collisions at $(v_\infty/c) = 0.10$ associated with the K-X-rays from lead (□) is compared to results of coupled channel calculations including multistep processes, performed by Soff et al. /14/. Also shown is the single δ-ray spectrum (Σ) (data from /17/).

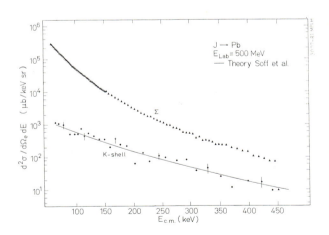

### Figure 5b

The same as in 5a for J → Pb collisions at $(v_\infty/c) = 0.09$.

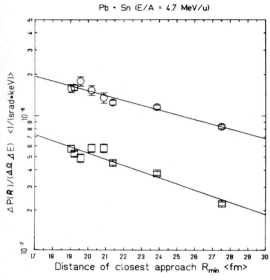

Pb + Sn (E/A = 4.7 MeV/u)

Distance of closest approach $R_{min}$ <fm>

Figure 6a

$\delta$-ray emission probability for Pb-Sn collisions as a function of the distance of closest approach $R_{min}$ for fixed $\delta$-ray kinetic energies :  (O) 540 keV and ($\square$) 720 keV. The solid line represents a fit to the data with :

$$P \propto \exp[-\frac{2R_{min}(E_{kin}+|BE|)}{\hbar v_\infty}]$$

$(v_\infty/c) = 0.10$ (data from /17/).

A scaling rule for $1s\sigma$-excitation probabilities as a function of the distance of closest approach and the momentum transfer has been proposed and successfully applied by Bosch et al. /12/ considering results of X-ray measurements. The $\delta$-ray emission probability per energy interval is given by :

$$\frac{\Delta P(R_{min})}{\Delta E \Delta \Omega} = d_0^2(Z) \; (\frac{m_e c^2}{E_{kin}+m_e c^2})^{c(Z)} \; \exp[-\frac{m'R_{min}(E_{min}+|BE|)}{\hbar v_\infty}] \tag{9}$$

with :  $m'=2$ for heavy collision systems;

$(E_{min}+|BE|)/\hbar v_\infty = q_{min}$ - minimum momentum transfer required;

$\exp(-2R_{min}q_{min})$ is based on the scaling behaviour of the wave function $|\Psi|^2 \sim 1/R^2$;

$d_0^2(Z)\cdot((E_{kin}/m_e c^2)+1)^{-C(Z)}$ reflects the coupling strength to the continuum;

$C(Z)$ is supposed to be a smooth function of Z. For X-ray measurements $<E_{kin}> = 0$ is assumed.

The solid lines in Figure 6a represent a least square fit to the data using $P \propto \exp(-2 P_{min}q_{min})$. From the fitted value of $q_{min}$ one can extract the binding energy for known $E_{kin}$ :  For $E_{kin} = 720$ keV it gives $|BE| = 310$ keV compared with 320 keV from /15/. For $E_{kin} = 540$ keV one obtains a smaller value of $|BE| = 250$ keV which can be caused by contributions from the L-shell. No reasonable agreement is achieved, however, if Equation (9) is fitted to the absolute values of the data. Similar behaviour is observed for lower kinetic energy $\delta$-rays and larger distances of closest approach, shown in Figure 6b for I → Pb collisions at $(v_\infty/c) = 0.09$ /16/. The slope of the emission probability distribution exhibits an exponential fall-off with the distance of closest approach $R_{min}$, steeper for higher energies, according to $P \propto \exp(-2 R_{min}q_{min})$. The "average" binding energy of 60 keV can be extracted, indicating the dominance of contributions from the L-shell. Similar to the data for Pb-Sn collisions the absolute intensities cannot be reproduced by Equation (9) due to strong deviations from the proposed $(E_{kin}/m_0 c^2+1)^{-C(Z)}$ energy dependence.

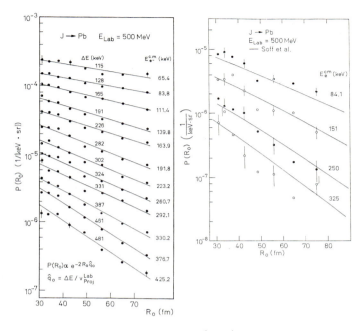

Figure 6b (Left)
δ-ray emission probability for I → Pb collisions as a function
of the distance of closest approach $R_O$ for fixed δ-ray kinetic
energies $E_e^{C.M.}$ (keV). The solid lines represent results of fit
to the slope of the data with an exponential function
$P \propto \exp[-2\,R_O\Delta E/\hbar v]$. Also indicated in the figure are the values
obtained from the fit for the energy transfer required : ΔE.

Figure 6c (Right)
Emission probability for δ-rays associated with Pb-K-X-rays
(K-shell, united atom) for I → Pb collisions as a function of
the distance of closest approach, $R_O$, for fixed δ-ray kinetic
energies $E_e^{CM}$ (keV) as indicated. The solid lines show results of
calculations performed by Soff et al. /14/.

In Figure 6c the dependence on the distance of closest approach of the emission
probabilities for electrons from the K-shell of the (I → Pb) united atom are displayed for four different δ-ray kinetic energies. Triple coincidences have been
performed, detecting δ-ray associated with particles scattered through an angle
and in additional coincidence with K-X-rays from lead /16/. The solid line represents coupled channel calculations, including multistep processes, performed by
Soff et al. /14/. Both slope and absolute intensities are in a good agreement
within the experimental errors. This emphasizes the rôle of multistep processes
/14/ for proper interpretation of measured absolute values. The scaling law from
Equation (9), however, reproduces well the behaviour of δ-ray emission probabilities (L-shell contributions included) as a function of the distance of closest
approach $R_{min}$ and the minimum momentum transfer required $q_{min}$.

## VERY HEAVY COLLISION SYSTEM

First measurements of high energy δ-rays from very heavy quasi-atoms were carried
out, as already mentioned, for $^{208}$Pb+$^{208}$Pb collisions at 4.7 MeV/u bombarding
energy /17/ - far below the Coulomb barrier. Albeit for symmetrical collision the
ion trajectory can be defined unambiguously only for scattering angles $\theta_{CM} = \pi/2$
and the contributions from the 1sσ-innermost state cannot be separated by a K-X-ray
coincidence requirement, the $^{208}$Pb+$^{208}$Pb system gives the unique opportunity to
study δ-ray spectra without any distortion due to nuclear de-excitation processes.
The only source of background - the decay of the Coulomb-excited 3$^-$ state
(2.614 MeV) via internal e$^+$-e$^-$-pair creation - is expected to be extremely weak,
as shown in measurements of positron yields /18/.

In Figure 7 the measured double-differential cross sections /17/ are compared with
results of coupled channel calculations performed by Soff et al. /14/, which include multistep excitation processes. Shown are the calculated total δ-ray distribution, representing a sum over all initial bound states up to 4sσ- and 4p$\frac{1}{2}$σ-
state, as well as contributions from the 1sσ- and 2p$\frac{1}{2}$σ-state. The experimental
data exceed the calculated values by a factor of about three. The slope of the
measured spectrum, however, is reproduced well. It should be noted in this context
that perturbation theory values /19/ also agree in the shape with the measured
spectrum, underestimating the data by more than a factor of 6. The pronounced
difference between the contributions from 1sσ- and 2p$\frac{1}{2}$σ-states, shown in Figure 7,
demonstrate again that the emission process depends strongly on the energy (or
momentum-) transfer $\Delta E = |BE| + E_{min}$ required and that the 1sσ-binding energy
cannot be neglected even at δ-ray kinetic energies as high as 1.7 MeV.

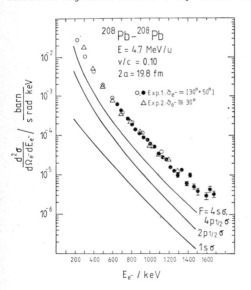

Figure 7
δ-ray spectra from $^{208}$Pb+$^{208}$Pb at bombarding energy 4.7 MeV/u.
Exp. 1 - "orange"-β-spectrometer
Exp. 2 - achromatic electron channel
The solid lines represent results of
coupled channel calculations performed
by Soff et al. /14/ for the total
spectrum (F = 4sσ, 4p$\frac{1}{2}$σ denotes the
Fermi surface) and for contributions
from the 1sσ- and 2p$\frac{1}{2}$σ-states.

The dependence of the emission probabilities on the ion scattering angle at
several fixed δ-ray kinetic energies between 470 keV and 1.8 MeV is given in
Figure 8. The experimental data represent the total emission probability - i.e.
contributions from several inner-shells are superimposed. The results of the
triple coincidence measurement - i.e. δ-ray-scattered particles and K-X-rays - do
not show, within the limits of error, any significant deviations from the distributions presented in Figure 8. From this one has to conclude that a large fraction

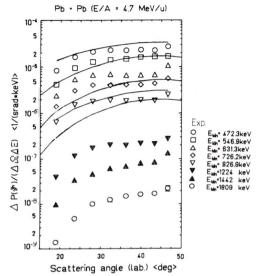

Pb + Pb (E/A = 4.7 MeV/u)

$\Delta P(\vartheta)/(\Delta\Omega\,\Delta E)$ $<1/(sr\cdot rad\cdot keV)>$

Scattering angle (lab.) <deg>

Exp.

| | |
|---|---|
| ○ | $E_{kin}$= 472.3keV |
| □ | $E_{kin}$= 546.9keV |
| △ | $E_{kin}$= 631.3keV |
| ◇ | $E_{kin}$= 726.2keV |
| ▽ | $E_{kin}$= 826.9keV |
| ▼ | $E_{kin}$=1224 keV |
| ▲ | $E_{kin}$=1442 keV |
| ○ | $E_{kin}$=1809 keV |

Figure 8
δ-ray emission probabilities as a function of the scattering angle of the detected ion for kinetic energies of the δ-electrons between 470 keV and 1.8 MeV as indicated. The solid lines represent theoretical calcu- lations of Soff et al. for following kinetic energies (from the top) : 400, 500, 700, 800, and 900 keV. The theoretical curves for 500, 700, and 900 keV are result of calculations for a triple coincidence - δ-ray- K-X-ray and scattered particles /14/.

of the K-X-rays observed is not associated with holes in the 1sσ- or 2p½σ-states but results from vacancies produced in higher shells (with ionization probabilities in the order of unity), which may couple to the 1sσ-orbital in the outgoing phase of the collision. This finding allows to compare the data in Fig. 8 with calculated impact parameter dependences both of the total as well as of the K-X-ray coincident δ-ray emission probabilities /14/. Since the scattering particles are identical, the emission probabilities for impact parameter b(θ) and those for (π-θ) have been weighted with the corresponding Rutherford cross section and folded. The agreement with the experimental data is surprisingly good, keeping in mind the factor of three difference between the total energy distributions presented in Figure 7. Apparently the δ-ray emission associated with particles scattered through small angles is underestimated by the theory, especially for higher δ-ray kinetic energies, at which the effects of rotational coupling (not included in the theory) can be neglected.

Under realistic assumptions for the energy transfer required, the slope of the data is reproduced by the scaling law from Equation (9) only in the low energy region. For higher δ-ray energies the calculated values exhibit a steeper increase with the increasing ion scattering angle. Further experiments are planned to study this un- expected behaviour, with additional emphasis of the dependence on the relative ion velocity. (The X-ray production probabilties as a function of the ion velocity ex- hibit at fixed impact parameter an unexpected maximum for $v_\infty/c$ = 0.10 /20/).

I would like to thank Prof. P. Armbruster, Dr. H. Backe, Dr. H. Bokemeyer, and Dr. G. Soff for many helpful discussions.

*) The major part of the experiments presented in this progress report have been carried out by a GSI-Heidelberg-Munich collaboration with following members : F. Bosch and C. Kozhuharov (GSI-Darmstadt); E. Berdermann, M. Clemente, and P. Kienle (Technische Universität München); F. Güttner, W. Koenig, B. Martin, B. Povh, H. Skapa, J. Soltani, and Th. Walcher (Max-Planck-Institut für Kern- physik, Heidelberg).

**) δ-ray distributions have been investigated before the early 1970's mostly in the low kinetic energy region (from eV to several keV), at very low bombarding energies (below several MeV), for very light projectiles (p or α) and/or projectile-target combinations (cf. e.c. Ref. 6 and further references therein).

REFERENCES

[1]  Pomeranchuk, I. and Smorodinsky, I., J. Phys. USSR 9 (1945) 97;
     Voronkov, V.V., and Kolesnikov, N.N., Sov. Phys. JETP 12 (1951) 136;
     Werner, F.W. and Wheeler, J.A., Phys. Rev. 109 (1958) 126; Pieper, W.
     and Greiner, W., Z. Physik 218 (1969) 327; Müller, B., Rafelski, J.,
     and Greiner W., Z. Phys. 257 (1972) 62; 257 (1972) 183, see also G. Soff,
     W. Greiner, W. Betz and B. Müller, Phys. Rev. A20 (1979) 169 and further
     references therein; Zeldovich, Y.B. and Popov, V.S., Sov. Phys. - Usp. 14
     (1972) 673; Rein, D., Z. Phys. 221 (1969) 423

[2]  Meyerhof, W.E. and Tjaulberg, Ann. Rev. Nucl. Sci. 27 (1977) 279;
     Mokler, P.H., Folkmann, F.: Structure and Collisions of Ions and Atoms,
     Sellin, I.A. (ed.) (Springer, Berlin 1978)

[3]  Meyerhof, W.E., Phys. Rev. Lett. 31 (1973) 1341; Demkov, Y.N., Sov. Phys.
     JETP, 18 (1969) 138

[4]  Kozhuharov, C., Kienle, P., Jakubaßa, D., Kleber, M., Phys. Rev. Lett. 39
     (1977) 540

[5]  Bosch, F., Krimm, H., Martin, B., Povh, B., Walcher, Th., Traxel, K.:
     Phys. Lett 78B (1978) 568

[6]  Merzbacher, E., Lewis, H.W., in: Flügge (ed.), Encyclopedia of Physics
     (Springer, Berlin 1958) Vol. XXXIV; Stolterfoht, N., in: Sellin, I.A.
     (ed.) Structure and Collisions of Ions and Atoms (Springer, Berlin 1978)

[7]  Kienle, P., Kleber, M., Kozhuharov, C., Martin, B., Povh, B., Schwalm, D.:
     Vorschlag zur Untersuchung elektromagnetischer Prozesse in hohen Feldern,
     Heidelberg, Munich 1975 (unpublished); Bosch, F., Krimm, H., Martin, B.,
     Traxel, K., Walcher, Th.: Vorschlag zur Spektroskopie von δ-Elektronen
     als Untersuchung der Quantenelektrodynamik überkritischer Felder, Heidel-
     berg 1977 (unpublished)

[8]  Bosch, F., Güttner, F., Martin, B., Meyer-Schützmeister, L., Kienle, P.,
     Koenig, W., Kozhuharov, C., Krimm, H., Povh, B., Scappa, H., Soltani, I.,
     Walcher, Th., in: XVIII International Winter Meeting on Nuclear Physics,
     Bormio, Italy, 21-26 Jan., 1980, University of Milan Ed. Supplemento n. 13;
     Güttner, F., Koenig, W., Martin, B., Povh, B., Scappa, H., Soltani, I.,
     Walcher, Th., Bosch, F., Kozhuharov, C., submitted to Z. Physik A

[9]  Moll, E., Kankeleit, E., Nukleonik 7 (1965) 180

[10] Müller, B. and Greiner, W., Z. Naturforschung 31a (1976) 1

[11] Bang J. and Hansteen, J.M., Kgl. Dan. Vid. Selsk. Mat.-Fys. Medd. 31
     (1959) No. 13; Armbruster, P., Behncke, H.-H., Hagmann, S., Liesen, D.,
     Folkmann, F., and Mokler, P.H., Z. Phys. A288 (1978) 277; Bang, J. and
     Hansteen, J.M., preprint Nordita 80/22.

[12] Müller, B., Soff, G., Greiner, W., Ceausescu, V., Z. Phys. A285 (1978) 27;
     Bosch, F., Liesen, D., Armbruster, P., Maor, D., Mokler, P.H., Schmidt-
     Böcking, H., and Schuch, R., Z. Phys. A296 (1980) 11

[13] Krimm, H., Diplomarbeit, Universität Heidelberg, 1978 (unpublished);
     Güttner, F., Diplomarbeit, Universität Heidelberg, 1979 (unpublished)

[14]  Soff, G., Reinhardt, J., Müller, B., and Greiner W., Z. Phys. <u>A294</u> (1980) 137; Soff, G., private communication

[15]  Fricke, B., Soff, G., GSI-Report T1-74

[16]  Güttner, F., Koenig, W., Martin, B., Povh, B., Skapa, H., Soltani, J., Walcher, Th., Kozhuharov, C., Kienle, P., Meyer-Schützmeister, L., European Conference on Atomic Physics, April 6-10, 1981; Kowalski, J., zu Putlitz, G., Weber, H.G., (Ed.) Book of Abstracts part II (Heidelberg 1981) 899; Skapa, H., Diplomarbeit, Universität Heidelberg, 1980 (unpublished)

[17]  Berdermann, E., Bokemeyer, H., Bosch, F., Clemente, M., Güttner, F., Kienle, P., Koenig, W., Kozhuharov, C., Krimm, H., Martin, B., Povh, B., Traxel, K., Walcher, Th., GSI-Jahresbericht 1978, p. 95; Berdermann, E., Bokemeyer, H., Bosch, F., Clemente, M., Güttner, F., Kienle, P., Kozhuharov, C., Krimm, H., Martin, B., Povh, B., Traxel, K., Walcher, Th., Verh. DPG (VI) 14 (1979) 492; Berdermann, E., Bosch, F., Bokemeyer, H., Clemente, H., Güttner, F., Kienle, P., Koenig, W., Kozhuharov, C., Martin, B., Povh, B., Tsertos, H., Walcher, Th., GSI-Scientific Report 1979, p. 100; Berdermann, E., Bokemeyer, H., Bosch, F., Clemente, M., Güttner, F., Kienle, P., Koenig, W., Kozhuharov, C., Martin, B., Povh, B., Tsertos, H., Walcher, Th., Verh. DPG (VI) 15 (1980) 1168

[18]  Backe, H., Handschug, L., Hessberger, F., Kankeleit, E., Richter, L., Weik, F., Willwater, R., Bokemeyer, H., Vincent, P., Nakayama, V., Greenberg, J.S., Phys. Rev. Lett. <u>40</u> (1978) 1443; Kozhuharov, C., Kienle, P., Berdermann, E., Bokemeyer, H., Greenberg, J.S., Nakayama, V., Vincent, P., Backe, H., Handschug, L., Kankeleit, E., Phys. Rev. Lett. <u>42</u> (1979) 376; Tsertos, H., Diplomarbeit, Technische Universität München, 1980 (unpublished)

[19]  Soff, G., Betz, W., Müller, B., Greiner, W., Merzbacher, E., Phys. Lett. <u>65A</u> (1978) 19; Jakubaßa, D.H., Z. Phys. <u>A293</u> (1979) 281

[20]  Liesen, D., Armbruster, P., Behncke, H.-H., Bosch, F., Hagmann, S., Mokler, P.H., Schmidt-Böcking, H., Schuch, R., in: Electronic and Atomic Collisions; Oda, N. and Takayanagi, K. (ed.) North-Holland, 1980, p. 337

PHYSICS OF ELECTRONIC AND ATOMIC COLLISIONS
S. Datz (editor)
© North-Holland Publishing Company, 1982

ELECTRON CAPTURE AND LOSS TO CONTINUUM

I. A. Sellin

University of Tennessee, Knoxville TN  37916, and
Oak Ridge National Laboratory, Oak Ridge TN 37830
U.S.A.

A sharp cusp in the velocity spectrum of electrons ejected in
ion-atom and ion-solid collisions arises from capture to low-
lying, projectile-centered continuum states for fast bare or
nearly bare projectiles; from loss to low-lying, projectile-
centered continuum states when loosely bound projectile electrons
are available; and from a combination of these processes, modified by solid state effects, when solid targets are employed.
The respective cusp shapes observed are only superficially
similar.  Differences in cusp shape and yield are compared and
contrasted for projectiles varying in energies from $\sim$ 100 keV/u
to 8500 keV/u traversing gaseous targets with $Z_T$ = 2-18; thin
solid polycrystalline targets (C, Al, Ag, Au); and <110>, <100>,
and <111> axial channels in Au.  For heavy ions the results of
both singles and coincidence experiments, wherein the electron
is detected in coincidence with an emergent ion of charge state
$q_e$ are reviewed.  Results from investigators at the Universities
of Aarhus, Freiburg, Frankfurt, Georgia, Nebraska, Sussex, Tennessee and Western Ontario; East Carolina, New York, and Texas
A & M Universities; AERE Harwell; Centro Atomico Bariloche; Oak
Ridge National Laboratory; and CEN-Saclay are mentioned.

I.  INTRODUCTION AND OVERVIEW

A sharp cusp in the velocity spectrum of electrons, ejected in ion-atom and ion-
solid collisions, is observed when the ejected electron velocity $\vec{v}_e$ matches that
of the emergent ion $\vec{v}$ in both speed and direction[1].  In ion-atom collisions, the
electrons primarily originate from capture to low-lying, projectile-centered continuum states (ECC) for fast bare or nearly bare projectiles, and primarily from
loss to those low-lying continuum states (ELC) when loosely bound projectile electrons are available.  Most investigators now agree[2,3] that ECC cusps are strongly
skewed toward *lower* velocities and exhibit full widths half maxima roughly proportional to v (neglecting target shell effects, which are sometimes strong)[4].

A close examination of recent ELC data from our laboratory shows that ELC cusps
are instead nearly symmetric, with widths independent of v in the velocity range
6-18 au, a result *not* predicted by recent theory[5].  Figure 1 provides a striking
illustration of the differences observed.  For $0^{7+}$ projectiles on Ar in Figure 2,
a cusp obtained in coincidence with $0^{8+}$, an ion that has lost one electron (dotted
line), is overlaid with a cusp containing all electrons not detected in coincidence with $0^{8+}$ (solid line).  This effectively resolves the ELC and ECC contributions as the data set for both cusps is identical, and the strong difference between ECC and ELC cusp shapes is directly demonstrated.

In contrast, "convoy" electron cusps produced in heavy ion-solid collisions at
MeV/u energies are slightly skewed toward *high* electron velocities, but exhibit
velocity-independent widths, very similar to ELC cusp widths.  (For protons,

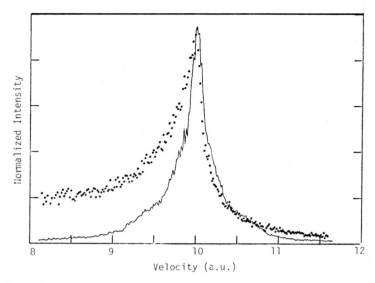

Figure 1.  Velocity spectrum of electrons emitted in the forward direction in collisions of 40 MeV $O^{3+}$ ions with Ar (solid line).  The spectrum shows a typical
ELC cusp, with low-energy projectile rest frame Auger structure superposed.  A
representative peak-normalized ECC spectrum (dots) for 40 MeV $O^{8+}$ ions traversing
Ar is superposed.

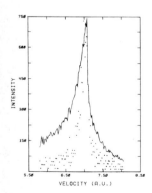

Figure 2.  Cusp for 20 MeV $O_8^{7+}$ projectiles on Ar obtained in coincidence with $O^{8+}$ (dotted line), overlaid
with cusp containing all electrons not detected in
coincidence with $O^{8+}$ (solid line).  The former is an
ELC spectrum while the latter represents an ECC spectrum.  Differences in ELC and ECC cusp shapes are
obvious.

however, typically at energies of a few hundred keV, other investigators[1,6] indicate a skew toward lower energies, though in these data, background treatment
issues tend to cloud the meaning of the skew observed.)  While the shape of the
convoy peaks is approximately independent of projectile Z, velocity, and target
material, we find that the yields in polycrystalline targets exhibit a strong dependence on projectile Z and velocity.  Attempts have been made to link convoy
electron production to binary ECC or ELC processes, sometimes at the last layer,
or alternatively to a solid-state wake-riding model[7], but our measured dependence
of cusp shape and yield on heavy projectile charge state and energy[8] are inconsistent with the predictions of available theories.  Figure 3 contrasts ECC with

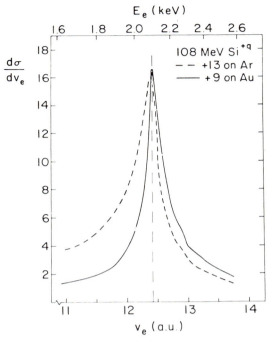

Figure 3. The differential production cross section $d\sigma/dv_e$ as a function of the electron velocity, $v_e$, of continuum electrons emerging near $0°$ with respect to the ion beam, for 108 MeV $Si^{13+}$ in Ar gas at 30 mTorr and for $Si^{9+}$ on a 100 $\mu g/cm^2$ Au foil. The electron energy scale appears at the top of the figure. The mean emergent charge from the solid target is also 13. From Laubert et al., Ref. 8.

convoy distribution line shapes. Figure 4 illustrates the lack of agreement of observed cusp widths with the predictions of ECC theory applied at or near the last atom layer[9], and with the predictions of so-called "wake-riding" theories[7]. These theories seek to explain the origin of the convoy electrons in terms of electrons trapped into an oscillatory electron density polarization potential trailing each projectile which are then liberated at the exit surface.

When coincidence with emergent ion charge state $q_e$ is required[10], ECC cusps can be sorted as to whether b = 0, 1, 2... additional bound-state captures occurred during the same collision which generated the continuum electron. Similarly, ELC cusps can be sorted as to how many additional electrons were lost[11]. The shapes observed are relatively independent of whether or not additional capture or loss events occurred. The yields (production cross sections) tend to mimic the beam velocity, projectile Z, and projectile charge q dependence of corresponding single and multiple electron capture and loss cross sections.

For convoy electron production in solids, cusp shapes are again found to be approximately independent of $q_e$[8,12]. More remarkably, for polycrystalline and randomly oriented monocrystalline targets, the yields are found to be independent of $q_e$, i.e., to mirror the unweighted statistical fraction of emergent ions of each charge state, even though there is an appreciable projectile Z dependence and reason to believe that the *observed* convoys (many are lost through scattering) originate well within one mean free path for charge changing of the exit surface. For well-channeled ions, however, the convoy yield is strongly suppressed[13,14], pointing to the necessity of close approach to an atomic string in the bulk as a necessary precursor[13] for convoy production.

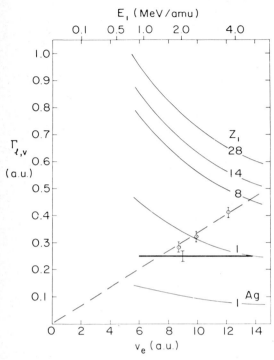

Figure 4. The full width at
half maximum of the longitudinal
electron velocity distribution,
$\Gamma_\ell$, in au, for continuum electrons as a function of the electron velocity $v_e$, in au. The
incident projectile energy, in
MeV/u, appears at the top of the
figure. The dashed line is the
ECC prediction. The solid lines
are the predictions of the WR
theory for an Al target with
the indicated projectiles. The
lowest solid curve marked Ag is
the prediction of WR theory for
protons incident on Ag. The
experimental data for solids
correspond to $\Gamma_\ell \simeq 0.25$ au and
are represented by the heavy
solid line. The open points
represent the gaseous target
(Ne and Ar) results for $O^{8+}$
and $Si^{14+}$. From Laubert,
Ref. 8.

Studies of ECC processes for bare ions on light atoms - H and He - especially at
the highest velocities are of particular importance. In recent articles[15]
Shakeshaft and Spruch have focused attention on an unusual opportunity to test
higher-order Born contributions to charge transfer amplitudes through study of ECC
cusp asymmetries. Moreover, strict scaling laws linking ECC studies using p, d,
and alpha beams with corresponding studies using higher Z bare projectiles should
exist, and when derived should be subjected to experimental test by comparing the
results for low Z projectiles with the results for higher Z bare projectiles in
the same gas targets. At sufficiently high ion velocities, the mean charge state
of an ion emergent from a solid also corresponds to $\bar{q}_e = Z$. Thus also in the
solid target case, scaling laws linking proton or alpha particle-induced convoy
electron yields to those for higher Z projectiles may be realizable.


## II.  SUMMARY OF PROGRESS CONCERNING ELECTRON CAPTURE TO THE CONTINUUM (ECC)

In this section and the two subsequent ones, we incorporate an earlier review[3] of
ECC, ELC, and convoy electron production, updating and amending it to take account
of more recent data and publications.

Electron capture to the continuum describes capture to projectile-centered continuum states, where the capture proceeds in analogy to electron transfer to
bound states, but the wave function which describes the motion of the electron
after the collision is instead a projectile-centered continuum wave function. The
phenomenon therefore represents a form of ionization, but one in which, for example, a plane wave description of the captured electron is completely inappropriate.
Rather, Coulomb waves centered on the projectile become a more appropriate

description.  Macek, in a series of publications with Eugene Rudd and others
dating back to 1970[16], makes the following analogy.  Ionization can be thought of
as the natural continuation of excitation to a sequence of orbits of ever-increas-
ing principal quantum number into the continuum.  The excitation cross sections
continue smoothly right through the ionization limit, provided an appropriate normalization of continuum states vis-a-vis excitation to high n Rydberg states per
unit bandwidth Δn is considered.  In like fashion, one may envision electron capture events accompanying an ion-atom encounter into a sequence of orbits of ever-
increasing principal quantum number n, whose production rate also continues
smoothly from the region of high Rydberg states just below the continuum into the
continuum.  Somehow, this process went experimentally undiscovered and theoretically neglected during the 50-odd years which have elapsed since the initial
development of the quantum theory.  Although quantum mechanical theories of excitation, ionization, and capture to bound states were worked out in the 1920's and
1930's, the electron capture contribution to ionization was somehow ignored.  That
it can sometimes be extremely important is illustrated by a 1978 paper by Shakeshaft[17], who finds that for certain energies ($\sim$ 40 keV), more than half the total
cross section for ionization of hydrogen by protons is accounted for by this process.  Understanding of the important interactions characterizing ion-atom, ion-
molecule, and ion-solid collisions is thereby changed significantly.

It appears that the discussion of the electron capture to continuum process initially arose in the course of attempting to explain orders of magnitude disagreements between experimental differential cross section data and Born approximation
calculations in the ionization of He and $H_2$ by proton impact at angles $\sim$ 10° with
respect to the forward direction.  According to Macek's 1970 article[18] on the
theory of the forward peak in the angular distributions of electrons ejected by
fast protons, it was Rudd who elaborated a suggestion by Oldham, Jr.[19] to explain
the forward peaking, by arguing that when the emitted electron is ejected at a
velocity $\vec{v}_e$ near the projectile velocity $\vec{v}$, some electrons are subjected to a
strong Coulomb drag in the forward direction before moving away from the proton
as free electrons.  Macek then devised a first-order approximation for the final
state wave function of the electron, proton, and residual ion which would describe
an electron being "carried along" by a proton.  He used an expansion of Faddeev's
equations incorporating Green's functions for three non-interacting particles. To
a first approximation Macek's expression for the ionization amplitude can be written:

$$a = a_{DI} + a_{ECC} - a_P,$$                                                          (1)

where a represents the total amplitude for electron ionization, $a_{DI}$ represents the
amplitude for direct ionization calculated in some approximation which does not
account for strong projectile distortion, $a_{ECC}$ is a "charge exchange to continuum"
amplitude which is dominant near $\vec{v}_e = \vec{v}$, and $a_P$ is a counter term arising because
the term with all outgoing particles described by plane waves has been counted
twice.  The final wave function of the electron in the DI term, for example, can
be represented as a Coulomb wave centered at the target.  If $\vec{v}_e \simeq \vec{v}$, then the
electron has large momentum with respect to the target, and a Coulomb wave centered at the target describing this electron differs little from a plane wave.
Therefore, the first and the third term approximately cancel.  The ECC term becomes large and dominates.  If $\vec{v}_e \gtrsim \vec{v}$, all three terms contribute and cancellation
is incomplete.  According to Macek[18], as elaborated by Duncan et al.[20] and Lucas
and Macek[20], interference effects are predicted to be observable when the phase
of the ECC amplitude in Eq. (1) is varying rapidly owing to the Coulomb distortion
of the final state wave function, especially for small (but non-zero) angles for
which the direct-ionization and charge-transfer amplitudes are comparable.  Apart
from the issue of interference effects, Macek's theory predicts an approximately
symmetric distribution of electrons in velocity space centered at $\vec{v}_e = \vec{v}$, whose
yield scales as $Z^3$.

In contrast, there exist several single amplitude theories of the forward peak which seek to describe the $\bar{v}_e = \bar{v}$ peak through some enhancement factor multiplying a smoothly varying amplitude for direct ionization away from the forward peak, calculated, for example, in Born approximation to first or second order. The best known of these are due to Salin[21] and to Dettmann, Harrison, and Lucas[9]. Working within the framework of a first Born approximation to ionization, Salin devised an approximate scheme for introducing projectile distortion of the continuum electron wave function over and above the more usual allowance for target distortion. As discussed by a recent comment by Ponce and Meckbach[22], the final state may be thought of as a two-center continuum molecular orbital. Salin's distorted wave approximation leads to one-center continuum atomic orbitals about the appropriate nucleus for electrons slow with respect to one of the nuclei, while describing the electron in the combined atom continuum when the electron is fast with respect to both heavy particles. As discussed by Rødbro and Andersen[23], Salin's theory predicts an asymmetric peak in the velocity spectrum of ejected electrons at low projectile velocities which tends toward symmetry at high velocities because of the underlying shape of the Born cross section, a prediction unfulfilled either in the data of Ref. 23 or in data from our laboratory. Obviously, no interference can occur in this or any other single amplitude theory.

The theory of Dettmann et al.[9] explicitly considers charge transfer to the continuum to be the dominant process in the production of $\bar{v}_e \simeq \bar{v}$ electrons, which are represented in terms of projectile-centered Coulomb waves. The structure developed for the T matrix describing the scattering has a very similar form to that for bound-state capture, except that a continuum state wave function replaces a bound-state wave function in the final state. A Coulomb wave normalization factor therefore arises which can be viewed as the origin of a divergence of the differential cross section at zero deg, which in turn gives rise to a cusp in the measured velocity distribution near zero deg after averaging over some finite analyzer acceptance angle. Using an approximation analogous to the Brinkman-Kramers approximation for bound-state capture, they obtained the result (in the notation of Ref. 15 except that $Z_B = Z$, $Z_T$ = target Z):

$$\frac{d\sigma_{BK}}{dv_e} (1s \rightarrow cont.) \simeq \frac{2^{17}}{5} Z_T^5 Z^3 \left(\frac{e^2}{\hbar v}\right)^{10} \left(\frac{1}{v^2}\right) \{[(v_e - v)^2 + (v\Theta_0)^2]^{1/2} \qquad (2)$$

$$- |v_e - v|\} \pi a_0^2 = 4\pi^2 Z F_{BK}(v) \left(\frac{e^2}{\hbar}\right) \{[(v_e - v)^2 + (v\Theta_0)^2]^{1/2} - |v_e - v|\},$$

where

$$F_{BK}(v) = \frac{2^{16}}{10\pi^2} \frac{Z^2 Z_T^5}{\left(\frac{e^2}{\hbar}\right)^3} \left(\frac{e^2}{\hbar}\right)^{12}. \qquad (3)$$

In an attempted further refinement, their second-order Born estimate of F leads to

$$F_{BK} \rightarrow F(v) \simeq F_{BK} \{0.3 + O(v^{-1})\}. \qquad (4)$$

The now familiar simple analytic cusp in the cross section at $v_e = v$ described by Eq. (2) essentially arises from the singularity in the Coulomb wave normalization factor $e^{\eta\pi/2} \Gamma(1 + i\eta)$, where the continuum Coulomb wave function can be written

$$\Psi_{\vec{k}}(\vec{r}) = (2\pi)^{-3/2} e^{\eta\pi/2} \Gamma(1 + i\eta) e^{i\vec{k}\cdot\vec{r}} {}_1F_1[-i\eta, 1, -i(kr + \vec{k}\cdot\vec{r})], \qquad (5)$$

where

$$k = |\vec{k}|, \quad \vec{k} = \left(\frac{m}{\hbar}\right)(\vec{v}_e - \vec{v}), \text{ and } \eta \equiv Z/(a_0 k).$$

Dettmann et al. made the approximation $\eta \gg 1$, or

$$e^{\eta\pi} |\Gamma(1 + i\eta)|^2 = 2\pi\eta/ (1 - e^{-2\pi\eta}) \simeq 2\pi\eta \equiv f(k). \qquad (6)$$

The form derived for the differential cross section is

$$\frac{d\sigma}{d\vec{v}_e} = f(k) F(v). \qquad (7)$$

If the doubly differential cross section in this form is integrated over a cone of half angle $\Theta_0$ centered on the forward direction, the simple result given in Eq. (2) is derived.

Here the assertion is that $f = f(k)$, *not* $f(\vec{k})$ - i.e., isotropy in the projectile rest frame is predicted. Put another way, as in the BK approximation to bound-state capture, capture to s states is predicted to dominate at high velocities. *The asymmetry of all ECC cusps measured in the work of our laboratory conflicts with this view.*

The theory of Dettmann et al. applies when the projectile velocity v is much greater than $v_T$, the initial velocity of the electron to be captured in the target (i.e., the approximation $v_T \simeq 0$ is made). *Our data exhibit strong target shell effects if the condition $v \gg v_T$ is not well satisfied.*

The pioneering experimental work on ECC was done by Crooks and Rudd[1] for fast protons on He. A process identified as ECC (CEC for charge exchange to the continuum, in their parlance) by Lucas and collaborators was discovered at nearly the same time by Harrison and Lucas[1] in experiments on fast protons emergent from C foils. Harrison and Lucas observed a cusp in the electron velocity distribution centered at $\vec{v}_e = \vec{v}$ for electrons ejected in the forward direction. *We would instead characterize these $\vec{v}_e = \vec{v}$ electrons as convoy electrons, whose origin we believe cannot be described by a simple ECC (CEC) mechanism at or near the exit surface* (at least not for the fast, heavy multiply ionized projectiles in the range 6-18 au used in the work of our laboratory).

At the time of the earlier review[3], Lucas had also continued work with various collaborators[1,6,9,14,24,25] for various H, He projectiles, mostly in thin solid and occasionally in gas targets, at projectile energies up to 1.2 MeV/u. Extensive work with both solid and gas targets and H, He projectiles had also been undertaken by Menendez, Duncan and co-workers[26] and with solid targets by Meckbach et al.[6] The latter authors prepared a summary article[6] which emphasized several controversial disagreements between the various data and the theoretical predictions discussed so far for ECC as of 1977.

For reasons of space limitation we shall not review here some most interesting work by Duncan and Menendez[26] on ELC by H$^-$ in Ar except in passing, as the final state interaction is then often between an electron and a neutral residual particle, and different physical principles might be expected to apply. The 1977 and 1979 papers of Menendez and Duncan explore these questions in detail.[26]

Chiu, McGowan, and Mitchell[27] extended the work of Meckbach et al.[6] to gas targets and found support for an ECC description in gas targets, in opposition to the alleged deficiencies of such a description for solid targets, as found in the latter work. Soon thereafter Steckelmacher et al.[6] criticized the interpretation given the data of Meckbach et al. on instrumental resolution and background treatment grounds, and argued for the validity of the ECC (CEC) model for the solid target case as well. They also argued against the rival, solid-state "wake-riding" description of convoy electron production in solid targets.[7] A still unresolved dispute arose shortly thereafter[27], with Chiu et al. maintaining not only the legitimacy of a background subtraction method used but also maintaining the

correctness of the independence of the cusp width observed for C targets originally asserted in the work of Meckbach et al.[6] Steckelmacher et al.[6] argued that
to the contrary, a linear $\propto v\Theta_0$ dependence of cusp width is observed for C targets
if full widths are measured at the $\sim$ 70% maximum points and interpreted this result as support for the ECC (CEC) model of convoy electron production of Dettmann
et al.[9] Heavy ion data from our laboratory do not resolve this dispute, though
for heavy ions we do observe v-independence of cusp width under conditions of much
smaller "background" than present in the experiments of Ref. 6 (see section on
convoy production, below).

In the ion-atom collision case, for 0.7 MeV/u $d^+$ projectiles on Ar we confirmed,
apart from a measurable asymmetry, that the cusp shapes agreed approximately with
those predicted by Dettmann et al.[9], as had been earlier established by Cranage
and Lucas.[24] However, Cranage and Lucas had subtracted a large background attributed to direct ionization.

While at the time we were puzzled to find much smaller backgrounds than those of
Cranage and Lucas, we believe subsequent results suggest that their early results
may have suffered from spurious backgrounds of unexplained origin now reduced or
absent with their more recently developed electron spectrometric apparatus.[28]

The question of whether and how to subtract backgrounds underneath the cusps is
fundamental. A convincing argument was made by Rudd and Macek[16] and subsequently
by many others that no unambiguous subtraction of electrons due to direct, target-
centered ionization processes from those due to projectile-centered ECC processes
can be made. Essentially, one argues that electrons reaching particular energy-
momentum continuum states are physically indistinguishable, and the decision as
to which mechanism operates to produce a specific electron cannot be made even in
principle.

In addition to published yield measurements using $C^{6+}$ and $O^{8+}$ projectiles, we also
carried out as yet unpublished yield measurements for deuterons and bare $Si^{14+}$
ions in Ar at the same velocities. A simple $Z^{2.2\pm0.2}$ power law derived from interpreting our data for $C^{6+}$ and $O^{8+}$ projectiles traversing Ar at velocities $\sim$ 6 to
12 au is found to correctly predict both the d + Ar and $Si^{14+}$ + Ar cross sections
observed within error bars. A $Z^3$ dependence characterizing theories of Macek and
of Dettmann et al. (in a hydrogenic target approximation) is not found. The dependence of the cusp yields on projectile velocity was also established. Least-
squares fits to the experimental exponent in $v^{-n}$ yielded n = 4.55 ± 0.19 and n =
4.65 ± 0.14 for $C^{6+}$ and $O^{8+}$, respectively, where standard errors in the fits are
indicated, much weaker than the $v^{-10}$ dependence of Eq. (2). In these experiments,
asymptotically high velocities had not been reached. These exponents are in fact
very similar to those fitted to the corresponding bound-state electron capture
cross sections measured by Macdonald and Martin[29] for the same projectile-target
combination at overlapping velocities.

At the time of the earlier review[3], Rødbro and Andersen[23] had submitted a paper
concerning ECC for 0.015 - 1.5 MeV/u $H^+$ in He, in Ne, Ar, and $H_2$. They studied
doubly differential ECC cross sections $d^2\sigma/dEd\Omega$ as well as yields, using a parallel plate electrostatic analyzer with energy resolution $\Delta E/E$ of 1% FWHM and a
collection cone of half angle $\Theta_0 \sim 6.3 \times 10^{-3}$ rad. A sharp maximum was found in
the cross sections for v $\sim$ 1.4 $v_T$, where $v_T$ is the velocity of the target electron captured. In accord with our results, the decrease of the cross section
with increasing proton velocities resembles that for capture to bound states in
the same collision system. Shell effects were seen for Ne and Ar, a result again
in good accord with findings for our higher Z, higher v data.[4,11] Rødbro and Andersen extracted a reduced cross section $\sigma_c$ from the data and compared it to cross
sections for capture to excited states measured by other investigators. It was
derived by extrapolating ECC cross sections to bound state capture cross sections
into Rydberg states across the ionization limit according to $\sigma_n = n^{-3}\sigma_c$ in accordance with the recipe originally proposed by Rudd and Macek.[16] Because agreement

is remarkably good (though there appears to be a ∿ 25% analysis error we comment upon below), we have applied the same analysis to our ECC and convoy production data with the results noted in section IV.

Unfortunately, Rødbro and Andersen chose to quote cross section data after subtraction of a smooth background as in Figure 5, which is taken from their paper. We have argued that background subtraction, apart from spurious background signals reaching the detector from unrelated or secondary processes, is improper. Fortunately, for the most part the errors introduced are seen to be modest over the greater part of the energy range used, so that their quoted data would appear still to have considerable value.

Both our ECC cross section scale and that of Rødbro and Andersen have been rendered absolute by calibrating the apparatus with respect to absolute cross section measurements for Ar L-shell Auger electron production in the range 158-212 eV by Stolterfoht, et al.[30] From the work of Rudd[31] it is known that these cross sections are nearly isotropic (to within ∿ 10%). There is excellent consistency of the cross section scales used by Rødbro and Andersen and in our laboratory; in fact, the agreement is fortuitously good, if the same background subtraction method is applied. In quoting results from our laboratory, however, we will make no such background subtraction, as we believe it is unjustified. The absolute cross sections presented by Rødbro and Andersen were estimated to be correct to better than 25%, and the relative errors to be better than 10%. We are inclined to view these error bars as optimistic, for two reasons: (a) The so-called "background" subtracted in their analysis should probably be included in the cross section; and (b) the singly differential cross section $d\sigma/d\Omega$ is averaged over a finite solid angle $\Delta\Omega$. If the electron velocity distribution is sharply peaked then a slight pointing error such that the acceptance cone of the analyzer is not exactly centered at 0° introduces an error in $d\sigma/d\Omega$. This error is larger if the electron velocity distribution is asymmetric. The asymmetry of ECC cusps is now well established and arises in part from a discontinuity in the doubly differential cross section for ECC when viewed in the projectile rest frame because contributions from higher partial waves (for example, p waves) become important. The errors in our absolute cross sections are estimated at 40%. Error bars, indicating relative errors, represent one standard deviation errors in the range of repeated measurements. They are typically much smaller than 40%. Figures 6 and 7 exhibit yields from the paper of Rødbro and Andersen for protons traversing He, Ne, Ar, and He as a function of projectile velocity.

We now compare in detail our heavy ion ECC data on gases with the proton impact data of Rødbro and Andersen. Because they use a questionable background subtraction procedure, in effect setting the signal equal to zero in the cusp wings,

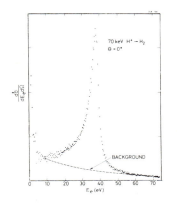

Figure 5. The doubly differential cross section for protons in $H_2$ from data as measured by Rødbro and Andersen. The fitted background under the forward peak is shown. When calculating the singly differential cross section, or forward peak yield, only the peak area above this background was evaluated.

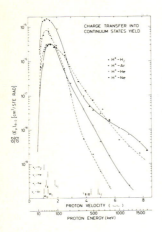

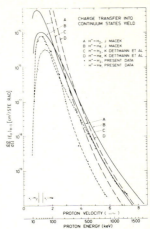

Figure 6.  Singly differential
cross sections (capture to continuum states yields) for He,
Ne, Ar, and $H_2$ targets versus
proton velocity.  From Ref. 23.

Figure 7.  Singly differential cross sections (capture to continuum states yields)
for He and $H_2$ targets versus proton velocity.  The experimental results are compared with calculations based on the
theories of Macek and Dettmann et al.
From Ref. 23.

Rødbro and Andersen avoided specifying the limits of integration on the cusp
region when extracting yields from their data.

From our data we have extracted absolute values for the differential cross section
$d\sigma/d\Omega$ for ECC by numerically integrating the cusp-shaped velocity distribution
over the velocity interval $v(1 \pm \alpha)$, where $\alpha$ is arbitrarily and conveniently set
to 0.04.  In effect, we proportionately adjust the limits of integration to match
the linear increase in $\Gamma = 3/2\ v\Theta_0$, according to the formula of Dettmann et al.[9]

We define the differential cross section for electron ejection into a small cone
of half angle $\Theta_0$ about the forward direction by

$$\left.\frac{d\sigma}{d\Omega}\right|_\alpha^{\Delta\Omega(\Theta_0)} \equiv \frac{1}{\Delta\Omega} \int_{\Delta\Omega(\Theta_0)} \int_{v(1\pm\alpha)} \frac{d^2\sigma}{d\Omega dv_e}\ dv_e\ d\Omega. \qquad (8)$$

This cross section obviously depends on the integration limits, i.e., on $\alpha$. If the
angular distribution of ECC electrons is cusp shaped about the forward direction
then $\left.\frac{d\sigma}{d\Omega}\right|_\alpha^{\Delta\Omega(\Theta_0)}$ also depends strongly on the acceptance angle of the electron
energy analyzer and increases as $\Theta_0$ is decreased.  When comparing results obtained
with different experimental setups, care must be taken that the integration limits
and the chosen solid angle $\Delta\Omega$ are noted.

As noted earlier, our uncertainties in the absolute values of $\left.\frac{d\sigma}{d\Omega}\right|_\alpha^{\Delta\Omega(\Theta_0)}$ are esti-
mated to be $\sim \pm40\%$, the major sources being the uncertainty in the absolute cross
section by which the instrument was calibrated ($\sim 20\%$), the uncertainty in the
background corrections for Ar L-auger yield measurements ($\sim 10\%$) and uncertainties
in the relative efficiency of the instrument.  In Table I we present our results
with and without the type of subtraction indicated by Rødbro and Andersen, together with results quoted from their paper.  A second comparable set of absolute

cross section data for ECC processes is that of Cranage and Lucas[24] for 300-1200 keV protons traversing $H_2$, He, Ne, and Ar targets. Their cusps were integrated over the constant velocity interval ($v \pm e^2/4\hbar$). Their acceptance angle $\Theta_0$ was 0.05 rad. Their results, without subtraction of large, slowly varying backgrounds, are also compared to ours in Table I. Differences of about $\pm 10\%$ are expected solely because of the different limits of integration used.

Following the discussion in the unpublished Ph.D. thesis (1979) of C. R. Vane, for 1.8-3.9 MeV/u $C^{6+}$, $O^{8+}$, and $Si^{14+}$ on He, Ne, and Ar, our ECC cross sections $\frac{d\sigma}{d\Omega}\Big|_{\alpha}$ $\frac{\Delta\Omega = 1.8 \text{ deg}}{\alpha = 0.04}$ are plotted vs. ion velocity in Figures 8, 9, and 10. For the same targets, curves for different projectile ions are nearly linear and parallel, depicting both a common velocity dependence and a velocity-independent $Z_1$ dependence.

For the $O^{8+}$-Ar and -He data the velocity dependences are found, through computer fitting of the data, to be well represented in this energy region by $v^{-4.3\pm0.3}$ and $v^{-8.4\pm0.6}$, respectively, but that for $O^{8+}$-Ne is less well represented by any simple power law. It is interesting to note that the approximate velocity dependences of the total cross sections for single electron capture for $O^{8+}$-He and -Ar are approximately the same as these. Macdonald and Martin[29] found total single capture dependences of $\sim v^{-3}$ for Ar and $v^{-8}$ for He for $O^{8+}$ velocities greater than $\sim 4.5$ au.

While the conditions for validity of the Born approximation are not well met for these many-electron targets, where inner-shell electrons may have orbital velocities in excess of the projectile velocity, a comparison with the predicted velocity dependence is revealing. The second Born theory of Dettmann et al.[9] yields a nearly target-independent, approximate $v^{-10}$ power law for total cross sections, while Macek's treatment with scaled hydrogenic target wave functions gives a $v^{-9.2}$ dependence for He targets.[23] The Ne and Ar target data obtained here have experimental velocity dependences significantly different from $v^{-10}$, but the He target velocity dependence is much nearer the theoretical value. This might be expected since the He electron velocity is less than one-sixth that of the 8.7 au projectiles, for example, and only 1s electrons can contribute, as is assumed in the development of the theories. However, $Ze^2/\hbar v$, the parameter usually considered important in determining the applicability of the Born approximation, is about one for the 8.7 au velocity projectile case, in clear violation of the assumption that it is much less than one.

TABLE I

ABSOLUTE CROSS SECTIONS FOR ECC FROM $H^+$ AND $D^+$ -- Ne AND Ar

| Projectile Ion | Energy (meV/u) | Velocity (au) | Target | $\left[\frac{d\sigma}{d\Omega}\right]_1$ | $\left[\frac{d\sigma}{d\Omega}\right]_2$ | $\left[\frac{d\sigma}{d\Omega}\right]_3$ | $\left[\frac{d\sigma}{d\Omega}\right]_4$ |
|---|---|---|---|---|---|---|---|
| | | | | ($\times 10^{-19}$ cm$^2$/Steradian) | | | |
| $D^+$ | 0.70 | 5.29 | Ar | 22.8 | 20.3 | 16.9 | 12.($H^+$) |
| $D^+$ | 0.70 | 5.29 | Ne | 25.2 | 25.0 | 18.0 | 16.($H^+$) |
| $H^+$ | 1.40 | 7.49 | Ar | 5.82 | 5.02 | 7. | |
| $H^+$ | 1.88 | 8.70 | Ar | 1.72 | 1.75 | | |
| $H^+$ | 2.00 | 8.94 | Ar | 1.3 | 1.3 | 3.1 | |

[1]Present measurement with yield integral and background subtraction identical to that of Reference 23.

[2]Source: Rodbro and Anderson, Reference 23 (1.88 and 2.0 MeV/u points extrapolated from 1.7 MeV/u)$\Theta_0$ = .36°

[3]Present measurement integrating over limits $v$ ($1 \pm \alpha$), $\alpha$ = 0.04; no background subtraction. $\Theta_0$ = 1.8°

[4]Source: Cranage and Lucas, Reference 24. No background subtraction.

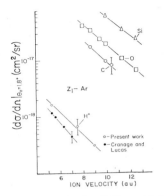

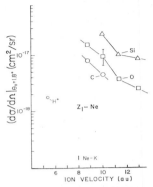

Figure 8.  ECC singly differential cross sections for $H^+$, $C^{6+}$, $O^{8+}$, and $Si^{14+}$ on Ar.  Velocity integration interval is $v(1-\alpha)$ to $v(1+\alpha)$, $\alpha = 0.04$.

Figure 9.  ECC singly differential cross sections for $H^+$, $C^{6+}$, $O^{8+}$, and $Si^{14+}$ on Ne.  Velocity integration interval is $v(1-\alpha)$ to $v(1+\alpha)$, $\alpha = 0.04$.

For the $O^{8+}$-Ne data displayed in Figure 9 there is an observable deviation from strict power law dependence in the region between 9 - 10 au.  It is assumed that this increased yield originates from Ne K-shell capture to the continuum and is phenomenon related to velocity matching of the projectile and target electrons, since the Ne K-shell orbital velocity is approximately 8 au, as labeled in Figure 9.  The increased yield arises because of a rapid rise in the low-energy side of the peak at ion velocities beyond $v = 8$ au.  It is also suspected that the slow velocity variation for Ar target data in the velocity region studied is in part due to the increased Ar L-shell contribution to continuum capture which shows an onset at about $v = 6$ au.[23]  The same onset or additional contribution is observed in $O^{8+}$-Ar *total* electron capture cross sections but at a slightly higher incident ion velocity.[29]

The Z dependence of the data displayed in Figures 8, 9, and 10 is graphically displayed in Figure 11, where it is seen that a simple power law satisfactorily

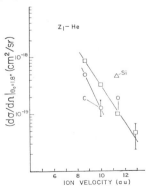

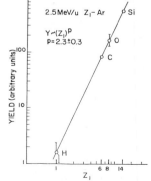

Figure 10.  ECC singly differential cross sections for $C^{6+}$, $O^{8+}$, and $Si^{14+}$ on He.  Velocity integration interval is $v(1-\alpha)$ to $v(1+\alpha)$, $\alpha = 0.04$.

Figure 11.  $Z_1$ dependence of singly differential cross section for 2.50 MeV/u $Z_1$-Ar.  $H^+$ point arrived at by extrapolation of data in Figure 8.

describes the Z-Ar data at 2.5 MeV/u for these projectiles. The point for Z = 1 on this graph is taken from an extrapolation of the $H^+$-Ar data in Figure 8 to 2.5 MeV/u (hence the relatively large uncertainty displayed in the figure for that data point). Representing the cross sections as proportional to $Z^p$, with p determined through a least-squares fitting of the data, gives p = 2.3 ± .3, a value not in accord with predicted $Z^3$ dependence given in most theories[9] (in a hydrogen-like target approximation). Variation of the limits of integration from $\alpha$ = 0.01 to 0.04 does not shift the value of p noticeably. Similarly, yields calculated for ECC peaks from $O^{8+}$ and $Si^{14+}$ traversing argon target gas, and from which direct ionization backgrounds have been subtracted, exhibit the same $\sim Z^{2.3}$ dependence.

As remarked earlier, the theories of Dettmann et al.[9] and of Macek[18] predict an approximately symmetric peak in the velocity distribution of ECC electrons emitted into a small cone about the forward direction.

About three years ago, a potentially important step forward was made by Shakeshaft and Spruch.[15] They suggested that the asymmetry of measured ECC cusps[2] might be the first experimental indication of the importance of including the second Born term in Born expansion charge transfer calculations, even though the impact velocities were appreciably below those required to assure dominance in total cross section. In the first Born approximation the shape of the cusp-like peak observed in the ECC velocity distribution is centered at $v_e$ = v and is symmetric about v, owing to a $\sim v^{-2\ell}$ dependence for ejected-electron partial waves. However, for the second Born term all partial waves are thought to have comparable importance. The first Born term depends only on the magnitude of the vector $(\vec{v}_e - \vec{v})$, implying an isotropic velocity distribution in the rest frame of the projectile characteristic of s-wave continuum states. Theoretically the asymmetry arises entirely from second Born terms, for which the differential cross section $d\sigma/dv$ is asymmetric under the transformation $(\vec{v}_e - \vec{v}) \rightarrow -(\vec{v}_e - \vec{v})$.

A counter-conjecture concerning the origin of the observed asymmetry is provided by Chan and Eichler[32], who note that retention of terms linear in $\Delta v_e/v = (|\vec{v}_e - \vec{v}|)/v$ beyond those incorporated in the first Born-Brinkman-Kramers (BK) approach originally used by Dettmann, Harrison, and Lucas produces a similar asymmetry. However, predictions of Refs. 15 and 32 concerning the projectile Z and v dependence of both shape and yield are very different, as is the predicted shape of the corresponding cusps.

The characteristic feature of the Shakeshaft and Spruch shape is the sheer drop on the high-velocity side of the peak. When convoluted with the instrument function, a drop is expected whose slope and width are essentially determined by the analyzer resolution function - a feature displayed by every ECC cusp we have ever observed for $C^{6+}$, $O^{8+}$, $Si^{14+}$, and $Ar^{18+}$ in He, Ne, and Ar at all velocities (5 - 18 au)! This property is not shared by the Chan and Eichler shape. Second, the CE shape is predicted to become symmetric at high v as $\sim 1/v$. There is no evidence in our data for any decline in asymmetry. On the contrary, for He the asymmetry (as defined below) is found to increase slightly in the range 15-18 au. For $C^{6+}$, and for $Si^{14+}$ on Ne, the asymmetry is an increasing function of v in the range 7 - 18 au, rising sharply as the beam velocity matches the neon K velocity. Third, the ECC asymptotic velocity dependence of $d\sigma/dv$ predicted by Dettmann, Harrison, and Lucas[9], when integrated over an appropriately scaled velocity region (e.g., $(1 - \alpha)v$ to $(1 + \alpha)v$ with $\alpha$ = 0.04) is $\sim v^{-10}$. This dependence coincides with our experimental results for $Ar^{18+}$ in He, which (over the range v = 15 - 18 au) scale as $\sim v^{-9.9}$, suggesting that the asymptotic velocity region has been reached (not so for either Ne of Ar targets, where the yields scale as $\sim v^{-7}$ over the same range). To test both conjectures, we measured ECC spectra for higher velocities (15 - 18 au) and for heavier bare projectiles ($Ar^{18+}$) than ever used heretofore in He, Ne, and Ar targets, anticipating shell effects in the latter two. Strictly, the theoretical results apply to hydrogenic targets and may need modification, even for He.

Typical data for $Ar^{18+}$ on He are displayed in Figure 12 overlaid with the best fits with the data obtained with a convoluted Shakeshaft-Spruch (SS) line shape (curve A), and a convoluted Chan-Eichler (CE) line shape (curve D). Data were acquired over a wider velocity range, but since detailed line-shape predictions apply only very near $v = v_e$, only the corresponding region is displayed. Convolution of parametrized theoretical line shapes (curves B and E) with the 1.7% FWHM measured apparatus function shown in curve H (from electron-gun calibration data) yielded the best fits (curves A and D). That the asymmetries observed are valid is demonstrated in curve G, which depicts the narrower, much more symmetric line shape (with Auger structure superposed) obtained for forward electron loss from 8.5 MeV/u $Ar^{13+}$ on He. Here only the incident beam was switched, with all other conditions unaltered.

Because the theoretical curves B and E apply to Z = 6 at 9 au in a hydrogenic target, our comparisons are only partially appropriate. However, these comparisons are sufficiently successful that a calculation for v = 15 - 20 au in He would be very worthwhile. Standard reduced $X^2$ tests, in addition to a deviation test to be described, exhibit a marked preference for the SS as opposed to the CE shapes. For the data of Figure 12 the fitted SS line shape yields $X^2 = 1.2 \pm 0.2$, whereas the CE shape yields $1.8 \pm 0.4$. At the same velocity, the analogous values are $X^2 = 6.5$ and 10 for Ne, and 8.9 and 10.8 for Ar. These values demonstrate the inappropriateness of a single-cusp fit to data we expect to be characterized by overlapping cusps of somewhat different width for each target shell. Curves C and F are derived from the deviation spectrum ($y_i - y_{fit}$ vs. i) corresponding to A and B. To extract trends from the large statistical scatter in the deviation spectrum, we have used a moving-average technique. Curves C and F represent ten-channel moving averages, smaller than but of the same order as the analyzer resolution (16 channels). The clear preference for the SS vs. the CE fit is exhibited by the large dip in curve F, which shows that the CE shape simply lacks the dropoff characterizing both the SS line shape and - by this test - the data. A moving average over fewer channels enhances the valley in the deviation spectrum (at the expense of scatter). In dozens of spectra acquired for He, Ne, and Ar, pronounced valleys were invariably observed for CE fits, and were not observed for SS fits. A direct experimental measure of the asymmetry is the ratio $(\Gamma_\ell - \Gamma_r)/(\Gamma_r + \Gamma_\ell)$, where $\Gamma_\ell$ and $\Gamma_r$ are the half widths (half maximum) of the cusp to the left and right of the peak. At 15 au, the measured values in He, Ne, and Ar are 0.28, 0.45, and 0.35, respectively. At 18 au, they are 0.35, 0.39, and 0.44, respectively. The range errors are $\pm 0.07$ in each case.

Though our shape and velocity-dependence data are much better in accord with Ref. 15 than with Ref. 32, other predictions of Ref. 15 are not observed. For example, the asymmetries observed in Ne and Ar are very similar for all bare projectiles for Z in the range 6 - 18, a finding not in accord with a predicted strongly Z-dependent asymmetry. Also, the yields scale at $\sim 1:200:500$ for 18 au $Ar^{18+}$ on He, Ne, and Ar, respectively, a dependence much weaker than a simple $Z_T^5$ dependence.[15] Here, the theoretical restriction to the case of asymptotic velocities in hydrogen may be limiting.

In order to provide some uniform comparative measure of cusp shape as a function of projectile velocity v and atomic number Z for various target gases, the half widths at half maximum for both the low-energy and the high-energy side of the cusp-shaped peaks were measured and corrected for analyzer-dependent errors. Results are displayed in Figures 13 and 14 for $O^{8+}$ and $Si^{14+}$, respectively, traversing He, Ne, and Ar. The half-widths of the high-energy side, $\Gamma_r$, are narrower than predicted by the analytic form of Dettmann et al.[9] and are observed to be independent of target and projectile Z, and nearly independent of v over the ranges measured. The small v dependence is fully consistent with the increase expected from the analyzer resolution alone.

The low-energy side half-widths, $\Gamma_\ell$, are wider for all cases than the corresponding $\Gamma_r$, in contradiction to predicted symmetric peak shapes in the theory of

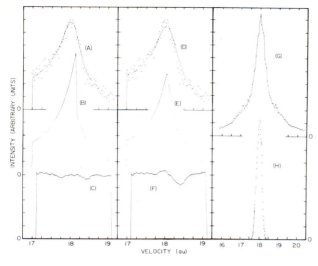

Figure 12. Comparison of the central portion of ECC cusps obtained for 18.1 au $Ar^{18+}$ on He with the overlaid convoluted line shapes read from Ref. 15 (A) and Ref. 32 (D). The respective best fit theoretical shapes (B) and (E), when convoluted with the measured apparatus function (curve H), produces fits (A) and (D). A narrower, more symmetric ELC spectrum for 8.5 MeV/u $Ar^{13+}$ on He is shown in (G). Curves (C) and (F) display ten-channel moving averages of the corresponding deviation spectra ($y_i - y_{fit}$ vs. i), as discussed in the text.

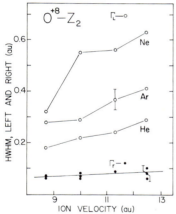

Figure 13. Half-widths at half-maxima on lower, $\Gamma_\ell$, and higher, $\Gamma_r$, energy sides of the forward ECC peaks for $Si^{14+}$-Ar and Ne.

Figure 14. Half-widths at half-maxima on lower, $\Gamma_\ell$, and higher, $\Gamma_r$, energy sides of the forward ECC peaks for $Si^{14+}$-Ar and Ne.

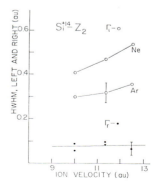

Dettmann et al.[9]  It is noted, however, that the He and Ar cusp half-widths, $\Gamma_\ell$,
do vary approximately linearly with ion velocity throughout the full range v =
8 - 12.5 au as predicted, while the half-width, $\Gamma_\ell$, for Ne target *does not*.  The
increase in $\Gamma_\ell$, from 0.29 au to 0.49 au in the interval v = 8.7 - 10 au has been
re-examined in greater detail with a 30° parallel plate electrostatic analyzer of
higher intrinsic energy and angular resolution than the spherical sector analyzer
employed for the bulk of this study.  The results are displayed in Figure 15,
where a sharp rise in $\Gamma_\ell$ is noted at v $\simeq$ 10 au.  The increase in $\Gamma_\ell$ occurs when
the projectile velocity nearly matches the Ne K-shell electron orbital velocity.
Similarly for 15 and 18.1 au $Ar^{18+}$ projectiles on He, Ne, and Ar targets the low-
energy side half-width $\Gamma_\ell$ increases approximately linearly with v for He and Ne
but rises more steeply for Ar targets, where for 18.1 au projectiles the ion velocity exceeds the Ar K-shell electron orbital velocity.  Therefore, ECC peak shapes
are found to be highly target-dependent even approaching $v_e$ = v for ion velocities
near the target electron orbital velocities.  However, comparing cusp shapes for
different projectiles but the same targets reveals little or no Z dependence in
the shape.

The step-function behavior of the ECC cusp indicated in Figure 12 can be given a
quite simple physical interpretation, as discussed recently by Lucas et al. in
their 1980 paper[6] on convoy electron production by 0.5 - 2.5 MeV protons in carbon
foils.  At such low velocities the role of higher partial waves present in higher-
order Born amplitudes is not an issue; indeed, for low to intermediate velocity
collisions, the presence of partial wave amplitudes $\ell > 0$ is not at all surprising.
One may express a general forward electron scattering amplitude in the projectile
rest frame as

$$f = f_s^0 + f_p^0 + f_d^0 \cdot \cdot \cdot, \tag{9}$$

where the superscripts remind us that at exactly $\Theta'$ = 0, where $\Theta'$ = the polar
angle of emission in this frame, only m = 0 states may be populated ($\vec{v} \times \vec{k}$ has no
corresponding component parallel to the beam for $\vec{k}$ along the forward direction).
Slightly away from the forward direction, other m values could be expected to
enter in.  Now odd-$\ell$ terms involve odd-order Legendre polynominals, i.e., odd
powers of cos $\Theta'$.  Taking $f_p$ as an example, cos $\Theta'$ changes from 1 to -1 as $\Theta'$
ranges from 0 to $\pi$.  Observation of electrons emitted into a small cone in the
laboratory system near $\Theta$ = 0 implies collection of electrons emitted near *both* $\Theta'$
= 0 and $\Theta'$ = $\pi$, since the relation between $\Theta$ and $\Theta'$ is double-valued.  Hence in
the forward direction, $f \simeq f_s + f_p \cdot \cdot \cdot \cdot$, and in the backward direction, $f \simeq$
$f_s - f_p \cdot \cdot \cdot \cdot$   Not only does one anticipate a step, but a measure of its size
directly indicates the relative importance of higher partial waves.

Still more recently, Macek et al.[33] have extended the discussion of Shakeshaft
and Spruch[15] to consider other evidence for second Born contributions to ECC by
positive ions, and provide explicit curves for the discontinuity of f at 0°, and
for gentler changes in f at 0.2° and 0.8°, calculated in second Born approximation
for $H^+$ - He collisions at 1.2 MeV.  Figure 16 displays their calculation.

For another type of experiment[26], Macek et al.[33] have closely analyzed measurements of the cusp peak position as a function of electron ejection angle.  They
predict that the peak position varies as $v_e$ = v cos $\Theta$ for the first Born theory,
but varies linearly with $\Theta$ as $\Theta \to 0$ if higher-order Born terms are important.
Their analysis[33] of earlier results[26] for 2 MeV $He^{++}$ on Ar provides strong backing
for the importance of higher Born terms in the scattering amplitude.

In the context of interpreting ECC asymmetries measured for 40 - 200 keV protons
in He, Meckbach, Nemirovsky, and Garibotti[34] have independently discussed both v-
dependent and higher partial wave contributions to asymmetries which they are able
to characterize in a general case, independent of a specific application.  The
general expression for $d\sigma/dv$ involves four generalized coefficients $B_0^{(0)}$, $B_1^{(0)}$,
$B_0^{(1)}$, and $B_1^{(1)}$,

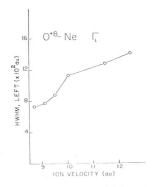

Figure 15. Half-widths at half-maxima on lower energy side of the forward $O^{8+}$-Ne ECC peaks.

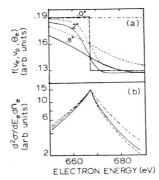

Figure 16. For the process $H^+ + He \rightarrow (H^+ + e) + He^+$ at incident energy 1.225 MeV. (a) Predictions for $f$ ($v_e$, $v_p$, $\Theta_e$) for second Born approximation (solid line), partial-wave expansion (dashed line), for laboratory scattering angles 0°, 0.2°, and 0.8°. The curves are normalized at $\Theta_e = 0$ and $E_e = 650$ eV. (b) The cross sections from first-order Faddeev-equation results are denoted by the dot-dashed line, the second Born results by the solid line, and measurement by the dashed line. The curves are normalized at the peak energy to measurements. See Ref. 33.

$$d\sigma/dv = (1/k)[B_0^{(0)}(v) + B_1^{(0)}(v)\cos \Theta' \ kB_0^{(0)}(v) + kB_1^{(1)}(v)\cos \Theta']. \qquad (10)$$

Their experimental ECC cusps are not only asymmetric, but the cusp shape varies considerably depending on the value of $\Theta_0$ chosen. Figure 17 displays a set of experimental ECC cusps observed for 191 keV p = He collisions, where the labels A through F correspond to $\Theta_0$ values of 2.5, 2.0, 1.5, 1.0, 0.5 and 0.17 deg, respec-

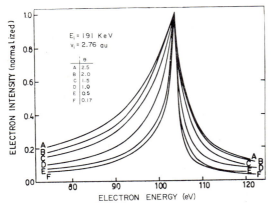

Figure 17. Set of experimental ECC cusps obtained at $E_j$ = 191 keV and for different acceptances contained A through F as described in the text. From Ref. 34.

tively. Best fits to the cusps, obtained using Eq.(10) broadened by finite apparatus energy and angular resolution, were obtained with the values of $B_0^{(0)}$, $B_0^{(1)}$ $B_1^{(0)}$, and $B_1^{(1)}$ qualitatively indicated in Figure 18 (the $U_j$'s are velocity-weighted apparatus-broadened averages over Legendre polynomials $P_j$).

From this discussion, it is evident that the simple Dett-mann formula, Eq. (2), which describes a symmetric cusp in both longitudinal and trans-verse k components, cannot begin to account for the longitudinal asymmetries observed, let alone the quali-tative changes in shape as

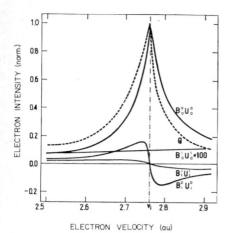

ELECTRON INTENSITY (norm.)

ELECTRON VELOCITY (au)

Figure 18.  Contributions of the
terms in Eq. (10) to the fitted
cusps.  Note that the cusp peak
is shifted to a velocity slightly
lower than v.

$\Theta_0$ is varied.  A clear task of the immediate future is detailed angular distribution measurements with milliradian resolution near the forward direction.  Perhaps
even interferences in d$\sigma$ with increasing angle caused by mixtures of higher partial waves will be revealed.  At sufficiently high asymptotic velocities, specific
partial wave contributions from particular higher-order Born terms may be isolated.

## III.  ELECTRON LOSS TO THE CONTINUUM (ELC)

When partially ionized projectiles undergo ion-atom collisions, it is now well
known[2,5,26] that a superficially similar peak in the velocity spectrum of electrons
emitted in the forward direction arises from projectile ionization, and that cross
sections for ELC dominate wherever loosely bound projectile electrons are available.

At the time of the 1978 review[3], Drepper and Briggs[5] had published a theory concerning electron loss in binary collisions using fast projectile ions as in He[+] on
He collisions in which the laboratory-frame velocity distribution of the forward
ejected electrons displays a sharp cusp in both energy and angle, centered near
$\vec{v}_e = \vec{v}$.  The theoretical cusp shape closely resembles that predicted in the theories of Dettmann, Harrison, and Lucas[9], Macek[18], and Salin[21] for the forward peak
in electron capture to continuum states (ECC) by bare ions for the very good reason that the same Coulomb factor (cf. Eq. (5)) appears in the differential cross
section for ELC as for ECC in their theory.  In particular, the FWHM is predicted
to scale as v$\Theta_0$.  As we shall see, our heavy ion data is in poor agreement with
this model.

Menendez et al.[26] performed the prototype charge-state-variation experiment, in
comparing electron spectra for 2 MeV He[+] and He[++] on Ar, finding no significant
differences in the cusp shapes.  Why there are strong differences between ECC and
ELC cusp shapes in our heavy ion data and only minor differences in the shapes observed by Menendez et al. is a mystery.  Menendez conjectures[26] that his observation angle of $\Theta = 2°$ (as opposed to 0 deg) may account for the similarities.

More puzzling still are two other studies of ELC vs. ECC shapes for He[+], He[++] on
He vs. H$_2$ collisions, one of which, by Lucas et al. (0.4 = 3.25 MeV, He target,
Ref. 35) confirms that the shapes are highly similar, and the second of which,
by Meckbach et al. (0.1 MeV, H$_2$ target, Ref. 36), yields a highly asymmetric ECC
shape for He[++] impact, but a narrower, more nearly symmetric shape for He[+] impact.

Figure 19 depicts a representative $He^+$ on $H_2$ spectrum acquired by Meckbach et al. Curiously, Lucas[37] reports finding that ELC cusps for $He^+$ on He are clearly skewed toward lower velocities, in the $He^+$ energy range noted above, but those observed using $H_2^+$ on $H_2$ are not. Figure 20 depicts representative spectra for $H_2^+$ on $H_2$ acquired by Lucas et al. They find the prediction[9] that $\Gamma = 3/2 \ v \ \Theta_0$ to be wrong. Their value[37] measured at $\Theta_0 = 0.05$ rad is $\Gamma \simeq 0.65 \pm .15$ au, and is independent of v in the range 0.15 - 1 MeV/u. Though the heavy ion data from our laboratory

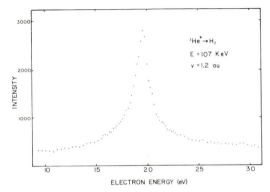

is not directly comparable to that for $He^+$, $He^{++}$ in He, $H_2$ or $H_2^+$ in $H_2$, Figure 2 does provide a most convincing demonstration that ECC and ELC line shapes are indeed very different, at least for the Z, v range in question. Moreover, $\Gamma$, the full-width half maximum of the ELC velocity distribution, has been shown to be surprisingly independent of v.[38]

Figure 19.  Spectrum of ELC plus ECC electrons for 107 keV $He^+$ ions in $H_2$ gas. From Ref. 34.

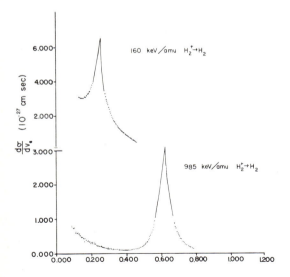

ELECTRON VELOCITY    (10 au)

Figure 20.  Differential cross section for electron ejection into a solid angle of $(0.1)^2$ steradian about the forward direction (M. Lucas and R. Cranage, private communication).

In Figure 21 are plotted $\Gamma$ and the ratio $\Gamma_L/\Gamma_R$ (a measure of the asymmetry) for cusps produced by 70 MeV $Si^{q+}$ (q = 6 - 14) traversing Ar and Ne. We note wide and asymmetric cusps when ECC is an important contributing process. For example, $\Gamma(Ar) = 0.46 \pm 0.02$ au and $\Gamma_L/\Gamma_R(Ar) = 2.3 \pm 0.2$ when q = 14. If only tightly bound 1s projectile electrons are carried into the collision (q = 13, 12), the still wide and asymmetric cusps indicate that a large fraction of the cusp electrons correspond to ECC. When more loosely bound 2s projectile electrons are present (q = 11, 10), the measured widths and asymmetries decrease, but only when the projectile ions carry some more loosely bound 2p electrons into the collision (q $\leq$ 9) do the cusp asymmetries disappear within experimental uncertainties. (All experimental uncertainties quoted are one standard deviation range errors from numerous independent measurements.) We conclude that for $Si^{q+}$ (q = 6 - 9) projectiles, the ELC cross section dominates the ECC section, and the measured cusps are dominated by ELC. For $O^{q+}$ (q = 4 - 8), Vane et al.[2] showed that the integrated cusp yield rises steeply when projectile L-shell electrons are present. Plots of

$\Gamma_L/\Gamma_R$ against projectile charge state ($0^{q+}$, q = 3 - 8) show that they become narrower and nearly symmetric whenever any projectile $2\ell$ electrons are present. Again we conclude that for $0^{q+}$ (q = 3 - 5) projectiles the measured cusps are dominated by ELC.

In Figure 22 the FWHM of representative ELC cusps for $Si^{9+}$ and $0^{5+}$ projectiles on Ne and Ar targets is plotted as a function of the ion velocity. For these projectiles, within experimental uncertainties, the FWHM of the ELC cusps are independent of projectile velocity. For $Si^{q+}$ (q = 6 - 9), $0^{q+}$ (q = 3 - 5) and $C^{q+}$ (q = 2, 3), we have further found only a weak dependence of the FWHM on projectile charge state (mostly due to broadening due to autoionization lines in the wings), and on the target gas (He, Ne, Ar). We summarize these results by stating that for Si, 0, and C projectiles in the velocity range 7 - 12.5 au traversing He, Ne, and Ar targets, the FWHM of the ELC cusps is $\Gamma$ = 0.26 ± 0.04 au, independent of q, v, and projectile Z (but dependent on the $\theta_0$ sampled by the apparatus). Using the same experimental set-up, Laubert et al.[8] found that the FWHM of convoy electron cusps measured when Si and 0 projectiles traverse solid targets (C, Al, Ag and Au) is independent of projectile nuclear charge, projectile charge state, velocity and target over any similar ranges and is summarized by $\Gamma$ = 0.25 ± 0.02 au, suggesting a close relationship between ELC and convoy electron production processes. We will discuss convoy electron spectrum and yield further in Section IV.

Before we compare experimental ELC results with theoretical predictions by Briggs and Drepper and by Day[5], we note that these treatments are all approximate evaluations of the full first Born approximation of Drepper and Briggs.[5] Their conclusions should be compared with data in the region $e^2/\hbar v$ << 1. The data represented here only marginally satisfy this criterion.

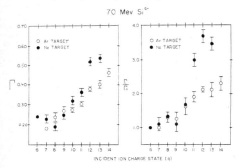

Figure 21. (a) Full width at half maximum, $\Gamma$, and (b) the ratio of left width to right width at half maximum, $\Gamma_L/\Gamma_R$, of the forward electron peak for 70 MeV $Si^{q+}$ (q = 6 - 14) traversing Ar and Ne as a function of incident ion charge q. When the ECC cross section dominates the ELC cross section (q ≥ 1), the Ne cusps become wider and more asymmetric than the Ar cusps, since for Ne targets the beam velocity exceeds the K-shell electron velocity (velocity matching criterion). Data points without error bars represent only one measurement.

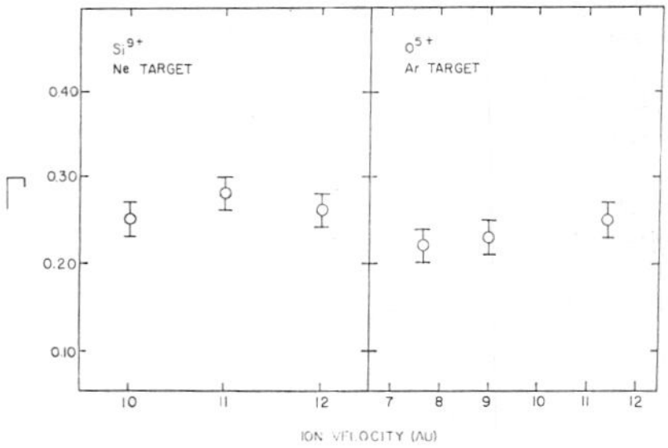

Figure 22. Full width at half maximum, $\Gamma$, of ELC cusps as a function of ion velocity $v_i$ for $Si^{9+}$ traversing Ne, and $0^{5+}$ traversing Ar. For $Si^{q+}$ (q = 6 - 9), $0^{q+}$ (q = 3 - 5) and $C^{q+}$ (q = 2 - 3) on He, Ne and Ar targets, $\Gamma$ is found to be independent of v within experimental error.

Briggs and Drepper derive an expression for the doubly differential ELC cross section for a one-electron projectile on a neutral N-electron target. Since electrons ejected into a narrow cone in the forward direction are those with projectile frame speeds $k \sim 0$, the major k dependence in the cross section is predicted to arise from the normalization factor in a Coulomb wave centered on the projectile. Similarly, the doubly differential ECC cross section for small k is predicted to be primarily determined by the projectile Coulomb potential, and not by the target.[9,18] Therefore, Briggs and Drepper derived an analytic expression describing the cusp shape of the ELC velocity distribution similar to that derived by Dettmann et al.[9] for ECC electrons. The FWHM of both the ELC and the ECC cusps is predicted to increase linearly with $v(\Gamma = 3/2 \ v\Theta_0)$. While Vane et al.[4] found that the FWHM of ECC cusps was roughly proportional to v (but note the asymmetry!), we find that the FWHM of the symmetric ELC cusps is independent of v under the same experimental conditions. On the other hand, the criterion $e^2/\hbar v \ll 1$ also applies to the theory of Dettmann et al., a condition only marginally satisfied by data we discuss.

Day[5] derives a form of the doubly differential ELC cross section in the plane wave Born approximation, including the possibility of variations of the cross section with respect to projectile frame polar angle. The FWHM of the cusp is then given by $\Gamma = 3/2 \ \Theta_0 v \ (1 + 3/2 \ \beta + \ldots)$, $-1 \leq \beta \leq 2$, where the anisotropy parameter $\beta$ is, in general, a function of v. The parameter $\beta$ allows $\Gamma(v)$ to depart from linear behavior. However, since $\Gamma$ is measured to be independent of projectile velocity, projectile nuclear charge Z, projectile charge state q and target in the velocity range 7 - 12.5 au, severe restrictions are placed on the velocity and $Z_1$ dependences of the asymmetry parameter $\beta$.

Beyond questions of ELC shape, there are questions of ELC yield, and the v dependence thereof. Briggs and Drepper[5] predicted that within the first Born approximation, the cross section for ejection of electrons into a small cone centered on the forward direction tends to a constant value at asymptotically high velocities, and quoted unpublished results of Strong and Lucas which appear to tend toward v independence for impact energies $\gtrsim$ 3 MeV. A large peak observed in the yield at $\sim$ 1 MeV impact energy was attributed to ECC. Absolute ELC cross sections[37] as measured by Lucas and collaborators for $H_2^+$ on $H_2$ using a solid angle of 0.01 sr about the forward direction, and integrating over background subtracted ELC cusps, are displayed in Figure 23. For some reason, no conspicuous peak corresponding to a large ECC contribution is seen as a function of v.

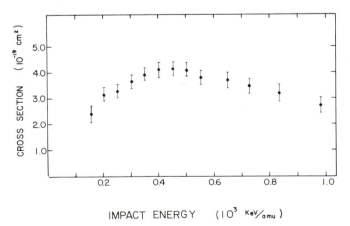

Figure 23. Cross section for electron loss in $H_2^+ \rightarrow H_2$ collisions into $(0.1)^2$ steradian about the forward direction and integrated over electron velocities 1/4 au either side of the peaks. The error bars do not include systematic errors. (M. Lucas and R. Cranage, private communication.)

CROSS SECTION ($10^{-19}$ cm$^2$)

IMPACT ENERGY ($10^3$ $^{KeV}/_{amu}$)

In our laboratory we have measured absolute ELC cross sections for Si, O, and C projectiles traversing Ar, Ne, and He targets. We have measured singles spectra when the projectiles carried L-shell electrons into the collision, and ELC spectra in coincidence with scattered ion charge state for electron loss from L- and K-shells. The ELC cross sections were determined by integrating the cusps over the velocity interval v ± 0.5 au. We have established that the ELC cross sections associated with single, double and triple electron loss, and the total ELC cross sections independent of how many additional electrons were lost, follow the same trends with energy and incident charge state as the electron loss cross sections measured by Macdonald and Martin.[29] Figure 24 compares the total ELC cross sections for $O^{9+}$ projectiles on Ar targets with the total electron loss cross sections for the same projectiles and targets from Ref. 29.

The symmetric maxima and minima appearing in the wings of ELC cusps originally misled us into seeking a description of the observed structure in terms of interference, which as noted above has been predicted but not found for ECC, and has not been predicted in ELC. Much of the interpretive discussion of this data in Refs. 3 and 39 is therefore obsolete. The real origin of the observed structures was revealed in new measurements at higher resolution, and was published very shortly thereafter, in Ref. 40. Since no single electron excitation can lead to autoionization lines in the energy ranges in question, autoionization was initially considered improbable as a source of the observed structures. Nevertheless, double-electron excitations proved to be responsible. Higher projectile energy and improved energy resolution showed that the structures found earlier form a Rydberg series of intense lines. These lines can be assigned to doubly excited autoionizing states of the projectile instead of the two-electron continuum states conjectured previously. We have collected spectra for $C^{2+}$ and $O^{4+}$ projectiles at energies between 0.5 and 1.56 MeV/u using Ar as the target gas. In all spectra at least three resolved lines have been observed in each wing of the cusp. A representative spectrum is shown in Figure 25. A series of lines is visible in both wings of this cusp. These lines originate from $(1s^2 2p\, n\ell)$ projectile autoionizing states.

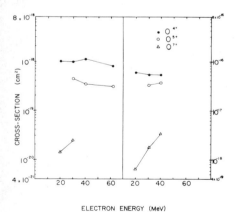

ELECTRON ENERGY (MeV)

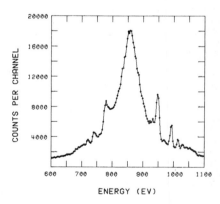

ENERGY (EV)

Figure 24. a) Total ELC cross sections for $O^{9+}$ projectiles on Ar targets. The integration limits are v ± 0.5 au and $\Theta_0 = 1.8°$. b) Total electron loss cross sections measured by Macdonald and Martin.[29]

Figure 25. Energy spectrum of electrons emitted in the forward direction in collisions of 25 MeV $O^{4+}$ on Ar. The spectrum shows a cusp-shaped peak produced mainly by projectile electron loss.

In the projectile rest frame the energies of these lines correspond to the following series of transitions: $(1s^2 2p \, n\ell) \rightarrow (1s^2 2s) + e^-(k, \ell')$. For doubly excited $C^{2+}$, projectile rest frame energies of $2.7 \pm 0.2$, $4.2 \pm 0.3$, $5.2 \pm 0.3$ eV, . . . were measured, corresponding to n = 5, 6, 7, . . . . . For $O^{4+}$, energies of $2.1 \pm 0.2$, $4.7 \pm 0.3$, $6.3 \pm 0.3$ eV, . . . were measured (n = 6, 7, 8, . . . .). The energies of the ejected electrons are accurately predicted by the simple formula $E = E_\infty - (13.6 \, q^2)/(n - \mu)^2$, where E is the energy in eV of the series limit, q is the charge of the residual ion, n is the principal quantum number of an electron in a Rydberg autoionizing state, and $\mu$ is the quantum defect. The values for E can be taken from atomic energy level tables[41], and are 8.00 eV for carbon and 11.98 eV for oxygen. The experimental E values can be made to agree with calculated values within error bars, for quantum defects $\mu = 0.2 \pm 0.1$ for carbon and $0.12 \pm 0.07$ for oxygen. Transitions from states with values of n < 5 for carbon and n < 6 for oxygen are energetically forbidden.

Assuming an isotropic angular distribution of the electrons in the projectile rest frame and estimating the transmission function of the analyzer, a crude estimate of cross sections for the production of autoionizing states was made. For 25 MeV $O^{4+}$ on Ar the total cross section corresponding to the state corresponding to the most intense line (n = 6) is $\sim 1 - 2 \times 10^{-19}$ cm$^2$. Summing over all lines up to the series limit, one gets a cross section which is about four times larger. A comparison with total projectile electron loss sections measured by Macdonald and Martin[29] shows that the fraction of electrons emitted in the autoionization channels discussed here is $\sim 1\%$ of the total electrons lost.

A comparison with the Si data obtained in earlier, lower resolution measurements strongly suggests that the structure there originates from the same kind of autoionization states. The lowest distinct structure corresponds to an energy in the projectile frame of about 4.3 eV. This energy was assigned to transitions from states with n = 9. In order to explain the similar structures observed with $Si^{11+}$ (and $Si^{12+}$), simultaneous single capture with excitation or double capture into excited states has to be considered. In fact, in all the collision processes discussed here, two electron transition processes are required to account for production of the observed states. The one- and two-electron capture processes apparently have high cross sections at the lower beam energies (less than 2.5 MeV/u), where strong structures were found for $Si^{11+}$ and $Si^{12+}$ projectiles, but have much smaller cross sections at energies above 3.5 MeV/u, where they were not discernible.

Thus, the analysis of electrons emitted into the forward direction permits the study of doubly excited, high Rydberg autoionizing states where low Auger energies (2 to 20 eV) can be very conveniently detected. For Be-like ions, there is a high probability for simultaneous excitation of both electrons to bound states (one high lying). For silicon, double electron capture events to excited states of the projectile, and events in which there is simultaneous capture and excitation involving two electrons were found to have high cross sections at energies below 2.5 MeV/amu. The single and double electron capture cross sections have comparable magnitude. A similar interpretation of our data has been given independently by Gersten and Cohen.[42] Earlier, Lucas and Harrison[42] had discussed similar autoionization electron production from 271 keV N ions traversing C foils.

Space limitations imposed by the editor and publisher regrettably make inclusion of
a detailed discussion of forward electron production in ion-solid collisions (convoy electron production) beyond that in the abstract and Section I impossible here.
The remaining two manuscript sections, IV. Forward Electron Ejection in Ion-Solid
Collisions: Convoy Electron Production, and V. Capture and Loss to Rydberg and
Super-Rydberg States in Gases and Solids, are available from the author. While the
work of a number of other investigators[43-55] in this fascinating field can thus
not be adequately represented here, a thorough discussion of their work is presented in Sections IV and V.

## ACKNOWLEDGEMENTS

A number of colleagues cited in the references to the work of our laboratory have
contributed greatly to the carrying out of the experiments described, aided by
generous help from staff at Oak Ridge National Laboratory, Lawrence Berkeley Laboratory, and Brookhaven National Laboratory. In the work described here, the contributions of C. R. Vane, S. B. Elston and M. Breinig have been the most substantial and sustained over the longest period of time. Substantial contributions have
also been made by our ORNL colleagues G. Alton and M. O. Krause. Finally, we would
like to acknowledge the energetic participation of our late colleague, Roman Laubert.
The work on atomic collisions in solids is the final work we were able to carry out
in collaboration with him before his death; we will sorely miss him.

This work was partially supported by the National Science Foundation; the Office
of Naval Research; the Fundamental Interactions Branch, Division of Chemical
Sciences, Office of Basic Energy Sciences, U. S. Department of Energy, under contract W-7405-eng-26; and the East Carolina Research Council.

## REFERENCES

[1] Crooks, G. B. and Rudd, M. E., Phys. Rev. Lett. 25 (1970) 1599; and Harrison,
K. G. and Lucas, M. W., Phys. Lett. 33A (1970) 142 and 35A (1971) 402.

[2] Vane, C. R., Sellin, I. A., Suter, M., Alton, G., Elston, S. B., Griffin,
P. M., and Thoe, R. S., Phys. Rev. Lett. 40 (1978) 1020; Vane, C. R., IEEE
Transactions on Nuclear Science MS-26, No. 1 (1979) 1078.

[3] Sellin, I. A., Journal de Physique Colloque, Supp. No. 2, 40 (1978) C1-225.

[4] Vane, C. R., Sellin, I. A., Suter, M., Elston, S. B., Alton, G. D. and Thoe,
R. S., in: Takayanagi, K. and Oda, N. (eds.), Proceedings of the Eleventh
International Conference on the Physics of Electronic and Atomic Collisions
(Society for Atomic Collision Research, Kyoto, 1979) 752.

[5] Drepper, F. and Briggs, J. S., J. Phys. B 9 (1976) 2063; Briggs, J. S. and
Drepper, F., J. Phys. B 11 (1978) 4033; and Day, M. H., J. Phys. B 13 (1980)
665.

[6] Meckbach, W., Chiu, K. C. R., Brongersma, H. H. and McGowan, J. W., J. Phys.
B 10 (1977) 3255, and references therein; Steckelmacher, W., Strong, R.,
Khan, M. N. and Lucas, M. W., J. Phys. B 11 (1978) 2711; and Lucas, M. W.,
Steckelmacher, W., Macek, J. and Potter, J. E., J. Phys. B 13 (1980) 4833.

[7] Brandt, W. and Ritchie, R. H., Phys. Lett. 62A (1977) 374; Neelavathi, V. N.,
Ritchie, R. H. and Brandt, W., Phys. Rev. Lett. 33 (1974) 302; Meckbach, W.,
Arista, N. and Brandt, W., Phys. Lett. 65A (1978) 113; and Day, M. H., Phys.
Rev. Lett. 44 (1980) 752.

[8]   Laubert, R., Sellin, I. A., Vane, C. R., Suter, M.,Elston, S. B., Alton,
       G. D. and Thoe, R. S., Phys. Rev. Lett. 41 (1978) 712; and Laubert, R.,
       Sellin, I. A., Vane, C. R., Suter, M., Elston, S. B., Alton, G. D., and
       Thoe, R. S., Nucl. Inst. and Meth. 170 (1980) 557.

[9]   Dettmann, K., Harrison, K. G. and Lucas, M. W., J. Phys. B 7 (1974) 269.

[10]  Vane, C. R., Sellin, I. A., Elston, S. B., Suter, M., Thoe, R. S., Alton,
       G. D., Berry, S. and Glass, G., Phys. Rev. Lett. 43 (1979) 1388.

[11]  Breinig, M., Elston, S. B., Huldt, S., Liljeby, L., Vane, C. R., Berry, S.
       D., Schauer, M. and Sellin, I. A., submitted to Phys. Rev. A.

[12]  Laubert, R., Huldt, S., Breinig, M., Liljeby, L., Elston, S. B., Thoe, R.
       S., and Sellin, I. A., J. Phys. B 14 (1981) 859.

[13]  Elston, S. B., Sellin, I. A., Breinig, M., Huldt, S., Liljeby, L., Thoe,
       R. S., Datz, S., Overbury, S. and Laubert, R., Phys. Rev. Lett. 46 (1981)
       321.

[14]  Dettmann, K., Khan, M. N. and Lucas, M., J. Phys. C 9 (1976) 1879.

[15]  Shakeshaft, R. and Spruch, L., Rev. Mod. Phys. 51 (1979) 369; and Phys. Rev.
       Lett. 41 (1978) 1037.

[16]  Rudd, M. E. and Macek, J., Case Studies in Atomic Physics 3 (1972) 125.

[17]  Shakeshaft, R., Phys. Rev. A 18 (1978) 1930.

[18]  Macek, J., Phys. Rev. A 1 (1970) 235.

[19]  Oldham, W. J. B., Jr., Phys. Rev. A 140 (1965) 1477, and 161 (1967) 1.

[20]  Duncan, M. M., Menendez, M. G., Eisele, F. L., and Macek, J., Phys. Rev. A
       15 (1977) 1785; and Lucas, M. W., in: Gemmell, D. S. (ed.), Proceedings of the
       Workshop on Physics with Fast Molecular-Ion Beams (Argonne National Labora-
       tory, August 1979) 291.

[21]  Salin, A., J. Phys. B 2 (1969) 631 and 1255; and J. Phys. B 5 (1972) 979.

[22]  Ponce, V. H. and Meckbach, W., to be published in Comments on Atomic and
       Molecular Physics.

[23]  Rødbro, M. and Andersen, F. D., J. Phys. B 12 (1979) 2883.

[24]  Cranage, R. W. and Lucas, M. W., J. Phys. B 9 (1976) 445.

[25]  Strong, R. and Lucas, M. W., Phys. Rev. Lett. 39 (1977) 1349.

[26]  Menendez, M. G. and Duncan, M. M., in:  Sellin, I. A. and Pegg, D. J. (eds.),
       Beam-Foil Spectroscopy (Plenum, New York, 1976), Vol. 2 p. 623; Menendez, M.
       G. and Duncan, M. M., Phys. Lett. 54A (1975) 409, Phys. Rev. A13 (1976) 566,
       Phys. Lett. 56A (1976) 177, and Phys. Rev. Lett. 40 (1978) 1642; Menendez,
       M. G., Duncan, M. M., Eisele, F. L. and Junker, B. R., Phys. Rev. A 15 (1977)
       80; Duncan, M. M. and Menendez, M. G., Phys. Rev. A 16 (1977) 1799, and 20
       (1979) 2327; and Menendez, M. G., private communication and to be published.

[27]  Chiu, K. C. R., McGowan, J. W. and Mitchell, L. B. A., J. Phys. B 11 (1978)
      L117; Chiu, K. C. R., Meckbach, W., Sanchez Sarmiento, G. and McGowan, J. W.,
      J. Phys. B 12 (1979) L147; and Steckelmacher, W. and Lucas, M. W., J. Phys.
      B 12 (1979) L152.

[28]  Steckelmacher, W. and Lucas, M. W., J. Phys. E 12 (1979) 961.

[29]  Macdonald, J. R. and Martin, F. W., Phys. Rev. A 4 (1971) 1965.

[30]  Stolterfoht, N., Schneider, D. and Ziem, P., Phys. Rev. A 10 (1974) 81.

[31]  Rudd, M. E., Phys. Rev. A 10 (1974) 518.

[32]  Chan, F. T. and Eichler, J., Phys. Rev. A 20 (1979) 367.

[33]  Macek, J., Potter, J. E., Duncan, M. M., Menendez, M. G., Lucas, M. W. and
      Steckelmacher, W., Phys. Rev. Lett. 46 (1981) 1571.

[34]  Meckbach, W., Nemirovsky, I. B. and Garibotti, C. R., private communication
      and to be published.

[35]  Lucas, M. W. and Strong, R., Bull. Am. Phys. Soc. 23 (1978) 1087.

[36]  Meckbach, W., private communication and to be published.

[37]  Lucas, M. W., private communication and to be published.

[38]  Breinig, M., Schauer, M., Sellin, I. A., Elston, S. B., Vane, C. R., Thoe,
      R. S. and Suter, M., J. Phys. B 14 (1981) L291.

[39]  Suter, M., Vane, C. R., Sellin, I. A., Elston, S. B., Alton, G. D., Thoe,
      R. S. and Laubert, R., Phys. Rev. Lett. 41 (1978) 399.

[40]  Suter, M., Vane, C. R., Elston, S. B., Alton, G. D., Griffin, P. M., Thoe,
      R. S., Williams, L., Sellin, I. A., and Laubert, R., Z. Phys. A 289 (1979)
      433.

[41]  Bashkin, S. and Stoner, J. O., Jr., Atomic Energy Levels and Grotian Diagrams
      (North Holland/American Elsevier, Amsterdam and New York, 1975).

[42]  Gersten, J. and Cohen, M., Phys. Rev. A 21 (1980) 1354; Lucas, M. W. and
      Harrison, K. G., J. Phys. B 5 (1972) L20; and Strong, R. and Lucas, M. W.,
      Phys. Rev. Lett. 39 (1977) 1349.

[43]  Datz, S., Nucl. Inst. and Meth. 132 (1976) 16; and Datz, S., Appleton, B. K.,
      Biggerstaff, J. A., Noggle, T. S. and Verbeek, H., Book of Abstracts, VIth
      International Conference on Atomic Collisions in Solids (Amsterdam, 1975) 162.

[44]  Dettmann, K., Khan, M. N. and Lucas, M. W., J. Phys. C 9 (1976) 1879; Khan,
      M. N. and Lucas, M. W., Book of Abstracts, VIth International Conference on
      Atomic Collisions in Solids (Amsterdam, 1975) 160; and Khan, M. N. and Lucas,
      M. W., Phys. Rev. B 19 (1979) 5578.

[45]  Ashley, J. C., Tong, C. J. and Ritchie, R. H., Surface Science 81 (1979) 409,
      and references therein.

[46] Hyajesh, A. R., Steckelmacher, W. and Lucas, M. W., J. Phys. C 11 (1978) 2917; Meggitt, B. T., Harrison, K. G. and Lucas, M. W., J. Phys. B 6 (1973) L362; and Lucas, M. W., private communication and to be published.

[47] Ponce, V. H., Gonzalez-Lepera, E., Meckbach, W. and Nemirovsky, I. B., private communication and submitted for publication.

[48] Latz, R., Astner, G., Frischkorn, H. J., Koschar, P., Pfennig, P., Schader, J. and Groeneveld, K.-O., presented at the Ninth International Conference on Atomic Collisions in Solids, Lyon, France, July, 1981, and to be published.

[49] Yamazaki, Y. and Oda, N., presented at the Ninth International Conference on Atomic Collisions in Solids, Lyon, France, July 1981, and to be published.

[50] Gladieux, A. and Chateau-Thierry, A., presented at the Ninth International Conference on Atomic Collisions in Solids, Lyon, France, July 1981, and to be published.

[51] Datz, S., Martin, F. W., Moak, C. D., Appleton, B. R., and Bridwell, L. B., in: Andersen, S. (ed), Atomic Collisions in Solids IV (Gordon & Breach Science Publishers, 1972) 87.

[52] Brown, M. D., Ellsworth, L. D., Guffey, J. A., Chiao, T., Pettus, E. W., Winters, L. M. and Macdonald, J. R., Phys. Rev. A 10 (1974) 1255.

[53] Brandt, W., Laubert, R., Mourino, M. and Schwarzschild, A., Phys. Lett. 30 (1973) 358.

[54] Guffey, P., Ellsworth, L. and Macdonald, J. R., Phys. Rev. A 15 (1977) 1863.

[55] Sellin, I. A., Breinig, M., Brandt, W. and Laubert, R., to be published in Nucl. Inst. and Meth. as part of the Proceedings of the Ninth International Conference on Atomic Collisions in Solids, Lyon, France, July 1981; Sellin, I. A., Elston, S. B., Breinig, M. and Huldt, S., to be published in Nucl. Inst. and Meth. as part of the Proceedings of the Sixth International Conference on Fast Ion Beam Spectroscopy, Laval University, Quebec, P.Q., August 1981; and Breinig, M., Elston, S. B., Huldt, S., Liljeby, L., Vane, C. R., Berry, S. D., Glass, G. A., Schauer, M., Sellin, I. A., Alton, G. D., Datz, S., Overbury, S., Laubert, R., and Suter, M., submitted for publication to Phys. Rev. A.

PHYSICS OF ELECTRONIC AND ATOMIC COLLISIONS
S. Datz (editor)
© North-Holland Publishing Company, 1982

COLLISIONS BETWEEN POSITIVE IONS

H.B. Gilbody

Department of Pure and Applied Physics
The Queen's University of Belfast
Belfast, United Kingdom

A report is given of recent experimental studies of charge
transfer and ionization in collisions between positive ions.
The results are considered in relation to current theoretical
descriptions.

INTRODUCTION

It is now two decades since methods based on the use of fast intersecting beams
were successfully developed for studies of collisions between charged particles.
In the first attempt to study collisions between positive ions, Guidini and his
collaborators (1965) obtained data of limited accuracy for $H_2^+$-$N_2^+$ collisions at keV
energies.  In subsequent work Brouillard and Delfosse (1967) described a merged
beam method for the study of $He^{2+}$-$He^+$ charge transfer collisions at low energies.
However, since 1977 reliable experimental data on charge transfer and ionization
for a number of different collision processes have become available.  Apart from
my own group in Belfast, groups in Newcastle (Dolder), Culham (Harrison) and
Louvain-la-Neuve (Brouillard) are now active in this field.

Much of the current interest in charge transfer and ionization in collisions
between positive ions stems from their relevance to controlled thermonuclear
fusion research (c.f. Gilbody 1979).  For example, in a Tokamak device, particle
loss and plasma cooling can occur through charge transfer processes of the type

$$H^+ + X^{q+} \rightarrow H + X^{(q+1)+}$$ (1)

involving collisions between hydrogen isotope fuel ions and impurity species.
Fast H atoms formed in this way escape from the magnetic confinement while the
charge state of X is increased leading to increased energy loss through electron-
ion recombination, line radiation and free-free bremsstrahlung.  Similar
processes involving $He^{2+}$ ions are also relevant to the effectiveness of $\alpha$ particle
heating.

There is also considerable interest in schemes to promote fusion by the heating
and inertial confinement of DT pellets by intense pulsed beams of GeV energy
heavy ions (c.f. Godlove 1979).  The design of suitable beam drivers may be
greatly influenced by ion-ion collisions within the beam.  These occur at
energies $\sim 0.5$ keV $amu^{-1}$ (Kim 1976) due to the lateral and transverse motion of
the beam particles.  If ions which change their charge as a result of charge
transfer

$$X^+ + X^+ \rightarrow X + X^{2+}$$ (2)

or ionization

$$X^+ + X^+ \rightarrow X^+ + X^{2+} + e$$ (3)

are lost from the beam, this may result in not only an unacceptable reduction in
beam intensity but in serious wall-effect secondary processes within the
accelerator/storage ring.

In this short report the experimental data obtained over the past few years will

be discussed in relation to current theoretical descriptions of charge transfer
and ionization.

EXPERIMENTAL APPROACH

The general principles underlying intersecting beam methods have been described
elsewhere (c.f. Dolder 1969) and only a short summary of the experimental
approach need be given here.

For the processes of the type

$$X^+ + Y^+ \rightarrow X + Y^{2+} \tag{4a}$$

$$\rightarrow X^+ + Y^{2+} + e \tag{4b}$$

most measurements so far have provided only cross sections $\sigma(Y^{2+})$ for the
production of $Y^{2+}$ ions from the combined processes of charge transfer and
ionization.   The main experimental problems arise from the low densities of the
primary ion beams and the comparatively small signal count rates in relation to
the background signals arising from the interaction of both beams with the
residual gas.   Efficient separation of the collision products formed in the beam
intersection region from the primary beam component (which may be up to $10^{12}$ times
larger) requires very carefully designed electrostatic or magnetic analyzers.

The angle at which the two beams intersect may be chosen to provide centre of mass
collision energies spanning a wide range.   In the measurements in Belfast which
cover c.m. energies within the range 38-400 keV, the beams are arranged to
intersect at $90^0$ while the Newcastle and Culham groups have used beams inclined at
angles of a few degrees to provide lower collision energies.   At Louvain-la-Neuve
a merged beam apparatus has provided data on the $He^{2+}$-$He^+$ system for interaction
energies down to 10 eV.

Fig 1a shows the essential features of the experimental arrangement used in our
measurements in Belfast.   The apparatus first described by Mitchell et al (1977)
has since undergone a number of improvements which have been described in our
subsequent papers.   A fast projectile beam $X^+$ of energy variable within the
range 60-600 keV is arranged to intersect at $90^0$ a $Y^+$ target beam of energy which
can be set within the range 4-15 keV.   Products of charge changing collisions
formed in each beam during transit through the background gas prior to the intersection region are effectively removed by the beam deflector plates $S_1$ and $S_2$.
The intersection region is screened from stray  particles and fields by plates
$E_1$-$E_4$.   The target beam defining arrangement T is designed to trap back-scattered
particles.   The base pressure in the intersection region is normally about
$4 \times 10^{-10}$ torr.

The $Y^{2+}$ collision products are separated from the $Y^+$ primary component of the beam
by the $45^0$ electrostatic analyzer $A_1$ followed by a second $90^0$ analyzer $A_2$
terminated by the Johnston multiplier $M_2$.   The transmission efficiency of the
combined analyzers is near 100% while the angular acceptance (up to $3.3^0$) of $M_2$ is
large enough to accommodate most scattered particles.   The $Y^+$ beam is measured as
a current by the Faraday cup $F_2$ while a similar cup $F_3$ is used for transmission
checks and for multiplier calibration.   Electrostatic deflection of the
projectile beam at $A_3$ allows the fast atom collision products X to be counted by
the Johnston multiplier $M_1$.

In the measurements of $\sigma(Y^{2+})$, both beams are modulated (by electrostatic
deflection) in a carefully programmed sequence of pulses synchronised with
periodic gating of two scalers receiving pulses from $M_2$ in the arrangement shown
in Fig 1b.   The difference in the count rate of the two scalers can then be
identified with the true $Y^{2+}$ signal and $\sigma(Y^{2+})$ is then obtained from the
expression

$$\sigma(Y^{2+}) = Se^2V_1V_2F/I_1I_2(V_1{}^2 + V_2{}^2)^{\frac{1}{2}} \tag{5}$$

where S is the $Y^{2+}$ signal count rate after allowing for the measured efficiency of detection, $I_1$, $I_2$, $V_1$ and $V_2$ are the currents and velocities of the projectile

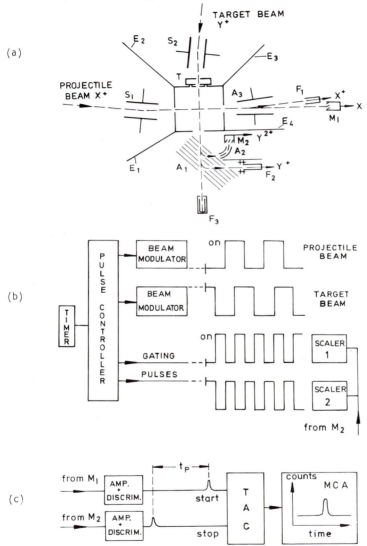

Fig 1. Schematic diagram of experimental arrangement used by Belfast group showing (a) main features of intersecting beam system (b) signal recovery scheme for measurements of $\sigma(Y^{2+})$ and (c) signal recovery scheme for measurements of $\sigma_c$

and target beams and F is the effective beam height (form factor).   F is
determined by scanning the beam profiles in the beam intersection region using
moveable slits used in conjunction with Faraday cups (not shown).   It is
important to demonstrate that cross sections are independent of beam intensities
and shapes and plots of S against $I_1 I_2/F$ are shown to be strictly linear for each
interaction energy.   It is worth emphasising that the cross sections determined
from (5) are absolute and do not rely on any other calibration or normalisation.
In our studies of $H^+$-$He^+$ collisions, the uncertainty in the absolute magnitude of
$\sigma(He^{2+})$ was assessed at ±6.5%.

Unlike other laboratories, we have also successfully used a coincidence technique
to obtain, so far for $H^+$-$He^+$ and $H^+$-$Li^+$ collisions, a separate measurement of the
cross section $\sigma_c$ for the charge transfer channel (4a).   In the first arrangement
(shown in Fig 1c), pulses from $M_1$ corresponding to the X atom signal provide the
start pulse for a time-to-amplitude converter whereas pulses from $M_2$ corresponding
to the $Y^{2+}$ signal provide the stop pulses.   A peak corresponding to the X-$Y^{2+}$
coincidences is then identified on a multi-channel analyzer.   In an alternative
arrangement, which reduces deadtime problems, dual coincidence units have been
used with appropriate delay lines to record pulses from $M_1$ and $M_2$ and produce
random-plus-true coincidences on one scaler and random only on a second scaler.
The charge transfer cross section $\sigma_c$ is then obtained from an expression similar
to (5) in which S is the coincidence count rate.   In our studies of $H^+$-$He^+$
collisions the uncertainty in our absolute values of $\sigma_c$ was assessed at 13.5%.
With a knowledge of $\sigma_c$, the cross section $\sigma_i$ for the ionization channel (4b) may
be obtained from the difference $\sigma_i = \sigma(Y^{2+}) - \sigma_c$.

The following Table summarises the experimental data available from groups in
Belfast (B), Culham (C), Louvain-la-Neuve (L) and Newcastle (N).

| Process | C.M. Energy (keV) | Measured cross section | Group | References |
|---|---|---|---|---|
| $H^+$+$He^+$ | 40-402 | $\sigma(He^{2+}),\sigma_c,\sigma_i$ | B | Mitchell et al (1977) Angel et al (1978a,b) |
| $H^+$+$He^+$ | 3-29 | $\sigma(He^{2+})$ | N | Peart et al (1977a) |
| $H^+$+$Li^+$ | 62-350 | $\sigma(Li^{2+}),\sigma_c,\sigma_i$ | B | Sewell et al (1980) |
| $H^+$+$C^+$ | 70-462 | $\sigma(C^{2+})$ | B | Neill et al (1981) |
| $H^+$+$N^+$ | 70-465 | $\sigma(N^{2+})$ | B | Neill et al (1981) |
| $H^+$+$Mg^+$ | 1-45 | $\sigma(Mg^{2+})$ | N | Peart et al (1977b) |
| $H^+$+$Ti^+$ | 1.5-18 | $\sigma(Ti^{2+})$ | C | Hobbis et al (1979) |
| $H^+$+$Fe^+$ | 1.5-18 | $\sigma(Fe^{2+})$ | C | Hobbis et al (1979) |
| $^3He^{2+}$+$^4He^+$ | 0.1-20 | $\sigma(^4He^{2+})$ | N | Peart and Dolder (1979) |
| $^4He^{2+}$+$^3He^+$ | 0.01-1.7 | $\sigma(^3He^{2+})$ | L | Jognaux et al (1978) |
| $Li^+$+$Li^+$ | 19-88 | $\sigma(Li^{2+})$ | N | Peart et al (1981) |
| $Na^+$+$Na^+$ | 19-88 | $\sigma(Na^{2+})$ | N | Peart et al (1981) |
| $K^+$+$K^+$ | 19-88 | $\sigma(K^{2+})$ | N | Peart et al (1981) |
| $Cs^+$+$Cs^+$ | 40-280 | $\sigma(Cs^{2+})$ | B | Dunn et al (1979) |
| $Cs^+$+$Cs^+$ | 19-88 | $\sigma(Cs^{2+}),\sigma_c$ | N | Peart et al (1981 |
| $Rb^+$+$Rb^+$ | 19-88 | $\sigma(Rb^{2+})$ | N | Peart et al (1981) |
| $Xe^+$+$Xe^+$ | 38-303 | $\sigma(Xe^{2+})$ | B | Angel et al (1980 |
| $Ar^+$+$Ar^+$ | 50-370 | $\sigma(Ar^{2+})$ | B | Angel et al (1980) |

RESULTS

Results for the reaction

$$H^+ + He^+ \rightarrow H + He^{2+} \tag{6a}$$

$$\rightarrow H^+ + He^{2+} + e \tag{6b}$$

are shown in Figs 2 and 3. Our cross sections $\sigma(He^{2+})$ for $He^{2+}$ production, $\sigma_c$ for charge transfer and $\sigma_i = (\sigma(He^{2+})-\sigma_c)$ for ionization measured in the range 40-380 keV (Angel et al 1978a,b) are seen to be in excellent accord with the values of $\sigma(He^{2+})$ measured by Peart et al (1977a) in the range 3-29 keV. While charge transfer provides the main contribution to $\sigma(He^{2+})$ at the lower collision energies, the ionization cross section increases with energy until $\sigma_i \approx \sigma(He^{2+})$ at c.m. energies above about 200 keV.

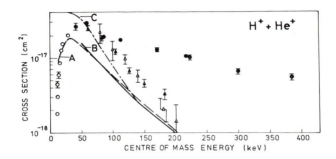

Fig 2.   Cross sections $\sigma_c$ for charge transfer in $H^+$-$He^+$ collisions: ▲ , Angel et al (1978b)(90% confidence limits); △ , Olson (1978) CTMC theory (99% confidence limits); Curve A, lower limit to $\sigma_c$ based on theoretical data of Rapp (1974); Curve B, Sinha and Sil (1979); Curve C, Belkić (1980). Cross sections $\sigma(He^{2+})$ for $He^{2+}$ formation: ● , Angel et al (1978a); O , Peart et al (1977a).

In Fig 2 cross sections $\sigma_c$ for the charge transfer process (6a) calculated by Olson (1978) using the classical trajectory Monte-Carlo (CTMC) method are seen to be in reasonable accord with our experimental values within the combined statistical uncertainties. A Coulomb-Born approximation has been used by Sinha and Sil (1979) in which allowance is made for capture into s and p states of H for principal quantum numbers $1 \leq n \leq 6$. This provides capture cross sections about a factor of two smaller than our experimental values and not greatly different from those for the inverse process

$$He^{2+} + H(1s) \rightarrow He^+(1s) + H^+ \tag{7}$$

calculated by Rapp (1974) using a multi-state impact parameter method. On the basis of time-reversal-invariance, cross sections for (7) would be expected to give a lower limit estimate of $\sigma_c$ for (6a)(which involves all bound states of H) at the same c.m. energy. The continuum distorted wave method used by Belkić (1980) leads to values of $\sigma_c$ (also shown in Fig 2) which are in generally poor accord both in magnitude and energy dependence with the experimental data. For c.m. energies in the range 1.6-14 keV Winter et al (1980) have used a coupled-molecular-state approach to obtain values of $\sigma_c$ (not shown) which are in excellent accord with experiment.

A theoretical estimate of the ionization cross section $\sigma_i$ of (6b) for $H^+$+$He^+$ collisions may be obtained by scaling the cross sections for $H^+$-H collisions based on the Born approximation (Bates and Griffing 1953). Calculations by Bates and Boyd (1962a) indicate that for interactions of a bare nucleus with a hydrogenic ion, the effects of Coulomb repulsion in ionization are negligible except in a region on the low energy side of the cross section maximum.    At the higher impact energies they suggest that cross sections for a process

$$A^{Z_p+} + B^{(Z_t-1)+} \rightarrow A^{Z_p+} + B^{Z_t+} + e \qquad (8)$$

characterised by projectile and target atomic numbers $Z_p$ and $Z_t$ may be obtained from the Born cross section for $H^+$-H collisions by multiplying the cross section scale by $Z_p^2/Z_t^4$ and the projectile energy scale by $M_pZ_t^2$ where $M_p$ is the projectile mass.

Cross sections for (6b) obtained by scaling in this way are in generally poor accord (Fig 3) with our experimental values of $\sigma_i = (\sigma(He^{2+}) - \sigma_c)$.    At the highest c.m. energy of 386 keV the experimental value is only 79±6% of the value scaled from the $H^+$-H Born cross section at the equivalent energy of 121 keV $H^+$ incident on stationary H atoms.    However, this discrepancy is not surprising in view of recent high precision measurements in this laboratory (Shah and Gilbody 1981) of $\sigma_i$ for $H^+$+H collisions which, when normalised to the Born approximation at 1500 keV, are only about 83% of the Born value at 121 keV.    In addition, Shah and Gilbody (1981) through studies of ionization in $He^{2+}$-H collisions, have confirmed the inadequacy of the Born scaling procedure at such low collision energies.

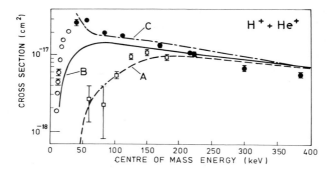

Fig 3.    Cross sections $\sigma_i$ for ionization in $H^+$-$He^+$
          collisions:  $\square$ , Angel et al (1978b);  Curve A,
          scaled from CTMC results of Banks et al (1976);
          Curve B, scaled from Born approximation (Bates
          and Griffing 1953);  Curve C, Belkić (1980:
          Cross sections $\sigma(He^{2+})$ for $He^{2+}$ formation:
          $\bullet$ , Angel et al (1978a); $\circ$, Peart et al (1977a).

Olson (1978) using the CTMC method has calculated ionization cross sections for (6b) which agree (within the statistical uncertainties) with scaled CTMC values for $H^+$-H collisions (Banks et al 1976) shown in Fig 3.    While these cross sections agree closely with the Born values at the highest impact energies, they provide much better agreement with our experimental values of $\sigma_i$ at the lower impact energies.    Values of $\sigma_i$ calculated by Belkić (1980) using the continuum distorted wave method are in very poor accord with experiment.

The $^3He^{2+}$-$^4He^+$ interaction has been studied experimentally by Peart and Dolder (1979) at c.m. energies in the range 0.1-20 keV.    Measured cross sections for $He^{2+}$ formation are believed to be primarily due to the resonant charge transfer process

$$^3He^{2+} + {}^4He^+(1s) \rightarrow {}^3He^+(1s) + {}^4He^{2+} \qquad (9)$$

Similar measurements for the $^4He^{2+}$-$^3He^+$ interaction have been carried out by

Jognaux et al (1978). Unfortunately in both experiments, the angular acceptance of the collision product analyzers and detectors was too small to accommodate all the scattered He$^{2+}$ particles so that precise comparison with theoretical predictions is precluded. Cross sections for charge transfer have been calculated by Bates and Boyd (1962b) and Dickinson and Hardie (1979) using perturbed stationary state methods. Cross sections measured by Peart and Dolder are only about 50% and 70% of the theoretical values at 106 keV and 20.5 keV respectively. However, Dickinson and Hardie (1979), by considering the differential cross sections, have applied a correction to their total cross sections appropriate to the angular acceptance of the apparatus of Peart and Dolder. Agreement with the cross sections measured by Peart and Dolder is then within 18%. Below 60 ev these corrected theoretical values show a similar degree of accord with the results of Jognaux et al (1978) (whose apparatus provided a larger angular acceptance), but are roughly 40% larger in the range 100-1700 eV.

Fig 4.    Data for H$^+$ - Li$^+$
                collisions

$\sigma(Li^{2+})$:  ●, Sewell et al (1980);

**O**, Olson (1980)

$\sigma_i$ :  ▲ , Sewell et al (1980);

△ ,  Olson (1980);

Curve A,  scaled from Merzbacher
                and Lewis (1958);

Curve B,  scaled from Bell and
                Kingston (1969)

$\sigma_c$ :  ■ , Sewell et al (1980);

◪ ,  Bogdanov et al (1965);

□ ,  Olson (1980);

Curve M,  Mapleton et al (1975);

Curve BS, Banyard and
                Shirtcliffe (1979);

Curve BK, Todd (1981)

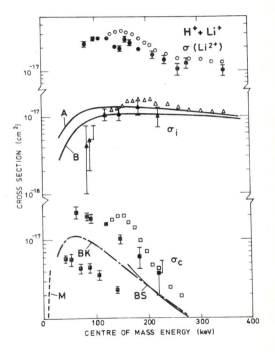

In this laboratory (Sewell et al 1980) we have measured cross sections $\sigma(Li^{2+})$ for Li$^{2+}$ formation, $\sigma_c$ for charge transfer and $\sigma_i = (\sigma(Li^{2+}) - \sigma_c)$ for ionization in H$^+$-Li$^+$ collisions. These results (Fig 4) show that as the interaction energy decreases, charge transfer makes an increasing contribution to $\sigma(Li^{2+})$ and becomes dominant below about 140 keV. At the higher impact energies values of $\sigma_i$ are approaching $\sigma(Li^{2+})$. Values of $\sigma_i$ calculated by Olson (cited by Sewell et al 1980) using the CTMC method are in reasonable accord with our data and, at the higher impact energies, with values (curves A and B) based on the Born approximation. The results in Curve A were obtained by scaling the K shell ionization cross sections of Merzbacher and Lewis (1958) calculated using simplified wave functions. This procedure is equivalent to the classical scaling of ionization cross sections described by Vriens (1969) and for isoelectronic systems by

Thomson (1912).    Curve B was obtained by scaling the ionization cross sections for H+-He collisions calculated by Bell and Kingston (1969) using accurate wavefunctions.

Values of $\sigma_c$ for H+-Li+ collisions calculated by Olson (cited by Sewell et al 1980), while in good accord with experiment (Fig 4) at high collision energies, increases more rapidly with decreasing energy and peak at a much higher energy. Values of $\sigma_c$ calculated by Banyard and Shirtcliffe (1979) using a continuum-distorted-wave method and by Todd (1980) using the Brinkman-Kramers approximation are approximately a factor of two smaller than experiment.    In Fig 4 are also shown low energy theoretical estimates of $\sigma_c$ obtained by Mapleton et al (1975) using a two-state atomic expansion method.    In addition, it is interesting to note that experimental values obtained by Bogdanov et al (1965) from a study of the scattering of protons in a lithium arc, while about four times smaller, exhibit in energy dependence similar to our experimental values.

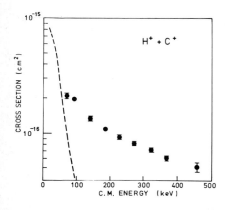

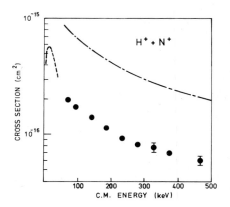

Fig 5.    Data for H+ - C+ collisions.

●  ; $\sigma(C^{2+})$,  Neill et al (1981).
---- ; Cross section for inverse charge
        transfer process, Goffe et al
        (1979).

Fig 6.    Data for H+ - N+ collisions.

●  ; $\sigma(N^{2+})$, Neill et al (1981).
—·—·; $\sigma_i$, Peach (1971).
- - -; Cross section for inverse
        charge transfer process,
        Phaneuf et al (1979).

Figs 5 and 6 show our measured cross sections (Neill et al 1981) $\sigma(C^{2+})$ and $\sigma(N^{2+})$ for H+-C+ and H+-N+ collisions within the range 70-465 keV.    These measurements were carried out with C+ and N+ ions believed to be predominantly in the ground state.    In the absence of a separate measurement, an estimate of the charge transfer contribution to the measured cross sections is provided by those for the corresponding reverse process

$$X^{2+} + H \rightarrow X^+ + H^+ \tag{10}$$

provided that the excited state channels are not greatly different.    Cross sections for C²+ impact (Goffe et al 1979) and for N²+ impact (Phaneuf et al 1979) when plotted at the same c.m. energy (Figs 5 and 6) indicate that the charge transfer contributions to $\sigma(C^{2+})$ and $\sigma(N^{2+})$ become large at the low impact energies.    Cross sections for proton impact ionization of N+ calculated by Peach (1971) using the Born approximation, are seen (Fig 6) to considerably exceed measured values of $\sigma(N^{2+})$.

Figs 7 and 8 show cross sections $\sigma(Ti^{2+})$ and $\sigma(Mg^{2+})$ for $H^+$-$Ti^+$ and $H^+$-$Mg^+$ collisions measured by Hobbis et al (1979) and Peart et al (1977b) in the c.m. ranges 1.5-18 keV and 1-45 keV respectively. If, as seems likely, the measured cross sections are dominated by the charge transfer process which involves small energy defects of 0.03 eV and -1.4 eV respectively for the H(1s) formation channel, close agreement with cross sections for the inverse process (10) might be expected if ground state species are predominant. Cross sections for (10) measured in this laboratory (McCullough et al 1979) for $Ti^{2+}$-H and $Mg^{2+}$-H collisions and theoretical estimates by Bates et al (1964) for $Mg^{2+}$-H collisions over an overlapping range of centre of mass energies are seen to be in excellent agreement with the corresponding ion-ion cross sections.

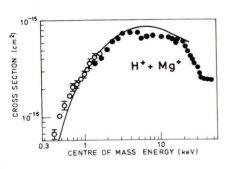

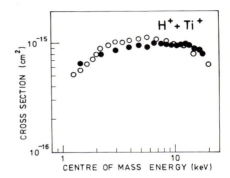

Fig 7.   Data for $H^+$ - $Mg^+$ collisions.    Fig 8.   Data for $H^+$ - $Ti^+$ collisions.

$\sigma(Mg^{2+})$: ●, Peart et al (1977b).    $\sigma(Ti^{2+})$: ●, Hobbis et al (1979).
Cross section for inverse charge    Cross section for inverse charge
transfer: ○, McCullough et al (1979);    transfer: ○, McCullough et al (1979).
———, Bates et al (1964).

Hobbis et al (1979) have also obtained cross sections for $Fe^{2+}$ production in $H^+$-$Fe^+$ collisions at c.m. energies within the range 1.5-18 keV. Above 6 keV, cross sections are comparable in shape and magnitude with those for $H^+$-$Ti^+$ and $H^+$-$Mg^+$ collisions. However below 6 keV, $\sigma(Fe^{2+})$ increases with decreasing collision energy to a value of about $1.5 \times 10^{-15} cm^2$ at 1.5 keV. This low energy behaviour is surprising on the basis of the 16.2 eV energy defect of the charge transfer channel assuming ground state species. Unfortunately there are no data for the inverse $Fe^{2+}$-H charge transfer process for comparison.

At energies where measured cross sections $\sigma(Y^{2+})$ for proton impact are dominated by ionization it is interesting to examine the extent to which the available experimental data can be described in terms of a general scaling relation. For isoelectronic systems A and B classical scaling based on the Thomson (1912) theory indicates that ionization cross sections $\sigma_i(A)$ and $\sigma_i(B)$ are related by the expression

$$\sigma_i(A)/\sigma_i(B) = |U(B)/U(A)|^2$$

for corresponding reduced energies where U(A) and U(B) are the ionization energies of A and B. In the more general scaling law described by Vriens (1969), the ionization cross sections $\sigma_i$ should scale to a universal form when expressed as $\bar{\sigma}_i = U^2\sigma_i/R^2Z_p^2n$ where $Z_p$ is the projectile charge, U is the ionization energy, R is the Rydberg constant and n is the number of electrons responsible for the process. The projectile energy $E_p$ is scaled by the relation $E = E_p/\lambda U$ where $\lambda$ is the projectile mass expressed in units of electron mass. Fig 9 shows experimental data for $H^+$ in $He^+$, $Li^+$, $C^+$ and $N^+$ scaled in this way (taking n as the number of outer shell electrons) together with measured ionization cross sections for $H^+$ in H and He and for $He^{2+}$ in H. It will be seen that agreement between the scaled values is generally well within a factor of two.

Fig 9.   Classical scaling of
         ionization cross sections.

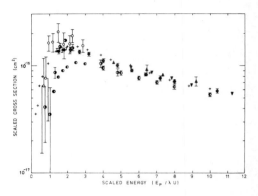

▼  $H^+ + C^+$ (Neill et al 1981)

▲  $H^+ + N^+$ (Neill et al 1981)

○  $H^+ + Li^+$ (Sewell et al 1980)

◑  $H^+ + He^+$ (Angel et al 1978b)

●  $H^+ + H$ (Shah and Gilbody 1981)

◐  $He^{2+} + H$ (Shah and Gilbody 1981)

+  $H^+ + He$ (de Heer et al 1966,
             Hooper et al 1962)

Now let us consider collisions between multi-electron ions.  We have measured
(Dunn et al 1979), Angel et al 1980) cross sections $\sigma(X^{2+})$ for the combined
processes

$$X^+ + X^+ \rightarrow X + X^{2+} \qquad \text{charge transfer} \qquad (11a)$$
$$\rightarrow X^{m+} + X^{2+} + me \qquad \text{ionization for } m \geqslant 1 \qquad (11b)$$

(where m is unspecified) in $Cs^+$-$Cs^+$, $Xe^+$-$Xe^+$ and $Ar^+$-$Ar^+$ collisions within the
c.m. energy range 19-280 keV.  In parallel measurements (unpublished) we
have also obtained charge transfer cross sections $\sigma_{21}$ for the inverse charge
transfer process

$$X^{2+} + X \rightarrow X^+ + X^+ \qquad (12)$$

which for $Cs^{2+}$, $Xe^{2+}$ and $Ar^{2+}$ ions are roughly 100, 10 and 3 times the
corresponding values of $\sigma(X^{2+})$ over the same c.m. energy range.  It is therefore
evident that, in these cases, $\sigma_{21}$ does not provide a reasonable lower-limit
estimate of $\sigma(X^{2+})$.

In Fig 10 our cross sections $\sigma(Cs^{2+})$ for $Cs^{2+}$ production in $Cs^+$-$Cs^+$ collisions
(Dunn et al 1979) measured in the c.m. range 40-280 keV are shown together with
the recent data of Peart et al (1981) measured in the c.m. range 19-88 keV.  The
two sets of data are in generally good accord except for our two lowest energy
values which are considerably larger than those of Peart et al.  This discrepancy
requires further examination especially since Peart et al claim, as a result of
careful checks, that the smaller angular acceptance of their collision product
detector was adequate to record essentially all the scattered collision products.

Theoretical studies of processes of the type (11) are difficult since they involve
the coupling between the nuclear and electronic motions.  A molecular approach
has been applied to slow $Cs^+$-$Cs^+$ collisions by Das et al (1978), Olson (1978) and
Olson and Liu (1981).  The latter have used the Fano-Lichten molecular orbital
description to show that collisions will be dominated by ionization rather than
charge transfer.  This takes place through a pseudo-crossing of the $5p\sigma_u$ and
$6s\sigma_g$ states at an internuclear separation of about 2.1 $a_0$.  Here the dominant
mechanisms are:- $Cs^+ + Cs^+ \rightarrow |Cs^{2+}(5p\sigma_g{}^2 6s\sigma_g{}^2)| \rightarrow Cs^{+*}(6s^2) + Cs^+$ $(13a)$

$$\rightarrow Cs^{+*}(6s) + Cs^{+*}(6s) \qquad (13b)$$

where (13a) leads to the ionization channel (11b) through autoionization of the
doubly excited product ion and (13b) can also lead to a small contribution to
(11b) via molecular ionization.  On the basis of this model and by reference to
known data for the Ar-Ar system, Olson and Liu estimate that $\sigma(Cs^{2+}) \approx 3 \times 10^{-16} cm^2$
at 50 keV a value which compares favourably with the experimental values in
Fig 10.  A rough experimental estimate of $\sigma_c$ for (11a) by Peart et al (1981) in

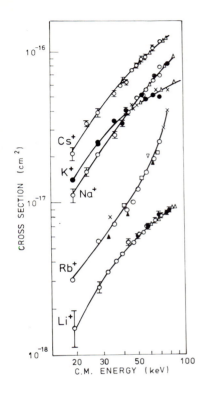

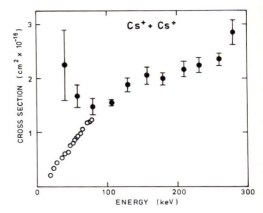

Fig 10. Cross sections for $Cs^{2+}$ formation
          in $Cs^+$-$Cs^+$ collisions.

●    Dunn et al (1979)
○    Peart et al (1981)

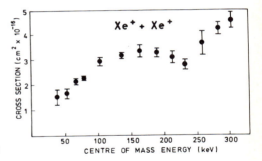

Fig 11.   Cross sections for $X^{2+}$
          formation in $X^+$-$X^+$
          collisions for alkali ions
          (Peart et al 1981).

Fig 12. Cross sections for $Xe^{2+}$ formation
          in $Xe^+$-$Xe^+$ collisions
          (Angel et al 1980).

the c.m. range 28-47 keV is at least an order magnitude smaller than $\sigma(Cs^{2+})$ and
is thus in general accord with the theoretical model.

Fig 11 shows cross sections $\sigma(X^{2+})$ for $X^+$-$X^+$ collisions measured recently by
Peart et al (1981) for all the alkali ions in the c.m. range 19-88 keV.   Cross
sections do not increase systematically with the mass of the ions but all show a
monotonic increase with increasing energy.

Fig 12 shows our measured cross sections $\sigma(Xe^{2+})$ for $Xe^+$-$Xe^+$ collisions.   In
this case it might be expected that the charge transfer contribution would be
greater than for the closed-shell $Cs^+$-$Cs^+$ system.   However it will be seen that
the values of $\sigma(Xe^{2+})$ and $\sigma(Cs^{2+})$ are comparable in magnitude.   For the $Ba^+$-$Ba^+$
system, where each ion has a loosely bound electron, Sramek et al (1980) have
used a classical trajectory method with basic functions obtained from a multi-
configuration valence-bond method to estimate charge transfer cross sections of
up to $2.4 \times 10^{-15} cm^2$ in the c.m. range 25-500 keV.

REFERENCES

Angel,G.C., Dunn, K.F., Neill, P.A. and Gilbody, H.B.  J.Phys.B 13(1980) L391.
Angel,G.C., Dunn, K.F., Sewell, E.C. and Gilbody, H.B. J.Phys.B 11(1978a)  L49.
Angel,G.C., Sewell, E.C., Dunn, K.F. and Gilbody, H.B. J.Phys.B 11(1978b) L297.
Banks, D., Barnes, K.S. and Wilson, J.McB.  J.Phys.B 9 (1976) L141.
Banyard, K.E. and Shirtcliffe, G.W.  J.Phys.B 12(1979) 3247.
Bates, D.R. and Boyd, A.H.  Proc.Phys.Soc. 79(1962a) 710.
Bates, D.R. and Boyd, A.H.  Proc.Phys.Soc. 80(1962b) 1301.
Bates, D.R. and Griffing, G.  Proc.Phys.Soc. A66(1958) 961.
Bates, D.R., Johnston, H.C. and Stewart, A.L.  Proc.Phys.Soc. 84(1964) 517.
Belkić, Dž.  J.Phys.B 13(1980) L589.
Bell, K.L. and Kingston, A.E.  J.Phys.B 2 (1969) 653.
Bogdanov, G.F., Karkhov, A.N. and Kucheryaev, Ya. A.  Atomnaya Energiya 19 (1965)
    1316.
Brouillard, F. and Delfosse, J.M.  Abstracts Proc. 5th ICPEAC Lemingrad (1967) 159.
Das, G., Raffenetti, R.C. and Kim, Y-K.  Argonne Nat. Lab. Report (1978)
    ANL-79-41 p.195.
Dickinson, A.S. and Hardie, D.J.W.  J.Phys.B 12 (1979) 4147.
Dolder, K.T.  Case Studies in Atomic Collision Physics (1969) Vol 1 p.250.
Dunn, K.F., Angel, G.C. and Gilbody, H.B.  J.Phys.B 12 (1979) L623 and
    unpublished data.
Gilbody, H.B.  Advances in Atomic and Molecular Physics 15 (1979) p.293.
Godlove, T.F.  IEEE Trans. Nucl. Sci. 26 (1979) 2997.
Goffe, T.V., Shah, M.B. and Gilbody, H.B.  J.Phys.B 12 (1979) 3763.
Guidini, J., Manus, C., Sinda, T. and Watel, G.  Abstracts Proc. 4th ICPEAC
    Quebec (1965) 450.
de Heer, F.J., Schutten, J. and Moustafa, H.  Physica 32 (1966) 1766.
Hobbis, D.A., Nicholson, P., Harrison, M.F.A. and Montague, R.G. (1979) private
    communication.
Hooper, J.W., Harmer, D.S., Martin, D.W. and McDaniel, E.W.  Phys.Rev. 125 (1962)
    2000.
Jognaux, A., Brouillard, F. and Szucs, S.  J.Phys.B 11 (1978) L669.
Kim, Y-K.  1976 U.S. Dept. of Commerce Report LBL-5543 (1976) p.11.
McCullough, R.W., Nutt, W.L. and Gilbody, H.B.  J.Phys.B 12 (1979) 4159.
Mapleton, R.A., Schneeberger, M.F. and Steele, C.A.  USAF Report (1975)
    AF-CRL-TR-75-0053.
Merzbacher, E. and Lewis, H.W.  Handb. Phys. 34 (1958) 166.
Mitchell, J.B.A., Dunn, K.F., Angel, G.C., Browning, R. and Gilbody, H.B.
    J.Phys.B 10 (1979) 1897.
Neill, P.A., Angel, G.C., Dunn, K.F. and Gilbody, H.B.  (1981) unpublished.
Olson, R.E.  J.Phys.B 11 (1978) L227.
Olson, R.E.  Argonne Nat. Lab. Report (1978) ANL-79-41 p.195.
Olson, R.E. and Liu, B.  J.Phys. B 14 (1981) L279.
Peach, G.  J.Phys. B 4 (1971) 1670 and private communication.
Peart, B. and Dolder, K.T.  J.Phys.B 12 (1979) 4155.
Peart, B., Forrest, R.A. and Dolder, K.T.  J.Phys.B 14 (1981) 1655 and private
    communication.
Peart, B., Gee, D.M. and Dolder, K.T.  J.Phys. B 10 (1977b) 2683.
Peart, B., Grey, R. and Dolder, K.T.  J.Phys.B 10 (1977a) 2675.
Phaneuf, R.A., Meyer, F.W. and McKnight, R.H.  Phys. Rev. A17 (1979) 534.
Rapp, D.  J. Chem. Phys. 61 (1974) 3777.
Sewell, E.C., Angel, G.C., Dunn, K.F. and Gilbody, H.B.  J.Phys.B 13 (1980) 2269.
Shah, M.B. and Gilbody, H.B.  J.Phys.B (1981) - in press.
Sinha, C. and Sil, N.C.  Phys. Lett. 71A (1979) 201.
Sramek, S.J., Macek, J.H. and Gallup, G.A. Phys.Rev. A 22 (1980) 1467.
Thomson, J.J.  Phil. Mag 23 (1912) 449.
Todd, N.  (1981)  private communication.
Vriens, L.  Case Studies in Atomic Collision Physics (1969) Vol 1 p.337.
Winter, T.G., Hatton, G.J. and Lane, N.F.  Phys. Rev. A 22 (1980) 930.

PHYSICS OF ELECTRONIC AND ATOMIC COLLISIONS
S. Datz (editor)
© North-Holland Publishing Company, 1982

INTRODUCTION TO THE INVITED PAPERS COMMEMORATING V. ČERMÁK

Andrew J. Yencha
Department of Physics
State University of New York at Albany
Albany, New York, 12222, U.S.A.

This session of invited papers, which includes the paper on Penning ionization by Prof. Niehaus and the paper on collisions of Rydberg atoms by Dr. Gounand, is being held to commemorate the scientific achievements and contributions of the late Vladimír Čermák who died on January 4, 1980, in Prague, Czechoslovakia. Dr. Čermák's career was one of considerable depth and breadth--the type of career in science we all strive to achieve. Two areas where Dr. Čermák made major contributions are represented by the two topics of this session--namely Penning Ionization and Collisions of Rydberg Atoms. Since I'm sure our invited speakers will touch on some of the scientific aspects of Dr. Čermák's life, I should like to say a few words about Vladimír Čermák, the person.

Anyone who had ever met Vladimír knows how much he loved life and his fellow man and how in tune he was with nature. One of his major pastimes in Czechoslovakia was to walk for hours with his wife Věra in the hills surrounding Prague and in the Šumava mountains in southern Bohemia where he was born. He took great delight in observing nature from the standpoint of being a true participant of nature. In leisure moments he often talked of his sojourns through the woods and forest of Czechoslovakia, and his home was full of pictures of animals. I believe the fox was his favorite.

In all the world, there were three special places which Vladimír loved the most. The first was his birthplace in southern Bohemia. The second was Boulder, Colorado, where he twice was a visiting scientist in the 1960's at the Joint Institute of Laboratory Astrophysics. The third was Paris where he spent a year as a Visiting Professor at the University of Paris-Sud. He took great delight in speaking French, and once while I was in his laboratory in Prague he received a telephone call from Orsay from Jean Durup and others gathered in his laboratory. I believe Jean Futrell was one of those who spoke to him as well. When he hung up the phone after his rather lengthy conversations in French with a number of people, Vladimír was truly beaming with joy at having spoken to so many of his French colleagues and friends. It therefore was fitting that the last ICPEAC Vladimír attended was the one held in Paris in 1977. For those of you who were there and met him, I'm sure you can attest to the gleam in his eye and love in his heart for having had the opportunity to attend that particular ICPEAC meeting.

PHYSICS OF ELECTRONIC AND ATOMIC COLLISIONS
S. Datz (editor)
© North-Holland Publishing Company, 1982

PENNING IONIZATION

A. Niehaus

Fysisch Laboratorium der Rijksuniversiteit Utrecht,
Princetonplein 5
3584 CC Utrecht
The Netherlands

The influence of the work of V. Cermak on the development of the
field of Penning ionization is demonstrated by discussing
a selection of his contributions in the frame of our present
knowledge. It is further tried to characterize present research
on Penning ionization by discussing a selection of most recent
contributions.

INTRODUCTION

The developement of research on ionizing collisions of excited particles with
atoms and molecules into an active field of collision physics, now well known as
the field of Penning ionization, has to a very large extent been determined by
the important contributions of V. Cermak, who died on 4 January 1980. This review
lecture on Penning ionization is devoted to him, in appreciation of his work and
in admiration of his continuous effort to serve science.

We want to select a few examples showing in which way, and how strongly, Cermak
initiated and influenced research in the field of Penning ionization. In the
early sixties he was mainly involved in mass spectrometric  studies of the decomposition of polyatomic molecular ions formed in collisions with ions or neutral
excited particles (1-4). These studies were carried out to investigate the uni-
molecular decomposition at certain internal energies. In their paper "Ionizing
reactions of noble gas atoms in metastable states with polyatomic molecules" in
(1965) (2) Cermak and Herman point out for the first time the characteristics
and the possible value of Penning ionization electron spectroscopy (PIES): "the
process of ionization results in releasing of only one electron; the excitation
energy transferred to the molecular ion is given by

$$E_e = E_n - IP - E_K \qquad (1)$$

($E_n$ – electronic excitation energy of the impacting particle, IP – ionization
potential of the molecule, $E_K$ – kinetic energy of the released electron), and
depends therefore on the energy of the released electron. This in principle makes
it possible to determine directly the distribution of the excitation energy of
the molecular ion by measuring the distribution of kinetic energy of the released
electron. (In experiments with molecular beams, the data on the angular distribution of electron velocities may be obtained too)".In 1966 (5) Cermak reports
the first successful application of (PIES) and presents spectra for a number of
molecules. These papers may be regarded as the origin of the development of the
field of (PIES), a method which has most contributed to our present knowledge
of the various aspects of Penning ionization reactions.

From his first "Retarding-potential measurements of the kinetic energy of electrons released in Penning ionization" (5), which were performed with a rather
low energy resolution as compared to his later measurements, as well as compared
to measurements of other groups, Cermak was able to draw the conclusion that the
excited molecular states were populated "probably by a Frank-Condon transition",

a result that later has been confirmed in numerous studies for a large number of
molecules ionized by rare gas metastables.

Viewing back after many years of research on Penning ionization we may say that
essential details regarding the basic mechanisms of Penning – and associative
ionization have already been correctly formulated in the early papers of Cermak.
It is perhaps characteristic that these ideas were put forward on the basis of
rather indirect evidence. In their paper "Associative ionization in mixtures of
carbon monoxide with sodium and patassium and the mechanism of associative ionization reactions" (6) Herman and Cermak propose (1966) – based on mass spectrometric measurements of associative ions and their appearance potentials – two
possible mechanisms of associative ionization of a metal atoms Me with an excited particle $R_m^*$ :

$$R_m^* + Me \rightarrow RMe^+ + e^-  \qquad (2)$$

They distinguish between two cases, (a) and (b), depending on whether the asymptotic electronic energy of $R_m^* + Me$ is below the energy of $R + Me^+$ (case (a)), or
above (case (b)), and propose potential curve diagrams describing the corresponding mechanisms. The diagrams are reproduced in Fig. 1(a) and (b). The mechanisms

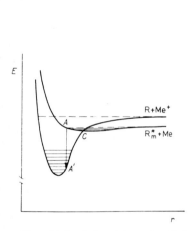
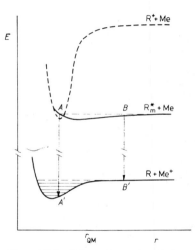

Fig. 1(a): Schematic potential curve        Fig. 1(b): Schematic potential curve
diagram describing associative ioni-        diagram describing associative ionization case (a) (see text).                 zation case (b) (see text).

(Figures reproduced from Herman and Cermak (1966).)

are described as follows: (a) "During a central collision of $R^*$ and Me the internuclear separation diminishes until the turning point A is reached. As a potential curve $R^*$ - Me lies in this region above the potential curve of the ion, a
spontaneous relasing of an electron can occur and the system passes in a Franck-
Condon transition to the potential curve of the ion ..." (b) "The central and
noncentral collisions lead to the releasing of an electron of the kinetic energy
given by the vertical separation of the potential energy curves $R^*$ - Me and
$R - Me^+$, and to the Franck-Condon transitions to the $R - Me^+$ curve. If the transition occurs above the potential minimum of the lower curve and the relative
kinetic energy of the colliding particles can be stored in the molecular ion, a
diatomic molecular ion $RMe^+$ in the ground state is formed. If this is not the
case (in Fig. 1b for larger than $r_{QM}$ roughly), there are two products of the

reaction besides the electron Me$^+$ and R (Penning Ionization)". Herman and Cermak conclude by saying: "It is evident that a direct proof, whether the suggested mechanism is correct, can be provided by the measurement of the energy of the electron released in the ionization process. Such an electron spectroscopy would be very valuable for our better understanding of the processes of Penning and associative ionization".

Later experimental investigations using the method of (PIES) have indeed provided direct evidence for the essential correctness of the proposed mechanisms and the later formulated two state optical potential description of Penning ionization may be regarded as their theoretical version.

Cermak has always been especially interested in the phenomena arising in Penning ionization of polyatomic molecules. His systematic studies using the method of high resolution (PIES) (e.g. 7-13) have yielded numerous results that now form a basis for further work, even in other fields. For instance, the observed propensity for ionization of certain molecular orbitals in a Penning ionization allows to obtain information on the influence of condensation on these orbitals, by comparison of Penning electron spectra from molecules condensed on cold surfaces with spectra from the same molecules in the gas phase.

We have, of course, not been able to give a full account of Cermaks work here. We hope, however, with the examples given, to have indicated the importance of his work for the developement of the field of Penning ionization during the last 15 years.


REMARKS ON THE PRESENT STATE OF THE FIELD

Review lectures and progress reports on Penning ionization have been presented at this conference continuously (14-16). Considering the literature of the last ten years, one observes that research in this field has gradually entered a new "phase". In the early seventies most experimental and theoretical studies were aimed at a clarification of the ionization process itself. In more recent years, however, after the basic mechanism and the theoretical description of Penning ionization in its simplest form could be regarded as established, the emphasis has changed: the Penning ionization process is now, generally speaking, predominantly used to obtain information on collisions, in an equivalent way, as, for instance, the processes of photon emission or electron scattering are used since a long time to obtain information on the basic quantities that can be probed by them. This can especially be recognized in the following discussion of the advances of the last two years.

To facilitate this discussion we first very shortly describe the present state of our knowledge on Penning ionization.

The theoretical description is based on a two state potential curve model that relies on the Born-Oppenheimer separation of nuclear - and electronic coordinates. The initial electronic state of the Penning system - consisting of the excited particle $A^*$ and the target particle B is described by the complex potential $\tilde{V}_*(R) = V_*(R) - i\Gamma(R)/2$. The imaginary part accounts for the decay by electron emission into the final electronic state $V_+(R)$ dissociating into $A + B^+$. In this description the Penning transition $A^* + B \rightarrow A + B^+ + e^-$ is necessarily vertical. Experimentally observable quantities - such as ionization cross section, scattering cross section, and electron energy distributions, can be calculated from the theoretical quantities $\tilde{V}_*(R)$ and $V_+(R)$, either by exact numerical integration of the Schrödinger equation for the heavy particle motion, or by semiclassical and classical methods. In this form the description is of course limited to the simplest case of a Penning reaction, namely to a case where (i) $A^*$ and B are atoms, (ii) $\tilde{V}_*(R)$ and $V_+(R)$ are non degenerate in the whole region of accessible distances R (no crossings), (iii) the relative collision velocity is low enough

to exclude diabatic mixing with the other molecular states. Nevertheless, so far
in none of the cases where a rather stringent test of the theory was possible by
comparison with experiments (e.g. 17, 18) significant inconsistencies were recognized. Whereby it should be pointed out that these tests only pertained to the
description of the "collision part" of the problem and not to the functions
$\tilde{V}_*(R)$ and $V_+(R)$. Although some "ab initio" determinations of these functions have
been performed, up until present they must be considered as model function due
to insufficient accuracy.

Predominantly, Penning systems with $A^*$ being a metastable rare gas atom have been
studied experimentally. From these studies much information on $\tilde{V}_*(R)$ and $V_+(R)$,
as well as on branching ratios in cases where more than one final state potential
curve $V_+(R)$ are accessible, has been obtained. In addition, deviations from the
two state behaviour due to diabatic mixing among different initial state curves
$\tilde{V}_*(R)$ have been reported. For molecular targets (M) it is found that in most
cases the population of vibrational states within one electronic state is very
similar to the population arising for photoionization, with slight deviations due
to interaction with the projectile particle (A) in the initial or final state of
the Penning reactions.

This very short and necessarily incomplete account of our present knowledge may
suffice to serve as background for the discussion of new developments. The main
problems investigated presently will become evident in that discussion.

VELOCITY DEPENDENCE OF TOTAL IONIZATION CROSS SECTIONS FOR $He^*(2^3S)$-Ar AND
$He^*(2^1S)$-Ar

In a recent paper Burdenski et al (19) report integral total and integral ionization cross sections for the systems $He^*(2^3S)$-Ar and $He^*(2^1S)$-Ar in the c.m.
collision energy range 2-230 meV. As compared to earlier data (20, 21), that were
restricted to energies > 20 meV, this constitutes a considerable advance, since
around 20 meV the ionization cross sections exhibit a minimum which has been predicted before, on the basis of calculations with empirically derived potentials
$\tilde{V}_*(R)$, but not been observed. The results of Burdenski et al (19) therefore allow
a more reliable determination of $\tilde{V}_*(R)$ than was possible before. The experiment
was performed as follows.

The helium beam effuses from a multichannel array which is operated at room temperature and at liquid nitrogen and liquid hydrogen temperatures. The atoms are
then excited by electron impact, pass a region where the $He^*(2^1S)$-component can
be removed by quenching, and are finally velocity selected by means of a mechanical selector, before entering the scattering chamber containing the Ar-gas.
This chamber is cooled to liquid nitrogen temperature. Effective relative cross
sections - pertaining to an averaging over 5% beam velocity spread - for total
scattering $Q_{eff}$, and for ionization $\sigma_{eff}$ are measured as a function of beam velocity. The results obtained are reproduced in Fig. 2. The absolute scale is determined by theoretical calculations. These calculations are exact numerical
solutions of the radial Schrödinger equation for a complex $\tilde{V}_*(R)$, the parameters
of which are fixed to reproduce the shape of the observed velocity dependence.
In the case of $He^*(2^3S)$-Ar it turned out that the real part of $\tilde{V}_*(R)$ determined
by Siska (22), from differential elastic cross section data, together with a
slightly modified function $\Gamma(R)$ give the satisfactory agreement seen in Fig. 2.
With the unmodified $\tilde{V}_*(R)$ of ref. (22) a significant deviation from the measured
shape of the cross section curve was observed at low velocities. On the other
hand the modified $\tilde{V}_*(R)$ reproduces the measured $Q_{eff}$ and the earlier differential
cross section data (22, 23, 24) as well. Although there always remains the question of uniqueness of a potential derived by fixing parameters in a model function,
it is very probable that the potential derived by Burdenski et al (19) for the
most thoroughly studied Penning system $He^*(2^3S)$-Ar is improved as compared to the
earlier determined ones. In the case of $He^*(2^1S)$-Ar Burdenski et al (19) find

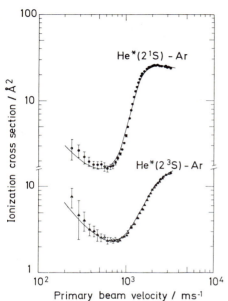

Fig. 2: Velocity dependence of effective total ionization cross sections. ( | ) : experimental data (——) : calculations using augmented optical potentials (reproduced from Burdenski et al (1981)

that their own measurements of $\sigma_{eff}$ and $Q_{eff}$ can well be reproduced (see fig. 2) in calculations with a function $\tilde{V}_*(R)$, whose real part is slightly different from the real part of a potential obtained from differential scattering data (25), but whose imaginary part is rather drastically changed. They invoke a function $\Gamma(R)$ that saturates at $R_c = 3.14$ Å in order to be able to reproduce the observed decrease of $\sigma_{eff}$ at high collision energies. This new potential, however, leads to calculated differential elastic cross sections that deviate significantly from experimental data (23, 26-29). The theoretical absolute values of cross sections for He$^*$($2^3$S)-Ar in Fig. 2 agree well with the earlier experimental values obtained from a calibration of relative measurements (20) to destruction rate constants measured by the flowing afterglow method (30). In case of He$^*$($2^1$S)-Ar the absolute values of Fig. 2 are about 10% higher than the earlier ones (20), but still within the limits of possible error of the absolute rate constants (30).

An important result of the work of Burdenski et al (19) is the confirmation of a peculiar hump in the real part of $\tilde{V}_*(R)$ for He$^*$($2^1$S)-Ar. This hump was proposed on the basis of differential elastic cross section data (24). As a general explanation for the existence of such a hump in He$^*$($2^1$S)-target interaction potentials, adiabatic mixing with the He$^*$($2^1$P)-target channel was proposed (29, 31). One-electron model potential calculations by Siska (32) lend further support for this explanation.

PENNING IONIZATION OF Ar, Kr AND Xe BY Ne$^*$($3s$ $^3P_{2,0}$)

While experimental data on Penning systems involving He$^*$($2^1$S, $2^3$S) metastables can often be analyzed successfully on the basis of the two state model involving one single complex potential for the initial state, obvious problems arise for Penning systems with Ne$^*$($2p^5 3s$) $^3P_2$ and $^3P_0$. These have energies of 16.619 eV and 16.715 eV, respectively, and the resonance state $^3P_1$ lies in between with an energy of 16.671. The most important problems to be anticipated are the following: (i) possible mixing of the potential curves, dissociating into these three states, during the Penning collision, (ii) possible splitting into different potential curves according to different values of the projection of the total electronic angular momentum ($\Omega$) at small distances, and (iii) different and possibly complicated imaginary parts for these potentials. Some progress towards a solution of these problems has been made in recent articles by Gregor and Siska (33), and by Hotop et al (34).

Gregor and Siska (33) carried out differential elastic scattering measurements at selected collision energies in the thermal region, using two supersonic beams. The Ne-metastables are formed by electron impact leading to a ratio of 5:1 for

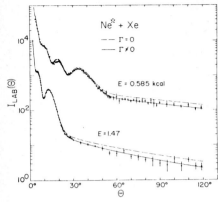

Fig. 3: Differential elastic cross sections.
(—) calculation with optical potential
(--) calculation with imaginary part
$\Gamma(R) = 0$.
(reproduced from Gregor et al (1981).)

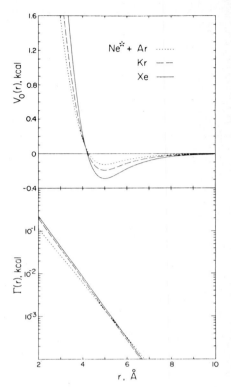

Fig. 4: Derived optical potentials for Ne* + Ar, Kr, Xe; real parts in upper panel, resonance widths $\Gamma(R)$ in lower.
(reproduced from Gregor et al (1981).)

$^3P_2$ and $^3P_0$ metastables in the beam. No state selection was applied. As an example cross section data for Ne* – Xe are reproduced in Fig. 3. In analyzing their data Gregor and Siska assume a single complex potential, i.e. they assume (i) the $^3P_2$-target-, and $^3P_0$-target potentials to be identical, and (ii) the $^3P_2$-target potential to be negligibly split according to the three different $\Omega$-values. Since the Ne*-cross section data show less structure due to ionization than corresponding He*-data, the authors find it further necessary to use earlier measured relative velocity dependent ionization cross section (35, 20, 36) as a constraint on the complex potential. The fit calculations in which parameters of the model-complex potential are determined are carried out using a partial wave expansion with JWKB complex phase shifts. The resulting real and imaginary parts of the potentials for Ne*/Ar, Kr, Xe are reproduced in Fig. 4. With these potentials both, the measured differential cross section data, and the relative ionization cross sections are well reproduced. The solid line through the measured points of fig. 3 demonstrates this for Ne* – Xe. The broken line is calculated with the real potential alone, showing indeed only a small influence of ionization on the differential elastic cross section. Gregor and Siska (33) compare the van der Waals parameters of the real part of their single Ne* – rare gas potentials with those of known Na-rare gas potentials (37), and find very close agreement (see table 1). They point out that this is evidence of the dominance of the role of the Ne*-3s electron for long range interactions, the details of the ionic core of Ne* being of little importance, and they conclude further that this characteristic of Ne$^*$-interactions supports the one-potential model used in their analysis.

The results and conclusions of Gregor and Siska (33) mentioned so far are partly supported and partly supplemented by results of recent work of Hotop et al (34) who studied Ne$^*$($^3P_2$, $^3P_0$)-rare gas systems by the method of Penning ionization electron spectroscopy with state selected Ne$^*$-metastables. State selection of a mixed Ne$^*$($^3P_2$, $^3P_0$)-beam of a near-Maxwellian velocity distribution at an effective temperature in the range 350-400°K is achieved by optical excitation with a narrow band dye laser, and electron energy distributions are measured with a 127°-cylindrical condenser of resolution 18-20 meV.

For each Ne$^*$ – X system (X: rare gas) four distributions are observed, corresponding to the two metastable states Ne$^*$($^3P_2$) and Ne$^*$($^3P_0$), and the two fine structure states X$^+$($^2P_{3/2}$) and X$^+$($^2P_{1/2}$) of the rare gas ions formed. As an example the distributions for Xe are reproduced in Fig. 5. The small, narrow peaks visible

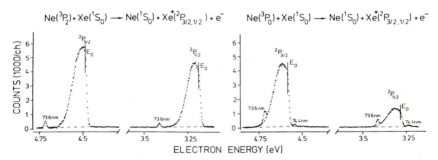

Fig. 5: Penning ionization electron spectra for the systems Ne$^*$($^3P_2$) + Xe, and Ne$^*$($^3P_0$) + Xe. (reproduced from Hotop et al (1981).)

in all distributions are due to photoionization of Xe by Ne-resonance photons and allow an accurate absolute energy calibration. The energy $E_o$ indicated in each distribution is the energy the Penning electron would have in case of a transition at infinite distance, and is given by $E_o = E^*$ – IP, with $E^*$ the excitation energy of the metastable and IP the energy needed to form the corresponding ion state of the rare gas from the ground state. Hotop et al (34) extract from the measured distributions values for the following quantities:

(i)   final state branching ratios $\sigma(^2P_{3/2})/\sigma(^2P_{1/2})$. These are directly given by the area under the corresponding distributions.

(ii)  total cross section ratios $\sigma(^3P_2)/\sigma(^3P_0)$ for the two Ne-metastables. Values are here obtained from the corresponding distributions making use of the known flux ratio of 5:1 for Ne$^*$($^3P_2$) and Ne$^*$($^3P_0$).

(iii) well depths (D$^*$) for the Ne$^*$-rare gas interaction potentials. D$^*$-values are obtained from the approximate formula (39)

$$D^* \approx E_o - \varepsilon_* - V_+(R_e^*) \qquad (3)$$

with $R_e^*$ the equilibrium distance of $V_*(R)$, and $\varepsilon_*$ the energy position of the low energy edge of the measured distributions. According to the semiclassical prescription due to Miller (40) $\varepsilon_*$ is the energy at which the intensity of the low energy peak of the distribution has dropped to 44% of its peak value. $V_+(R_e^*)$ is estimated to be $\approx$ – 6 meV. The authors point out that the D$^*$-values obtained in this way are to be considered tentative.

The experimental data on the Ne$^*$-rare gas systems obtained by Hotop et al (34) and by Gregor and Siska (33) are summarized in table 1. Also given are well depth

| Collision system | $\frac{\sigma(^2P_{3/2})}{\sigma(^2P_{1/2})}$ a | $\frac{\sigma(^3P_2)}{\sigma(^3P_0)}$ a | $D^*$ (meV) a | $D^*$ (meV) b | $R_e^*$(Å) b | Collision system | $D$(meV) c | $R_e$(Å) c |
|---|---|---|---|---|---|---|---|---|
| Ne$^3P_2$ + Ar | 1.51 (7) | 0.81 (8) | ~ 5 | 5.48 | 5.0 | Na + Ar | 5.31 | 5.01 |
| Ne$^3P_0$ + Ar | 3.94 (24) | | ~ 4 | | | | | |
| Ne$^3P_2$ + Kr | 1.34 (6) | 0.82 (6) | ~ 7 | 8.35 | 5.0 | Na + Kr | 8.68 | 4.96 |
| Ne$^3P_0$ + Kr | 3.83 (20) | | ~ 5 | | | | | |
| Ne$^3P_2$ + Xe | 1.25 (6) | 0.86 (6) | ~17 | 12.5 | 5.0 | Na + Xe | 13.0 | 5.06 |
| Ne$^3P_0$ + Xe | 3.54 (14) | | ~15 | | | | | |

(a) ref. 34; (b) ref. 33; (c) ref. 37

Table 1: Experimental data on the Ne$^*$-rare gas Penning systems (left part of the table). $D^*$ and $R_e^*$ are well depth and equilibrium distance of the interaction potential. For comparison, well depth and equilibrium distance for corresponding Na-rare gas potential (right side).

and equilibrium distance for Na-rare gas potentials. The close agreement between Ne$^*$-rare gas potentials, and Na-rare gas potentials has been pointed out in both ref. (33) and (34). An important result is that the tentative values of the potentials for Ne$^*$($^3P_2$)- and Ne$^*$($^3P_0$) interaction potentials are equal (34). This supports the single potential model of ref. (33). Further support to this model is given by the observed similarity of the electron distributions for Ne($^3P_2$) and Ne($^3P_0$) systems (see fig. 5), which indicates that no significant $\Omega$-splitting occurs in the accessible region of distances for $^3P_2$ rare gas potentials. While these findings are well explained by assuming that the interactions with Ne$^*$($2p^5$ 3s) are dominated by the 3s-electron, with the core structure being of little importance, the strongly non statistical branching ratios observed by Hotop et al (34) seem to indicate that the $2p^5$-core orientation during the collision plays an important role for the imaginary part of the interaction. This has been discussed in a recent theoretical paper by Morgner (41).

Morgner develops a theoretical description that allows to interprete the two final state branching ratios $R_f(^3P_2)$ and $R_f(^3P_0)$, and the total ionization cross section ratio for Ne$^*$($^3P_2$) and Ne$^*$($^3P_0$), $R_{tot}$ (see table 1), in terms of transition amplitudes $u_{m'm}$, where m' is the quantum number characterizing the orientation of the rare gas ion 2p-hole in the final state of the collision system Ne$^*$- X$^+$($np^5$), and m the same quantum number for the Ne$^*$($2p^5$3s) 2p-hole in the Ne$^*$ - X system before ionization. The three observed quantities $R_f(^3P_2)$, $R_f(^3P_0)$, and $R_{tot}$ are expressed in terms of the relative values of five matrix elements $u_{00}$, $u_{11}$, $u_{10}$, $u_{01}$, and $u_{-11}$.

Morgner points out that he derived the expressions under the assumption that the ionizing transition proceeds via the "exchange mechanism" (38), i.e. an electron from the target is transferred to the empty 2p-Ne$^*$ orbital, accompanied by the ejection of the 3s Neon electron. For this mechanism he finds, by considering theoretical expressions of the matrix elements $u_{m'm}$, that their values should decrease as $|m' - m|$ increases. In his analysis of the Ne$^*$ - Ar data Morgner sets $u_{-11} = 0$. Choosing the normalization $u_{00} = 1$, he then determines the three remaining numbers $u_{01}$, $u_{10}$, $u_{11}$ from the three experimental quantities. He finds $u_{10}$ and $u_{11}$ to be approximately zero, within the errors given by the experimental uncertainties, and $u_{01} = - 0.171 \pm 0.085$. Recalling the definition of the matrix

elements $u_{m'm}$ given above, this result means that in the ionization of Ar by Ne[*]
only Ne-Ar[+*] systems with $\Sigma$-symmetry are formed with significant probability, the
main contribution ($u_{00}$ = 1) being caused by transfer of a $3p\sigma$ electron of Ar to
the $2p^5(\sigma)$-hole of Ne[*],and a smaller contribution ($u_{01}$ = 0.171 ± 0.085) by transfer
of a $3p\sigma$ electron of Ar to the $2p^5(\pi)$-hole of Ne[*]. The latter contribution leads
to ejection of an electron of $\pi$-symmetry with respect to the internuclear axis.

Morgner points out that the result is not quite definite because it is based on
the, albeit very probable, assumption that $u_{-11} \approx 0$. Further support for the
"selection rule" - predominant formation of the ionized system in a $\Sigma$-state in
Penning ionization of rare gases - is given by the result of an analysis of the
He[*]$(2^3S)$-Ar system (42), where the same selection rule is found to hold. As Morgner points out (41), a final proof for Ne[*]-rare gas systems can be obtained by
performing experiments with oriented Ne[*] metastables.

The described analysis of Morgner (41), that led to a convincing interpretation
of the results of Hotop et al (34) in terms of the $\Sigma$-selection rule, was based
on the assumption of the "exchange mechanism" (38) for Penning ionization. On the
other hand Gregor and Siska (33) conclude from a discussion of the distance dependence of the imaginary part of the Ne[*]-rare gas potentials determined by them,
that the so called "radiative mechanism" (38) is dominant. We would like to point
out that the branching ratios observed by Cermak (8) for the population of excited states of Hg[+] in Penning ionization by Ne[*], which he found to be very different from the ones for photoionization, is strong evidence against the dominance
of the radiative mechanism in Penning systems involving Ne[*].


STUDIES OF ASSOCIATIVE IONIZATION

Cermak (6) has proposed two mechanisms for associative ionization in collisions
of excited atoms with other species. These are the mechanisms described in the
introduction as cases (a) and (b), and shown in Fig. 1. The validity of the proposed mechanism for case (b) - where the initial state lies in the ionization
continuum at large R was demonstrated rather early by electron spectroscopy (38).
For case (a) - where the initial state lies below the continuum at large R, and
has to cross into it in order to make possible associative ionization by the
mechanism proposed by Cermak (6) - also an alternative mechanism has been discussed (43). In rather recent studies, however, it was verified for X[*] - H systems
(X[*] = Ar[*], Kr[*], Xe[*]) by electron spectroscopy that indeed the mechanism proposed
by Cermak is dominant (44-46). We discuss now a very recent electron spectroscopic
study of associative ionization in thermal Ne[*]$(^3P_2)$ - H, and Ar[*]$(^3P_2)$ - H collisions by Lorenzen et al (47). The first system belongs to case (b) and the second
one to case (a).

Metastables were produced in a cold cathode discharge ($\sim$ 350°K),and H-atoms in a
radio frequency discharge ($\sim$ 300°K). In the case of Ne[*] the mixed beam, containing the $(^3P_2)$ - and $(^3P_0)$ metastables in a ratio of about 5:1, was state selected
by optical excitation (48, 49). The electrons were analyzed in a 127° cylindrical
condensor at a resolution of 20 meV. The spectra were corrected for energy dependent transmission and, if necessary, for contributions from other target
species than H. As example we reproduce in Fig. 6 the spectrum for the system
Ar[*]$(^3P_2)$ - H as obtained by Lorenzen et al (47).

In associative ionization processes, here

$$X^* + H \to XH^+(v, J) + e^-(\varepsilon),\qquad(4)$$

the electron distribution $P(\varepsilon)$ directly images the population distribution $P(v, J)$
of the rovibronic states of the associative ion because of energy conservation:

$$\varepsilon(v, J) = E_o + E_{rel} + U^{+*}(v, J)\qquad(5)$$

with $E_o$ the excitation energy of $X^*$ minus ionization potential of H, $U^+(v, J)$ the binding energy of the (v, J)-state of $XH^+$ relative to the dissociation limit $X + H^+$, and $E_{rel}$ the relative collision energy in the initial state (47). The spectra of Lorenzen et al show for the first time resolved rotational structure, as seen in Fig. 6. They demonstrate that from the energy position of individual

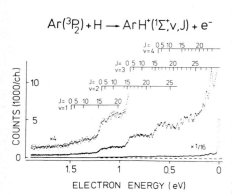

$$Ar(^3P_2) + H \longrightarrow ArH^+(^1\Sigma^+, v, J) + e^-$$

Fig. 6: Electron spectrum due to associative ionization in $Ar^*(^3P_2)+H$ collision, showing rovibronic structure. (reproduced from Lorenzen et al (1980).)

rovibronic lines $\varepsilon(v, J)$ detailed information on the $XH^+$-potential can be obtained. Starting with an analytical modified Morse potential, as determined by Weise et al (50), they assign on the basis of quantum mechanical calculations individual lines, and determine in this way the potential. They determine the well depths $D_e(NeH^+, X^1\Sigma^+) = 2.27$ (3) eV, and $D_e(ArH^+, ^1\Sigma^+) = 4.02$ (3) eV. They further point out that an evaluation of the distributions P(v,J) for a known $XH^+$-potential will lead to very detailed information on the complex $X^*$-H-potential.

With their state selected $Ne^*$-beam Lorenzen et al are able to distinguish spectra for $Ne^*(^3P_2)$ and $Ne^*(^3P_0)$. The $Ne(^3P_2)$-spectrum is several eV wide and in essential agreement with the spectrum of Morgner (51), apart from additional structure. This spectrum is made up of a part, with $\varepsilon > E_o + E_{rel}$, that corresponds to formation of stable $NeH^+$, and another part, with $\varepsilon < E_o + E_{rel}$, corresponding to Penning ionization. The ratio of the latter to the first part is about 2. The existence of an extended low energy part indicates a deep well of $Ne^*(^3P_2)$-H potential. From the position of the low energy edge of this part Morgner (51) determined a well depth of 2.55 eV. The $Ne(^3P_0)$-spectrum is very narrow ($\approx$ 50 meV) and centered at $E_o$, and the intensity corresponds to an ionization cross section about 60 times smaller than that for $Ne(^3P_2)$ (47). Expecting an analogous behaviour for $Ar^*$-H, Lorenzen et al ascribe the observed broad spectrum to $Ar^*(^3P_2)$-H interactions.

For $Ar^*(^3P_2)$-H the value $E_o$ is negative by about 2 eV so that always $\varepsilon > E_o + E_{rel}$ and only associative ionization occurs. The shape of P(v, J) is not discussed by Lorenzen et al. Although it is very similar to the one of $Ne^*(^3P_2)$-H in the region $\varepsilon > E_o + E_{rel}$, which indicates the mechanism of ionization to be the type (a) proposed by Cermak (6), it is possible that the observed enhancement of population of the (v, J)-states corresponding to $\varepsilon \gtrsim 0$ is due to the other proposed mechanism (43), where it is assumed that first the $X^*$-H system adiabatically changes to curves dissociating to $X - H^{**}$ ($H^{**}$: Rydberg states) followed by loss of the loosely bound electron caused by diabatic coupling to the continuum $XH^+$. This may be expected in view of the close correspondence of the two processes (i) decay of a molecular system by emission of a slow electron into the continuum, and (ii) transition into a molecular Rydberg state below the same continuum (52-54).

Very detailed information on the process of associative ionization can also be obtained by optical spectroscopy of the associative ions formed. This has recently been demonstrated by Hartmann and Winn who analyzed the visible fluorescence arising from reaction of $Ar^*(^3P_2)$-metastables with Ca-atoms in a flowing afterglow (55). Due to the low ionization energy of Ca not only the ground state $Ca^+(4s^2S_{1/2})$ but also excited states $Ca^+(3d\,^2D_{5/2,\,3/2})$ and $Ca^+(4p\,^2P_{3/2,\,1/2})$ can be formed, the latter ones being observable through the allowed transitions to the ground state. To the red of these $Ca^+$-lines, a molecular band system was

found that was attributed to the $(A^2\Pi \to X^2\Sigma^+)$ transition of CaAr$^+$.
On the red side this band system changes over into an undulatory continuum ascribed to transitions into the vibrational continuum of $X^2\Sigma^+$. Using model potentials containing, in addition to the known polarizability of Ar, two free parameters, the measured spectrum is simulated in fit calculations performed either by a semiclassical method or by numerical integration of the radial Schrödinger equation. In this way rather accurate values for the parameters are obtained. Hartmann and Winn (55) report a well depth of 0.124 (12) eV, and an equilibrium distance of 2.8 Å, for $X^2\Sigma^+$ and a well depth of 0.607 (12) eV, and an equilibrium distance of 2.6 Å, for $A^2\Pi$. Further, absolute rate constants for population of the observed levels are measured, and, by comparison of the intensity of Ca$^+$(4 p $^2P_{3/2,}$ $_{1/2}$)-lines with the intensity of the $(A^2\Pi \to X^2\Sigma^+)$-band, separate partial Penning- and associative ionization cross sections are obtained. These are 33,7 Å for Penning ionization into Ca$^+$(4p $^2P_{3/2,\ 1/2}$) by Ar$^*$($^3P_2$), and 5.5 Å$^2$ for associative ionization.

Another interesting result of Hartmann and Winn (55) is the strongly non statistic population of the two fine structure states of Ca$^+$(4p $^2P_{3/2,\ 1/2}$). They find a ratio of (J = 3/2):(J = 1/2)= 5:1. As possible explanation the authors propose a propensity to conserve total angular momentum in the ionization by Ar$^*$($^2P_2$).

DETERMINATION OF INITIAL STATE INTERACTION POTENTIALS BY PENNING IONIZATION
SPECTROSCOPY

Since the first electron spectroscopic studies of Penning systems that showed the influence of the initial state potential curve on the observed electron spectra (56), it has been the aim of many studies to determine these potentials by analyzing Penning spectra. The most extensive study of this type was carried out by Morgner and Niehaus (57) for He($2^3S$)/H. Since the electron spectrum reflects the difference potential $V_*(R) - V_+(R)$ rather than $V_*(R)$ alone, in principle $V_+(R)$ must be known in order to determine $V_*(R)$. Another requirement for a reliable determination of $V_*(R)$ is that its well depth must be large compared to the spread of relative collision energies and also compared to the resolution of the electron detector. Especially favourable systems in this context are those involving an open shell atom as target, because it forms a strong binding with the excited atom.

In a very recent paper Lorenzen et al (58) report on the systems, Ne($^3P_2$) and Ne($^3P_0$) with Li, Na and K. The electron spectra for Ne$^*$($^3P_2$) are found to be broad with a pronounced low energy peak that is shifted towards low energies as compared to $E_o$ (difference between excitation energy of Ne$^*$ and ionization energy of the alkali), while the spectra for Ne$^*$($^3P_0$) are narrow and close to $E_o$. As an example in Fig. 7 their spectrum for Ne($^3P_2$)- Na($^2S_{1/2}$) is shown. Well depths are obtained using relation (3), whereby the edge position $\varepsilon_*$ is determined by fitting the low energy peak and a clearly visible second peak with a semiclassical expression (59) that approximately describes the distribution in this region of electron energies corresponding to transitions around the minimum of the difference potential. In addition to $\varepsilon_*$ this evaluation yields the curvature of the difference potential at the minimum. The quantity $V_+(R_o)$ of relation (3) which contributes only a small

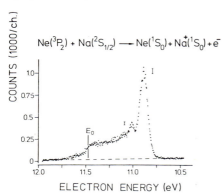

$$Ne(^3P_2) + Na(^2S_{1/2}) \longrightarrow Ne(^1S_0) + N\overset{+}{a}(^1S_0) + e^-$$

COUNTS (1000/ch.)

ELECTRON ENERGY (eV)

Fig. 7: Penning ionization electron spectrum for Ne$^*$($^3P_2$) + Na($^2S_{1/2}$) (reproduced by Lorenzen et al (1981).)

correction, is obtained from theoretical potential curves $V_+(R)$ for these systems, using the great similarity between $Ne^*(^3P_2)$ and Na regarding their binding properties to determine $R_e$.

The well depths obtained by Lorenzen et al (58) are $D(^3P_2 - Li) = 798$ (22) meV, $D(^3P_2 - Na) = 672$ (18) meV, and $D(^3P_2 - K) = 561$ (18) meV, for $Ne^*(^3P_2)$, and $D(^3P_0 - Li) = 35$ (28) meV, $D(^3P_0 - Na) = 48$ (20) meV, and $D(^3P_0 - K) = 32$ (20) meV, for $Ne^*(^3P_0)$. In addition, equilibrium distances, curvatures at these distances, and zero crossing distances for these potentials are reported. All quantities for $Ne(^3P_2)$-Alkali potentials are found to be very similar to the ones of the corresponding Na-alkali $^1\Sigma$-potentials. As an example we quote the ratios of well depths given by Lorenzen et al (58), for Li, Na and K targets, respectively: 0.94 (3), 0.906 (27), 0.859 (28).

By comparison of integrated intensities of $^3P_0$ - and $^3P_2$ spectra the following cross section ratios, $\sigma(^3P_2)/\sigma(^3P_0)$, are found for Li, Na and K, respectively: 11.9 (24), 10.7 (13), and 15.7 (31). The large cross section ratios are expected from the potential curves (e.g. 60).


PENNING ION - PENNING ELECTRON COINCIDENCE STUDIES

Penning ionization reactions are much more complicated for molecules than for atoms. This is mainly due to the large number of additional degrees of freedom, and to the close spacing of electronic states of Penning systems involving molecules. Our knowledge on such systems from experimental methods as described in the previous sections therefore is much less complete than for atoms. A discussion of recent work may be found in two recent articles (16, 61) and in references cited therein. As has been demonstrated by Münzer and Niehaus (62) and by Goy et al (63) a new method consisting of the coincident detection of energy analyzed Penning electrons and mass selected ions formed opens new possibilities to study Penning systems involving molecular targets. Preliminary results on the $He^*(2^3S)/H_2$-system, studied by Münzer and Niehaus, have already been discussed by Hotop (16). We therefore report here only on the work of Goy et al who have investigated the $He(2^3S)/NO_2$ system.

In order to achieve a sufficient coincidence counting rate an integral electron spectrometer of large acceptance angle ($\sim 1\%$ of $4\pi$) and somewhat lower energy resolution ($\sim 2\%$ of the electron energy) is used. The mass selection of ions $(NO_2^+, NO^+, O^+)$ is performed by time of flight analysis. For the purpose of interpretation both, Penning ion - Penning electron, and photo ion-photoelectron coincidence (integral) spectra are measured. These two spectra, measured with 584 Å photons and $He(2^3S)$-metastables, are reproduced in Fig. 8(a) and (b), respectively.

One advantage of these coincident spectra is that they are free of background. This allows to identify and to evaluate broad features which in non coincident spectra usually have been interpreted as a continuous background due to scattered electrons.

We can only sketch here the course of argumentation of Goy et al (63) that leads to a rather complete and unambiguous determination of the $He(2^3S)$-$NO_2$ Penning reaction. Based on assignments of electronic states obtained by Brundle et al (64) from their 584 Å-photoelectron spectra, information on the fragmentation of $NO_2^+$ at certain internal energies in the various electronic states is derived from the spectrum in Fig. 8(a). Most important for the interpretation of the Penning process is the finding that $NO_2^+$ is stable only if either formed in the electronic ground state $(^1A_2)$, or in the first three levels of the vibrational bending mode in the first electronically excited $NO_2^+(^3B_2)$-state. The instability of $NO_2^+$ in higher vibrational levels of the $(^3B_2)$-state and in other electronic states is evident through the absence of an increase of the coincidence count rate $e^-(NO_2^+)$ below 8.2 eV in Fig. 8(a). The sharp rise corresponding to population of the first $(^3B_2)$-

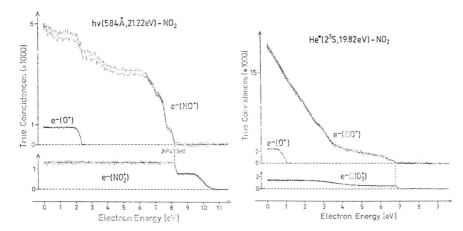

Fig. 8(a): Photoionization of $NO_2$. Spectra of electrons which are coincident with the ions $NO_2^+$, $NO^+$, $O^+$. (reproduced from Morgner (1981).)

Fig. 8(b): Penning ionization of $NO_2$. Spectra of electrons which are coincident with $NO_2^+$, $NO^+$, $O^+$. (reproduced from Morgner (1981).)

state is also observed in the $e^-$ ($NO_2^+$) Penning spectrum.
It lies at the position given by $E_o$ and proves that the Penning transition occurs from a "weak-interaction" initial state. The same is true for the sharp rise in the $e^-$ ($O^+$)-intensity due to population of the second ($^3B_2$)-state of $NO_2$. On the other hand, the rise corresponding to population of the ($^1A_1$)-state of $NO_2$ is absent around $E_o$ in the $e^-$ ($NO_2^+$)-intensity of the Penning spectrum. Instead an increase of intensity is observed between 5 and 2 eV. This increase is due to formation of $NO_2^+$ in the ($^1A_1$) and ($^3B_2$) state, and therefore corresponds to a deep minimum in the initial state potential from which the transition occurs. Approximate evaluation of this feature on the basis of relation (3) yields a minimum of $D_e$ = (4.8 + 0.1, - 0.3) eV. Thus there exist two initial state potential curves from which Penning transitions occur (i) a weak interaction potential that populates triplet final states, and (ii) a strong interaction curve that populates ($^1A_1$) and ($^3B_2$) states. Since He* ($2^3S$) - $NO_2(^2A_1)$ can form a quartet and a doublet curve, Goy et al (63) identify the weak interaction curve as the quartet curve, and the strong interaction curve as the doublet curve, because in this way all observed transitions are spin allowed. They further remark that the absence of any structure due to spin forbidden transitions is a most direct proof for the validity of the spin conservation rule. The very strong attraction on the doublet initial state potential curve is ascribed to an adiabatic rearrangement at a "curve crossing" with the doublet ionic curve of $He^+$ ($^2S$) - $NO_2^-(^1A_1)$ configuration.


FINAL REMARKS

The field of investigations of processes occurring in collisions of excited species with atoms and molecules has grown enormously during the last decade. In order to be able to carry out a discussion in some detail we have selected those topics that in our opinion are most closely connected to the work of Cermak, i.e. the "classic" field of Penning ionization. We will now try to shortly indicate new developements in the wider field.

In a number of recent papers spectroscopy of electrons ejected in metastable surface collisions has been applied to obtain information on the electronic structure of clean or adsorbate covered surfaces (e.g. 65-69). One of the two possible
mechanisms invoked to describe electron ejection is the so called Auger deexcitation process (AD) which is equivalent to Penning ionization in the gas phase. This
means that the (AD)-electron spectrum directly reflects the electronic properties
of the outermost surface layer, in the same way as Penning ionization spectroscopy does for gas phase particles. Interpretation of (AD)-spectra have         in
the past been hampered by the presence of the other possible process leading to
electron ejection, the so called resonance ionization Auger neutralization (RI +
AN) process, in which first the excited electron of the metastable is transferred
to empty states above the Fermi level of the solid, and secondly the projectile
ion is neutralized by transfer of a valence electron, accompanied by an ejection
of another valence electron from the solid. Evaluation of spectra due to this
process requires self deconvolution of the density distribution of target states
(69) and therefore yields less direct information. Conrad et al (68,70) have demonstrated that under certain conditions the (RI + AN) can be suppressed or its
contribution be distinguished from the (AD)-contribution. Thus it appears possible
to successfully apply Penning electron spectroscopy to the study of surfaces.
The (RI + AN)-process is also suppressed if several layers of molecules are condensed on (cold) surfaces. This has been demonstrated in a series of publications
of Kuchitsu and coworkers (e.g. 71). By comparison of the Penning spectra from
condensed molecules with corresponding spectra obtained for gas phase molecules,
information on the forces responsible for condensation can thus be obtained.

Valuable information on the Penning process has been obtained in the past from measurements of the destruction of metastables in collisions, by monitoring the metastable density in the presence of some target gas. Since Penning ionization is often
the only destruction process the absolute destruction rates obtained allow to
derive effective absolute Penning cross sections. Absolute cross sections derived
in this way from flowing afterglow measurements are still accepted to be the most reliable ones (e.g. 30). In addition, since measurements are not so time consuming
and the method very versatile, it is possible to perform rather comprehensive
studies. In a recent paper Wren and Setser report results obtained by the flowing
afterglow technique for the quenching of the metastable states of Ne, Ar, and Kr
by Hg-atoms (72). They discuss these results in terms of interaction potentials
and corresponding models for Penning ionization. Ueno et al (73) use the
$N_2^+(B^2\Sigma_u^+ \rightarrow X^2\Sigma_g^+)$ optical emission of $N_2^+$ formed in collisions with He$^*$($2^3$S) as a
monitor signal for the He($2^3$S) concentration. By adding target gas and by measu-
ring the $N_2^+$-emission after electron pulse excitation they determine destruction
rate constants for a large number of He($2^3$S)/target systems.

While considerable information on ionizing collision processes involving excited
atoms in long lived metastable states, and in long lived high Rydberg states
(see f.i. 74) has accumulated, rather little is known on systems involving short
lived excited states. Investigations of such systems have now become possible by
application of lasers. Bussert et al (75) reported the first measurements of electron spectra arising in collisions of laser excited Ne$^*$($2p^5 3p\ ^3D_3$) atoms with
several target gases. They create the Ne$^*$-atom in certain states of alignment and
polarisation with respect to the relative collision velocity, and demonstrate
that this influences in a characteristic way the observed electron spectra. It
appears that a whole new field in Penning electron spectroscopy opens up when
oriented excited atoms are used. The first successful beam study of ionizing
collisions of laser excited He in the short lived ($3^1$P)-state has been reported
by Pesnelle et al (76). In the crossing region of a velocity selected He-meta-
stable beam with a Ne-gas target He$^*$ ($3^1$P) is created by laser excitation from the
He($2^1$S)-state with an efficiency of close to 100%, and the velocity dependence
of associative ionization leading to HeNe$^+$-formation is measured.

Since in the Penning process charged particles are formed that are easily detected, this process is a sensitive means to study the interaction of strong laser

fields with a collision system. Weiner and Dingels (77) conclude from measurements of the $Na_2^-$ and $Na^+$ intensity as a function of laser power in Na/Na collisions that they have observed laser assisted Penning ionization in collisions of two $Na^*(3p)$ atoms.

Finally we would like to point out that Penning ionization is also an important process in collisions of ground state particles at somewhat higher collision energies, where the quasimolecule may be excited to states that can decay by electron ejection (59), and in collisions at rather low energies between ions with atoms or molecules where after adiabatic rearrangement the system may decay via a Penning process (53). Further, the spontaneous ionization process occurring in slow collisions with highly charged ions by a direct mechanism of the type $A^{q+} + B \rightarrow A^{(q-1)+} + B^{++} + e^-$ is very closely related to the Penning process and can be described within the same theoretical framework (e.g. 54). In all these spontaneous ionization processes of a collision system the ejected electron serves as a sensitive probe yielding very direct information on the collision itself.

ACKNOWLEDGEMENT

It is a pleasure to thank H. Hotop and H. Morgner for valuable suggestions regarding the organization of this manuscript, and for communicating material prior to publication.

REFERENCES

[ 1] Cermak, V. and Herman, Z., Collection Czech. Chem. Comm. 27 (1962) 406.
[ 2] Cermak, V. and Herman, Z., Collection Czech. Chem. Comm. 30 (1965) 169.
[ 3] Herman, Z. and Cermak, V., Collection Czech. Chem. Comm. 30 (1965) 2114.
[ 4] Herman, Z. and Cermak, V., Collection Czech. Chem. Comm. 33 (1968) 468.
[ 5] Cermak, V., J. Chem. Phys. 44 (1966) 3781.
[ 6] Herman, Z. and Cermak, V., Collection Czech. Chem. Comm. 31 (1966) 649.
[ 7] Cermak, V., Collection Czech. Chem. Comm. 33 (1968) 2739.
[ 8] Cermak, V. and Herman, Z., Chem. Phys. Lett. 2 (1968) 353.
[ 9] Cermak, V. and Ozenne, J.B., J. Mass Spectrom. Ion Phys. 7 (1971) 399.
[ 10] Cermak, V. and Yencha, A.J., J. Electron Spectrosc. Relat. Phenom. 8 (1979) 109.
[ 11] Cermak, V., J. Electron Spectrosc. Relat. Phenom. 9 (1976) 419.
[ 12] Cermak, V., Spirko, V. and Yencha, A.Y., J. Electron Spectrosc. Rel. Phenom. 8 (1976) 339
[ 13] Cermak, V. and Yencha, A.J., J. Electron Spectrosc. Relat. Phenom. 11 (1977) 67.
[ 14] Niehaus, A. in "The Physics of Electronic and Atomic Collisions", Proc. VIII ICPEAC, ed. by B.C. Cobic and M.V. Kurepa, Institute of Physics, Beograd (1973), p. 649.
[ 15] Manus, C., Pesnelle, A. and Watel, G. in "The Physics of Electronic and Atomic Collisions", Proc. X ICPEAC, ed. G. Watel, North Holland, 1978, p. 525.
[ 16] Hotop, H. in "The Physics of Electronic and Atomic Collisions", Proc. XI ICPEAC, ed. by N. Oda and K. Takayanagi, North Holland, 1980, p.271.
[ 17] Hickman, A.P. and Morgner, H., J. Phys. B9 (1976) 1765.
[ 18] Morgner, H. and Niehaus, A., J. Phys. B12 (1979) 1805.
[ 19] Burdenski, S., Feltgen, R., Lichtenfeld, F. and Pauly, H., Chem. Phys. Lett. 78 (1981) 296.
[ 20] Illenberger, E. and Niehaus, A., Z. Physik B20 (1975) 33
[ 21] Pesnelle, A., Watel, G. and Manus, C., J. Chem. Phys. 62 (1975) 3590.
[ 22] Siska, P.E., Chem. Phys. Lett. 63 (1979) 25.
[ 23] Chen, C.H., Haberland, H. and Lee, Y.T., J. Chem. Phys. 61 (1974) 3095.
[ 24] Brutschy, B., Haberland, H. and Schmidt, K., J. Phys. B9 (1976) 2693.
[ 25] Werner, F.J., Staatsexamensarbeit, Univ. Freiburg (1979), unpublished.

[26]  Brutschy , B., Haberland, H., Morgner, H. and Schmidt, K., Phys. Rev. Lett.
      36 (1976) 1299.
[27]  Haberland, H. and Schmidt, K., J. Phys. B10 (1977) 695.
[28]  Jordan, R.M., Martin, D.W. and Siska, P.E., J. Chem. Phys. 67 (1977) 3392.
[29]  Martin, D.W., Gregor, R.W., Jordan, R.M. and Siska, P.E., J. Chem. Phys.
      69 (1978) 2833.
[30]  Schmeltekopf, A.L., Fehsenfeld, F.C., J. Chem. Phys. 53 (1970) 3173.
[31]  Siska, P.E., Chem. Phys. Lett. 63 (1979) 25.
[32]  Siska, P.E., J. Chem. Phys. 71 (1979) 3942.
[33]  Gregor, R.W. and Siska, P.E., J. Chem. Phys. 74 (1981) 1078.
[34]  Hotop, H., Lorenzen, J. and Zastrow, A., J. Electron Spectrosc. Relat.
      Phenom. 23 (in press).
[35]  Tang, S.Y., Marcus, A.B. and Muschlitz, E.E., Jr., J. Chem. Phys. 56 (1972)
      566.
[36]  Neynaber, R.H. and Tang, S.Y., J. Chem. Phys. 70 (1979) 4272.
[37]  Buck, V. and Pauly, H., Z. Physik 208 (1968) 208, 390.
[38]  Hotop, H. and Niehaus, A., Z. Physik 228 (1969) 68.
[39]  Hotop, H. and Niehaus, A., Z. Physik 238 (1970) 452.
[40]  Miller, W.H., J. Chem. Phys. 52 (1970) 3563.
[41]  Morgner, H., J. Phys. B (to be published).
[42]  Hoffmann, V. and Morgner, H., J. Phys. B12 (1979) 2857.
[43]  Berry, R.S. in C. Schlier, Ed., Proceedings of the International School of
      Physics "Enrico Fermi", Course 44, Academic, New York, 1970, p. 193.
[44]  Morgner, H. and Niehaus, A., "Electronic and Atomic Collisions", in J.S. Ris-
      ley and R. Geballe, eds., Abstracts IXth ICPEAC, Seattle, 1975, p. 1073.
[45]  Niehaus, A., in "Physics of Ionized Gases", Ed. B. Navinsek, J. Stefan Insti-
      tute, Ljubljana, 1976), p. 143.
[46]  Morgner, H., PhD. Thesis Univ. Freiburg, 1976.
[47]  Lorenzen, J., Hotop, H., Ruf, M.W., and Morgner, H., Z. Phys. A297 (1980) 19.
[48]  Dunning, F.B., Cook, T.B., West, W.S. and Stebbings, R.F., Rev. Scient.
      Instr. 46 (1975) 1072.
[49]  Hotop, H. and Zastrow, A., abstracts of papers Xth ICPEAC, Paris. 1977,p.306.
[50]  Weise, H.P., Mittmann, H.V., Ding, A., Henglein, A., Z. Naturf. 26a (1971)
      1122.
[51]  Morgner, H., J. Phys. B12 (1979) 2171.
[52]  Miller, W.H. and Morgner, H., J. Chem. Phys. 67 (1977) 4923.
[53]  Hultzsch, W., Kronast, W., Niehaus, A. and Ruf, M.W., J. Phys. B12 (1979)1821.
[54]  Niehaus, A., Comments Atom. Mol. Phys. 9 (1980) 153.
[55]  Hartmann, D.C. and Winn, J.S., J. Chem. Phys. 74 (1981) 4320.
[56]  Fuchs, V. and Niehaus, A., Phys. Rev. Lett. 21 (1968) 1136.
[57]  Morgner, H. and Niehaus, A., J. Phys. B12 (1979) 1805.
[58]  Lorenzen, J., Hotop, H. and Ruf, M.W., Z. Physik (to be published).
[59]  Gerber, G. and Niehaus, A., J. Phys. B9 (1976) 123.
[60]  Niehaus, A., Ber. Bunsenges. Phys. Chemie, 77 (1973) 632.
[61]  Niehaus, A. in "The Excited State in Chemical Physics" (Part 2), Ed. J.W.
      McGowan, John Wiley (1981) p. 399.
[62]  Münzer, A. and Niehaus A., J. Electron Spectrosc. Relat. Phenom 23 (in press).
[63]  Goy, W., Kohls, V. and Morgner, H., J. Electron Spectrosc. Rel. Phenom. 23
      (in press).
[64]  Brundle, C.R., Neumann, D., Price, W.C., Evans, D., Potts, A.W., Streets,
      D.G., J. Chem. Phys. 53 (1970) 705.
[65]  Boiziau, C., Garot, C., Nuvolone, R. and Roussel, J., J. Physique Letters
      39 (1978) L339.
[66]  Johnson, P.D. and Delchar, R.A., Surface Sci. 77 (1978) 400.
[67]  Wang, S.W. and Ertl, G., Surface Sci 93 (1980) L75.
[68]  Conrad, H., Ertl, G., Küppers, J., Wang, S.W., Gerard, K. and Haberland, H.,
      Phys. Rev. Lett. 42 (1979) 1082.
[69]  Boiziau,C., Garot, C., Nuvolone, R. and Roussel, J., Surface Sci 91 (1981)
      313.
[70]  Conrad, H., Ertl, G., Küppers, J. and Sesselman, W., Surface Sci 100 (1980)
      L 461.

[71] Yencha, A.J., Kubota, H., Fukuyama, T., Kondow, T. and Kuchitsu, K., J. Electron Spectrosc. Relat. Phenom. 23 (to be published).

[72] Wren, D.J. and Setser, D.W., J. Chem. Phys. 74 (1981) 2331.

[73] Ueno, T., Yokoyama, A., Takao, S. and Hatano, Y., Chem. Phys. 45 (1980) 261

[74] Stebbings, R.F., in "The Physics of Electronic and Atomic Collisions", Proc. X ICPEAC, ed. G. Watel, North Holland, 1978, p. 549.

[75] Bussert, W., Hotop, H., Lorenzen, J., Ruf, M.W., Abstracts of papers, XI ICPEAC, Kyoto, (1979).

[76] Pesnelle, A., Runge, S., Sevin, D., Wolffer, N. and Watel, G., J. Phys. B14 (1981) 1827.

[77] Weiner, J. and Polak-Dingels, P., J. Chem. Phys. 74 (1981) 508.

PHYSICS OF ELECTRONIC AND ATOMIC COLLISIONS
S. Datz (editor)
© North-Holland Publishing Company, 1982

COLLISIONS OF RYDBERG ATOMS

F. Gounand[†]
CEN Saclay, 91191 Gif sur Yvette Cedex, France

Rydberg atoms are like hydrogen in their essential properties, because when the principal quantum number n is high enough the outer electron is influenced mainly by the charge of the core and not by its structure. Table I gives the typical, and rather unusual, orders of magnitude of some properties of Rydberg atoms as well as the corresponding scaling laws. In that table n has to be understood as being the effective quantum number (defined by $E = R_y/n^2$, where E is the binding energy and $R_y$ the Rydberg constant).

Table 1
Some Properties of the Rydberg Atoms

| Property | Scaling | Typical Value (n ~ 25) |
|---|---|---|
| Radius | $n^2$ | ~ 300 Å |
| Geometric Cross–Sections | $n^4$ | ~ $10^5$ Å$^2$ |
| Binding Energy | $1/n^2$ | ~ 200 cm$^{-1}$ |
| Separation of Adjacent n Levels | $1/n^3$ | ~ 10 cm$^{-1}$ |
| Lifetime | $n^3$ | ~ 15 μs |
| Ionizing Field $E_c$ | $1/n^4$ | ~ 800 V/cm |

Because of their large size and the weak binding energy of the outer electron Rydberg atoms are expected to be extremely sensitive to any kind of external perturbation, in particular that resulting from the interaction with another atom, a molecule, or a charged particle. Accordingly their collisional behavior should differ considerably from the one observed for low-lying states, thus justifying the interest of physicists.

The large size of the Rydberg atom suggests that in the collision of a Rydberg atom A with a perturber P the outer electron and the ionic core may behave as separate scatterers during the collision, allowing simplified theoretical treatment (Fig. 1). One may even speculate whether only one of the two interactions (i.e., $e^-$–P or $A^+$–P) has to be taken into account for a given process. This is one of the crucial questions which physicists have tried to answer in the past few years. Note also that one can hope that Rydberg atom collisions may provide useful information on both $e^-$–P or $A^+$–P scattering at very low energies, data that should be difficult to obtain by other techniques.

255

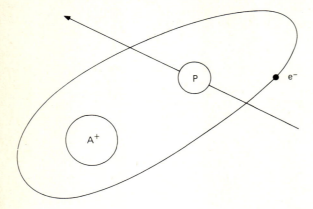

Fig. 1. Collision of a Rydberg atom A* with a particle P.

Experimental studies of collisions involving Rydberg atoms were started 50 years ago by pressure shift measurements of Rydberg spectral lines.[1]   Fermi, in a famous paper,[2] explained the observed results and laid the foundation of the theoretical treatment of Rydberg atom collisions.   In the last decade, due to the development of new experimental techniques, there has been a strong revival of interest in this topic.   Many aspects of collisions of Rydberg atoms have been studied.

An important aspect in the study of Rydberg collisions is the large number of the energetically allowed reactions (even at thermal energies).   Most frequently several processes occur concurrently, each contributing noticeably to the total depopulation (quenching) of the initial Rydberg state.   Table II lists some of these processes.   The last column indicates the type of particle for which extensive experimental results have been presented in Kyoto.[3]   (At: atoms, M: molecules, C: charged particles)

Table II
Some Collisional Processes Involving Rydberg Atoms

| Reaction | Perturber | |
|---|---|---|
| $A(n,\ell) + P \rightarrow A(n,\ell') + P$ | $\ell$ changing | At, M |
| $A(n,\ell) + P \rightarrow A(n',\ell') + P$ | n changing | At, M |
| $A(n,\ell) + P \rightarrow A(\neq n,\ell)$ | quenching | At, M |
| $A(n,\ell) + P \rightarrow A^+ + P + e^-$ | | |
| $\rightarrow A^+ + P^-$ | collisional ionization | M |
| $\rightarrow A^{P+} + e^-$ | | |

More practical reasons have also prompted the studies of collisions involving Rydberg atoms. For example, they play an important role in astrophysical and laboratory plasmas. Rydberg collisions may also affect the overall efficiency of laser isotope separation. Rydberg atoms also offer the promise of sensitive far infrared detection.

In the last ICPEAC a symposium on Rydberg states was held and we refer the interested reader to the corresponding five lectures[3] for extensive references to works appearing before 1978. This progress report will be organized as follows. We first briefly review the experimental techniques. In the second part we present a brief survey of the works that have appeared since the last ICPEAC which are along the lines presented at the Kyoto symposium. In the third part we focus more extensively on three new lines of experimental research, namely:

- Collisions in the presence of a DC electric field
- Ion-Rydberg atom collisions
- Resonant Rydberg-Rydberg collisions.

The paper ends with some concluding remarks.


I.  Brief Survey of Experimental Techniques

In an ideal experiment atoms are produced in a well-defined Rydberg state and interact with a target. Then the various (charged or neutral) end products have to be identified, detected, and the corresponding cross sections (or rate constants) have to be measured.

Selective excitation of Rydberg levels is most commonly achieved by pulsed dye lasers, due to their high power, their flexibility, and because they are especially suitable for time-resolved detection. Wavelengths from UV to near IR can be obtained with good spectral resolution ($\sim .5$ cm$^{-1}$) allowing selective excitation of levels up to $n \sim 45$ to be routinely performed. The one or multistep excitation takes place either from the ground state or from a metastable state (populated by appropriate techniques).

Neutral end products are usually detected either by time-resolved fluorescence or by selective field ionization (SFI). The first method has been extensively used in cell experiments. It is especially well-suited for the measurements of the quenching cross section of Rydberg levels. However, due to its rapidly decreasing efficiency as n increases, it is only tractable up to moderately high n values ($n \lesssim 20$). The SFI technique takes advantage of the fact that the outer electron, due to its low binding energy, can be pulled off by a modest electric field. Classically, for hydrogenic systems the required ionizing field is about $3.2 \times 10^8 n^{-4}$ V/cm. In fact the dynamics of the ionizing process is complex, but field ionization has been studied in detail.[4,5] The technique, specific to Rydberg state studies, has been proven simple, efficient and selective if careful analysis of the SFI signal is achieved. A typical field ionization apparatus is shown in Fig. 2. A pulsed electric field (of $\sim 5$ kV/cm for $n \sim 20$) is applied to the interactive region where the collision takes place. The resulting ions are extracted and detected by an electron multiplier. The method has been mainly used in beam apparatus although cell measurements are possible. Finally, if one excepts the line shape studies performed by Doppler-free two-photon spectroscopy[6] or photon echoes technique,[7] the SFI method is the only one used for collision studies at high n values. The study of ionic products can be achieved by the classical methods of charged particle detection.  The mass selection of the ions is most useful.    The reader interested in the details of the experimental techniques is referred to Ref. [8].

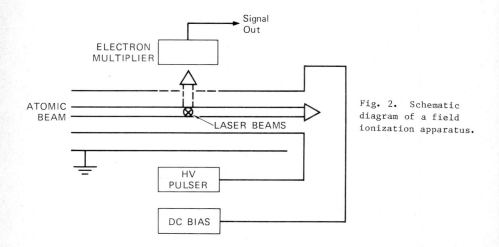

Fig. 2.  Schematic diagram of a field ionization apparatus.

## II.   Recent Progress in the Lines Already Presented in the Kyoto Conference

The most widely investigated processes are listed in Table II.  It has been shown that most of the observed results can be explained by only taking into account the $e^-$perturber interaction, although in some cases effects ascribable to the core are observed.  The resulting cross sections are much less or of the order of magnitude (in the case of a strong $e^-$-P interaction) of the geometric cross section.   The n dependence of Rydberg atom-atom cross sections usually shows a maximum for some $n_{max}$ value before a decrease is observed.[9]  The $n_{max}$ and $\sigma_{nmax}$ values as well as the whole shape of the  $\sigma(n)$  curve strongly depends, as expected, on the nature of the perturber and especially of the characteristics of the $e^-$-P scattering at low energies.  We present now a brief survey of the most recent experimental and theoretical works along that line.

### A.   Experimental Works

Hugon et al.[10] have studied the depopulation of highly excited Rb(ns) states (n ~ 25) induced by He atoms.  The measurements were performed in a cell by a time-resolved SFI technique.   The cross sections obtained  (~ $10^2$ $Å^2$)  agree reasonably well with a scaling formula derived by Hickman,[11] taking into account only the $e^-$-P interaction.  Similar experiments on Na(ns) states have also been performed by Boulmer et al.[2]  Line-shape studies of both broadening and shift due to alkali$^*$-alkali[6,13-16] and alkali$^*$-rare gas[17,18] collisions have been reported for high n values.  In some cases, as expected from the Fermi theory, effects of the ionic core-perturber interaction are observed.  These measurements at high n values provide results of special interest for a valuable comparison of the existing theories because the high n values range is especially appropriate for simple theoretical calculation.

Collisional ionization[19] in atom\*-atom collisions has been reported[20] for the Rb(9s)-Rb(5s) pair. The experiment used a mass-spectrometer allowing a detailed study of the end products. The reported cross section for associative ionization agrees well with the theoretical work of Mihajlov and Janev.[21] No experimental evidence for the formation of negative Rb⁻ ion has been obtained. This type of study, although of great theoretical and practical interest, is difficult, because the cross sections of the reactions leading to neutral end products are usually much larger than those corresponding to ionizing processes.

Various molecular targets have been investigated.[22-26] The results are usually in good agreement with the so-called free electron model.[3,8] Evidence for resonant energy transfer from electronic to vibrational energy in $Na^*-CH_4$ and $Na^*-CD_4$ collisions has been reported.[27] The resonance of the width is about 50 cm⁻¹. Moreover some selection rules for this type of process have been obtained and explained using a simple model.

## B. Theoretical Works

The validity of the approach taking into account only the $e^--P$ interaction has been widely discussed from different points of view[11,28-32] (involved energy defects, strength of the $e^--P$ interaction, excitation of the Rydberg atom). In particular the calculation of correction terms to the pure impulse formulation performed by Hahn[33] have led to good agreement with experimental data for the $\ell$-mixing process even at low (n ∼ 10) n values for highly polarizable targets. The influence of the ionic core-perturber interaction has also been investigated by various authors.[28-29,34] The recent works of Matsusawa[35] and Hickman[34] indicated that this influence is negligible compared to the $e^--P$ interaction for the angular mixing process at high n values. The obtention of simple scaling laws,[11,30,33] useful at high n values, have also been reported. These asymptotic formulae are clearly of great interest for astrophysicists as well as for people involved in plasma modeling. Some of them have already received an experimental confirmation.[10] Note finally that the free electron model has been able to explain the oscillations observed in the self broadening and shift of the alkali spectral lines.[15,36,37]

## III. A. Collisions in the Presence of a DC Electric Field

Two groups have reported recently the study of such processes.[38,39] The two experiments giving qualitatively the same results we will describe more extensively the one performed by Slusher et al.[39] This type of experiment is important from a practical point of view because under many circumstances in which collisions are thought to be important an external field is present (plasma physics, fusion experiments, astrophysics...). The studied process is the already widely investigated $\ell$-mixing process,[8,9] i.e., the collisional transfer of population from an initially populated n,$\ell$ state to all the nearly degenerate $\ell$ states of the same n, i.e., for example, in that particular case

$$Xe(31f) + Xe \rightarrow Xe(31; \ell \geqslant 4) + Xe \tag{1}$$

Fig. 3 shows the energy diagram of Xe in the field region of interest here. This figure displays the first effect of an applied static electric field: the energy levels are split by the field so that the zero field energy separations between the 31f state and the (31, $\ell \geqslant 4$) states noticeably vary when a modest DC field is present (in fact $\ell$ is no longer a good quantum number in the presence of the

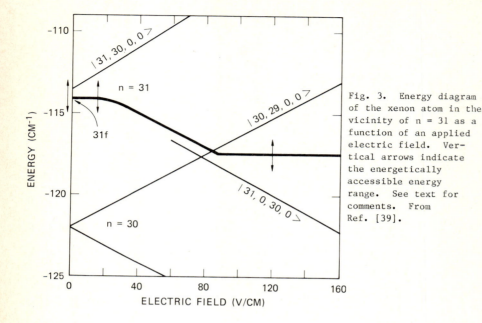

Fig. 3. Energy diagram of the xenon atom in the vicinity of n = 31 as a function of an applied electric field. Vertical arrows indicate the energetically accessible energy range. See text for comments. From Ref. [39].

field but $\ell$ states are adiabatically correlated to the Stark states labelled, as usual, by their parabolic quantum numbers). The energy defect between initial and final states being an important parameter in low-energy collisions a variation in the cross-section should occur when the field is applied. A second effect of the field is to strongly modify the shape of the wavefunction;[40] in Stark states the electron is localized on the positive or negative z side of the atom depending upon whether the energy shift is red or blue. In zero field the wavefunction exhibits a much more symmetric form. Thus if the cross section is related to the overlap integral between the initial and the final states, as expected for short range interaction, it should also exhibit some variation with the application of a DC field. Both indicated effects should lead to two observations: a change in the value of the cross section and the appearance of some selection rules.

The experimental apparatus has already been widely described elsewhere.[32] The obtained results are shown in Fig. 4 and can be explained qualitatively in the following way by considering only the variation of the energy levels due to the presence of the field. If one assumes that the energy transferable in the collision is of the order of 1 cm$^{-1}$ (this figure can be obtained by considering energy and momentum conservation for the collision between the quasi free Rydberg electron and the ground state Xe atom) Fig. 3 shows that in zero field all the (31,$\ell \geqslant 4$) states are accessible leading to the broad $P_1$ feature (for details on the field ionization process we refer the reader to Ref. [4, 5, 8]). In a field of 15 V/cm the arrow in Fig. 3 shows that only the lower members of the n = 31 Stark manifolds are energetically accessible leading to the observed reduction of the width of the $P_1$ feature. When E = 120 V/cm the Stark shifted state lies in the region where both the n = 31 and 30 manifolds overlap allowing

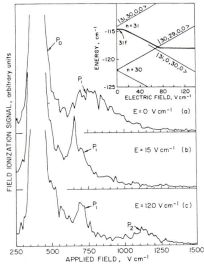

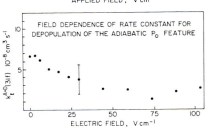

Fig. 4. SFI data obtained in Xe(31f)-Xe collisions for three different values of the DC electric field. The lower part of the figure shows the E dependence of the rate for process (1) deduced from the evolution of the $P_0$ adiabatic feature (the error bar shown is tentative). From Ref. [39].

a population transfer to some lower members of the n = 31 manifold as well as to some upper members of the n = 30 manifold. This explains the observation of a second feature $P_2$. In the absence of any detailed calculations as well as in view of the obvious crudeness of the above discussion, one cannot rule out the possibility of explaining, at least partly, the observed results by the overlap effect mentioned earlier.

The rate constant for depopulation of the adiabatic $P_0$ feature as a function of the applied field is also shown in Fig. 4. It exhibits a clear decrease. Thus the two expected effects of the field (variation in the cross section and in the distribution of the final states) have been demonstrated.

B. Rydberg Atom-Ion Collisions

The atom[*]-ion collisions are of great interest in plasma physics and astrophysics. The basic difference between atom[*]-ion and atom[*]-atom collisions is that in the atom[*]-ion case the involved interaction is long range (e[−]-ion). Thus the cross sections are not limited by the geometric size of the Rydberg atoms and huge cross sections can be expected.

Although the field has been very active for a long time from a theoretical point
of view,[41-44] state-resolved studies of ion-Rydberg collisions only appeared
recently.   Burniaux et al.[45] have reported absolute cross sections for the $e^-$
capture reaction:

$$He^{++} + H^*(n) \rightarrow He^+ + H^+ \quad (8 \leqslant n \leqslant 24) \tag{2}$$

More recently Kim and Meyer[46] have measured cross sections for $e^-$ removal in
$N^{3+} + H^*(n)$ collisions.   These two works have confirmed that ionization is
dominant at high reduced velocity $\tilde{v}$ (i.e., the ratio of the incoming ion
velocity to the velocity of the Rydberg $e^-$) while capture takes place at low $\tilde{v}$
values, as expected from classical trajectory Monte-Carlo calculations (CTMC).[44]

We will focus now on the recent experiment performed by MacAdam et al. on the
$\ell$ mixing processes induced by ion collisions.[47,48]   The process is the same as
the one discussed in III B, namely,

$$Na(nd) + X^+ \rightarrow Na(n,\ell \geqslant 3) + X^+ \quad (X^+ = He^+, Ne^+, Ar^+) \tag{3}$$

The reaction is studied in the matching velocity region (i.e., $\tilde{v} \sim 1$), contrary
to the case of atom*-atom collisions for which the collision velocities are at
most 1% as great as the orbital velocities.   The apparatus is shown in Fig. 5.   A
time-resolved SFI detection technique is used to analyze the end products.

The ion beam (F, A, $A_1$ X) is extracted from a discharge source, focussed by ion
optics (E, $L_1$, $L_2$) and velocity selected (W).   In the collision region it crossed
a Na thermal beam and two laser beams (not shown in the Figure) that create the
Na(nd) states (by the usual two-step photoexcitation procedure).   $P_1$ and $P_2$ are
the plates used for the SFI detection, the SFI ions being measured by an $e^-$-
multiplier (EM).   The incident ion beam current is measured by the Faraday cup
FC.   Slits A6 and A7 allow the determination of the beam profile.   Deflection
plates (D) are used for switching the beam from the collision region into a
second Faraday cup (FR).

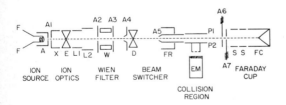

Fig. 5.  Apparatus for the
study of Rydberg atom-ion
collisions.  See text.  From
Ref. [48].

Fig. 6 shows time-resolved SFI signals for Na(28d) exposed to weak and strong
fluxes of 1 keV $Ne^+$ ions.   The peak centered at channel 8 has an area roughly
proportional to the Na(28d) population while the one centered at channel 18 comes
from the field-ionization of the collisionally populated Na(28; $\ell \geqslant 3$)
states.   For zero ion current only the peak centered at channel 8 occurs.   Note
that the total area under each of the three curves has approximately a constant
value.   Fig. 7 shows the n dependence of the fractional decrease R in area of the
signal related to the Na(nd) state (peak at channel 8 in Fig. 6) induced by the
ion beam.   (The R value can be shown to be proportional to the $\ell$-mixing cross

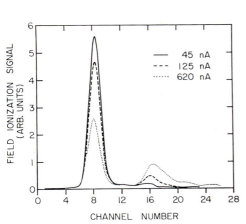

Fig. 6. Time resolved SFI signals from Na(28d) state for three different currents of 1 keV Ne$^+$ beam. From Ref. [48].

Fig. 7. n dependence of the $\ell$-change signals (see text) for Ne$^+$(1 keV) − Na(nd) collisions. The solid line is a power law least−squares fit to the data. The dashed line shows the theoretical result of Ref. [49] normalized at n = 25. From Ref. [48].

section.) The dashed line shows the theoretical result of Percival and Richards[49] normalized at n = 15. The agreement between theory and experiment is excellent; the $\ell$-mixing cross section scales like n$^5$ in the matching velocity region ($\tilde{v} \simeq 1$). A careful analysis of the geometry of the collision region as well as of the SFI signals allows the authors to put some of their data on an absolute basis. For example the $\ell$-changing cross section for Na(28d) colliding with 1 keV Ne$^+$ ions is reported to be $\sigma$(28d) = 5.25 10$^8$ Å$^2$, in good agreement with theoretical calculations. Note the huge size of the cross section which is about 10$^3$ times larger than the geometric cross section and corresponds to a typical impact parameter of the order of 1 μm! Note finally that the authors found some evidence that under single collision conditions the final $\ell$ value is not far from 3 indicating that dipole-type transitions probably dominate the process as suggested by the large impact parameter, in agreement with the theoretical results. In Table III we compare the main features of the $\ell$-mixing cross sections induced either by ion or by neutral atoms.

Table III
Some Features of the $\ell$-mixing Process Induced either by Ions or by Atoms
(for n $\lesssim$ 30)

| Property | Ion<br>($\tilde{v} \sim 1$) | Atom<br>$\tilde{v} \leqslant 10^{-2}$ |
|---|---|---|
| Interaction | Long-range | Short-range |
| n Dependence | $n^5$ | $n_{max}$ |
| Nature of the target | Independent<br>(1 keV Ne$^+$ = 2 keV Ar$^+$) | Strongly depends on<br>$e^-$-P scattering |
| Magnitude | $\sim 10^3 \, \sigma_{geom}$ | $\lll \sigma_{geom}$ |

## C.   Resonant Rydberg-Rydberg Collisions

Atomic Rydberg levels provide a unique possibility to study resonant collisions
for two reasons:

- Their energy separations may be easily changed in discrete steps by changing the n or $\ell$ quantum numbers.
- A continuous tuning of the energy separations is also possible by applying modest magnetic or electric fields.

The first possibility has already been used to study resonant electronic to
vibrational[27] energy transfer from Na$^*$ to $CH_4$ or $CD_4$ and electronic to rotational
energy transfer[50] from Xe$^*$ to $NH_3$.  The resonance widths reported are 50 cm$^{-1}$ and
6 cm$^{-1}$ respectively.  Resonant charge transfer collisions are also under current
investigation.[51]  But the second possibility (continuous tuning) offers the
possibility of observing very sharp resonances.  Such kinds of sharp resonances
has recently been reported by Safinya et al. for Na$^*$-Na$^*$ collisions.[52]

Fig. 8 shows an energy map of the Na$^*$ states in the vicinity of the 20s level.
In a field of $\sim$ 200 V/cm the 20s state lies just midway between the 20p and 19p
levels.  In fact as shown by Fig. 8 there are four values of E that give rise to
exact resonances due to the splitting of the $|m|$ = 0 and 1 sublevels of the p
states.  It is observed that for the resonant values of the electric field the
population in the s state is rapidly transferred to the p states.  The
experimental apparatus has been described elsewhere.[4]  The SFI method is used for
the detection of the population transfer.  An example of the results obtained is
shown in Fig. 9.  The field ionization voltage was set so that the population of
the upper p state was recorded as a function of the DC electric field.  Note that
the background signal (due to the black-body radiative induced population transfer[53]) is linear in the laser power while the resonant collisional signal is
quadratic.  By comparing the resonant collisional signal with the total number of
excited atoms the authors deduced the cross sections for the process

$$Na(ns) + Na(ns) \rightarrow Na(np) + Na((n-1)p); \quad 16 < n < 27 \qquad (4)$$

Various experimental tests and theoretical estimates were performed in order to
ensure that superradiance or related cooperative phenomena[54,55] cannot be
responsible for the observed signals.  Fig. 10 shows the n dependence of the
cross sections for process (4) as well as the n dependence of the collision time

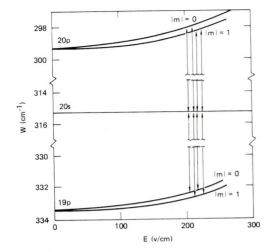

Fig. 8. Energy levels for the 19p, 20s, and 20p sodium states in an electric field. The collisional resonances are shown by the arrows. From Ref. [52].

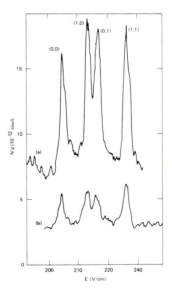

Fig. 9. Signal from the 20p state, after population of the 20s state, versus dc electric field at two exciting laser powers. (a) 1.0 and (b) 0.4. The collisional resonance signals are labelled by the $|m|$ values of the upper and lower p states (see Fig. 8). From Ref. [52].

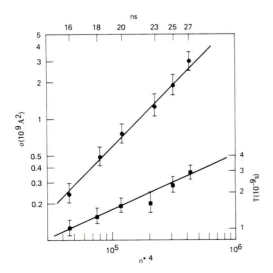

Fig. 10. Observed resonant cross sections (•) and colli- sion times (■) plotted versus $n^{*4}$ on logarithmic scales. The values of the ns states are given at the top for reference. From Ref. [52].

as deduced from the observed resonance widths. The measured cross sections are of the order of $10^9$ Å$^2$, corresponding to typical impact parameters of about 2 μm, (!), and scale as $n^{*4}$ (where $n^*$ is the effective quantum number). These huge values (the largest ever observed for atom-atom collisions) reflect the resonant behavior of the process; the non-resonant collision cross sections for reaction (4) are expected from CTMC calculations to be about 2 to 3 orders of magnitude smaller than the resonant ones. The collision times scale as $n^{*2}$ and are of the order of 2 ns. This unusually long duration may allow to introduce perturbations in mid-collision.

The observed cross sections and widths can be compared with calculations based on the interaction of the dipole moments of the two colliding atoms. Several approaches for dipole-dipole collisions are available.[56-58] The cross section at resonance is found to be given (in atomic units), to about an order of magnitude, by

$$\sigma \sim \frac{\mu_1 \mu_2}{v} \sim \frac{n^{*4}}{v} \sim \frac{\sigma_{geom}}{v} \qquad (5)$$

where $\mu_1$ and $\mu_2$ are the ns-np and ns $-$ (n $-$ 1)p dipole moments which both are of the order of $n^{*2}$ and v is the collision velocity. The collision time T is given by

$$T \sim \frac{\sqrt{\sigma}}{v} \sim \frac{n^{*2}}{v^{3/2}} \qquad (6)$$

Both formulae (5) and (6) exhibit the observed n dependences. Numerical evaluations give $\sigma \sim 3 \times 10^8$ Å$^2$ and T $\sim$ 1 ns for n = 18 in good, somewhat fortuitous, agreement with the observed values (Fig. 10). Note that $\sigma$ is larger than the geometrical cross section by a factor of about 1/v. As for ion-Rydberg cross sections no turn-over in the n dependence of the cross section is observed, contrary to the case of atom$^*$-atom collisions.

IV.  Final Remarks

The three examples we have extensively discussed display some of the most interesting features of collisions involving Rydberg atoms:

-    The importance of the e$^-$-P interaction. In particular when this interaction is long range huge cross sections can be obtained.
-    The application of an external field can noticeably modify the features of a given collisional process.
-    Rydberg atoms provide unique capabilities for studying resonant collisions.

Clearly more experimental as well as theoretical work is needed along these lines. For example, the possibility of extracting quantitative information on the e$^-$-P (or A$^+$-P) scattering at very low energies has certainly not been fully explored. The long collision time observed in resonant Rydberg-Rydberg collisions should also open new areas for the experimentalist. The possibility of inducing selection rules for a collisional process should also deserve further attention. Finally one can be sure that the rather unusual properties of Rydberg atoms will allow in the near future the opening of other new lines of research in collision physics.

[†]Visiting scientist at SRI International.

Acknowledgements:

The author is grateful to T. F. Gallagher and R. Kachru for helpful discussions and comments. It is a pleasure to acknowledge the warm hospitality of all the members of the Molecular Physics Laboratory at SRI International during the preparation of this manuscript. This work was partly supported by the ONR Physics Division under Contract No. N00014-79-C-0202.

References

1. Amaldi, E., and Segre, E., Nuovo Cimento 11, 145 (1934).
2. Fermi, E., Nuovo Cimento 11, 157 (1934).
3. Electronic and Atomic Collisions; Oda, N., and Takayanagi, K., editors. Invited papers and progress reports, XI ICPEAC, Kyoto (1979), pp. 457-506.
4. Gallagher, T., F., Humphrey, L. M., Cooke, W. E., Hill, R. M., and Edelstein, S. A., Phys. Rev. A 16, 1098 (1977).
5. Jeys, T. H., Foltz, G. W., Smith, K. A., Beiting, E. J., Kellert, F. G., Dunning, F. B., and Stebbings, R. F., Phys. Rev. Lett. 44, 390 (1980).
6. Weber, K. H. and Niemax, K., Opt. Comm. 28, 317 (1979).
7. Flusberg, A., Kachru, R., Mossberg, T. W., and Hartmann, S. R., Phys. Rev. A 19, 1607 (1979).
8. Rydberg Atoms, Dunning, F. B., and Stebbings, R. F., editors. Cambridge University Press, to be published (1982).
9. See for example: Gallagher, T. F., Edelstein, S. A., and Hill, R. M., Phys. Rev. A 15, 1945 (1977). Also Hugon, M., Gounand, F., Fournier, P. R., and Berlande, J., J. Phys. B 12, 2707 (1979).
10. Hugon, M., Fournier, P. R., and DePrunelé, E., J. Phys. B, to be published.
11. Hickman, A. P., Phys. Rev. A 23, 87 (1981).
12. Boulmer, J., Delepch, J. F., Gauthier, J. C., and Safinya, K. A., to be published.
13. Weber, R. H., and Niemax, K., Opt. Comm. 31, 52 (1979).
14. Stoicheff, B. P., and Weinberger, E., Phys. Rev. Lett. 44, 733 (1980).
15. Stoicheff, B. P. Thompson, D. C., and Weinberger, E., Proceedings of the 5th International Conference on Spectral Line Shapes; Seidel, J., and Wende, B., editors, Berlin (1980).
16. Niemax, K., and Weber, K. H., in [15].
17. Kachru, R., Mossberg, T. W., and Hartmann, S. R. Phys. Rev. A 21, 1124 (1980).
18. Brillet, W. L., and Gallagher, A., Phys. Rev. A 22, 1012 (1980).
19. Klucharev, A. N., Lazarenko, A. V., and Vrijnovic, V., J. Phys. B 13 1143 (1980). Also Dimicoli, I., and Botter, R., J. Chem. Phys. 74, 2346 and 2355 (1981).
20. Cheret, M., Spielfiedel, A., Durand, R., and Deloche, R., submitted to J. Phys. B (1981).
21. Mihajlov, A. A., and Janev, R. K., submitted to Phys. Rev. A (1981). Also Janev, R. K., and Mihajlov, A. A., Phys. Rev. A 20, 1890 (1979), ibid A 21, 819 (1980).
22. Kellert, F. G., Smith, K. A., Rundel, R. D., Dunning, F. B., and Stebbings, R. F., J. Chem. Phys. 72, 3179 (1980).
23. Kellert, F. G., Higgs, C., Smith, K. A., Hildebrandt, G. F., Dunning, F. B., and Stebbings, R. F., J. Chem. Phys. 72, 6312 (1980).
24. Higgs, C., Smith, K. A., Dunning, F. B., and Stebbings, R. F., submitted to Phys. Rev. A (1980).
25. Hitachi, A., King, T. A., Kubota, S., and Doke, T., Phys. Rev. A 22, 863 (1980).
26. Safinya, K. A., and Gallagher, T. F., Phys. Rev. A 22, 1588 (1980).
27. Gallagher, T. F., Ruff, G. A., and Safinya, K. A., Phys. Rev. A 22, 843 (1980).
28. Flannery, M. R., J. Phys. B 13, L657 (1980).

29.  Flannery, M. R., Phys. Rev. A 22, 2408 (1980).
30.  Matsusawa, M., J. Phys. B 12, 3743 (1979).  See also Matsusawa, M. in [8].
31.  Sasano, K., and Matsusawa, M., J. Phys. B 14, L91 (1981).
32.  Cheng, L. Y., and Van Regemorter, H., J. Phys. B, (to be published).
33.  Hahn, Y., J. Phys. B 14, 985 (1981).
34.  Hickman, A. P. submitted to J. Phys. B (1981).
35.  Matsusawa, M., J. Phys. B (to be published).
36.  Kaulakis, B. P., Presnyakov, L. P., and Serapinas, P. D., Zh. Eksp. Teo. Fiz.
     30, 60 (1979) (J.E.T.P. Lett. 30, 53 (1980)).
37.  Matsusawa, M. private communication.
38.  Gounand, F., Gallagher, T. F., Safinya, K. A., and Sandner, W., p. 1113,
     Abstracts of Contributed Papers, XII ICPEAC, Datz, S., Editor, Gatlinburg
     (1981).
39.  Slusher, M., Higgs, C., Smith, K. A., Dunning, F. B., and Stebbings, R. F.,
     p. 1111, same issue as [38].
40.  Kleppner, D., Littman, M. G., and Zimmerman, M. L., in [8].
41.  Percival, I. C., and Richards, D., Adv. Atom. Mol. Phys. 11, 1 (1975) and
     references therein.
42.  Olson, R. E., J. Phys. B 13, 483 (1980).
43.  Olson, R. E., and MacKellar, A. D., Phys. Rev. Lett. 46, 1451 (1981).
44.  Hickman, A. P., Olson, R. E. and Pascale, J. in [8].
45.  Burniaux, M., Brouillard, F., Jognaux, A., Govers, T. R., and Szuy, S., J.
     Phys. B 12, 2421 (1977).
46.  Kim, H. J., and Meyer, F. W., Phys. Rev. Lett. 44, 1047 (1980).
47.  MacAdam, K. B., Crosby, D. A., and Rolfes, R., Phys. Rev. Lett. 44, 980
     (1980).
48.  MacAdam, K. B., Rolfes, R. and Crosby, D. A., submitted to Phys. Rev. A
     (1981).
49.  Percival, I. C., and Richards, D. R., J. Phys. B 10, 1497 (1977).
50.  Smith, K. A., Kellert, F. G. Rundel, R. D., Dunning, F. B., and Stebbings,
     R. F., Phys. Rev. Lett. 40, 1362 (1978).
51.  MacAdam, K. B., submitted to Comments on Atomic and Molecular Physics,
     (1981).  See also Ref. [38], p. 1115 (1981).
52.  Safinya, K. A., Delpech, J. F., Gounand, F., Sandner, W., and Gallagher,
     T. F., Phys. Rev. Lett. (to be published 1981).
53.  Gallagher, T. F., and Cooke, W. E., Phys. Rev. Lett. 42, 835 (1979).
54.  Gross, M., Fabre, C., Goy, P., Haroche, S., and Raimond, J. M., Phys. Rev.
     Lett. 43, 343 (1979).
55.  Gounand, F., Hugon, M., Fournier, P. R., and Berlande, J., J. Phys. B 12, 547
     (1979).
56.  Mott, N. F., and Massey, H.S.W., The Theory of Atomic Collisions, Clarendon
     Press, Oxford (1950).
57.  Purcell, E. M., Astrophys. J. 116, 457 (1952).
58.  Anderson, P. W., Phys. Rev. 76, 647 (1949).

PHYSICS OF ELECTRONIC AND ATOMIC COLLISIONS
S. Datz (editor)
© North-Holland Publishing Company, 1982

THEORY OF COLLISIONAL ELECTRON DETACHMENT

Alexander Z. Devdariani

Department of Optics
Leningrad State University
Leningrad
USSR

It can be seen from the experimental data that
the zero range potential model is valid for the
description of the negative ions destruction in
collisions with atoms. The problem of calculating
the escaped electrons spectrum and detachment
cross section is discussed. Special attention is
given to the calculation of the cross section
near the reaction threshold. The influence of
quasimolecular terms crossing on the detached
electrons spectra is considered.

INTRODUCTION

In quantum mechanics the problem of two discrete energy levels approaching one another  during the variation of the external  parameters has been considered in detail. After certain schematization
this problem may be reduced to the solution of the system of two
ordinary differential equations. The problems considered in this
paper are related to the formation of the electron in collisions.
The destruction of negative ions (NI) in collisions is an example
of such processes

$$A + B^- \longrightarrow A + B + e \tag{1}$$

$$A + B^- \longrightarrow AB + e \tag{2}$$

Since after the collision the electron state belongs to the continuum the problem being considered concerns the interaction of a
single level with the infinite number of other levels. Even after
the schematization such problems result in partial differential
equations. The final theoretical problem of the investigation of
the detachment processes (1) – (2) consists in the calculation of
the cross sections (full and differential) and the spectrum of the
electrons formed.

THE DEVELOPMENT OF ZERO RANGE POTENTIAL MODEL (ZRPM)

The model which describes the process (1) – (2) was proposed by
Demkov /1/. It reduces to the consideration of the electron behaviour in the range where the discrete quasimolecular level crosses
the boundary of a continuum at $R = Ro$. As the atom approaches
the NI in collisions the binding energy  of an additional electron
diminishes, and its wave function area becomes greater than that
of a nonzero effective potential. Therefore the wave function

269

satisfies the Schrödinger equation for a free particle with a zero
centrifugal member.(Here we deal with the case in which the wave
function of a loosely bound electron approaches the spherically
symmetric one when $R \rightarrow R_0$). The existence of the effective potential well of variable depth may be taken into account, if we
assume the logarithmic derivative to be the function of time
(zero range potential model). Thus the problem reduced to the
equation

$$i\, \frac{\partial \psi}{\partial t} = -\frac{1}{2}\frac{\partial^2 \psi}{\partial z^2}$$ (3)

with boundary condition

$$\frac{1}{\psi}\frac{\partial \psi}{\partial z}\Big|_{z=0} = f(t)$$ (4)

$f < 0$ corresponds to the bound state of the electron, $f = 0$ - to
the crossing of the descrete level with the continuum. In the investigation of NI destruction in collisions the initial condition corresponds to the bound state of the electron for $t \rightarrow \infty$,
so its binding energy is $E = -f2/2$. For the arbitrary form of
$f(t)$ the solution presents certain mathematical difficulties,
therefore the problem has been solved in [1] for $f = +\gamma t$ and
the form of the escaping electrons distribution function has been
obtained for the detachment probability value $P_d \approx 1$,

$$W(k) = \frac{2k^2}{\gamma}\, exp\left(-\frac{2k^3}{3\gamma}\right)$$ (5)

k - the escaping electron impulse.

To determine the spectrum for intermediate cases $0 < P < 1$ and to
find the value of $P_d$ itself for various parameters of the problem
it is useful to consider the solution of (3) - (4) at square approximation for

$$f(t) = -\alpha t^2 + \beta, (\alpha > 0,\ \beta \leq 0).$$

Such an approximation is a good one for the real $f(t)$ in the
crossing area and at the same time the analytical solution is
still possible. Such a problem has been analytically solved in
[2]. It turned out that the amplitude of electron distribution in
the continuum could be described satisfactorily by the formula

$$\psi(k) = \frac{k}{[\alpha(-ik+\beta)]^{1/4}}\, exp\, i\left[\frac{2}{5\sqrt{\alpha}}(-ik+\beta)^{5/2} - \frac{2}{3}\frac{\beta}{\sqrt{\alpha}}(-ik+\beta)^{3/2} + \frac{\pi}{4}\right]$$ (6)

The spectrum and the detachment probability (see Figs. 1, 2)
depend on the parameter $\lambda = \beta^{1/5}/\alpha$ . As in Demkov model the initial part of the spectrum is proportional to k2; however, the
exponent governing the distribution decay for large values of k
varies from k2 for $\lambda < -1.8$ through k5/2 for$\lambda \approx 0$ to k3 for$\lambda > 2$
and is described by formula (6) [2]. Such spectrum  behaviour
may be checked up experimentally by observing the electrons
spectrum for the fixed scattering angle between the atoms becouse
$\lambda$ is related in a certain way to the collision energy E and the
impact parameter $\rho$ (see below).

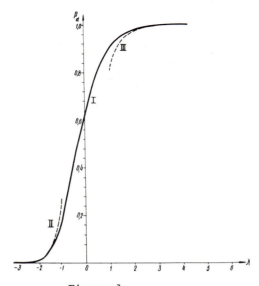

Figure 1
The detachment probability
1 - data *[2]*, 2 - data */7/* 3 - data */1/*

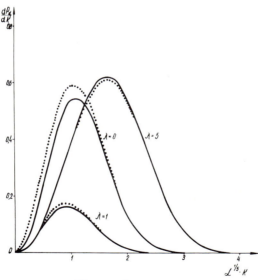

Figure 2
The detached electrons spectrum
Solid curve - the computation,
dashed curve - formula (6)

## THE DETACHED ELECTRONS SPECTRUM

Bydin was the first to observe the electron spectra of NI destroyed during the collision of J-, Br-, Cl- with atoms of some inert
gases /3/. The most interesting systems for checking ZRPM are
those of H- - He, Ne, Ar since it is experimentally shown
(Lam et al. /4/, Risley /5/) that  in this case the electron
distribution is isotropic. It was these pairs which were experimentally investigated in detail by Esaulov et al. /6/ for the
case of the fixed impact parameter and of the energy interval
0.08 - 20 kev. In Fig. 3 (from /6/) the comparison of the experi-

Figure 3
Detachment peak for H- - He collisions at E = 250 ev.
The figure  shows the theoretical convoluted
spectrum (·) and experiment  ( - )  /6/

mental data with formula (6) is shown for the case when the collision kinematics was taken into account and $\lambda$ was assumed to be
$\simeq$ 0 (the turning point for the atomic radial motion is close to
the point where the descrete level crosses the boundary of continuum). The dependence of the maximum position W(k) on the collision energy also agrees well. The agreement which is established
in /6/ makes it possible to state that for the systems H- - He,
Ne, Ar the destruction of NI in the energy interval investigated
is satisfactorily described by ZRPM.

Summing up the discussion of the spectra we present the formula
derived by Chaplik /7/ for those adiabatic collisions which retain
the adiabatic character at any value of t.

$$\varphi(k) = g \frac{k}{2^{1/4}} \frac{1}{\left[\frac{d\, f(t_k)}{dt}\right]^{1/2}} exp\left[i\frac{k^2}{2}t_k + \frac{i}{2}\int^{t_k} f^2(t)dt\right] \quad (7)$$

where $t_k$ is determined as the root of the equation:

$$-\frac{f^2(t_k)}{2} = \frac{k^2}{2} \quad (8)$$

g – a constant, which is not determined in /7/.

It is shown in /8/ that formula ( 7 ) really descibes all the known spectra which have been analytically investigated for the problem (3) – (4) and $g = 2^{3/4} exp\left(i \frac{\pi}{4}\right)$.

ELECTRON DETACHMENT CROSS SECTION

Let us now discuss the problem of calculating the NI destruction cross section. As it follows from Risley's review /9/ the bulk of experimental data on destruction cross sections at nonsymmetric collisions may be represented as in Fig. 4. In range 1 below the

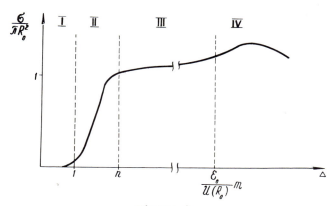

Figure 4

The schematic dependence of the detachment cross section of NI during the collision with the atoms, $\mathcal{E}_o$ – the binding energy of electron in NI, $m$ is the reduced mass of the colliding particles

reaction threshikd the transitions are essentially of quantum, tunneling character, in range 11 (near the threshold) the cross section can be described by the formula given in the works by Bydin, Dukelsky /10/ and Mason, Vander s lice /11/

$$\frac{\sigma'}{\pi R_o^2} = \left(1 - \frac{1}{\Delta}\right), \qquad \Delta = \frac{E}{u^-(R_o)} \qquad (9)$$

The derivation of this formula is based on the assumption that when the distance R between the colliding particles reaches the critical value of Ro we have $P_d = 1$ and when R > Ro then $P_d = 0$.

In range 111 the cross section increases slowly, and at last in range 1V, where the colliding atoms velocity is much greater than that of loosely bound electrons we can ignore the binding energy and consider the NI destruction process to be the consequence of the elastic scattering of a free electron on the projectile. For range 1V the cross section was calculated by Lopantseva and Firsov /12/.

Let us show that the cross section calculation for ranges 1 – 11 – 111 may be based on ZRPM. The dependence $P_d(\lambda)$ can be obtained from Fig. 1, thus the calculation of the destruction cross section:

$$\sigma = 2\pi \int_0^\infty P_d\,(\rho)\,d\rho \qquad (10)$$

reduced to establishing the connection between $\lambda$ and collision parameters. If we know the terms U(R) and U⁻(R) for the systems AB and AB⁻ respectively we can define for $R > R_0$

$$f(R) = -\sqrt{2\,(u - u^-)} \qquad (11)$$

Then the classical trajectory R(t) in the potential U⁻ and the real dependence f(t) can be defined. Further the question arises as to what extent the real f(t) can be approximated by a parabola. In this case the coincidence of the time moments $t_f$ when the collision becomes nonadiabatic for the real and parabolic f(t) may be taken as a criterion. The investigation at various reasonable analytical approximations to f(R), for example

$$f(R) = -\sqrt{2\,\mathcal{E}_0}\left[1 - exp\left(-\tfrac{\bar{\bar{z}}}{\xi}(R^2 - R_0^2)\right)\right] \qquad (12)$$

shows /13/ that the real f(t) can be well approximated by a parabola for collision velocity

$$v \le \frac{\mathcal{E}_0}{4\sqrt{\xi}} \qquad (13)$$

$\mathcal{E}_0$ – the binding energy of electron in NI.

Using the data of Olson, Lia /14/ for example, we can see that the parabolic approximation is valid for the energies $E \le 10$ ev, i.e. in regions 1-11 near the reaction threshold. In this case the parameter $\lambda$ for $U^-(R \approx R_0) = Ae^{-\gamma R}$ is

$$\lambda = \frac{\Theta}{\Delta^{1/5}}\left(1 - \frac{1}{\Delta} - \frac{\rho^2}{R_0^2}\right) \qquad (14)$$

where

$$\Theta = \left|\frac{df}{dr}\right|_{R=R_0}^{4/5} \frac{m^{1/5}}{2\,U^-(R_0)}\left(\frac{2}{v}\right)^{6/5} \qquad (15)$$

m – the reduced mass of the colliding atoms.

The destruction cross section is

$$\sigma = \pi R_0^2\,\frac{\Delta^{1/5}}{\Theta}\int_{-\infty}^{\lambda(\rho=0)} P_d\,(\lambda)\,d\lambda \qquad (16)$$

The integral in (16) can be analytically calculated for various regions of parameters $\Theta$, $\Delta$. The dependence $\sigma(\Theta, \Delta)$ (see Fig. 5) clearly represents the experimental situation. Assuming that $P_d(\lambda) = 0$ when $\lambda < 0$ and $P(\lambda) = 1$ when $\lambda > 0$ we obtain (9) from (16) for the region near the reaction threshold.

For $\Delta \gg 1$ (region 111) formula (16) transforms continuously

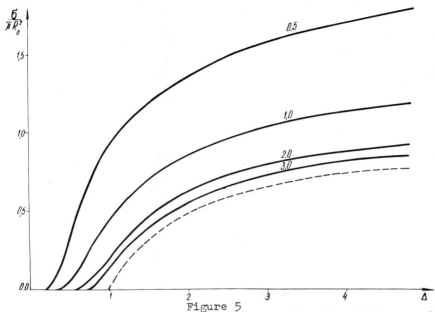

Figure 5

The detachment cross section near the reaction threshold according to formula (16). Figures – the value of parameter $\theta$ , dashed line – cross section according to formula (9)

into the formulas of /15/ which have been derived for the case of the rectilinear flight. In region lll the cross section increases of El/5, and the mechanism of this phenomenon can be qualitatively related not only to the increasing of the impact parameter region, where $0 < P < 1$, but also to the fact that $P_d$ ($\rho$ = Ro) $> 0.5$. In region lll the condition (13) is no longer valid, tf for the parabola is smaller than tf for the real f (tf is "forgetting point"), so the cross section calculated in /15/ is smaller than the experimental one. The significance of tf was first noted in the paper by Herzenberg, Ojha /16/ where this fact was taken into account in the H⁻ – He cross section calculation and thus the agreement with the experiment became essentially better.

Besides the numerical solution is also possible for the problem (3) – (4) with the real f(t). This problem has been solved by Gauyacq /17, 18/, the delicate distinction in the isotopic phenomena for H⁻ – He and H– – Ne systems being illustrated by the calculations (the physical meaning of the isotopic phenomena in the NI destruction is discussed in /19/).

Good agreement of the calculation with the experimental data Champion et al. /20/ gives us an additional argument in favour of ZRPM.

THE ACCOUNT OF THE QUANTUM MECHANICAL CHARACTER OF THE ATOMS MOTION

In the problems considered above the electron detachment and the
atoms motion were analysed independently by of introducing the
trajectory R(t). This assumption is invalid when the collision
energy values are near the reaction threshold, $\Delta \simeq 1$. Thus, in
case of the numerical solution of problem (3) – (4) it becomes
difficult to choose the potential for the R(t) calculation when
$R < Ro$.  It should be noted that for the case of transitions
between discrete levels the problem of choosing the trajectory
is solved in /21/. For the problem of the NI destruction in collisions followed by the s-electron ejection the problem of choosing
the trajectory can be analysed using the solution of equation /22/:

$$\left[ -\frac{1}{2m}\frac{\partial^2}{\partial R^2} - \frac{1}{2}\frac{\partial^2}{\partial z^2} + \beta\,(R_o - R)\right]\psi = E\,\psi \qquad (17)$$

with the boundary condition:

$$\left(\frac{1}{\psi}\,\frac{\partial \psi}{\partial z}\right)_{z=0} = \alpha\,(R_o - R) \qquad (18)$$

Problem (17) – (18) generalizes the problem (3) – (4) by taking
into account the quantum character of the atomic motion. Two cases
are to be distinguished: 1) the detachment is caused by the crossing of the AB⁻ system term with the A – B atoms repulsion term
(the first problem, Fig. 6);  2) the detachment is caused by the

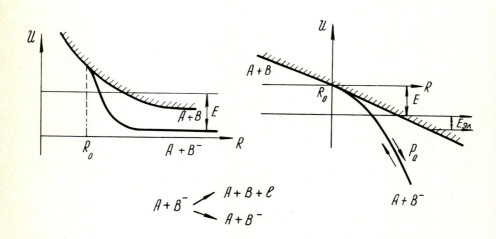

Figure 6
The real and the model schemes of the terms for the first
problem (see the text)

crossing of AB⁻ system term with the A – B atoms attraction term (the second problem, Fig. 7).

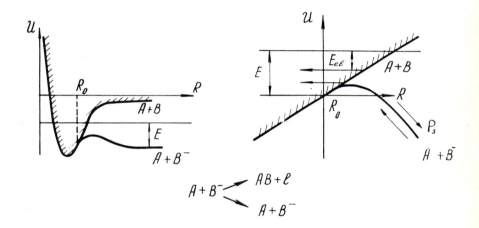

Figure 7
The real and the model schemes of the terms for the second problem (see the text)

The second problem may be used for the analysis of the reaction (2). The problems being discussed, have been solved in papers /8, 23/, the case of horizontal position of the continuous spectrum boundary being considered in /24/.

Among the main results of /8, 23/ the following should be mentioned the electronic spectra for problems 1 and 2 turn out to be related

$$W_2\,(\kappa) = W_1\,(\kappa)\left(1 - P_{d_1}\right)\exp\left[\widetilde{\pi}\,(\eta - \delta^2)\right], \tag{19}$$

$$\delta^2 = \frac{\beta m^{1/4}}{\alpha^{3/2}} \quad , \quad \eta = \frac{2Em}{\alpha} \tag{20}$$

and

$$P_{d_2} = \frac{P_{d_1}}{P_{d_1} + exp\left[-\widetilde{\pi}(\eta - \delta^2)\right]} \tag{21}$$

The physical meaning of the last inequality is quite clear. It is related to the fact that the discrete level in problem 1 is localized farther from the boundary of the continuum than that in problem 2. In problem 1 the detachment is of tunneling character only if $E < 0$, but in problem 2 the tunneling occurs even when $E < \beta^2 / 2\alpha^2$

In problem 1 the amplitude of the transition to the continuum states satisfies the equation:

$$\frac{d^2 \varphi_1}{dz^2} - \frac{3}{z} \frac{d\varphi_1}{dz} + \left( -\delta^2 z^4 - 2i\delta^{3/2} z^3 + \eta z^2 + \frac{3}{z^2} \right) \varphi_1 = 0 \tag{22}$$

where
$$\kappa = z \frac{\beta^{1/2}}{\alpha^{1/4} m^{1/8}}$$

and for nearly all values of parameters $\eta$, $\delta$ the function $\varphi_1(k)$ can be approximately described by the formula of JWKB-approach for equation (22):

$$\varphi_1(z) \approx \frac{\sqrt{2} \, \delta^{3/4} z}{(\delta^2 z^2 + 2i\delta^{3/2} z - \eta)^{1/4}} \exp\left( i \int_{\widetilde{z}_1}^{-iz} z' \sqrt{\delta^2 z'^2 + 2\delta^{3/2} z' + \eta} \; dz \right), \tag{23}$$

$\widetilde{z}_1$ – the root of the integrand in (23). It is to be noted that formula (23) may be derived by the generalization of the known Landau method $/25/$ in the case of transitions to the continuum

$$\left| \varphi_1(\kappa) \right| \approx \exp\left[ -\mathcal{I}m \left( \int_{x_1}^{x_0} \sqrt{2\left(E - \frac{\kappa^2}{2} - U(x)\right)} dx - \int_{x_2}^{x_0} \sqrt{2(E - U^-(x))} dx \right) \right] \tag{24}$$

where $x_{1,2}$ – the turning point of radial motion of the atoms in the potential fields U and U$^-$, $x_0$ – the stationary phase point, which is determined by the condition

$$-\frac{\kappa^2}{2} = U(x_0) - U^-(x_0) \tag{25}$$

The detachment probability

$$P_{d_1} = \int_0^\infty \left| \varphi_1(\kappa) \right|^2 d\kappa \tag{26}$$

depends on the two parameters $\delta$, $\eta$. The approximate expressions for this value are derived for three regions of the parameters $\delta$, $\eta$ for the intermediate cases $Pd$ being obtained by the numerical integration. To calculate the detachment cross section we should relate $\eta$ and $\delta$ to the impact parameters $\varrho$, $E$. It appears that:

$$\delta = \frac{\beta m^{1/4}}{\alpha^{3/2}} \left( 1 + 2 \frac{\varrho^2}{R_0^2} \frac{E}{R_0 \beta} \right), \quad \eta = \frac{2Em}{\alpha} \left( 1 - \frac{1}{\Delta} - \frac{\varrho^2}{R_0^2} \right). \tag{27}$$

therefore the integral (10) reduces to the integral over the ray in $\eta - \delta$ plane and can be easily calculated for the parameters of any value. It should be emphasized that in this approach the values Ro, U-(Ro), $\alpha$ , $\beta$ are used as the initial parameters; the trajectory R($t$) is not calculated.

CONCLUSION

The formulation of any collision model in terms of S-matrix properties reduces to determining a certain scheme of S-matrix poles location and of their motion. In the problems considered above the electron was supposed to be in the virtual state (the account of one S-matrix pole on the imaginary axis). Such a mechanism results in cross section increasing. On the other hand, the cross section decreases as the collision velocity increases provided the collision process is governed by the decay of the quasistationary state (two symmetric poles). Roughly speaking it occurs due to the diminishing of the collision time. It was stated in /15/ that full consideration of the problem taking into account the transformation of a virtual level into quasistationary one does not cancel the conclusion that the tunneling contribution increases if the collision velocity increases since the quasistationary state forms at R < Ro.

At a considerable distance from the continuum boundary we deal with the quasistationary states only; such states may interact with one another both directly and via the continuum. In the problems concerning NI quasistationary terms appear, for example, in the consideration of the charge exchange $O_2 - O^-$ (Mathis, Snow /26/). The theory of such transitions developed in /27/ and the influence of the interaction on the shape of electronic spectra is considered in /28/.

ACKNOWLEDGEMENTS

The author wishes to thank prof. Yu.N.Demkov for the help and the stimulating discussion of the problems considered in the present paper. The author also thanks prof. N.P.Penkin for supporting the investigation and dr. V.M.Borodin for discussion.

REFERENCES:

/1/ Demkov, Yu.N., Electron detachment in slow collisions between negative ions and atoms, Zh. Eksp. Teor. Fiz. 46 (1964) 1126-1136.

/2/ Devdariani, A.Z., Electron transitions from bound state to continuous spectrum, Teor. Mat. Fiz. 11 (1972) 213-225.

/3/ Bydin, Yu.F., The energy Spectrum of electrons released in collisions of negative halogen ions with the atoms of inert gases, in: Abst. papers V ICPEAC (Nauka, Leningrad, 1967), 218-220.

/4/ Lam, S.K., Delos, J.B., Champion, R.L. and Doverspike, L.D., Electron detachment in low-energy collisions of $H^-$ and $D^-$ with He, P.R. A 9 (1974) 1828-1839.

/5/ Risley, J.S., The negative hydrogen ion and its behavior in
atomic collisions, in: zu Putlitz, G. (ed.), Atomic Physics
(Plenum Press, New York and London, 1975).

/6/ Esaulov, V., Dhuicq, D. and Gauyacq, J.P., Differential
study of H- - inert-gas collisions,  J.Phys. B: Atom. Molec.
Phys.  11 (1978) 1049-1067.

/7/ Chaplik, A.V., Quantum transitions to a continuous spectrum
due to adiabatic perturbations , Zh. Eksp. Teor. Fiz. 45
(1963) 1518-1522.

/8/ Devdariani, A.Z., Destructions of negative ions in slow
collisions with atoms, Ph.D .  Thesis, Dept. of Physics,
Leningrad State Univ. (1972).

/9/ Risley, J.S., A review of negative ion  detachment cross
sections, in: Oda, N.,and Takayanagi, K. (eds.), Electronic
and Atomic Collisions (North-Holland, Amsterdam, 1980).

/10/ Bydin, Yu.F. and Dukelsky, V.M., The detachment of an
electron  in collisions negative halogen ions with atoms of
inert gases and the hydrogen molecules, Zh. Eksp. Teor. Fiz.
31 (1956) 568-578.

/11/ Mason, E.A. and Vanderslice J.T., Interection energy and
scattering cross sections of H- Ions in helium,  J. Chem.
Phys. 28 (1958) 253-257.

/12/ Lopantseva, G.B. and Firsov, O.B., Breakup of fast negative
ions in inert gases,  Sov. Phys. - JETP 23 (1966) 648-650.

/13/ Devdariani, A.Z., Distructions of negative ions in the
collisions with atoms near the threshold of the reaction,
Zh. Eksp.  Teor. Fiz., to be published.

/14/ Olson, R.E. and Liu, B., Interections of H and H⁻ with He
and Ne, Phys. Rev. A 22 (1980) 1389-1394.

/15/ Devdariani, A.Z., The cross-section of the destruction of the
negative ion in collision with an atom, Sov. Phys. - Tech.
Phys. 18 (1973) 255-258.

/16/ Herzenberg, A. and Ojha, P.,  Electron detachment in collisions of negative ions: H⁻ + He, Phys. Rev. A 20 (1979)
1905-1914.

/17/ Gauyacq, J.P., Electron  detachment in low-energy H⁻ - Ne
collisions: evidence of a purely dynamical phenomena,
J. Phys. B 13 (1980) L 501-504.

/18/ Gauyacq, J.P., Theoretical study of the detachment process
in the zero-range potential approximation. H- - He collisions,
J. Phys. B: Atom. Molec. Phys. 13 (1980) 4417-4439.

/19/ Smirnov, B.M., Negative ions (Atomizdat, Moscow, 1978).

/20/ Champion, R.L., Doverspike, L.D. and Lam, S.K., Electron
detachment from negative ions: The effects of isotopic
substitution, Phys. Rev. A 13 (1976) 617-621.

/21/ Bykhovsky, V.K., Nikitin, E.E. and Ovchinnikova, M.Ya.,
Probability of a nonadiabatic transition near the turning
point, Zh. Eksp. Teor. Fiz. 47 (1964) 750-756.

/22/ Devdariani A.Z., Electron detachment on slow collisions
of negative ions with atoms. The quantum mechanical treatment
of atomic motion, in: Abst. papers Vl ICPEAC (Cambridge,
USA, 1969) 550-551.

/23/ Devdariani, A.Z., Demkov, Yu.N., The destruction of negative
ions near the reaction threshold, in: Abst. papers Vlll
ICPEAC (Beograd, 1973) 840-841.

/24/ Devdariani, A.Z., Demkov, Yu.N., Destruction and formation
of negative ions in slow collisions with atoms, Teor. Mat.
Fiz. 21 (1974) 74-85.

/25/ Landau, L.D., Lifschiz, E.M., Kvantovaja Mehanika (Nauka,
Moscow, 1974).

/26/ Mathis, R.F. and Snow, W.R., Charge transfer between $O^-$
ions and $O_2$ molecules in the ground state and singlet delta
excited state, J. Chem. Phys. 61 (1979) 4274-4278.

/27/ Devdariani, A.Z., Ostrovsky, V.N. and Sebyakin, Yu.N.,
Crossing of quasistationary levels, Sov. Phys. JETP
44 (1976) 477-482.

/28/ Devdariani, A.Z., Ostrovsky, V.N. and Sebyakin, Yu.N.,
Characteristics of electron and photon spectra associated
with interection between quasistationary terms, Sov. Phys.
JETP 49 (1979) 266-273.

PHYSICS OF ELECTRONIC AND ATOMIC COLLISIONS
S. Datz (editor)
© North-Holland Publishing Company, 1982

COLLISIONAL ELECTRON DETACHMENT OF $Cl^-$ ON
DIATOMIC MOLECULES

B. K. Annis and S. Datz

Oak Ridge National Laboratory
Oak Ridge, Tennessee   37830
U.S.A.

Recent experimental results for collisional electron detachment of $Cl^-$ by $H_2/D_2$, $N_2$, $O_2$, NO, and CO are discussed.  The
emphasis is on angular distributions and energy loss measurements for laboratory energies of a few hundred eV.  Evidence
for the possibility of bound excited states of $N_2Cl$ and COCl
and the role of target negative ion resonant states is
presented.

INTRODUCTION

Much of the collisional electron detachment work which has been published since
the last ICPEAC has been concerned with the collisions of negative atomic ions
and atomic targets, and the intricacies of this work could well be the subject of
a progress report.  However, it will have to suffice to say that the question of
appropriate theoretical analysis of these supposedly simply systems is not yet
fully resolved.[1]  Despite the difficulties or perhaps because of them, there has
been a shift in emphasis--at least on the part of experimentalists--to molecular
targets, and much of this work has only just become available at this meeting.  A
synthesis of all the new material is beyond the scope of this report which instead
will be concerned primarily with our work at Oak Ridge National Laboratory on
detachment of $Cl^-$ by the common atmospheric diatomic targets at sub-keV projectile energies as well as the closely related work of Champion, Doverspike, and
co-workers at the College of William and Mary.

The interest in the detachment behavior of $Cl^-$ is due in part to the high electron affinity of chlorine.  As a result, the orbital velocity of the attached
electron is greater than the relative collision velocity of a typical collision
pair until projectile energies on the order of a few kilovolts are reached.
Consequently, at lower energies a quasi-molecular description ought to be applicable.  At the opposite extreme of the electron affinity scale, $H^-$ for example,
the orbital velocity of the extra electron will often be less than typical collision velocities and so the scattering may more closely resemble that of a free
electron.  A considerable variation in the collisional detachment for arbitrary
systems is thus to be expected.

Three types of detachment processes are of interest for the systems we have
studied.  These are

$$Cl^- + AB \rightarrow Cl + AB + e \qquad (1)$$

$$Cl^- + AB \rightarrow Cl + AB* + e \qquad (2)$$

$$Cl^- + AB \rightarrow Cl + AB^- \qquad (3)$$

Reaction (1) is the so-called "simple" or "direct" detachment process and corresponds to the direct emission of the electron from the projectile accompanied only

by rotational/vibrational excitation of the target. This process is most analogous to detachment by an atomic target. As indicated above, the best theoretical
description for this type of detachment is still an open matter.[1] However, at the
relatively low collision velocities of the present study the local complex potential model[2] should provide a useful framework for discussing detachment collisions. This same model has been successfully applied to the related processes of
Penning ionization[3] and associative/dissociative attachment,[4] and the physical
picture resulting from this model is diagrammed in Fig. 1.

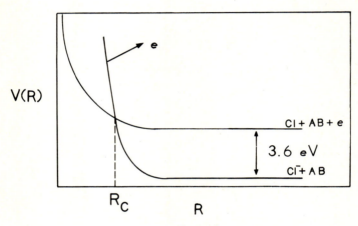

Figure 1
Complex Potential Model

The potential curve appropriate to the ion-neutral interaction is assumed to
cross that for a neutral-neutral interaction at some value of the internuclear
separation $R_c$. The latter potential curve actually represents the lower limit of
a continuum of states corresponding to the heavy particles and a free electron.
For a sufficiently small impact parameter the colliding particles will penetrate
to values of R such that $R < R_c$, but in this region there will be a finite probability that the electron will be ejected and carry off an amount of kinetic
energy equal to the difference in the two potentials. In principle, at a given
value of R, an electron may be emitted on the inward part of the trajectory or
remain attached until the turning point is reached and be emitted on the way back
out. In either case the K.E. of the electron will be the same but the heavy particles will experience different trajectories and structure may appear in the
differential cross section. Figure 1 shows potential curves which are purely
repulsive with a crossing point corresponding to a value of the energy exceeding
the electron affinity of Cl. This is justified by the recent results of
Doverspike, et al.,[5] who established that the thresholds for several of the collision pairs of interest to us do indeed exceed the electron affinity by 1-4 eV.

Reaction (2) can be incorporated in this picture, at least in a qualitative sense,
since there will also be a potential curve appropriate to the interaction of neutral chlorine and the target molecule in an electronically excited state. The
trajectories of the scattered heavy particles will then reflect the relative
probabilities of electron emission due to the crossings with the ground and
excited state potentials.

Reaction (3), the charge transfer process, is complicated by the fact that for the diatomic targets so far considered the negative molecular product ion is typically a resonant state with a lifetime of $10^{-15}$ to $10^{-10}$ sec. Evidence for the formation of such a resonant state was obtained by Risley[6] in an analysis of the electron spectrum produced by collisions of $H^-$ and $N_2$; however, the degree to which this channel was involved in the detachment collisions was not clear. One of the objectives of our study was to determine if there was evidence for such an occurrence in other systems.

The total cross section data (along with the results for related rare gas[7]) for some diatomic[5] targets obtained by the William and Mary group has been plotted in Fig. 2 as a function of relative kinetic energy. For the low energies shown, no particular structure is evident and the limited amount of data at higher energies indicate that the cross sections slowly increase until energies in the keV range are reached. As an example, at 14 keV the detachment cross section for $Cl^-/CO$ increases to about 45 $A_0^2$ and does not decrease until approximately 50 keV.[8] No particular pattern is as yet obvious from the threshold results which established values in the range $4\frac{1}{2}$ to $5\frac{1}{2}$ eV for $H_2$, $D_2$, and $O_2$ and $7-7\frac{1}{2}$ eV for He, Ar, $N_2$, and CO.

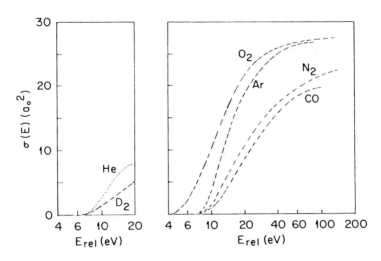

Figure 2
Total Detachment Cross Sections of $Cl^-$ in Various Gases

The initial effort at Oak Ridge to characterize detachment from $Cl^-$ by diatomic targets was a determination of the energy losses of the neutralized chlorine product resulting from collisions with $D_2$ and $H_2$.[9] The experiment involved the use of time-of-flight methods. The complexities of energy loss involving molecular targets are more readily put in perspective by comparison with a closely related rare gas target. In this case the kinematic effects are identical in both systems and do not obscure such a comparison. The energy loss spectra for $Cl$ resulting from a collision with He at a scattering angle of 1° and projectile energy of 350 eV is shown in Fig. 3.

For the He target only "simple" detachment is observed with an energy loss of 4 eV which corresponds to the minimum endothermicity of 3.6 eV (E.A. of Cl) plus a few tenths of an eV kinetic energy of the ejected electron. The energy loss spectrum for a $D_2$ target at the same energy and scattering angle is shown in Fig. 4.

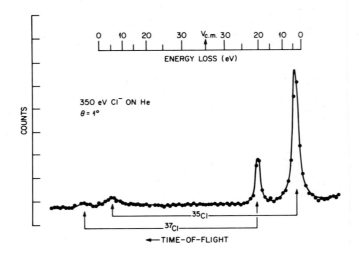

Figure 3
Energy Loss Spectrum on Detachment Collisions with He;
Peaks at left are caused by back scattered (in C.M.) particles

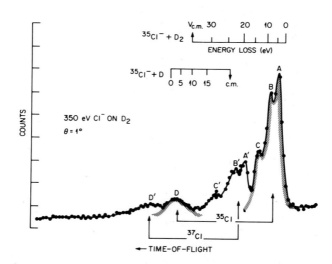

Figure 4
Energy Loss Spectrum of Cl$^0$ Formed by Detachment with D$_2$

Although the dominant peak (A) in the spectrum is essentially identical to that for He, two additional channels labelled B and C are readily apparent.  The energy loss of peak B corresponds to dissociative attachment of $D_2$, i.e.,

$$Cl^- + D_2 \rightarrow Cl + D + D^-$$

or

$$Cl + D + D + e^-$$

Peak C is ascribed to an electron transfer to the $^2\Sigma_g^+$ resonance of $H_2^-$ which is a repulsive state and also ultimately leads to dissociation.  Peak D corresponds to the backscattered contribution arising from simple detachment and is the equivalent to Peak A for small impact parameter collisions.  No isotope effects were observed in this study so that the preceeding comments also apply to an $H_2$ target.

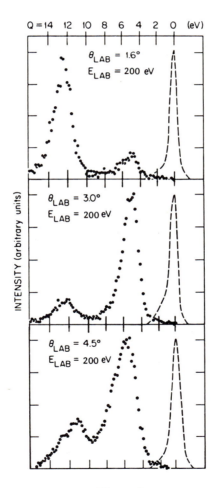

Figure 5
Energy Absorbed (Q) in Detachment Collisions with $N_2$ Target

In the $Cl^-$ and $H_2$ system the maximum scattering angle is ∿3° and differential
measurements are difficult. ·As the mass of the target molecule is increased, the
kinematic constraints are relaxed and it was possible to measure energy losses
and scattered intensity as a function of angle.  Such measurements required an
apparatus with provision for scanning a wider angular range than was available
with that used in the hydrogen experiments.  Consequently a second apparatus which
incorporated basically the same ion source and time-of-flight techniques was
developed.  The angular resolution was 0.5°, a scattering cell was used, and beam
energies were 150, 200, and 300 eV.

A sample of the data[10] for a nitrogen target for a laboratory beam energy of 200
eV is shown in Fig. 5 for various angles.  An illustration of the apparatus energy
resolution is provided by the spectra for the elastic scattering of $Cl^-$ by argon
which are shown as dashed peaks.  Two energy loss peaks in the $N_2$ data are clearly
resolved which correspond to losses of approximately 6 and 12 eV.  As with $D_2$ and
He, it is useful to make a comparison between $N_2$ and Ar.  The kinematics are
similar for the latter targets but unlike $N_2$ the only feature in the detachment
energy loss spectrum for an argon target was a peak centered in the range of 4-4½
eV.  At an angle of 4.5° the full width at half maximum of the argon peak was
less than half that shown for $N_2$.

The remarkable feature of this data is that at the lowest scattering angles a
high energy loss channel dominates the detachment process.  Although scattering
data on related systems is rather limited there is no indication that high energy
loss low angle scattering is to be expected.  For example, Inouye, et al.,[11] have
done energy loss measurements on the isoelectronic system $K^+ + N_2$ and Fernandez,
et al.,[12] have obtained energy loss data for $Ar^+ + N_2$.  In both experiments
energy loss peaks which, when the E.A. of Cl is taken into account, correspond
roughly  to our 12 eV peak, but this peak was present only at larger values of
$(\tau \equiv E\theta)$ while the low angle scattering was dominated by quasi-elastic processes.
In addition, Champion and Doverspike[10] measured the inelastic losses associated
with the non-detaching channel for $Cl^- + N_2$ and found no evidence for a high-loss
peak.  Consequently, it would appear that there is something unique about the
detachment channel.

The most probable energy loss (Q) for several beam energies are plotted as a
function of $\tau$ in Fig. 6.  The 12 eV loss channel varies with $\tau$ in a manner markedly different from that for the 6 eV channel.  Furthermore, if the $\tau$ dependence of
the high energy loss channel for the $K^+/N_2$ and $Ar^+/N_2$ systems were plotted in this
manner, in both cases the Q value would show an increase with $\tau$ rather than the
decrease shown in Fig. 6.  The energy losses appropriate to detachment accompanied
by target excitation are shown on the right-hand ordinate.  The most likely candi-
dates for the 12 eV channel appear to be excitation of the $a^1\Pi_g$ and the $a'^1\Sigma_u^-$
states of $N_2$ or charge transfer to the $N_2^-$ resonances labelled "a" and "a'".[13]
The peculiar $\tau$ dependence of this channel strongly suggests that more than one
electronic excitation is involved since excitation of a single level is expected
to increase with scattering angle as vibrational-rotational excitations of the
excited state occur.

Formation of the autodetaching state of $Cl^-$ observed by Cunningham and Edwards[14]
is discounted because the energy losses at high $\tau$ are too small, and at lower $\tau$
values we found no evidence for this channel with other targets (with the exception of CO).  In addition, separate experiments were carried out with $F^-$ and the
same 12 eV energy loss peak appeared.  Excitation of the analogous auto-detaching
states of $F^-$ should require an energy loss of approximately 16 eV.

The low energy loss channel (5-7 eV) shown in Fig. 6 is appropriate (within the
experimental uncertainty) to the formation of the $^2\Pi_g$ resonant state of $N_2^-$.  The
relative differential cross sections for each of the two channels are given in
Fig. 7.  The low-loss channel can be seen to dominate the scattering in the

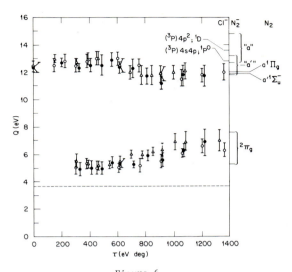

Figure 6
Energy Loss, Q, in Detachment Collisions of $(Cl^- + N_2)$ vs $\tau$

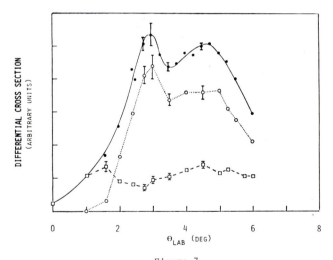

Figure 7
Differential Cross Section for $Cl^0$ Formed in Detachment Collisions with $N_2$ at 200 eV (Lab). Low-loss Channel, ○; high-loss channel, □.

intermediate ranges of angles and to pass through a local minimum at 650-700 eV deg. The high-loss channel also has a local minimum at about 500 eV-deg. Similar structure was also observed at a nominal beam energy of 150 eV[10] while at 300 eV only a hint of the minimum in the high-loss channel was observed.

Energy loss measurements with CO as a target produced results qualitatively similar to those for $N_2$, i.e., two energy loss peaks were clearly resolved and within
experimental uncertainty the Q vs τ behavior of the low loss channel was found to
be indistinguishable from that shown in Fig. 6.  As was the case for $N_2$, this
energy loss is also commensurate with charge transfer to a $^2\Pi$ resonance of $CO^-$.
In addition, the Q values for the high loss channel were observed to be ~11.5 eV
at low τ and unlike $N_2$ increased slowly to about 13 eV.  CO has fewer excited
states than $N_2$ and no resonances in the appropriate energy range.  The 11.5 eV
threshold appears to be most appropriate to excitation of the A $^1\Pi$ state of CO.
As with $N_2$, the energy loss spectra obtained in some $F^-$/CO experiments lead us to
conclude that target rather than projectile excitation dominates the scattering.

The relative angular differential data for a 200 eV ($Cl^-$) beam in CO is shown in
Fig. 8.  Like $N_2$, CO targets produce significant high-loss, low-angle scattering.
However, the angular intensity variation for both channels is more extreme for CO.
The same type of structure was also found in the data for beam energies of 150
and 300 eV.

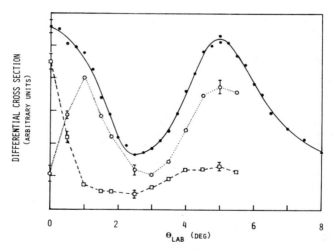

Figure 8
Differential Cross Section for $Cl^0$ from Collisions of $Cl^-$ with CO at 200 eV (lab.)

Representative loss spectra for NO and $O_2$ targets at beam energies of 200 eV (lab.)
are shown in Fig. 9.  Unlike the $N_2$ and CO data, the results for NO and $O_2$ show no
indication that high energy loss channels play a role in the low angle scattering.
In the case of NO, as the scattering angle is increased it is apparent that another
channel with an energy loss of about 12 eV does begin to contribute to detachment.

Both the magnitude of the most probable Q value for the dominant channel in the
NO scattering (5-7 eV) and the τ dependence of Q closely resembles that for $N_2$
and CO; consequently there is no evidence for charge transfer to the stable ground
vibrational state of $NO^-(^3\Sigma^-)$.  Evidence for $^1\Delta$ and $^1\Sigma^+$ resonant states of $NO^-$ in
the range 0.7 ∿ 2.5 eV have been found in electron impact experiments by Tronc.
et al.,[15] and charge transfer from $Cl^-$ to these states might contribute to our
observed energy losses.

The most probable Q value for detachment by $O_2$ at low scattering angles is about

4 eV (Fig. 9). Charge transfer to a bound state of $O_2^-$ would give rise to an energy loss in the 3.2 to 3.6 eV range; hence, if the $X^2\Pi_g$ state of $O_2^-$ is involved, our measurements indicate that it would have to be through an autodetaching vibrational level. The scattered intensity at a laboratory angle of 1.0° is quite low and this is responsible for the ragged appearance of the high energy loss side of this spectrum. However, as the scattering angle is increased to 4.0°, intensity is no longer a problem and it is quite clear that a second channel opens. It is not clear as to whether or not the process giving rise to the 4 eV peak at 1.0° is the same process causing the peak of ~4.2 eV at 4.0°. At the latter scattering angle we also observed an energy loss of about 4.2 eV with an argon target; consequently, in this case the energy loss measurements are insufficient to unequivocally identify the detachment channel.

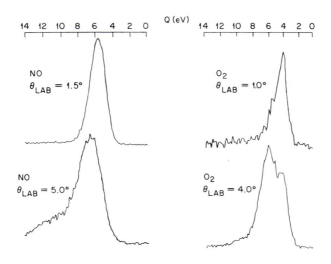

Figure 9
Energy Loss Spectra for NO and $O_2$ at 200 eV (lab.)

Despite the structure in the energy loss measurements for $O_2$, the angular differential cross section proved to be featureless (Fig. 10) in constrast to that observed for the other targets. The differential cross section for a laboratory energy of 200 eV for $Cl^-$ on NO is also shown in Fig. 10. Although it is difficult to unambiguously separate the relative contribution of the high-loss channel which the TOF spectra indicate is present at larger scattering angles, comparison of the spectra for NO and $O_2$ in the Q range of 8-14 eV suggests that the shoulder at about 7° in the NO data of Fig. 10 can be interpreted as structure in the low-loss channel.

DISCUSSION

The data presented here are sufficient to demonstrate that collisional electron detachment by the simple diatomic molecules $H_2$, $N_2$, NO, $O_2$, and CO proceeds in a complicated and highly specific manner. Within the framework of the local complex potential model it is apparent that several potential surfaces and electron ejection probability functions are necessary for each molecule in order to account for the observations. In general this material is not yet available; however, in

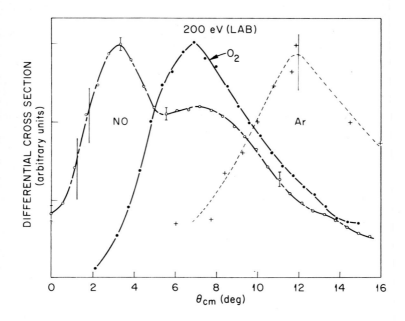

Figure 10
Differential Cross Sections for NO, ○, 0 , •; and Ar, +, at 2000 eV (lab.)

the case of Cl⁻ + NO and Cl⁻ + Ar there is some information on the potential that
may be useful in providing an insight into the difference in differential detachment cross sections for these collision pairs. As may be seen in Fig. 10 where a
differential cross section for Cl⁻ + Ar collisions based on the work of Fayeton,
et al.[16] has been plotted, the detachment scattering for NO occurs at substantially smaller angles than that with an argon target.

NO and Cl form a stable, bent molecule which has a well depth of 1.6 eV[17] along
the N-Cl axis, and by fitting a Morse potential to some of the spectroscopic
data[18] we have generated the NOCl curve shown in Fig. 11. First approximations
to the Cl⁻ + NO potential are provided by the potentials for the isoelectronic
systems K⁺ + NO and Ar + NO. The former was obtained from elastic scattering[19]
and the latter comes from the electron gas calculations of Nielson, et al.[20]
Appropriate projections of the potentials along the N-Cl axis have been used and
it is apparent from Fig. 11 that the potentials for the incoming part of the
trajectory intersect the NOCl curve near the minimum in the attractive well. This
is in marked contrast to the potentials for Cl⁻ + Ar and Cl + Ar as calculated by
Olson and Liu[21] and also shown in Fig. 11. These potentials are purely repulsive
and cross at about 6.5 eV. With these potentials we can now make an estimate of
the threshold or minimum scattering angle for a detachment collision. In such a
case the electron is ejected with zero kinetic energy at the crossing point which
also serves as the turning point in the heavy particle trajectory. The resulting
threshold angles are given by vertical lines in Fig. 10. The threshold angle for
Cl⁻ with Ar proves to be an overestimate, but the threshold data[7] suggest
that the crossing point of the potential should be reduced by 3 eV which should
bring the threshold angle into better agreement with the experimental data. However, the point is that these approximate calculations strongly suggest that in
order to further reduce the scattering angle to the range appropriate to Cl⁻/NO

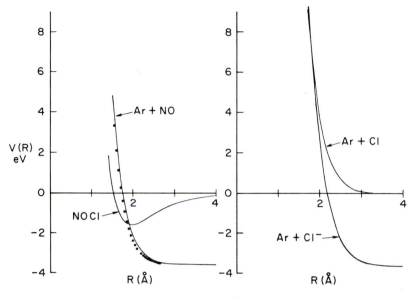

Figure 11
Potentials. ($\cdots$ shows $K^+ + NO.^{18}$)

it is necessary to have curve crossing in an attractive region of the exit channel. The energy loss measurements near the forward direction with an NO target
show that an endothermicity of some 5 eV is involved whereas in the above calculation the appropriate value is 3.6 eV. If the threshold process corresponds to
"simple" detachment accompanied by vibrational-rotational excitation then the
additional energy loss should further reduce the threshold angle to give even
better agreement with the observations. As discussed above, the energy losses are
also in accord with formation of $NO^-$ resonances. If the threshold process corresponds to charge transfer, then the forward scattering implies that at some
angle of attack there must be a minimum in the $Cl + NO^-$ surface.

If we now reconsider the observations of the low-angle, high-loss scattering with
$N_2$ and CO, the same type of threshold angle calculation suggests that a bound excited state such as that sketched in Fig. 12 ought to be able to account for the
low angle scattering. The "black box" area shown in Fig. 12 indicates that there
is too little information to order the curves in this region. Since the outer
shell of $O^-$ is isoelectronic with neutral chlorine, we have included in Fig. 12
a simplified representation of $N_2O^-$ bent at an angle of 125°. This is based on
the adiabatic correlation diagram of Krauss, et al.[22] (A similar diagram is also
appropriate for $[CO_2^-]^*$ which is the analog of $[COCl]^*$.) Since no stable ground
of $N_2Cl$ is known to exist, consideration of Fig. 12 suggests the possibility that
low energy associative detachment collisions could populate an excited state which
might then radiatively dissociate in a manner analogous to the rare gas-halogen
excimers. A second generation of experiments at lower energies involving spectroscopic product analysis and/or or a search for backscattered metastable product
should help to determine the nature of the excited state.

One of the objectives of this survey was to determine the evidence for the involvement of the negative ion resonant states of the targets. Detachment by charge

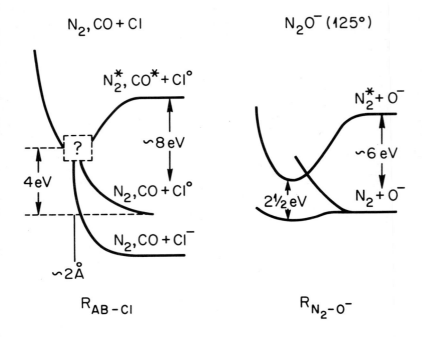

Figure 12
Schematic Potential Curves

transfer should result in an energy loss rather different from that for "simple" detachment. For $N_2$, CO, and NO, the observed energy losses are consistent with charge transfer. However, there do not appear to be any unambiguous arguments for discounting the possibility that the energy losses are in part attributable to "simple" detachment accompanied by vibrational-rotational excitation. This is in contrast to the results of Tuan and Esaulov[23] for detachment collisions of H⁻ and $N_2$ and CO where definite structure which could be attributed to the charge transfer and "simple" detachment channels is apparent. In the case of the more tightly bound Cl⁻, experiemtns involving measurement of the electron spectra might resolve the matter.

This research was sponsored by the Division of Chemical Sciences, Office of Basic Energy Sciences, U.S. Department of Energy, under contract W-7405-eng-26 with Union Carbide Corporation.

REFERENCES

[1]  de Vreugd, C., Wijnaendts van Resandt, R. W., Delos, J. B., and Los, J., Chem. Phys. [to be published).

[2]  Mizuno, J., and Chen, J.C.Y., Phys. Rev. A4 (1971) 1500–1516.

[3]  Miller, W. H., J. Chem. Phys. 52 (1970) 3563–3572; Bieniek, R. J., Phys. Rev. A18 (1978) 392–413.

[4]  Chen, J.C.Y., in: Advances in Radiation Chemistry VI, Curtan, M., and Magee, J. L. (eds.) (1968) 245–376.

[5]  Doverspike, L. D., Smith, B. T., and Champion, R. L., Phys. Rev. A22 (1980) 393–398.

[6]  Risley, J. A., Phys. Rev. A16 (1977) 2346–2351.

[7]  Smith, B. T., Edwards, III, W. R., Doverspike, L. D., and Champion, R. L., Phys. Rev. A18 (1978) 945–954.

[8]  Il'in, R. N. Sakharov, V. I., and Serenkov, I. T., Abstracts of Contributed Papers, XII ICPEAC, Gatlinburg (1981) 535–536.

[9]  Cheung, J. T., and Datz, S., J. Chem. Phys. 73 (1980) 3159–3165.

[10]  Annis, B. K., Datz, S., Champion, R. L., and Doverspike, L. D., Phys. Rev. Lett. 45 (1980) 1554–1557.

[11]  Inouye, H., Niurao, K., and Sato, Y., J. Chem. Phys. 64 (1976) 1250–1251.

[12]  Fernandez, S. M., Eriksen, F. J., Bray, A. V., and Pollack, E., Phys. Rev. A12 (1975) 1252–1260.

[13]  Schulz, G. J., Rev. Mod. Phys. 45 (1973) 423–486.

[14]  Cunningham, D. L., and Edwards, A. K., Phys. Rev. A8 (1973) 2960–2964.

[15]  Tronc, M., Huetz, A., Landau, M., Pichou, F., and Reinhardt, J., J. Phys. B: Atom. Molec. Phys. 8 (1975) 1160–1169.

[16]  Fayeton, J., Dhuicq, D., and Barat, M., J. Phys. B: Atom. Molec. Phys. 11 (1978) 1267–1281. The data used in Fig. 10 was actually obtained with a beam energy of 1000 eV, but our experience has been that such data can be scaled to different energies.

[17]  Darwent, B. de B., Natl. Stand. Ref. Data, Natl. Bur. Std. 31 (1970).

[18]  Botschwina, P., Haertner, H., and Sawodny, W., Chem. Phys. Lett. 74 (1980) 156–159.

[19]  Amdur, I., Jordan, J. E., Fung, L.W.-M., Hermans, L.J.F., Johnson, S. E., and Hance, R. L., J. Chem. Phys. 59 (1973) 5329–5332.

[20]  Nielsen, G. C., Parker, G. A., and Pack, R. T., J. Chem. Phys. 66 (1977) 1396–1401.

[21]  Olson, R. E., and Liu, B., Phys. Rev. A17 (1978) 1568–1574.

[22]   Krauss, M., Hopper, D. G., Fortune, P. J., Wahl, A. C., and Tiernan, T. O.,
       Potential Energy Surfaces for Air Triatomics, VI.  ARL TR 75-0202,
       Aerospace Res. Laboratories (1975).

[23]   Tuan, V. N., and Esaulov, V. A., Abstracts of Contributed Papers, XII
       ICPEAC, Gatlinburg (1981) 531-532.

PHYSICS OF ELECTRONIC AND ATOMIC COLLISIONS
S. Datz (editor)
© North-Holland Publishing Company, 1982

MUONIUM FORMATION AND REACTIVITY
IN THE GAS PHASE

Donald G. Fleming

Dept. of Chemistry and TRIUMF, University of
British Columbia, Vancouver, B.C. V6T 1W5
Canada.

The muonium atom is chemically identical to hydrogen but
only 1/9 the mass. Hence, studying the interactions of
muonium with its environment provides the possibility
for an unprecedented extension of the most sensitive end
of the isotopic mass scale in describing collision
phenomena. Results from the current gas phase experimental
programme at TRIUMF dealing with muonium formation and its
reactivity are discussed, with particular reference to their
theoretical importance.

INTRODUCTION

The positive muon ($\mu^+$), one of natures "elementary particles", is produced 100%
longitudinally polarized in the decay of positive pions ($\pi^+ \rightarrow \mu^+ \nu_\mu$), ultimately
decaying with a mean life of 2.2 $\mu$s into a positron and two neutrinos ($\mu^+ \rightarrow e^+ \nu_e \bar{\nu}_\mu$).
In this decay process, because of the definite helicity of neutrinos, the $e^+$ exits
preferentially along the muon spin, a fact responsible for the first measurements
of parity violation in the decay sequence $\pi^+ \rightarrow \mu^+ \rightarrow e^+$ [1] and which today, some
25 years later, is responsible for the now blossoming field of $\mu$SR (Muon Spin Rotation or Relaxation or Resonance) [2]. From the point of view of its physical-
chemical interactions with matter though, the $\mu^+$ can be regarded simply as an
ultralight proton ($M_\mu = 1/9 M_p$); in like manner, the muonium atom (Mu = $\mu^+ e^-$) is
regarded as an ultralight isotope of hydrogen. Muonium has often been regarded
as an "exotic" atom and, in particular, the hyperfine interaction between $\mu^+$ and
$e^-$ spins has provided one of the most important tests to date of the predictions
of QED [3]. As expected for isotopes, the hyperfine interaction frequencies in
muonium ($\nu_0$ = 4463.3 MHz) and hydrogen ($\nu_0$ = 1420.4 MHz) differ from each other
only by their respective gyromagnetic ratios ($g_\mu/g_p$ = 3.183). In this context,
it should be noted that positronium ($e^+ e^-$), another exotic atom, has no nucleus
and cannot properly be regarded as an isotope of hydrogen. Although perhaps seemingly less "exotic", the factor of 9 difference in mass between Mu and H is
unprecedented and provides the possibility of very sensitive tests of the importance of mass effects in the interactions of these very simple atomic systems with
their environments.

Muonium is being actively studied, primarily in condensed media but also in gases,
at the world's "meson factories": TRIUMF (Vancouver, Canada), LAMPF (Los Alamos,
USA) and SIN (Villigen, Switzerland), as well as at older facilities; at CERN
(Geneva, Switzerland) and JINR (Dubna, USSR), with a new programme now also
operational at KEK (Tsukuba, Japan). In the solid state, the $\mu^+ - e^-$ hyperfine
interaction can be highly perturbed by the lattice with the result that both
"normal" and "anomalous" muonium states corresponding to isotropic and anisotropic
interactions have now been identified in a number of solids, notably Si, Ge,
diamond and quartz. In liquids, muonium chemistry is being pursued both from the

point of view of muonium formation and reaction kinetics, as well as the characterization of muonic ($\mu^+$) radicals. Gases, however, remain the ideal medium for theoretically interpreting the interactions of the muon and of muonium since, particularly at low pressures, one is free of the complicating many body effects associated with condensed media. This article will be concerned exclusively with muonium formation and its reactivity in the gas phase.

The earliest study of muonium formation and muonium chemistry in gases was the work of Mobley et al. [4], using conventional "backward" muons of typically 125 MeV/c momentum and hence necessitating the use of high pressure ($\sim$ 40 atm) targets). The prime focus of this work was a study of the chemical and physical (spin exchange) interactions of muonium in a longitudinal magnetic field. Although certainly representing a significant beginning to gas phase muonium studies, these data are characterized by poor statistics and in some cases questionable interpretation [5]. Later work by Stambaugh et al. [6] and also by Barnett et al. [7] concentrated on a study of muonium formation in gases, again using relatively high energy muons and, in the case of ref. 6, high pressure targets. The utility of "surface" muons in the study of gases is self evident. In a surface muon beam, muons originate from $\pi^+$ decays on the surface of the production target and hence are essentially monochromatic (4.1 MeV) and 100% polarized [8]. After traversing counters and beam line and target windows, the $\mu^+$ enters the gas with $\lesssim$ 2.5 MeV which means it will stop in a distance $\sim$ 40 cm in a gas (eg., Ar) at 1 atm. pressure. The first gas phase muonium chemistry and atomic physics studies using this type of beam were carried out at the now defunct 184" cyclotron at Lawrence Berkeley Laboratory [9,10]. This effort is now being carried on at the TRIUMF cyclotron. Current areas of interest will be highlighted in the discussion to follow.

## $\mu^+$ POLARIZATION IN MUONIUM

In pion decay, since the $\nu_\mu$ has definite negative helicity while the $\pi^+$ itself has spin zero, the $\mu^+$ in the decay process $\pi^+ \to \mu^+ \nu_\mu$ is forced by (angular) momentum conservation to have negative helicity also. Thus the $\mu^+$ is borne (with 4.1 MeV kinetic energy) 100% longitudinally polarized in the rest frame of the pion and this polarization is maintained as the $\mu^+$ slows down in matter from several MeV to near thermal energies, <u>until</u> the onset of muonium formation. Since the $\mu^+$ is initially polarized ($\alpha_\mu$) but the captured electron is not ($\alpha_e, \beta_e$) muonium forms initially in parallel ($|\alpha_\mu \alpha_e>$) and antiparallel ($|\alpha_\mu \beta_e>$) states with equal probability. In zero and weak longitudinal magnetic fields, the parallel (triplet) state is an eigenstate of the (isotropic) Zeeman hamiltonian,

$$H = g_e \beta_e \overline{S}_e \cdot \overline{B} - g_\mu \beta_\mu \overline{I}_\mu \cdot \overline{B} + A \overline{S}_e \cdot \overline{I}_\mu \qquad (1)$$

but the antiparallel state is not, oscillating instead between $|\alpha_\mu \beta_e> = 1/\sqrt{2}[|1 0> + |0 0>]$ and $|\beta_\mu \alpha_e> = 1/\sqrt{2}[|1 0> - |0 0>]$ at the hyperfine frequency $\nu_0$ = 4463.3 MHz ( $FM_F$ represent the usual hyperfine coupled quantum numbers). Consequently, the formation of antiparallel ("singlet") muonium for sufficiently long times effectively depolarizes the positive muon, since experimental time resolutions are typically $\gtrsim$1 ns, much greater than $1/\nu_0$ = 0.22 ns. The solutions to Eqn. (1) are given in the familiar field dependent energy level (Breit Rabi) diagram in Figure 1.

In a <u>transverse</u> magnetic field (i.e., perpendicular to the $\mu^+$ spin), neither the parallel nor antiparallel states are eigenstates but these can be expressed in terms of the eigenstates of the Zeeman hamiltonian by a suitable transformation [11-13]. The time dependence of the $\mu^+$ polarization in muonium then has the form

$$P_\mu(t) = 1/4[(1+\delta)(e^{i\nu 12t} + e^{-i\nu 34t}) + (1-\delta)(e^{i\nu 23t} + e^{i\nu 14t})] \qquad (2)$$

where the $\nu_{ij}$'s are defined in Fig. 1 and $\delta = x/\sqrt{1+x^2}$; x is the dimensionless ratio $H/H_0$, where H is the applied field and $H_0$ = 1585 G is the contact field of the

$\mu^+$ at the electron (compare 503 G for the H atom). In moderate magnetic fields, H $\lesssim$ 200 G, $\delta \approx$ 0 and there are four allowed ($\Delta M = \pm 1$) transition frequencies; but, both $\nu_{14}$ and $\nu_{34}$ are comparable to $\nu_0$(4463 MHz) and hence not resolvable with the $\lesssim$ 1ns time resolution of a typical $\mu$SR experiment. Consequently, there are only two observable Mu frequencies, $\nu_{12}$ and $\nu_{23}$. In much weaker magnetic fields, B $\lesssim$ 10 G, the frequencies $\nu_{12}$ and $\nu_{23}$ coalesce into a single frequency ($\nu_{Mu}$ = 1.39xH(G)MHz) characteristic of coherent Larmor precession of "triplet" muonium. Classically, the loss of "singlet" muonium can be identified with its zero spin and hence zero precession.

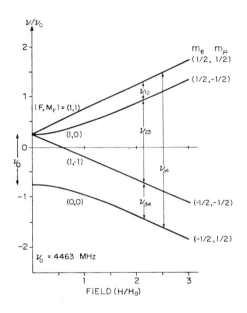

*Figure 1: Breit-Rabi diagram for the energy levels of muonium in a magnetic field, plotted in terms of the dimensionless variable H/Ho, where Ho = 1585G for muonium.*

## THE EXPERIMENTAL MSR SIGNAL

As described above, in a weak transverse magnetic field, triplet Mu will precess with a characteristic Larmor frequency (13.9 MHz at 10 G). If a counter system is fixed in the plane of precession, then since the probability of $e^+$ detection is maximal along the $\mu^+$ spin direction, each time the muon spin (in muonium) sweeps past the counter system there will be an enhanced probability for detecting the decay positron. Hence, a plot of number of detected positrons N(t) vs. time is expected to exhibit oscillatory behaviour. A typical time histogram of this nature is shown in Figure 2 (top) for $\mu^+$ stopped in $N_2$ gas at 1 atm. pressure. The data in Fig. 2 can be fit to the functional form

$$N(t) = N_0\ e^{-t/\tau_\mu}\ [1 + S(t)] + BG \qquad (3)$$

where $N_0$ is a normalization, $\tau_\mu$ = 2.197 $\mu$s is the muon lifetime, BG is a time independent background term and S(t) is the "Muonium Spin Rotation (MSR)" signal of interest. It has the form

$$S(t) = A_{Mu}\ e^{-\lambda t}(\cos \omega_{Mu}t + \phi_{Mu}) + A_\mu(\cos \omega_\mu t - \phi_\mu) \qquad (4)$$

where $A_{Mu}$, $\omega_{Mu}$, $\phi_{Mu}$ and $A_\mu$, $\omega_\mu$, $\phi_\mu$ are the amplitudes, frequencies and initial phases for the muonium and diamagnetic $\mu^+$ fractions, respectively. In addition,

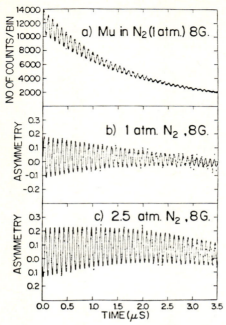

*Figure 2.* *Muonium Spin Rotation (MSR) in pure $N_2$ in an 8G transverse magnetic field. The top spectrum gives the time histogram N(t) at 1 atm. pressure while the middle spectrum gives the corresponding signal of interest, S(t). There is a noticeable background relaxation, $\lambda_Q$. The bottom of the figure gives the signal S(t) in 2.5 atm. $N_2$ moderator.*

Eqn. (4) contains the transverse relaxation rate $\lambda={}^1/T_2$, defined in analogy with NMR or ESR, corresponding to the interaction of the Mu atom with its environment. An example signal S(t) is also shown in Fig. 2 (middle) corresponding to the N(t) spectrum immediately above; the bottom part of the same figure shows S(t) for muons in $N_2$ at 2.5 atm. pressure. Note the pronounced increase in the amplitude $A_{Mu}$ at the higher pressure. The slow modulation apparent in Fig. 2 is due to a $\sim 20\%$ $\mu^+$ signal which precesses 103 times slower than Mu in the same field. For further details of the technique, consult Refs. 10 - 15.

An experiment consists of stopping a (surface) $\mu^+$ beam in the gas, collecting a histogram of events and analyzing it to yield the parameters of interest; notably $A_{Mu}$, $A_\mu$ and $\lambda$. The relaxation of the signal, $\lambda$, can be due to either chemical reaction of the Mu atom or to spin interactions with paramagnetic molecules. It is important to note that there can only be one Mu atom in the reaction system at a time; hence, the study of Mu reaction rates is particularly clean, free of competing reactions which often plague similar studies of H atom interactions. As such, the relaxation rate, $\lambda$, can be interpreted in terms of a (pseudo) first order rate constant,

$$\lambda = \lambda_0 + \lambda_c + \lambda_d = \lambda_0 + k[X] \qquad (5)$$

where $\lambda_0$ is the background value mentioned above, $\lambda_c$ corresponds to chemical reaction and $\lambda_d$ to Mu depolarization via electron spin exchange, while [X] represents the concentration of added reactant. In practice, effects ascribed to $\lambda_c$ and $\lambda_d$ are separable, as described in more detail below.

PRESSURE DEPENDENT MUONIUM FORMATION

In its slowing down process, the $\mu^+$ basically passes through three energy/time domains [14-16]. Most of its energy is lost in the first domain by the usual Bethe-Bloch type of ionization processes in which the $\mu^+$ slows from an incident energy of $\sim 2.5$ MeV to about 30 keV in a time of $\sim 10$ ns at 1 atm. pressure. Muonium formation then ensues via a series of change exchange cycles with modera- tor M, $\mu^+ + M \rightleftharpoons Mu + M^+$ with cross sections $\sigma_{10}$ (Mu formation) and $\sigma_{01}$ ($\mu^+$ formation again). In this charge exchange regime, the $\mu^+$ slows to an energy of $\sim 20$-30 eV during a time $\gtrsim 0.1$ ns at 1 atm pressure, based on extrapolations of available proton charge exchange data [17,18]. Although a total of something like 100 charge exchange cycles are expected [15,17] it is probably only the last few cycles which are important in deciding the fate of the $\mu^+$ as either "free" muon or as the muonium atom itself. If the time between collisions is long enough (com- pared to $1/\nu_0 = 0.22$ nsec), then appreciable depolarization due to antiparallel muonium formation can occur. Such an effect should be inversely proportional to the moderator pressure and is clearly revealed by the data. After emerging from the charge exchange regime, the $\mu^+/Mu$ ensemble subsequently thermalizes in the gas via elastic and inelastic collision processes. In 1 atm Ar, e.g., this takes $\sim 10$ ns $(\gg 1/\nu_0)$ so that 50% of the muonium formed is effectively depolarized at observation times.

The fractions of muonium ($f_{Mu}$) and of free muon ($f_\mu$) which thermalize in the gas can be obtained from the relations

$$f_{Mu} = \frac{2A_{Mu}}{A_{Tot}} \quad \text{and} \quad f_\mu = \frac{A_\mu}{A_{Tot}} \quad (6)$$

where $A_{Tot} = 2A_{Mu} + A_\mu$ and $A_{Mu}$, $A_\mu$ are the measured amplitudes for muonium and muon formation, respectively, as described earlier. The muonium amplitude has been multiplied by two to account for the unobserved antiparallel fraction. De- fined in this way, $f_{Mu} + f_\mu = 1$. The muon amplitude is in fact most likely due to the formation of $\mu^+$ molecular ions. The evidence for this will not be discussed here but lies in the fact that the muon signal relaxes noticeably upon the addi- tion of an impurity gas (eg, Xe in Ne) due to thermal muonium formation [14,15]. The neutral fraction $f_{Mu}$ can be compared with the corresponding fraction $f_H$ esti- mated from proton change exchange cross sections [15,18]. Representative data for the noble gases and for $N_2$ as a function of pressure are compared with the H atom fractions in Table I. This table also gives the "absolute asymmetries, $A_{abs}$", in the last column, which are expressed as a percentage fraction of the maximum $A_\mu$ obtained in Al under identical conditions (the asymmetry itself can be a function of beam line tune, counter efficiencies and particularly degrader thickness in the positron telescopes).

It is clear from the data in Table I that the measured amplitudes (and hence abso- lute asymmetries) for $\mu^+$ and Mu formation are strongly pressure dependent but the relative fractions $f_\mu$ and $f_{Mu}$ are not. That the fractions found should be pres- sure independent is intuitively reasonable: the time between collisions will change with pressure but not the number of collisions and/or cross sections in a given energy regime. Indeed, there is good agreement between $f_{Mu}$ and the corres- ponding $f_H$ value extrapolated from proton change exchange studies. Evidence that the $\mu^+/Mu$ asymmetry can change markedly with pressure can easily be seen in the case of $N_2$ in Fig. 1 and is a general feature in all gases studied [14,15]. Similar results in kr have been reported in a separate study at LAMPF [19].

Our first interpretation of such a pressure dependent asymmetry was that it was likely due to an extended stopping distribution of the $\mu^+/Mu$ ensemble at low pressures [14,16]. In this case, there will be a solid angle effect which must also include the effects of a distributed initial phase, tending to reduce the value of the expected asymmetry. This effect certainly occurs but Monte Carlo calculations show that it can be only about a 20% effect, at most [15], whereas there are pressure effects (on $A_{abs}$) in Table I of a factor of two and more. This is clearly a real effect and, as mentioned above, can be understood in terms of the increased time spent as singlet muonium during the charge exchange regime.

Table I : Pressure Dependent Amplitudes and Fractions for Muon and Muonium Thermalization in $N_2$ and the Noble Gases

| Gas | P(atm) | $A_\mu$ a) | $A_{Mu}$ | $A_{Tot}$ | $f_\mu(\%)$ | $f_{Mu}(\%)$ | $F_H(\%)$ b) | $A_{abs}(\%)$ |
|-----|--------|-----------|----------|-----------|-------------|--------------|--------------|---------------|
| He  | 1.2 | .12 | 0 | .12 | 100 | 0 | | 41 |
|     | 2.8 | .14 | 0 | .14 | 100 | 0 | 10 | 63 |
|     | 3.0 | .14 | 0 | .14 | 100 | 0 | | 63 |
| Ne  | 0.40 | .06 | .005 | .07 | 86 | 14 | | 26 |
|     | 1.2 | .17 | .005 | .18 | 94 | 6 | 10 | 43 |
|     | 1.6 | .24 | 0 | .24 | 100 | 0 | | 62 |
|     | 2.0 | .29 | .025 | .34 | 86 | 14 | | 83 |
| Ar  | 0.6 | .02 | .038 | .10 | 20 | 80 | | 44 |
|     | 1.1 | .06 | .10 | .26 | 25 | 75 | 80 | 76 |
|     | 2.5 | .10 | .13 | .36 | 28 | 72 | | 92 |
| Kr  | 0.40 | 0 | .040 | .08 | 0 | 100 | | 41 |
|     | 0.65 | 0 | .086 | .17 | 0 | 100 | 100 | 55 |
|     | 0.90 | 0 | .12 | .24 | 0 | 100 | | 71 |
| Xe  | 0.40 | 0 | .05 | .10 | 0 | 100 | 100 | 42 |
|     | 0.65 | 0 | .089 | .18 | 0 | 100 | | 63 |
| $N_2$ | 1.0 | .05 | .12 | .29 | 16 | 84 | 80 | 95 |
|     | 2.5 | .08 | .17 | .42 | 18 | 82 | | 100 |

a) some $\mu^+$ will scatter into the walls of the target vessel, particularly at low pressures. The values for $A_\mu$ given here have been corrected for this "wall signal" and are estimated to be accurate to ± .02

b) Thermalized neutral H atom fractions estimated from proton charge exchange data of Ref.18.

## MUONIUM REACTIVITY IN THE GAS PHASE

As noted above there are basically two types of reaction processes which can contribute to a relaxation of the MSR signal, chemical reaction and spin exchange. In a chemical reaction, e.g., $Mu+F_2 \rightarrow MuF+F$, the $\mu^+$ is placed in a diamagnetic environment in which it precesses (incoherently) 103 times slower than in muonium itself. In a spin exchange reaction, e.g. $Mu(\uparrow)+NO(\downarrow) \rightarrow Mu(\downarrow)+NO(\uparrow)$, the $\mu^+e^-$ hyperfine interaction is intrinsically changed which concomitantly reduces the $\mu^+$ polarization (in the case of "spinflip") hence causing a relaxation of the signal. To date bimolecular rate constants (Eqn. (5)) for the gas phase chemical reactions $Mu+Br_2[10]$, $Mu+Cl_2[20]$, $Mu+F_2[21]$, Mu+HBr, $Mu+C_2H_4$ and $Mu+H_2$ have been measured, in the temperature range $\sim$300-$\sim$500K for $Cl_2$, $F_2$, HBr and $C_2H_4$ and 300-850K for $Mu+H_2$. The spin exchange reactions $Mu+O_2$ and Mu+NO have been measured in the temperature range 295-478K; an example plot of $\lambda_{vs}[X]$ from Eqn. (5) is shown in Fig. 3 for Mu+NO at two different temperatures. The bimolecular rate constants k are obtained immediately from the slopes of such plots. The data is summarized in Table II. The last two columns give the 300K ratio of $k_{Mu}/k_H$ and $k_H/k_D$ where $k_H$ and $k_D$ are the rate constants for the analogous H and D atom reactions.

Chemical Reactions

In a bulk kinetic experiment such as the present ones, the bimolecular (thermal) rate constant k(T) is related to the reaction cross section by

$$k(T) = \left(\frac{8}{\pi\mu}\right)^{1/2} \left(\frac{1}{k_\beta T}\right)^{3/2} \int \sigma(E) E e^{-E/k_\beta T} dE \qquad (7)$$

where $\mu$ is the reduced mass of the colliding partners and $\sigma(E)$ in their energy dependent reaction cross section. The functional form of $\sigma(E)$ must be known before Eqn. (7) can be evaluated, either from an explicit evaluation of the reaction dynamics (as, e.g., in Refs. 28 - 30) or from some model dependent calculation. The usual "threshold" models typically give the expected Arrhenius dependence for the rate constant

$$k(T) = \sigma_0 \left(\frac{8k_B T}{\pi\mu}\right)^{1/2} e^{-E_a/k_B T} \qquad (8)$$

where $\sigma_0$ is some effective hard sphere collision cross section and $E_a$ is the (threshold) activation energy. An example Arrhenius plot giving results for the Mu + $C_2H_4$ reaction is shown in Fig. 4. Activation energies for all reactions studied to date are given in Table II.

Table II:   Mu, H, D Rate Constants, $k(10^{-11}$ cc.molec$^{-1}$s$^{-1}$)

| Molecule | $k_{Mu}$ [a) | $E_a^{Mu}$(eV) | $E_a^{H}$(eV) | $k_{Mu}/k_H$ [a) b) | $k_H/k_D$ [a) |
|---|---|---|---|---|---|
| $F_2$ | 2.33±0.16 | .040±.009 | .089±.003 | ~6 - 14 | |
| $Cl_2$ | 8.47±0.33 | .061±.009 | .052±.004 | 3 - 5 | 1.5 |
| $Br_2$ | 39.8 ±5.0 | - | .078±.017 | 4 - 6 | 1.5 |
| HBr [c) | 1.56±.03 | .050±.015 | .11 ±.005 | 2.5- 4.5 | 1.8 |
| $C_2H_4$ [c) | 0.57±.05 | .050±.012 | .091±.002 | 5 - 6 | 1.4 |
| $H_2$ [c) | $(6\pm2)\times10^{-3}$ | 0.55±.06 | 0.35±.03 | .055 | ~0.4 [d) |
| $O_2$ [e) | 26.0±3.0 | - | - | ~2.9 | |
| NO | 31.0±4.0 | - | - | ~2.9 | |

a)   at 300K, except $H_2$ at 800K

b)   Unfortunately, many of the H atom valves reported in the literature differ considerably from each other, which frustrates the important comparison to be made with Mu. This uncertainty is reflected in the range given for the ratio $k_{Mu}/k_H$. Recent values for $k_H$ can be found in the following references: H+F$_2$[22], H+Cl$_2$[23], H+B$_2$[23], H+HB$_2$[24] and H+C$_2$H$_4$[25]. See also discussion in Ref. 12.

c)   preliminary data from TRIUMF

d)   from H+D$_2$ and D+H$_2$ data of Ref. 26, at 800K

e)   Spin exchange reactions. $k_H$ values calculated from reported cross sections. Ref. 27.

----------------------------------------------------------------

The important point of note here is, regardless of the sophistication with which Eqn. (7) is evaluated, there is a "trivial" kinetic isotope effect expected which is just a manifestation of the mean collision velocity. In the case of Mu and H, which differ by a factor of about 9 in mass, $k_{Mu}/k_H = 2.9$, all other things being equal, whereas D and H should give a ratio $k_H/k_D = 1.4$. It is the deviations from these ratios which are interesting and indicative of dynamic isotope effects affecting $\sigma(E)$ itself. From Table II one can see that many of the reactions of Mu are "interesting" in that the ratio $k_{Mu}/k_H$ differs considerably from 2.9, particularly for Mu+F$_2$, Mu+C$_2$H$_4$ and Mu+H$_2$, even allowing for a considerable uncertainty in the H atom numbers;  on the other hand, $k_H/k_D$ shows no such dramatic behaviour.

In a chemical reaction there are basically two ingredients affecting $\sigma(E)$ and hence $k(T)$:  the potential energy surface describing the intermolecular interactions themselves and the explicit calculation of the collision dynamics occurring on this surface. The potential surface is always quantum mechanical in nature (although sometimes more empirical than theoretical) but the reaction dynamics are often calculated from classical trajectories particularly when 3D calculations are being done. In general, these two effects are not separable. Both quantum-mechanical and classical reaction calculations on an accurately known surface (the Liu-Siegbahn $H_3$ surface) are currently being carried out for the isotope variations of $H+H_2$ [28,30] so that for the first time the reaction rate theory can be "tested" independent of the potential energy surface itself. In this regard, the present measurements of $Mu+H_2$ provide important possibly even crucial data because the isotope effect in comparing $H, D+H_2$ is so small [31]. Indeed, at 800K, $k_{Mu}/k_H$=.055 and this value is really in excellent agreement with exact thermally averaged state-to-state 1D quantum calculations of Connor et al. [28]; 3D calculations which one would expect to be important in the case of these light molecules are currently underway [28,30].

Thermally averaged rate constants can often most easily be understood within the framework of quasiclassical transition state theory [29,30], the textbook example of which postulates some intermediate or "transition state" complex at the saddle point of the potential surface. In this picture the Arrhenius activation energy is the difference between the (vibrationally adiabatic) zero point energy levels of the target molecule and the activated complex. Since Mu is so much lighter than hydrogen, the zero point level of the Mu substituted complex will be considerably higher than the H one giving $Ea^{Mu}>Ea^H$ and, hence $k_{Mu}<k_H$ is expected, contrary to the results in Table II in every case except $Mu+H_2$. The $Mu+H_2$ results can in fact be qualitatively understood on this basis although it should be noted that the reaction is highly endothermic (0.31 eV from the v=0 state).

An effect which predicts just the opposite trend in rate constants is quantum tunneling, long thought to be important in chemical reactions [32] but one which has always been difficult to establish due to the small mass difference between the most common isotopes, hydrogen and deuterium. Clearly muonium with its large difference in mass compared to hydrogen should be a far more sensitive probe of the importance of tunneling in chemical reactions than anthing else, a feeling which provided considerable motivation for initiating our programme in gas phase muonium chemistry [10]. Detailed 1D calculations showing the importance of tunneling in a chemical reaction have been carried out by Connor and co-workers for the isotopic ractions of $H+F_2$ using a semiempirical LEPS surface [28,29]. In chemical reactions, tunneling is much more of a "dynamic" concept than the static one common, e.g., in nuclear alpha decay. In dynamic tunneling, collisions which are energetically allowed occur quantum mechanically but are classically forbidden, not because of an energy barrier but because of the reaction dynamics. In the case of $Mu+F_2$, tunneling is found to dominate the reaction at 300K and is in fact the only isotope to exhibit appreciable static tunneling; the predicted rate constant ratio at 300K is 6.6, in excellent agreement with what is reported in Table II based on the H atom numbers of Ref. 33 but considerably lower than the value of 14 found based on more recent H atom measurements [22]; this discrepancy may indicate an incorrect surface topology.

In the $F_2$ reactions, the potential barrier is "early" in the reaction plane so that translational energy is far more important than vibrational energy in determining chemical reactivity and hence the full impact of the tunneling correction for the very light Mu atom comes into play. An immediate manifestation of the importance of tunneling can be found in the experimental activation energies of Table II. The $Mu+F_2$ value of .04 eV is half of that of the corresponding $H+F_2$ reaction, indicating a preferred reaction path for the muonium reaction which is not apparent, e.g., in the $Cl_2$ reaction. Another way to appreciate this point is to consider the Tolman definition of the activation energy, which follows from Eqn. (8) according to

$$Ea=-k_B \frac{d[\ln k(T)]}{d[1/T]} = \langle E^*\rangle - 3/2 k_B T \qquad (9)$$

In Eqn. (9), $\langle E^*\rangle$ represents the average energy of those molecules that lead to

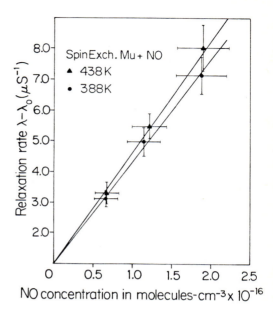

*Figure 3:* *The relaxation rate $\lambda_d$ for muonium spin exchange reactions with NO at 438K and 478K vs. NO concentration, calculated from PV=nRT. The bimolecular rate constants $k_d$ are obtained from the slopes.*

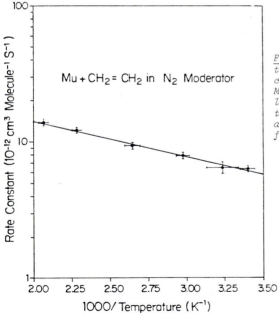

*Figure 4:* *An Arrhenius plot of the rate constants $k(T)$ for the chemical (addition) reaction of Muonium with ethylene in about 1 atm. $N_2$ moderator, in the temperature range 294-484K. The activation energy $E_a$ is obtained from the slope.*

reaction and hence depends on the specific nature of the reaction dynamics through the cross section $\sigma(E)$ in Eqn. (7); for Mu, $<E^*>$ is smaller than for H due to its enhanced tunneling capability and hence $E_a$ is also smaller. It is interesting to note that in the reaction of $Mu+C_2H_4$, both the rate constant and activation energy ratios of the analogous H atom reactions are very similar to that of $Mu+F_2$, possibly indicating that the barrier is also early on the $H+C_2H_4$ potential surface. This is another example of the type of information that can be gained from studying (gas phase) muonium chemistry. The isotopic utility of muonium in establishing viable potential energy surfaces for chemical reactivity has been discussed by Jakubetz with reference to the $H+Cl_2$ reaction [34].

Spin Exchange Reactions

As noted already, the spin exchange reactions of muonium with NO and $O_2$ have been measured in the temperature range 295 - 478K at moderator pressures of 1.0 and 2.5 atmospheres [5,16]. These reactions had previously been studied by Mobley et al. [4] at relatively high moderator pressures (∿40 atm) using a longitudinal field technique. Although a noble effort the data are generally of poor quality and moreover possible background contributions due to termolecular chemical reactions of the type $Mu+O_2 \xrightarrow{M} MuO_2$ in moderator M were ignored, rendering the subsequent interpretation questionnable. Such effects could well be significant at the large pressures used in Ref. 4, but are completely negligible at the low (∿1 atm) pressures used in the present study (see Ref. 5 for further discussion). Consequently, in collisions with paramagnetic $O_2(S=1)$ or $NO(S=1/2)$, the relaxation of the MSR signal can be ascribed to $\mu^+$ depolarization via electron spin exchange yielding depolarization rate constants $k_d$ in the manner described earlier (Figs. 2 and 3). These rate constants exhibit a temperature dependence consistent with $T^{\frac{1}{2}}$ [15,16].

Although some energy dependence to the spin exchange cross section $\sigma_{SE}$ may be anticipated, this is not expected to be a large effect in the temperature range of interest. Indeed, if $\sigma_{SE}$ is energy independent than equation (7) yields the usual "hard sphere" result consistent with the data ($E_a=0$ in Eqn. (8)). Hence, experimental depolarization rate constants $\sigma_d$ can be found directly from the measured $k_d$ by the relationship $k_d=\sigma_d\bar{v}$, where $\bar{v}=\sqrt{8k_BT/\pi\mu}$ is the mean collision velocity. The spin exchange cross sections of interest are related to $\sigma_d$ by a statistical factor defined by $\sigma_d=\frac{1}{2}f\sigma_{SE}$, where in a transverse field, f=3/4 for Mu+NO and f=8/9 for $Mu+O_2$ [15]. The spin exchange cross sections so obtained are found to be constant within errors. Results for the $Mu+O_2$ reaction are plotted along with the theoretical calculations of Aquilanti et al. [35] in Fig. 5. Temperature (and pressure) averaged values for $O_2$ and NO are compared with the corresponding H atom numbers [27] in Table III.

Table III:   Spin Exchange Cross Sections[a]  for Mu(H)+$O_2$,NO

| Molecule | $\sigma_{SE}(H)$ | $\sigma_{SE}(Mu)$ |
|----------|------------------|-------------------|
| $O_2$    | 9.0±1.0          | 9.2±0.5           |
| NO       | 11.2±1.8         | 10.3±0.4          |
| H        | 23.1±2.8         | 25[b]             |

a)  all given in units of $10^{-16}$ $cm^2$

b)  theoretical calculation from Ref. 45.

The H atom numbers are also plotted in Fig. 5. For reference, Table III also contains the experimental numbers for H+H spin exchange near room temperature [36] and the theoretical calculations for the corresponding Mu+H spin exchange reaction [37].

There are essentially three interesting features contained in the results of Table III: i) the $\sigma_{SE}$ for both H and Mu are essentially temperature independent in the temperature range ∿300-∿500K ii) there is no isotope effect, in marked con-

trast to the chemical reactions discussed above iii) $\sigma_{SE}$ for NO is consistently
higher than that for $O_2$, whereas both are about half of the H+H cross section.

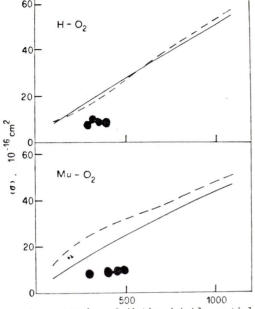

*Figure 5: The theoretical calcu-
lations of Aquilanti, Grossi and
Lagana [35] of spin exchange cross
sections $\sigma_{SE}$ for the H+$O_2$ (top) and
Mu+$O_2$ (bottom) reactions, giving
results for the SI (dotted line)
and OFD (solid line) interactions.
The experimental cross sections
are shown as solid circles: the
H atom results are from Ref. 27,
the Mu atom results from the
present TRIUMF work.*

For the scattering of distinguishable particles, the electron spin exchange cross
section is given by the partial wave sum

$$\sigma_{SE} = \frac{\pi}{k^2} \sum_{\ell} (2\ell+1)\sin^2\Delta_{\ell} \qquad (10)$$

where $\Delta_{\ell}$ is the difference between "singlet" and "triplet" phase shifts in the case
of Mu(H)+NO(S=1/2) or the difference between "doublet" and "quartet" phase shifts
in the case of Mu(H)+$O_2$(S=1). Explicit calculations of Eqn.(10) have been carried
out by Shizgal [37] comparing the Mu+H and H+H spin exchange reactions using the
kolos-Wolinwiencz potential for $H_2$ (see also Ref. 38). In this case the interaction potential is purely spherical and accurately known and the calculated results
naturally give excellent agreement with H+H experiments [36] (the Mu+H experiment
is a difficult one and has not yet been done). At low energies ($\lesssim 10^{-2}$eV) resonances dominate the calculated Mu+H spin exchange cross sections but the thermally
averaged cross sections in the region $\sim$300-500K exhibit a smooth behaviour with
little or no isotope effect. Cross sections for the spin exchange reactions of
Mu and H with $O_2$ have been carried out by Aguilanti et al. [35] but in this case
the interaction potential is rather poorly known compared to the H+H case. Moreover, rotational excitation of the $O_2$ molecule is no doubt important so that the
interaction is no longer spherically symmetric, necessitating some approximations
in the evaluation of the phase shifts in Eqn.(10). Two approximations have been
reported in Ref. 35; a spherical interaction (SI) approximation (which might be
expected to yield results qualitatively similar to H+H) and an Oriented Frame
Decoupling (OFD) approximation, which is similar to a sudden approximation. These
calculations for both H and Mu+$O_2$ are compared with the experimental data in Fig 5.
It is interesting to note that in the case of H+$O_2$ (top) there is little difference in the calculated $\sigma_{SE}$ between the SI and OFD interactions, but there is a
rather dramatic difference in the case of Mu+$O_2$ (bottom); the OFD calculation
gives considerably better agreement with experiment, although still about a factor
of 2 too high. In this regard, as with the chemical reactions discussed previously, it appears that Mu is considerably more sensitive to details of the interaction

than H is. Although the temperature dependence predicted by the theory is not seen
in the data,it should be noted that over the range $\sim$300-500K the experimental error bars can be accommodated by the theoretical trend. The ratio of $\sigma_{SE}(H)/\sigma_{SE}(Mu)$
is, however, in good agreement with the experimental results. Considering the
largely phenomenological interaction potential of Ref. 35, the level of agreement
between experiment and (OFD) theory is not too bad.

## SUMMARY AND PROSPECTUS

Despite its "exotic" beginnings, the muonium atom can in a very real sense be regarded as the ultimate light isotope of hydrogen, extending the most sensitive end
of the isotopic mass scale from, at most, a factor of 3 in comparing H and T to a
factor of as much as 27 in comparing Mu and T. Insofar as the study of mass effects is of interest in atomic and chemical physics then the study of the interactions of muonium with its environment provides the possibility of sensitive experimental tests of the theory of collision processes, particularly in chemical reaction dynamics. Moreover, the experimental technique demands that there be no
more than one Mu atom in the system at a time so that one can study the interaction
of muonium free of complicating "2nd order" effects which can obscure similar
studies in H atom chemistry.

Results in the temperature range 300-500K have been obtained for the chemical
reactions Mu+$F_2$, $Cl_2$, $Br_2$, $H_2$, HBr and $C_2H_4$, revealing a wide variation in isotope
effects in comparison with the analogous H atom reaction. In terms of the bimolecular rate constants k, the ratio $k_{Mu}/k_H$ varies from about 1/20 in the $H_2$ reaction
(at 800K) to about 10 in the $F_2$ reaction. Both quasi-classical (zero point) and
quantum mechanical (tunneling) effects play important roles, the former dominating
the Mu+$H_2$ reaction and the latter dominating the Mu+$F_2$ reaction. The spin exchange reactions Mu+$O_2$ and Mu=NO have also been studied revealing no isotope effect
other than a simple "kinetic" one ($k_{Mu}/k_H$=2.9) in the temperature range $\sim$300-500K.
These results are in qualitative agreement with current theoretical calculations.

Future experiments will involve extending our measurements to both lower and higher
temperatures. There is at the moment relatively little Mu atom data available
with which to compare with a much vaster array of H atom data and so part of our
programme must be to make routine measurements on a variety of reaction systems.
New directions also beckon. The spin exchange interaction of Mu with the alkalai
metal vapors as well as with the H atom itself would be very interesting as well
as providing challenging experimental problems. Also, the measurement of chemical
reaction rates from vibrationally excited molecules (e.g. Mu+$H_2$(v=1)) would be an
extremely valuable extension of the technique and one that could have a major impact on reaction rate theory. It may even be possible to carry out single photon
counting of the deexcitation of vibrationally excited muonium containing molecules.
And, of course, it would be nice to be able to produce a muonium beam.

Finally, I would like to sincerely thank my (graduate student) coworkers, Dr.
David M. Garner and Dr.Randall J. Mikula, without whose enthusiasm, tireless
efforts and good "plumbing" much of what is reported above might never
have happened.

## REFERENCES

[1]  Garwin, R.L., Lederman, L.M., Weinrich, M., Phys. Rev. <u>105</u> (1957) 1415;
     Friedman, J.I., Telegdi, V.L., ibid, 1681.

[2]  See proceedings of 1st and 2nd Int'l Conferences on Muon Spin Rotation;
     Rorschach, Switzerland, 1978, in Hyperfine Interactions, Vol. <u>6</u> (1979);
     Vancouver, Canada, 1980, in Hyperfine Interactions, Vol. <u>8</u> (1981).

[3]  D.E. Casperson et al., Phys. Rev. Letts. <u>38</u> (1977) 956.

[4] Mobley, R.M., Ph.D. Thesis, Yale University, 1967; R.M. Mobley et al., J. Chem. Phys. 44 (1966) 4354.

[5] Fleming, D.G., Mikula, R.J. and Garner, D.M., J. Chem. Phys. 73 (1980) 2751.

[6] Stambaugh, R.D., Ph.D. Thesis, Yale University, 1974; R.D. Stambaugh et al. Phys. Rev. Letts. 33 (1974) 568.

[7] B.A. Barnett et al., Phys. Rev. A11 (1975) 39.

[8] Brewer, J.H., Hyp. Int. 8 (1981) 831; Pifer, A.E., Bowen, T., Kendall, K.R., Nucl. Inst. Methods 135 (1976) 39.

[9] Brewer, J.H. and Fleming, D.G., et al., Abstracts of IXth IXPEAC, Seattle, 1975, Vol. 1, p. 157,159.

[10] Fleming, D.G. et al., J. Chem. Phys. 64 (1976) 1281.

[11] D.G. Fleming et al., A.C.S., Advances in Chemistry 175 (1979) 279.

[12] Garner, D.M., Ph.D. Thesis, University of British Columbia, 1979.

[13] Brewer, J.H., Crowe, K.M., Gygax, F.N. and Schenck, A., Muon Physics, Academic Press, New York, 1975, V.W. Hughes and C.S. Wu, eds., Vol.III,p.3.

[14] Fleming, D.G. and Mikula, R.J. et al., Hyp. Int. 6 (1979) 379, 8 (1981) 307.

[15] Mikula, R.J., Ph.D. Thesis, University of British Columbia, 1981.

[16] Fleming, D.G., Mikula, R.J. and Garner, D.M.,Hyp. Int. 9 (1981) 207; ibid, 8 (1981) 337.

[17] Allison, S.K., Rev. Mod. Phys. 30 (1958) 1137.

[18] Tawara, H. and Russek, A., Rev. Mod. Phys. 45 (1973) 178.

[19] Bolton, P.R. et al., Bull. Am. Phys. Soc. 24 (1979) 675.

[20] Fleming, D.G. et al., Chem. Phys. Letts. 48 (1977) 393.

[21] Garner, D.M., Fleming, D.G. and Brewer, J.H., Chem. Phys. Letts. 55 (1978) 163.

[22] Homan, K.J., Schweinfurth, H., and Warnatz, J., Ber. Bunsenges - Phys. Chem. 81 (1977) 724.

[23] Jaffe, S. and Clyne, M.A.A., J. Chem. Soc. Far. Trans. 2, 77 (1981) 531.

[24] Jourdain, J.L., Le Bras, G. and Combourieu, J., Chem. Phys. Letts. 78 (1981) 483.

[25] Sugawara, K., Okazaki, K. and Sato, S., Chem. Phys. Letts. 78 (1981) 259.

[26] Westenberg, A.A. and De Haas, N., J. Chem. Phys., 47 (1967) 1393.

[27] Anderle, M., et al., Phys. Rev. 23A (1981) 34 ; Gordon, E.B. et al., JETP Letts. 17 (1973) 395.

[28] Connor, J.N.L., Hyp. Int. 8 (1981) 423.

[29] Connor, J.N.L., Jakubetz, W. and Laganà, A., J. Phys. Chem., 83 (1979) 73; Connor, J.N.L., Jakubetz, W. and Manz, J., Chem. Phys. Letts. 45 (1977) 265.

[30] Garrett, B.C. and Truhlar, D.G., J. Chem. Phys. 72 (1980) 3460; Truhlar, D.G. and Connor, J.N.L., private communication and to be published.

[31] Sun, J.D. et al., Phys. Rev. Letts. 44 (1980) 1211.

[32] Johnston, H.S., Gas Phase Reaction Rate Theory, The Ronald Press Company, N.Y. 1966.

[33] Rabideau, S.W., Hecht, H.G. and Lewis, W.B., J. Magn. Reson. 6 (1972) 384.

[34] Jakubetz, W., Hyp.Int. 6 (1979) 387; J. Am. Chem. Soc. 101 (1979) 298.

[35] Aquilanti, V., Grossi, G. and Laganà, A., Hyp. Int. 8 (1981) 347.

[36] Desaintfuscien, M. and Audoin, C., Phys. Rev. A13 (1976) 2070.

[37] Shizgal, B., J. Phys. B. Atom. Molec. Phys., 12 (1979) 3611; Allison, A.C., Phys. Rev. 5A (1972) 2695.

[38] Berlinsky, A.J. and Shizgal, B., Can. J. Phys. 58 (1980) 881.

PHYSICS OF ELECTRONIC AND ATOMIC COLLISIONS
S. Datz (editor)
© North-Holland Publishing Company, 1982

LOW AND INTERMEDIATE ENERGY POSITRON-GAS COLLISIONS

T.S. Stein and W.E. Kauppila

Department of Physics and Astronomy
Wayne State University
Detroit, Michigan 48202
U.S.A.

This paper reports on progress in positron-gas cross section measurements. Total cross section ($Q_T$) measurements for 0.3 to 1000 eV $e^+$-He, Ne, Ar, Kr, Xe, $H_2$, $N_2$, $CO_2$, $O_2$, $CH_4$, and CO collisions are summarized. Interesting features in the $Q_T$ curves such as Ramsauer-Townsend effects and onsets of positronium (Ps) formation are discussed. Comparisons of $e^+$ and $e^-$ $Q_T$ measurements are presented to illustrate some of the interesting differences and similarities in the scattering of these projectiles from various atoms and molecules. Recent $e^+$ scattering investigations that go beyond $Q_T$ experiments are discussed including differential and inelastic scattering, Ps formation, atomic excitation, and resonance searches.

INTRODUCTION

Low energy $e^+$ beams suitable for direct measurements of total scattering cross sections for $e^+$-atom collisions were first developed and used for that purpose[1] in 1972. Such measurements are of interest because they involve interactions of antimatter with matter, and also because they can help provide a better understanding of the scattering of electrons by atoms and molecules, a subject of great importance to many different fields of science and technology. Comparisons between $e^+$- and $e^-$-atom (molecule) scattering reveal some interesting differences and similarities. The static interaction is attractive for electrons and repulsive for positrons, while the polarization interaction is attractive for both projectiles. The exchange interaction contributes only to $e^-$ scattering. The combined effect of the static and polarization interactions is that they add to each other in $e^-$ scattering whereas there is a tendency toward cancellation in $e^+$ scattering. This results in smaller scattering cross sections, in general, for positrons than for electrons at low energies. As the projectile energy is increased, the polarization and exchange interactions eventually become negligible, compared with the static interaction, which has the same magnitude for positrons and electrons. This results in a merging of the corresponding $e^+$ and $e^-$ scattering cross sections at sufficiently high projectile energies. Two scattering processes which occur for positrons (but not for electrons) are annihilation and positronium (Ps) formation (real and virtual). Annihilation is not expected[2] to be a significant effect in $e^+$ scattering except at near-thermal energies. Ps formation, on the other hand, has been found to be a significant process in $e^+$-beam-gas collision studies.

EXPERIMENTS FOR MEASUREMENTS OF POSITRON-ATOM (MOLECULE) TOTAL CROSS SECTIONS

One of the major difficulties encountered in $e^+$-atom cross section measurements is associated with producing sufficiently intense $e^+$ beams of well-defined energy. Table 1 summarizes the characteristics of the $e^+$ beams used by various groups that have reported $e^+$ $Q_T$ measurements in the first decade of activity in this area. The experiment of Costello et al.[1] utilized a 55 MeV electron linear accelerator

Table 1.  Beam characteristics for $e^+$ total scattering experiments

| Laboratory | Source | Scattering region | Energy analysis | Energy range (eV) | $\Delta E^{FWHM}$ (eV) | Detector | Detected $I_o$ (#/sec) |
|---|---|---|---|---|---|---|---|
| Gulf GA (1972)[a] | PP | $B_\parallel$ | TOF | 1-26 | 1-2 | 2γ | - |
| London (1972)[b] | $^{22}$Na | $B_\parallel$ | TOF | 2-1000 | ~1 | γ | few |
| Toronto (1973)[c] | $^{22}$Na | $B_\parallel$ | 90°E | 4-300 | ~1 | 2γ | < 1 |
| Swansea (1975)[d] | $^{22}$Na | $B_\perp$ | $B_\perp$ | 13-1000 | 0.14E | 2γ | few |
| Detroit (1976)[e] | $^{11}$C | $B_\parallel$ | $E_R$ | 0.3-800 | < 0.1 | CEM | > 100 |
| Texas (1977)[f] | $^{22}$Na | $B_\parallel$ | TOF | 2-50 | ~1 | 2γ | few |
| Bielefeld (1978)[g] | $^{22}$Na | ~FF | TOF | 1-6 | 0.4 (@6eV) | CEM | < 1 |

[a] Costello et al., Ref. 1

[b] Canter et al., Ref. 3

[c] Jaduszliwer and Paul, Ref. 4

[d] Dutton et al., Ref. 5

[e] Kauppila et al., Ref. 9

[f] Burciaga et al., Ref. 6

[g] Wilson, Ref. 7; Sinapius et al., Ref. 8

to produce positrons by pair production (PP). All of the other laboratories[3-8] use commercially-available $^{22}$Na as $e^+$ sources, except for the Detroit group[9] which uses the proton beam of a 4.75 MeV Van de Graaff accelerator to produce an $^{11}$C $e^+$ source by the reaction $^{11}$B(p,n)$^{11}$C. A variety of backscattering and transmission moderators have been used to obtain low energy positrons with relatively narrow energy distributions from the high energy, broad energy width fluxes resulting from $\beta^+$-decay. Griffith and Heyland[10] have reviewed the various techniques which have been utilized to obtain the low energy $e^+$ beams used in cross section measurements. As a result of the relatively low intensities of the $e^+$ beams used in scattering experiments, most of the positron groups use a **rather** long scattering region and an axial magnetic field (sometimes curved) to transport the $e^+$ beam through that region. The laboratories which do not use an axial magnetic field in the scattering region are Swansea,[5] where the positrons move along a circular path in a transverse magnetic field (a Ramsauer type of system), and Bielefeld,[7,8] where the positrons move in a field-free (FF) scattering region (except for a single axial magnetic focussing lens). The methods used to analyze the energy of the $e^+$ beams are time of flight (TOF),[1,3,6-8] a 90° electrostatic analyzer (90°E),[4] a transverse magnetic field with beam defining apertures ($B_\perp$),[5] and a retarding electrostatic field ($E_R$).[9] In all cases, except for Toronto, the method of energy analysis also provides some discrimination against the detection of positrons which have been scattered at small angles. The Toronto group[4] has relied on the use of collimators in their scattering region, variations in their axial magnetic field strength, and Monte Carlo calculations in order to take small-angle elastic scattering into account, and they have used a retarding electrostatic field to discriminate against inelastically scattered positrons. The $e^+$ energies used in $Q_T$ measurements range from 0.3 eV at Detroit[11] to 1000 eV at Swansea[12] and London[13]. The energy widths, $\Delta E^{FWHM}$ (full-width at half-maximum) of the slow $e^+$ beams used in scattering experiments are typically about 1 eV or more, except for the Detroit energy width of less than 0.1 eV.[9] The three methods used for detecting positrons are to (1) observe the two coincident annihilation gamma rays (2γ)[1,4-6] using two NaI scintillation counters, (2) observe one or both annihilation gamma rays (γ)[3] using a single NaI well counter, and (3) use a Channeltron electron multiplier (CEM)[7-9]. The detected primary beam currents ($I_o$) range from less than 1/sec to more than 100/sec.

The basic experimental method used by all of the groups which have measured $Q_T$, is to study the attenuation of the $e^+$ beam as it passes through a gas scattering region. Under "ideal" experimental conditions, $Q_T$ can be obtained from the expression,

$$I = I_o e^{-nQ_T L}$$

where $I_o$ is the detected beam intensity with no gas in the scattering region, I is the detected beam intensity **with** gas of number density, n, in the scattering region, and L is the path length of the $e^+$ beam through the scattering region. If any scattered positrons manage to reach the detector during the measurement of I, this will result in measured cross sections being too low.

LOW ENERGY POSITRON-INERT GAS TOTAL CROSS SECTION MEASUREMENTS

$Q_T$ measurements[1,8,11,12,14-18] for $e^+$-He collisions are shown in Fig. 1 along with the results of several theoretical calculations.[19-25] All of these measurements

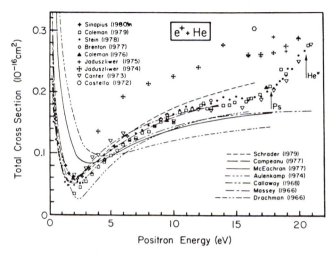

Fig. 1. $e^+$-He total cross sections at low energies. The experimental results (in chronological order), represented by symbols, are from Refs. 1, 14-17, 12, 11, 18, and 8. The theoretical results, represented by curves, are from Refs. 19-25. The "n" refers to normalized results.

are absolute except the normalized (n) values of Sinapius et al.[8] The values reported by Sinapius et al.[8] are the results of Wilson[7] corrected for scattering in the moderator and accelerator regions of their apparatus, which lowered Wilson's results by an average of 18%. Most of the measurements[8,11,12,14,17,18] lie in a fairly narrow band in Fig. 1. One of the interesting qualitative features is the observation of a Ramsauer-Townsend effect (a minimum in $Q_T$) by Stein et al.[11] and Sinapius et al.[8] near 2 eV with a steeply rising cross section at the lowest energies. All of the theoretical results in Fig. 1 indicate the existence of a cross section minimum. Cross section minima such as this were first observed by Ramsauer[26-28] and Townsend and Bailey[29] in the 1920's for low energy $e^-$-Ar, Kr, and Xe collisions. These minima were so deep, it appeared that the target gas was nearly transparent to the projectile electrons. The Ramsauer-Townsend minima arise from quantum mechanical effects associated with a net attractive interaction between the projectile and the target atoms. Another interesting feature of most of the experimental data in Fig. 1 is the noticeable increase in $Q_T$ as the energy is increased above the Ps formation threshold, which shows up most clearly in the narrow energy-width measurements of Stein it al.[11]

The $e^+$ experimental $Q_T$ results[8,9,11,14,17,18,30-35] for Ne, Ar, Kr, and Xe are displayed in Fig. 2 along with the results of several theories.[20,25,36-40] In a qualitative sense, the shapes of the theoretical curves at low energies are quite similar to some of the experimental results. Stein et al.[11] observe a rather deep Ramsauer-Townsend effect in Ne near 0.6 eV. Kauppila et al.[9] observe a shallow cross section minimum in Ar near 2 eV. There are dramatic increases in the measured $Q_T$ values above the Ps formation thresholds for all the inert gases, illustrated most clearly in the narrow energy-width measurements made in Detroit.[9,11,30] The theoretical cross section curves which extend above the Ps formation thresholds in Figs. 1 and 2 do not account for inelastic scattering, and thus are not expected to show the abrupt increases near the Ps formation thresholds which show up in the experimental results.

The qualitative features of the low energy $e^+$ $Q_T$ curves for the inert gases are summarized in Fig. 3 with the measurements of Stein et al.[11] for He and Ne,

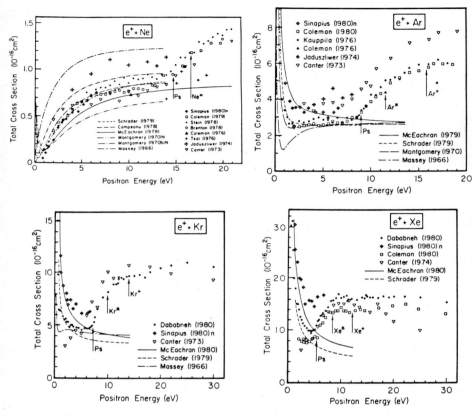

Fig. 2.  e⁺-Ne, Ar, Kr, and Xe low energy $Q_T$ results.  Ne experimental results are from Refs. 14,31,32,17,33,11,18, and 8.  Ne theoretical results are from Refs. 20, 36,36,37,38, and 25.  Ar experimental results are from Refs. 14,31,17,9,34, and 8. Ar theoretical results are from Refs. 20,36,25, and 39.  Kr experimental results are from Refs. 14,8, and 30.  Kr theoretical results are from Refs. 20,25, and 40. Xe experimental results are from Refs. 35,34,8, and 30.  Xe theoretical results are from Refs. 25 and 40.  UN refers to the (2p-d) unnormalized results and N refers to the (2s-p)+(2p-d) normalized results of Ref. 36.

Kauppila et al.[9] for Ar, and Dababneh et al.[30] for Kr, and Xe.  For comparison, the corresponding e⁻ $Q_T$ curves are also provided in Fig. 3, where the results of Ramsauer[26-28] (for Ne, Ar, Kr, and Xe) and Milloy and Crompton[41] (for He) are used for energies below a few eV, and the results of Kauppila et al.[9,42] (for Ar and He), Stein et al.[11] (for Ne) and Dababneh et al.[30] (for Kr and Xe) are displayed at the higher energies.  It is interesting to note that the situation regarding the existence of Ramsauer-Townsend effects in the inert gases is nearly reversed for positrons compared with electrons in the sense that positrons exhibit Ramsauer-Townsend minima only for the lighter inert gases (He, Ne), and possibly a shallow minimum for Ar, whereas electrons exhibit Ramsauer-Townsend minima only for the heavier inert gases, Ar, Kr, and Xe.  The e⁺ curves in Fig. 3 can be used to obtain estimates of the Ps formation cross sections ($Q_{Ps}$) (crosshatched regions) for e⁺ energies between the thresholds for Ps formation and atomic excitation, assuming that

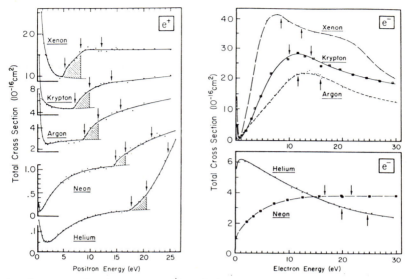

Fig. 3.  $Q_T$ curves for low energy $e^+$- and $e^-$-inert gas scattering.  The arrows refer to inelastic thresholds;  Ps formation, atomic excitation and ionization in
the $e^+$ case;  atomic excitation and ionization in the $e^-$ case.

the elastic scattering cross sections are smoothly varying as the $e^+$ energy increases through the Ps formation thresholds.  (For a further discussion of $Q_{Ps}$, refer
to the section on "Inelastic Scattering Investigations".)  The $e^+$-He curve in Fig.3
indicates a noticeable increase in the slope at the atomic excitation threshold energy, which was first pointed out by Coleman et al.[43] in an analysis of the data
of Canter et al.[14]

It is not a simple matter to ascertain which $e^+$ $Q_T$ measurements are the most reliable.  Two recent review articles[10,44] discuss errors in several of the experiments
represented in Figs. 1 and 2.  Although experimental groups will often make estimates of potential systematic errors in their experiments, the estimated magnitudes
of such errors may be incorrect, and there may be systematic errors which are overlooked which are as large or larger than those which are painstakingly considered.
An approach initiated by the Detroit group[9,11,30] is to make the corresponding measurements for each gas with electrons and positrons using the same experimental approach and system.  An advantage of this approach is that most of the potential systematic errors should equally affect the $e^-$ and $e^+$ measurements.  The Detroit $e^-$-
He results[42] are within a few percent of several other sets of experimental results[41,45-47] and theoretical calculations.[21,48-50]  The Detroit $e^-$ measurements for
the other inert gases[9,11,30] have been compared with other $e^-$ measurements and they
also appear to be quite reliable.  The Bielefeld group[8] has also made absolute $e^-$-
inert gas $Q_T$ measurements which are in quite good agreement with the Detroit results.[9,11,30]

It is informative to examine the $Q_T$ results for low energy $e^+$-inert gas collisions,
keeping in mind the $e^-$-inert gas results referred to above.  For the $e^+$-He case,
Fig. 1 indicates that there are several calculations[20,21,24,25] that agree quite
well with the experiments in the vicinity of the Ramsauer-Townsend minimum.  The
variational calculation of Campeanu and Humberston[24] and an exchange adiabatic

approximation calculation of Massey et al.[20] remain in quite good agreement with
several of the experiments up to the highest energies of overlap. Since the calculation of Campeanu and Humberston is likely to be the most reliable of all the
above calculations, it is of interest to consider comparisons between this theory
and the various experiments. Campeanu and Humberston have devoted considerable attention[24,51,52] to trying to explain the discrepancy between their calculations and
the measurements made in Detroit[11,53] and Texas[6,18] for energies below 6 eV, where
the experiments are lower. Humberston[51] contends that the Detroit and Texas results are both too low below 6 eV due to the neglect of positrons scattered through
small angles and that the measurements of Canter et al.[14] which are claimed to a-
gree with Campeanu and Humberston[24] below 6 eV, are more accurate in the vicinity
of 2 eV. Humberston[51] deduces that at 2 eV, the discrepancy between the results of
Campeanu and Humberston[24] and those of Stein et al.[11] could be explained by an angular discrimination of 12°, while an angular discrimination of 20° could explain
the discrepancy with the results of Burciaga et al.[6] In a reassessment of the experiment of Canter et al.[14] by Griffith et al.,[54] it is estimated that a "realistic
upper limit" for the angular discrimination of the experiment of Canter et al.[14]
is 10° rather than 35° as was initially suggested by Canter et al.[3]

In order to determine if differing angular discriminations could provide a consistent explanation for the discrepancies between the results of Campeanu and Humberston[24] and the various experimental results, we have calculated the percent error
introduced into the total elastic scattering cross section ($Q_{el}$) for e$^+$-He and e$^-$-
He collisions due to various values of angular discrimination at various energies.
The results for positrons, summarized in Fig. 4, were obtained using the s-wave
phase shifts of Campeanu and Humberston,[24] the p-wave phase shifts of Humberston
and Campeanu,[52] the lowest set of d-wave phase shifts of Drachman,[55] and the higher phase shifts (up to $\ell=20$) from the formula of O'Malley et al.,[56]

$$\delta_\ell = \pi k^2 P / ((2\ell+3)(2\ell+1)(2\ell-1)) a_o$$

where P is the static dipole polarizability ($1.383 a_o^3$) of the target atom (He) and
k is the wave number of the projectile. The corresponding percent errors for e$^-$-
He collisions, also shown in Fig. 4, were obtained using the s-, p-, and d-wave
phase shifts of Callaway et al.[21] (from their "EP" calculation) and the higher
phase shifts (up to $\ell=20$) from the formula of O'Malley et al.[56] given above. Fig. 4
clearly illustrates that the percent error in $Q_{el}$ (and therefore in $Q_T$) for a particular angular discrimination reaches a maximum for e$^+$-He collisions in the region of the Ramsauer-Townsend minimum (about 2 eV). An experiment with a large value for the angular discrimination could thus be expected to show a deeper Ramsauer-

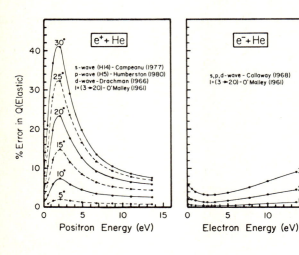

Fig. 4. Percent errors
in measured e$^+$-He and e$^-$-
He total cross sections
due to the neglect of
small angle elastic
scattering for various
angular discriminations.

Townsend effect than actually exists. The angular discrimination in the experiment of Stein et al.[11] was initially estimated to be about 13° at 2 eV. Fig. 4 indicates that an angular discrimination of about 17° could explain the difference between the Detroit results[11] in e[+]-He and the theoretical results of Campeanu and Humberston[24] near 2 eV in Fig. 1. A more recent and more complete analysis of the angular discrimination for low energy positrons by Dababneh et al.[30] using the same apparatus as that used by Stein et al.[11] gave an angular discrimination of 15-20° for e[+] energies between 1 and 20 eV, which is consistent with the 17° value required to explain the difference between the measurements of Detroit[11] and the values of Campeanu and Humberston[24] at 2 eV. The results of Canter et al.[14] between 3.5 and 6 eV are measurably higher than the curve of Campeanu and Humberston[24], and this is inconsistent with any explanation based on angular discrimination, alone. Coleman et al.[18] make no estimate of their angular discrimination, but have taken precautions to minimize the effect of small angle scattering. Of all the measurements, the corrected, normalized, values reported by Sinapius et al.[8] (who estimate an angular discrimination of 7°) are in the best agreement with Campeanu and Humberston between 1 and 6 eV. A major part of the discrepancy between the e[+]-He measurements of Sinapius et al.[8] and Stein et al.[11] could be due to the differing angular discriminations of the respective experiments.

An observation which at first glance seems inconsistent with an angular discrimination argument is that the measurements of Stein et al.[11] for e[+]-He are measurably higher than the curve of Campeanu and Humberston[24] shown in Fig. 1 above 7 eV. However, we found that the curve in Fig. 1 (obtained from Fig. 1 of Ref. 24) was noticeably lower at energies above the Ramsauer-Townsend minimum than one which we obtained using the s-wave phase shifts of Campeanu and Humberston,[24] the p-wave phase shifts of Humberston and Campeanu,[52] the lowest set of d-wave phase shifts of Drachman,[55] and higher phase shifts (up to $\ell$=20) from the formula of O'Malley[56] referred to above, with only part of the discrepancy between the two theoretical curves being accounted for by the inclusion of the higher partial waves. The resulting theoretical curve is shown in Fig. 5 along with the measurements of Canter et al.[14] and Stein et al.[11] The percent error estimates shown in Fig. 4 indicate

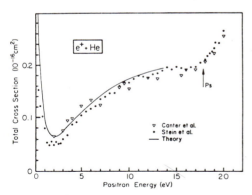

that a 15-20° angular discrimination (as estimated by Dababneh et al.[30]) would result in a remarkable consistency between the theoretical results in Fig. 5 and the measurements of Stein et al.[11] not only at 2 eV, but also for all energies above 2 eV. The results of Canter et al.[14] on the other hand, are inconsistent with the theoretical curve, being lower than the curve between 7 and 13.6 eV and measurably higher than the curve between 3.5 and 6 eV. It is interesting to note from the percent error curves in Fig. 4 that a 15-20° angular discrimination for low energy e[-]-He $Q_T$ measurements would introduce less than a 5% error in those measurements from 0 to 13.6 eV in contrast

Fig. 5. e[+]-He total cross sections at low energy. The experimental results are from Refs. 14 and 11. The theoretical curve is a "corrected" version of the Campeanu-Humberston-Drachman curve in Fig. 1.

to the e[+]-He case, which suggests that the excellent agreement of the Detroit e[-]-He results with several other experiments and theories may be an indication that systematic errors in the Detroit measurements, other than that due to angular discrimination, are relatively small. We have used calculations of phase shifts available in the literature to make percent error (in $Q_{el}$) estimates for various angular discriminations and projectile energies for positrons colliding with the heavier inert gases in order to search for some consistent patterns in the measurements for these gases. There are some intriguing consistencies which emerge in these studies. With the exception of Ne, the measurements of Sinapius et al.[8] are consistently higher than those of Detroit[9,11,30] in each of the heavier inert gases, as was the case for He. Preliminary estimates of percent errors in $Q_{el}$ due to various angular discriminations, indicate that a significant part of the discrepancies between the measurements of Sinapius et al.[8] and Detroit[9,30] in Ar, Kr, and Xe could be related to the differing angular discriminations of the respective experiments as was the case for He. Our percent error estimates for the $e^+$-Ne case indicate that the differing angular discriminations of the Bielefeld and Detroit experiments would not introduce a noticeable discrepancy between their measurements above 2 eV. Fig. 2 indicates that the Bielefeld results are in reasonably good agreement with the Detroit results above 2 eV, but average lower at the lowest energies. If the Detroit results should be shifted upward in energy by $\sim$0.2 eV (as discussed in the section on "Resonance Searches"), this would make the agreement between Bielefeld and Detroit even better. The measurements of London[14] for the heavier inert gases do not appear to fit any consistent pattern in relation to the other experiments. The results of the Texas group[6,18] for the heavier inert gases are in rather good agreement with Detroit[9,11] for Ne and Ar, and with London[35] for Xe.

INTERMEDIATE ENERGY $e^+$-INERT GAS TOTAL CROSS SECTION MEASUREMENTS

Experimental results[12-14,16-18,57,58] for the inert gases at intermediate energies are shown in Figs. 6 and 7 along with several theoretical calculations.[59-66] The measurements reported by Griffith et al.[13] were obtained by normalization of their results to the measurements of Coleman et al.[17] between 30 and 100 eV. The other measurements in Figs. 6 and 7 are absolute. Above 50 eV, the measurements of Toronto,[16,32] Swansea[12,33] and Detroit[57] are in good agreement (generally within 10% of each other) for He, Ne, and Ar. The normalized measurements of Griffith et al.[13] are also in good agreement with the other results for He and Ne, except from 200-400 eV for He where Griffith et al.[13] are 10-15% lower. For Ar, the measurements of Griffith et al.[13] are about 10% higher than Tsai et al.[32] and Kauppila et al.[57] below 100 eV, and more than 10% lower than Tsai et al.,[32] Kauppila et al.,[57] and Brenton et al.[33] above 200 eV. In all cases, the higher energy results of Canter et al.[14,35] and Coleman et al.[17] are lower than the other measurements, which has been attributed to their inadequate discrimination against

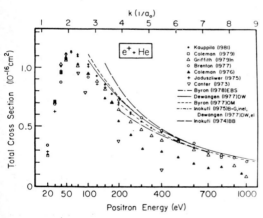

Fig. 6. $e^+$-He total cross sections at intermediate energies. The experimental results are from Refs. 14,16,17,12,13,18, and 57. The theoretical results are from Refs. (59 and 60), (61 and 63),62,63, and 64. The "n" refers to normalized results. The other code letters refer to the type of calculation.

small angle scattering. On the basis of these comparisons it seems that the most reliable measurements for energies above 50 eV are those of Toronto,[16,32] Swansea,[12,33] and Detroit.[57,58] The Detroit group[57,58] has also made the corresponding $e^-$ measurements on the same gases which are in remarkable agreement with the recent measurements of Blaauw et al.[47] and Wagenaar and de Heer.[67] The only significant differences between the $e^-$ measurements of these two laboratories could be explained by their differing angular discriminations for elastic scattering, estimated to be less than 1° for the Amsterdam group and typically 5-6° for Detroit. For positrons at energies above 100 eV, the estimated angular discrimination for the Detroit work

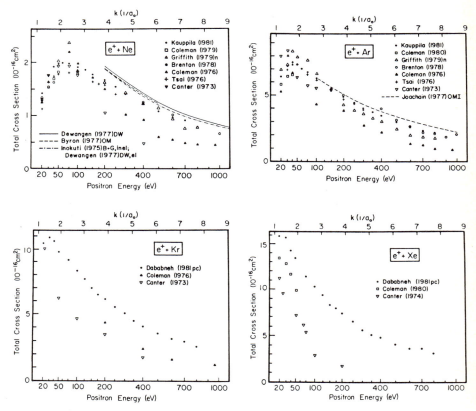

Fig. 7. $e^+$-Ne, Ar, Kr, and Xe total cross sections at intermediate energies. The Ne experimental results are from Refs. 14,32,17,33,13,18, and 57. Ne theoretical results are from Refs.(63, 61 and 65),62, and 63. Ar experimental results are from Refs. 14,32,17,33,13,34, and 57. Ar theoretical results are from Ref. 66. Kr experimental results are from Refs. 14,17, and 58. Xe experimental results are from Refs. 35,34, and 58.

is generally 6-8° for He, Ne, and Ar[57] and 8-10° for Kr and Xe[58] so that the measurements made in Detroit are subject to corrections which would increase their values by an amount depending on the nature of the differential elastic scattering of positrons, and the contribution of elastic scattering to $Q_T$.

Using the $e^+$-He differential elastic cross sections calculated by Byron and Joachain[62] (using an optical model formalism), Kauppila et al.[57] estimate that their $e^+$-He measurements may be an average of 1% too low due to the neglect of small angle elastic scattering from 100-500 eV. In surveying the various $e^+$-He theoretical results, Kauppila et al.[57] found that by adding the inelastic cross sections that can be calculated from the Bethe theory with an additional "gamma" term (related to the number of electrons in the target atom) of Inokuti et al.[61] and the elastic scattering cross sections calculated by Dewangen and Walters[63] using a distorted wave second Born approximation, $Q_T$ values (designated B+G, DW in Fig. 6) are obtained which agree to within 2% of their measurements above 200 eV. Brenton et al.[12] are in good agreement with this composite theory up to their highest energy,

1000 eV.  The theoretical calculations for Ne and Ar, shown in Fig. 7 are reasonably close to the measurements but do not merge with them at the  highest energies of overlap.

POSITRON- AND ELECTRON-INERT GAS $Q_T$ COMPARISON MEASUREMENTS

Comparisons made by Kauppila et al.[57] and Dababneh et al.[58] of measurements of $e^+$- and $e^-$-inert gas total cross sections up to 800 eV are shown in Fig. 8.  In these comparisons, the Detroit measurements[9,11,30,57,58] have been used for all $e^+$ energies and for $e^-$ energies above 2 eV.  For $e^-$ energies below 2 eV, the measurements of Milloy and Crompton[41] for He, Salop and Nakano[68] for Ne, Golden and Bandel[69] for Ar, and Ramsauer[27,28] for Kr and Xe are used.  Since from 2 to 800 eV, the $e^+$ and $e^-$ measurements of Detroit shown in Fig. 8 have been made with the same experimental apparatus and technique, most of the potential systematic errors should equally affect the $e^+$ and $e^-$ measurements.  The partial neglect of small angle elastic

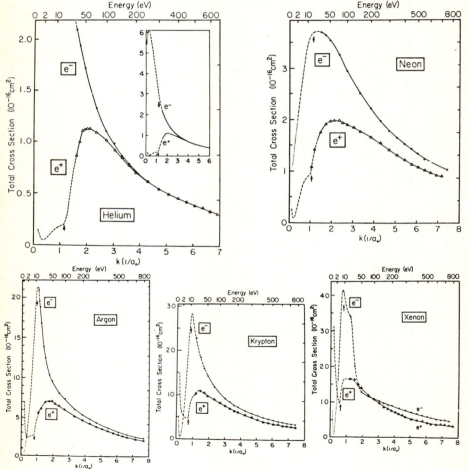

Fig. 8.  $e^+$- and $e^-$-He, Ne, Ar, Kr, and Xe total cross section comparisons.  The lowest inelastic thresholds for each projectile are indicated by arrows.

scattering, on the other hand, is a source of error which in general, does not affect the $e^+$ and $e^-$ measurements equally, since it depends on the angular discrimination (estimated to be roughly $5^\circ$ for all $e^-$ energies and ranging from $15-20^\circ$ at
low energies to $6-10^\circ$ at higher energies for positrons) and on the differential
elastic cross sections.

The $e^{\pm}$-He comparison in Fig. 8 provides a striking illustration of some of the differences and similarities in $e^+$ and $e^-$ scattering. At low energies the $e^+$ cross
section is about two orders of magnitude smaller than the $e^-$ cross section. This
is consistent with the fact that the polarization and static interactions are both
attractive in the $e^-$ case whereas there is a tendency toward cancellation of these
interactions in the $e^+$ case. In sharp contrast to the vastly different cross sections at low energies, there is an observed merging (to within 2%) of the $e^-$ and $e^+$
results above 200 eV. Kauppila et al.[57] estimate that the maximum amounts by which
their cross section measurements could be too low due to the neglect of small angle
elastic scattering are 2% for electrons and 1% for positrons. The merging of the
$Q_T$ curves isn't expected to occur at such low energies. The $e^+$ and $e^-$ distorted-
wave second Born approximation (DW) calculations of Dewangen and Walters[63] do not
merge (to within 2%) until 2000 eV. The composite (B+G,DW) calculations of Inokuti
et al.[61] and Dewangen and Walters referred to in the prior section, merges (to within 2%) at 1000 eV. At 200 eV, the DW calculations for $e^-$ are 21% higher than the
corresponding $e^+$ calculations while there is a 14% difference for the B+G,DW composite theory and a 21% difference for the eikonal Born series (EBS) calculations
of Byron.[64] The comparison $e^{\pm}$ measurements of the Detroit group[57,58] for Ne, Ar, Kr
and Xe shown in Fig. 8 do not indicate any merging of the cross sections up to the
highest energies studied. Xe exhibits a somewhat anomalous behavior between 40 and
90 eV, where the measured $e^+$ cross sections are larger than the corresponding $e^-$
cross sections.

POSITRON-MOLECULE TOTAL CROSS SECTIONS

$Q_T$ measurements[10,70-73] for $e^+$-$H_2$, $N_2$, $CO_2$, $O_2$, CO, and $CH_4$ at energies from 0.5 to
750 eV are shown in Figs. 9 and 10 along with a few available theoretical re-
sults.[74-77] A preliminary report[78] of the Detroit results was made for energies up to
50 eV for $H_2$ and up to 25 eV for $N_2$ and $CO_2$. Earlier measurements by Coleman et
al.[70] for $H_2$ are not included because they have been remeasured and reported by
Griffith and Heyland.[10] The measurements of Charlton et al. are normalized.[71] The
measurements for $H_2$ shown in Fig. 9 indicate the existence of a broad minimum below 9 eV and a dramatic increase near the Ps formation threshold. Another process
(besides Ps formation) which could contribute to the large increase is dissociation
of $H_2$, which becomes energetically favorable near 8.8 eV (the $H_2$ dissociation energy is 4.48 eV). A fixed-nuclei approximation calculation of the elastic scattering
cross section by Hara[74] is in quite good agreement with the measurements. An adiabatic nuclei approximation calculation by Baille et al.[75] is somewhat lower than
the measurements but similar in shape. The total cross section measurements of
Hoffman et al.[73] for $N_2$ also show a broad minimum with increasing cross sections at
the lowest energies, and near the Ps formation threshold. The measurements of
Coleman et al.[70] are generally lower than Hoffman et al. and do not indicate an increase at the lowest energies. For the $e^+$-$CO_2$ case, the measurements of Coleman
et al.[72] are much lower (20-40%) than the results of Hoffman et al.[73] Just above
the Ps formation threshold, an interesting bump appears in the results of Hoffman
et al.[73]

Comparison measurements by Hoffman et al.[73] of $e^{\pm}$-$H_2$ and $e^{\pm}$-$N_2$ total cross sections
are shown in Fig. 11. The $e^{\pm}$-$H_2$ results are merged for energies above 200 eV, but
the merging could be fortuitous if there is appreciable small angle elastic scattering for $H_2$. The $e^-$-$N_2$ results remain above the $e^+$-$N_2$ results up to the highest
energies studied.

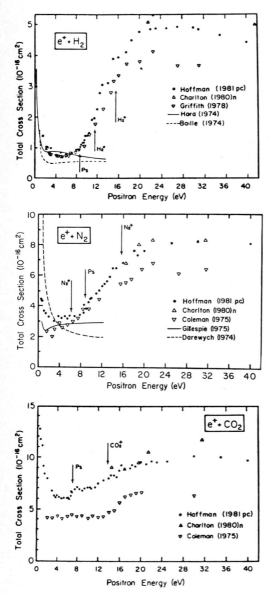

DIFFERENTIAL SCATTERING CROSS SECTIONS

Coleman and McNutt[79] have recently reported measurements of differential cross sections for the elastic scattering of 2-9 eV positrons by Ar for angles from 20-60°. In their experiment, slow positrons pass through a small gas cell (1 cm long) and travel a relatively large distance ($\sim$25 cm) through an evacuated straight flight tube in a strong axial magnetic field to a detector (Channeltron electron multiplier). The larger the angle through which an $e^+$ is scattered in the gas cell, the longer its TOF will be in the axial magnetic field. By alternately admitting gas into and then evacuating the gas cell, TOF histograms such as those shown in Fig. 12 are obtained. A "tail" can be observed on the long-time side of the "gas" TOF spectrum which is associated with detected positrons which have undergone forward elastic scattering. The "difference" spectrum shown in Fig. 12 is obtained by subtracting an appropriately adjusted "vacuum" TOF spectrum from the "gas" spectrum. Differential cross sections can be obtained by correlating the "difference" TOF spectrum with various angles of forward elastic scattering. The differential cross sections measured by Coleman and McNutt for $e^+$-Ar collisions are compared with the calculations of Schrader[25] (solid lines) and "scaled-down" calculations of McEachran et al.[39] (broken lines) in Fig. 13. The agreement between experiment and theory is rather good.

INELASTIC SCATTERING INVESTIGATIONS

Using $Q_T$ curves such as those illustrated in Fig. 3, Ps formation cross sections ($Q_{Ps}$) at the lowest excitation threshold energies for the inert gases have been estimated by Stein et al.[11], Kauppila et al.[9],

Fig. 9.  $e^+$-$H_2$, $N_2$, and $CO_2$ total cross sections at low energies. The $H_2$ experimental results are from Refs. 10,71, and 73. The $H_2$ theoretical results are from Refs. 75 and 74. The $N_2$ experimental results are from Refs. 70,71, and 73. The $N_2$ theoretical results are from Refs. 76 and 77. The $CO_2$ experimental results are from Refs. 72,71, and 73. The "n" refers to normalized results.

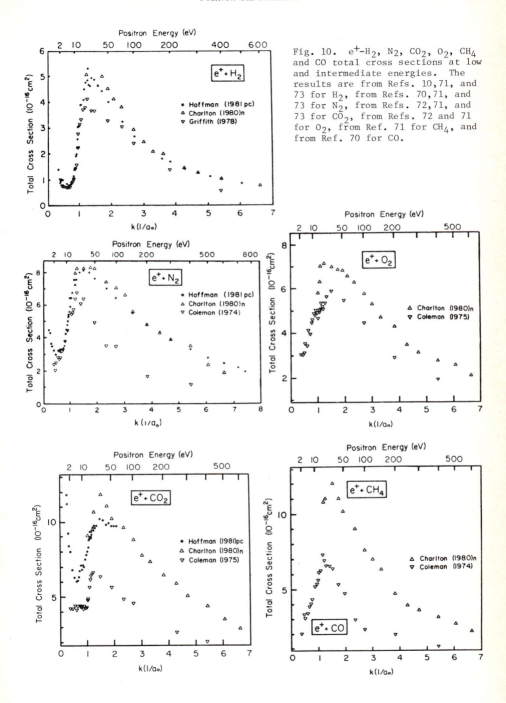

Fig. 10. $e^+$–$H_2$, $N_2$, $CO_2$, $O_2$, $CH_4$ and CO total cross sections at low and intermediate energies. The results are from Refs. 10,71, and 73 for $H_2$, from Refs. 70,71, and 73 for $N_2$, from Refs. 72,71, and 73 for $CO_2$, from Refs. 72 and 71 for $O_2$, from Ref. 71 for $CH_4$, and from Ref. 70 for CO.

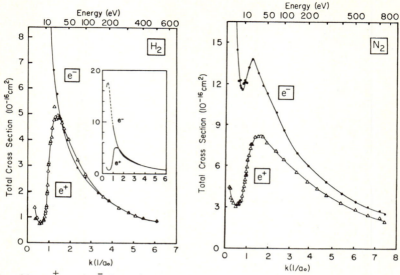

Fig. 11.   $e^+$- and $e^-$-$H_2$ and $N_2$ total cross section comparison measurements.

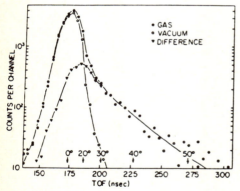

Fig. 12.   TOF histograms with and without Ar in the gas cell (from Coleman and McNutt (Ref. 79)).

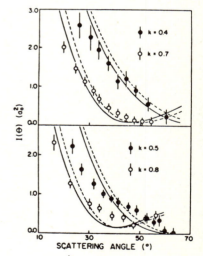

Fig. 13.   $e^+$-Ar results for $I(\theta)$ vs. $\theta$ at different $e^+$ mean energies. The error bars represent statistical standard deviations. (From Coleman and McNutt (Ref. 79)).

and Dababneh et al.[30]   The results for $Q_{Ps}$ in units of $10^{-16}$ cm² are: 0.07 for He at 20.6 eV, 0.20 for Ne at 16.6 eV, 2.0 for Ar at 11.5 eV, 3.5 for Kr at 9.9 eV, and 8 for Xe at 8.3 eV.   These estimates assume a smooth extrapolation of the elastic scattering cross sections from below the Ps formation thresholds.   For Xe, the slope of the total cross section curve is varying rapidly below the formation threshold, resulting in a larger uncertainty in the estimate of $Q_{Ps}$.   The extrapolated curve for Xe in Fig. 3 would give a value for $Q_{Ps}$

of $6.2 \times 10^{-16}$ cm$^2$ rather than the estimated $8 \times 10^{-16}$ cm$^2$ of Dababneh et al.[30] The estimates of $Q_{Ps}$ for He by Coleman et al.[72] based on the total cross section data of Canter et al.[35] give $(0.018 \pm 0.009) \times 10^{-16}$ cm$^2$ at 18 eV and $(0.062 \pm 0.026) \times 10^{-16}$ cm$^2$ at 20 eV, which seem consistent with the estimate of Stein et al.[11] at 20.6 eV. Coleman et al.[43] have used $e^+$-lifetime measurements to make lower limit estimates for $Q_{Ps}$ of $0.14 \times 10^{-16}$ cm$^2$ at 24.5 eV for He, $0.22 \times 10^{-16}$ cm$^2$ at 21.6 eV in Ne, and $0.88 \times 10^{-16}$ cm$^2$ at 15.7 eV for Ar. These lower-limit estimates seem to be consistent with the extrapolated estimates of the Detroit group[9,11,30] since the lifetime based estimates do seem to be lower.

Direct measurements of the energy dependence of the ortho-positronium formation cross section ($Q_{O-Ps}$) in He, Ar, H$_2$, and CH$_4$ have recently been measured by Charlton et al.[80] by passing a slow $e^+$ beam through a scattering chamber and counting triple coincidences from the 3$\gamma$ decay of O-Ps. Their results are shown in Fig. 14. In all four cases, the value of $Q_{O-Ps}$ peaks at relatively low energies, being about 3 eV above the respective Ps formation thresholds for both Ar and He. Based on their measurements, Charlton et al.[80] estimate the ratio $Q_{O-Ps}(Ar)/Q_{O-Ps}(He)$ is equal to 6.5. The peaks for Ar and He occur close to the respective lowest excitation thresholds for those gases (see Fig. 3), so we can estimate the above ratio from the Detroit total cross section results[9,11] and we obtain in this way a value of about 29 which is much larger than the ratio obtained by the London group.

Griffith et al.[81] have used their TOF system with a localized scattering region and a long flight tube to investigate intermediate energy $e^+$-He inelastic scattering. Fig. 15(a) shows two TOF spectra obtained by Griffith et al.[81] for 100 eV positrons when the scattering cell is alternately evacuated and partially filled with He. A secondary peak is observed in the "gas-in" spectrum which is associated with detected positrons that have been inelastically scattered in the forward direction, resulting in an increased TOF. Fig. 15(b) shows a "difference" spectrum of scattered positrons obtained from the spectra in Fig. 15(a). (The TOF in Fig. 15 increases from right to left). Some general information on inelastic scattering is obtained by Griffith et al.[81] by studying the shape and position of the secondary peak as a function of $e^+$ energy. The scattered positrons which have ionized atoms could possess a continuous range of energies which would result in a wider peak, whereas in excitation collisions, which involve two particles of greatly different masses, the energy losses experienced by the incident $e^+$ would be discrete. From studies of such secondary peaks, Griffith et al.[81] have deduced that most of the inelastic scattering (other than Ps formation which would not contribute to the secondary peak) at intermediate energies is due to ionization.

Coleman and Hutton[82] have obtained lower bounds on total excitation cross sections for 23-31 eV $e^+$-He collisions using a TOF technique. In this energy range, well-defined secondary peaks were observed in the TOF spectra corresponding to positrons which have lost 20.6 eV of energy. A TOF spectrum for an incident $e^+$ energy of 25.8 eV is shown in Fig. 16. At incident $e^+$ energies above 30 eV, a secondary peak associated with ionization overlaps the excitation peak, making it difficult to assign excitation cross sections. The secondary peak corresponding to 20.6 eV energy loss indicates that in the projectile energy range from 23 to 31 eV, the total excitation cross section is dominated by excitation of the $2^1S$ state and that there is appreciable small angle scattering associated with this excitation process. Fig. 17 shows the cross sections for the excitation of He (followed by forward scattering) deduced by Coleman and Hutton.

RESONANCE SEARCHES

Stein et al.[83] are currently using their narrow energy width ($<0.1$ eV) $e^+$ beam in a transmission experiment to search for $e^+$ scattering resonances. Fig. 18 shows measurements of the transmitted beam current versus the voltage applied to the $e^+$ source over 1.0 V ranges centered near the Ps formation threshold (9.0 eV) and the lowest atomic excitation threshold (11.5 eV) for Ar. These data were taken at

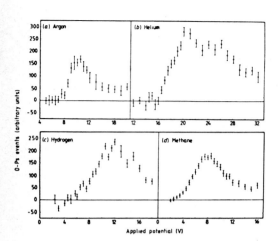

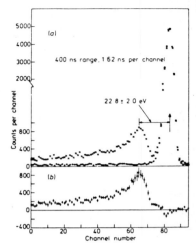

Fig. 14. The number of 0-Ps events, in arbitrary units, plotted against the applied accelerating voltage for each gas. (from Charlton et al. (Ref. 80)).

Fig. 15. TOF spectra of 100 eV positrons. (a) Closed circles: vacuum spectrum. Crosses: spectrum after scattering in He. (b) Spectrum of scattered positrons from the difference between the spectra in (a). (from Griffith et al. (Ref. 81)).

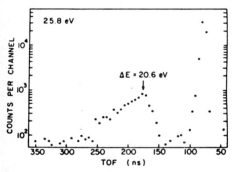

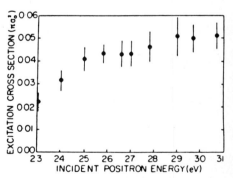

Fig. 16. TOF spectrum showing primary (unscattered) and secondary (excitation) peaks for positrons of mean incident energy 25.8 eV colliding with He. (from Coleman and Hutton (Ref. 82)).

Fig. 17. Cross sections for the excitation of He by $e^+$ impact followed by scattering into the forward direction. The error bars represent statistical and systematic uncertainties. (from Coleman and Hutton (Ref. 82)).

25 mV intervals with the primary beam attenuated by about 50%. In the data for Ar shown in Fig. 18, there is no convincing evidence of a resonance (which would be expected to manifest itself as a relatively narrow structure in the transmitted beam current). There is also no evidence of resonances in preliminary searches with He and $H_2$. However there is still useful information that is obtained from these studies. An abrupt change in the slope of the transmitted current curve appears in Fig. 18(a) which should correspond to the abrupt increase in the total cross section at the Ps formation threshold (see Fig. 3). The slope change in

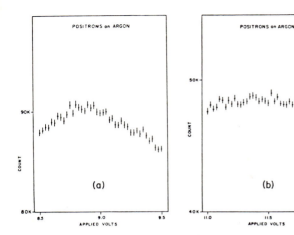

Fig. 18. Transmitted e+ beam currents in Ar versus voltage applied to the e+ source. The error bars represent statistical uncertainties.

Fig. 18(a) occurs at about 8.8 V, whereas the Ps formation threshold is known to be 9.0 eV. This implies that an applied voltage of 8.8 V corresponds to an e+ energy of 9.0 eV in the scattering region. The total cross section curves in Fig. 3, taken by the Detroit group and plotted versus applied voltage, provide additional evidence that their results should possibly be shifted to higher energies by approximately 0.2 eV, since those curves appear to be starting their abrupt increases a few tenths of an eV below the Ps formation thresholds. The curve in Fig. 18(b) does not show an appreciable change in its slope as the energy is increased through the atomic excitation threshold (11.5 eV) and this is supported by the corresponding $Q_T$ curve for Ar in Fig. 3, which is quite smooth between 11 and 12 eV.

ACKNOWLEDGEMENTS

We would like to thank Mr. Kevin Hoffman, Mr. Paul Felcyn, and Mr. Diab Jerius for their helpful assistance during the preparation of this manuscript. We acknowledge, with gratitude, the support of the National Science Foundation for our research program.

REFERENCES

[1] Costello, D.G., Groce, D.E., Herring, D.F., and McGowan, J. Wm., Can. J. Phys. 50 (1972) 23.
[2] Massey, H.S.W., Physics Today 29 (No. 3) (1976) 42.
[3] Canter, K.F., Coleman, P.G., Griffith, T.C., and Heyland, G.R., J. Phys. B 5 (1972) L167.
[4] Jaduszliwer, B. and Paul, D.A.L., Can. J. Phys. 51 (1973) 1565.
[5] Dutton, J., Harris, F.M., and Jones, R.A., J. Phys. B 8 (1975) L65.
[6] Burciaga, J.R., Coleman, P.G., Diana, L.M., and McNutt, J.D., J. Phys. B 10 (1977) L569.
[7] Wilson, W.G., J. Phys. B 11 (1978) L629.
[8] Sinapius, G., Raith, W., and Wilson, W.G., J. Phys. B 13 (1980) 4079.
[9] Kauppila, W.E., Stein, T.S., and Jesion, G., Phys. Rev. Letters 36 (1976) 580.
[10] Griffith, T.C. and Heyland, G.R., Physics Reports 39 (1978) 169.
[11] Stein, T.S., Kauppila, W.E., Pol, V., Smart, J.H., and Jesion, G., Phys. Rev. A 17 (1978) 1600.
[12] Brenton, A.G., Dutton, J., Harris, F.M., Jones, R.A., and Lewis, D.M., J. Phys. B 10 (1977) 2699.

[13] Griffith, T.C., Heyland, G.R., Lines, K.S., and Twomey, T.R., Appl. Phys. 19 (1979) 431.

[14] Canter, K.F., Coleman, P.G., Griffith, T.C., and Heyland, G.R., J. Phys. B 6 (1973) L201.

[15] Jaduszliwer, B. and Paul, D.A.L., Can. J. Phys. 52 (1974) 1047.

[16] Jaduszliwer, B., Nakashima, A., and Paul, D.A.L., Can. J. Phys. 53 (1975) 962.

[17] Coleman, P.G., Griffith, T.C., Heyland, G.R., and Twomey, T.R., Appl. Phys. 11 (1976) 321.

[18] Coleman, P.G., McNutt, J.D., Diana, L.M., and Burciaga, J.R., Phys. Rev. A 20 (1979) 145.

[19] Drachman, R.J., Phys. Rev. 144 (1966) 25.

[20] Massey, H.S.W., Lawson, J., Thompson, D.G., Quantum Theory of Atoms, Molecules, Solid State (Academic Press, New York, 1966) p.203.

[21] Callaway, J., LaBahn, R.W., Pu, R.T., and Duxler, W.M., Phys. Rev. 168 (1968) 12.

[22] Aulenkamp, H., Heiss, P., and Wichmann, E., Z. Phys. 268 (1974) 213.

[23] McEachran, R.P., Morgan, D.L., Ryman, A.G., and Stauffer, A.D., J. Phys. B 10 (1977) 663.

[24] Campeanu, R.I. and Humberston, J.W., J. Phys. B 10 (1977) L153.

[25] Schrader, D.M., Phys. Rev. 20 (1979) 918.

[26] Ramsauer, C., Ann. Physik (Leipzig) 66 (1921) 546.

[27] Ramsauer, C., Ann. Physik (Leipzig) 72 (1923) 345.

[28] Ramsauer, C. and Kollath, R., Ann. Physik (Leipzig) 3 (1929) 536.

[29] Townsend, J.S. and Bailey, V.A., Philos. Mag. 43 (1922) 593.

[30] Dababneh, M.S., Kauppila, W.E., Downing, J.P., Laperriere, F., Pol, V., Smart, J.H., and Stein, T.S., Phys. Rev. A 22 (1980) 1872.

[31] Jaduszliwer, B., and Paul, D.A.L., Can. J. Phys. 52 (1974) 272.

[32] Tsai, J.-S., Lebow, L., and Paul, D.A.L., Can. J. Phys. 54 (1976) 1741.

[33] Brenton, A.G, Dutton, J., and Harris, F.M., J. Phys. B 11 (1978) L15.

[34] Coleman, P.G., McNutt, J.D., Diana, L.M., and Hutton, J.T., Phys. Rev. A 22 (1980) 2290.

[35] Canter, K.F., Coleman, P.G., Griffith, T.C., and Heyland, G.R., Appl. Phys. 3 (1974) 249.

[36] Montgomery, R.E. and LaBahn, R.W., Can. J. Phys. 48 (1970) 1288. Total cross section values were provided by R.E. Montgomery and R.W. LaBahn (private communication).

[37] McEachran, R.P., Ryman, A.G., and Stauffer, A.D., J. Phys. B 11 (1978) 551.

[38] Campeanu, R.I., and Dubau, J., J. Phys. B 11 (1978) L567.

[39] McEachran, R.P., Ryman, A.G., and Stauffer, A.D., J. Phys. B 12 (1979) 1031.

[40] McEachran, R.P., Stauffer, A.D., and Campbell, L.E.M., J. Phys. B 13 (1980) 1281.

[41] Milloy, H.B. and Crompton, R.W., Phys. Rev. A 15 (1977) 1847.

[42] Kauppila, W.E., Stein, T.S., Jesion, G., Dababneh, M.S., and Pol, V., Rev. Sci. Instr. 48 (1977) 822.

[43] Coleman, P.G., Griffith, T.C., Heyland, G.R., and Killeen, T.L., J. Phys. B 8 (1975) L185.

[44] Griffith, T.C., Advances in Atomic and Molecular Physics, Vol. 15 (Academic Press, New York, 1979) p. 135.

[45] Kennerly, R.E. and Bonham, R.A., Phys. Rev. A 17 (1978) 1844.

[46] Charlton, M., Griffith, T.C., Heyland, G.R., and Twomey, T.R., J. Phys. B 13 (1980) L239.

[47] Blaauw, H.J., Wagenaar, R.W., Barends, D.H., and de Heer, F.J., J. Phys. B 13 (1980) 359.

[48] Yau, A.W., McEachran, R.P., and Stauffer, A.D., J. Phys. B 11 (1978) 2907.

[49] Nesbet, R.K., J. Phys. B 12 (1979) L243.

[50] Fon, W.C., Berrington, K.A., and Hibbert, A., J. Phys. B 14 (1981) 307.

[51] Humberston, J.W., J. Phys. B 11 (1978) L343.

[52] Humberston, J.W. and Campeanu, R.I., J. Phys. B 13 (1980) 4907.

[53] Kauppila, W.E., Stein, T.S., Pol, V., and Jesion, G., Proc. 4th Int. Conf. on Positron Annihilation, (Helsingor Abstracts, 1976) p. 25. Also see Stein, T.S., Kauppila, W.E., Pol, V., Jesion G., and Smart, J.H., Proc. 10th ICPEAC

(Commissariat a L'Energie Atomique, Paris, 1977) p. 804.

[54] Griffith, T.C., Heyland, G.R., Lines, K.S., and Twomey, T.R., J. Phys. B 11 (1978) L635.

[55] Drachman, R.J., Phys. Rev. 144 (1966) 25.

[56] O'Malley, T.F., Spruch, L., and Rosenberg, L., J. Math. Phys. 2 (1961) 491.

[57] Kauppila, W.E., Stein, T.S., Smart, J.H., Dababneh, M.S., Ho, Y.K., Downing, J.P., and Pol, V., Phys. Rev. A 24 (1981) 725.

[58] Dababneh, M.S., Hsieh, Y.-F., Kauppila, W.E., Pol, V., and Stein, T.S., private communication.

[59] Inokuti, M. and McDowell, M.R.C., J. Phys. B 7 (1974) 2382.

[60] Inokuti, M., Kim, Y.-K., and Platzman, R.L., Phys. Rev. 164 (1967) 55.

[61] Inokuti, M., Saxon, R.P., and Dehmer, J.L., Int. J. Radiat. Phys. Chem. 7 (1975) 109; Kim, Y.-K. and Inokuti, M., Phys. Rev. A 3 (1971) 665.

[62] Byron Jr., F.W. and Joachain, C.J., Phys. Rev. A 15 (1977) 128.

[63] Dewangen, D.P. and Walters, H.R.J., J. Phys. B 10 (1977) 637.

[64] Byron Jr., F.W., Phys. Rev. A 17 (1978) 170.

[65] Saxon, R.P., Phys. Rev. A 8 (1973) 839.

[66] Joachain, C.J., Vanderpoorten, R., Winters, K.H., and Byron Jr., F.W., J. Phys. B 10 (1977) 227.

[67] Wagenaar, R.W. and de Heer, F.J., J. Phys. B 13 (1980) 3855.

[68] Salop, A. and Nakano, H.H., Phys. Rev. A 2 (1970) 127.

[69] Golden, D.E. and Bandel, H.W., Phys. Rev. 149 (1966) 58.

[70] Coleman, P.G., Griffith, T.C., and Heyland, G.R., Appl. Phys. 4 (1974) 89.

[71] Charlton, M., Griffith, T.C., Heyland, G.R., and Wright, G.L., J. Phys. B 13 (1980) L353.

[72] Coleman, P.G., Griffith, T.C., Heyland, G.R., and Killeen, T.L., Atomic Physics 4 (Plenum Publishing Corp., New York, 1975) p. 355.

[73] Hoffman, K.R., Dababneh, M.S., Hsieh, Y.-F., Kauppila, W.E., Pol. V, Smart, J.H., and Stein, T.S., private communication.

[74] Hara, S., J. Phys. B 7 (1974) 1748.

[75] Baille, P., Darewych, J.W., and Lodge, J.G., Can. J. Phys. 52 (1974) 667.

[76] Darewych, J.W., and Baille, P., J. Phys. B 7 (1974) L1.

[77] Gillespie, E.S. and Thompson, D.G., J. Phys. B 8 (1975) 2858.

[78] Kauppila, W.E., Stein, T.S., Smart, J.H., and Pol, V., Abstracts of Papers of ICPEAC-X (Commissariat a L'Energie Atomique, Paris, 1977) p. 826.

[79] Coleman, P.G. and McNutt, J.D., Phys. Rev. Lett. 42 (1979) 1130.

[80] Charlton, M., Griffith, T.C., Heyland, G.R., Lines, K.S., and Wright, G.L., J. Phys. B 13 (1980) L757.

[81] Griffith, T.C., Heyland, G.R., Lines, K.S., and Twomey, T.R., J. Phys. B 12 (1979) L747.

[82] Coleman, P.G. and Hutton, J.T., Phys. Rev. Lett. 45 (1980) 2017.

[83] Stein, T.S, Laperriere, F., Dababneh, M.S., Hsieh, Y.-F., Pol, V. and Kauppila, W.E., to be published in the book of abstracts of contributed papers for ICPEAC-XII.

PHYSICS OF ELECTRONIC AND ATOMIC COLLISIONS
S. Datz (editor)
© North-Holland Publishing Company, 1982


ROTATIONAL RAINBOWS


Dieter Beck

Fakultät für Physik
Universität Bielefeld
Bielefeld
Germany.


With classical rigid shell scattering as a heuristic vehicle
the origin of rotational rainbows in molecular scattering
and their information on the (repulsive) anisotropy of the
intermolecular potential are indicated. Present observations
on this new probe to the interaction are discussed. First results of realistic, ab initio based theory compare well with
experiment.


I. INTRODUCTION

Rotational excitation is a ubiquitous result of molecular collisions at thermal
energy and above. It is an elementary process. But to relate it to its cause, the
anisotropy of the molecular interaction, is a long way.

Its direction has been shown by Curtiss [1], Takayanagi [2] and Dalgarno [3] and
their collaborators on the part of theory. Bernstein [4] and Toennies [5] and their
groups began to set experimental standards. Since then the formidable, very-many-
coupled-channels problem of the scattering theory has been battled; a wealth of approximation methods is the result, with the delicate decoupling procedures of
Pack [6], Kouri and McGuire [7] among them. Quantum chemistry has advanced to useful data on anisotropic potentials of many systems [8]. And the experiments have
been improved in range of energy and angle, in detection sensitivity and resolving
power. Single rotational transitions can now be resolved by conventional translation energy loss methods as well as by enormously more selective pump- and-probe
laser techniques [9 – 14].

These and other advances have been important steps in establishing the desired link
of rotational excitation and the molecular anisotropy. They also suggested the matter to be of intrinsic complexity, difficult to validate, and eventually necessitating laborious detail to obtain significant results. This may be greatly alleviated
by the finding of a characteristic structure in rotationally inelastic scattering
and the proof of its intimate relation to the anisotropy. Thomas termed the structure rotational rainbows [15, 16]. He was the first to get some specific notion of
its existence although many authors have seen traces of it before, mainly in calculations, but did not spell it out, while others considered somewhat analogous, but
different phenomena [17, 18]. Schinke [19] and, a bit later, Bowman [20] calculated
and clearly discussed the "angular vista" of the effect. At about this time Schepper, Ross and the author reported the first clear observation of rotational rainbows [21]. This made their possibly dominant nature and sensitive relation to the
anisotropy plainly apparent.

To present further progress a simple model is used in the following Sec. II to indicate how rotational rainbows come about. Sec. III is devoted to observations. More realistic, theoretical results as well as the first comparisons of ab initio
theory and experiment are sampled in Sec. IV. Sec. V concludes with a summary.

331

## II. THE HEURISTIC APPROACH: CLASSICAL, HARD SHELL SCATTERING

The term rotational rainbow suggests a *classical* turning point effect as is familar
in isotropic potential scattering for the deflection angle $\vartheta$ as a function of the
impact parameter b, if the potential has an inflection point. Consider a structureless atom and a rigid rotor diatom interacting by an anisotropic potential. Let the
molecules have random orientation and vanishing magnitude of angular momentum before collision, j = 0. This special case is both simple and conveys maximum information. It will suffice for the most as it is closely approximated or even obtained in current experiments. With dashes designating "after collision" assume further that j' or equivalently (by energy conservation) the relative recoil velocity
v' of collision partners is measured in addition to $\vartheta$. With two measured out-variables (j' or v' and $\vartheta$) as a function of three distributed in-variables of the scattering (b and molecular orientation $\beta$, $\alpha$) the general form in which eventual turning point behaviour affects the cross section is more complex than for elastic
scattering. We leave it to the debate of experts [22] and, instead, seek to simplify further by the following assertion. For rotational rainbow scattering to occur
three conditions are expected to be sufficient:
(i) Collisions must be rotationally sudden (initially and finally);
(ii) their outcome must be dominated by repulsive interaction, and
(iii) they must be of high (exciting and deflecting) action, i.e. typically trans-
ferring $\Delta j \gg 1$.
(iii) implies that the energy transfer be "fine-grained", j' may be considered qua-
si-continuous, indeed, essentially a classical variable. These conditions are compatible. They may often be well approximated in experiments on neutral, non-reactive atom-diatom systems at and above thermal energy and at sufficiently large deflection angles to improve the sudden and guarantee the repulsive nature of the collision. Rotational rainbows may really be a rather common feature of molecular collisions, except perhaps for very nearly spherical or hydrogenic molecules with large rotational quanta which will fail on (iii).

The sufficient character of the conditions simply means that the full story of rotational rainbows is not known yet. So far all clear observations comply with them.
But they really derive from classical, rigid or hard shell scattering which may be
considered the high action and sudden limit representation of repulsive collision
dynamics. This simple model, at least qualitatively, predicts all clearly recognized, classical features of rotational rainbows to date; and it suggests those beyond the reach of classical mechanics: Interferences. As it is analytically tractable to a near explicit expression for the cross section, it also yields instructive insight into how rotational rainbows come about. Let us, therefore, first in-
spect its results [23].

For atom-diatom collision partners the shell is reasonably of cylindrical symmetry, with its figure axis
coinciding with the internuclear axis and rigidly connected to it. Let X be measured along the figure axis
from the molecular center of mass as the origin, as is
indicated in Fig. 1. The shape of the shell is conveniently given by its (squared) contour over the figure axis F(X). The surface is then defined to be a rigid shell by assuming (a) instantaneous interaction
upon impact at "contact position" $\vec{R}$ (with respect to
the molecular center of mass) so that (b) the momentum transfer $\vec{\Delta p}$ has the direction of the outward normal unit vector $\hat{n}(\vec{R})$ of the shell at $\vec{R}$,

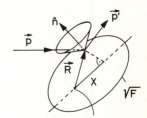

**Molecular Center of Mass**

$$\vec{\Delta p} \equiv \vec{p}' - \vec{p} = \Delta p \, \hat{n} \qquad (1)$$

and (c) angular momentum and energy are conserved
which yields, respectively,

Fig. 1

$$\vec{\Delta j} = -\Delta p \left( \vec{R} \times \hat{n} \right) , \qquad \Delta p = -\frac{2p \left( \hat{n} \cdot \hat{p} \right)}{1 + \frac{\mu}{I} \left( \vec{R} \times \hat{n} \right)^2} \tag{2}$$

$\mu$: Reduced atom-diatom mass,
I: Molecular moment of inertia.

Here $\Delta p$ is given for the case at hand, $j = 0$. Obviously, (1) and (2) determine the outcome of the collision completely for any given initial scattering asymptote. The out-variables of interest may be calculated, i.e. $j'$, $\vartheta$ in laser experiments or $v'$ $\vartheta$ conforming to the translation energy loss method. In the latter case the *reduced* recoil velocity $u^{::} = p'/p = v'/v$ will turn out to be more convenient and hence be used instead of $v'$. Both variable choices are, of course, equivalent; either one may be turned into the other by energy conservation. We omit the calculation of out-variables. But it does show two noteworthy features of rigid shell scattering: (a) To contribute to the intensity at a given outcome $j'$ or $u^{::}$ and $\vartheta$ projectiles must come in with fixed X and fixed angle of incidence with respect to $\tilde{n}$, i.e. on a cone as that shown in Fig. 1. This assigns a certain (somewhat involved) manifold of sets of impact parameter b and molecular orientation $\beta$, $\alpha$ to any given outcome. Such kind of order may be worth knowing. For more realistic (repulsive) interactions it will hold approximately. (b) Expressed in reduced out-variables like $u^{::}$ hard shell scattering (for $j = 0$) is no longer explicitly dependent on collision energy $E = (\mu/2)v^2$. Alternatively, the energy transfer $\Delta E$ is strictly proportional to E. Whenever initial rotation is small, this may be useful as an indicative that the scattering in a real situation approximates the rigid shell limit. Implementing the general, classical prescription for the double differential cross section (averaged over molecular orientation for consistence with the assumption $j = 0$) the analysis ends up with the result

$$J\left( u^{::}, \vartheta \,|\, j = 0 \right) = \frac{u^{::2} \left( 1 - u^{::} \cos \vartheta \right)^2}{\left( 1 - 2 u^{::} \cos \vartheta + u^{::2} \right)^3} \sum_{i=1}^{4(2)} \left| \frac{\left( F + \frac{1}{4} F'^2 \right)^{\frac{3}{2}}}{B'} \right|_{X_i} \tag{3a}$$

Here dashes designate differentiation with respect to X,

$$B \equiv \frac{\mu}{I} \left( \vec{R} \times \hat{n} \right)^2 = \frac{\mu}{I} \cdot \frac{F \left( \frac{1}{2} F' + X \right)^2}{F + \frac{1}{4} F'^2} = B(X) , \tag{3b}$$

a shell function, conveniently expressed as B(X) and related to the out-variables by

$$B = \frac{1 - u^{::2}}{1 - 2u^{::} \cos \vartheta + u^{::2}} . \tag{3c}$$

As expected the cross section as a function of $u^{::}$ is independent of energy. This form is simply converted to $J(j'; \vartheta | j = 0; E)$ by multiplication with the $j'$ to $u^{::}$ Jacobian obtained from energy conservation.

Eqs. (3) are an exact result in the classical and rigid shell approximations. (3a) expresses the nature of rotational rainbows as compactly as one can probably get to: The cross section exhibits (integrable) singularities if $u^{::}$, $\vartheta$ correspond to eventual turning points of the shell function B(X).

The use of these results is graphically indicated in Fig. 2 which is adapted to observed K - CO scattering [21]. Let the shell be a simple, prolate ellipsoid of revolution with semi-axes c and a (<c) and with its center of symmetry displaced by a small distance $X_0$ from the molecular center of mass to crudely simulate the heteronuclear diatom. This is shown in the upper center of Fig. 2. With the shell, i.e. F(X) given B(X) is obtained from (3b) in the lower center, exhibiting two unequal maxima $B_{1,2}$ and a minimum $B_0 = 0$. For any given $\vartheta$ the range of B generated by the shell is "reflected" onto $u^{::}$ by the $B(u^{::}, \vartheta)$ relation (3c) in the right

Fig. 2

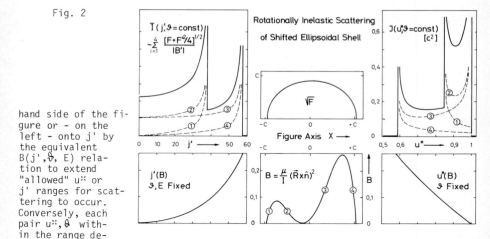

hand side of the fi-
gure or - on the
left - onto j' by
the equivalent
$B(j', \vartheta, E)$ relation to extend
"allowed" $u^{*}$ or
j' ranges for scat-
tering to occur.
Conversely, each
pair $u^{*}, \vartheta$ within the range de-
termines a unique B and, from B(X), four or two contributing values X at which the
shell form dependent sum of the cross section (3a) is evaluated. At any fixed $\vartheta$
the cross section as a function of $u^{*}$, thus, displays three singularities corresponding to the extrema $B_r$ of B(X). As a function of j' the singularity at elastic
scattering is transformed into a flat dependence by the j' to $u^{*}$ Jacobian multiplying $J(u^{*})$. The singularities r = 1,2, which appear in either form of J are the
rotational rainbows of the "shifted" ellipsoidal shell. The contributions to the
scattering from branches 1, 2 of B(X) are expected to lead to a "supernumerary"
interference if the superposition principle were taken into account. From eq. (2)
and note (a) above it may be inferred that at each $\vartheta$ they correspond to equal j'
component in a given spatial direction. This also holds for branches 3 and 4, but
with reversed sign of the component. The two pairs add incoherently.

For the shift $X_0 = 0$, simulating a homonuclear diatom, the B(X) maxima are of
equal height and the rainbow singularities coincide at a single rainbow recoil ve-
locity $u^{*}_{r=1}(\vartheta)$ or $j'_{r=1}(\vartheta, E)$. Fig. 3 shows the scattering for this case in the $\vartheta$,
v' domain. The cross section is non-vanishing on the shaded area. As exemplified
for $\vartheta = 60^0$ and $180^0$ each cut $\vartheta$ = const displays two singularities bounding the
allowed range,
one at elastic
scattering,
v' = v (which
will be trans-
formed away in a
representation
on j'), the other
the rainbow sin-
gularity at
$v'_{r=1}(\vartheta) \neq 1$.

Fig. 3

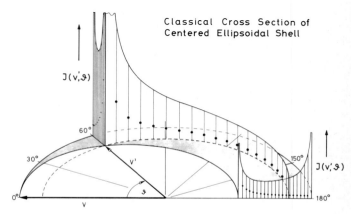

Classical Cross Section of
Centered Ellipsoidal Shell

For this recoil velocity, v, and $\vartheta$ given the value $B_1$ of the two equal maxima of B(X) is readily obtained from (3c). Use of (3b) yields

$$B_1 = \frac{\mu}{I}(c - a)^2 , \qquad (4)$$

the expected, absolute measure of the anisotropy of the ellipsoidal shell. In this sense the width of the allowed $u^{\ast}$ or j' interval of rotationally inelastic, rigid shell scattering at any given $\vartheta$ directly and sensitively tells the anisotropy of the shell. Within this range the form of the cross section reflects all the shell geometry in a strict point-to-point if not one-to-one correspondence.

If the scattering is scanned at fixed $u^{\ast}$ or j' as a function of angle, a third cut at v' = 0.84 v shows what is to be expected. The abrupt onset upon entering the allowed regime forms a halo about the forward direction. The angular width of the halo increases with the inelasticity according to eq. (3c) with $B = B_1$ (this determines the allowed, i.e. shaded regime in Fig. 3). In all cuts the dots separate (twice) the contributions expected to give rise to interference in a description going beyond classical mechanics.

As given the hard shell analysis permits use of any shell geometry of cylindrical symmetry. For 8-shaped geometries it needs (angle dependent) correction for shadowing and multiple contacts. The latter type of correction must also be considered in cases of excessive final rotation. The generally sufficient one-contact version given here has been studied in two dimensions by Bosanac to detail the relation of model parameters to the allowed regime and various other approximations [24]. It has also been extended to allow for finite initial rotation [25]. Except for the most super-elastic event the intensity at given $u^{\ast}$ or j' (=0), $\vartheta$ is found to be averaged over extended parts of the shell. The analytic form of the persisting, but shifted and "smeared" rainbow singularities is interestingly the same as that expected for r-dependent potentials [26]. Strong quenching of the supernumerary interference may be inferred.

This ends the classical and hard shell account of rotational rainbows. It brings out some, perhaps all of the *simple* features of rotationally inelastic scattering in its strong coupling limit, its order, its information, and the non-trivial turning point behaviour involved which has nothing to do with the r-dependence or an attractive well of the interaction.

## III. OBSERVATIONS

Let us next see what is actually observed. Fig. 4 shows an early sample obtained by Schepper et al. [21]. A conventional translation energy loss method is used. K atoms (1) are prepared by the seeding technique to raise and reasonably define collision energy. Molecules (2) come in with low initial rotation, $\bar{j} = 2.5 \pm 1$, by isentropic expansion of the crossed beam. Excitation is detected by velocity selection of scattered K atoms. The final j' remain unresolved. Studying rotational rainbows this is not really a serious drawback. As we have seen, the message of this phenomenon, by definition, resides in *many* ("fine-grained") quantized channels. The scattering of K by $N_2$ and CO is scanned at a fixed, large angle in the center-of-mass system (CMS) as a function of reduced recoil velocity. The collision energy is substantially above thermal to render eventual attractive well action negligible. The expected properties of rotational rainbows are clearly seen. The scattering is sharply restricted to allowed $u^{\ast}$ intervals with falloff flanks almost entirely determined by resolving power. For the inversion symmetric K - $N_2$ potential it sets in with a strong maximum at $u^{\ast} = 1$, elastic scattering, and breaks off with a second maximum at the rainbow velocity of about 0.75. For the heteronuclear system there are three maxima, the less and more inelastic being the rotational rainbows for O - and C - end scattering off CO, respectively. At the indicated energy and $100^0 \leqslant \vartheta \leqslant 180^0$ the (weak) angle dependence of the allowed $u^{\ast}$ interval was examined and found to be compatible with the prediction of the hard

shell model for a fixed ellipsoidal anisotropy (c - a) = 0.27 A for
K - $N_2$ and (c - a) = 0.29 A for
K - CO with a shift $X_0$ = 0.12 A.
The near equality of the anisotropies, obtained from very different cross sections, is a reasonable result. At $\vartheta$ = 150° and
0.34 < E < 1.24 eV the energy
dependence of the allowed $u^x$
interval was studied. It is *not*
constant as would be expected
for   a shell of fixed geometry.
Instead, it increases with E
(differently so for $N_2$ and CO)
to approach limiting values which
correspond to anisotropies near
the quoted (c - a), exceeding
them by at most 10%. The authors
argue that at the large observation angle this increase may, indeed, reflect a locally variable
anisotropy which increases and
finally levels off as the potential is probed farther in. This
conjecture invites further analysis. Contour maps of anisotropic potentials simply extracted
from the energy dependence of
rotational rainbow scattering
- an attractive aspect. Some verification of the rainbow nature
of data as in Fig. 4 has been se-
cured by isotopic substitution
[27]. Replacing natural CO by

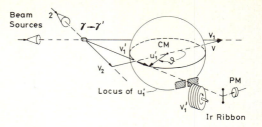

Final j'Unresolved.
Average Initial j Low by Isentropic Expansion.

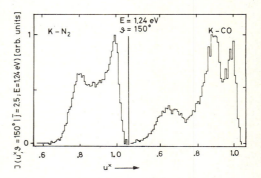

Fig. 4 showing rotational rainbows in
K - $N_2$ and K - CO scattering.

$C^{18}O$ shifts the molecular center of mass by $\Delta X_0$ = 0.032 Å with respect to the unaltered force field. The two CO rotational rainbows are clearly found to respond to
this change by coming farther apart. An ellipsoidal shell analysis of the observations yields: To within experimental error (c - a) remains unchanged by isotopic
substitution, while $X_0$ does increase by $\Delta X_0$ = 0.024 Å. The authors give a reasonable argument that even the minute deviation of 0.008 Å of the model result from
reality may be significant and due to the erroneous attempt to fit some K - CO
equipotential of $C_{\infty v}$ symmetry by a shifted, inversion symmetric shell. Whether this
will prove to be correct or not, any of the numbers quoted impressively documents
the sensitive dependence of rotational rainbows on the anisotropy.

Another example of the effect was obtained by Kinsey, Pritchard and their collaborators at M.I.T. on the system $Na_2$ - Ar [13a]. As indicated in the upper part of
Fig. 5 they employed a highly selective laser technique and, thus, were able to resolve single rotational level-to-level transitions. A first laser optically pumps
and thereby labels a specific j in a chopped mode. A second probe laser beam shines
into the scattering volume. It is directed antiparallel to the relative velocity of
collision partners and is set to induce fluorescence if the desired, final level j'
is collisionally populated *without* deflection. It is then slightly detuned so that,
because of the Doppler shift, the molecules deflected by some angle will absorb to
give laser-induced fluorescence. Note that this is a direct CMS detection catching
*all* azimuthal angles. The price to be payed for the resulting, fancy luminosity of
the scheme is a deterioration of angular resolving power as $\vartheta \to 0, \pi$. With the laser method scanning at fixed j' as a function of $\vartheta$ is naturally preferred. The results (for initial j = 7) show the strong angular vista of the single rainbow maximum of the inversion symmetric $Na_2$ - Ar potential. They also clearly demonstrate

the halo of increasing angular width with increasing j'. At larger (energy) transfer the cut into the allowed regime occurs later - as expected. These experiments use lower collision energy and extend to smaller angles than those on K - $N_2$ and CO. At an estimated, attractive well depth $\varepsilon \lesssim 1$ meV and the small rotation constant $B_e = 1.92 \times 10^{-5}$ eV of $Na_2$ collisions are, however, sudden and for $\vartheta > 10^0$ also predominantly repulsive.

The same authors also observed simultaneous rotational and vibrational excitation with $\Delta v = 1$, energetically corresponding to a $7 \to 32$ rotational transition [13b]. Perhaps not unexpected, the cross sections retain the typical rainbow shape, with maxima of $\Delta j$ transitions at or very near the angles of the corresponding vibrationally elastic process. At the collision energy the slow $Na_2$ vibration is about to loose its adiabatic character. Therefore, the observation angle (via collision duration) strongly affects the vibrational transition probability. Accordingly, small $\Delta j$ cross sections, peaking at small angles, are drastically smaller when vibration is simultaneuously excited. For large $\Delta j$, large angle

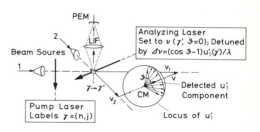

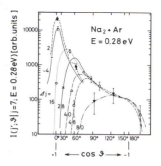

Direct CMS Detection by Doppler Shift; Specific to Detection of All Azimuthal Angles. State $\gamma'$ and $\vartheta$. Angular Resolving Power Poor at $\vartheta \sim 0, \pi$.

Fig. 5. Angular cuts of the $Na_2$ - Ar rotational rainbow.

peaking, the difference attenuates to a factor of $\sim 2$ in favour of the vibrationally elastic process. These observations are in full, qualitative accord with an JOS calculation based on an ab initio potential for the system $Na_2$ - He by Müller and Schinke [28]. Vib-rot excitation in these systems, thus, appears to walk along in rotational rainbow fashion, but under the hat of vibrational adiabaticity. It is noteworthy that its makeup under strongly non-adiabatic, i.e. curve-crossing conditions is altogether different [29].

More (and the very first) observations on pure rotational level-to-level differential cross sections are due to Bergmann and the group in Kaiserslautern. They also use a pump-and-probe laser technique, but probe Doppler-free in the scattered beam, see Fig. 6. This trades signal for better angular resolving power and (unwanted) double valued kinematics. They established rotational rainbow scattering for $Na_2$ - He battling rather unfavourable kinematical conditions [12a]. A spot check on $Na_2$ - Ar slightly deviates from the M.I.T. data, but clearly shows better angular resolving power at $\vartheta \sim 40^0$ [12b]. The system $Na_2$ - Ne was more extensively studied. A set of $j \to 28$ transitions is of similar nature as the data of Fig. 5, better resolved but more restricted in $\Delta j$ [12b]. Another set is shown in Fig. 6 [12c]. Again scanning $\vartheta$ small $\Delta j$ imply a huge rainbow maximum at small angles, partly within the unscattered beam. But with j or j' = 0 the hard shell model suggests and the JOS approximation predicts that the supernumerary interference should be there and, indeed, it is. The lines in the diagram give the result of an JOS calculation using an ab initio potential. The experimental data are normalized to the calculated at a single point, the maximum of the $0 \to 6$ transition. The amount of agreement is astounding.

The systems for which clear observations of rotational rainbows have been reported up to this summer are summarized in Tab. I. The $D_2$ - CO data are of Buck's laboratory in Göttingen [30]. Superimposed on a single $D_2$ transition they show recoil velocity distributions of the expected angular behaviour of a CO rotational rainbow. The as yet unpublished K - $O_2(\tilde{a}^4A")$ and K - $CO_2$ data were measured in Bielefeld. The latter are the first observation on a triatomic (linear) molecule. Conspicuous structure has also been reported in $Li^+$ - CO scattering [31]. It has not been included in Tab. I as its origin is unclear. It may, however, represent an aspect of rotational rainbows under the influence of considerable attractive interaction. In the second column $u^{\ast}$, $\vartheta$ means j' unresolved, $u^{\ast}$ scanned at constant $\vartheta$; j', $\vartheta$ indicates resolved j', $\vartheta$ scanned at j' = const. As a preliminary evaluation the third column shows the ellipsoidal hard shell anisotropy for those systems where neglect of attractive well influence appears justified. Some of the quoted errors are but weakly

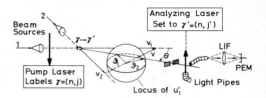

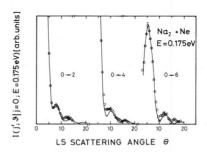

Kinematics Double Valued.
Comparison with IOS Calculation
on ab initio Potential.

Fig. 6 demonstrating the expected supernumerary oscillations of rotational rainbows in $Na_2$ - Ne.

related to the quality of the original data, but reflect the author's ignorance of experimental detail. The last two columns give collision energy and angular range of the experiments to enable estimates of the potential value corresponding to the anisotropy. Qualitatively the data show much reasonable consistence. Their quantitative significance awaits future check [41].

| System | Mode of Observation | (c-a) [Å] | E [eV] | $\vartheta$ |
|---|---|---|---|---|
| K - $N_2$ | $u^{\ast}$, $\vartheta$ | 0.27 ± 0.02 | 1.24 | 150° |
| K - CO | $\bar{u}^{\ast}$, $\vartheta$ | 0.29 ± 0.03 | 1.24 | 150° |
| K - $O_2$ | $\bar{u}^{\ast}$, $\vartheta$ | 0.35 ± 0.03 | 1.36 | 140° |
| K - $CO_2$ | $\bar{u}^{\ast}$, $\vartheta$ | 0.7 ± 0.07 | 0.52 | 160° |
| He-$Na_2$ | $\bar{j}'$, $\vartheta$ | 1.1 ± 0.3 | 0.09 | >100° |
| Ne-$Na_2$ | $\bar{j}'$, $\vartheta$ | 0.85 ± 0.15 | 0.175 | ∿ 60° |
| Ar-$Na_2$ | j', $\vartheta$ | 0.93 ± 0.15 | 0.28 | ∿ 70° |
| $D_2$-CO | $u^{\ast}$, $\vartheta$ | --- | 0.087 | ∿ 90° |

Tab. I Systems for which rotational rainbow scattering has been reported.

## IV. A QUANTITATIVE APPROACH: AB INITIO BASED IOS THEORY

Progress on a more realistic description of rotational rainbows than hard shell scattering is currently been reviewed in the literature [32]. So we may be a bit sketchy on it. It concentrates on two aspects. One is to spot, validate and adapt the appropriate scattering theory approximation. The other is to combine it with a reliable potential calculation and see whether this accounts for observation.

Regarding the first aspect it is suggested by the above conditions (i) and (iii) that the infinite-order-sudden approximation (IOSA) [33] is a suitable candidate. Let the previous atom-rigid rotor diatom model system interact by a potential $V(R,\gamma)$ with R the distance of the atom from the diatom center of mass and $\gamma$ the angle between the molecular internuclear axis and R. The $0 \to j'$ scattering amplitude in IOSA is

$$f(j'; \vartheta \mid 0; R_o) = \frac{i}{2} \cdot \frac{(-1)^{j'}}{(R_j, R_o)^{\frac{1}{2}}} \sum_{\ell} (2\ell + 1) P_\ell(\cos \vartheta) T^\ell(j' \mid 0) \qquad (5)$$

with the transition matrix element

$$T^\ell(j' \mid 0) = \frac{1}{2}(2j' + 1)^{\frac{1}{2}} \int_0^\pi d\gamma \, \sin\gamma \left(1 - e^{2i\eta_\ell(\gamma)}\right) P_{j'}(\cos \gamma) \qquad (6)$$

$$R_j^2 = \frac{2\mu}{\hbar^2}\left(E - B_e j(j + 1)\right).$$

The rotational excitation problem is effectively reduced to that of elastic scattering with the phase shifts $\eta_1(\gamma)$ calculated for all, in each case fixed atom diatom orientations $\gamma$. For finite initial rotation the IOSA accounts by its "factorisation" formula [34]

$$I(j'; \vartheta \mid j; R_o) = \left(\frac{R_o}{R_j}\right)^2 \sum_{j''} C^2(j, j'', j' \mid 0,0,0) I(j'; \vartheta \mid 0; R_o) \qquad (7)$$

$$C \neq 0 \quad \text{for} \quad |j - j''| \leqslant j' \leqslant j + j'', \quad \text{Clebsch-Gordan coefficient.}$$

The IOSA was used in the first finding of the angular orientation oscillations [19, 20]. On the basis of model potentials adapted to experimental collision partners it was, then, rapidly shown to also reproduce the main features of early observations, e.g. those of Fig. 4 [35, 36]. In these calculations the j' dependence of the cross section at $\vartheta =$ const was examined to find the rotational rainbow supernumerary interference and recognize the angular orientation oscillations as the angular cuts of the supernumeraries [35]. In accord with the high action condition (iii) semiclassical arguments have been applied to the IOS scattering amplitude and have, indeed, been indispensable to the general understanding of rotational rainbows [19, 20, 35, 36]. The key finding [20] is a two-dimensional generalisation of what is known from simple isotropic potential scattering: The IOS amplitude, eqs.(5), (6), has stationary phase points at which

$$2\frac{\partial \eta_\ell(\ell,\gamma)}{\partial \ell} = \mp \vartheta, \qquad 2\frac{\partial \eta_\ell(\ell,\gamma)}{\partial \gamma} = \mp \left(j' + \frac{1}{2}\right)$$

are simultaneously satisfied; its magnitude is ruled by the determinant

$$\begin{vmatrix} 2\frac{\partial^2 \eta}{\partial \ell^2} & 2\frac{\partial^2 \eta}{\partial \ell \partial \gamma} \\ 2\frac{\partial^2 \eta}{\partial \gamma \partial \ell} & 2\frac{\partial^2 \eta}{\partial \gamma^2} \end{vmatrix} = \frac{\partial(\vartheta, j')}{\partial(\ell, \gamma)}$$

which in turn is closely related to the Jacobian appearing in the classical cross section and for the rigid shell condenses into B', eq. (3a). Korsch and Schinke worked out an Airy-uniform, semiclassical approximation to the IOSA for repulsive potentials [37]. It approximates the IOSA exceedingly well even for the least favourable conditions of Tab. I, i.e. for $Na_2$ - He. Based on this and a further separability assumption Schinke has proposed a cross section-to-potential inversion procedure applicable in the strong and repulsive coupling limit [38]. The IOSA and

and its semiclassical version have also been used to compare the scattering of an ab initio Na$_2$ - He potential to that of an adapted hard shell [39]. This clarifies the quantitative merits of the latter which should not be overestimated [40].

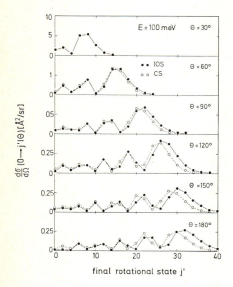

final rotational state j'

Fig. 7   IOSA and CS cross sections of Na$_2$ - Ne.

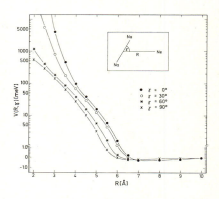

Fig. 8   Ab initio CI potential of Na$_2$ - Ne.

Some of the results mentioned in the last paragraph are illustrated in Fig. 7 by later work of Schinke, Müller and Meyer on Na$_2$ - Ne [8d]. The full dots represent 0 → j' cross sections generated in IOSA by an ab initio CI potential. It is seen in Fig. 8 to be considerably anisotropic, but of negligible attraction. Recalling cuts $\vartheta$ = const of Fig. 3 (transformed to j' instead of v') hard shell as well as semiclassical, Airy-type features are clearly apparent in the quantum cross sections of Fig. 7: The single classical rainbow singularity of the inversion symmetric potential is turned into a finite, main maximum which *gradually* decays into the "dark", classically forbidden j' range; on its "bright" side it resolves into the supernumerary interference oscillating about an average which is flat as j' → 0; as the angle increases, the (average) cross section falls off in an expanding, allowed j' range. Obviously, an angular cut at some fixed j' will again produce an oscillatory pattern, the orientation oscillations, which, expectedly, form a forward halo of angular width increasing with j'. The weighted sum in eq.(7) of the (2j+1) finite 0 → j" cross sections of a j" dependence as in Fig. 7 qualitatively explains the IOSA statement on the effect of finite, initial rotation: The main rainbow maximum is not severely distorted; its j' position is nearly independent of j, but the supernumerary oscillation is strongly quenched [8c, d, 36]. This is consistent with the observations, particularly with those of Fig. 6.

The IOSA violates both angular momentum and energy conservation. The open circles in Fig. 7 show the result of a (much more laborious) coupled states (CS) calculation which does not suffer from the latter defect. It is seen that, because of its "energy sudden" assumption, the IOSA increasingly overestimates large transfer. In Na$_2$ - Ne the energy transferred at the main maximum and $\vartheta$ = π amounts to about 20% of collision energy. Such and larger fractional energy transfer is common to most systems of Tab. I. Corrections [35, 42] and some quantitative estimate of its re-

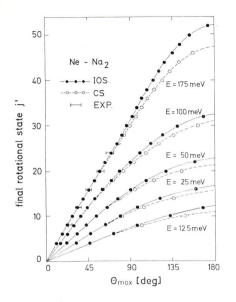

Fig. 9 Angle dependence of the allowed j' range for scattering to occur in $Na_2$-Ne. Experimental results are shown as error bars for E=0.175 eV.

liability with respect to the other defect would appear to be necessary in order to generally exploit the sensitivity of the described effect.

Finally, turning to comparison with experiment the feature most clearly observed is the width of the allowed j' interval and its dependence on angle. This is shown in Fig. 9 for several collision energies by plotting j' as a function of the angle $\vartheta_{max}$ at which the main maximum of the $0 \rightarrow j'$ cross section appears. E = 0.175 eV is the energy of Bergmann's experiments. His results, corrected for finite initial rotation, are entered as error bars. Considering that there is no adjustable parameter in the figure the agreement exceeds expectation. Details of reasonable agreement have also been obtained for He-$Na_2$ [8c].

## V. SUMMARY AND ACKNOWLEDGEMENT.

The finding of rotational rainbows has uncovered some simple and general elements of strong-coupling, rotational excitation in atom-diatom systems. Classical, rigid shell scattering conveys their essence. The infinite-order-sudden approximation promises to provide a tractable, quantitative account. A more complete and general, quantitative estimate of its error is needed. Combined with a reliable evaluation procedure the measurement of rotational rainbows would appear to be the royal road to vaguely known, (repulsive) molecular anisotropies. To be efficient the relation of properties of the potential to those of the scattering will require more attention. Attractive potential influence [43] and systems of more complicated nuclear geometry merit further study.

Thanks are due to K.Bergmann, U.Buck, H.J.Korsch, D.E.Pritchard, and R.Schinke who have made their work available prior to publication. On the own part the working company of U.Ross and W.Schepper has been indispensable, exciting and cheerful.

## VI. REFERENCES AND FOOTNOTES.

[1] Curtiss,C.F. and Adler,F.T., J.Chem.Phys. 20 (1952) 2045 and many following papers, in particular loc.cit. 49 (1968) 1952.
[2] Takayanagi,K., Prog.Theor.Phys. 11 (1954) 557; Prog.Theor.Phys.(Kyoto)Suppl. 25 (1963) 40; Adv.At.Mol.Phys. 1 (1965) 149; review.
[3] Arthurs,A.M. and Dalgarno,A., Proc.Roy.Soc. London A 256 (1960) 540.
[4] Blythe,A.R., Grosser,A.E. and Bernstein,R.B., J.Chem.Phys. 41 (1964) 1917.
[5] Toennies,J.P., Z.Physik 182 (1965) 257.
[6] Pack,R.T., J.Chem.Phys. 60 (1974) 633.
[7] McGuire,P. and Kouri,D.J., J.Chem.Phys. 60 (1974) 2488.
[8] For some recent examples see
    (a) Flower,D.R., Launay,J.M., Kochanski,E. and Prisette,J., J.Chem.Phys. 37 (1979) 355;
    (b) Habitz,P., Chem.Phys. 54 (1980) 131;

(c) Schinke,R., Müller,W.,      Meyer,W. and McGuire,P., J.Chem.Phys.74 (1981) 3916;

(d) Schinke,R., Müller,W. and Meyer,W., J.Chem.Phys.Dec.1981.

[9]   Faubel,M. and Toennies,J.P., Adv.At.Mol.Phys. 13 (1977), 226; review; translational energy loss method, mainly on ions.

[10]  (a) Gentry,W.R. in Oda,N. and Takayanagi,K. (eds.) Electronic and Atomic Collisions (North-Holland, Amsterdam 1980);

      (b) Andres,J., Buck,U., Huisken,F., Schleusener,J. and Torello,F., J.Chem. Phys. 73 (1980) 5620;
      translation energy loss method on hydrogenic neutrals.

[11]  Dagdigian,P.J. and Wilcomb,B.E. J.Chem.Phys. 72 (1980) 6462; laser method.

[12]  (a) Bergmann,K., Hefter,U. and Witt,J., J.Chem.Phys. 72 (1980) 4777;

      (b) Bergmann,K., Hefter,U., Mattheus,A. and Witt,J., Chem.Phys.Lett. 78 (1981) 61;

      (c) Hefter,U., Jones,P.L., Mattheus,A., Witt.J. and Bergmann,K., Phys.Rev. Lett. 49 (1981) 915; laser method.

[13]  (a) Serri,J.A., Morales,A., Moskowitz,E., Pritchard,D.E., Becker,C.H. and Kinsey,J.L., J.Chem.Phys. 72 (1980) 6304;

      (b) Serri,J.A., Becker,C.H., Elbel,M.B., Kinsey,J.L., Moskowitz,W.P. and Pritchard,D.E., J.Chem.Phys. 74 (1981) 5116; laser method.

[14]  Faubel,M., Kohl,K.H. and Toennies,J.P., J.Chem.Phys. 73 (1980) 2506; translation energy loss method on non-hydrogenic neutrals.

[15]  Thomas,L.D., J Chem.Phys. 67 (1977) 5224.

[16]  Other terms such as bulge effect and halo scattering have also been used.

[17]  Miller,W.H., Adv.Chem.Phys. 25 (1974) 69.

[18]  Gentry,W.R., J.Chem.Phys. 60 (1974) 2547.

[19]  Schinke,R., Chem.Phys. 34 (1978) 65.

[20]  Bowman,J.M., Chem.Phys.Lett. 62 (1979) 309.

[21]  Schepper,W., Ross,U. and Beck,D., Z.Physik A 290 (1979) 131.

[22]  Thomas,L.D., J.Chem.Phys. 73 (1980) 5905;
      Bowman,J.L. and Lee,K.T., J.Chem.Phys. 74 (1981) 2664.

[23]  Beck,D., Ross,U. and Schepper,W., Z.Physik A 293 (1979) 107.

[24]  Bosanac,S., Phys.Rev.A 22 (1980) 2617.

[25]  Beck,D., Ross,U. and Schepper,W., Z.Physik A 299 (1981) 97.

[26]  Korsch,H.J. and Richards,D., J.Phys. B 14 (1981) 1973.

[27]  Beck,D., Ross,U. and Schepper,W. Phys.Rev. A 19 (1979) 2173.

[28]  Müller,W. and Schinke,R., J.Chem.Phys. 75 (1981) 1219.

[29]  Ross,U., Schepper,W. and Beck,D., to be published; results on $K-O_2$, $CO_2$.

[30]  Andres,J. Buck,U., Meyer,H. and Launay,J.M., J.Chem.Phys. submitted.

[31]  Eastes,W., Ross,U., and Toennies,J.P., Chem.Phys. 39 (1979) 407.

[32]  Schinke,R. and Bowman,J.M. in: Bowmann,J.M.(ed.), Molecular Collision Dynamics, Topics in Current Physics (Springer, to be published).

[33]  Secrest,D., J.Chem.Phys. 62 (1975) 710;
      Hunter,L.W., J.Chem.Phys. 62 (1975) 2855.

[34]  Goldflam,R., Green,S. and Kouri,D.J., J.Chem.Phys. 67 (1977) 4149;
      Goldflam,R., Kouri,D.J. and Green,S., J.Chem.Phys. 67 (1977) 5661.

[35]  Schinke,R. and McGuire,P., J.Chem.Phys. 71 (1979) 4201.

[36]  Schinke,R., J Chem.Phys. 72 (1980) 1120.

[37]  Korsch,H.J. and Schinke,R., J.Chem.Phys. 73 (1980) 1222.

[38]  Schinke,R., J.Chem.Phys. 73 (1980) 6117.

[39]  Korsch,H.J. and Schinke,R., J.Chem.Phys. 1981.

[40]  Alexander,M.H. and Dagdigian,P.J., J.Chem.Phys. 73 (1980) 1233.

[41]  Fig.8 provides a first, preliminary check for $Na_2-Ne$. At the observation conditions the equipotential of closest approach is $V\sim 0.04$ eV. Its "effective, ellipsoidal" anisotrpy  is read off Fig. 8 to be 0.7±0.05 Å , within error limits agreeing with the (c-a) of the table!

[42]  Korsch,H.J. and Schinke,R., J.Chem.Phys. 65 (1976) 1462.

[43]  Bentley,J., J.Chem.Phys.73 (1980) 4708.

PHYSICS OF ELECTRONIC AND ATOMIC COLLISIONS
S. Datz (editor)
© North-Holland Publishing Company, 1982

# LOW ENERGY ION MOLECULE COLLISIONS

Kazuo Takayanagi

Institute of Space and Astronautical Science
Komaba 4-6-1, Meguro-ku
Tokyo 153   JAPAN

In low-energy ion-molecule collisions, the long-range orientation-dependent forces, such as the ion-dipole and ion-
quadrupole interactions, often play the principal role, not
only in the rotational excitations of the molecule, but also
in the determination of the transport cross sections.   The
Langevin cross section, which provides us with an upper
bound to the ion-molecule reaction cross section at low
energies, is also modified considerably by these orientation-
dependent forces.   In this report, previous theoretical
works are briefly reviewed and the recently proposed PRS
(Perturbed Rotational State) approach is described to some
details.   The PRS approach has been able to handle both the
rotational transitions and the transport and reaction cross
sections, as long as the long-range interactions are of the
primary importance.

## INTRODUCTION AND HISTORICAL BACKGROUND

A neutral molecule usually has a dipole moment or a quadrupole moment  or both, and
also the anisotropy in the polarizability.   Therefore, the ion-neutral molecule
system is characterized by the long-range orientation-dependent forces.   In the
low-velocity ion-molecule collisions, these forces induce (i) rotational transition
in molecule and (ii) distortion of the relative motion.   The transport cross sections, such as the momentum-transfer cross section, will be affected by these interactions.   The Langevin cross section, which provides us with an upper bound to the
ion-molecule reaction cross section at low velocities, is also modified by these
orientation-dependent forces.

During the past decade or so, an enormous number of papers on inelastic molecular
collisions have been published in scientific journals.   In most of them, however,
the long-range orientation-dependent forces play only a minor role.   This is because the short-range forces are more important in rotational transitions in $H_2$,
the most frequently studied molecule, and in vibrational transitions in many molecules.   However, in low-temperature gases, such as the cold interstellar molecular
clouds, the long-range ion-molecule interactions cannot be neglected.   Here,  the
rotational excitation is the major inelastic process and the ion-molecule reaction
is the most important type of chemical reactions in gas phase.   These collision
processes are affected considerably by the long-range interactions.

Most of the interstellar molecules are detected by radio telescopes in rotational
emission lines.   Therefore, it is interesting to study collisional excitation of
rotation in low temperature gases ($T < 100$ K).   It is expected that polar molecules
are excited efficiently by ion collisions.   Thus, rotational excitations in ion-
polar molecule collisions have been the subject of theoretical studies by many
authors.   Some of the representative approximation methods so far applied include
IP (Impact-Parameter)-Born [1][2], IP-Unitarized Born [3][4], IP-Sudden [4-6],

IP-CC(Close-Coupling) [4][7], Wavemechanical CC[8], and various decoupling approximation methods [9].    Because of the astrophysical interest, almost all the calculations in the sub-eV region were done for one particular system, namely

$$H^+ + CN(j=0) \rightarrow H^+ + CN(j=1).\tag{1}$$

These theoretical calculations are based on the expansion of the rotational wave
function in terms of the spherical harmonics, i.e., the eigenfunctions of the free
rotation of the linear molecule.    In the low-velocity collisions, however, the
molecular rotation adjusts itself almost adiabatically, particularly when the molecule is under the long-range orientation-dependent forces.    It is likely, therefore, that the use of the adiabatic basis functions will give a more appropriate
approach to the problem under consideration.    The present author introduced the
PRS (Perturbed Rotational State) approach at a US-Japan Seminar on Molecular Energy
Transfer (March 22-24, 1976 in Tokyo).    Later we have developed this approach and
applied it to the ion-polar molecule collisions [10-14], to the ion-nonpolar molecule collisions with a large quadrupole interaction [15], and also to the collisions between two polar molecules [16].    Further discussions on the PRS approach
will be given in the following sections of this report.

A chemical reaction between neutral molecules usually requires a large amount of
activation energy, so that its rate constant decreases rapidly when the gas temperature decreases.    For exothermic ion-neutral reactions, however , the rate constant
usually remains finite down to extremely low temperatures.    Chemistry of the iono-
spheric D region ($T \sim 200K$) and the interstellar clouds already mentioned is,
therefore, determined predominantly by ion-neutral reactions.    A small number of
ions in these gases play important roles as catalyzers.

The large rate of ion-neutral reactions is usually explained by the Langevin theory
where the polarization force plays the major role.    The polarization force is the
interaction between the electric dipole induced on the neutral molecule and the ion
charge.    For a large intermolecular separation R, the isotropic part of the interaction potential has the well-known form

$$V \sim - \alpha(Ze)^2/2R^4,\tag{2}$$

where $\alpha$ is the polarizability of the molecule and Ze the charge of the ion.    When
the centrifugal potential $E(b/R)^2$, where E is the collision energy and b the impact
parameter, is added to (2), the resulting effective potential has a barrier.    The
barrier height becomes equal to E for a critical impact parameter $b = b_c$.    If the
impact parameter is larger than $b_c$, the barrier is higher than E and the colliding
pair will have no chance to come to a close distance except due to the tunnel
effect which has a very small probability.    For $b < b_c$, on the other hand, the
colliding particles pass over the barrier and they are accelerated and hit together.
We call this violent collision a "hitting" process, although some people call it
a "capture" or "orbiting" collision.    [The term "capture" is not appropriate
unless a collision complex is definitely formed.]    Very often the exothermic reaction cross sections inferred from laboratory experiments are close to the Langevin
cross section which is given by

$$\sigma_L = \pi b_c^2 = \pi(2\alpha Z^2 e^2/E)^{1/2}\tag{3}$$

That is, in many systems, a violent collision leads to a reaction with a probability
comparable with unity.    It is noted that $v \cdot \sigma_L$, v being the collision velocity, is
independent of v, so that the Langevin rate constant is temperature independent.
When the neutral molecule has a large dipole moment, however, observed rate constant sometimes exceeds the Langevin value $v \cdot \sigma_L$.    Therefore, it is important to
study the influence of the dipole interaction on the ion-molecule collision processes.

Although there are various methods [1-9] to study the rotational excitations in molecular collisions, none of these methods (except the PRS approach [12]) have been applied to study the modification of the Langevin theory. The latter problem has been studied by other groups of people in entirely different fashions. Among the early studies, classical trajectory calculations by Dugan and his colleagues [17][18] are important contributions in this field. This type of numerical work, however, requires a considerable time on computation. Furthermore, it is rather difficult to derive conclusions of wide applicability out of these calculations. More recently, various average potential approaches have been proposed. The ADO (Average Dipole Orientation) and AQO (Average Quadrupole Orientation) models have been introduced by Su and Bowers [19-22]. Thermodynamic average potential approach has been used by Barker and Ridge [23] and modified later by Turulski and Forys [24] and by Celli, Weddle and Ridge [25].

In the ADO theory, the ion-dipole interaction is averaged over the molecular orientation with a classically-determined probability. An effective spherically-symmetric potential is thus obtained and used to study the relative motion of the colliding pair. The theory treats the rotational motion of the molecule "in an average sense" and "it is not intended to represent microscopic phenomena in any fashion" [22]. The AQO theory is the same as the ADO theory except the dipole interaction is replaced with the quadrupole interaction [21]. Barker and Ridge [23] took average of the ion-dipole interaction over the Boltzmann distribution and used the resulting temperature-dependent potential to study the orbiting. Turulski and Forys [24] and Celli et al [25] proposed to use the average free energy instead of the average energy for the effective potential. In the low pressure gas, there is little chance for the colliding pair to exchange energy with the heat reservoir (other molecules in the gas) during collision, so that the validity of the use of a thermodynamic average potential is doubtful. Furthermore, it is not certain whether the cross section of a particular kind calculated for an average potential should be close to the average of cross sections calculated for different potentials.

Still another approach, which is basically a kind of TST (Transition State Theory) has been proposed. The TST was originally designed [26] for neutral-neutral reactions which are characterized by the presence of the activation energy. Here the reacting system passes through a well-defined point (transition state) on a potential hypersurface in the configuration space. In ion-molecule reactions, however, there is no activation energy. Thus some modification of the TST is necessary as Bates [27] has pointed out. One has to calculate the flux of the representative particle passing across a surface dividing the reactants and products in the configuration space. Very recently, Chesnavich, Su and Bowers [28] have introduced a variational method to estimate an upper bound to the rate of surface crossing.

Although these theories give results in more or less good agreement with experiment at room temperature, these theories are all based on the classical mechanics and they do not take account of the discreteness of the rotational energy levels, which should be important in the low-temperature gases.

Dugan and Magee [17] and Hyatt and Stanton [29] took account of the discreteness of the rotational levels and applied the theory of second-order Stark effect to obtain the adiabatic potential in the ion-polar molecule collisions:

$$\varepsilon_{jm}(R) \sim j(j+1)B + \left(\frac{DZe}{R^2}\right)^2 \frac{j(j+1) - 3m^2}{2j(j+1)(2j-1)(2j+3)B}, \qquad (4)$$

where D is the dipole moment and B the rotational constant (in energy unit) of the molecule, and j, m are the quantum numbers for the rotational angular momentum and its projection on the intermolecular distance vector $\vec{R}$. Possibility of transitions between the adiabatic curves has been neglected in these works. The second-order Stark effect (4) was used also by Bottcher [7] in his discussions of orbiting collisions.

OUTLINE OF THE PRS THEORY

In order to avoid unnecessary complications in the equations, we shall assume that
the interaction is the pure dipole interaction and the molecule is linear and
electronically in a $^1\Sigma$ state.    Extension to the other cases is straightforward.
The PRS functions and the potential curves are determined by solving the eigenvalue
problem for each fixed value of R.

$$[ H_{rot} + V ] \chi = \varepsilon(R) \chi, \tag{5}$$

$$V = \frac{DZe}{R^2} \cos\theta', \tag{6}$$

where $H_{rot}$ is the Hamiltonian for the free rotation of the molecule of which the
eigenvalue and eigenfunction are, respectively, $j(j+1)B$ and $Y_{jm}(\theta', \phi')$.    The
quantization axis has been taken along $\vec{R}$ rather than the space-fixed z-axis.

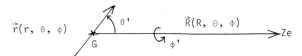

$\vec{r}(r, \theta, \phi)$     $\theta'$     $\vec{R}(R, \Theta, \Phi)$ $\longrightarrow Ze$
                    G      $\phi'$

Fig. 1    Coordinate system.   G is the center-of-mass
          the molecule.

To solve (5), $\chi$ is expanded as

$$\chi = \sum_\ell X_{\ell m}(R) Y_{\ell m}(\theta', \phi'). \tag{7}$$

Since the interaction (6) does not depend on $\phi'$, there is no m-mixing.    The coefficients X's satisfy

$$[ \ell(\ell+1) - u ] X_{\ell m} + \frac{1}{x^2} \sum_{\ell'} < \ell m |\cos \theta'| \ell'm > X_{\ell'm} = 0, \tag{8}$$

where the reduced (dimentionless) quantities

$$\varepsilon / B = u, \qquad R / (DZe/B)^{1/2} = x \tag{9}$$

have been introduced.    The eigenvalue u's are determined by solving the secular
equation as usual.    The asymptotic form of $\varepsilon(R)$ is, of course, identical with (4).
It is noted that $\varepsilon_{j,m} = \varepsilon_{j,-m}$.    The equation (8) does not explicitly depend on
D and B.    Therefore, the solution of (8) is universal.    Furthermore, we can
assume without losing generality that all the solutions X's are real.    The PRS
$\chi$'s are specified by their asymptotic forms:

$$\chi_{jm}(\theta', \phi'; R) \xrightarrow{R \to \infty} Y_{jm}(\theta', \phi'). \tag{10}$$

The rotational wave function $\psi$, which is a solution of

$$i\hbar \frac{\partial\psi}{\partial t} = [ H_{rot} + \frac{DZe}{R^2} \cos \theta' ] \psi, \tag{11}$$

is now expanded as

$$\psi = \sum_{jm} C_{jm}^{j_o m_o}(\tau) \chi_{jm}(\theta', \phi'; R(\tau)) \exp[-i\int_{-\infty}^{\tau} u_{jm}(\tau')d\tau'], \tag{12}$$

where the reduced time $Bt/\hbar = \tau$ has been introduced. $j_0m_0$ specify the initial rotational state. We determine the coefficients C's by solving a set of coupled differential equations, which are obtained by substituting (12) into (11). The coupling matrix elements are of the following form:

$$<jm|d/dt|j'm'> = \frac{d\vec{R}}{dt} \cdot \int \chi_{jm}^* \nabla_R \chi_{j'm'} \sin\theta' \, d\theta' \, d\phi'. \tag{13}$$

For distant collisions, we can assume that the relative motion is a rectilinear motion with a constant velocity. Then the reduced distance x is related to the reduced time $\tau$ by

$$x^2 = p^2 + \beta^2 \tau^2, \tag{14}$$

where

$$p = b/(DZe/B)^{1/2} \quad \text{and} \quad \beta = \hbar v/(DZeB)^{1/2} \tag{15}$$

are the reduced impact parameter and the reduced velocity, respectively. It can be shown easily that the probability P of any particular transition is a function of p and $\beta$ only. The excitation (or de-excitation) cross section will be given by

$$\sigma = 2\pi \int_0^\infty P(p, \beta) \, b \, db = 2\pi \frac{DZe}{B} \int_0^\infty P(p, \beta) \, p \, dp. \tag{16}$$

Thus the reduced cross section $\sigma/(DZe/B)$ is a function of $\beta$ only. This scaling law holds as far as the distant encounters dominate in the excitation in the ion-polar molecule collision.

When the impact parameter is not sufficiently large, it is desirable to take account of the trajectory bending. Sakimoto [11] has proposed to use the follow-ing energy and angular-momentum relations:

$$\frac{M}{2}(dR/dt)^2 + \frac{MR^2}{2}(\omega_{x'}^2 + \omega_{y'}^2) + <\psi(t)|H_{rot} + V|\psi(t)>$$

$$= \text{the total energy of the system}, \tag{17a}$$

$$MR^2\omega_{x'} + <\psi(t)|j_{x'}|\psi(t)> = J_{x'}, \tag{17b}$$

$$MR^2\omega_{y'} + <\psi(t)|j_{y'}|\psi(t)> = J_{y'}, \tag{17c}$$

where $J_{x'}$ and $J_{y'}$ are the x' and y' components of the total angular momentum $\vec{J}$ of the system, while $j_{x'}$ and $j_{y'}$ are the molecular angular momentum operators. The coordinates (x', y', z') are the rotating frame with z' always along the vector $\vec{R}$. $\omega$ denotes the angular velocity of the rotating frame relative to the space-fixed frame. The components $J_x$, $J_y$, and $J_z$ of $\vec{J}$ in the space-fixed frame are individually conserved in the classical approximation. The components $J_{x'}$ and $J_{y'}$ in (17) are related to $J_x$, etc. by

$$J_{x'} = -J_z \sin\Theta + (J_x \cos\Phi + J_y \sin\Phi)\cos\Theta, \tag{18a}$$

$$J_{y'} = J_y \cos\Phi - J_x \sin\Phi. \tag{18b}$$

Here, $\Theta, \Phi$ are the polar angles of $\vec{R}$ in the space-fixed frame where the z-axis is chosen along the direction of the ion incidence. If the wave function $\psi(t)$ is given at each time t, we can determine the relative motion by means of (17), (18). In this way, the classical trajectory can be determined simultaneously with the rotational state of the molecule. Application of this prescription to the ion-molecule reactions and to the ion transport in molecular gases will be discussed in a later section.

It is straightforward to include, in addition to the dipole interaction, the quadrupole and the polarization interactions in the theory. The interaction V in (6) is then replaced by

$$V = -\frac{\alpha(Ze)^2}{2R^4} + \frac{DZe}{R^2}\cos\theta' + \left(\frac{QZe}{R^3} - \frac{\alpha'(Ze)^2}{2R^4}\right) P_2(\cos\theta'), \qquad (19)$$

where Q is the quadrupole moment, $\alpha$ and $\alpha'$ are the spherical and anisotropic parts of the polarizability of the molecule and $P_2$ is the Legendre polynomial. For non-polar linear molecule in a $^1\Sigma$ state, the asymptotic form of the adiabatic potential curves can be obtained easily by the standard perturbation theory as

$$\varepsilon_{jm} = j(j+1)B - \frac{\alpha(Ze)^2}{2R^4} + \left(\frac{QZe}{R^3} - \frac{\alpha'(Ze)^2}{2R^4}\right)\frac{j(j+1) - 3m^2}{(2j-1)(2j+3)}$$

$$+ \text{ terms with } \left(\frac{QZe}{R^3} - \frac{\alpha'(Ze)^2}{2R^4}\right)^2 + \cdots \qquad (20)$$

Depending on the sign of QZ, the quadrupole and the anisotropic polarization terms are additive or subtractive, so that the resulting potential curves are different. Even when the polarization interaction can be neglected, the terms of odd power in QZ and the even power terms are mutually additive or subtractive according to the sign of the product QZ. For pure dipole interaction, the dipole moment D appears in (4) and in the higher-order terms only in the even powers, so that D → - D does not result in any change in the potential curves.

For the pure quadrupole interaction, the adiabatic potential $\varepsilon_{jm}(R)$ and the PRS functions are represented by the universal functions, except for scaling factors, if we introduce the following reduced (dimensionless) quantities: (Z = +1 is assumed for simplicity)

$$x = R / (|Q|e/B)^{1/3}, \quad u(x) = \varepsilon(R)/B. \qquad (21)$$

So far, we have assumed that the molecule is linear. If the molecule is a symmetric top, the linear Stark effect comes in. The rotational state of the molecule is now specified by the three quantum numbers J, K, and M. [The J here is not the same as the previously used notation for the angular momentum of the whole system.] When K = 0, the molecule behaves as a linear molecule with j = J, m = M. When K ≠ 0, the dipole moment is not smeared out by the rotation, so that there is a strong orientation-dependent interaction between the molecule and the ion. Because of this linear Stark effect, the adiabatic potential curves for K ≠ 0 are quite different from those for K = 0. Namely, when K ≠ 0, the degeneracy of the states with M and -M is removed, and the splitting of the levels with different values of M becomes appreciable at distances much larger than when K = 0. The quantum number K represents the rotation of the molecule around its symmetry axis. Since the dipole moment of the molecule is along the symmetry axis, the dipole interaction alone cannot change the value of K in collision. A study by Sakimoto of low-energy ion collisions with asymmetric rotors is now in progress.

The PRS approach so far described is based on the semiclassical (impact parameter) approximation. There is some ambiguity in determining the classical trajectory for the relative motion in the semiclassical theories. Although a fully quantum-mechanical PRS theory has been proposed [30], it has been applied to only a limited type of scattering problems. It is desirable to extend and improve the theory further.

ROTATIONAL EXCITATIONS

Fig. 2 illustrates the rotational exciation probability in the ion-linear polar molecule collision as a function of the reduced impact parameter p. The curves represent either $P(j=0 \rightarrow 1)$ or $P(0 \rightarrow 2)$ multiplied by p, so that the area under each curve is proportional to the excitation cross section (16) for the process. Since the p × P curves will never exceed the straight line for p × 1.0, we can estimate the upper limit to the contribution from the innermost region where, because of the slow convergence of the calculation, we often stop calculating the transition probability. For $H^+ + CN$, the reduced velocity β = 1.133, 0.654, 0.463 correspond to the collision energy of 0.12, 0.04 and 0.02 eV, respectively. The solid lines are the results of straight-path calculations, while the small circles for β = 0.654 represent the results of the bent-trajectory calculation based on (17), (18). In the latter calculations, the reduced mass used is that of $H^+ + CN$ and the spherical part of the polarization force has been included in

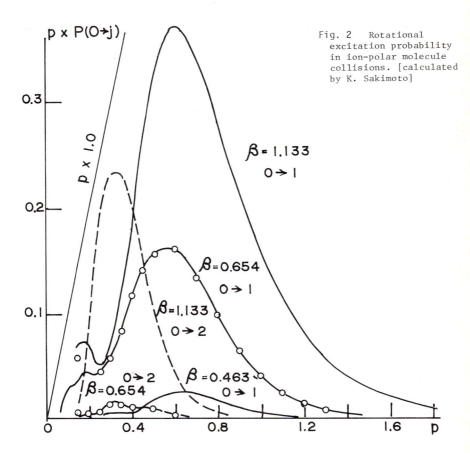

Fig. 2    Rotational excitation probability in ion–polar molecule collisions. [calculated by K. Sakimoto]

determining the trajectory.   The two calculations agree very well except in the
innermost region.   It is noted that one unit of the reduced distance $\sqrt{(De/B)}$   in
this system is 136 Å.   The main contribution to the excitation cross section comes
from the region of p > 0.2, or b > 27 Å.   Because of the large impact parameter,
the trajectory does not deviate much from the straight line.   For the same reason,
the pure dipole interaction is the only important part of the interaction in this
system.   It is noted further that the straight-path results are universal.
Namely, the main part of the curves in Fig. 2 can be used for any other ion-linear
polar molecule collisions, provided that one unit of the reduced distance (=
$\sqrt{(De/B)}$ ) is not too small.   In Table 1, the value of  $\sqrt{(De/B)}$ is listed for some
molecules.   The unit distance represents the intermolecular separation where the
magnitude of the dipole interaction $De/R^2$ becomes equal to B.   The constant B
gives a measure of the rotational level spacings.   If the interaction energy is
much smaller than the level spacings, the molecular rotation will not be perturbed
appreciably by the collision partner.   Thus, roughly speaking, $R = \sqrt{(De/B)}$ is the
effective range of the dipole interaction.

DePristo and Alexander [8] have applied the
fully wavemechanical CC method to the rotational excitation in $H^+$ + CN collisions.
The total angular momentum quantum number L
in such a calculation may be transformed into the classical impact parameter by the relation b = (L + 1/2)/k, where k is the wave
number of the relative motion.   Using this
relation, the direct comparison between their
calculation and the PRS results becomes
possible.   It has been found that the PRS
results agree very well with the more laborious wavemechanical CC calculations [10],
[11].   The PRS results have been compared
favorably also with the results of the IP-CC
calculations where a sufficient number of the
basis functions are included to assure the
convergence of the result [14].

Table 1

| molecule | $\sqrt{(De/B)}$ | β *) |
|----------|-----------------|------|
| CsF  | 1018 Å | 0.68 |
| SiO  | 322    | 0.55 |
| CO   | 37.5   | 1.8  |
| HF   | 46     | 0.13 |
| HCl  | 51     | 0.24 |
| CN   | 136    | 0.50 |
| HCN  | 219    | 0.40 |
| OCS  | 292    | 2.1  |
| $N_2O$ | 96   | 3.5  |

*) at 0.03 eV. Reduced mass of the
system is assumed to be 1 amu.

The effective range of the dipole interaction in the rotational transition is
$\sqrt{(De/B)}$ as mentioned earlier, while the level spacings are of the order of B.
Therefore, the Massey's adiabaticity condition for the collision velocity v is

$$\sqrt{(De/B)} \cdot B \; / \; \hbar v \gg 1. \qquad\qquad (22)$$

It is easy to see that this condition is nothing but  β << 1.   Namely, the
rotational transition probabilities will be very small  when β  becomes  much
less than unity.   That this is the case is clearly seen in Fig. 3, where the cross
sections  $\sigma(0 \to 1)$ and  $\sigma(0 \to 2)$ are shown as functions of  β.   Table 1 shows
the value of β at 0.03 eV.   When the reduced mass of the colliding pair is 1 amu,
β is larger than unity in some cases, but usually less than unity.   If the collision partner is a heavy ion, β will be much less, so that the near-adiabatic condition will hold.

The IP-CC calculations by Jamieson, Kalaghan and Dalgarno [4] and the wavemechanical CC calculations by DePristo and Alexander [8] have given the excitation cross
section $\sigma(0 \to 1)$ which, when plotted as a  function of  β, has a maximum in the
region of β = 2 ∿ 3, beyond which the cross section decreases [14].

For symmetric top molecules, the cross section  $\sigma(J,K,M \to J'K,M')$ considerably
depends on M and M'.   Sakimoto [14] has found, however,

$$(1/3) \sum_{MM'} \sigma(1,1,M \to 2,1,M') \simeq (1/3) \sum_{MM'} \sigma(1,0,M \to 2,0,M'). \qquad (23)$$

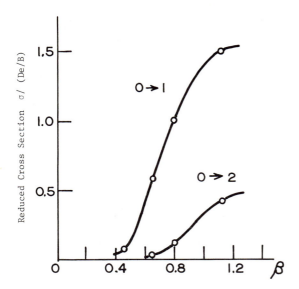

Fig.3  Rotational excitation cross sections
for ion-polar molecule collisions. Open
circles are the results of the bent-trajectory
calculation, while the solid lines were obtained for straight path [11].

Namely, the average cross
section does not depend
much on the K quantum number at least in this calculation.

For quadrupolar molecules,
the calculations are more
or less the same, except
that the selection rule is
now $\Delta j = 0, \pm 2, \pm 4, \ldots$ as
long as the quadrupole plus
polarization interactions
alone are included in the
calculation.    Curves for
$p \times P(0 \to 2)$ vs $p$ similar
to Fig. 2 for ion-dipole
collisions have been ob-
tained for $CO_2{}^+ + CO_2$
collision [15].    The unit
length in this problem is

$$(|Q|e/B)^{1/3}$$

(cf. (21) ).    In Table 2,
one unit of the reduced
distance is listed for
some quadrupolar molecules.
Also shown are the values
of the reduced velocity
for the pure quadrupole
interaction

$$\beta' = \hbar v / (|Q|eB^2)^{1/3} \quad (24)$$

at 0.03 eV.    According to the reference 15, the main contribution to the excitation cross section comes from the outer region of $p > 0.2$.    For $CO_2$, this corresponds to the impact parameter $b > 13$ Å.    Therefore, the asymptotic form of the
interaction is sufficiently accurate as the intermolecular potential and the assumption of the straight path  is permissible.

For $N_2$, these assumptions are still not so
bad, but for $H_2$, all the important contribution to the cross section comes from
close encounters, so that both the short-
range interactions and the trajectory
bending are expected to be important.

APPLICATIONS OF THE PRS THEORY TO CHEMICAL
REACTIONS AND TRANSPORT PHENOMENA

Sakimoto [11] has proposed a prescription
to determine the classical trajectories by
(17) and (18).    Using his approach, we
can calculate the hitting cross section,
i.e., an upper bound to the reaction cross
section.    The method also gives us the
differential cross section in the collision
process, from which we can calculate the
momentum-transfer and other transport cross sections.

Table 2

| molecule | $(|Q|e/B)^{1/3}$ | $\beta'$ *) |
|---|---|---|
| $H_2$ | 6.4 Å | 0.33 |
| $N_2$ | 26.4 | 2.4 |
| $O_2$ | 18.7 | 4.7 |
| CO **) | 31.6 | 2.1 |
| $CO_2$ | 64.3 | 5.1 |
| $C_2H_2$ | 52.9 | 2.1 |

*) at E = 0.03eV.  Reduced mass of
the system is assumed to be 1 amu.
**) weakly polar.

For each given value of the collision energy, we find out the critical impact parameter $b_c$, such that for $b > b_c$ the colliding particles are prevented by the barrier of the effective potential from close encounter, while for $b < b_c$ the colliding pair reach to a small, specified distance. The hitting cross section, which is an extension of the Langevin cross section to the polar or the quadrupolar interaction, is then given by $\pi b_c^2$.

For a given value of impact parameter b and the initial rotational state $j_0 m_0$, the transition amplitude C's (cf. (12)) and the scattering angle $\Theta_{sc}$ in the nonrotating, center-of-mass frame are determined. If we neglect the interference effect between the trajectories giving the same value of $\Theta_{sc}$, the momentum-transfer cross section can be calculated by the formula

$$\sigma_m(j_0) = \frac{2\pi}{2j_0+1} \sum_{jmm_0} \int_0^\infty \left(1 - \frac{k_j}{k_{j_0}} \cos \Theta_{sc}\right) \left|C_{j\ m}^{j_0 m_0}\right|^2 b\ db. \quad (25)$$

In the reference 12, we calculated the hitting cross section $\sigma_{hit}$ for the three systems: $H^+ + NH_3$, $NH_3^+ + NH_3$ and $C^+ + HCN$. The initial rotational state of the neutral molecule was assumed either $j_0 = 0$ or $j_0 = 1$. When the reduced cross section $\sigma_{hit}/(De/B)$ was plotted against the reduced energy $E/B$, it was found that all the three systems gave almost the same results. The explanation for this finding is as follows: First of all, at the thermal collision energy of 0.03 eV, the reduced velocity $\beta$ is often less than unity (Table 1) even when the reduced mass of the colliding pair is 1 amu. If the reduced mass is larger, $\beta$ is smaller still. Therefore, the molecular rotation is almost adiabatic in many cases, so that the transition probability is small. This is for $j_0 = 0 \rightarrow 1$ transition. For transitions between the higher levels, the energy separations are larger so that the transition probabilities for these are expected to be smaller. When the intermolecular distance decreases, some of the sublevels for a given j may have crossings with sublevels for $j \pm 1$. The crossings usually take place between the levels with considerably different values of m quantum number. Since the matrix element (13) vanishes for such a combination of states, the curve crossing does not enhance the excitation probability.

It must be noted here that the sublevels belonging to the same $j_0$ are degenerate at infinite separation of the molecules. Furthermore, we are describing the molecular rotation in the rotating frame. In the laboratory frame, a molecule in one of the unperturbed rotational state will remain in the same state if the collision is really a distant one where the intermolecular interaction is negligible. In the rotating frame, this gives rise to an apparent mixing between the sublevels. This apparent transitions must be taken account for collisions of large p. However, in $\sigma_{hit}$ and $\sigma_m$ the main contribution comes from the region of much smaller p where the trajectory bending is appreciable. Here the separations of the potential curves are considerable even among the sublevels with the same j. Furthermore, the angular velocity of the relative motion is small for small p except in a narrow portion of the trajectory near the closest point, so that the rotational coupling which is responsible for the apparent transitions between the sublevels is not effective. For these reasons, we can neglect the coupling between the different potential curves, to a good approximation, when we calculate $\sigma_{hit}$ and $\sigma_m$ at thermal energies. Under this decoupling approximation, it is easy to show that for the pure dipole interaction the reduced cross sections $\sigma_{hit}/(De/B)$ and $\sigma_m/(De/B)$ are the functions of $E/B$ and p and independent of the reduced mass. This scaling law explains why we have obtained almost the same results for different systems when the reduced cross sections are plotted against the reduced energy $E/B$.

With the decoupling approximation, it can be shown that in low temperature gases (and thus for low $\beta$), the calculated cross sections depend considerably on the initial rotational state of the molecule. The smaller the rotational quantum number j, the larger the hitting cross section. Some examples are shown in Table 3. In these cases, the dipole interaction has a large influence on the cross

sections.   This is important in the interstellar clouds where most of the molecules are in the ground or low-lying states of rotation.  As j increases, the
molecule rotates more and more rapidly, so that the dipole interaction is smeared
out and the spherical polarization interaction becomes relatively more important.
Under this situation, the calculated hitting cross section approaches the Langevin
value.

Table 3.   Reaction rate constants ($10^{-9} cm^3/s$) [12][31]

| reactions | The PRS value: $<v \cdot \sigma(hit)>$ | | | | | | Other values |
| | $j = 0$ | | | $j = 1$ | | | |
| T(K)=20 | 60 | 100 | 20 | 60 | 100 | | ref. |
|---|---|---|---|---|---|---|---|
| $H^+ + OH \rightarrow OH^+ + H$ | 31 | 25 | 21 | 14 | 14 | 13 | 1.0 | [32] |
| $C^+ + CH \rightarrow$ products | 12 | 8.9 | 7.7 | 5.3 | 5.3 | 5.0 | 3.0 | [32] |
| $C^+ + SiO \rightarrow Si^+ + CO$ | 35 | 23 | 18 | 29 | 21 | 17 | 1.0 | [33] |
| $H_3^+ + CN \rightarrow HCN^+ + H_2$ | 28 | 19 | 15 | 22 | 16 | 13 | 2.0 | [32] |
| $H_3^+ + OH \rightarrow H_2O^+ + H_2$ | 19 | 15 | 13 | 8.4 | 8.3 | 7.9 | 2.0 | [32] |

Recently, Sakimoto [34] has performed more calculations of $\sigma_{hit}$ and $\sigma_m$ for the
ion-polar molecule collisions and reached some more or less general conclusions:
(i) For the individual sublevels within a few lowest states, the decoupling approximation (neglecting the transitions between the adiabatic potential curves) is
good for calculating both $\sigma_{hit}$ and $\sigma_m$.
(ii) Even for higher rotational states, the adiabatic approximation is not bad.
In many cases, the reduced velocity  β is small as compared with unity at thermal
energies, so that the inelastic process (change in j) can be neglected.   He
studied the system $Li^+ + HCl$ by neglecting inelastic channels and taking only sublevels within j = 4 state into account.   (The rotational Boltzmann distribution for
HCl has the maximum near j = 4 at room temperature.)   For larger impact parameter,
the apparent transition is appreciable, and the initial m=4 state ends up at m= -4
(no net change in the space-fixed frame).   For smaller impact parameter, however,
the state m=4 (initial state) is dominantly populated throughout collision.  Since
the impact parameters which primarily contribute to the cross sections are rather
small, we can assume that the collision is nearly adiabatic in calculating $\sigma_{hit}$
and  $\sigma_m$, reconfirming the conclusion already mentioned.
(iii) When  j>>1 but $\alpha Bj^2/D^2 <<1$, he has found that the reduced cross section
$\sigma_{hit}/(De/B)$ is a numerical factor times $1/(j + 1/2)\sqrt{(E/B)}$.   Thus

$$\sigma_{hit} \propto \frac{De}{\sqrt{(E \cdot E_{rot})}} ,$$

where $E_{rot} = B(j + 1/2)^2$ is the rotational energy of the molecule.   This relation
indicates that $\sigma_{hit}$ is independent of B.   The same conclusion has been  reached
earlier by Dugan and Magee [17] from their classical numerical calculations and by
Chesnavich et al [28] in the TST approach.   For a few lowest j values, the quantum
mechanical effects become important and the above statement is no longer applicable.
Sakimoto has also found that $(j + 1/2)\sigma_m$ is also nearly independent of j when j is
not very small.
(iv) As an example of the symmetric-top molecule, the system $H^+ + NH_3(J,K)$ has been
studied for (J,K) = (4,0) and (5,5).   These two states have almost the same rotational energy in this molecule.   At room temperature, the hitting cross sections
for these states are not much different.   However, at lower collision energies,
the difference becomes considerable.   This is very important in the interstellar
chemistry.   The hitting rate (and thus the reaction rate) for the symmetric-top
molecule with K ≠ 0 is much larger than for the linear or K = 0 molecules and its
value depends on K.   K-dependence is found also for $\sigma_m$ not only at lower temperatures but also at room temperature.
(v) The collision frequency, which is closely related to the momentum-transfer cross

section, has been calculated for the system $CH_2F^+ + CH_3F$ (symmetric-top). It is found that states with $K \neq 0$ enhance the momentum-transfer cross section. By including this effect, the theoretical collision frequency agrees with the experimental values obtained by Buttrill [35] within 10 %.

It is my pleasure to thank Mr. K. Sakimoto for providing me with numerical data from hiscalculations, for reminding me of some important references and for various discussions on many parts of the preliminary manuscript.

REFERENCES

[1] A.E.E.Rogers and A.H.Barrett, Astrophys. J. 151, 163 (1968).
[2] W.M.Goss and G.B.Field, Astrophys. J. 151, 177 (1968).
[3] K.Takayanagi and Y.Itikawa, Publ. Astron. Soc. Japan 20, 376 (1968).
[4] M.J.Jamieson, P.M.Kalaghan and A.Dalgarno, J. Phys. B(Atom. Mol. Phys.) 8, 2140 (1975).
[5] M.J.Jamieson, Chem. Phys. Letters 37, 191 (1976).
[6] A.E.DePristo, S.D.Augustin, R.Ramswamy and H.Rabitz, J. Chem. Phys. 71, 850 (1979).
[7] C.Bottcher, Chem. Phys. Letters 66, 126 (1979).
[8] A.E.DePristo and M.H.Alexander, J. Phys. B(Atom. Mol. Phys.) 9, 2713 (1976).
[9] see, for instance, A.S.Dickinson, Computer Phys. Commun. 17, 51 (1979).
[10] K.Takayanagi, J. Phys. Soc. Japan 45, 976 (1978).
[11] K.Sakimoto, J. Phys. Soc. Japan 48, 1683 (1980).
[12] K.Sakimoto and K.Takayanagi, J. Phys. Soc. Japan 48, 2076 (1980).
[13] K.Takayanagi, Comments Atom. Mol. Phys. 9, 143 (1980).
[14] K.Sakimoto, J. Phys. Soc. Japan 50, 1668 (1981).
[15] K.Takayanagi, Atomic Collision Research in Japan - Progress Report - No.7, p.42 (1981); XII ICPEAC, Book of Abstracts of Contributed Papers, p.921 (1981).
[16] M.Hashi, S.Tsuchiya and K.Takayanagi, J. Phys. Soc. Japan 49, 1486 (1980).
[17] J.V.Dugan,Jr. and J.L.Magee, J. Chem. Phys. 47, 3103 (1967).
[18] J.V.Dugan,Jr., Chem. Phys. Letters 21, 476 (1973) and references therein.
[19] T.Su and M.T.Bowers, J. Chem. Phys. 58, 3027 (1973).
[20] T.Su and M.T.Bowers, Intern. J. Mass Spectry. & Ion Phys. 12, 347 (1973).
[21] T.Su and M.T.Bowers, Intern. J. Mass Spectry. & Ion Phys. 17, 309 (1975).
[22] M.T.Bowers and T.Su, in "Interaction Between Ions and Molecules", ed. by P. Ausloos (Plenum Press, New York) p.163 (1975).
[23] R.A.Barker and D.P.Ridge, J. Chem. Phys. 64, 4411 (1976).
[24] J.Turulski and M.Forys, J. Phys. Chem. 83, 2815 (1979).
[25] F.Celli, G.Weddle and D.P.Ridge, J. Chem. Phys. 73, 801 (1980).
[26] S.Glasston, K.J.Laidler and H.Eyring, "Theory of Rate Processes", (McGraw-Hill Book Company, Inc., New York, 1941).
[27] D.R.Bates, Proc. Roy. Soc. A360, 1 (1978).
[28] W.J.Chesnavich, T.Su and M.T.Bowers, J. Chem. Phys. 72, 2641 (1980).
[29] D.Hyatt and L.Stanton, Proc. Roy. Soc. A318, 107 (1970).
[30] K.Sakimoto, "Quantum-mechanical PRS theory - New coupled-state approach for rotational excitation", ISAS Research Note No.125 (1980), unpublished.
[31] K.Sakimoto, "Ion-polar molecule reaction rates in interstellar clouds", ISAS Research Note No.102 (1980), unpublished.
[32] E.Herbst and W.Klemperer, Astrophys. J. 185, 505 (1973).
[33] W.Langer, Astrophys. J. 206, 699 (1976).
[34] K.Sakimoto, "Ion transports in polar gases", to be published.
[35] S.E.Buttrill, J. Chem. Phys. 58, 656 (1973).

PHYSICS OF ELECTRONIC AND ATOMIC COLLISIONS
S. Datz (editor)
© North-Holland Publishing Company, 1982

ULTRAHIGH-RESOLUTION ION ENERGY-LOSS SPECTROSCOPY

Nobuo Kobayashi

Department of Physics
Tokyo Metropolitan University
Setagaya-ku, Tokyo
JAPAN

Principles for achieving an energy resolution of better
than 10 meV in ion energy-loss spectroscopy are considered.
The spectrometer used in our laboratory is described. The
ultimate energy resolution of the apparatus as good as 7
meV F.W.H.M. was achieved in the medium energy region.
The experimental results of rotational excitation of $H_2$
and vibrational excitation of some simple molecules by $Li^+$
impact are presented. Preliminary measurements of fine-
structure and vibronic transitions in the collisions of
rare gas ions with atoms and molecules are also reported.

1. INTRODUCTION

In the last two decades, study of atomic collision physics has been made a great
progress with the advance of a translational energy spectroscopy of particle beams.
During this period many workers have concentrated their efforts on the improvement
in energy resolution of spectrometers. As a consequence of the improvement in the
resolving power in energy, many exciting works discovering new phenomena have been
performed. However, even now, it is not so easy to get an energy resolution of
better than 10 meV. Especially in the case of charged particle scattering, experiment with such excellent resolution has not been made as yet because of great
difficulties in generating a monoenergitic beam. Therefore, in particle beam
spectroscopy, we call an energy resolution to be better than 10 meV "ultrahigh
resolution".

The development of electron spectroscopy using electrostatic condencers has made
great contributions to the advance of particle beam spectroscopy. In the case of
electron spectroscopy, absence of magnetic fields is the necessary condition to
achieve a high energy resolution because of very small mass of an electron. The
earth magnetic field must be compensated as low as 0.1 mG by using μ-metal shields
and Hermholtz coils [1]. Furthermore, large electron current well defined in
energy is required because that the electron is scattered in a very wide angle.
Even though, the energy resolution of the apparatus is improved up to ultrahigh,
it is not sure whether electron collision experiments are able to be made with
such high resolution from lack of the intensity.

Contrary to that, an ion has relatively large mass and, then, the effect of earth
magnetic field is not very serious compared with the case of electrons. Furthermore, scattered ions are concentrated within a relatively narrow angle. Therefore, it is natural to expect that ultrahigh-resolution ion energy-loss spectroscopy may be attained by using electrostatic condencers without great difficulties.

From the physical view point, vibrational and rotational freedoms play important
roles in ion-molecule collisions including charge transfer and ion-molecule reactions. Vibrationally and rotationally state-resolved measurements are necessary
for full understanding of ion-molecule collisions. During several years, ion

355

energy-loss measurements for vibrational and rotational excitation of ground state
molecules have become active subjects.  Such experiments provide useful informations about ion-molecule interactions, especially anisotropic parts of interaction
potentials.  Most of the experiments are concentrated in very low energy region
and differential cross sections are mainly studied.

On the other hand, only a few vibrationally and rotationally state-resolved ion
energy-loss measurements have been made in medium energy region.  As will be
discussed in the next section, the ultrahigh energy resolution in the ion scattering experiments is expected to be easily obtained at the medium energy than
very low energy.  Our attention has been mainly directed to the energy region
above several ten eV.  Up to present, the ultimate energy resolution of the
apparatus as good as 7 meV has been achieved in our laboratory.

The present report deals with the consideration on the principle for getting an
ultrahigh-resolution ion energy-loss spectroscopy and measurements carried out in
our laboratory.

## 2.  PRINCIPLE CONSIDERATION FOR ULTRAHIGH RESOLUTION ION SPECTROSCOPY

While large mass of an ion is expected to be advantageous in getting the ultrahigh
resolution, translational energies of scattered ions are broadened because of
large momentum transfers in collisions between particles of almost equal mass.
Therefore, the ultrahigh-resolution ion energy-loss spectroscopy is only possible
in very low energy region or at a very small scattering angle.

Detailed consideration on the broadening of line shape in ion energy-loss spectra
was presented by Lorents and Conklin [2].  The observed line width, $\delta E$, in the
spectra arises from three sources, the energy resolution of the apparatus, $\delta Eap.$,
thermal motion of target particles, $\delta Eth$, and angular spread, $\delta Ean.$  Assuming
that three sources are independent, the over all width of the line shape is given
by

$$\delta E = [(\delta Eap.)^2 + (\delta Eth.)^2 + (\delta Ean.)^2]^{1/2}. \tag{1}$$

All of these sources must be reduced in order to perform high energy resolution
ion scattering experiments.  The first and second terms can be reduced by the
improvement in experimental technique.  For example, the effect of the thermal
motion is able to be reduced by cooling the target particles or using a supersonic
nozzle beam. For the simplification of the discussion, $Eap. = Eth. = 0$ is supposed
here to be achieved.

The relation between the excitation energy Q and the translational energy $E_1$ of
an ion scattered at laboratory angle $\Theta$ is exactly determined from energy and
momentum conservation laws.  Assuming that target particles are at rest, we obtain

$$Q = 2\gamma(E_0E_1)^{1/2}\cos\theta + (1 - \gamma)E_0 - (1 + \gamma)E_1 , \tag{2}$$

where $\gamma = M_1/M_2$ is the mass ratio of the incident to target particle and $E_0$ is the
incident energy.  When the incident energy is much larger than the excitation
energy and the scattering angle is very small, the energy difference of the
incident particle before and after the collision is given by a very simple form as

$$\Delta E = (E_0 - E_1) = \gamma E_0\theta^2 + Q. \tag{3}$$

This relation is very useful and means that all information on collisions is , in
principle, possible to be obtained.  The first term on the light hand side of this
equation corresponds to the energy change of the incident particle due to the
recoil of the target particle.  In the case of ion collision experiments, angular
resolution of the analyzer is, then, important to receive a high energy resolution.
The angular spread in the scattered ions which accepted by the detector of a
finite angular resolution affects to the broadening of the line shape of the
spectrum.  Assuming angular independent cross sections over the acceptance angle
of the analyzer, this energy spread is given by

$$\delta E_{an.} = 2\gamma E_0\theta\delta\theta, \tag{4}$$

where $\delta\theta$ is the angular resolution of the analyzer. The $\delta E_{an.}$ is zero at $\theta = 0°$, but this becomes very large with increase of scattering angle except for very low collision energies. From the eq.(4), as the necessary condition to accomplish ultrahigh ion energy-loss measurement, $\delta E < 10$ meV, we obtain

$$E_0\theta < (1/2\gamma)(0.01/\delta\theta), \tag{5}$$

where $E_0$ is in eV units. Taking a practical value, $\delta\theta(\text{F.W.H.M.}) = 0.4°$, this condition is shown in Fig.1. Only when the collision energy and the scattering angle are the values under the curve shown in the figure, the ultrahigh resolution is able to be received.

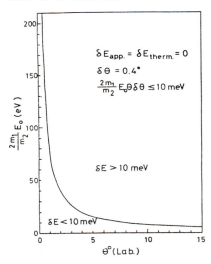

Fig.1 The necessary condition to accomplish ultrahigh resolution ion energy-loss measurements in the collision between incident and target particles of mass $m_1$ and $m_2$. Only when the collision energy $E_0$ and the scattering angle in laboratory system $\theta$ are the values under the curve, the energy resolution better than 10 meV is possible to be achieved.

The collision energy $E_0$ of the ordinate is, of course, different for different collision system. When the target is hydrogen molecules, the incident energy corresponding to the value 100 eV of the ordinate become 100 eV, 14 eV and 2.5 eV for the incident ions $H^+$, $Li^+$ and $Ar^+$, respectively. At a glance, the ultrahigh resolution ion energy-loss spectroscopy is restricted in a very narrow region concerning the collision energy and the scattering angle and seems to be rather impracticable. Many objects of study are, however, possible even in such narrow region.

## 2.1. LOW ENERGY REGION

At first, let us consider the very low energy region near the abscissa. In this region, differential cross section measurements are main subjects. Some rotationally state-resolved experiments have already been carried out. Toennies and coworkers have first succeeded in obtaining rotationally state-resolved energy-loss spectra of $H_2$ by impact of $Li^+$ and $H^+$ with a time-of-flight spectrometer [3]. After that Herman and Linder have studied the rotational excitation in the system $H^+$ - $H_2$ [4]. In the energy region below several eV, ion-molecule reactions become important processes. Teloy and coworkers have performed very interesting experiments for the reaction $H^+ + D_2 \longrightarrow D^+ + HD$. This work was reported at Xth ICPEAC [5]. Besides these rotationally state-resolved works, many vibrationally state-resolved measurements have been made. Such studies for rotational and vibrational excitation in ion-molecule scattering provide important information to decision of

potential surfaces. The detailed comparison between state-resolved differential
scattering measurements and fully *ab initio* calculation in the system $H^+$ - $H_2$ were
made by Linder and coworkers and reported at the last ICPEAC [6].

From the experimental view point, it seems to be very difficult to obtain an
ultrahigh energy resolution in the differential cross section measurements. The
energy resolution of the apparatus is, in fact, added to the energy spread due to
the angular spread of scattered ions. Over all resolution of a few ten meV has
been only just achieved up to present. Therefore, the rotationally state-resolved
measurement is limited to hydrogen molecules. When the target is $H_2$, incident ion
species are also restricted to one of light mass such as proton and lithium. If
anyone intend to make differential scattering measurements of the rotational
excitation of $H_2$ in the collision with ions of larger mass than lithium, beam
experiment must be carried out at a collision energy lower than 1 eV.

## 2.2. FORWARD SCATTERING REGION

The ultrahigh energy resolution is also possible to be achieved in the region near
the ordinate. In this region, while the experiments must be made at nearly zero
scattering angle, we are free from the restriction on the collision energy. This
is convenient for accomplishing ultrahigh-resolution ion-scattering measurements.
The scattered ions are concentrated within a very narrow angle at a slightly high
collision energy. Herrero and Doering have succeeded in state-resolved measurements for the vibrational excitation of $H_2$ by proton impact in the energy range
from 5 eV to 1500 eV [7]. They found that most of scattered ions associated with
the vibrational excitation are concentrated within an acceptance angle of the
analyzer settled at zero degree. Then, they were able to obtain absolute integral
cross sections above 100 eV. Since integral cross sections are the fundamental
values in atomic collision physics, this is considered to be advantageous. In
addition to that, the energy spread due to the angular spread of the scattered
particles vanishes as far as the analyzer with an appropriate angular resolution
is settled at zero degree. Therefore, if the velocity distribution of the target
molecules can be reduced, the resolution of an apparatus is the only source for
broadening of line widths in loss spectra. This seems to be a great advantage to
accomplish ultrahigh-resolution ion energy-loss measurements. Therefore, our
attention has mainly directed to the energy range above several ten eV.

## 3. EXPERIMENTALS

A schematic diagram of the ion energy-loss spectrometer at Tokyo Metropolitan
University is shown in Fig.2. The set-up of our apparatus is very simple, and no
particular means to compensate the earth magnetic field are adopted. The spectrometer consists of an ion source, an energy selector, a collision chamber, an
energy analyzer and an ion detector. The ion source, the energy selector and the
ion accelerating lense system are settled on a rotatable table.

The energy selector and analyzer are electrostatic hemispherical condencers with
real enetrance and exit apertures. The mean radii of them are both 75 mm, and the
entrance and exit apertures, which are made of molybdenum, is 0.5 mm in diameter.
The geometrical energy resolution is 1/300 of the transmission energy. The
distance between inner and outer electrodes of the condencer is 30 mm. Aluminum
alloy has been chosen as a material for both selector and analyzer in order to
reduce the weight of large condencers. Usually aluminum alloy is not used as a
material for the selector, since its surface is oxidizable and homogeneous surface
potential is not ensured. No particular means for antioxidization is not adopted
and the parts are only cleaned by organic solvent before setting in the vacuum
vessel. However, we have never experienced such trouble.

When the energy spectra are measured at nearly zero degree angle as the present
case, the intensity of the primary ions passing through the collision cell without
scattering and elastically scattered ions is much stronger than that of

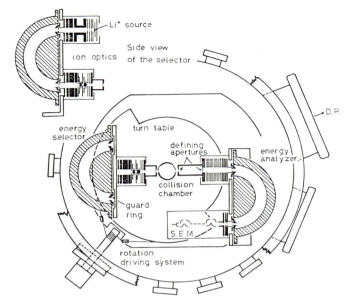

Fig.2   Schematic diagram of the ion energy-loss spectrometer used.

inelastically scattered ones.  The small inelastic peaks must be examined in the
presence of large primary peak, so that the sharp cuts of the tails of the peak
profile is the most important.  An electrostatic condencer has an aberration for
the beams entering the condencer with divergence against the central orbit.   In
the case of hemispherical one, the aberration is the first order, and the F.W.H.M.
is given by

$$\Delta E(1/2) = Et[(d/2R) + (\alpha^2/2)], \qquad (6)$$

where Et is the transmition energy in the analyzer, $\alpha$ is the divergence angle at
the entrance aperture.  Even though the divergence angle is only several degree,
the second term comes to the same magnitude of the gemetrical resolution.   Furthermore, the divergence of the beam entering the analyzer affects to prolong the tail
of the peak profile on the energy loss side of the spectrum.   The parallel beam
entrance to the analyzer is, in my opinion, the most important for achieving the
ultrahigh resolution ion-energy loss spectroscopy.

From this reason, we adopted a nearly uniform field as deceleration potential for
ions entering the analyzer.  The ions scattered from the target molecules in the
collision chamber pass through two angular defining apertures, acceptance angle of
which is ± 0.45°.  The ions are decelerated to the transmission energy of the
analyzer by this field.  If this field is exactly uniform, the ion beam diverge
considerably because of a high deceleration rate.  Therefore, the deceleration
potential is slightly modified from the uniformity by applying appropriate potentials, which is empirically determined, to some plates of deceleration lense.

At first stage of this study, we adopted Li$^+$ ion as the projectile by the following reasons. 1) Li$^+$ can be produced by thermionic emission, so that the velocity
distribution of the ions extracted from the source is relatively narrow.  This
must be a merit to reduce the background count rate.  2) Since the ionization
potential of Li is much smaller than those of target molecules and Li$^+$ has a

He-like structure, the exchange interaction is expected to be negligible compared with the electrostatic long range interactions. If so, this must be advantageous to the comparison between experiments and theories, because the interaction potential is simply given.

The whole sytems is housed in all metal bell-jar of 600 mm in inner diameter, and pumped by a 1200 l/sec oil-diffusion pump. The ultimate pressure is $1 \times 10^{-8}$ Torr and typical background pressure is $3 \times 10^{-7}$ Torr when the scattering gas of about $2 \times 10^{-4}$ Torr is introduced. This ultrahigh vacuum provides stable operation and a long life time of several months.

Up to present, we have succeeded in achieving the ultimate instrumental energy resolution as low as 7 meV. When the collision energy is below 200 eV, an energy resolution of better than 10 meV is routinely obtained. With this excellent energy resolution, the typical ion beam current is still of the order of $10^{-14}$ A. This beam intensity is enough high to forward scattering experiments.

## 4. RESULTS

### 4.1. VIBRATIONAL AND ROTATIONAL EXCITATION OF SIMPLE MOLECULES BY Li$^+$ IMPACT

So far we have studied vibrational excitation of simple molecules, $H_2$, $D_2$, $N_2$, $O_2$: Raman active [8], CO, NO, $CO_2$, $H_2O$, $CH_4$: Infrared active [9], and rotational excitation of $H_2$ [10] by Li$^+$ impact in the energy range from 40 eV to 1500 eV.

### A. ENERGY LOSS SPECTRA

As an example of collisionally induced vibrational transitions, a typical energy-loss spectrum for 203 eV Li$^+$ incident on $CO_2$ is shown in Fig.3. This spectrum was obtained by setting the analyzer at zero degree. The transmission energies in the selector and analyzer were both about 3 eV. The target gas was kept at room temperature. The molecule $CO_2$ is well known as infrared active one. The bending (83 meV) and asymmetric stretching (291 meV) modes are infrared active, whereas symmetric stretching mode (171 meV) is Raman active. The peaks corresponding to

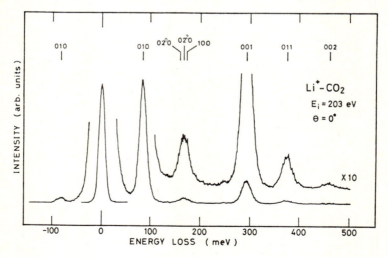

Fig.3  Typical energy-loss spectrum of 203 eV Li$^+$ incident on $CO_2$. The peak at -83 meV corresponds to the super elastic transition from (010) to (000).

the excitation of harmonics of the bending and asymmetic stretching modes and coupled modes of them are clearly observed as well as those of fundamentals.  On the left hand side of the primary peak, super elastic peak corresponding to the deexcitation from (010) to (000) is also clearly separated inspite of that initial abundance of the (010) state in the target molecules is only about 4 %.

As examples of energy-loss spectra measured with ultrahigh energy resolution Fig. 4 shows the rotational transitions of normal-$H_2$ and para-$H_2$ by 103 eV $Li^+$ incident. These spectra were obtained on almost the same conditions to the case of vibrational excitation of $CO_2$.  By best adjusting ion optics, the F.W.H.M. of 9.5 meV for the primary beam, which means instrumental energy resolution, were received. In the figures, the tails of the primary peak profiles measured without target gas in the collision chamber are shown by the dotted lines.  It can be seen that the tails of the primary peakds are sharply cut.  This indicates that decelerated ion beam successfully enters into the analyzer without divergence.

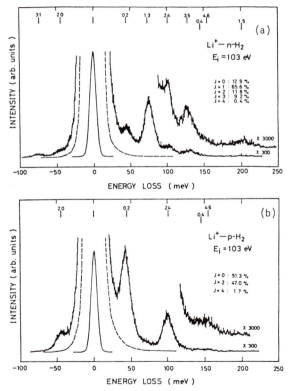

Fig.4  Typical energy-loss spectra for

(a) $Li^+$ - normal $H_2$

(b) $Li^+$ - para $H_2$

scattering at 103 eV and 0°. The F.W.H.M. of the primary peaks are 9.5 meV.  The tails of the primary peaks without target gas are shown by the broken curves.  The target gas was kept at room temperature.  The initial rotational state populations calculated according to the Boltzman distribution at 293 K are also shown.

In the case of n-$H_2$, which consists of ortho-$H_2$ and para-$H_2$ at the ratio of 3:1, the initial rotational states are mainly populated from J=0 to J=3 according to the Boltzman distribution.  The initial abundance of the states are shown in the figures.  In the spectrum the transitions with $\Delta J=2$ from these states are dominant. The transition with $\Delta J=4$, J=1 → 5, is weakly observed.  The spectrum of para-$H_2$ is rather simple since only the initial rotational states with even number are allowed.  In this spectrum the transition with $\Delta J=4$ is not clearly observed because of overlapping with J=4 → 6 transition.

In both spectra, most of all peaks are clearly separated. Especially the transition J=0 → 2 of n-H$_2$ is well separated from the primary peak, even though the
abundance of J=0 is only 13 % and the excitation energy is 43.9 meV. As far as we
know, this is first successful observation, using an nergy loss spectrometer, of
J= 0 → 2 excitation of hydrogen molecules which are at room temperature.

The F.W.H.M. of the inelastic peaks are about 16 meV, even though that of the
primary peak is 9.5 meV. This broadening must be due to the thermal motion of the
target molecules, as mentioned above. Therefore, the disagreement of the tails of
the primary peaks with and without target gas is attributed to the effect of
elastic collisions. If the velocity distribution can be reduced by cooling the
target gas, these peaks must be much clearly separated.

B.   CROSS SECTION MEASUREMENTS

Excitation cross sections were obtained from the relative intensities of the
inelastically scattered beams to the primary beam in an energy loss spectrum.
The angular distributions of the scattered ions associated with the excitation
were almost the same as that of the primary beam above several hundred eV in all
the case studied. Therefore, above these energies, almost all scattered ions
associated with vibrational or rotational excitation are expected to be accepted
with the detector by setting the analyzer at zero degree. We were, then, able to
determine  the integral cross sections from the energy-loss spectra obtained at a
scattering angle of zero degree.

As the examples, the measured cross sections for the vibrational excitation of CO
and N$_2$, which are very similar in molecular constants except for the fact that the
former is infrared and the latter is Raman active, are shown in Fig.5. The
intergral cross sections were obtained above 400 eV and 300 eV for CO and N$_2$,
respectively. Below these energies, measured cross sections are interpreted as
partial ones for the forward scattering within the acceptance angle of the analyzer, ± 0.45°.

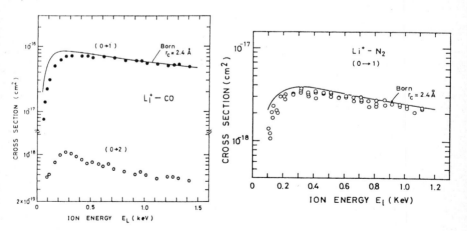

Fig.5  Measured cross sections for the vibrational excitation of CO,
infrared active, and N$_2$, Raman active molecules. The solid curves
are cross sections calculated by the first order Born approximation
considering only charge-dipole and charge-induced dipole potential
with conventional cut-off function for CO and N$_2$, respectively.

The magnitudes of the cross sections for the excitation of infrared active mode are much larger than those for Raman active one. While the cross sections for the excitation of harmonics of CO could be obtained, no peak corresponding to the v=0 → 2 transition could be observed in the case of $N_2$. The cross section of v=0 → 2 transition of $N_2$ was suggested to be smaller than $10^{-20}$ cm$^2$. Another characteristic feature is seen in Fig.5(a). The magnitude of the cross sections for the excitation of harmonics $\sigma(0 \to 2)$ are much smaller than those of fundamentals $\sigma(0 \to 1)$. Furthermore, $\sigma(0 \to 2)$ reaches the maximum at lower energy and has stronger energy dependence than $\sigma(0 \to 1)$. This characteristic energy dependence is also observed in the excitation of harmonics and coupled modes of other infrared active molecules.

As the excitation occurs strongly forward scattering and as Li$^+$ has a He-like structure, the dominant interactions are considered to be long range electrostatic forces, for example, polarization force and charge-dipole potential. The measured cross sections for the excitation of fundamentals are compared with first Born approximation considering only the charge-induced dipole and charge-dipole interactions with conventional cut-off function for the excitation of Raman active and infrared active molecules, respectively. The Born cross sections using the same cut-off factor, 2.4 Å, which is estimated from experimental and thoretical potentials, are shown in Fig.5 by solid curves. The agreements between experimental and theoretical cross sections are fairly good. Such excellent agreements are obtained for all molecules with only one exception of oxygen molecules. This fact indicates that the long range electrostatic forces are dominant interactions in Li$^+$ impact vibrational excitation in the medium energy region.

Contrary to that, in the case of the excitation of harmonics or coupled modes, stronger energy dependences of the measured cross sections were not be able to be explained by first order Born approximation. Iwamatsu et al. [11] calculated the cross sections of CO and NO with impact parameter method on the assumption that the transitions to the higher levels occur step by step through intermediate levels. The charge-dipole potentials with an appropriate cut-off factor were adopted in the same way to our Born calculations. Their calculation successfully reproduced the energy dependences as well as the magnitudes of the measured cross sections. This fact implies that the excitation of harmonics and coupled modes by Li$^+$ impact at the medium energy is dominated by the second order perturbation through the intermediate state.

The ultrahigh energy resolution of our ion energy-loss spectrometer and use of normal- and para-$H_2$ made comprehensive state-resolved study of the rotational

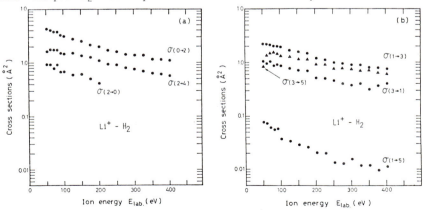

Fig.6  Measured cross sections for the rotational excitation of $H_2$ by Li$^+$ impact.

excitation of hydrogen molecules possible in the energy range of 40 eV to 400 eV.
The integral cross sections for the transitions, J=0 → 2, J=1 → 3, J=2 → 4, J=3 →
5, J=2 → 0, J=3 → 1 and J=1 → 5, were obtained above 150 eV.  In Fig.6(a) and (b),
the measured cross sections of the rotational excitation from even J and odd J are
shown, respectively.  Above 100 eV, the energy dependence of the cross sections of
each transition is almost the same except for the J=1 → 5 transition.  The
magnitudes of the cross sections of the excitation and de-excitation with ΔJ=2
have simple initial J dependence.  In the case of excitation, the cross sections
$\sigma(J \rightarrow J + 2)$, decrease with increase of J.  On the other hand, $\sigma(J \rightarrow J - 2)$,
increases with increase of J.  These initial J dependence of the ΔJ=2 transitions
at 203 eV are shown in Fig.7.

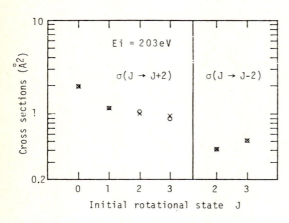

Fig.7  Dependence on the
initial rotational-state
numbers of the cross sections
with ΔJ=2 transitions.

O: experimental values taken
from Fig.6.

X: theoretical initial J
dependence predicted by eq.(7).

This result was semi-quantitatively compared with the first order perturbation
theory.  An interaction potential of an ion and a homonuclear molecule, within the
rigid rotator approximation, can be expanded by Legendre polinomials, $P_n(\cos\theta)$,
where θ is the angle measured from the molecular axis.  In that case, the rotational excitation with ΔJ=2 transition correlates to the anisotropic part with $P_2$
symmetry in the interaction potential.  As mentioned in the case of vibrational
excitation, long range electrostatic forces are considered to be dominant
interaction.  Therefore, in the present case, anisotropic parts of the cahrge-
quadrupole and charge-induced dipole interactions must be dominant.  Within the
first Born approximation taking account of these interactions, the leading terms
for determining the initial rotational state dependence of the cross sections with
ΔJ=2 transitions are given as follows [12]

$$\sigma \propto (J+2)(J+1)/(2J+3)(2J+1) \qquad \text{for excitation and} \qquad (7a)$$

$$\sigma \propto J(J-1)/(2J+1)(2J-1) \qquad \text{for de-excitation.} \qquad (7b)$$

These initial rotational state dependence predicted by the first Born approximation is shown in Fig.4.  The magnitude of the theoretical cross sections are
normalized to the experimental one at J=0.  Agreement between the experimental
and theoretical initial J dependences of the cross sections is fairly good.  This
means that the detailed balancing is well held and also suggests that the first
order perturbation theory can explain the ΔJ=2 transition in $Li^+$ - $H_2$ collisions
at the medium energy, even though the detailed calculation has not been made.

The cross sections for the ΔJ=4 transitions, $\sigma(J=1 \rightarrow 5)$, show stronger energy
dependence than those of the ΔJ=2 transitions.  On the analogy of the excitation
of harmonic oscillations mentioned above, this may be considered to the result
from the effect of the second order perturbation, to say step by step transition
through intermediate ΔJ=2 states.

## 4.2. FINE-STRUCTURE AND VIBRONIC TRANSITIONS

An electronic excitation of an incident ion or a target particle or the both frequently occurs at the medium energy as well as vibrational and rotational excitation. For the study of such electronic excitation, especially for the process in which metastable states are involved, the ion energy-loss spectroscopy must be also a powerful tool.

Very recentry, $Li^+$ ion source of our spectrometer has been changed with an electron impact type. A Nier type ion source and $90°$ sector mass selector made of ferrite magnets of 30 mm mean radii have been placed in front of the entrance aperture of the energy selector. By this modification, state-resolved measurements for the collisions of various ion species have been made possible. Preliminary experiments for the scattering of rare gas ions, including doubly charged ions, with atoms and molecules have been made. The results for the fine-structure transitions in $Ar^+(^2P_j)$ - Ar collisions are presented at this conference. While Johnson[13] has predicted the fine-structure transitions in this system, no direct observation has not been made for a long time. As far as we know, this measurement is the first successful observation of direct fine-structure transitions.

As an example, Fig.8 shows an energy-loss spectrum for $Kr^+$ - $O_2$ scattering at the collision energy 1008 eV and the scattering angle of $0°$. This spectrum is quite different from those for $Li^+$ scattering. In the spectrum, fine-structure transitions of doublet states of $Kr^+$ and vibronic excitation of $O_2$ are clearly observed as well as pure vibrational excitation of $O_2$. The prominent feature in this spectrum is the peak at 0.32 eV. This peak corresponds to the simultanious

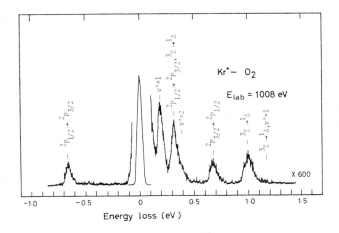

Fig.8 Typical energy-loss spectrum for $Kr^+$ - $O_2$ scattering at 1008 eV and $0°$.

transitions of the projectile and the target,

$$Kr^+(^2P_{1/2}) + O_2(^3\Sigma, v=0) \rightarrow Kr^+(^2P_{3/2}) + O_2(^1\Delta, v'=0). \qquad (9)$$

We have measured energy-loss spectra for $Kr^+$ - CO scattering. In that case the coupled transition of the fine-structure transitions of the projectile and vibrational transitions of target molecules,

$$Kr^+(^2P_{1/2}) + CO('\Sigma, v=0) \rightarrow Kr^+(^2P_{3/2}) + CO('\Sigma, v'=1) + 0.400 \text{ eV}, \qquad (10)$$

has been also observed. These facts suggest that the fine-structures of the projectile and target particles also play important roles in ion scattering at medium energies.

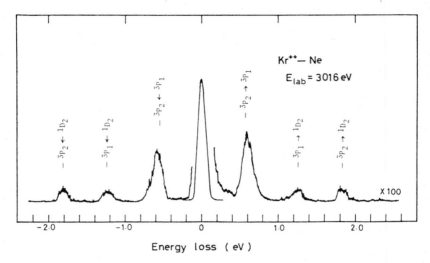

Fig.9  An energy-loss spectrum of 3016 eV Kr$^{++}$ incident on Ne
obtained at a scattering angle of 0°.  Spin non-conservative
transitions are observed as well as fine-structure transitions.
The states $^3P_1$ and $^3P_0$ of which energy difference is 0.115 eV
are not separated.

In Fig.9, an energy-loss spectrum for the scattering of Kr$^{++}$ - Ne is shown.  The
fine-structure transitions among triplet states of Kr$^{++}$, $^3P_2 \to {}^3P_{1,0}$ and spin
non-conservative transition, $^3P_{1,0} \to {}^1D_2$ and $^3P_2 \to {}^1D_2$, and reverse of them are
observed, even though $^3P_1$ and $^3P_0$ are not resolved at present.  This sepectrum
indicates that spin-nonconservative transitions also quite frequently occur in ion
collisions at a medium energy.  Such simultaneous transitions in the projectile
and target particles and spin-nonconservative transitions have already been
observed by Moore [14] in the system N$^+$ - O$_2$, Kr, Xe with an ion-energy-loss
spectrometer at a few keV.  The energy resolution of his apparatus was not enough
good to resolve vibrational and fine-structure transitions.

The collisionally induced electronic and vibronic excitation, of course, can be
measured by optical methods with much high resolving power in energy.  However,
the excitation processes involving metastable states can not be observed.
Furthermore, it is impossible to distinguish direct excitation from such simultaneous transitions of the target and projectiles.  The ion energy-loss spectroscopy
is expected to provide more precise information for understanding ion-molecule
collisions as well as ion-atom collisions.

FUTURE PROSPECTS

Final goarl of our work is placed in performing fully state-resolved ion energy-
loss measurements for ion-molecule scattering.  In that sense, we just stand at a
starting point of our work.  Even though fully state-resolved experiment is
presently limited to hydrogen molecules, we are optimistic about the future of
ultrahigh resolution ion spectroscopy in the medium energy region.  In the near
future, the energy resolution near one meV will be achieved.  At that time fully
state-resolved ion energy-loss measurements including vibronic-rotational excitation of molecules and transitions to metastable states will be made for the
various molecules bisides hydrogen molecules.

ACKNOWLEGEMENTS

I am very much indebted to Prof. Y. Kaneko for many helpful discussions and for his continuous encouragement during the course of this work. I am also greate-fully appreciate to the collaboration of Dr. Y. Itho without whose efforts most of the work described herein would not have been performed. The energy-loss spectrometer used in this work constructed by Grat-in-aid from the Mitsubishi Foundation. This work was supported in part by a Grant-in-aid from the Ministry of Education, Science and Culture.

REFERENCES

1) K.K.Jung: *Electronic and Atomic Collisions* (North Holland, Amsterdam, 1980) p.787.
2) D.C.Lorents and G.M.Conklin: J. Phys.B5 950 (1972).
3) H.E.Bergh, M.Faubel and J.P.Toennies: Faraday Disc. Chem. Soc. 55 203 (1973), K.Rudolph and J.P.Toennies: J. Chem. Phys. 65 4483 (1976).
4) V.Herman and F.Linder: J. Phys. B11 493 (1978).
5) E.Teloy: *Electronic and Atomic Collisions* (North Holland, Amsterdam, 1978) p.591.
6) F.Linder: *Electronic and Atomic Collisions* (North Holland, Amsterdam 1980) p.531.
7) F.A.Herrero and J.P.Doering: Phys. Rev. A5 702 (1972).
8) N.Kobayashi, Y.Itoh and Y.Kaneko: J. Phys. Soc. Jpn. 46 208 (1979).
9) N.Kobayashi, Y.Itoh and Y.Kaneko: J. Phys. Soc. Jpn. 45 617 (1978), Y.Itoh, N. Kobayashi and Y.Kaneko: *ibid.* 46 1399 (1979).
10) Y.Itoh, N.Kobayashi and Y. Kaneko: J. Phys. Soc. Jpn. 46 1041 (1979), Y.Itoh, N.Kobayashi and Y.Kaneko: J. Phys. B14 679 (1981).
11) M.Iwamatsu, Y.Onodera, Y.Itoh, N.Kobayashi and Y.Kaneko: Chem. Phys. Lett. 77 585 (1981).
12) E.Gerjuoy and S.Stein: Phys. Rev. 97 1671 (1955).
13) R.E.Johnson: J. Phys. B3 539 (1970).
14) J.H.Moore Jr.: Phys. Rev. A8 2359 (1973), J.H.Moore Jr.: *ibid.* A10 724 (1974).

PHYSICS OF ELECTRONIC AND ATOMIC COLLISIONS
S. Datz (editor)
© North-Holland Publishing Company, 1982

SUBPICOSECOND EXPERIMENTS ON MOLECULAR VIBRATION

Aart W. Kleyn[†]
FOM Instituut voor Atoom-en Molecuulfysica, Amsterdam Wgm, The Netherlands
and
Keith T. Gillen[*]
Molecular Physics Laboratory, SRI International
Menlo Park, California 94025

Vibrational motion on the subpicosecond time scale of a transient collision
intermediate has been observed through its striking effects upon differential and
total cross sections in inelastic and reactive processes in atom- and ion-
molecule collisions. Transient charge transfer processes that form a short-lived
diatomic ion and then return to the original potential energy surface are shown
to be efficient paths to high vibrational excitation of molecular targets.
Similar effects should be observable in a large number of collision systems.

## Introduction

A precise knowledge of the potential energy surface for a triatomic system could,
in principle, be used to predict the detailed distribution of product states for
inelastic or reactive processes, that probe the triatomic surface as a transient
intermediate during a collision. In contrast, even when detailed state-to-state
rates are available, it is far more difficult to extract quantitative information
on the relevant potential surfaces from measurements of inelastic and reactive
processes. The critical significance of the collision intermediate justifies
strenuous experimental efforts to improve our view of its fleeting existence.

One approach to examining a collision intermediate is to prepare it directly in a
well-defined state or energy level and observe its subsequent evolution.
Examples using photoexcitation of neutrals and ions are given in the chapters by
A. Giusti-Suzor[1] and P. C. Cosby.[2] These "half-collision" experiments can
often analyze decay channels and surface couplings with great precision assuming
sufficient spectroscopic information is available.

An alternate and quite difficult approach involves the examination of a transient
collision intermediate by direct spectroscopic observation of the very short-
lived colliding system (see, e.g., chapter by J. Polanyi[3]) or by optical pumping of the system while it is reacting (see, e.g., chapter by A. Gallagher[4]).
These experiments, closely related to spectroscopic line broadening, are still in
an exploratory stage and must eventually address the difficult problem of determining the dependence of emission line widths upon nuclear coordinates in the
transient intermediate.

In contrast, our approach to examining the collision intermediate uses more traditional scattering techniques, but adjusts experimental conditions so that the
transient intermediate itself provides an internal clock whose timing can be
observed through its profound effects upon the resultant product distributions.

[†]Visiting scientist 1980-81 at IBM Research, San Jose, CA 95193, U.S.A.

[*]Research support from NSF and ONR are gratefully acknowledged.

The internal clock is the vibrational motion of a transient ion, formed as part of the short-lived intermediate.  We will demonstrate for a few relatively simple systems involving charge transfer intermediate states that very detailed information about the collision dynamics can be obtained.  In some cases vibrations of the transient intermediate on the subpicosecond time scale can be seen experimentally.  The vibrational motion of this transient species yields the most dramatic results in carefully chosen systems when the collision velocity is adjusted to make the collision time comparable to the vibrational time.

Atom-Atom Ion-Pair Formation

Most of the work described in this report will involve neutral atom-molecule collisions that are strongly influenced by charge transfer to ion-pair states during the collision process.  It is therefore useful first to consider a related, but simpler, atom-atom interaction, that of Na with I.  Shown in Figure 1 are the lowest $^1\Sigma^+$ diabatic covalent and ionic potential curves for NaI.  At large distances R, the ion-pair state is asymptotically higher than the covalent state by $\Delta E = IP(Na) - EA(I)$, where IP is the ionization potential of Na and EA is the electron affinity of the I atom.  The ion-pair state decreases in energy as R decreases and crosses the covalent state at $R = R_c$.  In the simplest approximation, considering only the coulombic term of $V_{ion}$, $R_c = 14.4/\Delta E$, with $\Delta E$ in eV and $R_c$ in Å.

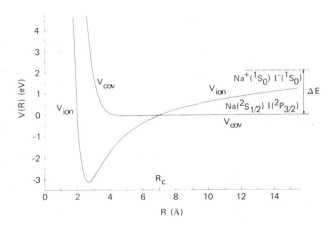

Figure 1.  Diabatic Na-I potentials.

Since the $^1\Sigma^+$ covalent curve is the same symmetry as the ion-pair curve, there will be an avoided crossing in the corresponding adiabatic curves at $R = R_c$.  For thermal energy Na and I atoms approaching each other along the $^1\Sigma^+$ curve (1/8 of the collisions statistically), an electron will jump from Na to I in the vicinity of $R_c$ as the adiabatic state abruptly changes from a covalent to an ionic character.  After the inner ionic region is visited, the outgoing trajectory will once again cross $R_c$ and the electron will return to the $Na^+$.

As one raises the collision energy, one expects eventually that a fraction of the collisions will proceed diabatically through the crossing region at $R_c$ and that for collision energies $E_o > \Delta E$, the ion-pair product channel $Na^+ + I^-$ would be observed. Formation of ions indeed has been observed in both total and differential cross section experiments.[5,6]

The probability for ion-pair formation depends, of course, on the strength of the coupling between the two interacting surfaces. In the simplest Landau-Zener (L-Z) description (see, for example, refs. 7, 8), the probability P for diabatic behavior at a crossing through the sphere of radius $R_c$ is

$$P = \exp\left(-\frac{2\pi H_{12}^2}{v_R |S|}\right), \qquad (1)$$

where $H_{12}$ is the coupling matrix element, $v_R$ is the radial velocity at $R = R_c$, and $S = \frac{d}{dR}(V_{cov} - V_{ion})$ evaluated at $R = R_c$. Faist and Levine[9] have obtained excellent fits to the Na + I scattering data by using improved adiabatic potentials and a slight modification of the crossing distance. Nonetheless the simple diabatic L-Z model is a reasonable first approximation to the scattering process.

## Atom-Molecule Systems: $O_2$ Target

Collisions of fast alkali atoms with $O_2$ molecules are influenced in an analogous way to Figure 1 by the crossing between the covalent incoming potential and strongly attractive ionic potentials. Necessarily the three atom system is vastly more complicated by the extra degrees of freedom, and the greater uncertainties in the knowledge of the potential surfaces and their interactions. The simple potential curves of Na + I are replaced by multidimensional potential surfaces in the triatomic system. Nonetheless, a carefully chosen set of physically reasonable simplifications allows us to generate an interesting classical picture of the collision dynamics.

The basic assumption is an uncoupling of the internal motion of the target molecule from the relative translational motion. To apply this assumption, we must exclude impact parameters that explore the inner repulsive core of the interaction potential. This neatly excludes the collisions resulting in the normal impulsive mechanisms for inelastic excitation transfer.

Consider a Cs atom approaching an $O_2$ molecule on a covalent surface. Assume that the O-O distance r is the classical equilibrium separation of $O_2$, $r_e$. At a $Cs$-$O_2$ distance, $R_c$, corresponding to a crossing with the ionic surface, an electron can jump to form $Cs^+ + O_2^-$. In the spirit of our simple picture, assume a vertical transition, an instantaneous electron jump that forms $O_2^-$ with an unchanged bond length, $r_e$. Clearly then, the crossing distance $R_c$ is given by $R_c = 14.4/[IP(Cs) - EA_v(O_2)]$ where $EA_v$ is the vertical electron affinity of $O_2$ at $r = r_e$. Since the equilibrium internuclear separation of $O_2^-$ is larger than that of $O_2$, the nascent $O_2^-$ begins its existence in a compressed state on its repulsive inner wall (see the $O_2$, $O_2^-$ potentials in Figure 2). As the $Cs^+$ and $O_2^-$ ions explore the attractive region inside $R_c$, the $O_2^-$ starts to expand, thus initiating an $O_2^-$ vibration. It is clear from Figure 2 that as the $O_2^-$ goes through one complete classical vibration, the vertical electron affinity (which depends on the instantaneous value of r) first increases, then decreases with a period identical to that of the $O_2^-$ vibration. Since $R_c$ depends on the instantaneous value of $EA_v(r)$ through

$$R_c = 14.4/\{IP(Cs) - EA_v[r_{O_2^-}(t)]\}, \qquad (2)$$

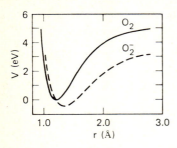

Figure 2.   Schematic $O_2$ and $O_2^-$ ground state
potential curves.

this crossing radius must vibrate in and out with the same period as the $O_2^-$.
Hence the position of the second crossing with the covalent surface depends
critically on the relative timing of the translational motion and the $O_2^-$ vibration.   Since the coupling $H_{12}$ decreases exponentially with $R_c$,[8,10] the branching probability P at this second crossing is also a very sensitive function of
the relative timing.   Depending on collision velocity and impact parameter, the
transient ion-pair intermediate may have a second crossing with the covalent
surface either in an expanded geometry that favors diabatic behavior or in a
compressed geometry that favors adiabatic behavior in the coupling region.

Figure 3 shows schematically $Cs + O_2$ collisions at $E_o = 35.5$ eV and various
impact parameters.   Each trajectory (drawn as a straight line for simplicity of
presentation) is assumed to convert to $Cs^+ + O_2^-$ ion-pairs at the initial
crossing at $R = R_c$.   At that instant ($t = 0$) the crossing radius $R_c$ starts
vibrating with the $O_2^-$ period in accord with Equation (2).   The calculated radius
$R_c'$ at the second crossing point (a function of b) is indicated in Figure 3.   The
lower trajectories labelled 1, 2, 3 have impact parameters that correspond to
reaching the second crossing after 1, 2, or 3 full vibrations of the $O_2^-$ (and of
$R_c'$).   These trajectories would tend to favor adiabatic passage through the second
crossing region yielding reneutralized products.   In contrast the upper trajectories have impact parameters that would reach the second crossing after ~ 1.5
or 2.5 $O_2^-$ vibrations and would yield larger diabatic probabilities for ion-pair
products.   Clearly the branching ratio would have an impact parameter dependence
that might show up as oscillations in the differential cross sections for this
system.

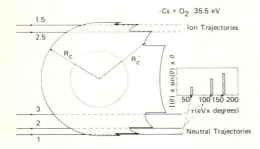

Figure 3.   Vibrational timing
effects on $R_c'$ for $Cs + O_2$
collisions.   The straight line
trajectories (at bottom) favor-
ing reneutralization would
suffer quite different attractive deflections during the time
spent on the ionic intermediate
surface.   The insert at the
right indicates angular positions and relative intensities
for three reneutralization peaks
predicted from classical trajectory calculations.

Figure 4 shows the angular distribution of neutral scattering in various collision systems.[11]   The 35.5 eV Cs + $O_2$ data indeed has distinct structure that is
consistent with the suggested vibrational timing arguments in Figure 3.  This can
be compared to the nearly flat distribution, associated with purely repulsive
potentials, evident for Cs + Ar scattering also in Fig. 4.  At higher collision
velocities, the structure for Cs + $O_2$ changes dramatically since the reneutralizing trajectories shift to smaller impact parameters and the number of observable vibrational oscillations decreases as the collision time decreases.  The
value of $R_c'$ not only influences the branching probability for reneutralization,
but also strongly affects the extent of deflection since $R_c'$ is the distance at
which the coulombic force is "switched off"; its variation implies a related
variation in the total time that the attractive force operates.

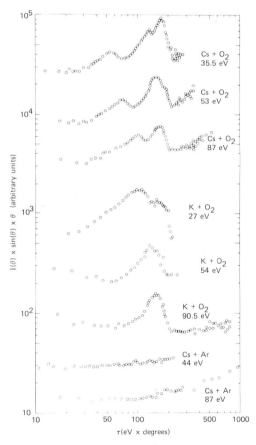

Figure 4.  Neutral product angular
distributions plotted vs. reduced
angle $\tau = E\theta$ for several collision
systems at various beam energies.

The energy dependence of the total
cross section for ion-pair formation
in the Cs + $O_2$ system[12] is shown
in Fig. 5.  Although integrating the
oscillating ion-pair formation
probability over all impact
parameters at each energy might be
expected to wash out any strong
oscillations in the velocity
dependence of the total cross
section, one still observes these
oscillations.  In fact, the interpretation of oscillatory total cross
section data for ion-pair formation
in Cs + $O_2$ and K + $O_2$ collisions[13]
yielded the first dramatic
indication that multiple transient
molecular vibrations could be used
as an internal clock for the
collision intermediate.

Returning to the differential cross
section measurements in Figure 4, we
note that the lighter K atom
explores a higher velocity regime
than Cs in the same energy range.
In fact, the reneutralization
structure simplifies considerably
for K + $O_2$ at 90 eV.  In this case
the time spent in the inner ionic
region is somewhat shorter than the
time for one $O_2^-$ vibration for all
impact parameters larger than the
extent of the repulsive inner
core.  The single peak is a rainbow
feature in the reneutralization
angular distribution.  Figure 6
shows a similar reneutralization
rainbow in the scattering of
metastable Ar atoms by $O_2$ at

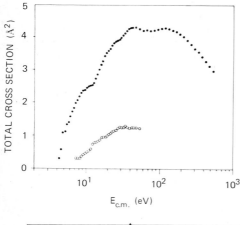

Figure 5.   Total ionization cross
section for Cs + $O_2$ collisions.
Closed circles show negative ion
products; open circles--electrons.

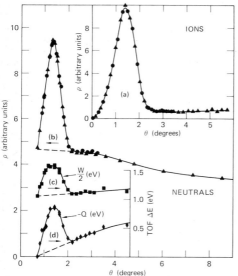

Figure 6.   $Ar^*$ + $O_2$ scattering
at $E_o$ + 114 eV.   (a) Product
$Ar^+$ angular distribution
$\rho = \theta \sin \theta \, d\sigma/d\omega$ vs. $\theta$.
(b) $Ar^*$ neutral angular
distribution.   (c) Half-width
of the product $Ar^*$ energy
profile vs. $\theta$.   (d) Average
$Ar^*$ c.m. energy loss vs.
laboratory $\theta$.

E = 114 eV,[14]   $Ar^*$ + $O_2 \rightarrow Ar^+ + O_2^- \rightarrow Ar^* + O_2$.   The corresponding ion-pair
rainbow for $Ar^*$ + $O_2 \rightarrow Ar^{2+} + O_2^-$ is also shown.   The neutral rainbow is located
at slightly smaller angles because the coulombic attraction is cut off at $R_c'$ when
reneutralization occurs.   The similarities of the $Ar^*$ and K reactions with $O_2$ are
not surprising and are consistent with many other comparison studies between
metastable rare gas and alkali reactions.[15-17]

Another important consequence of the vibrational timing upon the reneutralization
products is the very large vibrational excitation of $O_2$ that can result.   A
stretched $O_2^-$ would reneutralize vertically to a vibrationally excited $O_2$ as can
be seen in Figure 2.

Indeed, large vibrational $O_2$ excitation for scattering at angles in the rainbow region is observed by time-of-flight (TOF) measurements of scattered $Ar^{*}$[14] (and in the analogous $K + O_2$ system[18,19]). When the TOF inelasticity Q and its measured half-width $\frac{W}{2}$ are plotted vs. $\theta$ (Fig. 6), an isolated highly-inelastic feature appears precisely in the rainbow region built upon a second feature whose inelasticity slowly rises with $\theta$. The second feature is associated with the more common impulsive inelastic scattering on the repulsive inner wall of the covalent potential. This covalent contribution can be seen in the nearly flat background that underlies the sharp rainbow peak in the neutral angular distribution.

## Atom-Molecule Systems: Halogen Targets

In a series of experiments at the FOM institute, reported at the Paris ICPEAC[20] and reviewed elsewhere,[8,21] the effects on ion-pair formation of bond stretching in alkali atom halogen molecule collisions were studied. The results have since been confirmed in work by other groups.[22,23] Reactions of alkalis and $Ar^{*}$ with halogens have analogous mechanisms to those described for $O_2$ targets. However, there are great differences in the details of the potential surfaces and the resulting product distributions. First, since the vibration times of the heavy halogen negative ions are a factor of 5-10 larger than that of $O_2^-$, the halogen negative ion will only undergo a fraction of a vibration before the second crossing is reached. Second, the larger vertical electron affinity of the halogens and their much more rapid increase with r imply a larger $R_c$ and a very rapidly increasing $R_c(t)$. Hence in many cases, an adiabatic first passage through $R_c$ will yield such a rapid increase with t of $R_c$ that the second crossing will be almost entirely diabatic. Hence the ion-pair rainbow (as in Fig. 6) will have no accompanying neutral rainbow from reneutralization of the second crossing.

Yet strong rainbows in the neutral scattering have been observed and have been attributed to another consequence of bond stretching, vibrationally-induced third crossings. Consider the covalent route to the ionic surface with electron transfer occurring at the second crossing of the sphere of radius $R_c$. As the ions continue to larger distances, the $Br_2^-$ bond will start to stretch and $R_c$ will start to follow the departing ions. Since the ultimate increase in $R_c$ for halogens is very large, the expanding crossing radius will soon overtake the outgoing $K^+$ ion. This results in a third crossing created by the intramolecular motion of the $Br_2^-$. The situation is shown schematically in Fig. 7 for $K + Br_2$ at 20 eV.

$K + Br_2$   20 eV

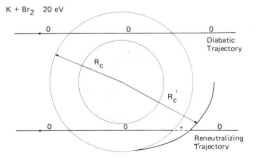

Figure 7. Simplified picture of a collision of a fast K atom and a fixed $Br_2$ molecule. The repulsive part of the potential is indicated by the sphere. Two types of trajectories leading to neutral products are shown. Whenever the system is neutral an O is inserted in the trajectory; the ionized state is indicated with a +.

If reneutralization occurs, $r \gtrsim r_e(Br_2)$ and consequently vibrational excitation will take place. This vibrational excitation indeed has been observed by Kleyn et al.[24] Since the coulomb force is only active for a very short time at the end of the collision the scattering angles are small. Thus the first effect of a third crossing occurring for covalent reneutralization is that anomalous vibrational excitation can be observed at very small scattering angles. The second effect of covalent reneutralization is that it leads to a vibrationally induced rainbow. The origin of this rainbow can be seen by referring to Fig. 7. At $t = 0$, $R'_c$ starts increasing, eventually overtaking the departing ion. For large b (near $\bar{R}_c$), the ion trajectory has a very small component in the R direction, is overtaken quickly by the expanding sphere, and suffers little attractive deflection. As b decreases, the time spent on the coulombic surface (before the third reneutralizing crossing) increases. However, the resulting increase in the attractive interaction with decreasing b is eventually compensated by the fact that the attractive force has an ever smaller component perpendicular to the trajectory. Hence a maximum attraction can occur in the deflection function for impact parameters larger than the repulsive core. This yields a rainbow that is shaped more by the bond-stretching dynamics than by the R dependence of the ion-pair potential. This rainbow feature occurs at very small scattering angles. An example (Fig. 8) is shown for collisions of $Ar^*$ with $I_2$ at 60 eV.[25] The cross section at angles beyond the rainbow region is flat, and due to underlying contributions from repulsive scattering. Similar behavior has been found for $K + Br_2$ collisions.[24]

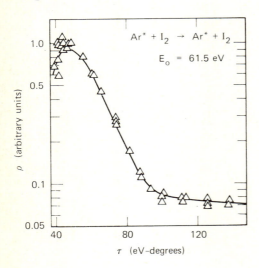

Figure 8.  Neutral product angular distribution $\rho = \theta \sin \theta \frac{d\sigma}{d\omega}$ plotted vs. reduced angle $\tau = E\cdot\theta$ for $Ar^* + I_2$ at 61.5 eV.

## Theoretical Models and Extension to Other Systems

The simple pictorial models described here can be made more quantitative by using classical trajectory surface hopping calculations on diabatic potentials, while still assuming an uncoupling between the r and R motions on the two potential surfaces. The calculational techniques and their application to ion-pair formation in collisions of alkali atoms and $Ar^*$ with halogen molecules have been described in detail (see, e.g., 8, 17, 24, 26). These calculational techniques have been extended to the systems considered here, both in their simplest form and with extensions to simulate adiabatic effects important for impact parameters near $R_c$. All of the important observed features--structure in the neutral angular distributions with $O_2$ targets, large $O_2$ excitations mediated by the ion-pair surface, oscillations in the total ion-pair formation cross section, cova-lent reneutralization rainbows in the scattering by halogen molecules--appear in the calculations.

The discussion of vibrational motion so far has been completely classical. This approach gives satisfactory qualitative explanations for the processes described above. If one is interested in the description of final vibrational state dis-tributions obtained using state specific detection, better methods are required. One method, the moving wavepacket approach introduced by Gislason et al.[27,28] has been very successful in explaining final vibrational state distributions from quenching experiments and is similar in spirit to the work discussed so far. In the moving wavepacket method, the time evolution of the vibrational wavefunction is represented by the evolution of a wavepacket on the respective potential sur-faces. When the potential surfaces are assumed additive in R and r the method is easy to use. The easiest and mathematically most elegant version of the method employs diabatic potential surfaces and the wavepacket is projected from one molecular potential to another at each adiabatic passing of a crossing. The expression for the final vibrational state population can be given in closed form.

The diabatic model has been used to explain energy-dependent oscillations in the final vibrational state distribution of $N_2(C^3\Pi_u)$ excited in the quenching of $Ar^*$ by ground state $N_2$.[27,29] The results are shown in Fig. 9. The experimental ratio ($v = 1/v = 0$) oscillates as a function of energy. The model calculations show that the oscillations are due to several vibrations of the intermediate $N_2$ molecule, which obtained some $N_2^-$ character from a higher $Ar^+ + N_2^-$ surface. The molecular vibrations during the collisions are reflected in the oscillatory motion of the wavepacket in its potential. The adiabatic version of the moving wavepacket model has been used to explain quenching of excited Na by $N_2$ and CO and very good agreement between theory and experiments has been obtained.[28]

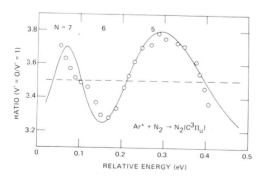

Figure 9. Population ratio ($v'=0/v'=1$) for the $N_2(C^3\Pi_u)$ product resulting from quenching of $Ar^*$ by $N_2$ as a function of energy. The circles are the experimental points; the solid line is the theoretical result. The number N refers to the number of half vibrations of the transient $N_2$ molecules.

The moving wavepacket model is an intermediate between the simple classical approach mainly used to explain this kind of atom-molecule collisions and more complete quantal studies. Klomp and Los[12] solved the close coupled equations for some limiting cases and demonstrated the potential importance of non-vertical transitions at low energies. Hickman,[30] using a classical path method and the Magnus approximation, observed oscillations in his calculated cross section for ion-pair formation in Cs + $O_2$ collisions. Recently, Becker and Saxon[31] performed close coupling calculations for inelastic K + $O_2$ collisions involving the ion-pair intermediate. These calculations confirm the simple ideas developed using the classical models. In addition, this kind of calculations gives much more insight into the detailed collision dynamics and serves as a standard with which the classical and semiclassical techniques can be compared and judged.

The effects of vibrational timing in the subpicosecond range have been mainly studied systematically in collisions of alkali or metastable rare gas atoms with simple molecules like halogens, $O_2$ and $N_2$. There is no reason to believe that these effects will not occur in many other collision systems. One obvious extension is to the study of ion-pair mediated effects in alkali collisions with larger molecules. Recently, the importance of "bond-bending" effects in the $NO_2^-$ ion[32] have been discussed in analogy to bond stretching in the halogens. Other recent experiments have indicated the possible involvement of similar vibrationally inelastic effects due to temporary charge transfer from $H^+$ or $H^-$ to $O_2$.[33,34] Collisionally-produced emissions in metastable rare gas-halogen reactions at near thermal energies seem to be similarly unaffected by bond stretching.[35] In addition, these effects should be observable in many excitation transfer or Penning ionization reactions.

## Summary

We have emphasized here the importance in atom-molecule collisions of vibrational excitation that is induced by involvement of another electronic state of the transient intermediate species. In contrast to the direct excitation of a vibrator by repulsive interactions on a single potential surface, this intermediate-induced excitation can yield high internal energies in large impact parameter collisions. Vibrational timing of the intermediate species on a subpicosecond time scale can be observed as structure in measurements of both differential and total cross sections.

There are several general requirements for direct observation of these effects. First, the collision must involve another electronic state of the diatomic molecule having a somewhat different geometry from that of the initial state. In an adiabatic description, the intermediate must access a region of the potential surface, where the molecular bond-strength is drastically modified from that of the isolated molecule. Systems with important low-lying ion-pair states satisfy this requirement. The second requirement for observation of vibrational timing is that the transfer from one diabatic state to the other occurs in a region of the potential surface where the repulsive forces are not important and do not significantly deform the molecular potentials. This is the case for many reactive systems that exhibit "harpooning" behavior, e.g., alkali and metastable rare gas collisions with halogens and $O_2$. This second requirement is a matter of convenience, since the general effects described here should be important even in systems where repulsive forces are significant in the region of electronic state interaction. A third requirement for clear observation of these vibrational effects is the need for the collision time to be not vastly different from the vibrational period of the intermediate diatomic. In general this is the case for collision energies from a fraction of an eV to several hundred eV, depending on the system.

We conclude that vibrational timing in the subpicosecond range is an important and neglected effect, the study of which can reveal the internal motion of transient states and thus provide detailed information on the dynamics of molecular collisions and the potential surfaces involved.

References

1.  A. Giusti-Suzor, XII ICPEAC, Gatlinburg Tenn. Invited Lect., Ed. S. Datz, North-Holland Publ., Amsterdam, 1981.
2.  P. C. Cosby, XII ICPEAC, Gatlinburg Tenn., Invited Lect., Ed. S. Datz, North-Holland Publ., Amsterdam, 1981.
3.  J. Polanyi, XII ICPEAC, Gatlinburg Tenn., Invited Lect., Ed. S. Datz, North-Holland Publ., Amsterdam, 1981.
4.  A. Gallagher, XII ICPEAC, Gatlinburg Tenn., Invited Lect., Ed. S. Datz, North-Holland Publ., Amsterdam, 1981.
5.  A.M.C. Moutinho, J. A. Aten, and J. Los, Physica (Utrecht) $\underline{53}$ (1973) 471.
6.  G. A. L. Delvigne and J. Los, Physica (Utrecht) $\underline{67}$ (1973) 166.
7.  A.P.M. Baede, Adv. Chem. Phys. $\underline{30}$ (1975) 463.
8.  J. Los and A. W. Kleyn, in: Alkali Halide Vapors, eds. P. Davidovits and D. L. McFadden (Academic Press, New York, 1979) p. 275.
9.  M. B. Faist, B. R. Johnson, and R. D. Levine, Chem. Phys. Lett. $\underline{32}$ (1975) 1; M. B. Faist and R. D. Levine, J. Chem. Phys. $\underline{64}$, (1976) 2953.
10.  R. E. Olson, F. T. Smith, and E. Bauer, Appl. Opt. $\underline{10}$ (1971) 1848.
11.  A. W. Kleyn, V. N. Khromov, and J. Los, J. Chem. Phys. $\underline{52}$ (1980) 5282; A. W. Kleyn, V. N. Khromov, and J. Los, Chem. Phys. $\underline{52}$ (1980) 65.
12.  U. C. Klomp and J. Los, to be published in Chem. Phys.
13.  A. W. Kleyn, M. M. Hubers, and J. Los, Chem. Phys. $\underline{34}$ (1978) 55.
14.  K. T. Gillen and T. M. Miller, Phys. Rev. Letters $\underline{45}$ (1980) 624.
15.  K. T. Gillen, T. D. Gaily, and D. C. Lorents, Chem. Phys. Letters $\underline{57}$ (1978) 192.
16.  D. W. Setser, T. D. Dreiling, H. C. Brashears, and J. H. Kolts, Faraday Discussions Chem. Soc. $\underline{67}$ (1978) 1613.
17.  A. P. Hickman and K. T. Gillen, J. Chem. Phys. $\underline{73}$ (1980) 3672.
18.  A. W. Kleyn, E. A. Gislason, and J. Los, Chem. Phys. $\underline{52}$ (1980) 81.
19.  C. E. Young, C. M. Sholeen, A. F. Wagner, A. E. Proctor, L. G. Pobo, and S. Wexler, J. Chem. Phys. $\underline{74}$ (1981) 1770.
20.  J. Los, X ICPEAC, Paris, Invited Lect., North Holland Publ., Amsterdam (1977), p. 617.
21.  A. W. Kleyn, E. A. Gislason, and J. Los, to be published in Phys. Reports.
22.  S. Wexler and E. K. Parks, Ann. Rev. Phys. Chem. $\underline{30}$ (1979) 179.
23.  K. Lacmann, Adv. Chem. Phys. $\underline{42}$, (1980) 513.
24.  A. W. Kleyn, E. A. Gislason, and J. Los, accepted for publ. in Chem. Phys.
25.  K. T. Gillen, unpublished results.
26.  J. A. Aten, G.E.H. Lanting, and J. Los, Chem. Phys. $\underline{19}$ (1977) 241.
27.  E. A. Gislason, A. W. Kleyn, and J. Los., Chem. Phys. Lett. $\underline{67}$ (1979) 252.
28.  E. A. Gislason, A. W. Kleyn, and J. Los, Chem. Phys. $\underline{59}$ (1981) 91.
29.  E. R. Cutshall and E. E. Muschlitz, J. Chem. Phys. $\underline{70}$ (1979) 3171.
30.  A. P. Hickman, J. Chem. Phys. $\underline{73}$ (1980) 4413.

31.  C. H. Becker and R. P. Saxon, accepted for publication in J. Chem. Phys.
32.  K. Kimura and K. Lacmann, Chem. Phys. Letters $\underline{70}$ (1980) 41.
33.  F. A. Gianturco, Uwe Gierz and J. Peter Toennies, J. Phys. B $\underline{14}$ (1981) 667.

34.  U. Hege and F. Linder, XII ICPEAC, Gatlinburg Tenn., 1981, Abstr. Contr.
     Papers, Ed. S. Datz, p. 945.
35.  M. S. de Vries and R. M. Martin, XII ICPEAC, Gatlinburg Tenn., 1981, Abstr.
     Contr. Papers, Ed. S. Datz, p. 949.

PHYSICS OF ELECTRONIC AND ATOMIC COLLISIONS
S. Datz (editor)
© North-Holland Publishing Company, 1982

APPLICATION OF QUANTUM DEFECT THEORY TO PREIONIZATION
PREDISSOCIATION AND DISSOCIATIVE RECOMBINATION

A. Giusti-Suzor

Laboratoire de Photophysique Moléculaire,
Université de Paris-Sud     91405 Orsay     France

The key features of the generalized multichannel quantum
defect theory are outlined and applications to two processes
involving both ionization and dissociation channels are
described : the competition between preionization and predissociation in the NO molecule, the dissociative recombination of the $H_2^+$ ion. In both cases, it is found that
apparent vibrational couplings between electronic continuum
and Rydberg series result actually mainly from an indirect
electrostatic interaction via the nuclear continuum.

INTRODUCTION

During the last years the frontier between collision and atomic or molecular
structure theories has become more and more permeable, in both ways. It is mainly
due to the developments of new experimental techniques (lasers, synchrotron
radiation ...) which allow a new insight into very excited spectral regions where
bound states a  continua are mixed. At the same time the increasing experimental
resolution yields detailed informations on the resonances arising in collision
cross sections (for example for dissociative recombination) due to temporary
capture into bound states. The theoretical treatments of these processes must
therefore deal simultaneously  and coherently with bound states and continua,
and often with different kinds of continua (ionization and dissociation). The
multichannel  quantum defect theory (MQDT) is based on this unified approach.

Only the key features of the theory will be outlined here, with references to
extensive developments in the literature. This report aims to describe the
application of MQDT to molecular processes involving simultaneously bound states,
electronic and nuclear continua, coupled all together by different types of interactions. First, we will study the competition between preionization and predissociation  for the decay of some Rydberg states in the NO molecule. Then we
will calculate the cross section for $H_2^+$ dissociative recombination, with
emphasis on its energy dependence. In  both cases comparison will be made with
recent experimental results which have motivated our theoretical studies.

THE KEY FEATURES OF MULTICHANNEL QUANTUM DEFECT THEORY

The original QDT[1] and MQDT[2] of Seaton proceeded by extrapolation of the properties of atomic Rydberg series to electron-atomic  ion collisions, the quantum
defect  becoming at positive energies the phase-shift  (in units of $\pi$) of the
scattered electron wavefunction. In this approach,  a "channel" is defined as the
set of the Rydberg states pertaining to a given series and the adjoigning  ionization continuum. Later this treatment was extended to molecular Rydberg series
by Fano[3], Jungen ant Atabek[4],  Jungen and Dill[5]. Recently, it has been
generalized[6] to long range potentials other than the Coulomb potential, that is

to "collision partners" other than electron-ion. This allowed in particular to treat molecular dissociation and predissociation problems[7], a channel consisting now in the set of the vibrational bound levels of a given potential and the adjoining dissociation continuum (see also the interesting "channel state representation" of Mies[8], closely related to MQDT). Finally, mixing of the two kinds of channels defined above yields a  uniform treatment of processes involving both electron-ion and atom-atom collisions (or half-collisions) such as dissociative recombination[9, 10], associative ionization (the inverse process) and molecular photoabsorption  above the first ionization and dissociation limits.

The basic concept allowing this unified approach is the distinction between short- and long- range interactions, which results in the subdivision of the configuration space into essentially two regions. The inner one, or "reaction zone", is characterized by strong many particle interactions, whereas in the external zone, each departing particle experiences only long range potential fields. Actually, QDT does not study the inner zone interactions, but retains only their effect on the wavefunctions in the external zone. It is cast in form of phase-shifts for the wavefunctions of the decoupled particles, and of mixing coefficients between the different channels, expressing that usually the coupling schemes are not the same inside and outside the reaction zone (frame transformations). These quantities serve as input parameters (equivalent to initial conditions at the boundary of the reaction zone) for MQDT calculations in the external region, where the total wavefunction is expressed as a linear superposition of long range potential eigenfunctions. Boundary conditions applied at infinity yield linear systems from which the coefficients of the superposition are determined, and then cross sections or spectral features are obtained. We stress that it is only at this stage that bound and continuum states are distinguished which correspond to closed or open channels respectively.

The inner zone parameters vary slowly with the "collision" energy on both sides of the threshold because they result from strong interactions involving exchange of large amounts of energy. Hence they may be obtained from (and used for) scattering as well as molecular structure studies, eventually with a scaling factor taking account of the density of states. They are determined either by ab-initio calculations, as for the case of $H_2$ in the second example below, or by analysis of experimental data, as for $N_2$ in our first example. This second "semi-empirical" approach is not a fitting procedure in which the parameters are varied until agreement is obtained between experimental and theoretical results : the inner zone parameters are extracted from experiments relative to energy ranges other than the process studied. Typically, spectroscopic observations in the discrete part of the spectra are used for studying more excited parts where bound states and continua coexist.

Let us finally precise the basic parameters necessary for a MQDT treatment of the two processes, vibronic preionization or capture, and electronic predissociation, involved in the forth coming examples.

i) Vibronic preionization concerns a set of Rydberg series converging towards the same electronic state of the ion, in different vibrational levels. Each series defines an "ionization channel"  characterized by its threshold. Since they correspond to the same electronic core, they all have a common electronic quantum defect $\mu(R)$ which is almost energy independant but may vary strongly with the internuclear distance R. Instead of introducing explicitely the terms of the Hamiltonian neglected in the Born-Oppenheimer approximation, which are responsible for the interchannel transitions, MQDT performs a frame transformation between short -and long- range vibrational basis sets[4, 5]. The resulting mixing coefficients are the integrals

$$\int \chi_v(R) \left\{ \begin{array}{c} \cos \pi \mu(R) \\ \\ \sin \pi \mu(R) \end{array} \right\} \chi_{v'}(R) \, dR \qquad (1)$$

where $\chi_v$ and $\chi_{v'}$ are ion vibrational wavefunctions. It is clear that the intensity of vibrational preionization depends on the variations of the quantum defect with R and that transitions with $|\Delta v| = 1$ are generally the strongest[11]. In the limit of $\mu$ constant with R (equivalent to a constant phase-shift for the continuum electron wavefunction) the integrals reduce to $\delta_{vv'}$ and vibrational preionization disappears : the effect of the core interactions on the external electron is independent of the nuclei position and no energy may be exchanged between the electron and the nuclei vibration.

ii) Electronic predissociation appears when a dissociative valence state, defining an open "dissociation channel", interacts with the Rydberg series. The corresponding inner zone parameters are the electrostatic couplings resulting from the Rydberg-valence configuration interaction. More precisely, they are reduced to a single almost energy independent quantity, owing to the normalization relation (12)

$$(n - \mu)^{3/2} \, V_n \simeq \text{constant} \qquad (2)$$

for the electronic couplings $V_n$ with the successive $n^{th}$ Rydberg states. The vibronic matrix elements obtained by vibrational averaging of this electronic quantity yield the interchannel mixing coefficients[10]. Note that this treatment is based on a diabatic representation of the molecular states. An adiabatic approach, in the MQDT framework, is also possible and is presently being developed (12).

For the processes initiated by photoabsorption, additional parameters are necessary : the electronic dipole moments D for transitions to the Rydberg and valence states. For the Rydberg series they reduce also to a single parameter, due to the relation, similar to (2)

$$(n - \mu)^{3/2} \, D_n \simeq \text{constant} . \qquad (3)$$

COMPETITION BETWEEN PREIONIZATION AND PREDISSOCIATION IN THE NO MOLECULE :

(in collaboration with Ch. JUNGEN)

### The experimental results

A high-quality photoionization spectrum of cold NO near threshold has been published recently by Ono, Linn, Prest, Ng and Miescher[14]. Comparison of this spectrum with high[15] and medium[16] resolution absorption spectra obtained earlier allowed unambiguous attribution of the observed photoionization peaks, and revealed two main features (see Figure 1).

i) Ionization is not the only decay mode for many of the resonant levels in this region, since the relative intensities of neighboring peaks are often quite different in the photoionization and photoabsorption spectra.

ii) Photoionization peaks corresponding to vibrational preionization via $\Delta v < -1$ transitions are present : for example the $5p\pi$ ($v = 3$) and $5p\pi$ ($v = 4$) resonances appear clearly on Figure 1, respectively above the $v = 0$ and $v = 1$

vibrational thresholds. They are typically as intense as peaks corresponding
to $\Delta v = -1$ transitions which are usually largely favored in the vibrational
preionization process.

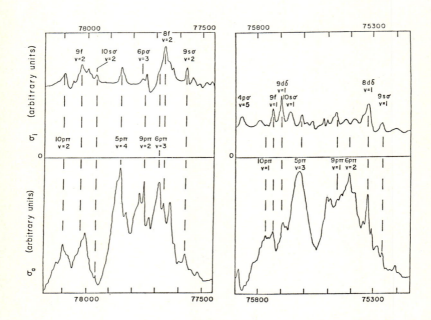

Figure 1
Details of   (a) the photoionization   spectrum   of   Ref. (14)
(b) the photoabsorption spectrum   of   Ref. (15)

The temperature is 150°K for (a), 78°K for (b) - The experimental resolution
is 9 cm$^{-1}$   for both spectra.

Point  i) is not surprising for the peaks assigned as (np$\pi$,v) levels of  $^2\Pi$
symmetry, since these levels lie above the dissociation limit of the B$^2\Pi$ valence
state which is known to cause strong perturbations in the np$\pi$ $^2\Pi$  Rydberg levels
below the dissociation limit. As an example, Figure 2 shows the successive
vibrational levels of the 5p$\pi$Q $^2\Pi$ state : the levels $v = 0$  and $v = 1$  are strongly
perturbed  by high bound levels of the B$^2\Pi$  state[17], the level $v = 2$ is predissociated  by the vibrational continuum of the same B$^2\Pi$ state, while the levels
$v = 3$ and $v = 4$ are both predissociated and preionized  since they lie above the
first ionization threshold.

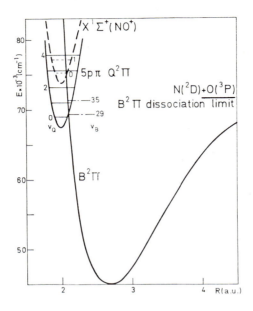

FIGURE 2

Example of $^2\Pi$ - $^2\Pi$ Rydberg valence interaction in NO

The MQDT treatment

From a deperturbation treatment of the precise spectroscopic data obtained in the discrete part of the absorption spectrum, Gallusser and Dressler[15] have extracted exactly the molecular parameters necessary to a diabatic MQDT treatment of the preionization - predissociation of the $p\pi$ $^2\Pi$ series : the electrostatic interaction between valence and Rydberg configurations, the electronic dipolar moments, and the quantum defect $\mu_{p\pi}$. The fact that the spectroscopic constants of the $NO^+$ ($X^1\Sigma^+$) ionic core and of the deperturbed ($np\pi$) Rydberg states are very similar indicates that $\mu_{p\pi}$ varies little with the internuclear distance[18] and it has been assumed to be constant in the calculations.

Our treatment included seven channels : six ionization channels defined by the thresholds v = 0 to 5 of the ion ground state, and the $B^2\Pi$ dissociation channel. The rotation was neglected in the MQDT calculations (we have verified that the widths of the peaks do not vary when the rotational quantum number J is changed) but in order to compare with observations we have computed the rotational profile of each band at 78°K, using a convolution procedure. We have also taken account of the experimental resolution.

## Results

Table 1 gives the widths obtained for the $5p\pi(v)$ peaks in the calculated photoabsorption cross section. They are also the widths of the corresponding peaks
in the photodissociation or photoionization cross sections. When taking account
of the rotational structure we reproduce quantitatively the observed widths, in
particular the alternation of broad and narrow individual lines due to the
oscillation of the overlaps between the Rydberg and dissociative state nuclear

Table 1  Calculated and observed peak widths of the
$5p\pi$ Rydberg levels of NO  $(cm^{-1})$

| level | calc. width | experimental resolution | calc. effective width [c] | obs. effective width | nature of effective observed width |
|---|---|---|---|---|---|
| $5p\pi, v=2$ | 7.6 | high[a] | 7.6 | $8^d$ | predissociation width |
|  |  | low[b] | – | – | – |
| $5p\pi, v=3$ | 24 | high | 61 | $\sim 50^d$ | } broadened rotational band contour |
|  |  | low | 61 | $\sim 69^e$ |  |
| $5p\pi, v=4$ | 0.9 | high | 0.9 | $1^d$ | preionization/predissociation width |
|  |  | low | 38 | $\sim 43^e$ | rotational band contour |

[a]  $0.1\ cm^{-1}$, corresponding to Ref. 16

[b]  $9\ cm^{-1}$, corresponding to Ref. 14 and Ref. 15.

[c]  rotational band structure $(78^\circ K)$ convoluted with the
     calculated level width from column 2 and with the appropriate
     experimental resolution width from column 3.

[d]  Ref. 16

[e]  estimated from the published spectrum of Ref. 15

wavefunctions. This confirms the continuity between the discrete and continuous
parts of the spectrum, since the molecular data used for the calculations come
from spectroscopic observations of the discrete levels. The shape of each individual line is almost Lorentzian (Figure 3). Convolution of the rotational
structure with both the individual line shape and the experimental resolution
yields theoretical band contours very close to the experimental profiles
(Figure 1) of the $5p\pi$ $(v=3)$ and $(v=4)$ peaks.

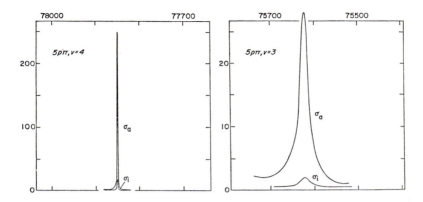

Figure 3

Calculated absorption ($\sigma_a$) and ionization ($\sigma_i$) cross sections (in Mb) near the 5p$\pi$ (v = 3) and 5p$\pi$ (v = 4)  Rydberg levels (infinite resolution, rotational structure neglected)

Coming now to the experimental features noted at the beginning, they can be understood in the light of our theoretical study

i) Predissociation is much more efficient than preionization for the decay of the superexcited (np$\pi$) $^2\Pi$ Rydberg states : the absorption cross sections in Figure 3 are the sum[+] of the small photoionization cross sections also drawn and of the much larger photodissociation cross sections, both being simultaneously calculated in our MQDT treatment. This explains the weakness of the corresponding peaks in the ionization spectrum in comparison with peaks of other symmetries, and also the marked steps at each vibrational threshold since the (p$\pi$) $^2\Pi$ states are predominent for the absorption in this part of the spectrum : usually, preionization resonances below the threshold shade off these steps in molecular photoionization.

ii) Nevertheless we find, in agreement with experiment, that preionization peaks do appear in the photoionization spectrum for these predissociated levels. No vibronic coupling enters in our calculations since the quantum defect has been assumed to be independent of the internuclear distance. Therefore, and this is the main result of our study, preionization of these levels is itself largely

---

[+]We neglect decay by fluorescence, much less rapid for these levels than preionization and predissociation.

(entirely in our calculations) induced by the <u>Rydberg - valence interaction</u> responsible for predissociation : the bound levels are coupled indirectly to the ionization continuum, via the dissociative channel (see Figure 4). This indirect process does not involve the overlaps between two bound vibrational wavefunctions of the ion as in the integrals (1), and thus the $|\Delta v| = 1$ transitions are not favored in comparison with $|\Delta v| > 1$ transitions (on the contrary, since the $|\Delta v| = 1$ transitions correspond to higher n values - n = 8, 9 instead of n = 5, 6 - and the electrostatic couplings $V_n$ decreases as $(n - \mu)^{-3/2}$ ).

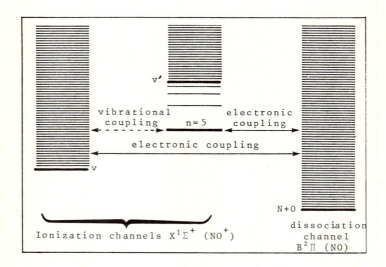

Figure 4

Schematic representation of the mechanism for $\Delta v < -1$ preionization

A similar process has been found in the $N_2$ molecule[19] connecting in this case vibrational and electronic preionizations : the electrostatic interaction between two continua (of same nature here) was responsible for an induced vibrational preionization, much stronger than usually.

We shall see now that such an indirect electronic coupling may also induce structures into dissociative recombination cross sections.

# THE DISSOCIATIVE RECOMBINATION OF $H_2^+$ WITH LOW ENERGY ELECTRONS

### (In collaboration with C. DERKITS and J.N. BARDSLEY)

The dissociative recombination of a molecular ion

$$AB^+ + e \rightarrow A + B$$

involves the same channels as the process of simultaneous ionization and dissociation : the interaction between electronic and nuclear continua is essential
and the enbedding of bound Rydberg levels in both continua (Figure 5) may lead
to resonant features in the cross sections. The only difference with the previous
process consists in the asymptotic conditions : here we have a true reactive
collision, the entrance channel being one of the open ionization channels
instead of the photon continuum, which will be omitted below (fluorescence
is still neglected here).

The $H_2^+$ dissociative recombination has been studied by several experimental
groups (20-22). The low energy, high resolution merged beam measurements of
Auerbach. et al.(21) revealed narrow window structures in the cross section.
However, attemps to reproduce them have failed and the main motivation of our
theoretical work was to assess the reality of these (or some of these) structures.

### The molecular data

The basic parameters necessary for the MQDT treatment(10) come here from ab-initio
calculations. For low energy electron incident upon ions in the first vibrational
levels of the ground state, the important dissociative state is the $(1\sigma_u)$  $^2\Sigma_g^+$
valence state. The choice of its potential curve and of the capture electronic
coupling, after calculations by Robb and Hazi, has been described by Bardsley(23).
The d waves ($\ell = 2$) play the major role in the capture and thus only the nd$\sigma$
Rydberg series have been considered. Seven ionization channels, corresponding
to the vibrational thresholds $v = 0$ to 6 of the ion ground state, have been included in the calculations. Their common quantum defect function $\mu$, deduced
from a deperturbation analysis of the adiabatic potential curves of Wolniewicz
and Dressler(24), is a slowly increasing function of the internuclear distance.

### Results and discussion

The cross sections obtained for various initial levels $v_0$ of the ions are similar
to that relative to $v_0 = 1$ shown on Figure 5a. The numerous resonances in form of
narrow dips correspond to temporary capture in bound Rydberg levels of the ionization channels  $v > v_0$  which are closed at low energy. The direct coupling
between these Rydberg levels and the entrance channel is purely vibronic and
very weak, specially for the levels with $v > v_0 + 1$  since the quantum defect
varies slowly with the internuclear distance. Therefore these bound levels play
a role in dissociative recombination essentially due to their underlineelectrostatic
coupling with the dissociative continuum(25), through the same indirect mechanism
as in the case of NO (see Figure 4). The window shape of the structures may be
compared with the windows sometimes observed in photoionization, corresponding
to states which are very weakly coupled to the photon continuum by play a role due
to  their interaction with the electron continuum (in our case, the photon and
electron continua are replaced by the electronic and nuclear continua respectively).

*A. Giusti-Suzor*

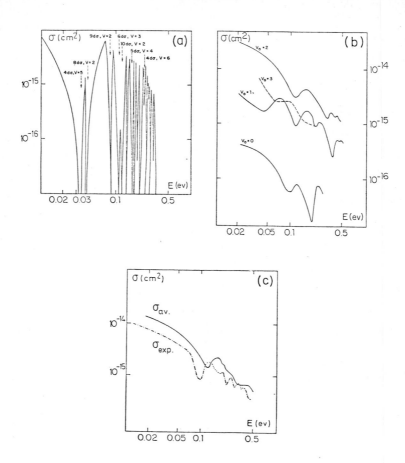

Figure 5:
Low energy cross sections for the dissociative recombinations of $H_2^+$ ions in various vibrational levels $v_0$

(a) calculated cross section for $v_0 = 1$

(b) calculated cross sections for $v_0 = 0, 1, 2$ (full lines) and $v_0 = 3$ (dotted line), convoluted with the experimental resolution 0.04 eV.

(c) $\sigma_{av}$ = averaged theoretical cross section ($v_0 \leqslant 2$) for the vibrational distribution (1:2:2) doe $v_0 = 0, 1, 2$.

$\sigma_{exp}$ = experimental cross section of Ref. 21, for $v_0 \leqslant 2$.

In order to compare the theoretical results with experiment, one has first to convolute the calculated cross sections with the experimental resolution which was 0.04 eV in the merged-beam experiment[21]. Figure 5b shows that most of the structures disappear giving rise to broad dips, especially a dip around 0.1 eV for $v_0$ = 0, 1, 2. It is due to the cumulative effect of four Rydberg levels, $5d\sigma$ ($v = v_0 + 3$), $6d\sigma$ ($v = v_0 + 2$), $8d\sigma$ and $9d\sigma$ ($v = v_0 + 1$). This dip occurs at lower energy for $v_0$ = 3 (and for $v_0 > 3$ ) due to the narrowing of the vibrational spacing. Then one must average the results over the vibrational distribution of $H_2^+$ ions. We assume that only vibrational levels $v_0 \leqslant 2$ are present and that they follow the threshold distribution of Van Busch and Dunn[26] in which the populations of the $v_0$ = 1 and 2 levels are almost equal and are twice that of the $v_0$ = 0 level. We obtain the averaged cross section represented by the full line in Figure 5c. The dashed line represents the experimental results of Auerbach et al.[21] when they remove at least partially the vibrational levels $v_0 > 2$ from their $H_2^+$ beam by adding Helium in their source. The small disagreement is probably due in part to the uncertainties in the vibrational distribution of the ions (the population of the level $v_0$ = 0 may be underestimated), in part to uncertainties in the relative velocity of the beams and in the theoretical molecular data. Nevertheless, we think that our calculations may explain the first dip near 0.1 eV observed in the cross section. On the contrary our calculations do not reproduce the structures at higher energies, the resonances of Figure 5a being there shaded off by the convolution with the experimental resolution.

If now vibrational levels $v_0 \geqslant 3$ are present in the ion beam, the dip near 0.1 eV will be much attenuated as may be deduced from the $v_0$ = 3 theoretical cross section in Figure 5b or from the "all $v_0$" experimental cross section[21]. This suggests an explanation of the experimental difficulties encountered for reproducing the structures in the low energy cross section : the $H_2^+$ ions are mixed with He atoms only in the canal coming from the discharge source, that is, during a time which is probably too short for the scavenging reaction to take place. Perhaps a selection of ions with $v_0 \leqslant 2$ was once exceptionally achieved.

The mechanisms which have been analyzed in this report put in light the relation between ionization and dissociation processes in molecules, not only in terms of competition for the decay of superexcited states : if the electronic and nuclear continua are coupled, they function like communicating vessels and cannot be treated separately. The generalized multichannel quantum defect formalism which treats them as different channels of a same collisional process is specially suitable for such a unified approach.

REFERENCES

1  Seaton, M.J., Mon. Not. R. Astron. Soc. 118, 504 (1958)
2  Seaton, M.J., Proc. Phys. Soc. 88, 811, 815 (1966)
3  Fano, U. Phys. Rev. A2, 353 (1970)
4  Jungen, Ch. and Atabek, O., J. Chem. Phys. 66, 5584 (1977)
5  Jungen, Ch. and Dill, D., J. Chem. Phys. 73, 3338 (1980)
6  Green, Ch., Fano, U., Strinati, G., Phy. Rev. A 19 1485 (1979) ;
   Fano, U., to be published in Comments Atom. Molec. Phys. (1981).

7   Colle, R., J. Chem. Phys. $\underline{77}$, 2910 (1981)

8   Mies, F.H., Mol. Phys. $\underline{41}$, 953, 973 (1980)

9   Lee, C.M., Phys. Rev. A$\underline{16}$, 109 (1977)

10  Giusti, A., J. Phys. B $\underline{13}$, 3867 (1980)

11  Berry, R.S., J. Chem. Phys. $\underline{45}$, 1228 (1966)

12  Bardsley, J.N., J. Phys. B$\underline{1}$, 365 (1968)

13  Raseev, G. and Le Rouzo, H., Book of Abstracts of contributed papers to the XII ICPEAC (Gattlinburg, July 1981).

14  Ono, Y., Linn , S.H., Prest, H.F., Ng, C.Y. and Miescher, E., J. Chem. Phys. $\underline{73}$, 4855 (1980)

15  Miescher, E., Lee, Y.T. and Gürtler, P., J. Chem. Phys. $\underline{68}$, 2753 (1978)

16  Miescher, E., Can. J. Phys. $\underline{54}$, 2074 (1976) ; Miescher, E. and Alberti, F., J. Phys. Chem., Ref. Data $\underline{5}$, 309 (1976).

17  Gallusser, R., Thesis, Eidgenössische Technische Hochschule, Zürich (1976) ; Gallusser, R. and Dressler, K., in preparation

18  Herzberg, G. and Jungen, Ch. J. Mol. Spectrosc. $\underline{41}$, 425 (1972)

19  Giusti-Suzor, A. and Lefebvre-Brion, H., Chem. Phys. Letters $\underline{76}$, 132 (1980) ; Tabché-Fouhailé, A., Ito, K., Fröhlich, H., Morin, P., Guyon, P.M. and Nenner, I., to be published in J. Chem. Phys. ; Lefebvre-Brion,H. and Giusti, A., Abstracts of the Workshop Electron Atom and Molecule Collisions, Bielefeld, May 1980 (Plenum Press, New York)

20  Peart, B. and Dolder, K.T., J. Phys. B $\underline{7}$, 236 (1974)

21  Auerbach, D., Cacak, R., Caudano, R., Gailly, T.D., Keysser, J., Mc Gavan, J.W., Mitchell, J.B.A. and Wilk, S.J., J. Phys. B $\underline{10}$, 3797 (1977)

22  Mathur, D. Khan, S.U. and Hasted, J.B., J. Phys. B$\underline{11}$, 3615 (1978)

23  Bardsley, J.N., Comments Atom. Mol. Phys. $\underline{10}$, 191 (1981)

24  Wolniewicz, L. and Dressler, K., J. Mol. Spectrosc. $\underline{67}$, 416 (1977)

25  O'Malley, T.F., J. Phys. B$\underline{14}$, 1229 (1981)

26  Von Busch, F. and Dunn, G.H., Phys. Rev. A$\underline{5}$, 1726 (1972)

PHYSICS OF ELECTRONIC AND ATOMIC COLLISIONS
S. Datz (editor)
© North-Holland Publishing Company, 1982

PROBING THE TRANSITION STATE
IN REACTIVE COLLISIONS

T. Carrington Jr.[a], J.C. Polanyi and R.J. Wolf

Department of Chemistry
University of Toronto
Toronto M5S 1A1, Ont.
Canada

There is lively interest in the possibility
of probing the < 1 ps intermediate in
reactive encounters.  Preliminary reports
of success, in both absorption and emission,
have appeared.  Since the emission was
examined over a range of wavelengths it
constitutes a primitive "spectroscopy of
the transition state".  In this paper we
use 3D Monte Carlo trajectories as a
means to simulating transition state spectra,
and conclude that this spectroscopy should
add materially to the clues that we have
regarding the molecular dynamics of the
simplest chemical reactions.

INTRODUCTION

When chemists denote a simple atom-molecule exchange reaction as,

$$A + BC \rightarrow AB + C$$

they imply the existence of a transitory species $ABC^{\ddagger}$; the
"transition state".  We use this term to denote all configurations
of the three particles intermediate between reagents, A + BC, and
products AB + C.

LIGHT ABSORPTION

There has been extensive discussion in recent years of the possibility of probing this transitory species by laser *irradiation*;
see ref. [1].  In one case evidence of success  has been reported [2].  The reaction was,

$$K + HgBr_2 \rightarrow KBr + HgBr \qquad (1)$$

Normally this reaction will only yield ground electronic state
$HgBr(X^2\Sigma^+)$, since the path to form HgBr* $(B^2\Sigma)$ is endothermic.  It
was found, however, that irradiation with intense laser pulses
at 595 nm gave a small but significant yield of HgBr*.  Since
595 nm radiation was not expected to be resonant with either
reagents or products, the observed effect appeared to be due to
absorption by the transition state, $KBrHgBr^{\ddagger}$.  Further work is in
progress [3].

LIGHT EMISSION

The lifetime of the reactive transition state $ABC^{\ddagger}$ will be at least as long as that of the unbound intermediate $AB^{\ddagger}$ in a collision

$$A + B \rightarrow AB^{\ddagger} \rightarrow A + B \qquad (2)$$

This type of "transition state" has been successfully observed spectroscopically; it evidences itself in collisions

$$A^{*} + B \rightarrow AB^{\ddagger *} \rightarrow A^{*} + B \qquad (3)$$

in the form of broad wings on the *emission* spectrum of A*,which are ascribed to emission from $AB^{\ddagger *}$ during the ~1 ps that A* and B have a repulsive interaction [4].  For example A may be $Na*(3^2P)$ and B an inert gas. Recently this type of study has been made for A* + BC, where A* was $Na*(3^2P)$ and BC was $N_2$ [5].  Figure 1 illustrates schematically the origin of red and blue wings on the A* resonance line, due to emission from $ABC^{\ddagger *}$ in a hypothetical collision A* + BC (non-reactive).

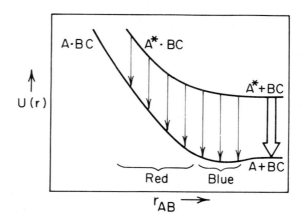

Figure 1
Schematic diagram of upper and lower optically-linked potential-energy curves, with arrows showing emission of the atomic line (A* → A, heavy arrow at right) and wing emission during the collision (A* · BC → A · BC).

It was this type of reasoning that led to the proposal for an alternative, and perhaps simpler, means of "probing the transition state" in chemical reactions [6].  The transition state $ABC^{\ddagger *}$ can be formed--and should therefore be observable in emission--when A*

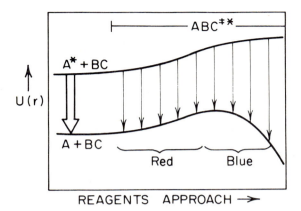

Figure 2
Schematic diagram of upper and lower optically-linked
potential-energy curves, with arrows showing emission of
the atomic line (A* → A, heavy arrow at left) and
emission by the reactive transition state as it proceeds
along the reagent approach coordinate(the emission is
ABC‡* → ABC‡).

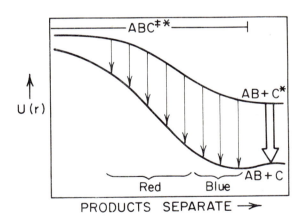

Figure 3
Schematic diagram of upper and lower optically-linked
potential-energy curves, with arrows showing emission
of the atomic line (C* → C, heavy arrow at right) and
emission by the reactive transition state as it proceeds along the coordinate of product separation (the
emission is ABC‡* → ABC‡).

collides with a reactive species BC,

$$A* + BC \rightarrow ABC^{\ddagger*} \rightarrow \text{Products} \qquad (4)$$

or when reagents A+BC are in the process of forming AB + C* as products,

$$A + BC \rightarrow ABC^{\ddagger*} \rightarrow AB + C* \qquad (5)$$

Reaction (4) is illustrated schematically in fig. 2, and reaction (5) is similarly illustrated in fig. 3. In both figs. 2 and 3 the emitting species is a reactive transition state $ABC^{\ddagger*}$; the difference is that in the former case (A* as reagent ) $ABC^{\ddagger*}$ can be regarded as A* in collision with BC along the approach coordinate, and in the latter case (C* as product) $ABC^{\ddagger*}$ comprises C* in collision with AB along the retreat coordinate (or "coordinate of separation"). Supposing that both A* and C* have optically allowed emissions, the oscillator strength for the case shown in fig. 2 can be depended on to be high when the transition state is along the approach coordinate, and in the latter case it will be high along the retreat coordinate.

The case of transition state emission for which experimental data exists is of the general type presented in fig. 3 [7]. The reaction between atomic fluorine and sodium dimer proceeds efficiently to form electronically excited atomic sodium in the $(3^2P)$ upper state of the D-line transition [8],

$$F + Na_2 \rightarrow FNaNa^{\ddagger*} \rightarrow NaF + Na*(3^2P) \qquad (6)$$

This reaction has been studied at reagent pressures of $10^{-4}$-$10^{-5}$ Torr in a crossed uncollimated "beam" apparatus [7]. Using a double monochromater to filter out the intense D-line emission (corresponding to the heavy downward arrow at the right of the schematic representation in Fig. 3) red- and blue-shifted wing emission was observed extending several hundred Å from the D-line, with an intensity of $\sim 10^{-6} \times I_D$ ($I_D$ is the D-line intensity) for a spectral slit width of 8.5 Å; the wings are depicted in fig. 1 of ref. [7].

In the following paragraphs we give preliminary results of a 3D Monte Carlo trajectory study in which we have generated a transition-state emission spectrum on the basis of assumed potential-energy hypersurfaces for a reaction with the energetics appropriate to reaction (6).

The general question we wish to address is the likely utility of this type of transition-state emission spectroscopy. Specifically we investigate the sensitivity of the far wing spectra to a) reagent collision energy, and (b) form of the reactive (ie upper) potential-energy surface.

The potential-energy surfaces (pes) used in the computations were of the LEPS (London Eyring Polanyi Sato [9]) type. These functions defined hypersurfaces $V(R_{AB}, R_{BC}, R_{AC})$, where V is the potential energy of electrons plus nuclei, and R are the subscripted internuclear separations for A + BC $\rightarrow$ AB + C. The LEPS potential provides a smooth interpolation between reagent (BC) and product (AB) Morse potentials. Some of the constants required in order

to specify the potential are normally obtained from the spectroscopy
of the relevant diatomics; the remainder constitute adjustable
parameters.  In the present work we used the following values
of the parameters required in order to define three LEPS equations,
namely two alternative upper pes termed $V^U(I)$ and $V^U(II)$, and a
common lower pes $V^L$:  $D_1=D_3=123 \cdot 68$ kcal/mole (all); $D_2=65 \cdot 37$,
$65 \cdot 37$, $16 \cdot 83$ kcal/mole; $\beta_1=\beta_3=1 \cdot 013$ Å$^{-1}$(all); $\beta_2=0 \cdot 435$, $0 \cdot 435$,
$0.857$ Å$^{-1}$; $R_1^0=R_2^0=1.926$ Å (all); $R_3^0=3.0789$ Å (all); $S_1=S_3=0.5$,
$0 \cdot 1667$, $0 \cdot 1667$; $S_2=0 \cdot 1667$(all).  These alternative constants
refer respectively to the three pes $V^U(I)$, $V^U(II)$, $V^L$.  The
symbols are defined in ref [10].

The first requirement to be met in choosing these constants was
(as noted above) that the energy release on $VU(I)$ and $V^U(II)$ correspond to F + Na$_2$ → NaF + Na* (3$^2$P), and that on $V^L$ to F + Na$_2$ →
NaF + Na(3$^2$S); see fig. 4.  The second requirement was that all

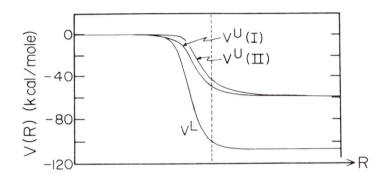

Figure 4
Two alternative (collinear) upper potential-
energy profiles, $V^U(I)$ and $V^U(II)$, and a (collinear)
lower potential-energy profile, $V^L$.  Energetics are
those for F + Na$_2$ → NaF + Na* (3$^2$P) on the upper
surface, and → NaF + Na(3$^2$S) on the lower.  The
broken vertical line indicates the position along
the reaction coordinate at which the old and
the new bonds are equally stretched from their
equilibrium separations

the surfaces be "attractive", ie that they release a substantial
fraction of their exothermicity as A and BC approach, since
these are 'harpooning' reactions [9,10].  That this requirement
is met is evident from fig. 4;  energy-release to the left of
the vertical broken line is largely along the coordinate of
approach, and that to the right is along the coordinate of
separation.  The final requirement was that the two alternative
upper surfaces, $V^U(I)$ and $V^U(II)$, differ by a moderate amount --

here again, see fig. 4. Using the definition given elsewhere
(e.g. ref [10]) the percentages of attractive energy release for
the three surfaces were $A_{\perp}=77, 61$ and 91% for $v^U(I)$, $v^U(II)$ and
$v^L$ respectively.

The model used in order to generate the transition state emission
spectra was similar in spirit to the quasi-static theory of line
broadening; ref. [4(a)]. However, where this model computed relative concentrations of emitters along one coordinate of
separation, $R_{AB}$, we were required to do so along three coordinates
of configuration-space; $R_{AB}$, $R_{BC}$, $R_{AC}$. The procedure used was to
run a batch of >1,000 reactive trajectories (with zero point
vibration, and rotation J=0) across the upper pes. As each
individual trajectory progressed across $v^U$ it was registered in
successive bins along the frequency coordinate $v$. The frequency
corresponding to each configuration was calculable, since $v^U-v^L$ was
known. The breadth of the frequency-bins was 20 cm$^{-1}$.

In order that a trajectory contribute to the emission intensity at
frequency $v$ it was necessary not only that the trajectory pass
through the corresponding region of $v^U-v^L$, but also that it emit.
The probability of emission at a given configuration was taken to
be proportional to the ratio of the rate of emission to the rate
of removal by propagation of the trajectory. The former quantity
was taken from experiment ($\tau=1.6 \times 10^{-8}$ s for Na*($3^2$P) $\rightarrow$ Na($3^2$S))
and the latter from the integration step size $\Delta t=1.2 \times 10^{-15}$s). This
was weighted by $v^3$ to obtain the relative probability of emission,
$P_{em}$, for each configuration.

Emission was deemed to occur with this small but finite probability at each integration-step, but the trajectory was continued
with an appropriately reduced weight. Trajectories were terminated at $R_{BC} > 11.5$ Å when $v^U-v^L$ was within 10cm$^{-1}$ (half a bin
width) of $v_D$, the frequency of the D-line. This theoretical
"resolution" was in approximate conformity with the experimental
spectral resolution [7]. All residual probability at the
terminal step (ammounting to ~0.9999 of the unit weight for each
trajectory) was ascribed to the D-line bin.

Though the concentration of emitters within an interval of configurations, and hence interval of frequencies, did not figure
explicitly in the foregoing, it figured implicitly. With the
stated $\Delta t$ a typical trajectory contributed many times to a
single bin in the spectrum. The number of such contributions was
inversely proportional to the local velocity -- just as the
concentration of emitters in any interval of configuration-space
would also be inversely proportional to the local velocity.

The computed far-wing spectra are illustrated in fig. 5. Reference
to the scale across the top of the figure ($\Delta v$(cm$^{-1}$)) shows that
the computed emission extends several thousand wave-numbers to
either side of the D-line, which is located at $\Delta v=0$. It should
be noted that in the experimental data of ref [7] the wings
recorded extended a few hundred wavenumbers from the D-line. In
this region of the nearer wing the present calculation gives wing
intensities relative to the D-line ($I_W/I_D$) ~$10^{-6}$, in approximate
accord with experiment. The far-wing pictured here has
$I_W/I_D \sim 10^{-7}$.

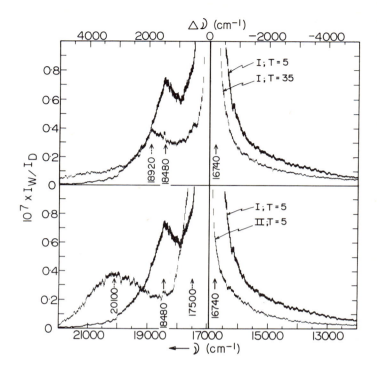

Figure 5

Computed transition state (i.e. wing) emission
intensities relative to the D-line emission
(labelled Δν=0 above). The top panel is for
one reactive pes at two collision energies,
whereas the bottom panel is for two alternative
reactive pes at a single collision energy (see text).

Both the computed nearer-wing and the computed far-wing show
interesting spectral features (not yet observed experimentally).
We have recorded those for the far-wing since they are (at least
for the present pes and collision energies) the most prominent.
In work now in progress [11] we interpret these features as
being due to the existance of favoured turning points for the
reactive trajectories proceeding across the upper pes; ie they
have an origin akin to the peaks in electronic emission intensity
in vibrationally-excited diatomic spectra at frequencies
corresponding to vibrational turning points. The existence
of these features should assist considerably in the interpretation

of the wing emission, ie in the correlation of $I_w(\nu)$ with the molecular dynamics.

We note three further aspects of the spectra recorded in fig. 5 which can be linked to the reaction dynamics. One is that the breadth of the wing emission and the location of the far-wing peak is sensitive to alteration in the reagent collision energy (see top panel; the heavy line refers to transition state emission at a collision energy T=5 kcals/mole, and the lighter line to T=35 kcal/mole). The far-wing peak in the blue shifts from 18480 to 18920 $cm^{-1}$. This is indicative of alteration in the characteristic path across the upper pes at enhanced collision energy [12].

The second favourable circumstance is that the wing-breadth and the location of the peak in the far wing is sensitive to a modest change in the form of the upper (reactive) pes from $V^U(I)$ to $V^U(II)$, with no change in $V^L$ or in T (see lower panel, fig. 5). The far-wing peak in the blue shifts from 18480 to 20100 $cm^{-1}$.

Finally we note that at the enhanced collision energy (upper panel, fig. 5) the far wing includes a significant contribution from frequencies shifted by $\Delta\nu \approx 4,000$ $cm^{-1}$. Examination of the collinear plots of energy differences, $V^U(I)-V^L$, shows that there are no collinear configurations for which $V^U(I)-V^L$ corresponds to $\Delta\nu$ as large as this. The explanation is that at enhanced collision energy there is a significantly increased likelihood of reaction through bent configurations $A \cdot B \cdot C$. This same dynamical effect has previously been inferred from the dependence of product rotational excitation on reagent collision energy in other reactive systems [12]. Here we have the possibility of observing bending in the transition state more directly.

CONCLUSION

It appears that spectroscopy of the transition state has the potential for furthering our understanding of reaction dynamics. Studies in state-to-state chemistry, whereby reagent and product attributes are specified and reactive cross sections measured, will continue to provide a rich lode of information from which categories of reaction dynamics can be deduced, but spectroscopy of the transition state should ultimately add to the clues at our disposal. The problem of reaction dynamics is so multivariant that a new category of information would represent a welcome development.

ACKNOWLEDGEMENTS

This work was supported by the Natural Sciences and Engineering Research Council of Canada (NSERC). T.C. thanks NSERC for the award of a Summer Studentship.

REFERENCES

[1]  Early work goes back to L.I. Gudzenko and S.I. Yakovlenko Sov. Phys. JETP 35 (1972) 877. For a review see T.F. George, I.H. Zimmerman, P.L. DeVries, J.M. Yuan, K.S. Lam, J.C. Bellum, H.W. Lee, M.S. Slutsky, and J.T. Liu, in Chemical

and Biochemical Applications of Lasers, ed. C.B. Moore
(Academic Press, New York, 1979) Vol. IV.  For a discussion
of the prospects for electronic excitation of transition
states by intense lasers, see J.C. Light and A. Altenberger-
Siczek, J. Chem. Phys. 70 (1979) 4108.

[2]   P. Hering, P.R. Brooks, R.F. Curl Jr., R.S. Judson and
      R.S. Lower, Phys. Rev. Lett. 44 (1980) 687.

[3]   P.R. Brooks.  Private Communication.

[4]   (a) R.E.M. Hedges, D.L. Drummond and A. Gallagher,
      Phys. Rev. A 6 (1972) 1519.  (b) W.P. West, P. Shuker and
      A. Gallagher, J. Chem. Phys. 68 (1978) 3864 and references
      therein.

[5]   A. Gallagher, Proc. XIIth ICPEAC  Conf., Gatlinberg, Tenn.
      USA, (North Holland, Amsterdam, 1981).

[6]   J.C. Polanyi, Faraday Disc. Chem. Soc. 67 (1979) 129.

[7]   P. Arrowsmith, F.E. Bartoszek, S.H.P. Bly, T. Carrington Jr.,
      P.E. Charters and J.C. Polanyi, J. Chem. Phys 73 (1980) 5895.

[8]   (a) D.O. Ham, J. Chem. Phys. 60 (1974) 1802.  (b) R.C. Olden-
      borg, J.L. Gole and R.N. Zare, J. Chem. Phys. 60 (1974) 4032.

[9]   P.J. Kuntz in Dynamics of Molecular Collisions, Part B, ed.
      W.H. Miller (Plenum Press, New York, 1976) Chapter 2, p. 53.

[10]  J.C. Polanyi and J.L. Schreiber in Physical Chemistry, An
      Advanced Treatise, Vol. VIA, Kinetics of Gas Reactions,
      eds. H. Eyring, D. Henderson, and W. Jost (Academic Press,
      New York, 1974) Chapter 6, p. 383.

[11]  J.C. Polanyi and R.J. Wolf, J. Chem. Phys. To be published.

[12]  A.M.G. Ding, L.J. Kirsch, D.S. Perry, J.C. Polanyi and
      J.L. Schreiber, Faraday Discuss. Chem. Soc. 55 (1973) 252.
      J.C. Polanyi and J.L. Schreiber, ibid. 62 (1977) 267.

a
  Present address:  Department of Chemistry, University of
  California, Berkeley, Cal. 94720, USA.

PHYSICS OF ELECTRONIC AND ATOMIC COLLISIONS
S. Datz (editor)
© North-Holland Publishing Company, 1982

THE ABSORPTION AND EMISSION OF RADIATION BY THE COLLISION COMPLEX

Alan Gallagher[*]

Joint Institute for Laboratory Astrophysics
National Bureau of Standards and University of Colorado
Boulder, Colorado   80309   U.S.A.

The absorption and emission of radiation by interacting atoms has been used for
many years to study atomic collisions.  Generally called line broadening, it
is actually a study of the diatomic collision complex.  It is also a form of
molecular spectroscopy, but it involves free rather than bound states, so it
requires the study of intensity information rather than the wavelengths of
bound-bound lines.

This field is now approaching a very exciting prospect of studying the collision
complex in reactive collisions.  It has long been recognized that the same ideas
and principles normally applied to two interacting atoms can also be applied to
an atom-molecule or triatomic interaction.  However, the complexity is very much
greater.  Thus, whereas line broadening by two colliding atoms has been studied
for many decades in many laboratories, triatomics are just beginning to be studied in only a few laboratories.  Yet this type of measurement holds great promise
as a powerful diagnostic of chemical reactions and other atom-molecule collision
processes.

I will first review the description of absorption and emission of radiation by
diatomics, then generalize to triatomics, discuss the new complications which
result, and end with an example under study in my laboratory.

(a) <u>Atom-Atom Collisions</u>.  The classical Franck-Condon principle (CFCP) leads
to great simplification in describing the absorption and emission of radiation by
a collision complex, or indeed by a bound molecule as well.  The CFCP predicts
that the photon energy is given by the difference in electronic state potential
energies V(R), independent of the nuclear velocity, i.e. when atoms at internuclear separation R make a radiative transition from electronic state j to k

$$h\nu = |V_k(R) - V_j(R)| \quad . \tag{1}$$

The total absorption or emission of radiation at frequency $\nu$ is thus given by
the total number of diatomic pairs at separation $R(\nu)$ times the radiative transition probability $\Gamma(R)$, where $R(\nu)$ is obtained from Eq. (1).  If atoms of type
B are in a thermal distribution about a potential $V_j(R)$ centered on an atom A,
then the probability of finding an atom B at $R \rightarrow R + dR$ from A is given by

$$P(R)dR = [B] \ 4\pi R^2 dR \ e^{-V_j(R)/kT} \quad , \tag{2}$$

where [B] refers to the density of species B.  Then to obtain the probability of
a spontaneous emission of frequency $\nu \rightarrow \nu + d\nu$ one uses $R(\nu)$ from Eq. (1) in Eq.
(2) and multiplies by the spontaneous emission rate $\Gamma(R)$ and the number of atoms

---

[*]Staff Member, Quantum Physics Division, National Bureau of Standards.

A in state j

$$I_{emiss}(\nu,T) = [A_j][B] \; 4\pi R(\nu)^2 \; \frac{dR(\nu)}{d\nu} \; \Gamma(R(\nu)) \; e^{-V_j(R(\nu))/kT} \; . \qquad (3)$$

If $R(\nu)$ is a monotonic function then Eq. (3) can be the basis of the following simple but powerful diagnostic technique for determining diatomic interactions $V_j(R)$: A measure of the T dependence of $I(\nu,T)$ yields $V_j(\nu)$, and $V_k(\nu)$ then follows from Eq. (1). To turn this into $V_j(R)$ and $V_k(R)$ requires $R(\nu)$, and this is obtained from the $T \to \infty$ limit of Eq. (3). Combining Eq. (3) with $I(\nu_o) = [A_j] \; \Gamma(\nu_o)$ yields

$$\frac{4\pi}{3} \{R^3(\nu) - R^3(\nu_1)\} = \int_{\nu_1}^{\nu} d\nu' \; \frac{I(\nu',\infty) \; \Gamma(\nu_o)}{I(\nu_o,\infty)[B] \; \Gamma(\nu')} \; . \qquad (4)$$

Here $\nu_o$ is the frequency of the j to k transition of atom A and $\Gamma(\nu_o)$ is the free atomic spontaneous-emission rate. If one can estimate or obtain $\Gamma(\nu_o)/\Gamma(\nu')$ from theory, then the right-hand side of Eq. (4) is obtained from the data. To obtain the function $R(\nu)$ from Eq. (4) one then chooses an $R(\nu_1)$ at the smallest R observed in the experiment. One then has $V_j(\nu)$, $V_k(\nu)$, and $R(\nu)$ so that the interaction potentials $V_j(R)$ and $V_k(R)$ are known. In Fig. 1 we have shown the rubidium-xenon A-X band and potentials as an example of these relationships.[1]

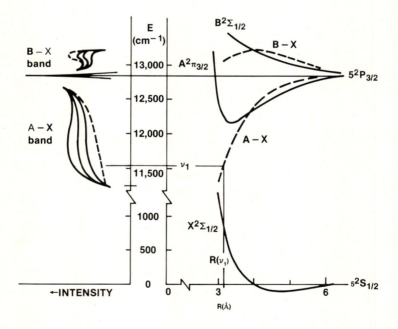

Fig. 1. The electronic energies of the Rb-Xe molecule are given on the right, and the RbXe emission spectrum, as a function of temperature, on the left. The lines at R = 3.2 Å and E = 11,550 cm$^{-1}$ intersect the (dashed) $V_A$-$V_X$ line, indicating the Eq. (1) relationship between R and $\nu$.

The absorption coefficient of the gas can replace the emission coefficient with minimal change in the above equations and procedures. Reference 2 is an example of such absorption measurements, and there are many examples of emission measurements, a few of which are referenced here.

The above procedure for virtually unique inversion of intensity data to yield adiabatic interaction potentials only works when a number of criteria are satisfied. First, it is necessary that each optical frequency $\nu$ be produced by only one pair of $V_j$-$V_k$; i.e., that there are no overlapping bands of populated levels. Otherwise Eq. (3) must be summed over bands and they are not easily separated. Note that this condition is satisfied in the case of Fig. 1 as the A–X and B–X bands appear on opposite sides of the atomic line. Next, the function $\nu(R)$ must be monotonic so that $R(\nu)$ is single valued. The B–X band in Fig. 1 is a case of a double valued $R(\nu)$; intensity data for this "satellite" case can often be reduced using some additional intensity information or plausible assumptions, but with a loss of accuracy and definiteness.[1,3] Next one needs theoretical information regarding $\Gamma(R)$. In the case where j–k is a strong atomic transition the variation in $\Gamma(R)$ is generally minimal, but there are many instances where $\Gamma(R)$ varies by orders of magnitude from the atomic value $\Gamma(\infty)$. Next, in order to use Eq. (3) the vapor must be in a thermal vibrational distribution in the initial state. This is almost always valid for absorption from ground states by a mixed gas, but it is often not valid for absorption or emission from an excited state. This is particularly true for excited states that have more than a few kT of binding, and care must be exercised to establish experimentally that the equilibrium limit is achieved.

In the case of colliding beams, beam-gas collisions, or other non-equilibrium situations, the Boltzmann distribution, $\exp(-V_j(R)/kT)$ in Eqs. (2) and (3) must be replaced with the orbit-averaged probability of occurrence of $R \to R + dR$. The general principle in some non-equilibrium cases has been described in Ref. 4.

Two other requirements are generally easily satisfied; the validity of the CFCP and a reasonable choice of $R(\nu_1)$. The CFCP is valid when the JWKB approximation is valid, i.e., when the deBroglie wavelength for the nuclei does change radically across one wavelength and $V_j(R)$ (and $V_k(R)$) is sufficiently separated from other states so that the nuclear motion does not cause nonadiabatic mixing; generally separations greater than 50 $cm^{-1}$ are sufficient. Finally, $R(\nu_1)$ can usually be obtained from ground-state collision or gas-transport data, from calculations, or from estimates of atomic sizes.

(b) Atom-Molecule Collisions. The CFCP applies to polyatomic collision complexes just as well as to diatomics, and it is not difficult to generalize the above picture to a collision of an atom with a diatomic molecule. A hypothetical example with the potentials depending only on internuclear-separations $R_1$ and $R_2$ is shown in Fig. 2.

The intensity in the triatomic case is given by the three-body probability distribution for position $R_1$, $R_2$, $\theta$ in the initial state j, again with $h\nu = |V_k(R_1,R_2,\theta) - V_j(R_1,R_2,\theta)|$. For a thermalized distribution this yields

$$I(\nu,T) \propto \Gamma(R_1,R_2,\theta) \frac{dVol(R_1,R_2,\theta)}{d\nu} e^{-V_j(R_1,R_2,\theta)/kT} , \qquad (5)$$

where Vol represents volume in configuration space.

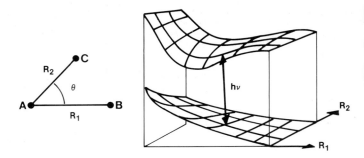

Fig. 2.  Hypothetical triatomic potential surfaces, for fixed $\theta$, with the Eq.
(5) relationship indicated by the line labeled $h\nu$.

However, the actual situation is now much more complicated, as very few of the
simplifying conditions of the diatomic case are now applicable.  First, the
$\nu(R_1,R_2,\theta)$ function will not be monotonic in all parameters and each value of
$\nu$ will generally arise from many regions of $R_1$, $R_2$, $\theta$ space.  Next, there will
be many more potential surfaces and these will frequently cross each other.  In
fact, chemical reactions and other processes of interest are often related to
such surface crossings.  Next, the transition moment will often undergo large
changes due to the changes in electronic structure and couplings that occur.
Finally there will be many causes to prevent the nuclei from attaining an equi-
librated distribution in coordinate space, or a classical-orbit distribution in
a beam experiment.

The situation need not be as bleak as indicated in the previous paragraph, as
will be shown by the following example, but if one wants to interpret observations the species and processes must be chosen with great care.  Furthermore,
theoretical guidance regarding the form of the three-body potential surfaces is
almost essential, and in all probability experiments must be viewed as a way of
testing and improving on the description of collisions that have been very thoroughly studied theoretically.

(c) The Example of Na-$N_2$.  As an example of an atom-molecule collision I will
discuss a process under study in our laboratory.  It has long been known that
collisions of $N_2$ with excited Na(3P) atoms quench the Na to the 3S state, resulting in vibrationally excited $N_2$.  The total quenching cross section has been
measured as well as the distribution of $N_2$ vibrational states which result.[5]  The
theory of this electronic-to-vibrational energy transfer process was originally
examined in a classic paper by Bauer et al.,[6] while Habitz[7] has recently published detailed calculations of the interactions.  Bauer et al. used a diabatic-
state picture and estimated the interactions, and they note that there are very
major uncertainties regarding these interactions.  Habitz, on the other hand, has
used an adiabatic-state description and carried out detailed model-calculations
of these potential surfaces.  The present measurements represent a relatively
direct test of these potentials, although their interpretation will also require
advances in the theory of collisional line shapes to allow for the presence of
electronic quenching.  We hope to improve understanding of the theory and of the
potentials with the current measurements, while taking advantage of these calculations that are already available.

The collisional quenching is believed to occur when the Na-$N_2$ separation R is
about 3 Å, while the two N atoms are separated by r ~ 1 Å.  Thus Bauer et al.

simplified the three-body potential surfaces by neglecting the angle $\theta$ between $\underset{\sim}{r}$ and $\underset{\sim}{R}$. In Fig. 3 we show the diabatic potentials considered by Bauer et al. Here the upper potential surface of $Na^*$-$N_2$ is the initial state and the approach of $Na(3P)$ toward the $N_2(v'=0)$ occurs along this surface, from the rear toward the front of the figure, with $r \sim 1.1$ Å. The surface of $Na^+$ on $N_2^-$ in the $\pi_g$ resonance state is indicated by the heavy-line potential surface. This $N_2$ state is the $\sim2$ eV electron scattering resonance of $N_2$, and in this diabatic picture the formation of this $Na^+$-$N_2^-$ electronic state can occur as the Na reached $\sim3$ Å where these diabatic potential surfaces cross. If a Landau-Zener jump occurs, the $Na^+$ is then attracted toward the $N_2^-$ along this ionic surface, while the two N atoms begin a wider, displaced vibration in the $N_2^-$ potential well. This $N_2^-$ vibration to larger r allows non-zero Franck-Condon factors to vibrationally excited $N_2$, so that transitions to the lower, $Na(3S)+N_2$, potential surface with $v'' > 0$ are possible. The Na and $N_2$ separate (toward the rear of the figure) on this surface, leaving vibrationally excited $N_2$.

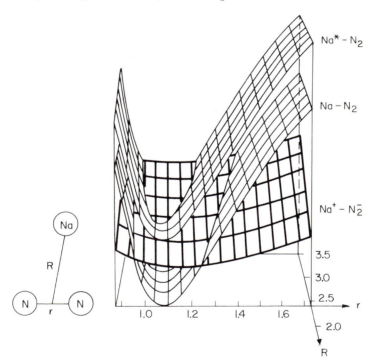

Fig. 3. The diabatic potential surfaces of Na-$N_2$ considered by Bauer et al., with dimensions in Å.

This diabatic picture offers valuable insight into the quenching process, but it can also be described using adiabatic potentials as Habitz has done. In addition to the greater accuracy of Habitz' potentials, the radiative transition occurs between the adiabatic potentials, and they are a more appropriate description here. The important features of these potentials are shown in Fig. 4, where we have shown the r,R dependence for the $\theta = 90°$ case that Habitz considers most important to the quenching process (the $Na^*$-$N_2$ interaction is most attractive

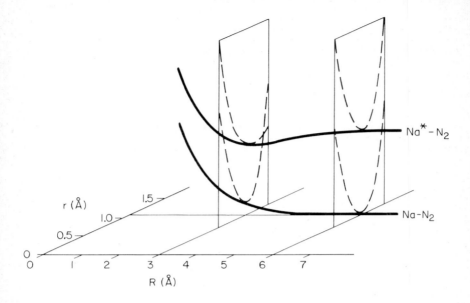

Fig. 4.   The Na*-N₂ (attractive branch) and Na-N₂ potential surfaces of Habitz
          for θ = 90°, shown diagrammatically as functions of r and R.   The R
          dependence of the potentials in the r = 1.1 Å plane, corresponding to
          the equilibrium separation of N₂, are shown as solid lines.   Their r
          dependence in the R = 3 Å and 6 Å planes are shown as dashed lines.

for this angle).   The mixing of the Na⁺ - N₂⁻ configuration into V*, plus other
terms, causes V* to increase only very slowly with increasing r for R ~ 3 Å and
θ = 90°, so that V and V* cross in the r ~ 1.2 Å, R ~ 3 Å region.   Thus the N
atoms tend to increase their separation as the Na* approaches 3 Å, and a major
component of the electronic quenching is expected in this crossing region.   This
is the same general region of quenching in the diabatic picture, and these two
pictures are also consistent in their general predictions regarding the nuclear
motion.   However, the Habitz calculation includes θ dependence and many more
contributions to the adiabatic potentials, and it is clearly preferable for
quantitative comparison to measurements.   Many of the bound vibrational states
of Na*-N₂ traverse the crossing region, so that they will be rapidly quenched
and the bound-state populations will be far below equilibrium values.   We do not
yet have a quantitative theory for the fluorescence spectrum expected under these
circumstances; instead we have initiated the measurements to see what occurs.

We have started measurements of two types of fluorescence spectra as a function
of cell temperature, and we would like to measure the spectrum for total absorption by Na-N₂ as well.   The experimental arrangement is shown diagrammatically
in Fig. 5.   In one measurement (type A) the laser wavelength is varied and the
D-line fluorescence intensity (5890 and 5896 Å) is measured.   This measures

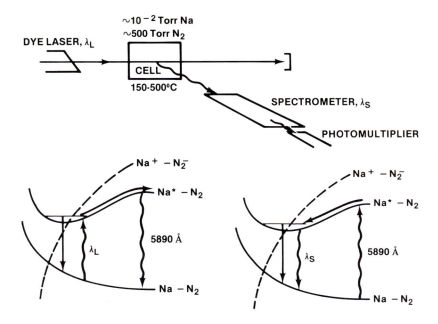

Fig. 5. The experimental arrangement, with the types of processes measured by tuning the laser on the spectrometer shown in the lower figures.

absorption by the $Na-N_2$ complex followed by separation to $Na^*$. In the second (B) measurement, the laser is fixed at 5890 Å, exciting $Na^*$, and we vary $\lambda_s$ to measure the spectrum radiated by the $Na^*-N_2$ complex. These two processes are indicated by arrows in Fig. 5. Our preliminary data are shown in Fig. 6. Note the strong red wing, consistent with the attractive $V^*$ of Habitz as well as with expected similarity at large R to Na-noble gas interactions. The blue wing is identified with a repulsive $Na^*-N_2$ branch, also consistent with Habitz, to the Na-noble gas case, and to preliminary measurements of fluorescence temperature dependence. Note the differences in spectral shape between the two types of measurements. The type A (laser tuning) spectrum becomes weaker with increasing wavelength relative to type B (fluorescence spectrum).

If the quenching did not occur, at the high $N_2$ densities of these data the $Na^*-N_2$ complex would attain an equilibrium distribution in r and R coordinates and the emission spectrum (type B) would be related to the absorption spectrum (type A) by a thermodynamic factor of $(\nu_0/\nu)^3 \exp\{-h(\nu-\nu_0)/kT\}$, where $\nu_0$ is the atomic Na transition frequency. Consequently, I have plotted the type A absorption spectrum times this factor as a short dashed line in Fig. 6. It is interesting that this agrees with the emission spectrum on the blue wing of the atomic transition, implying that quenching is relatively unimportant for this wing, or that it affects both types of spectra in the same way. Habitz does not expect significant quenching from this state, but in any event only free collision states of $Na^*-N_2$ may be involved and it is reasonable to then expect a reciprocity between absorption and emission by the complex even with quenching.

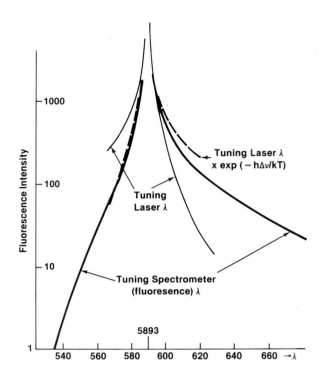

Fig. 6.  Measured fluorescence intensity as a function of laser wavelength and spectrometer wavelength, for $N_2$ density of ~$2 \times 10^{19}$ cm$^{-3}$, Na density of ~$10^{12}$ cm$^{-3}$, and T $\cong$ 500 K.

On the red wing, on the other hand, the type A spectrum multiplied by the thermodynamic factor does not agree with the type B spectrum. These are preliminary data, and this conclusion must be checked, but it is not unexpected. The type A and B fluorescence experiments produce and sample different distributions of bound, quasi-bound, and free collision states, so that noticeable differences in quenching rates may be expected between them. In addition, major differences are expected between these two spectra (corrected by thermodynamic factors) and total absorption, due to the rapid quenching expected for the bound and quasi-bound states that overlap the quenching region. Temperature dependences of these three types of spectra should also be very informative. We do not yet have quantitative predictions for these spectra, but I believe that when we do this detailed spectral information will yield a very definitive test of our understanding of this electronic-to-vibrational energy transfer process.

I have described the absorption and emission of radiation by this Na-$N_2$ collision complex as an example of a large class of triatomic collision processes that can be observed in unique detail by these techniques. Just as in diatomic systems these techniques complement the measurements of cross sections as tests of theoretical understanding. We and other laboratories are just beginning to apply these radiative techniques to triatomic collision complexes. It has now been demonstrated that these radiation processes can be measured, but the best analysis procedures are not yet apparent. Nonetheless, I am convinced that this very detailed probe directly into the collision complex will ultimately yield many valuable insights into reactive and inelastic collision processes.

The author thanks W. Kamke and B. Kamke who helped obtain the data. This work was supported in part by the National Science Foundation through Grant No. PHY79-04928 to the University of Colorado.

REFERENCES

[1] Drummond, D. L. and Gallagher, A., J. Chem. Phys. 60 (1974) 3426.
[2] Castex, M. C., J. Chem. Phys. 66 (1977) 3854.
[3] Bras, N. and Bousquet, C., J. Physique 42 (1981) 215.
[4] York, G., Scheps, R., and Gallagher, A., J. Chem.Phys. 63 (1975) 1052.
[5] Hertel, I. V., The excited state in chemical physics II, Adv. Chem. Phys. 45 (1981) 341.
[6] Bauer, E., Fisher, E. R., and Gilmore, F. R., J. Chem. Phys. 51 (1969) 4173.
[7] Habitz, P., Chem. Phys. 54 (1980) 131.

PHYSICS OF ELECTRONIC AND ATOMIC COLLISIONS
S. Datz (editor)
© North-Holland Publishing Company, 1982

WHAT'S REALLY HAPPENING IN LASER-INDUCED
PROCESSES IN COLLISIONS

R. Stephen Berry
Department of Chemistry and
The James Franck Institute
The University of Chicago
Chicago, Illinois  60637
U.S.A.

The experiments are reviewed in which a colliding
pair of particles undergoes a process specifically
as a consequence of being coupled to a radiation
field during collision.  Experiments are classified
and the results are surveyed, with some attention
to how unambiguously they can be categorized.
Some suggestions of future experiments are also
reviewed.

INTRODUCTION

This is a review, primarily of the experimental work,and a superfi-
cial overview of atomic and molecular processes that occur when a
pair of colliding particles are bathed in the field of a light
wave during their collision.  This class of such processes is
potentially large, far larger than the variety that has thus far
been observed.  They include absorption, emission and inelastic and
elastic scattering of radiation by collision complexes, accompanied
by any of the elastic, inelastic, ionizing, charge transfer, dissociative, associative or rearrangement collision processes that have
been the traditional subjects of the ICPEAC's.  The crucial characteristic of all the processes concerning us here is that their
occurrence is entirely dependent on or very sensitive to the
presence of the radation field during the collision.  We shall refer
to such processes as radiation-coupled collisions.

Radiation-coupled collisions (RCC's) can be classified according
to the radiation process -- absorptive, emissive, Rayleigh scattering
or Raman scattering -- and according to the types of collision process, for example according to which degrees of freedom of the
initial and final states are free and which are bound.  Moreover
these processes can also be classified according to the number of
photons involved and how many are resonant; for example the transfer
of energy from one atom to another accompanied by emission of one
photon by the collision coupled, first suggested by Gudzenko and
Yakovlenko[1], is a 1-$\phi$ emissive excitation transfer RCC.  Note that
the nomenclature I am using here does not include any photons used
prior to the collision to prepare the collision partners in excited
states, nor does it include photons emitted well after the collision,
even though they may be used as the indicators of the states in
which products emerged from the collision.

The RCC processes observed thus far are 1-$\phi$ electronic energy
transfer by dipole-dipole inelastic collisions[2-4] and by 1-$\phi$ dipole-
quadrupole collisions,[5,6] 1-$\phi$ emissive collisions,[7-12] 1-$\phi$ electron
transfer between a neutral atom 'and a positive ion,[13] 1-$\phi$ associative

and 2-ϕ Penning ionization,[14-18] 1-ϕ absorptive collisions[19-21] ,
Rayleigh and Raman scattering[22],1-ϕ emissive collisions[27], and
perhaps 1-ϕ associative rearrangement collisions.[28]  Our discussion
will review the status of each of these in turn and then address
some points of interpretation and of possible future experiments,
particularly to study other RCC processes.  (We omit x-ray processes
from quasimolecules on the arbitrary ground that their natural line
widths are larger than the inverse of a typical collision duration.)

We shall not try to review all the theoretical work on RCC's.  Such
reviews were given by Weiner[29-31] recently.  Our bibliography cites
a few theoretical papers published prior to the writing of those
reviews that now appear relevant to this topic, and several more
recent theoretical papers[32-44].  However, we shall discuss the
contents of the theoretical works only insofar as they are directly
relevant to particular experiments.  It is notable that the number
of theoretical papers on RCC's is at least a factor of three higher
than the number of experimental papers, at this time.  The simplest
explanation appears to be that the experiments are very difficult,
especially because many of them require strong laser fields, $10^4$W/
$cm^2$ or more, which can make all sorts of processes happen, other
than the desired one.

Weiner pointed out[30] that there are several ways to approach the
RCC conceptually.  The most widely-used is a perturbation  picture
in which the collision partners are treated as separated, interacting
atoms (or molecules), each perturbed by the radiation field.  A
powerful extension of this approach, which could be exploited systematically but has only been used in a few instances, treats the
colliding partners as a compound system.  This formalism uses
quantization appropriate to the compound state rather than that of
the separated fragments, when the collision partners are close
together.  We shall see a qualitative application of this approach
later.

The second approach, that is the most general and perhaps the most
satisfying in principle,is the dressed-state approach in which the
interacting collision partners are in eigenstate of the Hamiltonian
of matter-plus-radiation at all times. This approach has the apparent
disadvantage of being significantly more difficult to use as a computational tool.  However one calculation based on the dressed-states
method has been presented,[43] for the analysis of curve-crossing and
dissociation of diatomic molecules in intense fields.

The third approach is particularly attractive when spectroscopic
methods are used to detect the presence of RCC's, and to analyze
what goes on during such a collision.  This is the application of
concepts of collision broadening shifts of spectral lines.  It has
been used effectively to analyze energy-exchange RCC's.

ABSORPTIVE COLLISIONS

The first observations of processes properly called radiation-coupled
collisions were absorptive collisions.  These had been seen for many
years in the spectra of metal vapors in rare gases.[45-48]  For purposes
of low-resolution spectroscopy, little distinction was made between
processes occurring in collisions and processes involving van der
Waals molecules of rare gas atoms with metal atoms.  These very
weakly bound molecules have dissociation energies comparable to or
less than the mean thermal energies of typical collisions and, in

some instances suffer collisions frequently enough to broaden the structure and obscure individual rotational lines of bound molecules. The interpretations of features in the wings of atomic lines -- the "satellite lines" -- were, for many years, given in terms of the theory of line shapes:[32,49,50] The transient proximity of a rare gas atom to an alkali or other metal atom perturbs the metal, shifting both its ground and excited states, giving rise to a feature that might peak in intensity, and the peak could be either to the long or short wavelength side of the "parent" line. The existence and position of a peak is a function of the density of states, both the bound and continuum states, and of the variation of the optical transition probability with the internuclear distance of the metal and its perturber. In the simplest picture, with transitions only at classical turning points, the peak intensity occurs at the frequency corresponding to a vertical transition to the upper potential curve from that point on the lower curve where the product of the Boltzmann distribution of collision energies and the transition probability is a maximum. Refinements allow for transitions at any distance, and for interference between transition amplitudes at different distances. In all the processes described in the older literature, the excited compound species may be bound or free; if it is free, it dissociates to give only a single excited species (the metal atom) and a rare gas atom in its ground state.

A recent development in absorptive RCC's is the observation of pair absorption by colliding species capable of more than van der Waals bonding. The systems Ba+Ba and Ba+Tl were reported by White et al.,[20] and the alkali pairs Cs+Cs, Rb+Rb and Cs+Rb were recently reported by Hotop and Niemax[21]. In these experiments, a single photon excites both collision partners, producing two free excited atoms:

$$2Ba(6s^2\ {}^1S_0) + \hbar\omega(339.4nm) \rightarrow Ba(6s5p\ {}^1P_1) + Ba(6s5d\ {}^1D_2).$$

With densities of $7.9 \times 10^{16}$ for Ba and $1.4 \times 10^{17}$ for Tl, they observed absorption by colliding atoms to give

$$Ba(6s^2\ {}^1S_0) + Tl(6p^2P_{\frac{1}{2}}) + \hbar\omega(386.7) \rightarrow Ba(6s6p\ {}^1P_1) + Tl(6p^2\ P_{3/2}).$$

The absorption coefficients for these two processes at $10^{17}cm^{-3}$ are approximately 2.9% $cm^{-1}$ and 2.3% $cm^{-1}$, respectively. The absorption curves are asymmetric with maxima at the energies corresponding to separated atoms, and widths (fwhm) of about 15 $cm^{-1}$.

The electric dipole selection rules for such excitation require the transition to change the parity of the compound system. Because the parity of the weakly interacting pair is given by the product of parities of the states of the separated atoms, the dipole selection rule for absorptive RCC's is that one atom changes $L \rightarrow L \pm 1$ and the other atom undergoes $L' \rightarrow L'$ or $L' \pm 2$ (strictly, $L' \pm 2n$). Thus, for example, the transitions Cs(6s)+Rb(5s)→Cs*(5d)+Rb*(5p), Cs(6s)+Cs(6s)→Cs*(6p)+Cs*(5d) and Rb(5s)+Rb(5s)→Rb*(5p)+Rb*(6d or 8s) are observed but Cs(6s)+Rb(5s)→Cs*(6p)+Rb*(5p) is at best very weak. This last is allowed as a quadrupole transition but not as a dipole one.

EMISSIVE COLLISIONS

Emissive collisions in which one partner is initially excited have been studied in the same context as their counterpart absorptive collisions.[45,46,51] Three examples have been reported in which a

colliding pair of atoms, both in excited states, emits a single
photon to release all the excitation of both.[6,11,12] For the dipole-
allowed process[11] barium vapor in a heat pipe oven, at a density of
$10^{16}$ atoms/cm$^2$ in a buffer gas of Ar or Xe, was pumped optically by
a pulsed dye laser (7 nsec duration) at 661.4 nm to induce two-
photon excitation, Ba$(6s^2) \rightarrow$ Ba$*(6s6d^1D)$. The Ba$*(^1D)$ decayed radiatively to the desired Ba$*(6s6p^1P_1^\circ)$ and Ba$(6s5d^1D_2)$ levels. These
could be produced in densities of $2.6 \times 10^{14}$ and $5.3 \times 10^{14}$ cm$^{-3}$, respectively. The signal of the process

$$Ba*(6s5p^1P_1^\circ) + Ba*(6s5d^1D_2) \rightarrow 2Ba(6s^2 {}^1S_0) + \hbar\omega$$

was the light emitted at approximately 339.4 nm. This emission is
broad, relative to atomic lines: its fwhm is of order 3.5 nm
(ca 20cm$^{-1}$) and the peak is decidedly degraded to short wavelengths.
This suggests that the largest contribution to the lineshape, the
contribution from large impact parameters, corresponds to the longest
wavelength transitions, i.e. that the upper and lower potential
curves move apart as the collision partners approach.

The effective cross section for this emissive collision, based on the
experimentally determined populations of barium in the two excited
states, the cell temperature, and parameters of the experiment, is
$\sigma_{emis} \cong 2.6 \times 10^{-20}$ cm$^2$. The cross section for stimulated emission,
the reverse process, is inferred by White et al. to be $\sigma_{stim.ab.} \cong$
$3.4 \times 10^{-24}$ P/A cm$^2$, where the laser power per unit area is given in
watts/cm$^2$, consistent with a calculated value of $1.9 \times 10^{-24}$ P/A.

The cross section for the dipole-quadrupole process for $2Ba(5s5d^1D) \rightarrow$
$2Ba(5s^2 {}^1S) + \hbar\omega$ (438.6nm) is $3.4 \times 10^{-23}$ cm$^2$, with a factor of 7 uncertainty, for a spontaneous emissive RCC[6]. This value seems large for
a quadrupole process in light of the value for the pure dipole-
allowed process. We shall return to this question later.

The heteronuclear system

$$Ba*(6s6p^1P_1) + Tl*(6p^2P_{3/2}) \rightarrow Ba(6s^2 {}^1S_0) + Tl(6p^2P_{\frac{1}{2}}) + \hbar\omega (386.7nm)$$

shows emission[12] with a line shape precisely the same as its absorption curve, when the excited atoms are present simply at their equilibrium concentrations, in this case at temperatures of order 1700C.
Here is a particularly simple, elegant example of an RCC with no
strong laser field at all.

ELECTRONIC ENERGY TRANSFER

The first laser-induced RCC process to be studied was electronic
energy transfer from an excited atom of one kind to another, $A* + B + \hbar\omega \rightarrow A + B*$. Falcone et al.[2] observed the processes

$$Sr(5s5p {}^1P) + Ca(4s^2 {}^1S) + \hbar\omega (497.7nm) \rightarrow Sr(5s^2 {}^1S) + Ca(4p^2 {}^1S)$$

and

$$Sr(5s5p^1P) + Ca(4s^2 {}^1S) + \hbar\omega (471.1nm) \rightarrow Sr(5s^2 {}^1S) + Ca(4s5d^1D).$$

The initial excited 7p state of Sr was produced by pumping with a
flashlamp-pumped dye laser tuned 50cm$^{-1}$ to the red of the 5s→5p
resonance line. The density of excited strontium, measured by
fluorescence, was typically $3 \times 10^{13}$ cm$^{-3}$ and that of the ground-
state Ca, $6 \times 10^{15}$ cm$^3$. The excited calcium population was established

from the intensity of the $Ca(4p^2\ ^1S) \rightarrow (4s4p\ ^1p)$ emission at 551.3 nm and the 5d→4p emission at 519.0 nm. The signals were measured as functions of the wavelength of the "transfer" radiation, the radiation absorbed during the collision that makes the energy transfer possible. The 497.7 nm signal is asymmetric toward long wavelengths, with a width (fwhm) of $14cm^{-1}$. The velocity averaged cross section was predicted to be $2 \times 10^{-17}cm^2$ at a transfer power density of $5 \times 10^5$ w/cm$^2$; the measured cross section was $9 \times 10^{-18}cm^2$

The Stanford group continued[4] to study electronic energy transfer in the Sr-Ca system, using a precisely resonant 40 psec laser pulse at 460.7 nm to generate $Sr(5s5p\ ^1P°)$ and another intense 40 psec pulse at the transfer frequency, delayed by 5 nsec, to effect the excitation of $Ca(4p^2\ ^1S)$ during collisions. They also studied the process $Ca(4s4p\ ^1P)+Sr(5s^2\ ^1S)+\hbar\omega \rightarrow Ca(4s^2\ ^1S)+Sr(6s6d\ ^1D_2)$, by the same method. The cross section for production of Ca* is almost $10^{-13}cm^2$ when the power density is of order $3 \times 10^9$w/cm$^2$. The maximum in the radiation-induced energy transfer occurs at a frequency precisely equal to the difference between the electronic energy of the incident collision partners and that of the free products; there is no indication that kinetic energy plays any role in the transfer.

The system Eu-Sr was studied by Cahuzac and Toschek[3] in a similar manner. Europium atoms were excited from this $6s^2\ ^8S_{7/2}$ ground state to the $6s5p\ ^8P$ level by a nitrogen-pumped dye laser (460 nm, 4 nsec pulse duration). The excited europium atoms collided with $Sr(5s^2\ ^1S_0)$ atoms in the presence of a delayed 4 nsec pulse of light with tunable wavelength in the region of 658 nm, providing the doubly-excited $Sr(Sp^2\ ^1D_2)$ state. This system has three states corresponding to the $6s6p\ ^8P$ level, with J=9/2, 7/2 and 5/2. Any one of the three may be excited and, with red light of the proper frequency, may excite strontium in an RCC to its $^7P$ state, which is again detected by its emission. The state with J=9/2 gives rise to a single peak degraded to longer wavelengths; excitation to the state J=5/2 gives rise to a single asymmetric peak degraded to shorter wavelengths, and excitation to the state with J=7/2 gives rise to a broad emission degraded to short wavelengths but with a secondary maximum about 0.7 nm to the blue of its 651.07 nm maximum. This behavior fits well with the interpretation[30] that the potential curves involved when J=9/2 is excited are furthest apart when the distance R between Eu and Ca is large; that the potential curves of the states with J=5/2 are closest together when R goes to infinity, and the curves involved with $Eu(^8P_{7/2})$ is excited have a separation that first grows as R diminishes from infinity and then decreases again. That is, the J=9/2 case has the lower state, based on $Eu(^8P_{9/2})+Sr(^1S_0)$, exhibiting stronger repulsion than the upper state, based on dissociating to $Eu(^8S)+Sr$ $(5p^2\ ^1D)$. The J=5/2 case has the potential of the lower state less steep and less repulsive than that of the upper state and the J=7/2 case appears to have a potential minimum forthe atoms in the lower $Eu(^8P_{7/2})+Sr(^1S_0)$ state.

The measured cross section for the J=9/2 case was reported as $1.2 \times 10^{-20}(P/A)\ cm^2$ with P/A in watts/cm$^2$. Neglecting the near degeneracy, Cahuzac and Toschek[3] calculate $4.5 \times 10^{-20}(P/A)\ cm^2$.

In all these systems, there may be simpler two-photon excitations followed by collisional energy transfer, and pairs of one-photon excitations followed by collisional energy transfer "up-pumping" the energy acceptor.  By delaying the transfer radiation and then taking signal only in an interval including transfer pulse, the effects of these other processes, not RCC's, can be minimized.

Green et al.[5] have reported a dipole quadrupole process in the Ca-Sr system,

$$Sr*(5s5p\ ^1P_1)+Ca(4s^2\ ^1S_0)+\hbar\omega(510.7\ nm)\rightarrow Sr(5s^2\ ^1S_0)+Ca(3d4p\ ^1F_3).$$

A dye laser pumped by a Nd:YAG oscillator-amplifier excited the Sr to its initial $^1P$ state with pulses of 40 psec.  The transfer laser supplied 40 psec pulses delayed by 5 nsec.  The $^1F_3$ state of Ca was detected by its emission at 534.9 nm to give Ca$(4s3d\ ^1D_2)$. The cross section is $3\times10^{-14}$cm$^2$ at a power of $7\times10^9$W/cm$^2$ and is linear in power below this level.

Recently, White has reported[6] a process he interprets as a radiation-induced dipole-quadrupole excitation transfer RCC.  This involves the reaction $2Ba(6s^2\ ^1S_0)+\hbar\omega(438.6\ nm)\rightarrow2Ba(5s5d\ ^1D_2)$.  The "trans-fer" radiation at 438.6 nm is resonant with the final state of the two excited Ba atoms.  The process differs from the inverse of the emissive collision described in the previous section because the one excited barium atom was in a $^1P$ state and the other, in a $^1D$; here both are in $^1D$ states.  The line shape is again asymmetric toward short wavelengths, both in absorption and in emissive RCC's. The cross section for the process is estimated[5] as about $1.8\times10^{-25}$ (P/A) cm$^2$.

White interpreted this process as absorption by one Ba atom of single 438.6 nm or 22790 cm$^{-1}$ photon to give a virtual state dominated by the $6s6p\ ^1P_1$ state which lies 4730 cm$^{-1}$ below the proposed virtual state.  Then the Ba atom in its virtual state in collision with a Ba atom in its ground state "splits" the 22790 cm$^{-1}$ and transfers exactly half that energy, leaving both atoms in their $5s5d\ ^1D_2$ states.

An alternative description would classify this particular process as a 1-$\phi$ absorptive RCC, like those of Ref. 20.  In this picture, the photon at 22790 cm$^{-1}$ is absorbed by the two Ba atoms in colli-sion, producing a weakly dissociative state that correlates with two separated Ba$(^1D_2)$ atoms.  That the spontaneous emissive RCC is observed for this system and has the same line shape as the absorptive collision is consistent with this compound state picture.  If the isolated atom were truly coupled strongly to the radiation field so that the picture of an isolated atom in a virtual state were the correct inital state, one would expect some distortion by the radiation field of the potential curve of the excited state. If this were the case, the line shape for the spontaneous emissive RCC would differ from that of the absorptive or energy transfer RCC, but the line shape for a stimulated emissive RCC would look like that of the absorptive RCC.

For low power densities, all these energy transfer and absorptive processes have probabilities or cross sections proportional directly to the power.  For dipole-allowed processes at power levels above about $10^9$ or $10^{10}$ watts/cm$^2$, this dependence falls off to proportion-ality to $(P/A)^{\frac{1}{2}}$, as predicted by Gudzenko and Yakovlenko[1].

## CHARGE TRANSFER

Continuing their intensive study of the Ca-Sr system, the Stanford group[13] produced electron transfer RCC's by the process

$$Ca^+(4s\ ^2S_{\frac{1}{2}})+Sr(5s^2\ ^1S_0)+h\omega \rightarrow Ca(4s^2\ ^1S_0)+Sr^{+*}(5p\ ^1P_{3/2}).$$

The process is endoergic by 2.6ev in the absence of the radiation. The Ca$^+$ ions were produced by 2-$\phi$ excitation of Ca(4s$^2$ $^1$S) to Ca*(4s 4d$^1$D$_2$) and ionization of this state by a third photon of the same wavelength as the first two, all by a 40 psec pulse of several megawatts from a tunable dye laser pumped by a Nd-YAG oscillator-amplifier. A second 40 ps pulse from an independently tunable dye laser arrived 5 ns later to provide the pulse for the RCC. The excited strontium ions were detected by their emission at 407.8 nm due to the 5p→5s transition.

The peak in the curve of radiation-induced electron transfer occurs at 471.5 nm, not at 473.0 nm, the energy corresponding to exact resonance for the separated atoms. The implication of this is that the process does involve some exchange of kinetic and electronic energy, that is, that the transfer takes place in a region of Ca$^+$-Sr separation where the potential curve of at least one of the two relevant states is not horizontal. However the observed line is too narrow and symmetric and too close to 473.0 nm to fit the Landau-Zener curve crossing models proposed for this process.[33,34,52] The suggestion that other methods[53-55] be used is reminiscent of the proposal that electron transfer in near-resonant ion-ion neutralization also would be best treated by a formalism other than the Landau-Zener.[56,57]

The absolute cross section for radiation-induced electron transfer could not be determined very accurately. However Green et al. did manage a rough estimate of about 5x10$^{-15}$cm$^2$ at the power level of the transfer laser, about 10$^9$W/cm$^2$.

## PENNING AND ASSOCIATIVE IONIZATION

The first reported experiments on radiative-coupled Penning and associative ionization collisions[14] dealt with Li+Li. To date, only alkali atom-atom systems have been studied in this context.[15-18,31] With Li, the two processes are

$$2Li*(2p)+\hbar\omega \rightarrow Li^++Li(2s)+e \text{ (radiative-coupled Penning ionization)}$$

and

$$2Li^*(2p)+\hbar\omega \rightarrow Li_2^++e \text{ (radiative-coupled associative ionization)}.$$

The lithium system was chosen for the first experiments because the energy of 2Li*(2p) is insufficient to produce Li$_2^+$ + e, much less Li$^+$ + Li(2s) + e. The Li*(2p) is produced by absorption of resonance-frequency radiation from a tunable nitrogen-pumped dye laser. Ionization by 3-$\phi$ absorption at the frequency of the lithium resonance line is of course possible, both with Li atoms and Li$_2$ molecules. However by using two lasers, one at relatively low power tuned to the resonance line and the other at higher power but tuned off the resonant frequency by 250 to 1750cm$^{-1}$, the true ionizing RCC's could be distinguished unambiguously. The Li atoms were generated by colliding beams from opposing effusive ovens, producing concentrations of about 2x10$^{12}$cm$^{-3}$in the region intersecting the two opposing laser beams. The power in the laser beams was kept below 10$^5$W/cm$^2$. Detection was done by monitoring the mass-analyzed ion signal. The cross section for associative ionization was estimated to be about $5 \times 10^{-20} cm^2$ under typical operating conditions, or about $5 \times 10^{-25}$ (P/A) $cm^2$ with $10^5 W/cm^2$ of radiation. Incidentally, von Hellfeld et al. pointed out a large isotope effect in the relative cross section for associative ionization. The process discriminates very much in favor of the formation of $^6Li^7Li^+$ at the expense of $^6Li_2^+$ and $^7Li_2^+$. Penning ionization in the $Li_2$ system is the first true laser-induced 2-$\phi$ process to be observed.

The more complex situation with Na+Na has now been addressed.[15-17] Weiner and his associates[15,16] carried out experiments similar to those for Li+Li, with two opposed effusive beams of Na giving a density of about $3 \times 10^9$ atoms/$cm^3$ in the collision region. The intensity of the radiation field was about $3 \times 10^7 W/cm^2$, a hundredfold higher than in the experiments with Li. As with those experiments, the excited state, here the Na*(3p) state, was prepared with radiation from a dye laser tuned to the resonant $3s \to 3p$ frequency. With only one laser, both $Na^+$ and $Na_2^+$ ions appear at the frequency of the Na D lines, but at no other frequencies. However with one laser tuned to the frequency of one of the D lines and the other laser tuned over the range of about 570-615 nm, the intensities of $Na_2^+$ and $Na^+$ show strongly structured spectra. The spectrum of $Na_2^+$, especially, has very marked structure with intervals of about 120 $cm^{-1}$ and widths of about $20 cm^{-1}$, strongly suggestive of vibrational structure associated with the newly formed $Na_2^+$. The process is $Na(3s+\hbar\omega_1 \to Na^*(3p)$ followed by the RCC process $Na^*(3p)+Na(3s)+\hbar\omega_2 \to Na_2^+ + e$.

Both the $Na^+$ and $Na_2^+$ spectra also show sharp peaks due to processes that do not involve radiation-coupled collisions. These include

a)  $Na(3s\,^2S)+2\hbar\omega \to Na^*(4d\,^2D)$,  $Na^*(4d)+Na^*(3p) \to Na_2^+ + e$;

b)  $Na(3s\,^2S)+\hbar\omega \to Na^*(5s\,^2S)$,  $Na^*(5s\,^2S) + Na^*(3p\,^2P) \to Na_2^+ + e$;

c)  $Na(3s\,^2S)+\hbar\omega_1 \to Na^*(3p\,^2P_{3/2})$,  $Na^*(3p\,^2P_{3/2})+\hbar\omega_2 \to Na^*(5s\,^2S)$,

   $Na^*(5s\,^2S) + Na^*(3p\,^2P) \to Na_2^+ + e$.

The appearance of the corresponding peaks in the $Na_2^+$ spectrum and the $Na^+$ spectra from Penning ionization are all due to the absorption of a third photon, carrying the system to an energy where one electron and the vibrational motion may be in the continuum. The cross section for Penning ionization of Na was estimated[16] to be about $10^{-17} cm^2$ at a power of $6 \times 10^4 W/cm^2$.

While Weiner and Polak-Dingels[16] distinguished $Na_2^+$ formation by Penning ionization RCC's and $Na^+$ similarly by associative ionization RCC's at the frequencies of the NaD lines with power levels of order $10^4 W/cm^2$, Roussel et al.[17] observed the direct 2-$\phi$ Penning ionization RCC's with a single laser whose intensity is $10^5 W/cm^2$, over the range 580-615 nm. Detection and mass analysis of the ions was accomplished by single time-of-flight mass spectrometry. The spectrum is very rich; one can distinguish peaks in the $Na_2^+$ spectrum corresponding to $2Na(3s)+4\hbar\omega \to 2Na^*(3p)$, followed by $2Na^*(3p) \to Na_2^+ +e$, at both of the D-lines, and $2Na(3s)+4\hbar\omega \to 2Na^*(5s)$ followed by $2Na^*(5s) \to Na_2^+ +e$. These are of course not RCC processes. However the major part of the spectrum is a rich pattern that appears to correspond, peak for peak, with that reported for the two-laser system.[15] (Roussel et al. seem to feel the spectra are not nearly identical as I do.) This means that the factors responsible for the vibrational structure are approximately independent of whether the process occurs via Na*(3p)+Na(3s)+ $\hbar\omega$ or by 2Na(3s) + $2\hbar\omega$.

Strictly, we expect differences in these two processes which will be apparent as variations in the relative intensities of different pulses. The radiation coupling of Na*(3p) + Na(3s) + $\hbar\omega$ must surely be stronger at large internuclear distances than that of 2Na(3s) + $2\hbar\omega'$, when $\omega'$ is well off the atomic resonance and the energy of $2\hbar\omega'$ is the same as that of ($\hbar\omega$ + the 3s→3p excitation). Furthermore we expect the first process to occur predominantly at internuclear distances that make the nuclear kinetic energy of Na*(3p)+Na(3s) approximately equal to that of $Na_2^+$ in any of its vibrational states, while the 2Na(3s)+$2\hbar\omega'$ process is expected to occur at internuclear distances at which the kinetic energy of the ground state atoms is approximately equal to that of the vibrating $Na_2^+$ molecule. These conditions are direct carry-overs of the application of the Franck-Condon principle for both position and momentum, as it has been applied in ordinary Penning and associative ionization. The potential curves for excited alkali atoms are becoming well known, as at least one contributed paper in this Conference shows, so it will soon be possible, perhaps it is even possible now, to compute the relative intensities of vibrational peaks in the two-frequency, 2-$\phi$ spectrum[15] and in the single-frequency, 2-$\phi$ spectrum, [17] just from Franck-Condon factors alone. Then of course one will want to include the variation with internuclear distance in the radiative coupling strength to the final electronic state.

The most recent system for which Penning ionization RCC's have been studied is the mixed alkali, Li+Na.[18,31] Here, Na*(3p) + Li*(3p) lacks 0.2eV of the energy required to form $NaLi^+$ + e, and Na(3s) + Li*(3d) lacks 0.27 eV to form the same products. Both of these have been observed by 2-$\phi$ resonant excitation of the initial states followed (during the same laser pulse) by radiation-coupled associative ionization.

MOLECULAR REARRANGEMENT

On radiation-coupled collision yielding light as the system undergoes a rearrangement is the process $F+Na_2 \rightarrow [FNa_2]^* \rightarrow NaF+Na$ in which light is emitted by the compound state. This process, observed very recently by Arrowsmith et al.[27], is described in some detail in this volume by J. Polanyi, so it will not be discussed here.

Weiner suggested[42] that a chlorine atom might be set free in the RCC $Hg+Cl_2+\hbar\omega \rightarrow HgCl^*+Cl$, in the field of an ArF laser at $10^9$ W/cm . While this process has not been reported, a related halogen transfer was described by Hering et al.[28]:

$K + HgBr_2+\hbar\omega (ca595nm) \rightarrow KBr+HgBr^*$.

The emission of radiation of about 500 nm was taken as the signature of the HgBr*. The authors have subsequently discovered that there is a possible artifact via the reaction of traces of $K_2$ with $HgBr_2$ to give 2KBr + Hg. If the one of the KBr molecules were to hold nearly all the exothermicity of the reaction and absorb the incident laser photon, KBr could be produced in a fluorescent state.[53]

However the intensity of the emission is quite compatible with the expected collision lifetime of K with $HgBr_2$ and with the expected radiative lifetime (tens of nsec) of HgBr*. At this time, Hering et al. are attempting to determine whether the emission they observe is due to KBr* or HgBr*.

ELECTRON-ATOM COLLISIONS

Electrons in collision with atoms have been shown to couple with radiation. Andrick and Langhans[58,59] and Langhans[60] have studied the energy distribution of electrons backscattered (160°) from argon in the presence of the 10.6μm radiation from a $CO_2$ laser. The electrons undergo free-free transitions, either absorbing or emitting one photon of 117meV from the laser field. For electrons with energies between about 5 and 16eV but not at energies of scattering resonances, the quantity $\gamma^2$, the ratio of the differential RCC, per unit radiation field intensity, to the differential elastic cross section, is about $5\times10^{-8}$ $cm^2$/W.[59] Moreover $\gamma^2$ rises linearly with electron energy, as theory predicts. At the energies of scattering resonances, the effect seems to be considerably enhanced.[60] However it is difficult at present to make quantitative statements about this enhancement because the resonances in question at 11.098 and 11.270eV, are considerably narrower than the resolution of the electron spectrometer.

Other RCC's of electrons with atoms have been reported. Weingartshofer et al.[61] have described multiphoton processes analogous to the 1-φ transitions described by Andrick and Langhans. Langedam and van der Wiel[62] have presented data on free-free transitions between electron scattering resonances.

FURTHER REMARKS

We conclude with a few remarks regarding directions that might prove fruitful in the study of radiative-coupled-collisions.

First, I would reemphasize the desirability of using the compound state picture to interpret the processes occuring in RCC's. The separate-particle model appears to be adequate in many cases, but because of its stronger symmetry constraints, processes may be predicted to be highly improbable from the separate-particle view that are clearly allowed in the lower symmetry of the compound state. Moreover the additional information regarding the permissible kinds of coupling that go to make up the possible compound states may give significant insight. An example is in the case of $Ba(6s6p\ ^1P) + Ba(6s\ 5d\ ^1D) \rightarrow 2Ba(6s^2\ ^1S) + \hbar\omega$.

The manifold of states of $Ba_2$ that can be constructed from $Ba(^1P)$ and $Ba(^1D)$ is easy to write out. It includes two $^1\Sigma_u$ states and four doubly-degenerate $^1\Pi_u$ state, the only states strongly allowed by elastic dipole transitions from the ground $^1\Sigma_g$ state. The molecular orbital configurations of the most important of these states are:

A) $\ldots (\sigma_g 6s)^2 (\sigma_g 6p)(\sigma_u 5d)\ ^1\Sigma_u$ ,

B) $\ldots (\sigma_g 6s)^2 (\sigma_g 6p)(\pi_u 5d)\ ^1\Pi_u$ and

C) $\ldots (\sigma_g 6s)^2 (\sigma_u 6p)(\pi_g 5d)\ ^1\Pi_u$ .

The first of these must mix to some degree with the configuration

$$A') \ldots (\sigma_g 6s)^2 (\sigma_g 6p)(\sigma_u 6s) \; ^1\Sigma_u \; ,$$

which corresponds to orbital polarization by hybridization of the $\sigma_u 5d$ with the $\sigma_u 6s$ orbital. The picture that emerges from the compound state view of the process is then

$$Ba(^1P) + Ba(^1\Sigma) \rightarrow [Ba_2^* (\sigma_g 6s)(\sigma_u 5d + 6s) \; ^1\Sigma_u] \rightarrow Ba_2(\sigma_g 6s)^2 (\sigma_u 6s)^2 + \hbar\omega$$

via the $\sigma_g 6p \rightarrow \sigma_u 6s$ transition. The $Ba_2$ molecule disociates immediately, of course. This is not the only reasonable possibility. The
excited $Ba_2^*$ could have one electron in the $\pi_g 6p$ orbital, and emit
radiation from the $\pi_g 6p \rightarrow \sigma_u 6s$ transition. Polarization studies
would tell us the contribution of each of these.

The process described by White[6] as dipole-quadrupole may be interpreted as an allowed process in a compound state representation.
Bringing two $Ba(5s5d^1D_2)$ atoms together, one can generate several
states; among these, particularly the $\ldots (\sigma_g 5s)^2 (\sigma_g 5d)(\sigma_u 5d) \; ^1\Sigma_u$
and the $\ldots (\sigma_g 5s)^2 (\delta_g 5d)(\sigma_u 5d) ^1\Delta_u$ states are of interest here. The
orbital $\sigma_u 5d$ is in reality an antibonding $\sigma_u$ mixture of 5d and 5s,
as in the above case. An electron may move by dipole radiation
between this orbital and either the $\sigma_g 5d$ or the $\delta_g 5d$. This
mechanism offers a dipole-allowed path in the compound state that
is not available for the separated atoms.

Next, there are several sorts of experiments that might be carried
out with the subject at its present stage.

a) The electron exchange RCC deserves to be studied to compare
cases in which the transfer can occur over only a narrow range
of internuclear distances and cases in which the transfer may
occur over a long range. The need to look at this issue was
pointed out in connection with the experiments[13] and again in a
theoretical analysis by Copeland and Tang.[55]

b) The probability of energy transfer and electron transfer
should be studied as functions of the energy mismatch of the
transfer radiation and the relative energy of the colliding
particles, to determine how effectively kinetic energy may be
converted to electronic energy in such processes. The velocities of the outgoing particles might be determined as well.

c) The role of nuclear kinetic energy can be studied even more
effectively in Penning and associative ionization RCC's, by
measuring the energy of the electrons. In time, it will be
useful to measure the angular distribution of these electrons
as functions of the intensity of the transfer laser, to study the
field's polarization of the orbitals of the compound state.

d) The possibility of studying bound final states is now a clear
possibility, after the experiments of Polak-Dingles et al.[16]
and Roussel et al.[17] A first theoretical description of the
spectroscopy of such states was very recently given.[44]

e) Weiner[43] has proposed studying the effect of laser fields
on ion-pair forming collisions, based on the dressed-state
theoretical model of Lau. That seems a particularly attractive direction for studying processes that have no net

consumption of radiation energy.

f)   According to the analysis of Agre and Rapoport,[40] collisions that would be simple elastic processes stopped by their
centrifugal barriers may show enhanced resonances with states
inside the centrifugal barriers, in the presence of a suitable
laser field.

g)   Lau[35-39] has developed a theoretical model for radiation-
induced dissociation and predissociation.  These processes may
be considered half-collisions, or the inverses of RCC's whose
final states are bound.  Such processes are surely ripe for
experimental study now.

h)   Just as we have seen electron transfer RCC's, we might
expect analogous proton transfer processes.  These are probably
easier to effect than the heavy particle transfers that have
been discussed.  They could be achieved in ion-neutral collisions
such as $ArH^+ + Ca + \hbar\omega \rightarrow Ar + CaH^{+*}$, or in neutral-neutral collisions
yielding ion pairs:  $HCl + NH_3 + \hbar\omega \rightarrow Cl^- + NH_4^+$.  In such cases,
the interesting comparison is of course with the process unassis-
ted by a radiation field.

i)   The last example suggest the rich potential for RCC's involving molecular collision partners, which has been started with
the work of Arrowsmith et al.[27] and Hering et al.[28]

The subject has many possibilities and the types and numbers of
experiments done thus far are small.  The principal limitation seems
to be the cost of laser systems that can supply $10^5$ to $10^3 W/cm^2$ of
tunable power, particularly if it is to be supplied in 40 psec
pulses.  However we have seen examples[12,21,46-48,51] in which no
lasers were required, so the study of radiation-coupled collisions
is still open to experiments on modest budgets.

Acknowledgements.  This discussion was prepared at the Aspen Center
for Physics, for whose hospitality I am most grateful.

REFERENCES

[1]    Gudzenko, L.I. and Yakovlenko, S.I., Zh. Eksp. Teor. Fiz. 62
       (1972) 1686-1694; transl. Sov. Phys. - JETP 35 (1972) 877-881.

[2]    Falcone, R.W., Green, W.R., White, J.C., Young, J.F. and
       Harris, S.F., Phys. Rev. A 15 (1977) 1333-1335.

[3]    Cahuzak, Ph. and Toschek, P.E., Phys. Rev. Lett 42 (1978)
       1087-1090.

[4]    Green, W.R., Lukasik, J., Willison, J.R., Wright, M.D., Young,
       J.F. and Harris, S.E., Phys. Rev. Lett. 42 (1979) 970-973.

[5]    Green, W.R., Wright, M.D., Lukasik, J., Young, J.F. and Harris,
       S.E., Opt. Lett. 4 (1979) 265-267.

[6]    White, J.C., Phys. Rev. A 23 (1981) 1698-1702.

[7]    Lochte-Holtgreven, W., Naturwiss.  38 (1951) 258-159; see
       R.S. Berry, Chem. Revs. 69 (1969) 533-542 for references to
       subsequent work on radiative attachment of electrons.

[8]    van Thiel, M., Seery, D.J. and Britton, D. J. Phys. Chem. 69
       (1965) 834-842.

[9]    Carabetta, R.A. and Palmer, H.B., J. Chem. Phys. 45 (1967)
       1325.

[10]   Bates, D.R. Case Stud. Atom. Phy. 4 (1975) 59-85:  a review
       of radiative recombination of electrons.

[11]   White, J.C., Zdasiuk, G.A., Young, G.F. and Harris, S.E.,
       Phys. Rev. Lett. 43 (1979) 1709-1712.

[12]   Falcone, R.W. and Zdasiuk, G.A., Opt. Lett. 5 (1980) 365-367.

[13]   Green, W.R., Wright, M.D., Young, J.F. and Harris, S.E., Phys.
       Rev. Lett. 43 (1979) 120-123.

[14]   V. Hellfeld, A., Caddick, J. and Weiner, J., Phys. Rev. Lett.
       40 (1978) 1365-1373.

[15]   Polak-Dingels, P., Delpech, J.-F. and Weiner, J., Phys. Rev.
       Lett. 44 (1980) 1663-1666.

[16]   Weiner, J. and Polak-Dingels, P., J. Chem. Phys. 74 (1981)
       508-511.

[17]   Roussel, F., Carré, B., Breger, P., and Spiess, G., J. Phys.B
       14 (1981) L313-L319.

[18]   Polak-Dingels, P., Keller, J., Gauthier, J.C., Bras, N. and
       Weiner, J., Phys. Rev. A (in press).

[19]   Watanabe, R. and Welsh, H.L. Phys. Rev. Lett. 13 (1964) 810.

[20]   White, J.C., Zdasiuk, G.A., Young, G.F. and Harris, S.E.,
       Opt. Lett. 4 (1979) 137-139.

426 *R.S. Berry*

[21]  Hotop, R. and Niemax, K., J. Phys. B 13 (1980) L93-L99.

[22]  McTague, J.P. and Birnbaum, G., Phys. Rev. Lett. 21 (1968) 661-663.

[23]  McTague, J.P. and Birnbaum, G., Phys. Rev. A3 (1971) 1376-1383.

[24]  Barocchi, F., P. Mazzinghi and Zoppi, M., Phys. Rev. Lett. 41 (1978) 1785-1787.

[25]  Proffitt, M.H. and Frommhold, L., Phys. Rev. Lett. 42 (1979) 1473-1475.

[26]  Frommhold, L. and Proffitt, M.H., Phys. Rev. A 21 (1980) 1249-1255.

[27]  Arrowsmith, P., Bartoszek, F.E., Bly, J.H.P., Carrington, T., Charters, P.E. and Polanyi, J.C., J. Chem. Phys. 73 (1980) 5895-5897.

[28]  Hering, P., Brooks, P.R., Curl, R.F., Jr., Judson, R.S. and Lowe, R.S., Phys. Rev. Lett. 44 (1980) 687-690.

[29]  Weiner, J. Penning and associative ionization produced by intense optical fields, in:  Oda, N. and Takayanagi, K. (eds.), Electronic and Atomic Collisions (North-Holland, Amsterdam, 1980).

[30]  Weiner, J., Inelastic collision processes in the presence of intense optical fields, Lectures of the Summer School "Synamique Reactionelle dès Etats Excités," Les Houches, June, 1979.

[31]  Weiner, J., Collisioned ionization in the presence of intense optical radiation, presented at the "Meeting on photon-assisted collisions," Institute di Chimica Quantistica ed Energetica Molecolare, Pisa, Italy, June, 1981.

[32]  Lisitsa, V.S. and Yakovlenko, S.I., Zh. Eksp. Teor. Fiz. 66 (1974) 1550-1559; transl. Sov. Phys. - JETP 39 (1974) 759-763.

[33]  Vitlina, R.Z., Chaplik, A.V. and Éntin, M.V., Zh. Eks. Teor. Fiz. 67 (1974) 1667-1673; transl. Sov. Phys. - JETP 40 (1975) 829-832.

[34]  Dubov, V.S., Gudzenko, L.I., Gurvich, L.V. and Yakovlenko, S.I., Chem. Phys. Lett. 45 (1977) 330-333.

[35]  Lau, A.M.F., Phys. Rev. A 16 (1977) 1535-1542.

[36]  Nayfeh, M.H. and Payne, M.G., Phys. Rev. A 17 (1978) 1695-1706.

[37]  Lau, A.M.F., Phys. Rev. A 19 (1979) 1117-1131.

[38]  Lau, A.M.F., Phys. Rev. Lett. 43 (1979) 1009-1012.

[39]  Lau, A.M.F., Phys. Rev. A 22 (1980) 614-617.

[40]  Agre, M.Ya. and Rapoport, L.P., Opt. Spektrosk. 48 (1980)
1023-1026; transl. Opt. Spectrosc. (USSR) 48 (1980) 560-561.

[41]  Weiner, J., J. Chem. Phys. 72 (1980) 2856-2860.

[42]  Weiner, J., J. Chem. Phys. 72 (1980) 5731-5732.

[43]  Weiner, J., Chem. Phys. Lett. 76 (1980) 241-244.

[44]  Hutchinson, M. and George, T.F., Phys. Lett. 82A (1981) 119-
122.

[45]  Ch'en, A.J. and Takeo, M., Revs. Mod. Phys. 29 (1957) 29-93.

[46]  Hedges, R.E.M., Drummond, D.L. and Gallagher, A., Phys. Rev.
A 6 (1972) 1519-1544.

[47]  Happer, W., Moe, G. and Tam, A.C., Phys. Lett. 54A (1975)
405-406.

[48]  Moe, G., Tam, A.C. and Happer, W., Phys. Rev. A 14 (1976)
349-358.

[49]  Breen, R.G., Revs. Mod. Phys. 29 (1957) 94-143.

[50]  Gallagher, A. and Holstein, T., Phys. Rev. A 16 (1977) 2413-
2431.

[51]  Tam, A., Moe, G., Park, W. and Happer, W., Phys. Rev. Lett.
35 (1975) 85-87.

[52]  Gudzenko, L.I. and Yakovlenko, S.I., Zh. Tekh. Fiz. 45 (1975)
234-236; transl. Sov. Phys. Tech. Phys. 20 (1975) 150-153.

[53]  Rapp, D. and Francis, W.E., Chem. J. Phys. 37 (1962) 2631-2641.

[54]  Copeland, D.A. and Tang, C.L., J. Chem. Phys. 65 (1976) 3161-
3171.

[55]  Copeland, D.A. and Tang, C.L., J. Chem. Phys. 66 (1977) 5126-
5129.

[56]  Olson, R.E., Combust. Flame 30 (1977) 243-249.

[57]  Berry, R.S., J. Chim. Phys. 77 (1980) 759-768.

[58]  Andrick, D. and Langhans, L., J. Phys. B9 (1976) L459-461.

[59]  Andrick, D. and Langhans, L., J. Phys. B11 (1978) 2355-2360.

[60]  Langhans, L., J. Phys. B11 (1978) 2361-2366.

[61]  Weingartshofer, A., Holmes, J., Caudle, G., Clarke, E. and
Krüger, H., Phys. Rev. Lett. 39 (1977) 259-270.

[62]  Langendam, P.J.K., and van der Wiel, M.J., J. Phys. B11 (1978)
3603-3612.

[63]  Curl, R.F., private communication, June, 1981.

PHYSICS OF ELECTRONIC AND ATOMIC COLLISIONS
S. Datz (editor)
© North-Holland Publishing Company, 1982

OPENING REMARKS FOR THE FANO SYMPOSIUM

Joseph Macek

Behlen Laboratory of Physics
The University of Nebraska
Lincoln, Nebraska 68588-0111

I am happy to welcome all of you here tonight for this symposium in honor of
Professor Fano. Since Ugo has participated in this conference from its inception
24 years ago and has contributed key insights to every subject represented here,
it is indeed appropriate to honor him in this way. Because his contributions have
touched most of us in our own specialized areas, it is impossible to recognize all
of this work in four talks. We are fortunate, however, to have as our speakers
people whose own research encompasses major areas where Professor Fano has contri--
buted essential concepts.

One of the recurring themes of Professor Fano's work has been the role of electron
correlation in atomic structure and the aggregation of matter from the basic atomic
constituants. Fano recognized that such correlation was the key to understanding
the structure of atomic states with excitation energies in what is now know as the
vacuum ultraviolet. In this energy region atomic electrons are frequently asymptotically free and there are few sharp features that can be related to specific
aspects of the electron dynamics. One such aspect is the Cooper minimum. Charac-
teristically, Fano noted that an aspect of these minima, discussed by Mike Seaton,
would relate to electron spin polarization. Prof. Dr. J. Kessler will speak on
this effect, now known as the Fano effect.

With the rapid development in the early 60's of experimental techniques, such as
vacuum ultraviolet, electron and Auger spectroscopy to study atomic states in the
spectral range on the high energy side of the ultraviolet, the field was ripe for
the introduction of appropriate theoretical concepts. The R-matrix represents one
such concept exploited by Ugo and his co-workers. Phil Burke, who has contributed
much to this subject and to this conference, will speak on this concept.

The formation of molecules represents the first step in the aggregation of atoms.
Simple diatomic molecules share many of the spectral features of atoms, but
nuclear motion introduces added features requiring new concepts. By joining the
quantum defect theory and the idea of frame transformations, Fano and his students
showed how the effects of nuclear and electronic motion could be disentangled to
extract information on molecular states. Perhaps no one has carried out this program to a higher degree of perfection than Dr. Christian Jungen, our third speaker.

While it is appropriate to emphasize Professor Fano's scientific contributions
here we must not overlook another accomplishment of major importance for the future
of this conference. I refer here to the school of atomic theory that has formed
around him at the University of Chicago. The many fine young theorists he has
brought into the field represents another lasting contribution. Professor Tony
Starace is a member of that school. We will review, appropriately enough, electron
correlations and will bring us up to date on the work of the Fano school in this
area.

While the four talks conclude our symposium one must recognize that the topics
discussed, despite their breadth, do not touch on many of Fano's contributions to

the science represented at ICPEAC.  The Fano-Lichten model, alignment and orientation, and no doubt other topics, are omitted.  This is done for obvious reasons of time and organization, but I am confident that other speakers at this conference, in other symposia and progress reports, will take up whatever is omitted here, such is the influence of Fano's work on the scientific program of this community.

PHYSICS OF ELECTRONIC AND ATOMIC COLLISIONS
S. Datz (editor)
© North-Holland Publishing Company, 1982

# NEW PERSPECTIVES ON ELECTRON CORRELATIONS

Anthony F. Starace
Behlen Laboratory of Physics
The University of Nebraska
Lincoln, Nebraska  68588-0111
U.S.A.

Two formulations for describing electronic excitations developed
by Fano and his school are reviewed.  The transition matrix
formulation of atomic photoionization has provided new perspectives on the well-known random phase approximation and on its
relation to configuration interaction.  The hyperspherical
coordinate approach to two electron states and collision processes
has provided new perspectives on the classification of two-
electron excitation channels and on the evolution of highly
excited electronic states.

## I.  INTRODUCTION

Ugo Fano and his school have made many contributions to atomic theory.  A common
feature of these many contributions is that they are "appropriate" formulations
for particular physical problems.  That is, they are formulations in which the key
features of experimental observations stand out in an obvious way.  Not surprisingly, given such an appropriate formulation, one is often provided with new perspectives.  We shall concern ourselves here with some of the new perspectives that
have been provided on electron correlations.

Specifically, we shall review in this paper two such appropriate formulations of
Ugo Fano and his school which have in common that they treat effects of two electron excitations in particularly convenient ways.  The first, the atomic transition
matrix formulation for one-electron transition processes — in particular, for the
single electron photoionization process — is designed to treat the effects of
*virtual* pairs of excited electrons in addition to the usual final state interactions treated, for example, in a Hartree-Fock or close-coupling formulation.  The
atomic transition matrix formulation has provided new perspectives on the photoionization process and on the Random Phase Approximation (RPA) for describing this
process.  In particular, it has provided a new relation between the RPA and the
more usual configuration interaction theory; and it has provided a more general
definition of the RPA which permits one to treat open-shell atom photoionization
processes in the same way as closed-shell atom photoionization processes.

The second appropriate formulation to be discussed here is the hyperspherical coordinate approach for describing the joint motion of two electrons and in particular
*real* two electron excitation processes.  This approach has shown that the radial
and angular coordinates, in the six dimensional space appropriate for two electron
systems, are approximately separable.  This separability has in turn led to a new
classification scheme for Rydberg series of two electron excitations.  Most impor-
tantly, analysis of the breakdown of this approximate separability provides a new
perspective on the evolution of two-electron excitations.

## II.  ATOMIC TRANSITION MATRIX FORMULATION

Theoretical understanding of closed-shell atom photoionization cross sections[1] has
been based on the Random Phase Approximation (RPA),[2-4] which implies that, in addition to the usual final state interactions, virtual excitations of pairs of valence

electrons have an important influence on these cross sections. The importance of the electron correlations included in the RPA have been confirmed for closed-shell atoms by other theoretical methods, especially the Many-Body Perturbation Theory[5] and the R-Matrix Theory.[6] On the whole, RPA calculations — or their equivalent — agree with most experimental photoionization cross sections for closed-shell atoms to within about 10%.

The formal definition of the RPA is that, within a Hartree-Fock (HF) basis, one includes all particle-hole interactions to infinite order.[2] In particular, one must seemingly include an infinite number of virtually excited electron pair excitations and de-excitations. Given that the RPA works beautifully in describing atomic photoionization, the only additional wish that atomic physicists might have is for a closer relation between the RPA and the more familiar configuration interaction picture.

## A. Closed-Shell Atom Treatment:  Relation of the RPA to Configuration Interaction

With this in mind, Chang and Fano[7] developed the transition matrix formulation for closed-shell atom photoionization processes. Initial application was made to photoionization of the outer subshell of argon, i.e.,

$$Ar3p^6(^1S) + \gamma \to Ar^+3p^5(^2P)\varepsilon\ell(^1P) \ ,\tag{1}$$

where dipole selection rules limit $\ell$ to the values 0 and 2. We shall use the process (1) to illustrate the theory. Chang and Fano[7] choose the following configurations for the initial and final states:

$$<f| \equiv <3p^5 \ \psi_{\varepsilon d} \ (^1P)|\tag{2a}$$

$$|i> \equiv |3p^6(^1S)> \ + \ |3p^4\Phi_a\Phi_b \ (^1S)>\tag{2b}$$

Thus the final state consists of a HF core of electrons plus an excited electron, with angular momentum 2 and kinetic energy $\varepsilon$, described by the unknown wavefunction $\psi_{\varepsilon d}(r)$. (Note that we ignore the s-channel in this discussion for simplicity; in any case, it is much weaker than the d-channel.) The initial state consists of a HF ground state $|3p^6(^1S)>$ plus a correlation term having two electrons excited out of the HF ground state. These two electrons are described by the unknown wavefunctions $\Phi_a(r)$ and $\Phi_b(r)$, which we shall presume have orbital angular momentum 2.

The dipole matrix element between the initial and final states in Eq. (2) is:

$$<f|r^{(1)}|i> \ = 2(<\psi_{\varepsilon d}|r|3p> \ - \ <3p|r|\phi> )\ .\tag{3}$$

Here the matrix elements on the right represent radial integrals over r, the factor 2 is an overall angular factor, and the function $\phi(r)$ is defined as the following linear combination of the unknown correlation functions $\Phi_a(r)$ and $\Phi_b(r)$:

$$\phi(r) \equiv C(\Phi_a(r)<\psi_{\varepsilon d}|\Phi_b> \ + \ \Phi_b(r)<\psi_{\varepsilon d}|\Phi_a> )\tag{4}$$

In Eq. (4), C represents an angular factor and the brackets represent overlap integrals of $\Phi_a$ and $\Phi_b$ with $\psi_{\varepsilon d}$. It is because of these overlaps that $\Phi_a$ and $\Phi_b$ are assumed to have orbital angular momentum 2. Clearly, since the dipole operator is a one electron operator, the dipole transition matrix element between the correlation term in Eq. (2b) and the final state in Eq. (2a) is non-zero only if one of the virtually excited electrons, represented by the correlation function $\Phi_a$ or $\Phi_b$, has a non-zero overlap with the excited electron function $\psi_{\varepsilon d}$. The other virtually excited electron, represented by $\Phi_b$ or $\Phi_a$, is de-excited by the dipole interaction to the 3p-subshell.

To completely determine the dipole matrix element between the initial and final states in Eq. (2) for a particular photon energy, then, one only needs to determine the unknown final state wavefunction $\psi_{\varepsilon d}(r)$ and the particular linear combination of the correlation functions $\Phi_a(r)$ and $\Phi_b(r)$ represented by $\phi(r)$ in Eq. (4). To determine $\psi_{\varepsilon d}(r)$ and $\phi(r)$, Chang and Fano[7] start from the equation of motion of the outer product $|f><i|$,

$$H|f><i| - |f><i|H = \hbar\omega \, |f><i| \quad , \tag{5}$$

where H is the exact Hamiltonian for the atom and $\hbar\omega$ is the photon energy. Eq. (5) is integrated analytically over N-1 radial coordinates to obtain a set of coupled differential equations in the Nth coordinate r for the unknown functions $\psi_{\varepsilon d}(r)$, $\Phi_a(r)$, and $\Phi_b(r)$. Upon performing these integrations, the first term on the left in Eq. (5) gives final state interactions, the second term on the left gives initial state interactions, and the right hand side gives the first order transition matrix,[7,8] from which the method gets its name. Chang and Fano then approximate the resulting equations by dropping all terms involving $\Phi_a$ and $\Phi_b$ which cannot be cast in terms of the linear combination $\phi(r)$ in Eq. (4). Further approximations are made, by dropping all but a few interactions involving $\phi(r)$, to obtain a coupled set of differential equations for $\psi_{\varepsilon d}(r)$ and $\phi(r)$ which can be shown[7] to be equivalent to the RPA.

The results of the transition matrix formulation for the photoionization of argon[9] are compared with experiment[10] in Fig. 1. Curves I show the length and velocity results obtained in the HF approximation which results from setting all interactions involving $\phi(r)$ equal to zero when calculating the final state wavefunction $\psi_{\varepsilon d}(r)$ in the transition matrix formulation. Curves II show the results obtained from solving the equations for $\psi_{\varepsilon d}$ and $\phi(r)$ in an uncoupled approximation. Finally, curves III show the results obtained by solving the coupled equations for $\psi_{\varepsilon d}(r)$ and $\phi(r)$. These latter results are in very good agreement with experiment[10] and are equivalent to results obtained in the RPA.[2] The difference between the transition matrix formulation and the RPA lies, however, in the closer relation of the former to a configuration interaction approach (cf. Eq. (2)). Indeed, the configuration interaction picture of Chang and Fano[7] has been confirmed directly by the multiconfiguration Hartree-Fock approach to photoionization of Swanson and Armstrong.[11]

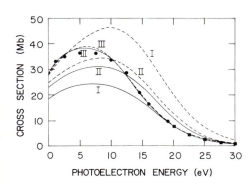

Fig. 1. Theoretical calculations of Chang[9] for the photoionization cross section of the 3p subshell of Ar. Dashed and solid lines give length and velocity results, respectively, in three levels of approximation discussed in the text. Experimentally measured values of the Ar cross section are indicated by the solid circles[10] and by the solid squares (Samson, unpublished) (from Ref. 9).

B.  Open-Shell Atom Generalization:  New Definition of the RPA

Theoretical understanding of the influence of electron correlations on the photoionization cross sections for open-shell atoms is less developed than for closed-shell atoms. This is due to the greater theoretical difficulty of dealing with atoms which are not spherically symmetric and which thus have a greater number of final state channels. In addition, theoretical approximation methods developed specifically for closed-shell atoms, such as the RPA, are not easily generalized to treat open-shell atoms. In particular, while several open-shell atom RPA theories have been developed,[12-16] these have been given in the form of matrix or integral equations which require the use of large numbers of basis functions for their solution; furthermore, these RPA theories differ from one another.

Recently, Starace and Shahabi[17] generalized the transition matrix formulation to treat open- or closed-shell atoms. In order to do so, they developed a graphical procedure for evaluating the integrations over N-1 coordinates in each term of the equation of motion, Eq. (5). The graphical method simplifies the treatment of antisymmetry and of angular momentum algebra and at the same time affords an insight into the physical processes involved similar to that afforded by many-body perturbation theory graphs. The graphical method is based on the graphical angular momentum algebra of Jucys et al.[18] and on the state vector graphs of Briggs.[19]

Starace and Shahabi[17] start by defining the initial $<i|$ and final $|f^\alpha>$ states in an open-shell atom photoionization process analogously to Chang and Fano's[7] states for closed-shell atoms, i.e.,

$$|f^\alpha> \equiv \sum_{\widetilde{LS}\ell} |n_0\ell_0^{q-1}(\widetilde{LS})\psi^\alpha_{(\widetilde{LS})\epsilon\ell} \, L_fS_f> \qquad (6a)$$

$$<i| \equiv <n_0\ell_0^q L_iS_i| \qquad (6b)$$

$$+ \sum_{\substack{\overline{LS}\ell_\phi \\ L_pS_p}} b(\overline{LS},L_pS_p,\ell_\phi) <n_0\ell_0^{q-2}(\overline{LS}) \, \Phi_a\Phi_b \, (L_pS_p)L_iS_i| \; .$$

Thus, the initial state is represented not only by the HF state of q electrons in the $n_0\ell_0$ subshell but also by correlation terms having two electrons excited out of the $n_0\ell_0$ subshell.

Using Eq. (6) for the form of the initial and final states, the graphical method permits the exact evaluation of all matrix elements in the equation of motion, Eq. (5), integrated over N-1 radial coordinates. Of these matrix elements, those involving the correlated part of the initial state are very complex. Starace and Shahabi[17] found, however, that if all interactions of the correlation functions $\Phi_a$ and $\Phi_b$ with the ionic core, $n_0\ell_0^{q-2}$, are approximated by requiring $\Phi_a$ and $\Phi_b$ to exchange zero orbital and spin angular mometum with the ionic core, then each interaction may be described in terms of linear combinations $\phi$ of $\Phi_a$ and $\Phi_b$, as in the closed-shell atom case. Furthermore when the weight factors for these particular interactions are calculated as if the ionic core contained a full subshell of electrons, instead of only q-2 electrons, then in the closed-shell atom case the resulting coupled differential equations are identical to those obtained algebraically by Chang and Fano, which are equivalent[7] to the RPA equations.

The approximation that the virtually excited electrons $\Phi_a$ and $\Phi_b$ exchange zero angular momentum with the ionic core provides then a set of equations equivalent to the RPA equations in the closed-shell atom case. It thus provides also a new definition of the RPA. Since this new definition is independent of whether the atom is initially open or closed, it therefore provides a new RPA-equivalent set of equations for open-shell atoms as well as a new understanding of the RPA.

In summary form, then, the key RPA-type approximations of the generalized transition matrix formulation for atomic photoionization are as follows:
  (a) The ground state $< i|$ should include configurations of the form
      $< n_0 \ell_0{}^{q-2} \Phi_a \Phi_b |$, in which two electrons are excited.
  (b) When $\Phi_a$ or $\Phi_b$ interact with the core $n_0 \ell_0{}^{q-2}$, one only includes the part of the interaction in which there is zero exchange of orbital and spin angular momenta.
  (c) The approximate interactions in (b) are calculated with a weight factor appropriate for a filled $n_0 \ell_0$ subshell.

These approximations give exactly the Chang-Fano equations[7] for closed-shell atoms and they give the close-coupling equations[20] when the correlation functions $\Phi_a$ and $\Phi_b$ are set equal to zero. The close connection to configuration interaction and the clear specification of the kinds of interactions included give a very good understanding of the RPA and its physical content.


III.  HYPERSPHERICAL COORDINATE APPROACH

While the use of hyperspherical coordinates to describe two-electron correlations is quite old,[21-28] it is only relatively recently that Macek[29] introduced a quasi-separable approximation in such coordinates which provided quantitatively accurate predictions of doubly-excited-state energies in He as well as a good description of two-electron dynamics. The initial success of the quasi-separable approximation stimulated Fano and his school to further develop the theory and to carry out numerous applications to doubly excited states[30-36] of He and H⁻ as well as to continuum processes[37-39] for He and for the e-H system. Most recently the hyperspherical coordinate approach has been extended successfully to treat atoms with more than two electrons.[40-44] While the quasi-separable approximation is only the zero order approximation to a general theory of atomic collisions,[45] it has provided new perspectives which we review here on electron correlations and on the electron excitation process.

A two electron wavefunction $\psi(\vec{r}_1, \vec{r}_2)$ is usually described by the six coordinates $r_1$, $r_2$, $\hat{r}_1$, and $\hat{r}_2$ of the two electrons. In hyperspherical coordinates the magnitudes of the individual coordinates, $r_1$ and $r_2$, are replaced by the hyperspherical radius $R \equiv (r_1{}^2 + r_2{}^2)^{1/2}$ and the hyperspherical angle $\alpha \equiv \arctan(r_2/r_1)$. Before summarizing the features of the Schrödinger equation in these coordinates let us look first at plots of approximate two-electron probabilities, $|\psi(R, \alpha, \hat{r}_1, \hat{r}_2)|^2$, in these coordinates. Fig. 2 shows contour plots[31,32] and Fig. 3 shows relief maps[36] for the probability distributions of the singly-excited state $1s2s$ $^1S$ and the doubly-excited state $2s^2$ $^1S$ of He. (Note that the wavefunctions are calculated in the approximation that each electron has an orbital angular momentum equal to zero in order to eliminate all dependence on the angular variables $\hat{r}_1$ and $\hat{r}_2$; since the angular dependence is trivial, these states are symmetric about $\alpha = \pi/4$, i.e., under interchange of $r_1$ and $r_2$.)

The most obvious distinguishing features of the two probability distributions is that that for the single excited state is largest along $\alpha \simeq 0$ and $\alpha \simeq \pi/2$ (implying one electron is much further from the nucleus than the other) while that for the doubly excited state is largest along $\alpha \simeq \pi/4$ (implying both electrons are comparably excited, i.e., $\alpha = \pi/4$ when $r_1 = r_2$). A second important feature is the behavior of the nodal lines for the two probability distributions. The 1s2s $^1$S state has a single nodal line along $R \simeq 2$, while the 2s$^2$ $^1$S state has two nodal lines along $\alpha \simeq$ constant: one along $5° \le \alpha \le 30°$ and the other along $60° \le \alpha \le 85°$. The fact that the pattern of nodal lines is approximately along the orthonormal grid of constant $R$ and constant $\alpha$ implies a quasi-separability of $R$ and $\alpha$ coordinates.

The nodal line pattern for a particular state serves also to classify the state.[36] The ground state of He, 1s$^2$ $^1$S, has a spherically symmetric probability distribution and is the first member of the singly-excited state channel 1sns $^1$S which converges to the He$^+$(n=1) threshold. The single node in $R$ for the state 1s2s $^1$S, shown in Figs. 2(a) and 3(a), characterizes it as the second member of the 1sns $^1$S channel. The state 2s$^2$ $^1$S, shown in Figs. 2(b) and 3(b), has no radial nodes. It is the first member of the Rydberg series 2sns $^1$S converging to

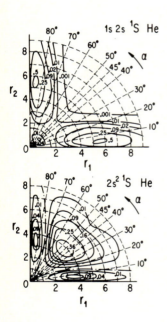

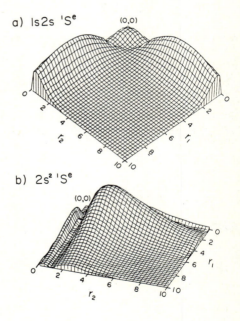

Fig. 2. Contour plot of the approximate probability distribution $|\psi(\vec{r}_1,\vec{r}_2)|^2$ for He. (a) 1s2s $^1$S (b) 2s$^2$ $^1$S. Solid Lines: lines of constant probability. Dot-Dash Lines: nodal lines. (From Ref. 32)

Fig. 3. Relief map of the approximate probability distribution $|\psi(\vec{r}_1,\vec{r}_2)|^2$ for He. (a) 1s2s $^1$S (b) 2s$^2$ $^1$S (From Ref. 36)

the He$^+$(n=2) threshold. The two nodes approximately along constant $\alpha$, symmetrical about $\alpha = \pi/4$, characterize 2s$^2$ $^1$S as a member of this second Rydberg channel.
Thus nodes in R characterize the excitation of a state within a channel while
nodes in $\alpha$ characterize the various channels.[36]

These features of the two-electron wavefunction $\psi(\vec{r}_1,\vec{r}_2)$, in the hyperspherical
coordinates R and $\alpha$, motivate the hyperspherical method of Macek,[29] who expands
$\psi(\vec{r}_1,\vec{r}_2)$ in terms of a complete set of adiabatic angle functions $\phi_\mu(R;\alpha,\hat{r}_1,\hat{r}_2)$:

$$\psi_E(R,\alpha,\hat{r}_1,\hat{r}_2) = (R^{5/2} \sin\alpha \, \cos\alpha)^{-1} \sum_\mu F_{\mu E}(R)\phi_\mu(R;\alpha,\hat{r}_1,\hat{r}_2) \tag{7}$$

The angle functions $\phi_\mu(R;\alpha,\hat{r}_1,\hat{r}_2)$ are dependent on the five angles $\alpha$, $\hat{r}_1$, and
$\hat{r}_2$ and are only parametrically dependent on R. They are the eigenstates of the
following differential equation,

$$\left[ \frac{-d^2}{d\alpha^2} + \frac{\ell_1^2}{\sin^2\alpha} + \frac{\ell_2^2}{\cos^2\alpha} - RC(\alpha,\theta_{12}) \right] \phi_\mu = -U_\mu(R)\phi_\mu \quad, \tag{8}$$

and the potentials $U_\mu(R)$ are the corresponding eigenvalues at each value of R.
In Eq. (8), the potential $-C(\alpha,\theta_{12})$ is proportional to the sum of the nuclear
and electrostatic potentials,

$$-C(\alpha,\theta_{12}) = -\frac{Ze^2}{\cos\alpha} - \frac{Ze^2}{\sin\alpha} + \frac{e^2}{(1 - \sin2\alpha \, \cos\theta_{12})^{1/2}} \tag{9}$$

Substituting Eq. (7) into the two-electron Schrödinger equation and using the
properties of the angle functions $\phi_\mu$ in Eq. (8), one obtains the following set of
coupled differential equations for the radial functions $F_{\mu E}(R)$:

$$\left( \frac{d^2}{dR^2} + \frac{U_\mu(R)+ \frac{1}{4}}{R^2} + 2E \right) F_{\mu E}(R) + \sum_{\mu'} W_{\mu,\mu'} F_{\mu'E}(R) = 0 \quad, \tag{10}$$

where

$$W_{\mu,\mu'} \equiv (\phi_\mu, \frac{\partial^2 \phi_{\mu'}}{\partial R^2}) + 2 (\phi_\mu, \frac{\partial \phi_{\mu'}}{\partial R}) \frac{\partial}{\partial R} \quad. \tag{11}$$

The brackets in Eq. (11) imply integration over angular variables only.

Each of the potentials $U_\mu(R)$ and its corresponding angular eigenfunction $\phi_\mu$
define the set of hyperspherical, two-electron channels $\mu$. These channels are
coupled through the radial derivative matrix elements $W_{\mu,\mu'}$ in Eq. (11). In
the separable — or adiabatic — approximation[29] one assumes that motion in $\alpha$
and motion in R are weakly coupled so that the derivatives of $\phi_\mu$ in Eq. (11) are
small enough that one may ignore the coupling terms $W_{\mu,\mu'}$. In this case, the
two electron wavefunction may be represented by a single term in the summation
in Eq. (7):

$$\psi_{\mu E}^{sep.}(R,\alpha,\hat{r}_1,\hat{r}_2) = (R^{5/2}\sin\alpha\,\cos\alpha)^{-1}\,F_{\mu E}(R)\phi_\mu(R;\alpha,\hat{r}_1,\hat{r}_2) \qquad (12)$$

Notice in Eq. (12) how all members of the channel $\mu$ have the same angular function $\phi_\mu$. Each state of excitation energy E within the channel $\mu$ is described by the radial function $F_{\mu E}$, which is calculated in the channel potential $U_\mu(R)$ using Eq. (10) and ignoring the coupling terms. Because each member of a Rydberg series of doubly excited states has the same angular function $\phi_\mu$ and has a radial function $F_{\mu E}(R)$ that is calculated in the same potential $U_\mu(R)$, the physical properties of states belonging to a particular channel $\mu$ are often immediately apparent upon examination of $U_\mu(R)$ and $\phi_\mu$. In what follows we illustrate the use of the potentials $U_\mu(R)$ to classify two-electron excitation channels. We then examine the numerical accuracy of the separable approximation. Next, we show how the variation of the angle functions $\phi_\mu$ with R provides a new perspective on the evolution of excitation processes. Lastly, we sketch very recent extensions of the hyperspherical coordinate approach to atoms having more than two electrons.

## A. Hyperspherical Classification of Two-Electron Excitation Channels

The first major success[29] of the separable approximation in hyperspherical co-ordinates was the classification and interpretation of the photoabsorption spectrum of He in the region of the doubly excited Rydberg states converging to the n = 2 threshold. In the usual classification scheme there should be three Rydberg series of such levels of comparable intensity: 2snp $^1P$, 2pnd $^1P$, and 2pns $^1P$. The experimental spectrum of Madden and Codling,[46] shown in Fig. 4, showed only one strong Rydberg series and one very weak Rydberg series. The third possible series was not observed. Cooper, Fano, and Prats[47] interpreted the relative intensities of the two observed series in terms of the so-called "+" and "-" series, $(2snp \pm 2pns)^1P$. The "+" series members are more intense than those of the "-" series because the corresponding wavefunctions of the "+" members have a much larger amplitude near the origin, allowing therefore a much larger overlap with the ground state. This scheme, however, does not explain the weakness of the 2pnd $^1P$ channel. Fig. 5, however, shows Macek's hyperspherical potentials $U_\mu(R)$ for the three channels $\mu$ converging to the n=2 state of He$^+$. One sees immediately that the three channels have vastly different centrifugal barriers near the origin, explaining the

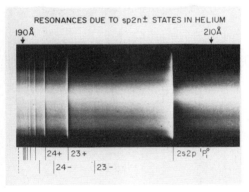

RESONANCES DUE TO sp2n± STATES IN HELIUM
190Å        210Å

|24+ |23+        |2s2p 'P₁°
|24- |23-

Fig. 4. Photoabsorption spectrum of He between 190 and 210 Å. The "+" and "-" series members are indicated below the spectrum. (From Ref. 46)

large intensity differences of the three allowed channels. Furthermore, the first two hyperspherical channels have the "+" and "-" characteristics predicted by Cooper et al.[47]

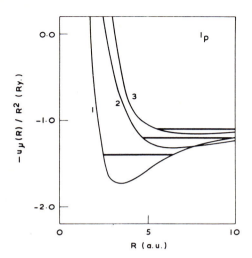

Fig. 5. Hyperspherical potential curves $U_\mu/R^2$ vs R for the three He doubly excited $^1$P channels converging to the n=2 state of He$^+$. (From Ref. 29).

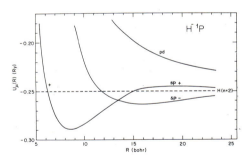

Fig. 6. Hyperspherical potential curves $U_\mu$ vs R for the three H$^-$ doubly excited $^1$P channels converging to the n=2 state of H. (From Ref. 33)

Similar work has been carried out for the doubly excited states of H$^-$ by Lin[33] and by Greene.[35] Fig. 6 shows Lin's[33] hyperspherical potentials for the three doubly excited Rydberg series converging to the n=2 state of H. The "+" channel is repulsive at large R and is not deep enough to support any bound states at small R. The repulsive barrier can, however, produce shape resonances in this channel above threshold. The "-" potential is attractive at large R and can support an infinity of Feshbach resonances. The "pd" channel is repulsive at all R values. Fig. 7 shows Greene's hyperspherical potentials for doubly excited channels converging to the n=3 state of H. In this case the "+" potential is always attractive and since its centrifugal barrier is weaker than those of the other channels, the "+" series is the most strongly excited from the ground state. Greene used this hyperspherical channel calculation and quantum defect theory (QDT) to interpret the resonances obtained by Hamm et al.[48] in the photodetachment spectrum of H$^-$ near the n=3 threshold as due to the "+" series resonances. The data and Greene's QDT fit are in excellent agreement, as shown in Fig. 8.

B. Numerical Results in the Separable Approximation

The separable approximation in hyperspherical coordinates thus provides a very accurate qualitative description of Rydberg series of doubly excited states in He and H$^-$. But how good are the quantitative predictions in this approximation? The answer depends on the excitation energy above the minimum in the hyperspherical potential $U_\mu$ of interest. For the lowest energy members calculated in the potentials $U_\mu(R)$, the separable approximation

energies are in excellent agreement with experiment and with other theoretical results; the separable approximation wavefunction may also be used confidently. However, higher energy members of a particular channel $\mu$ calculated in the potential $U_\mu(R)$ are increasingly too high in energy,[29] if bound, or have too negative phase shifts,[37-39] if unbound. This is not surprising since for higher excitation energies the coupling between the hyperspherical channels can no longer be ignored. Thus at present the separable approximation in the hyperspherical coordinate

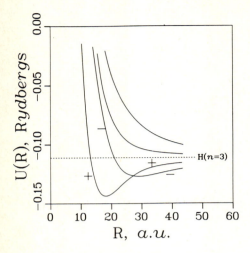

Fig. 7. Hyperspherical potential curves $U_\mu$ vs R for H$^-$ doubly excited $^1$P channels converging to the n=3 threshold of H. (From Ref. 35)

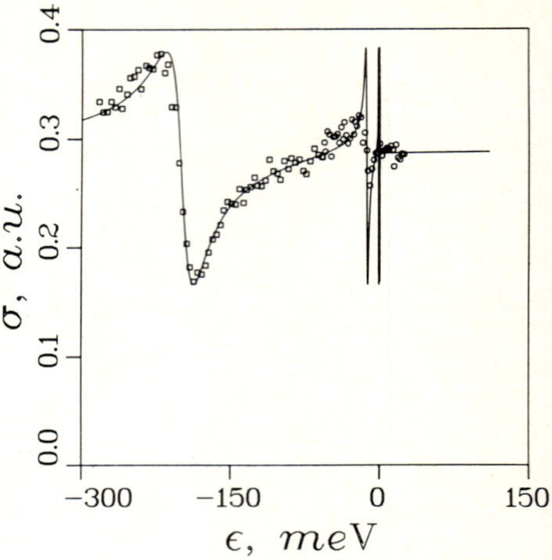

Fig. 8. Photodetachment cross section of H$^-$ near the n=3 threshold of H at $\epsilon = 0$. Data: Hamm et al.[48] Solid Line: QDT fit of Greene.[35] (From Ref. 35)

approach provides a very good initial approximation to the exact electron wave-function, but its systematic improvement for states of moderate and high excitation energy to provide state-of-the-art numerical predictions has only just begun[38,41] and remains a task for future research.

In what follows, we provide a few examples of the high level of accuracy to be expected from the separable approximation for the lowest excited states in particular hyperspherical potentials. Regarding level energies, Macek[29] calculated the He 2s2p($^1$P) resonance excitation energy as 68.138 eV as compared with the experimental value[46] of 60.135 ± .015 eV. Similarly, Miller and Starace[39] calculated the ground state energy of He as -2.895 a.u. as compared with the essentially exact non-relativistic theoretical value[49] of -2.904 a.u. Regarding phase shifts, Lin[37] calculated the e-H $^1$S phase shift at k = 0.1 to be 2.513 rad. as compared with the essentially exact theoretical value[50] of 2.553 rad.

Lastly, a recent calculation[39] of the photoionization cross section of He using separable approximation hyperspherical coordinate wavefunctions demonstrates the strengths and weaknesses of the method. The initial and final wavefunctions for the process

$$\text{He}(^1\text{S}) + \gamma \longrightarrow \text{He}^+ \; 1\text{s}(^2\text{S}) + \text{e}^-(^1\text{P}) \tag{13}$$

both have the form of Eq. (12). For the initial state, $\mu$ corresponds to the lowest $^1$S potential $U_\mu$(R), and for the final state, $\mu$ corresponds to the lowest $^1$P potential $U_\mu$(R). The photoionization cross section obtained using the separable approximation wavefunctions is shown in Fig. 9.

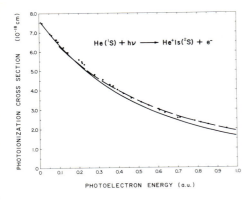

Fig. 9.  Photoionization cross section for He. <u>Full curve</u>: separable approximation (single channel) hyperspherical calculation of Miller and Starace (Ref. 39); <u>Dots</u>: Experimental results of Samson (Ref. 51); <u>Dashed Curve</u>: 1s-2s̄-2p̄ (four channel) close-coupling calculation of Jacobs (Ref. 52) (From Ref. 39).

Fig. 9 also shows the revised experimental results of Samson,[51] which have error bars of ±3%. The results lie within these error limits near threshold (for kinetic energies $0.0 \leqslant \varepsilon \leqslant 0.4$ a.u.) and in fact agree with experiment to within 1% at threshold. The hyperspherical results, however, are systematically lower than experiment above $\varepsilon = 0.4$ a.u. Of the many other theoretical calculations, we show the one with the best overall agreement with experiment: the four channel (1s-2s̄-2p̄) close-coupling calculation of Jacobs.[52] In comparison with the close-coupling results, the hyperspherical results are in better agreement with experiment below $\varepsilon = 0.2$ a.u. and are systematically lower above $\varepsilon = 0.2$ a.u.

## C.  Evolution of Two-Electron Excitations

The hyperspherical coordinate approach has not only been used to study stationary states, but also to understand qualitatively how a low-energy two electron state concentrated near the origin, upon receiving energy during a collision process, evolves to states of high excitation far from the origin. The key idea, stressed recently by Fano[53] and illustrated graphically by Lin,[36] is that such states describe motion along a potential ridge centered about the direction $\alpha = \pi/4$ (i.e., $r_1 = r_2$).

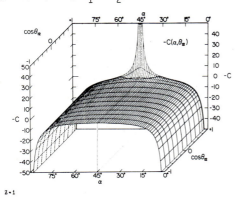

Fig. 10.  Relief map of the potential $-C(\alpha, \theta_{12})$ defined in Eq. (9) for $Z = 1$. (from Ref. 32).

Consider Eq. (8) for the channel functions $\phi_\mu(R; \alpha, \hat{r}_1, \hat{r}_2)$. The potential $-C(\alpha, \theta_{12})$, defined in Eq. (9), is shown in Fig. 10 for $Z = 1$. States having one electron more excited than the other, i.e., $r_2 \gg r_1$ or $r_1 \gg r_2$, have an angle function $\phi_\mu$ with maximum amplitude in the valleys of the potential in Fig. 10, near $\alpha = 0$ and $\alpha = \pi/2$. Comparably excited, doubly-excited states have $r_1 \sim r_2$ and thus the angle function $\phi_\mu$ for these states has maximum amplitude on the ridge of the potential in Fig. 10, near $\alpha = \pi/4$, and preferably near $\cos\theta_{12} = -1$ (i.e., on opposite sides of the nucleus). We consider

now the R-dependence of the .angle functions $\phi_\mu$.  Eq. (8) shows that the potential
-C is multiplied by R.  For large enough R, therefore, the potential -RC on the
ridge becomes equal to the eigenvalue $U_\mu(R)$.  At this "classical turning point"
the angle function $\phi_\mu$ has no more "kinetic energy" of motion in $\alpha$ on the ridge.
For larger R values, its amplitude on the ridge is exponentially damped and the
probability amplitude in the channel $\mu$ must retreat to the valleys of the potential in Fig. 10, implying that for such large R values $\mu$ describes states with one
electron more highly excited than the other.  Alternatively, the two electron
state on the ridge may "hop" to the next higher channel $\mu'$.  With a higher value
of $-U_{\mu'}(R)$, the two electron excitation could move to somewhat larger R along the
ridge since the difference between $-U_{\mu'}$ and the top of the potential ridge of -RC
would restore some positive "kinetic energy" of motion in $\alpha$.  Actually the vicinity of the classical turning point is propitious for such a transition to a higher
channel $\mu'$ since the coupling matrix elements (cf. Eq. (11) ) are largest precisely where the channel functions are changing most rapidly with R.

Lin[36] has shown graphically how the channel functions $\phi_\mu$ behave as functions of R.

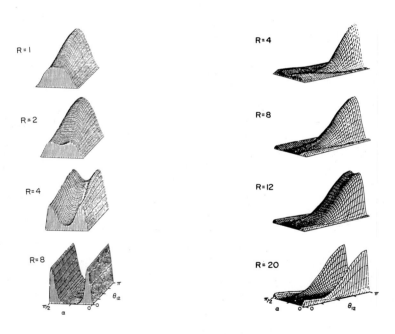

Fig. 11.  Plot of $|\phi_\mu(R;\alpha,\theta_{12})|^2$
vs. $\alpha$ and $\theta_{12}$ for various R values
for the first H$^-$ $^1$S hyperspherical
channel $\mu = 1$.  (from Ref. 36)

Fig. 12.  Plot of $|\phi_\mu(R;\alpha,\theta_{12})|^2$
vs. $\alpha$ and $\theta_{12}$ for various R values
for the second H$^-$ $^1$S hyperspherical
channel $\mu = 2$.  (from Ref. 36)

In Figs. 11 and 12 we show the $H^-(^1S)$ channel functions $\phi_\mu(R;\alpha,\theta_{12})$ for $\mu = 1$ and
$\mu = 2$ (i.e., the lowest two $^1S$ hyperspherical channels). In Fig. 11 one sees that
at $R = 1$ the charge distribution in the first channel is peaked about $\alpha = \pi/4$,
lying on the potential ridge. At $R = 4$, however, the charge distribution is va-
cating the ridge and moving to the valleys near $\alpha = 0$ and $\alpha = \pi/2$. By $R = 8$, $\mu = 1$
describes a channel with one electron much more highly excited than the other.
Fig. 12 shows the next higher hyperspherical channel function. Note that at $R = 4$,
precisely where $\mu = 1$ has a depression along the ridge, the $\mu = 2$ channel's charge
distribution has a maximum. This peak in $\mu = 2$ along the ridge progresses outward
to larger R values until at $R = 12$ a depression appears along the ridge. If two-
electron states in $\mu = 2$ are to move to larger R and remain comparably excited
they must hop again to the next higher hyperspherical channel, and so on.

This new perspective of two electron excitation states evolving toward large
radii R along a potential ridge has its origins in the Wannier-Peterkop-Rau[54]
analysis of electron impact ionization near threshold. Its application to
quantitative predictions of excitation cross sections remains a task for
future research.

### D.  Recent Extensions to Atoms Having More than Two Electrons

The new perspectives on electron correlations provided by the hyperspherical
coordinate approach are not limited to He and $H^-$. Indeed several applications
have already been made to heavier atoms. While detailed discussion of these
works is not possible in this short paper, we wish to alert the reader to this
recent progress. Firstly, Clark and Greene[40] have made a first attempt at a
hyperspherical description of the three electron systems Li and $H^{--}$. Greene,[41]
on the other hand, has treated Be and the alkaline earth atoms as two electron
systems, outside closed shells, in hyperspherical coordinates. The influence
of the inner closed shells on the outer electron pair is approximated by an
atomic potential. Watanabe[42] has gone still further in treating two electrons
in hyperspherical coordinates outside an open-shell ionic core. The influence
of the ionic core on the outer electrons is described by appropriate boundary
conditions on the hyperspherical wavefunction at the surface of the ionic core.
Both $K^-$ and $He^-$ have been treated in this way. Lastly, Lin[44] has presented a
method for generating basis sets of analytically determined, approximate hyperspherical wavefunctions. Such basis sets might be used to describe doubly
excited states of complex atoms in perturbation or scattering calculations.

### ACKNOWLEDGMENTS

The author wishes to thank U. Fano, C. D. Lin, C. H. Greene, and S. Watanabe
for providing results prior to publication. Research support of the National
Science Foundation under Grant No. PHY-8026055 is also gratefully acknowledged.

### REFERENCES

1.  Starace, A. F., "Trends in the Theory of Atomic Photoionization," Applied
    Optics 19, 4051 (1980); "Theory of Atomic Photoionization," in Handbuch
    der Physik 31, Mehlhorn, W. (ed.) (Springer-Verlag, Berlin, 1982).

2.  Amusia, M. Ya and Cheripkov, N. A., Case Studies in Atomic Physics 5, 47
    (1975).

3.  Wendin, G., in: Wuilleumier, F. J. (ed.) Photoionization and Other Probes
    of Many-Electron Interactions (Plenum Press, New York, 1976), p. 61.

4.   Johnson, W. R. and Lin, C. D., Phys. Rev. A 20, 964 (1979); Johnson, W. R. and Cheng, K. T., Phys. Rev. A 20, 978 (1979); Huang, K.-N., Johnson, W. R., and Cheng, K. T., Phys. Rev. Lett. 43, 1658 (1979).

5.   Kelly, H. P., in: Wuilleumier, F. J. (ed.), Photoionization and Other Probes of Many-Electron Interactions (Plenum Press, New York, 1976), p. 83.

6.   Burke, P. G. and Robb, W. D., Adv. Atomic Mol. Phys. 11, 143 (1975); Burke, P. G. and Taylor, K. T., J. Phys. B 8, 2620 (1975).

7.   Chang, T. N. and Fano, U., Phys. Rev. A 13, 263 (1976); Phys. Rev. A 13, 282 (1976).

8.   Löwdin, P. O., Phys. Rev. 97, 1474 (1955).

9.   Chang, T. N., Phys. Rev. 15, 2392 (1977).

10.  Samson, J. A. R., Adv. At. Mol. Phys. 2, 177 (1966).

11.  Swanson, J. R. and Armstrong, Jr., L., Phys. Rev. A 15, 661 (1977); Phys. Rev. A 16, 1117 (1977).

12.  Armstrong, Jr., L., J. Phys. B 7, 2320 (1974).

13.  Rowe, D. J. and Ngo-Trong, C., Rev. Mod. Phys. 47, 471 (1975).

14.  Dalgaard, E., J. Phys. B 8, 695 (1975).

15.  Starace, A. F. and Armstrong, Jr., L., Phys. Rev. A 13, 1850 (1976).

16.  Cherepkov, N. A. and Chernysheva, L. V., Phys. Letters 60A, 103 (1977); Izvestia Akademii Nauk SSSR 41, 2518 (1977) [Eng. Translation: Akademiia Nauk SSSR. Bulletin Physical Series (Allerton Press) 41, 47 (1977).]

17.  Starace, A. F. and Shahabi, S., Physica Scripta 21, 368 (1980); Phys. Rev. A (in press).

18.  Yutsis (Jucys), A. P., Levinson, I. B., and Vanagas, V. V., The Theory of Angular Momentum (Israel Program for Scientific Translations, Jerusalem, 1962).

19.
19.  Briggs, J. S., Rev. Mod. Phys. 43, 189 (1971).

20.  Smith, K., Henry, R. J. W., and Burke, P. G., Phys. Rev. 147, 21 (1966).

21.  Kemble, E. C., The Fundamental Principles of Quantum Mechanics with Elementary Applications, (Dover Publications, Inc. New York 1937) p. 210.

22.  Morse, P. M. and Feshbach, H., Methods of Theoretical Physics, Vol. II (McGraw-Hill Book Company, New York, 1953) pp. 1730ff.

23.  Fock, V., Izvest. Acad. Nauk USSR ser Fiz. 18, 161 (1954) [Eng. Transl.: Kong. Norske Videnskabers Selskabs Forh. 31, 138, 145 (1958).]

24.  Demkov, Y. N. and Ermolaev, A. M., Zh. Eksp. Teor. Fiz. 36, 896 (1959) [Sov. Phys. - JETP 36, 633 (1959)].

25.  Smith, F. T., Phys. Rev. 120, 1058 (1960).

26. Zickendraht, W., Annals of Physics 35, 18 (1965).

27. Frankowski, K. and Pekeris, C. L., Phys. Rev. 146, 46 (1966); Frankowski, K., Phys. Rev. 160, 1 (1967).

28. Macek, J. H., Phys. Rev. 160, 170 (1967).

29. Macek, J. H., J. Phys. B 2, 831 (1968).

30. Fano, U., Atomic Physics 1 (Plenum, New York, 1969), pp. 209-25.

31. Fano, U. and Lin, C. D. Atomic Physics 4 (Plenum, New York, 1975), pp. 47-70.

32. Lin, C. D., Phys. Rev. A 10, 1986 (1974).

33. Lin, C. D., Phys. Rev. Lett. 35, 1150 (1975); Phys. Rev. A 14, 30 (1976).

34. Klar, H., J. Phys. B. 7, L436 (1974); Klar, H. and Klar, M., J. Phys. B 13, 1057 (1980).

35. Greene, C. H., J. Phys. B 13, L39 (1980).

36. Lin, C. D., Phys. Rev. A (in press).

37. Lin, C. D., Phys. Rev. A 12, 493 (1975).

38. Klar, H. and Fano, U., Phys. Rev. Lett. 37, 1132 (1976); Klar, H., Phys. Rev. A 15, 1452; Klar, H. and Klar, M., Phys. Rev. A 17, 1007 (1978).

39. Miller, D. L. and Starace, A. F., J. Phys. B 13, L525 (1980).

40. Clark, C. W. and Greene, C. H., Phys. Rev. A 21, 1786 (1980).

41. Greene, C. H., Phys. Rev. A 23, 661 (1981).

42. Watanabe, S., Phys. Rev. A (in press).

43. Fano, U., Physica Scripta (in press).

44. Lin, C. D., Phys. Rev. A 23, 1585 (1981).

45. Fano, U., Phys. Rev. A (in press).

46. Madden, R. P. and Codling, K., Astrophys. J. 141, 364 (1965).

47. Cooper, J. W., Fano, U., and Prats, F., Phys. Rev. Lett. 10, 518 (1963).

48. Hamm, M. E., Hamm, R. W., Donahue, J., Gram, P. A. M., Pratt, J. C., Yates, M. A., Bolton, R. D., Clark, D. A., Bryant, H. C., Frost, C. A., and Smith, W. W., Phys. Rev. Lett. 43, 1715 (1979).

49. Pekeris, C. L., Phys. Rev. 112, 1649 (1958).

50. Schwartz, C., Phys. Rev. 124, 1468 (1961).

51. Samson, J. A. R., Phys. Reports 28C, 303 (1976).

52. Jacobs, V. L., Phys. Rev. A 3, 289.(1971).

53.   Fano, U., Phys. Rev. A <u>22</u>, 2660 (1980).

54.   Wannier, G., Phys. Rev. <u>90</u>, 817 (1953); Peterkop, R., J. Phys. B <u>4</u>, 513 (1971); Rau, A. R. P., Phys. Rev. A <u>4</u>, 207 (1971).

PHYSICS OF ELECTRONIC AND ATOMIC COLLISIONS
S. Datz (editor)
© North-Holland Publishing Company, 1982

R-MATRIX THEORY

P.G. Burke

Queen's University, Belfast, N. Ireland,
and
Daresbury Laboratory, Warrington, England

The development of the R-Matrix theory both as a conceptual
and as a computational tool for understanding a broad
range of atomic and molecular processes is reviewed. The
essential role of Fano in developing this concept and in
relating it to other approaches is stressed.

## 1. Introduction and Basic Concept

It gives me very great pleasure to present a paper at this symposium in honour of
Professor Ugo Fano. As our Chairman to-day Professor Macek has told us,
Professor Fano has contributed key insights into almost all areas represented at
this conference. In addition, many of us here to-day have been fortunate to have
been influenced and guided at initial stages of our research by Professor Fano.
In my own case the influence of Ugo Fano was critical in the initiation of my own
work on R-Matrix theory (Burke and Hibbert, 1969, Burke et al, 1971).

The realization that this approach, first introduced in nuclear physics by Wigner
(1946 a,b) and Wigner and Eisenbud (1947) and comprehensively reviewed by Lane and
Thomas (1958), could be used to describe atomic photoionization and electron atom
scattering first began to evolve in 1965 during an extremely stimulating summer
which I spent working with Ugo Fano, John Cooper and Joe Macek at the NBS in
Washington. We were trying at that time to understand the new and exciting resonance spectra being produced by Madden and Codling (1963, 1965) using radiation
from the 180 MeV synchrotron, and by Simpson (1964) Kuyatt et al (1965) and others
using a recently perfected high resolution electron spectrometer. It was clear
from this work that an understanding of processes such as atomic photoionization

$$ h\nu + A_i \rightarrow A_j^* \rightarrow A_k^+ + e^- \tag{1} $$

and electron-atom scattering

$$ e^- + A_i \rightarrow A_j^{-*} \rightarrow A_k + e^-, \tag{2} $$

where $A_j^*$ and $A_j^{-*}$ are intermediate resonance states, could be understood in terms
of an R-Matrix type theory.

The basic idea of R-Matrix theory is that configuration space describing the
electron-atom, ion or molecule complex is divided into two regions as illustrated
in figure 1.

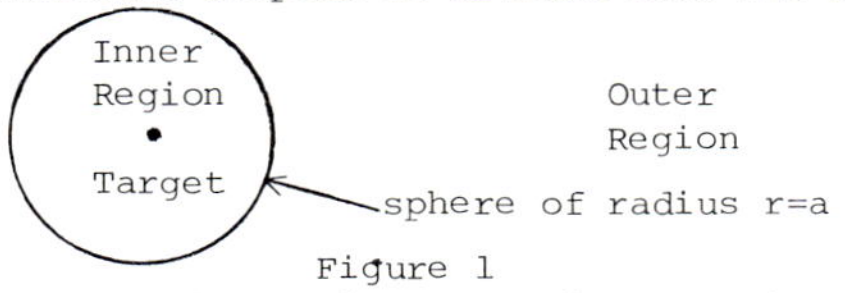

Figure 1
Division of Configuration Space in R-Matrix Theory

In the inner region the electron interaction with the target is strong, electron exchange and correlation effects are important and the intermediate complex behaves very much as a bound state.  Consequently bound state solution methods can be used in this region.  On the other hand, in the outer region, if the radius of the sphere is chosen to just envelope the charge distribution of the target, the electron interaction with the target is simple, and often analytic solutions are possible.  This occurs for example in electron-ion scattering when it is often possible to just include the Coulomb interaction of the scattered electron with the ion core.

The link between these two regions is provided by the R-Matrix which is defined by the equation

$$u_i(a) = \sum_j R_{ij}(E) \left( a \frac{du_j}{dr} - b_j u_j \right)_{r=a} \tag{3}$$

where $u_i(r)$ is the radial wavefunction describing the motion of the scattered electron in the i th channel, and the $b_j$ are parameters whose choice is discussed below.

The R-Matrix can be calculated as described in the next section by solving the pseudo-bound state problem in the inner region.  Alternatively use can be made of the fact that the R-Matrix is a meromorphic function of energy (with no threshold branch cuts) to parametrize it over a limited region.  This leads immediately, in the case of electron-ion scattering, to the Quantum Defect Theory introduced by Seaton (1958, 1966) and developed extensively by Fano (1970, 1981) and applied in many areas of atomic and molecular spectroscopy (e.g. Lu and Fano, 1970, Jungen, 1981).

The division of configuration space into two or more regions enables different representations or frames to be used in the different regions.  These frames are then connected by a unitary frame transformation of the R-Matrix on the boundary. The importance of this was emphasised for photodetachment near fine-structure thresholds by Rau and Fano (1971) and for electron molecule collisions near rotational and vibrational excitation thresholds by Chang and Fano (1972).  In both cases interactions which are weak compared with the strong electrostatic and exchange interactions in the inner region are neglected and are only included in the outer region.  In the case of photodetachment, the relativistic terms in the Hamiltonian are neglected in the inner region enabling LS coupling to be used, and in the case of electron molecule scattering the rotational and vibra- tional terms in the Hamiltonian are neglected enabling a body-fixed frame to be used.  In the latter case, this approximation is equivalent to the neglect of nuclear motion made in the Born-Oppenheimer approximation.  In the outer region these terms are included to yield the observed cross section to fine-structure or to rotational and vibrational levels of the target.

A key property of R-Matrix theory, as emphasised by Fano (1978), is that the energy spectrum of particles confined within the limited volume of the inner region is discrete, with a low density of levels.  Thus the number of configur- ations required to approximate the corresponding wavefunction may remain small over a limited but very important range of energies (say up to 30 eV).  By a suitable choice of the parameters $b_j$, Fano and Lee (1973) showed that each eigen energy of interest could be brought into coincidence with the total collision energy.  In this way, the corresponding eigenfunction accurately describes the collision complex occurring for example in the processes defined by equations (1) and (2).

2.   General Theory

The R-Matrix, for an electron scattering by an N-electron target, is calculated by solving the Schrödinger equation

$$(H - E) \ \Psi \ = \ 0 \tag{4}$$

in the inner region. If we expand $\Psi$ in this region in terms of a discrete basis satisfying arbitrary boundary conditions at $r = a$ then the Hamiltonian is not Hermitian owing to a non-zero surface term. This surface term can however be eliminated by introducing the Bloch operator $L_b$ (Bloch, 1957)

$$L_b \ = \ \sum_i \ | \ \Phi_i \ > \ \frac{1}{2} \ \delta(r - a) \ (\frac{d}{dr} - \frac{b_i}{r}) \ < \Phi_i | \tag{5}$$

where the $\Phi_i$ are channel functions formed from the target states coupled to the spin and angular variables of the scattered electron and the $b_i$ have already been introduced in equation (3). We now rewrite equation (4) as

$$(H + L_b - E) \ \Psi \ = \ L_b \ \Psi \tag{6}$$

which we can formally solve giving

$$\Psi \ = \ (H + L_b - E)^{-1} \ L_b \ \Psi \tag{7}$$

We now expand the Green's function $(H + L_b - E)^{-1}$ in the inner region in terms of a discrete basis defined by

$$\Psi_k \ = \ \mathscr{A} \sum_{ij} \ \Phi_i \ v_j(r) a_{ijk} \ + \ \sum_j \ \phi_j \ b_{jk} \tag{8}$$

where the $v_j(r)$ are radial basis orbitals representing the scattered electron which are non-zero on the boundary $r = a$, $\mathscr{A}$ is an antisymmetrization operator and the $\phi_j$ are $N + 1$ electron correlation functions which are insignificant by the boundary. The correlation functions are necessary to obtain accurate results for complex atoms, ions and molecules since, in the first expansion over channel functions, continuum channels are not included explicitly although often some pseudo-state channels are retained. We now determine the expansion coefficients $a_{ijk}$ and $b_{jk}$ by diagonalysing $H + L_b$ giving

$$< \ \Psi_k \ | H + L_b | \ \Psi_{k'} > \ = \ E_k \ \delta_{kk'} \tag{9}$$

where the integral is taken over the inner region. Since $H + L_b$ is Hermitian in this region the eigen energies $E_k$ are real. Expanding the Green's function in equation (7) then gives

$$| \ \Psi \ > \ = \ \frac{1}{2} \sum_{kj} \ \frac{| \ \Psi_k ><\ \Psi_k | \ \Phi_j >}{E_k - E} \ (\frac{d}{dr} - \frac{b_j}{r}) \ < \Phi_j | \ \Psi > \tag{10}$$

Projecting this equation onto the channel functions $\Phi_i$ and evaluating it on the boundary yields immediately our basic equation (3) where the radial wavefunctions

$$u_i(a) \ = \ < \ \Phi_i \ | \ \Psi >_{r=a} \tag{11}$$

and the R-Matrix

$$R_{ij}(E) \ = \ \frac{1}{2a} \sum_k \ \frac{W_{ik}(a) \ W_{jk}(a)}{E_k - E} \tag{12}$$

where we have introduced the surface amplitudes

$$W_{ik}(a) \ = \ \sum_j \ v_j(a) \ a_{ijk} \tag{13}$$

The main part of the work required to calculate the R-Matrix is the diagonalization of $H + L_b$ in equation (9). This needs to be carried out only once in order

to determine the R-Matrix at all energies.  The evaluation of the matrix elements
of $H+L_b$ is similar to a bound state calculation, the main differences are that
integrals involving the radial basis orbitals $V_j(r)$ are over a finite rather than
an infinite range, and the contribution from the Bloch operator must be included.
These differences are relatively trivial and therefore bound state codes can be
readily modified to carry out R-Matrix calculations.  This has proved particularly important in the case of molecules where the technology for carrying out
bound state calculations has been built up over many years.

The K-Matrix is obtained by matching to a complete set of solutions in the outer
region.  In this region electron exchange is by definition negligible and the
potential matrix has assumed its asymptotic form

$$V_{ij}(r) = \sum_{\lambda=1}^{\lambda_{max}} a_{ij}^{\lambda} r^{-\lambda-1} \tag{14}$$

The resultant coupled radial equations is then solved in this potential subject
to the usual boundary conditions

$$u_{ij}(r) \underset{r\to\infty}{\sim} k_i^{-\frac{1}{2}} (\sin\theta_i \delta_{ij} + \cos\theta_i K_{ij}) \quad \text{open channels}$$

$$u_{ij}(r) \underset{r\to\infty}{\sim} O(r^{-2}) \quad \text{closed channels} \tag{15}$$

where $\theta_i = k_i r - \frac{1}{2}\ell_i\pi$ for a neutral target, where $k_i$ and $\ell_i$ are the channel wave
numbers and angular momenta, and we have introduced a second index j on the
solution vector $u_i(r)$ to label the independent solutions.  Once the K-Matrix is
obtained it is a simple matter to calculate the T-Matrix and cross sections.

The choice of radial basis $V_i(r)$ and channel constants $b_i$ requires some discussion
at this point.  In the work on electron-atom and electron-ion scattering described
in detail by Burke and Robb (1975) it proved convenient to choose the $V_i(r)$ to be
numerical functions which were solutions of a radial wave equation with some
suitably chosen potential, and which satisfied fixed boundary conditions

$$\frac{a}{V_i(a)} \frac{dV_i}{dr}\Big|_{r=a} = b_i \tag{16}$$

where the $b_i$ were usually taken to be zero.  In this case, the Bloch operator
term vanishes identically, and thus need not be considered, however because all of
the basis orbitals satisfy the same boundary condition (16), expansion (12) is
slowly convergent and a correction for the omitted far-away poles suggested by
Buttle (1967) must be included.  A variational correction suggested by Zvijac et
al (1975) can also be applied.

In the work on electron-molecule scattering by Schneider (1975a, b, 1977),
Schneider and Hay (1976), Burke et al (1977), Buckley and Burke (1979) and Noble
et al (1981), either Gaussian or Slater type analytic orbitals were used to
represent the $V_i(r)$.  As emphasised above, this enables standard molecular bound
state codes to be used with little modification in the evaluation of equation (9).
Also since these orbitals satisfy arbitrary boundary conditions at $r = a$, the
Bloch operator term must be retained, however the Buttle correction is usually
not necessary.  The main limitation in the use of analytic rather than numerical
basis functions is the limited number that can be included in each channel before
linear dependence problems become important.  This limits the range of energies
over which the method is reliable.

In both of the methods described above, the logarithmic derivative of the wavefunction is not continuous on the boundary.  This difficulty can be overcome

using the eigenchannel method introduced in nuclear physics by Danos and Greiner (1966) and extended and applied in atomic physics by Fano and Lee (1973) and Lee (1974). In this method, the basis orbitals are first chosen to satisfy equation (16) for some arbitrary $b_i$, and the eigenenergies $E_k$, defined by equation (9), calculated. The $b_i$ are then adjusted and this whole process repeated iteratively until a complete set of eigenfunctions $\Psi_k$ satisfying

$$E_k \ (b_i) \ = \ E, \tag{17}$$

where E is the energy of interest, are found. It follows from equation (10) that the wavefunction in the inner region, corresponding to this energy can be expressed as a linear combination of these eigenfunctions. Further, since only the terms satisfying equation (17) contribute to equation (10), then the logarithmic derivative of the wavefunction is continuous on the boundary. For a given orbital basis this method gives higher accuracy than the previous methods since the convergence problems of expansion (12) are avoided. However this is at the expense of repeated diagonalizations of the Hamiltonian matrix at each energy rather than a single diagonalization for all energies.

The R-Matrix method using analytic basis orbitals is closely related to the Hulthen-Kohn variational method used extensively by Nesbet (1980) and collaborators in electron-atom scattering. The Hulthen-Kohn method starts from equation (8) but with the first expansion replaced by a sum over channel functions with either sine or cosine asymptotic forms. Substitution into the Hulthen-Kohn variational principle and carrying out the integrals, now over all space, gives a variational expansion for the K-Matrix defined by equation (15). It was pointed out by Oberoi and Nesbet (1973) that the Hulthen-Kohn variational method can be formulated as a hybrid method by incorporating the exact solution in the outer region, corresponding to the potential (14), into the variational basis. When this is done the hybrid variational method and the R-Matrix method become very similar in their computational requirements. At the present time the program packages developed for electron-atom scattering and electron-molecule scattering are being modified by Morgan (1981) to carry out Hulthen-Kohn calculations as well.

An important extension of the R-Matrix method to describe inelastic and reactive heavy particle collisions has been made by Light and Walker (1976) and has been reviewed by Light et al (1979). In this approach, the collision problem is reduced, in the usual way, to the solution of a set of coupled linear second-order differential equations in an appropriately chosen scattering coordinate. However, unlike the situation for electron scattering discussed above, the density of states in the scattering region is now often so high that it is no longer appropriate to use a single expansion in this region. Instead the scattering region is divided into a number of separate sub-regions by sub-dividing the range of the scattering coordinate, where each sub-region is connected to each of its neighbours by an R-Matrix. An independent expansion is then made in a discrete basis in each of these sub-regions and the Hamiltonian diagonalized in this basis. In this way, the R-Matrix can be propagated across each sub-region to where the interaction is negligible. The advantages of this R-Matrix propagation method are firstly that it is computationally very fast, secondly that it is very stable, even when closed channels are present, and finally that different finite representations or frames can be used in each sub-region enabling the characteristics of a rapidly changing wavefunction such as occurs in reactive scattering to be described. This approach has also recently been used in the outer region for electron scattering for situations when the potential in equation (14) is non-negligible out to very large values of r (Shimamura, 1977, Baluja and Burke, 1981).

3. Applications

In this section there is space to do no more than mention the wide range of

applications of R-Matrix theory that have been made. Without any attempt to give
a complete list of references, they are:

  (i)   Attachment energies of negative atomic ions. Work carried out by Le
        Dourneuf et al (1977) has given accurate results for C, N and O.

  (ii)  Electron-atom and electron-ion scattering. A general program written by
        Berrington et al (1974, 1978) has been very widely used for many atoms and
        ions.

(iii)  Photoionization of atoms and ions. Calculations using expansion (10) for
        the initial and final states have been carried out by Burke and Taylor
        (1975), Le Dourneuf et al (1975) and Combet Farnoux et al (1978) for
        complex atoms and ions.

  (iv)  Electron-molecule scattering. As well as the work on molecules mentioned
        in the previous section vibrational excitation and dissociative attachment
        has been included in the theory by Schneider et al (1979a) and applied to
        vibrational excitation of $N_2$ by Schneider et al (1979b).

  (v)   Free-free transitions for atoms and ions. The R-Matrix theory has been
        extended to free-free transitions by Bell et al (1977).

  (vi)  Dynamic dipole polarisabilities and two photon photoionization. The
        general theory for these processes has been developed by Shorer (1980).

 (vii)  Spectral line shifts. The R-Matrix method has been applied by Yamamota
        (1980) to obtain the spectral line-shifts in a helium plasma.

(viii)  Inelastic and reactive heavy particle collisions. The work of Light and
        Walker (1976) and Light et al (1979) has been discussed in the previous
        section.

## 4.  Future Directions

There are two areas which are under very active development at the moment. The
first is the inclusion of relativistic effects to enable processes involving
heavy atoms and ions to be studied. The second is the extension of the theory
to treat electron impact ionization.

Relativistic effects can be included in several different ways. For relatively
light atoms, when the relativistic terms in the Hamiltonian can be neglected in
the inner region, the frame transformation approach of Rau and Fano (1971),
already discussed in the first section, can be used. For heavier atoms where
these terms cannot be neglected in the inner region, they can be included in
the Hamiltonian in equation (9) within the Breit-Pauli approximation. Recently
Scott and Burke (1980) and Taylor and Scott (1981) have initiated such an approach
for electron scattering and photoionization. Lastly for very heavy atoms it is
necessary to use the Dirac Hamiltonian in equation (9) to describe the collision.
This was the approach considered by Chang (1975, 1977) and recently extended and
programmed by Norrington and Grant (1981). In conclusion, although a good start
has been made, many difficult theoretical and computational problems remain
before an arbitrary heavy atom can be treated and it is expected that this will be
a very active area of research in the future.

The extension of R-Matrix theory to treat electron-impact ionization raises a
fundamentally new problem, since two electrons now emerge from the inner region
in figure 1. So far no complete theory has been proposed although certain
limiting cases can be treated. The simplest is if the incident electron is fast
so that its interaction with the target can be treated by first-order perturbation theory. In this case the Coulomb-Born approximation can be used where the
wavefunctions describing the interaction of the slow electron with the target
atom or ion in both the initial and final states are expanded in an R-Matrix

basis as in equation (10). For a neutral target this approximation has been
considered by Robb et al (1975). If the incident electron is not fast, then
this approximation breaks down. However for high z ions the interaction between
the electrons in the outer region may often be small enough to be neglected,
although the interactions between the two electron and the ion in the inner
region is strong. In this case expansion (8) can be modified to include two
electrons in continuum orbitals and the Hamiltonian diagonalized in this basis
in the inner region. In the outer region Coulomb waves are then used to represent
the two outgoing electrons. This approximation is being considered by Taylor and
Taylor (1981). The final case is when the two outgoing electrons interact
strongly with each other and with the residual ion in the outer region. This is
the situation which gives rise to the well-known Wannier threshold law of ionization. Fano and Inokuti (1976) have given some preliminary consideration to this
problem.

They introduced the hyperspherical coordinates

$$R = \sqrt{r_1^2 + r_2^2} \, , \quad \alpha = \tan^{-1} \frac{r_1}{r_2} \qquad (18)$$

together with the angular variables $(\theta_1 \phi_1)$ and $(\theta_2 \phi_2)$ to describe the motion of
the two outgoing electrons. They then suggested than an outer region should be
defined in the $(r_1, r_2)$ plane defined by

$$r_1 > a \, , \quad r_2 > a \quad \text{and} \quad R > A \qquad (19)$$

where a and A are suitably chosen radii. So far no detailed calculations have
been carried out for this problem, but its fundamental importance in quantal
collision theory provides one of the most significant challenges for the future.

References

Baluja, K.L. and Burke, P.G., 1981, to be submitted to Comp. Phys. Commun.
Bell, K.L., Burke, P.G. and Kingston, A.E., 1977, J. Phys. B 10, 3117.
Berrington, K.A., Burke, P.G., Chang, J.J., Chivers, A.T., Robb, W.D. and
    Taylor, K.T., 1974, Comp. Phys. Commun. 8, 149.
Berrington, K.A., Burke, P.G., Le Dourneuf, M., Robb, W.D., Taylor, K.T., and
    Vo Ky Lan, 1978, Comp. Phys. Commun. 14, 367.
Bloch, C., 1957, Nucl. Phys. 4, 503.
Buckley, B.D. and Burke, P.G., 1979, in "Electron Molecule and Photon Molecule
    Collisions (Ed. T.N. Rescigno, V. McKoy and B.I. Schneider, Plenum Press,
    New York) p 133.
Burke, P.G. and Hibbert, A., 1969, Abstracts of VIth ICPEAC (MIT Press,
    Cambridge, Mass.) p 367.
Burke, P.G., Hibbert A. and Robb, W.D., 1971, J. Phys. B 4, 153.
Burke, P.G. Mackey, I. and Shimamura, I., 1977, J. Phys. B 10, 2497.
Burke, P.G. and Robb, W.D., 1975, Adv. in Atom. Molec. Phys. 11, 143.
Burke, P.G. and Taylor, K.T., 1975, J. Phys. B 8, 2620.
Buttle, P.J.A., 1967, Phys. Rev. 160, 719.
Chang, E.S. and Fano, U., 1972, Phys. Rev. A6, 173.
Chang, J.J., 1975, J. Phys. B. 8, 2327.
Chang, J.J., 1977, J. Phys. B. 10, 3335.
Combet Farnoux, F., Lamoureux, M. and Taylor, K.T., 1978, J. Phys. B11, 2855.
Danos, M. and Greiner, W., 1966, Phys. Rev. 146, 708.
Le Dourneuf, M., Vo Ky Lan and Burke, P.G., 1977, Comments on Atom. Molec. Phys.
    7, 1.
Le Dourneuf, M., Vo Ky Lan, Burke, P.G. and Taylor, K.T., 1975, J. Phys. B 8,
    2640.
Fano, U., 1970, Phys. Rev. A2, 353.

Fano, U., 1977, in "Electronic and Atomic Collisions" (Ed. G. Watel, North
  Holland Publishing Co.) p 271.
Fano, U., 1981, Comments on Atom. Molec. Phys. 10, 223.
Fano, U. and Inokuti, M., 1976, Argonne National Laboratory Report ANL-76-80.
Fano, U. and Lee, C.M., 1973, Phys. Rev. Letters 31, 1573.
Grant, I.P. and Norrington, P.H., 1981, Abstract at this Conference.
Jungen, C., 1981, Invited Paper at this Conference.
Kuyatt, C.E., Simpson, J.A. and Mielczarek, S.R., 1965, Phys. Rev. 138, A385.
Lane, A.M. and Thomas, R.G., 1958, Rev. Mod. Phys. 30, 257.
Lee, C.M., 1974, Phys. Rev. A10, 584.
Light, J.C. and Walker, R.B., 1976, J. Chem. Phys. 65, 4272.
Light, J.C., Walker, R.B., Stechal, E.B. and Schmalz, T.G., 1979, Comp. Phys.
  Commun. 17, 89.
Lu, K.T. and Fano, U., 1970, Phys. Rev. A2, 81.
Madden, R.P. and Codling, K., 1963, Phys. Rev. Lett. 10, 516.
Madden, R.P. and Codling K., 1965, Astrophys. J. 141, 364.
Morgan, L., 1981, private communication.
Nesbet, R.K., 1980, in "Variational Methods in Electron-Atom Scattering Theory"
  (Plenum Press, New York).
Noble, C.J., Burke, P.G. and Salvini, S., 1981, Abstract at this Conference.
Norrington, P.H. and Grant, I.P., 1981, J. Phys. B 14, L261.
Oberoi, R.S. and Nesbet, R.K., 1973, J. Comput. Phys. 12, 526.
Rau, A.R.P. and Fano, U., 1971, Phys. Rev. A4, 1751.
Robb, W.D., Rowntree, S.P. and Burnett, J., 1975, Phys. Rev. A11, 1193.
Schneider, B.I., 1975a, Chem. Phys. Lett. 31, 237.
Schneider, B.I., 1975b, Phys. Rev. A11, 1957.
Schneider, B.I., 1977, in "Electronic and Atomic Collisions" (Ed. G. Watel,
  North Holland Publishing Co.) p 257.
Schneider, B.I., Le Dourneuf, M. and Burke, P.G., 1977a, J. Phys. B 12, L365.
Schneider, B.I., Le Dourneuf, M, and Vo Ky Lan, 1977b, Phys. Rev. Lett. 43, 1926.
Schneider, B.I. and Hay, P.J., 1976, J. Phys. B 9, L165.
Scott, N.S. and Burke, P.G., 1980, J. Phys. B 13, 4299.
Seaton, M.J., 1958, Mon. Not. Roy. Astron. Soc. 118, 504.
Seaton, M.J., 1966, Proc. Phys. Soc. 88, 801.
Shimamura, I., 1977, in Reports on CECAM Workshop on Electron Molecule Scattering
  (Université Paris-Sud, Orsay, France).
Shorer, P., 1980, J. Phys. B 13, 2921.
Simpson, J.A., 1964, Rev. Sci. Instr. 35, 1698.
Taylor, K.T., and Scott, N.S., 1981, J. Phys. B 14, L237.
Taylor, K.T., and Taylor, A.J., 1981, private communication.
Wigner, E.P., 1946a, Phys. Rev. 70, 15.
Wigner, E.P., 1946b, Phys. Rev. 70, 606.
Wigner, E.P. and Eisenbud, L., 1947, Phys. Rev. 72, 29.
Yamamota, K., 1980, J. Phys. Soc. Japan 49, 730.
Zvijac, D.J., Heller, E.J. and Light, J.C., 1975, J. Phys. B 8, 1016.

PHYSICS OF ELECTRONIC AND ATOMIC COLLISIONS
S. Datz (editor)
© North-Holland Publishing Company, 1982

THE IMPACT OF FANO'S IDEAS ON MOLECULAR SPECTROSCOPY

Ch. JUNGEN

Laboratoire de Photophysique Moléculaire du C.N.R.S.[*]
Université de Paris-Sud
91405 - ORSAY - France.

Fano's contribution to molecular spectroscopy is reviewed.
Some recent applications of multichannel quantum defect
theory to molecular problems are discussed.

I. INTRODUCTION

Fano's direct involvement with molecular spectroscopy began in 1969, on an occa-
sion when Herzberg visited the University of Chicago, bringing with him a
highly resolved photographic recording of the photoabsorption spectrum of diatomic molecular hydrogen. This spectrum [1] shows several well-developed Rydberg
series which converge towards the various rotational ionization limits of $H_2$
(i.e. rotational levels of the vibrational-electronic ground state of $H_2^+$). In
the course of their evolution some of the series perturb each other strongly,
with the result that conspicuous periodic irregularities of level structure and
intensities appear, and extend beyond the ionization limit in the form of broad
asymmetric preionization (Beutler-Fano) profiles. The complicated structures
prevent straightforward extrapolation of the Rydberg series to their respective limits. Although Herzberg interpreted the observations qualitatively
correctly, he was not able at the time to extract a precise value of the ionization potential of molecular hydrogen -the initial goal of his study- in a simple
manner from the spectrum.

II. QUANTUM DEFECT THEORY AND FRAME TRANSFORMATIONS

Fano's attention was attracted by the aspects of molecular dynamics revealed by
the complex appearance of the spectrum. He was able to find an elegant solution
to the spectroscopic problem [2] by combining Seaton's multichannel quantum
defect theory [3] with the idea of frame transformations. (The latter concept
uses different reference frames simultaneously for the description of a given
particle, thereby taking account of different physical situations prevailing in
different regions of space). Fano thereby introduced the point of view of scattering theory into an important class of molecular spectroscopic problems.
Molecules are thus viewed in the context of their fragmentation channels, i.e.,
of the products (atoms or ions plus electrons) into which they may eventually
disintegrate or from which they may originally have been formed. A molecule
itself exists as a collision complex, when all particles of which it is composed
are assembled in a limited region of space, called the reaction zone. In this
zone their interactions are strong and their motions are complicated : it is
here, however, where the "fate" of the molecular complex is decided, i.e. where
the coupling between the different asymptotic channels originates. Excited
states observed by molecular spectroscopy often contain the key to an understanding of these short-range interactions because a dominating part of their probability amplitude lies inside this reaction zone. Outside this zone, the interactions

between the fragments are weaker and simpler, and local solutions to the
Schrödinger equation can in general be expressed analytically.

Seaton's multichannel quantum defect theory (MQDT) [3] does not deal explicitly
with the wavefunction inside the reaction zone ; rather, it replaces all of the
complex short-range interactions by their net effect on the outside part of the
radial function of a single escaping particle. This latter is usually an electron,
which may be weakly bound in a Rydberg state or may have sufficient energy to
escape. Indeed, a characteristic feature of MQDT is that it provides a unified
treatment of bound and continuum states in terms of the same parameters. Thus,
the concept of an ionization channel is extended to include the entire Rydberg
series which converges to that ionization threshold as well. Determination of
the total wavefunction then reduces to an application, at each energy, of
appropriate boundary conditions, simultaneously on the boundary of the reaction
zone ("core") and at infinity. A helpful circumstance thereby is that the quantum defects which represent the conditions on the surface of the core, vary
slowly with total energy. The reason is that the highly excited Rydberg electron
traverses the core boundary with high velocity, nearly independent of its slow
motion (bound or free) at large distance. The quantum defects therefore extra-
polate smoothly  through the ionization thresholds into the continuum region,
where they become the scattering phase shifts (to within a factor $\pi$). On the
other hand, observable quantities (absorption intensities, photoion currents) are
equally determined by the wavefunction for $r \to \infty$ where electron motion is slow
and depends critically on the energy (e.g. near a threshold) ; however this part
of the problem is handled analytically by MQDT.

As Fano realized, the concept of frame transformations becomes crucial in a
multichannel situation since on its path out to ionization the electron's potential energy changes drastically, from a sizeable fraction of an atomic unit to
near zero. The ratio of the electrostatic energy to the fine structure splittings
of the core thereby undergoes a corresponding change, in other words, the electron is subject to entirely different physical conditions close or far from the
core. If, in particular, the fine structure spacing of the isolated ion happens
to be of the order of $10^{-3}$ a.u. (which is the magnitude of rotational spacings
in molecules), one will find that several lower Rydberg members conform to one
fine structure pattern, characteristic of the complex, while all the higher
members conform to another, characteristic of the ion. (An example will be
discussed in more detail in sect. III). Obviously, there is then no one-to-one
correspondence between the ionization channels, defined by alternative levels
of the residual ion, and the short-range channels, represented by the lower part
of the Rydberg series and characterized, in contrast, by the fact that the ion
level pattern is dwarfed by the electronic potential energy. It is this experimental circumstance from which the idea of combining MQDT with frame transformation has sprung. Fano showed that the observed complexity of the series structures suggests a reformulation of quantum defect theory in which dynamical
behaviour, represented by the quantum defect parameters, is separated from purely
geometrical factors.

Fano's 1970 paper [2] has had consequences extending far beyond its original context. Numerous related papers have been published since, all of which adopt a
similar point of view, and deal not only with molecular ionization [4-6] but
also with dissociative recombination [7] , as well as with the spectroscopy of
rare gas [8] , alkaline earth [9] and group IV elements [10] , with negative
atomic ions [11] and *ab initio* evaluation of atomic parameters [12]. In the
following  I shall first outline those of Fano's original ideas which specifically concern the interplay of electron motion with molecular rotation, and then
indicate their more recent extension to include vibrational nuclear motion, to
the point where competition between alternative fragmentation channels (ioniza-

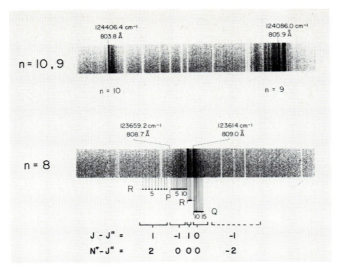

Figure 1
Members n = 8-10 of the Worley-Jenkins Rydberg series in $N_2$ (after Ref. 13).

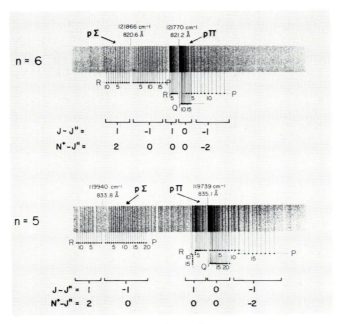

Figure 2
Members n = 5-6 of the Worley-Jenkins Rydberg series in $N_2$ (after Ref. 13).

tion and dissociation) can now be treated.

## III. ROTATION-ELECTRON COUPLING

Figs. 1 and 2 show small portions of the $N_2$ absorption spectrum observed by
Johns and Lepard [13] . This spectrum is in many ways similar to the $H_2$ spectrum
studied earlier by Herzberg and collaborators [1] . Each of the five bands shown
corresponds to excitation of a particular singlet np Rydberg state ; in an atom
each of these would consist of one single line. The molecular fine structure
components are primarily rotational and are grouped into branches corresponding
to the values J-J" = + 1, 0 and -1 of the change of total angular momentum occurring during the photoabsorption process (J" is the initial, J the final
total angular momentum). As n increases from 5 to 10 we see a striking transformation of the fine structure pattern which is characteristic of the transition
of the excited electron's angular motion from a molecule-fixed to a space-fixed
quantization frame. At low n the branches appear in two groups (molecular bands)
whose separation is the splitting of the atomic p state into molecular
p$\Sigma$ ($\ell$=1, $\Lambda$=0) and p$\Pi$ ($\ell$=1, $\Lambda$= $\pm$1) components ( $\Lambda$ is the orbital angular momentum component with respect to the molecular axis). As one expects the splitting
decreases with increasing n ; however we focus here on the simultaneous transformation of line spacings and branch intensities. Notice how the two "outer"
branches progressively lose intensity until they disappear altogether for n $\sim$ 10,
while the three central branches close up and stand out with identical structure.
Instead of labeling the branches according to the total angular momentum change
J - J", we can alternatively use $N^+$ - J" where $N^+$ is the angular momentum (exclu-
ding spin) of the ion core left behind by the Rydberg electron. (For $\ell$ = 1 $N^+$ is

restricted to the values $N^+ = J^+ \pm 1$ for parity $\pi = (-1)^J$ and to $N^+ = J$ for

$\pi = - (-1)^J$). We find (cf. Fig. 1) that the widely spaced weak ("forbidden")
branches have $N^+$ - J" = +2 and -2 while the narrow strong ("allowed") branches
all have $N^+$ - J" = 0. The replacement of the quantum number  $\Lambda$, characteristic
of the molecular complex, by the number $N^+$ referring to the isolated ion, is
thus evident.


The phenomenon is further discussed  with reference to Fig. 3. The left part
shows the Rydberg level and adjoining continuum component energies drawn for a
single J and parity value as functions of their position with respect to the
mean (atomic) energy. Thus the abscissa gives an enlarged view of the component
splittings seen on the ordinate. The pattern as a whole represents two ionization
channels and is intended to help visualize the transformation from the short-
range channels into the ionization channels. The curved dotted line indicates
the scaling law n$^{*-3}$ (n$^*$ = effective principal quantum number) according to
which small splittings in unperturbed Rydberg series behave : the change of
coupling regime shows up clearly by the fact that the splittings start to
deviate from the curve near n = 7 and tend instead towards the rotational level
splitting  E($N^+$ = J + 1) - E($N^+$ = J - 1) of the ion. Above the higher limit
($N^+$ = J + 1) the photoelectron groups corresponding to the two limits are
defined by the same constant energy difference. A horizontal dotted line marks
the energy near which the fine structure pattern changes.


Fano's realization was that this energy also defines a partitioning of the
space surrounding the core into two further zones, an outer zone B and an inner
zone A which also contains the core. These zones are illustrated in the right
part of Fig. 3, where the Coulomb potential which dominates the electron motion
outside the core is displayed : the boundary between zones A and B is drawn
corresponding to the classical outer turning point at the change-over energy.

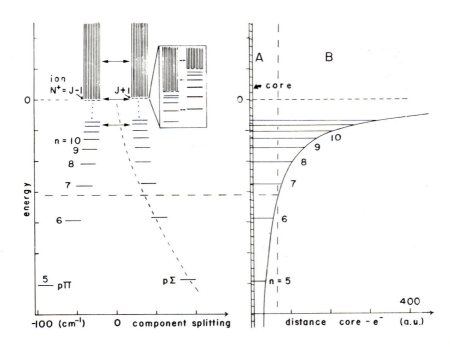

Figure 3
Left : energy level pattern of the Worley-Jenkins Rydberg series in
      $N_2$ (see text for details).
Right : Coulomb potential and Rydberg levels.

Zone A is characterized by vanishingly small rotational energy differences.
Thus the situation usually assumed in molecular spectroscopy prevails here,
i.e. the Born-Oppenheimer separation is valid and a molecule-fixed coordinate
frame may be employed. (Note that zone A pertaining to nuclear rotation extends
far beyond the core region where the anisotropic components of the core field
couple the electron to the molecular axis. E.g. the quadrupole field typically
drops to the magnitude of rotational splittings near $r \sim 10$ a.u. whereas zone
A extends to $r \sim 70$ to 100 a.u. for a first row molecule with $J \sim 5$ to 7).
The electronic wavefunction in region A can therefore be written for any energy
not too far above threshold in terms of only two parameters, namely the quantum
defects $\mu_{\Lambda + 0}$ and $\mu_{\Lambda = 1}$. These measure for each short-range channel the phase
difference with respect to a pure central Coulomb field acquired by the excited
electron inside the core. (The difference $\mu_{\Sigma} - \mu_{\Pi}$ is a measure of the anisotropic
interactions experienced at short range. E.G. the $\Sigma$-$\Pi$ splitting of the low
Rydberg members plotted in Fig. 3 is given directly by $(\mu_{\Sigma} - \mu_{\Pi})/ n^{*3}$ (a.u.)).

A MQD wavefunction valid in region B is expressed in the laboratory frame and
must also depend on the rotational level difference in the ion because this is
no longer negligible everywhere in zone B and even dominates at large r. An
important circumstance noted by Fano is that the two wavefunction expansions
constructed according to these principles are in fact equivalent throughout the

entire range from the surface of the core to the A/B boundary. The reason is that
the potential acting on the Rydberg electron remains Coulombic down to the core
surface, and is the same for all channels. Consequently, outside the core the
full molecular wavefunction depends everywhere only on the quantum defects
$\mu_\Sigma$ , $\mu_\Pi$ and on the rotational fine structure splitting $E(N^+ = J + 1)-E(N^+ = J-1)$.
In addition, the space-quantized wavefunction B must be re-expanded at short
range in molecule-fixed coordinates : the relevant 2 x 2 unitary matrix with
elements $< N^+|\Lambda >^{(\ell=1)}$ has been known for a long time [14].

It is instructive at this point to compare Seaton's and Fano's results for the
two-channel problem. As an example, the expression for discrete level positions
below the first ionization threshold is given below. Seaton obtained the important result that :

$$\tan \pi n_1^*(E) + K_{11}^{(s)} = \frac{K_{12}^{(s)}{}^2}{\tan \pi n_2^*(E) + K_{22}^{(s)}} , \tag{1}$$

where the $n_i^*$ , i = 1 and 2, are the effective principal quantum numbers given
by the Rydberg equation as functions of energy E and Rydberg limit $E_i$ by

$$E - E_i = - \frac{1}{2n_i^*{}^2(E)} \quad \text{(a.u.)} \tag{2}$$

Discrete energy levels occur whenever eqs. (1) and (2) are fulfilled simultaneously. $K^{(s)}$ is a reaction matrix which varies slowly with energy E (s stands
for "smooth" [15]) and represents the short range interactions between core and
electron. (For example, the element $K_{12}^{(s)}$ represents the coupling between the two
series ; when it is zero eq. (1) yields two independent series with constant
quantum defect). In Fano's 1970 formulation the $K^{(s)}$ matrix is further analyzed
according to

$$K_{ij}^{(s)} = \sum_\alpha < i|\alpha > \tan \pi\mu_\alpha < \alpha|j > , \tag{3}$$

where in our specific case $\alpha = \Lambda$ and $i = N^+$. Seaton's non-diagonal reaction
matrix containing three independent elements, is thereby reduced to the diagonal
set of two short-range quantum defects complemented by the analytically known
2 x 2 unitary frame transformation matrix whose elements depend on a single
rotation angle. The separation by eq. (3) of dynamical and geometrical factors
is physically satisfying. The role of zone A can be compared to that of a lens,
acting on the motion of the incoming electron in order to cause it to follow
the slow nuclear rotation before it actually hits the core and experiences the
short-range interactions. On the practical side, the reduction of the number of
dynamical parameters to be fitted (or calculated) is useful. Indeed, this
latter advantage has proved decisive in the application of MQDT to the coupling
of electron motion with nuclear vibrational motion where the number of channels
that must be considered is infinite.

Eqs. (1), (2) and (3) fully account for the level pattern shown in Fig. 3. (E.g. it can be shown easily that for small n, for which $n_1^* \sim n_2^*$, eqs. (1) and (3) yield two series with constant quantum defects $\mu_\Sigma$ and $\mu_\Pi$). The equations also account for local level perturbations occurring at high n between levels of different principal quantum number (indicated by an arrow in Fig. 3 but so far observed only in $H_2$ and $He_2$). Fano has also given expressions for the photo-ionization intensity profiles in the Beutler region, i.e. between the two thresholds (cf. inset in Fig. 3), as well as for the open continuum, i.e. above the second threshold. The expressions for photoionization cross sections involve, in addition to the quantum defects $\mu_\Sigma$ and $\mu_\Pi$ and ionization thresholds $E(N^+ = J - 1)$ and $E(N^+ = J + 1)$, two matrix elements for dipole absorption, $d_\Sigma$ and $d_\Pi$. These are classified according to the short-range channels $\Lambda$ because absorption takes place inside the core in region A. The relative magnitude of the $d_\Lambda$'s can be inferred from the relative strengths of the $N^+ - J'' = 0$ and $\pm 2$ branches at high n : for $N_2$ and $H_2$ the latter vanish implying that $d_\Sigma \simeq d_\Pi$ [1].

In the open continuum the rotational branching ratio of the ions produced [2] and the angular distribution of the photoelectrons [4] are features of interest. Interchannel coupling due to the anisotropic short-range interactions is documented both for $N_2$ and $H_2$ by the fact that the $N^+ - J'' = \pm 2$ branches show up weakly in the photoelectron spectrum. Table 1 compares the experimental data of Niehaus and Ruf [16] for $H_2$ with values predicted by MQDT on the basis of the quantum defects $\mu_\Sigma$ and $\mu_\Pi$ taken from the discrete spectrum. Allowance for vibrational motion has also been necessary in the calculations [17].

Table 1

Observed and calculated rotational branch intensities and asymmetry parameters for photoionization of $H_2$ (736 Å)

|  | Experiment<br>Niehaus and Ruf [16] | Theory<br>Raoult et al. [17] |
|---|---|---|
| $I_S/I_Q$ (90°) [a] | 0.041 ± 0.006 | 0.035 |
| $\beta_Q$ [b] | 1.95 ± 0.01 | 1.954 |
| $\beta_S$ | 0.45 ± 0.87 | 0.200 |
| $\beta_{QO}$ [c] | 1.93 ± 0.01 | 1.928 |

[a] intensity ratio of the $N^+ - J'' = 2$ and 0 rotational branches, observed by photoelectron detection at 90° with respect to the (unpolarized) light beam.

[b] the letters O, Q and S refer to the values $N^+ - J'' = 2$, 0 and -2, respectively.

[c] overlapped Q and O lines.

## IV. VIBRATION-ELECTRON COUPLING

An instructive way of introducing the subject of vibration-electron coupling is to consider two recent two-photon experiments performed on the $Na_2$ molecule. Carlson, Taylor and Schawlow [18], using their two-step "polarization labeling" technique, have reached numerous highly excited discrete Rydberg levels of $Na_2$ which they interpret in terms of excitation of a molecular d electron. Almost simultaneously, Leutwyler, Hofmann, Harri and Schumacher [19], using the sequential two-photon ionization (TPI) method, discovered an extensive vibrational preionization structure above threshold. The two experiments are related in that both involve, in the first step, the pumping of rotational levels of the intermediate B state.

It is tempting to apply the concept of short-range channels to these recent data, by suggesting that the data of Carlson et al. contain the key information on the short-range channels of the $Na_2$ system necessary to interpret the photoionization spectrum of Leutwyler et al., which thus far has resisted analysis. Carlson et al. presented their data in the form of a plot showing experimental dissociation energies $D_e$ of the $nd\Pi$ and $nd\Delta$ states of $Na_2$, with n ranging from 4 to 15.

A surprising result was that, for low n, $D_e$ was found to differ appreciably from the limiting value $D_e^+$ of the ion $Na_2^+$. This means that at short range the Rydberg electron contributes substantially to the chemical bond, i.e. "distorts" the $Na_2^+$ core considerably.

From these observations we can infer that the electronic potential energy curves $U_{nd\Lambda}(R)$ of the $Na_2$ Rydberg states and $U^+(R)$ of the ion $Na_2^+$ are not parallel to each other (R is the internuclear distance). The electronic Rydberg binding energy is therefore R-dependent and can be expressed by an R-dependent Rydberg equation as [1,20].

$$U_{n\Lambda}(R) = U^+(R) - \frac{1}{2(n - \mu_\Lambda(R))^2} \qquad (4)$$

The introduction of an R-dependent quantum defect function is the key to extension of MQDT to combined rotational/vibrational electronic coupling [1,5,21]. The short-range channels are now labeled more completely by $(R,\Lambda)$ while the ionization channels are labeled by $(v^+, N^+)$, where $v^+$ is the vibrational quantum number of the ion. A first step in the $Na_2$ problem would then be to extract the functions $\mu_\Lambda(R)$ from the data of Ref. 18. The $K^{(S)}$ matrix elements of eq. (1) may then be generalized to an integral over the vibrational motion :

$$K^{(s)}_{v^+,N^+;v^{+\prime},N^{+\prime}} = \sum_\Lambda <N^+|\Lambda> \int \chi_{v^+}(R) \tan \pi\mu_\Lambda(R) \chi_{v^{+\prime}}(R) \, dR < \Lambda|N^{+\prime}> \qquad (5)$$

Here the vibrational wavefunctions $\chi_{v^+}(R)$ of the ion act as the frame transformation matrix elements connecting the short-range $|R>$ channels to the asymptotic $|v^+>$ channels.

An extensive program of analysis has been conducted along these lines in recent years [5, 17] for the $H_2$ molecule and has yielded a quantitative account, not only of discrete level energies, but also of preionization line widths and shapes, partial vibrational photoionization cross sections as well as photoelectron angular distributions. As an example, Fig. 4 shows a small portion of the

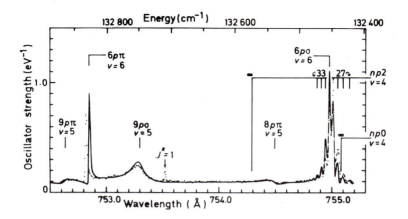

Figure 4

Preionization near the $v^+$ = 4, $N^+$ = 0 and 2 ionization thresholds in $H_2$ (78 °K ; J = 1, J" = 0). The observed (points from Ref. [22]) and calculated (full line after Ref. [23]) total photoionization oscillator strengths are shown as functions of wavelength.

photoionization spectrum of $H_2$ observed under high resolution by Dehmer and Chupka [22] . Preionization resonances of varying width and shape  can be seen, and include an interesting complex structure near 755 Å  which is due to the interplay between rotational and vibrational preionization. The good agreement between experiment and calculation [23] illustrates the power of the short-range- (or eigen-) channel concept.

## V. PREDISSOCIATION AND PREIONIZATION

The final subject to be mentioned here concerns the inclusion of dissociative channels in the MQDT treatment. As a consequence of the relatively low dissociation energy of the ion $H_2^+$ many of the higher Rydberg levels of $H_2$ lie above the dissociation limits of certain lower members  of the same series. For example, the 3pπ , v > 2 and 4pσ , v > 0 levels lie above the H(1s) + H(n=2) dissociation limit with which the 3pσ and 2pπ Rydberg states correlate. These levels are therefore predissociated. At higher energy, the 3pπ , v > 6 and 4pσ , v > 4 levels are subject to both predissociation and preionization.

The extension of MQDT to the simultaneous treatment of ionizative and dissociative fragmentation channels relies on the following considerations. Predissociation, like the ionization process itself, constitutes a breakdown of the Born-Oppenheimer approximation ; in other words, it occurs to the extent that the Rydberg electron in the predissociating  (e.g. 3pσ or 2pπ) state ventures beyond the boundary of region A. In other words, the process occurs when the electron is outside the core ($r > r_0$), in a region where MQDT provides explicit wavefunctions in terms of quantum defects. Further, the interactions leading to predissociation occur over a range of internuclear distances extending not far beyond the outer classical turning point $R_0$ of the predissociated level. It follows that the zone of configuration space relevant to the dissociation process (i.e. $r \geqslant r_0$ , $R \leqslant R_0$) is the same in which the treatment of vibration-ally bound levels, outlined in the preceding section, applies. It is therefore sufficient to solve the MQD equations (of the type of eq. (1)) in the finite range $0 < R \leqslant R_0$. For $R > R_0$ a single nuclear wavefunction describes the motion of the separating atoms fully ; this function emerges from the reaction zone at $R \cong R_0$ with a well-defined phase shift (relative to the phase of a free-particle wave). Indeed, the variation of this shift with energy becomes the main object of the calculation. For $R \leqslant R_0$ one needs a multichannel expansion of the wavefunction in terms of several vibrational-electronic components, owing to the vibration-electron coupling represented by the R-dependence of the quantum defects. Further, in order to match the wavefunctions for $R \leqslant R_0$ and $R > R_0$ , we must impose a common phase on all vibrational components present for $R \leqslant R_0$. The eigenvalue spectrum inside the reaction zone thereby becomes discrete even above the dissociation limit and can be obtained quite easily in terms of the functions $\mu_\Lambda(R)$. Note that this method again owes much to Fano, who, with

C.M. Lee, proposed a similar method [12] for the *ab initio* evaluation of atomic short-range interaction parameters.

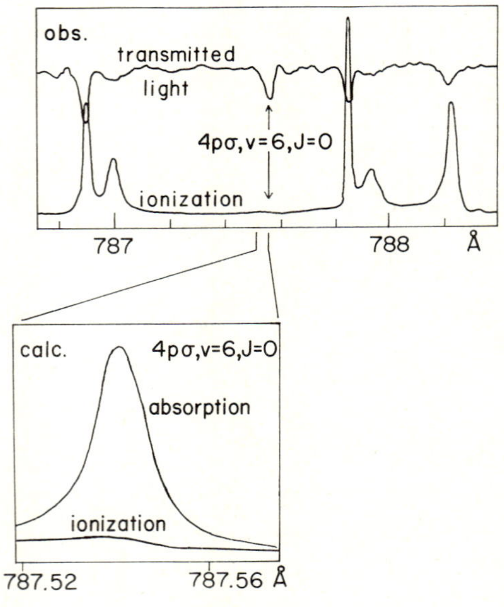

Figure 5

Competition of preionization and predissociation above the $v^+ = 1$ ionization threshold in $H_2$ (78 °K ; J=0,1,2 ; J"=0,1). Top : experimental spectra after Ref. [22] , showing the absence of the absorbing 4pσ , v=6, J=0 resonance in the ionization spectrum. Bottom : calculated absorption and ionization spectra near the 4pσ , v=6, J=0 resonance [24] . The difference between the intensity distributions shown corresponds to photodissociation.

As an example of the method, Fig. 5 illustrates the results of a calculation for the 4pσ , v=6, J=0 resonance of $H_2$ [24]. The 4pσ, v=6 level lies at an energy where the $H_2^+$ ($v^+$=0 and 1) + $e^-$ ionization channels as well as the H(1s) + H(n=2) dissociation channel are open. Dehmer and Chupka [22] have established for this level that predissociation prevails, based on the observation that the peak appears in the spectrum of the light transmitted by the $H_2$ sample but not in the spectrum signalled by production of $H_2^+$ ions (cf. Fig. 5). The calculation reproduces this behavior exactly.

The conclusion is that quantum defects, familiar to most physicists as electronic quantities relevant to the level spectra of atoms, have a broad significance also in molecular physics. In the example just presented they have served to account for predissociation, a process which involves nuclear motion and which is in fact close to a chemical reaction. Fano recognized this possibility very early.

## REFERENCES

[1] Herzberg, G. and Jungen, Ch., J. Mol. Spectrosc. 41 (1972) 425.
[2] Fano, U., Phys. Rev. A2 (1970) 353.
[3] Seaton, M.J., Proc. Phys. Soc. London 88 (1966) 801.
[4] Dill, D., Phys. Rev. A6 (1972) 160.
[5] Jungen, Ch. and Atabek, O., J. Chem. Phys. 66 (1977) 5584.
    Jungen, Ch. and Dill, D., J. Chem. Phys. 73 (1980) 3338.
    Raoult, M. and Jungen, Ch., J. Chem. Phys. 74 (1981) 3388.
[6] Ginter, D.S. and Ginter, M.L., J. Mol. Spectrosc. 82 (1980) 152.
[7] Giusti, A., J. Phys. B 13 (1980) 3867, see also Giusti, A., present volume.
[8] Lu, K.T. and Fano, U., Phys. Rev. A2 (1970) 81 ; Lu, K.T., Phys. Rev. A4 (1971) 579 ; Lee, C.M. and Lu, K.T., Phys. Rev. A8 (1973) 1241 ; Starace, A.F., J. Phys. B6 (1973) 76.
[9] Wynne, J.J. and Armstrong, J.A., Comments At. Mol. Phys. 8 (1979) 155 ; Armstrong, J.A., Wynne, J.J. and Esherick, P., J. Opt. Soc. Am. 69 (1979) 211 ; Brown, C.M. and Ginter, M.L., J. Opt. Soc. Am. 70 (1980) 87 ; and references cited therein.
[10]Brown, C.M., Tilford, S.G. and Ginter, M.L., J. Opt. Soc. Am. 67 (1977) 584, 607.
[11]Lee, C.M., Phys. Rev. A11 (1975) 1692.
[12]Fano, U. and Lee, C.M., Phys. Rev. Lett. 31 (1973) 1573 ; Lee, C.M., Phys. Rev. A10 (1974) 584.
[13]Johns, J.W.C. and Lepard, D.W., J. Mol. Spectrosc. 55 (1975) 374. See also Carroll, P.K., J. Chem. Phys. 58 (1973) 3597.
[14]Van Vleck, J.H., Rev. Mod. Phys. 23 (1951) 213.
[15]Fano, U., Phys. Rev. A17 (1978) 93.
[16]Niehaus, A. and Ruf, M.W., Chem. Phys. Lett. 11 (1971) 55 ; see also reanalysis by Chang, E.S., J. Phys. B11 (1978) L69.
[17]Raoult, M., Jungen, Ch. and Dill, D., J. Chim. Phys.-Chim. Biol. 77 (1980) 599.
[18]Carlson, N.W., Taylor, A.J. and Schawlow, A.L., Phys. Rev. Lett. 45 (1980) 18.
[19]Leutwyler, S., Herrmann, A., Wöste, L. and Schumacher, E., Chem. Phys. 48 (1980) 253.
[20]Mulliken, R.S., J. Am. Chem. Soc. 91 (1969) 4615.
[21]Chang, E.S., Dill, D. and Fano, U., in : Cobic, B.C. and Kurepa, M.V. (eds.), Abstracts of Papers, Proceedings of the Eight International Conference on the Physics of Electronic and Atomic Collisions (Institute of Physics, Belgrade, 1973), p. 536.

[22] Dehmer, P.M. and Chupka, W.A., J. Chem. Phys. 65 (1976) 2243.
[23] Jungen, Ch. and Raoult, M., Faraday Discuss. Chem. Soc. 71 (1981), in press.
[24] Jungen, Ch., to be published.


*    Laboratoire Associé à l'Université de Paris-Sud.

PHYSICS OF ELECTRONIC AND ATOMIC COLLISIONS
S. Datz (editor)
© North-Holland Publishing Company, 1982

THE FANO EFFECT AND ITS CONSEQUENCES

Joachim Kessler

Physikalisches Institut
Universität Münster
D-4400 Münster
West Germany

In 1969 Fano predicted that spin polarized photoelectrons can be ejected from unpolarized alkali
atoms, if circularly polarized light is used for
photoionization. This was an impetus for new developments in photoionization, culminating in the
insight that photoelectrons are usually polarized
no matter whether produced by circularly, linearly
or unpolarized light. Spin polarization of photoelectrons is a new observable which helps to completely understand the fundamental process of photoionization. The Fano effect also stimulated the development of efficient sources of polarized electrons.

## 1. The Fano Effect

When you read the recollections of M. Born[1] then you find that there
is one idea which he did not like very much: the idea, namely, that
people might regard him mainly as the inventor of the "Born approximation". He is, of course, definitely right since he made much
greater contributions to physics, even though at conferences like
the ICPEAC the Born approximation is quoted much more often than
his more fundamental work.

I can imagine that Ugo Fano's attitude towards the Fano effect is
similar. He himself seemed never to be so excited about the Fano
effect as other people, and this is quite intelligible to me when
I look at the wealth of his contributions to physics of which the
Fano effect is just one single feature.

On the other hand, Fano's idea how to produce spin-oriented or
polarized photoelectrons from unpolarized atoms opened up new
fields of research in physics. I will try to explain this in the
following and to mention in passing a few personal recollections
connected with the Fano effect.

For 40 years - from the late twenties to 1969 - physicists believed
that there is only one way to obtain polarized photoelectrons: to
start from a polarized target (for instance a polarized atomic beam
obtained by sending alkali atoms through a Stern Gerlach magnet) and
then to eject the oriented electrons by photoionization (cf. Fig. 1).
One thus produces polarized photoelectrons provided that the spins
of the electrons do not flip during photoionization. Mott and
Massey[2], in their famous book, give a semiclassical argument why
the spins should not flip; and indeed, this method works as has

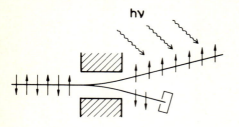

Fig. 1

Production of polarized
electrons by photoioni-
zation of polarized atoms.

been shown by the Yale group (W. Raith, M.S. Lubell and others)
who developed an efficient source of polarized electrons based
on this technique[3].

In this situation it was a great surprise when Ugo Fano[4] showed
that photoelectrons with high polarization can also be obtained
from unpolarized atoms. I had the great pleasure to witness the
moment when this idea occurred to him. It was at the Annual Meeting
of the American Physical Society on February 1, 1968, at Chicago.
I gave a review on the polarization effects in low-energy electron  scattering and tried to visualize why the polarization
maxima usually occur at cross-section minima. The first speaker
in the discussion was Ugo Fano. "Let me remind you", he said
"that the cross sections for photoionization of alkali atoms
have also such deep minima in their wavelength dependence, the
Cooper minima. Don't you think that at these minima there is a
favorable situation for producing polarized photoelectrons?"
This question was too difficult for me and I reacted with the
handwaving arguments you use when you cannot give a reasonable
answer, not knowing that Fano brought forth at this moment an
idea which would open new dimensions in photoionization. We had
some discussion after this session and in June 68 he sent me a
preprint in which his idea had been transformed into quantitative results. He predicted that one should obtain highly polarized photoelectrons if unpolarized cesium atoms are photoionized
by circularly polarized light.

This effect is caused by spin-orbit interaction which turns out
here (I put in Ugo's own words[5]) to be "a weak force with conspicuous effects". One of those conspicuous effects is the
following: There are certain wavelengths where the electron
spins do not retain their direction during the photoionization
process, but they flip with a great probability. Let me illustrate this by Fig. 2. It shows, qualitatively, the cross sections
for production of spin-up (e↑) and spin-down electrons (e↓) vs.
wavelength, if unpolarized alkali atoms are photoionized by circularly polarized light. The directions "up" and "down" are
defined here by the direction of the photon spins in the circularly polarized light which we assume as "up". The unpolarized
alkali beam can be considered to be a mixture of 50% spin-up
atoms (A↑) and 50% spin-down atoms (A↓). Let us pick out the
wavelength were $Q_e^{\downarrow}$ is zero. At this wavelength only e↑ are
produced. Since, on the other hand, half of the atoms in the

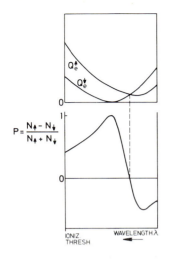

$$P = \frac{N_\uparrow - N_\downarrow}{N_\uparrow + N_\downarrow}$$

Fig. 2

Qualitative diagram of the photo-electron polarization P if unpolarized alkali atoms are photoionized by circularly polarized light. $Q_e^\uparrow$ and $Q_e^\downarrow$ are the cross sections for producing e↑ and e↓.

alkali beam have their spins "down", the electrons of all those atoms must flip their spins if they are photoionized. The old assumption that photoionization does not cause spin flips is therefore not true, though I must admit that I have picked out a wavelength here, where the spin-flip effect is particularly dramatic. But the spin flips are not the only reason why the photoelectrons become polarized. Another conspicuous effect of spin-orbit interaction is the following: The photoionization cross section depends on whether the photon spins of the circularly polarized light are parallel or antiparallel to the spins of the atomic electrons. At the right part of Fig. 2, for instance, the A↓ in the unpolarized alkali beam are photoionized with much larger probability than the A↑, so that in the end more e↓ than e↑ are produced even though also here one has quite a few spin flips into the ↑ direction. This dependence of the photoionization cross section on the relative spin orientation of photons and atoms has been directly verified by Lubell and Raith in the same year 1969 in which Fano's paper appeared[6].

So far I have only said that, with circularly polarized light, the cross sections for production of e↑ and e↓ differ from each other. This is, however, equivalent to saying that the photoelectrons have a polarization P, since the polarization is defined as the difference of the numbers of e↑ and e↓ produced, normalized to the total number of electrons: $P = (N_\uparrow - N_\downarrow)/N_\uparrow + N_\downarrow)$. Accordingly, from the cross sections mentioned one can immediately construct the polarization curve which is also shown in Fig. 2. Where $Q_e^\downarrow = 0$, P is 1, and where $Q_e^\downarrow > Q_e^\uparrow$ one has P < 0.

Incidentally, my discussion here did not strictly follow the line of Prof. Fano's original paper but I used a more phenomenological approach; this is mainly for reasons of time limitation and I hope, he will forgive me.

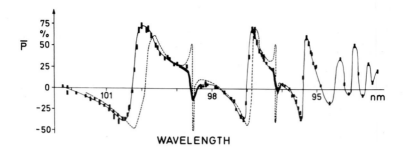

Fig. 3

Angle averaged polarization $\overline{P}$ of photoelectrons ejected by
circularly polarized light from xenon atoms. ——— experimental results, - - - - theoretical results[14]. For further
calculated results see refs. 15, 16.

When I had received Fano's preprint I did something very exceptional: I went to my graduate student J. Lorenz who had been working
for nearly two years at another experiment for his Ph.D. thesis
and said to him: "I need somebody for an exciting experiment. Would
you mind changing the subject of your thesis?" He agreed and after
a few months we had confirmed Fano's prediction: High spin polarization of the photoelectrons produced from free cesium atoms by
circularly polarized light. This would have been a very long title
for a paper and so we decided to choose the expression which
W. Raith and I used in our correspondence. In one of his letters
W. Raith had asked me: "How are your experiments with the Fano
effect going on?" So Lorenz and I called our paper[7] which appeared
in January 1970 "Experimental verification of the Fano effect" and
thus gave credit to the discoverer of this conspicuous effect. The
term "Fano effect" was immediately accepted by all the physicists
working along that line and is irreversibly integrated in the literature now.

## 2. The Consequences

That is all I can say here about the Fano effect as such; let me
now talk about its consequences. Although Ugo's original paper
concerned only free alkali atoms, the experiments were soon
extended to other substances. The next student who showed up
in my group was U. Heinzmann. He got the task to study photoemission by circularly polarized light from solid alkalis and
also found polarization of the photoelectrons which was, however,
much smaller[8]. Then we studied other atoms: Tl, noble gases, Ag,
Hg, other metals, whatever we touched yielded polarized photoelectrons, even certain molecules did[9-13]. Fig. 3 gives as an
example the polarization of photoelectrons produced by circularly
polarized light from xenon. The measurement where all the photoelectrons were observed independent of their direction of emission

has been made at the synchrotron in Bonn[10]. In the wavelength range used the photoionization cross section has a pronounced resonance structure due to autoionizing transitions. As you see, this resonance structure is also found in the polarization curve. Such autoionizing resonances in the polarization curve have first been predicted by Lee[14] in Fano's group in 1974. I am sorry that I do not have the time to explain how much such measurements help in classifying and understanding these resonances. This has been studied in detail by Heinzmann[16] who made not only his diploma thesis but also his Ph.D. and lecturer thesis with the Fano effect. So for him the consequences of the Fano effect were certainly important; but that is not the sort of consequences I plan to discuss here.

Let me instead mention the consequence of the Fano effect for the old dream of physicists, to have an efficient source of polarized electrons. When we started our measurements on the Fano effect one of my coworkers, K. Jost, was at Stanford in order to continue and finish a project I had started there together with R. Hofstadter: A continuous source of polarized electrons was to be constructed for a superconducting cw high-energy accelerator. The source was based on low-energy electron scattering. When I had read Fano's preprint I sent a message to Jost saying: if this Fano effect really works then you can give up our project. And indeed, there are several labs where notable currents of polarized electrons have been produced with the Fano effect. I mention Yale, the FOM Institute in Amsterdam, Bonn, and my group in Münster. Since our source yielded 81% polarization with currents up to 10 nA we felt that this is a Fantastic Apparatus for Notable Orientation of electron spins; that's why we call it the FANO source. Novel experiments became feasible by such sources, e.g. an exchange experiment at Yale[17] where polarized electrons were scattered from polarized atoms, and a so-called triple scattering experiment which we present at this Conference[18].

But that's not the end of the story of the sources. I mentioned before that in 1970 we had extracted polarized photoelectrons also from solid alkalis which were irradiated with circularly polarized light. In 1975 Pierce, Meyer, and Zürcher[19] in Zürich showed that by using GaAs instead of solid alkalis one obtains a polarization of about 40% with currents of several μA. And I must say that in the long and cumbersome history of polarized electron sources this was a real breakthrough. At least a dozen of laboratories have now installed such sources, and beautiful experiments have been made with them. I can mention only two of them: the measurement of parity nonconservation in deep inelastic electron scattering by Prescott and co-workers[20] and the study of surface magnetism by exchange scattering of polarized electrons made by Celotta, Pierce, and co-workers at NBS[21]. In order to avoid misunderstanding, let me say quite clearly that I do not want to claim that the GaAs source had not been found if Fano had not made his discovery. Solid-state physicists in Paris showed in the late sixties, that spin-oriented conduction electrons are produced in semiconducters which are irradiated with circularly polarized light[22]. This was definitely a first step into the direction of the GaAs source. Nevertheless, the field of polarized photoemission from nonmagnetic solids has certainly been much stimulated by all the activities engendered by Fano's idea.

| el. light | $P_z$ | $P_n$ | $P_p$ | $\bar{P}$ |
|---|---|---|---|---|
| σ | x | x | x | x |
| π | o | x | o | o |
| unpol. | o | x | o | o |

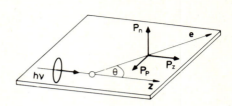

Fig. 4

Table: Polarization of photoelectrons ejected from unpolarized atoms for different polarization of incident light. x indicates the components different from zero. $\bar{P}$ is the polarization averaged over all directions of emission. Diagram: polarization components for circularly polarized light.

Let us now go on in the series of consequences. The question came up whether it is really necessary to use circularly polarized light in order to produce polarized photoelectrons. And not long after the first surprise, the Fano effect, came the second surprise when Cherepkov from Leningrad appeared at the Belgrade ICPEAC saying that also linearly polarized light produces polarized photoelectrons from unpolarized atoms[23]. In fact, even unpolarized light does it. The theoretical work done in the following years in Leningrad[24,25] and in Fano's group[14] showed that there is a wealth of polarization effects in photoionization. Fig. 4 gives a survey of the photoelectron polarization obtained with unpolarized atoms. The polarization of the photoelectrons observed along a certain direction can, in principle, have 3 components: $P_n$ perpendicular to the reaction plane, and $P_p$ and $P_z$ in the reaction plane (z axis = direction of light propagation). If circularly polarized light is used for photoionization, then all the components can exist and even the polarization $\bar{P}$ averaged over all directions of emission is different from zero. ( $\bar{P}$, by the way, is the polarization which has been observed in the Fano-effect experiments mentioned before.) If linearly polarized or unpolarized light is used, one obtains only polarization perpendicular to the reaction plane*). The other components disappear which is easily seen by simple symmetry arguments.

———————————

*) For a proper definition of the reaction plane in the cases of linearly and unpolarized radiation see, e.g., J. Kessler, Comments Atomic Mol. Phys. 10, 47 (1981) (The article contains a few trivial misprints, since proofs were not sent to the author)

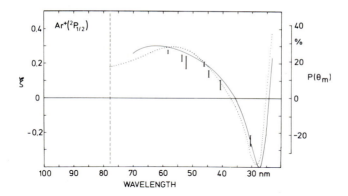

Fig. 5

Photoelectron polarization produced by unpolarized vuv light
from argon atoms, if the residual $Ar^+$ is left in the $^2P_{1/2}$
state. Angle of emission $\theta_m = 55°44'$. The right scale gives
the polarization (for a discussion of the parameter $\xi$ see
ref. 26). Broken line: ionisation threshold. ———— relativistic random phase approximation[29], ····· random phase
approximation with exchange[24].

The question is now: Do these polarization components which are
compatible with symmetry arguments really appear in experiment
and how large are they? This has been studied in the past few
years and I show you just two results which have been obtained
by G. Schönhense in Münster. Fig. 5 shows the polarization of
photoelectrons (right scale) ejected by unpolarized vuv light
from argon atoms which are of course unpolarized. Here, at
certain wavelengths, the photoelectrons emitted under 54°44'
(magic angle) have been spin analyzed in the case that the
residual ion is left in the $^2P_{1/2}$ state[26]. The polarization
is seen to be significant. Fig. 6 shows, for one wavelength
(58.43 nm), the complete angular dependence of the photoelectron polarization when linearly polarized light is used. The
target was xenon, the residual ion was left in the $^2P_{3/2}$ state[27].

These were only two curves. More results are available: Not only
ground states have been photoionized but also excited states,
where laser light instead of vuv radiation has been used[28];
calculations have not only been made in Chicago and Leningrad,
but also in Notre Dame where data of high accuracy have recently
been computed[29,30]. And all this work shows that the polarization
components given in Fig. 4 do exist and are significant. No
matter whether you use circularly, linearly or unpolarized
light you cannot help producing polarized photoelectrons! Or
to put it differently: Spin polarization of photoelectrons is

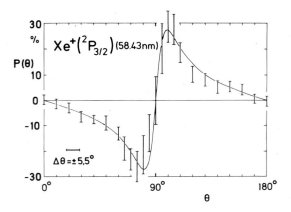

Fig. 6

Angular distribution of photoelectron polarization P($\theta$)
for the process Xe + h$\nu$(21.22eV) $\longrightarrow$ Xe$^+$($^2$P$_{3/2}$) + e$^-$.

not an exception, as physicists believed for a long time, but it
is the rule; you must be particularly lucky in order to produce
unpolarized photoelectrons!

This has the following important consequence: Previously, studies
of the photoeffect were restricted to observation of cross sections.
It was either the angle-integrated photoionization cross section $\sigma$
or - in more recent years - the angle-dependent differential cross
section $\sigma(\theta)$ that has been observed. From the latter observable one
obtains the characteristic parameter $\beta$ which yields deep insight
into the mechanism of photoionization. Today we have new observables, the components of the photoelectron polarization, whose
measurement yields new characteristic parameters besides $\beta$ and thus
opens new dimensions in the analysis of photoionization. The new
parameters yield further independent information on the photoeffect
which cannot be obtained by cross-section measurements. An analysis
which I cannot give here, has shown that polarization measurements
in conjunction with the cross-section measurements mentioned may
yield the complete set of parameters which are necessary for the
description of the photoionization process. This possibility to
completely understand photoionization within the framework of
current theory is certainly a very important consequence which
has been initiated by Fano's idea. I believe that there will be
great progress along this line, thanks to the wonderful cooperation which has developed in this field between the groups working
in the U.S., the Sovjet Union, and my country. Perhaps such a
cooperation is even the most important consequence of the Fano
effect.

I wonder, whether Ugo Fano saw all these consequences, when the
idea flashed through his mind on February 1, 1968.

REFERENCES:

1. M. Born, My Life: Recollections of a Nobel Laureate, Taylor and Francis Ltd., London 1978

2. N.F. Mott, H.S.W. Massey, The Theory of Atomic Collisions (Clarendon Press, Oxford 1965)

3. M.J. Alguard, J.E. Clendenin, R.D. Ehrlich, V.W. Hughes, J.S. Ladish, M.S. Lubell, K.P. Schüler, G. Baum, W. Raith, R.H. Miller, W. Lysenko, Nucl. Instrum. Meth. 163, 29 (1979)

4. U. Fano, Phys. Rev. 178, 131 (1969); Phys. Rev. 184, 250 (1969)

5. U. Fano, Comments Atomic Mol. Phys. II, 30 (1969)

6. M.S. Lubell, W. Raith, Phys. Rev. Letters 23, 211 (1969); G. Baum, M.S. Lubell, W. Raith, Phys. Rev. A 5, 1073 (1972)

7. J. Kessler, J. Lorenz, Phys. Rev. Letters 24, 87 (1970); U. Heinzmann, J. Kessler, J. Lorenz, Z. Physik 240, 42 (1970)

8. U. Heinzmann, K. Jost, J. Kessler, B. Ohnemus, Z. Physik 251, 354 (1972)

9. U. Heinzmann, H. Heuer, J. Kessler, Phys. Rev. Letters 34, 441 (1975)

10. U. Heinzmann, F. Schäfers, K. Thimm, A. Wolcke, J. Kessler, J. Phys. B: Atom. Molec. Phys. 12, L 679 (1979)

11. U. Heinzmann, F. Schäfers, J. Phys. B: Atom. Molec. Phys. 13, L 415 (1980)

12. U. Heinzmann, A. Wolcke, J. Kessler, J. Phys. B: Atom. Molec. Phys. 13, 3149 (1980)

13. U. Heinzmann, B. Osterheld, F. Schäfers, G. Schönhense, J. Phys. B: Atom. Molec. Phys. 14, L 79 (1980) and references therein

14. C.M. Lee, Phys. Rev. A 10, 1598 (1974)

15. W.R. Johnson, K.T. Cheng, K.-N. Huang, M. Le Dourneuf, Phys. Rev. A 22, 989 (1980)

16. U. Heinzmann, J. Phys. B: Atom. Molec. Phys. 13, 4353, 4367 (1980) and references therein

17. M.J. Alguard, V.W. Hughes, M.S. Lubell, P.F. Wainwright, Phys. Rev. Letters 39, 334 (1977)

18. O. Berger, J. Kessler, K.J. Kollath, R. Möllenkamp, W. Wübker, Phys. Rev. Letters 46, 768 (1981)

19. D.T. Pierce, F. Meier, P. Zürcher, Phys. Letters 51 A, 465 (1975)

20. C.Y. Prescott, W.B. Atwood, R.L.A. Cottrell, H. DeStaebler, E.L. Garwin, A. Gonidec, R.H. Miller, L.S. Rochester, T. Sato, D.J. Sherden, C.K. Sinclair, S. Stein, R.E. Taylor, J.E. Clendenin, V.W. Hughes, N. Sasao, K.P. Schüler, M.G. Borghini, K. Lübelsmeyer, W. Jentschke, Phys. Letters 77 B, 347 (1978)

21. R.J. Celotta, D.T. Pierce, G.-C. Wang, S.D. Bader, G.P. Felcher, Phys. Rev. Letters 43, 728 (1979)

22. R.R. Parsons, Phys. Rev. Letters 23, 1152 (1969)

23. N.A. Cherepkov, Phys. Letters 40 A, 119 (1972), Sov. Phys.-JETP 38, 463 (1974)

24. N.A. Cherepkov, J. Phys. B: Atom. Molec. Phys. 12, 1279 (1979)

25. N.A. Cherepkov, J. Phys. B: Atom. Molec. Phys. 13, L 689 (1980); 14, L 73 (1981) and references therein

26. U. Heinzmann, G. Schönhense, J. Kessler, J. Phys. B: Atom. Molec. Phys. 13, L 153 (1980)

27. G. Schönhense, Phys. Rev. Letters 44, 640 (1980)

28. H. Kaminski, J. Kessler, K.J. Kollath, Phys. Rev. Letters 45, 1161 (1980)

29. K.-N. Huang, W.R. Johnson, K.T. Cheng, Phys. Rev. Letters 43, 1658 (1979)

30. K.-N. Huang, Phys. Rev. A 22, 223 (1980) and references therein

PHYSICS OF ELECTRONIC AND ATOMIC COLLISIONS
S. Datz (editor)
© North-Holland Publishing Company, 1982

## USE OF HYPERSPHERICAL COORDINATES IN
## FEW-ELECTRON SYSTEMS

Hubert Klar
Fakultät für Physik
Universität Freiburg
Hermann-Herder-Str.3
7800 Freiburg/Germany

Correlated atomic wave functions in terms of hyperspherical coordinates are conveniently obtained diagonali-
sing the Hamiltonian in a basis of hyperspherical harmonics. The calculation of matrix elements of atomic
potentials between hyperspherical harmonics may be
simplified expanding the potentials into a series of
hyperspherical harmonics. The accurate construction of
this potential expansion is often nontrivial. This report presents a method for Coulomb interactions and a
method for non-Coulomb interactions, respectively, to
derive harmonic expansions on a hypersphere. The
present treatment is confined to two correlated electrons for the purpose of illustration; the generalisation to more particles is straightforward.

INTRODUCTION

During the last years considerable progress has been achieved in the investigation
of doubly excited states of two-electron systems solving the Schrödinger equation
using hyperspherical coordinates [1-8]. The method has been extended to three-
electron systems [9], to alcali earth atoms [10-11], to photoionisation [12], to
positron – H scattering [13], and recently also to the two-electron Dirac – Coulomb problem [14].

To simplify notations we restrict here ourselves to two-electron systems; the generalisation to more particles is straight forward [15]. Hyperspherical coordinates $(R,\omega)$ for two electrons are introduced in terms of the position vectors $\vec{r}_i$
relative to the nucleus by

$$\vec{r}_i = R \; \vec{f}_i \; (\omega) \qquad\qquad i = 1,2. \qquad\qquad (1)$$

The hyperradius R is given by $R = \sqrt{r_1^2 + r_2^2}$ and the symbol $\omega$ stands for a set of 5

angular coordinates [16-21]. The Hamiltonian reads then in a.u.

$$H = - \frac{1}{2} \; (R^{-5} \; \frac{\partial}{\partial R} \; R^5 \; \frac{\partial}{\partial R} - \frac{\Lambda^2}{R^2}) \; + \; V \; (R,\omega) \qquad\qquad (2)$$

where the squared grand angular momentum $\Lambda^2$ acts only in $\omega$. Solutions of the wave
equation are conveniently constructed expanding the wave function into a series
of hyperspherical harmonics $Y_{\lambda g} (\omega)$. These harmonics are eigenfunctions of $\Lambda^2$,

$$\Lambda^2 \; Y_{\lambda g} \; (\omega) \; = \; \lambda(\lambda+4) \; Y_{\lambda g} \qquad\qquad (3)$$

and g labels states degenerate $\lambda$. Along these lines one converts exactly the wave
equation into a one – dimensional system of equations in the variable R. This

conversion requires the calculation of matrix elements of the Hamiltonian (2) between harmonics,

$$H_{\lambda g \; \lambda' g'} = \int d\omega \; Y_{\lambda g} \; (\omega)^* H \; Y_{\lambda' g'} \; (\omega). \tag{4}$$

The accurate and fast evaluation of these matrix elements is often a non-trivial
problem for the potential part $V(R,\omega)$ because hyperspherical harmonics are complicated functions in several variables (3N-1 variables for N electrons), and
simple two-particle central potentials are non-central on the hypersphere. The
calculation of (4) is conveniently performed in two steps: (i) one expands the
potential into a series of hyperspherical harmonics

$$V \; (R,\omega) = \sum_{\lambda g} V_{\lambda g} \; (R) \; Y_{\lambda g} \; (\omega) \tag{5}$$

and (ii) one integrates a product of three harmonics over the hypersphere. The result of the second step is expressed in terms of $3 - \lambda$ coefficients. These coefficients are now known {22}, in special cases they have been used in Ref. 6-8.

The main subject of this report concerns the first step, eq. (5). We present in
the following an analytic method (A) for the expansion of Coulomb potentials {23}
into a series of hyperspherical harmonics. A more general method (B) applicable
to other potentials is derived below.

### HARMONIC EXPANSION OF ATOMIC POTENTIALS
#### A. COULOMB-INTERACTION

The two-electron Coulomb interaction

$$V = - \frac{Z}{r_1} - \frac{Z}{r_2} + \frac{1}{r_{12}} \tag{6}$$

satisfies the Poisson equation

$$(\Delta_1 + \Delta_2)V = 4\pi \{Z\delta(\vec{r}_1) + Z\delta(\vec{r}_2) - \delta(\vec{r}_1 - \vec{r}_2)\}. \tag{7}$$

Replacing there the single-electron coordinates $\vec{r}_1$, $\vec{r}_2$ by hyperspherical coordinates, see eq. (1), one easily derives from eq. (7)

$$V = - \frac{4\pi}{R} \int d\omega' \; G(\omega,\omega') \; \{Z\delta(\vec{f}_1(\omega')) + Z\delta(\vec{f}_2(\omega')) - \delta(\vec{f}_1(\omega') - \vec{f}_2(\omega'))\} \tag{8}$$

where the Greens function $G(\omega,\omega')$ is defined by

$$(\Lambda^2 + 3) \; G \; (\omega,\omega') \; = \; \delta^{(5)} (\omega - \omega') . \tag{9}$$

Since the harmonics $Y_{\lambda g}$ form a complete set of functions on the five-dimensional sphere the Greens function can be represented by the eigenfunction expansion

$$G \; (\omega,\omega') \; = \; \sum_{\lambda g} \frac{1}{\lambda(\lambda+4)+3} \; Y_{\lambda g} \; (\omega) \; Y_{\lambda g} \; (\omega')^* . \tag{10}$$

Substitution of eq. (10) into eq. (8) leads to the potential expansion coefficient

$$V_{\lambda g} \; (R) \; = \; - \; \frac{4\pi}{(\lambda+1) \; (\lambda+3)R} \; \times$$

$$\times \int d\omega \; Y_{\lambda g}(\omega)^* \{Z\delta(\vec{f}_1 (\omega))+Z\delta(\vec{f}_2(\omega)) \; - \; \delta(\vec{f}_1(\omega)-\vec{f}_2(\omega))\} . \tag{11}$$

The evaluation of the integrals in (11) is simple. Because of the rotational invariance of V only harmonics describing S-states occur. These harmonics depend only on $5 - 3 = 2$ body - fixed angles. The integrals have therefore the structure

$$\int d\omega \; Y_{\lambda g}^{L=0} \; (\omega) \; \delta(f_i(\omega)) \; = \; c \; Y_{\lambda g}^{L=0} \; (\omega_i)$$

where $\omega_i$ stands for the set of 2 angles at which $f_i=r_i/R$ is equal to zero. The constant c being independent of $\lambda$ and g depends on the choice of the functions $\vec{f}_i$ introduced in eq. (1).

## B. NON-COULOMB INTERACTION

We come now to the treatment of two correlated electrons in a many - electron system. Such a treatment is necessary for instance for the two outer electrons of negative alkali ions and for doubly excited states of alkali earth atoms.

We assume in the following that the Fourier transformation of an atomic potential $V(r_1)$ experienced by an electron at nuclear distance $r_1$ exists,

$$W(k_1) \; = \; (2\pi)^{-3/2} \int d\vec{r}_1 \; e^{-i\vec{k}_1\vec{r}_1} \; V(r_1) . \tag{12}$$

If the potential at large distances shows a Coulomb behaviour

$$\lim_{r\to\infty} rV(r) = Z \neq 0$$

one may handle the Coulomb part Z/r using method A described above and one applies the Fourier transform, eq. (12), to the remainder V(r) - Z/r.

In the sixdimensional momentum space of both electrons the Fourier representation of the potential $V(r_1)$ reads

$$V(r_1) = (2\pi)^{-3/2} \int d\vec{k}_1 \int d\vec{k}_2 \; e^{i(\vec{k}_1\vec{r}_1 + \vec{k}_2\vec{r}_2)} \; W(k_1) \; \delta^{(3)}(\vec{k}_2) \qquad (13)$$

with $W(k_1)$ given by eq. (12). It is now convenient to introduce hyperspherical coordinates in the momentum space; analogous to eq. (1) we write for the one-electron momenta

$$\vec{k}_i = K \, \vec{g}_i(\Omega) \qquad\qquad i = 1,2 \qquad\qquad (14)$$

with $K = \sqrt{k_1^2 + k_2^2}$ and $\Omega$ standing for a set of 5 hyperspherical angles. Substituting these momentum coordinates into eq. (13) we obtain

$$V(r_1) = (2\pi)^{-3/2} \int_0^\infty dK \; K^2 W(K) \int d\Omega e^{i\vec{K}\cdot\vec{R}} \; \delta^{(3)}(\vec{g}_2(\Omega)). \qquad (15)$$

The expansion of the potential into position space harmonics results now from the partial wave expansion of a plane wave in six dimensions {24}

$$e^{i\vec{K}\vec{R}} = \frac{4}{K^2R^2} \sum_{\lambda=0}^\infty i^\lambda (\lambda+2) \; J_{\lambda+2}(KR) \; C_\lambda^2(\hat{K}\cdot\hat{R}). \qquad (16)$$

Note that harmonics on an n-dimensional sphere can be expressed by Gegenbauer polynomials $C_\lambda^\nu$ with $\nu = \frac{1}{2}(n-2)$ {25}. We have therefore selected the index $\nu=2$ corresponding to $n=6$ for two electrons. Substituting eqs. (12) and (16) into eq. (15) and integrating over K {26} we find

$$V(r_1) = \frac{2}{\pi^2 R^2} \sum_\lambda i^\lambda \int_0^R dr \; r V(r) \sin\{(\lambda+2)\arcsin\frac{r}{R}\} \times$$

$$\times \int d\Omega \; C_\lambda^2(\hat{K}\cdot\hat{R}) \; \delta^{(3)}(\vec{g}_2(\Omega)). \qquad (17)$$

The expansion (5) is obtained using in eq. (17) the addition theorem for hyperspherical harmonics {25,27}

$$C_\lambda^2(\hat{K}\cdot\hat{R}) = n_\lambda \sum_g Y_{\lambda g}(\Omega)^* \; Y_{\lambda g}(\omega) \qquad (18)$$

where the summation runs over all harmonics degenerate in $\lambda$. The coefficient $n_\lambda$ depends on the dimensionality and has in our case the value[1] {27}

$$n_\lambda = \frac{2\pi^3}{\lambda+2} \qquad (19)$$

Note that eq. (17) contains only real quantities because for O(3) invariant potentials only even values of $\lambda$ occur.

The generalisation of this method to more complicated (spin-dependent, non-central, ...) interactions is straightforward starting from the appropriate Fourier representation corresponding to eq. (12).

Footnote:

1. The addition theorem given in Reference {25} contains a printing error. In equation (2) on page 243 one has to do the replacement

$$C^{(1/2)\rho}_{p}(1) \rightarrow C^{(1/2)\rho}_{n}(1) \ .$$

REFERENCES:

{1} Macek, J.H., Journ. Phys. B1 (1968) 831.

{2} Lin, C.D., Phys. Rev. A10 (1974) 1968.

{3} Lin, C.D., Phys. Rev. A14 (1976) 30.

{4} Fano, U. and Lin, C.D., Correlations of excited electrons, in: zu Putlitz, G., Weber, E.W. and Winnacher A. (eds.), Atomic Physics, Vol. 4 (Plenum, New York, 1975).

{5} Fano, U., Physics Today 29 (1976) 32.

{6} Klar, M., Korrelation in 2-Elektronen-Systemen, Ph.D. Thesis, Dept. of Physics, Freiburg Univ. (1978).

{7} Klar, H. and Klar, M., Phys. Rev. A17 (1978) 1077.

{8} Klar, H. and Klar, M., Journ. Phys. B13 (1980) 1057.

{9} Clark, C.W. and Greene, C.H., Phys. Rev. A21 (1980) 1786.

{10}Greene, C.H., Phys. Rev. A23 (1981) 661.

{11}Fano, U., Physica Scripta 23 (1981) in press.

{12}Miller, D.L. and Starace, A.F., Phys. Rev. A23 (1981) in press.

{13}Pelikan, E., priv. communication.

{14}Lawen, M., priv. communication.

{15}Knirk, D.L., Journ. Chem. Phys. 60 (1974) 66.

{16}Fock, V.A., Kong. Nors. Vidensk. Selsk. Forhandl. 31 (1958) 145.

{17}Smith, F.T., Journ. Math. Phys. 3 (1962) 375.

{18}Dragt, A., Journ. Math. Phys. 6 (1965) 533.

{19}Pustovalov, V.V. and Simonov, Yu.A., Sov. Phys. JETP 24 (1967) 230.

{20}Pustovalov, V.V. and Smorodinski, Ya.A., Sov. Journ. Nucl. Phys. 10 (1970) 729.

{21}Nyiri, Yu. and Smorodinski, Ya.A., Sov. Journ. Nucl. Phys. 12 (1971) 109.

{22}Aguila, F. del, Journ. Math. Phys. 21 (1980) 2327.

{23}Klar, H., Journ. Phys. B7 (1974) L 436.

{24}Watson,G.N., Theory of Bessel Functions (Univ. Press, Cambridge, 1962).

{25}Erdélyi, A., Magnus, W., Oberhettinger, F., Tricomi, F.G., Higher Transcen-
    dental Functions, Vol. 2 (Mc Graw-Hill, New York, 1953).

{26}Oberhettinger, F., Tabellen zur Fourier Transformation (Springer, Berlin, 1957).

{27}Lee, S.P., The Three-Nucleon Problem, Ph.D. Thesis, Dept. of Math., London
    Imperial College (1969).

PHYSICS OF ELECTRONIC AND ATOMIC COLLISIONS
S. Datz (editor)
© North-Holland Publishing Company, 1982

CLOSE-COUPLING METHODS FOR ION ATOM COLLISIONS

A. Salin

Laboratoire d'Astrophysique, Université de Bordeaux I
33405 Talence
FRANCE

I - INTRODUCTION

I shall deal in this report with the application of close coupling (C.C.) methods for ion-atom collisions when the energy of relative motion is above a few tens or hundred eV per atomic mass unit. Then, an infinite number of channels are usually openned and a large number of partial waves for the relative motion have to be taken into account. For the sake of simplicity, the discussion will be given in a semi-classical approach : most applications of the C.C. method use semi-classical (or "impact-parameter") methods. Furthermore, in the energy range of interest, all the important features of semi-classical approaches have their counterpart in a full quantum mechanical treatment. Hence we consider the time evolution of the electronic state, the evolution of the nuclei being described by a classical law of motion.

The essence of the C.C. method is an expansion of the electronic wave-function describing the electronic state onto a set of basis functions $\{ \phi_n (\underset{\sim}{r}, t, \lambda) \}$. Let us call $\Psi$ the exact solution of :

$$(H_{e\ell} - i \frac{d}{dt}\big|_{\underset{\sim}{r}}) \ \Psi \ (\underset{\sim}{r}, t) = 0 \tag{1}$$

with asymptotic conditions depending on the process considered. $H_{e\ell}$ is the electronic hamiltonian and $\underset{\sim}{r}$ is a set of electronic coordinates. We introduce the approximate wave-function :

$$\Psi_a \ (\underset{\sim}{r}, t) = \sum_{n=1}^{N} a_n (t) \ \Phi_n (\underset{\sim}{r}, t, \lambda) \tag{2}$$

The fact that N is finite is basic to the C.C. method. We have made explicit in the form of $\phi_n$ that the basis functions may be dependent on a set of parameters $\{ \lambda \}$. We have to answer two questions : how to choose the $\phi_n$ ? Which equations should the $a_n$ satisfy ? The two questions may not be independent. There has been a large number of answers in the litterature, which raises a third question : which criterium do we have to compare the various methods ?

We may classify the solutions in two different groups.

1°) If (and only if) the set $\{ \phi_n \}$ can be made complete, the expansion (2) converges to the exact solution as $N \to \infty$. A logical strategy, then, is to study the possible convergence of the

expansion. However the computation time increases very rapidly with
N, essentially due to the large number of matrix elements that have
to be computed. Consequently, it is advisable to choose the form of
$\phi_n$ that makes the computation of these matrix elements as fast as
possible. This approach is similar to the one followed in quantum
chemistry for the computation of molecular properties. It can be
illustrated by the work of Shakeshaft[1] and Morrisson and Öpik[2]
on one electron systems. It is surprising that no calculations have
been done along that line for many electron systems since, in principle, it is well suited to that case. In particular Gaussian type
orbitals (GTO) have the great advantage that matrix elements can be
evaluated analytically even when translation factors are included
in the expansion[3].

2°) The more popular approach is to choose the basis functions $\phi_n$
    "conveniently" in an attempt to minimize the number of terms.
    This is basically the idea behind the molecular approximation
    that we shall discuss later. An other possibility is to try to
    "optimize" the basis functions by introducing parameters deter-
    mined through variationnal methods.


II - DERIVATION OF C.C. EQUATIONS IN SEMI-CLASSICAL METHODS

Which equation shall we use to define $\Psi_a$? The exact wave function
satisfies various equivalent equations which may not be equivalent
in an approximate calculation. This arbitrariness has been often
overlooked. The usual procedure is to define P (t) as the projector
onto the space spanned by $\{\phi_n\}$, Q (t) = 1 - P (t) and impose :

$$P (t) (H - i \frac{\partial}{\partial t}) \Psi_a = 0 \qquad (3)$$

The exact solution is such that :

$$P (t) (H - i \frac{\partial}{\partial t}) P \Psi = - P (t) (H - i \frac{\partial}{\partial t}) Q (t) \Psi \qquad (4)$$

so that obviously $P \Psi \neq \Psi_a$.

The usual justification of (3) goes through a variationnal procedure[4]. Similarly, tentatives to define $\{\phi_n\}$ in the "best possible
way" have relied on variationnal methods (e.g. the Euler-Lagrange
optimization procedure of Cheshire[5]). Let us comment on these
approaches.

We have to determine some quantity q (e.g. the scattering amplitude).
One constructs a functionnal $\mathcal{L} (\Psi_T)$ and gets a variationnal principle
when $\mathcal{L} (\Psi_T)$ is stationnary for arbitrary variations of $\Psi_T$ around
the exact value $\Psi$.[†] Basic to the significance of this variationnal
principle are the two conditions :

    - the "trial" space spanned by $\{\Psi_T\}$ should include the
      exact solution

    - the exact value corresponds to stationnarity with respect
      to arbitrary variation of $\Psi_T$.

[†] This has the implication that $\mathcal{L}$ is stationnary around $\Psi$ if we make a variation
   with respect to a limited set of parameters. However the converse is not true.

These two conditions make the use of variationnal principle for C.C. calculations of doubtful value in most cases since <u>we have no guarantee that the points where $\mathcal{L}$ is stationnary</u> (there may be more than one) <u>yield the best value of q.</u>

These shortcomings do not apply to extremum principles on q itself. Well known examples are extremum principles on reaction matrix[18] or the Rayleigh-Ritz principle for bound states. In the latter case, for example, one knows that the value of the energy improves as it goes down. The fact that the evolution operator is unbound is at the root of the difficulty in the semi classical theory of collisions (or in the quantal theory when an infinite number of channels are open).

Details on the above discussion can be found in the papers by Storm, Shakeshaft and Spruch, Weglein[6] and references therein. Storm and Rapp[7] have calculated upper bounds on the error on the transition amplitude ("error functions") for proton-atomic hydrogen collisions and a sample of their results is given in figure 1. To my knowledge this is the only calculation of this type. As can be seen from the figure, much has still to be done in order to get useful bounds.

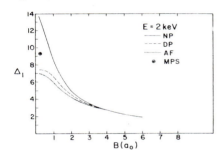

*Figure 1* : *Error function versus impact parameter for $H^+ - H(1s)$ collisions. Curve extracted from |7|. For all the cases considered here the error function is greater than one.*

In conclusion, variationnal principles should not be used as a smoke-screen that permits to avoid a discussion on the physics put into the approximate solution. For the moment, the only criterium we have to compare various methods is that $Q(t)$ can be made small, i.e. the expansion converges quickly.

III - Molecular expansion

The molecular expansion is a good example of how we try to accelerate the convergence of the expansion (1) by using physical intuition. The only study on the convergence of the molecular expansion is that of Winter and Lane[8] for the $He^{2+} - H$ system. More recently Hatton, Lane and Winter[9] have resumed their study with the introduction of plane wave translation factors[10]. They claim to obtain convergence more rapidly. This claim should be considered with caution since the completeness of the Bates - McCarroll expansion is doubtful[11]. However we have inevitably to face the "momentum transfer" problem basically connected with the molecular C.C. expansion. We shall not treat the problem here in details since it deserves large developments by itself and an extensive litterature is available (see e.g.[12] and references therein).

The molecular expansion is defined by :

$$\Psi_{mol} = \sum_n a_n (t) \, \chi_n (\underset{\sim}{r}, R) \, D_n (\underset{\sim}{r}, R) \tag{5}$$

in terms of the molecular function $\chi_n (\underset{\sim}{r}, R)$ defining a state of the electrons for a fixed internuclear distance R. The electron translation factor (ETF) $D_n$ is introduced, from the point of view of C.C. methods, to accelerate the convergence of the expansion (beside curing the method from     well known deficiencies). Of course, a bad choice of ETF may slow down the convergence. To get some hints on the suitable form of ETF, it is worthwhile to understand what is missing in the description of the collision when $D_n = 1$.

One generally justifies the molecular approximation by "thinking in configuration space" : the electrons adjust their motion to the ins- tantaneous position of the nuclei so that the electronic density in configuration space is very close to the electronic density of a state (or a superposition of a small number of states) defined for fixed nuclei (molecular wave function). This picture is completely incorrect in momentum space. The momentum distribution of the mole- cular wave functions has qualitatively the following shortcomings :

- it is independent of the reference frame for the momenta since it is defined for fixed nuclei.
- it is (by definition) independent of the collision velocity or any parameter characterising the collision.

Hence it cannot describe easily the evolution of the electronic momen- tum distribution in the course of the collision. For example the ini- tial state has a momentum distribution centered about the value $q_0$ related to the translation of the electron with the atom. Therefore we have to reconcile the intuition behind the molecular approximation with a correct picture of the momentum density as the collision goes on. A very attractive way of doing this has been proposed by Ponce[13].

Consider for simplicity a one electron system. The exact solution $\Psi$ satisfies the continuity equation :

$$Re \, \overset{\rightarrow}{\nabla} (i \, \Psi \, \overset{\rightarrow}{\nabla} \, \Psi^{\bigstar}) - \frac{\partial}{\partial t} \mid \Psi \mid^2 = 0 \tag{6}$$

which states that for every point in C.S. the rate of change of the electron density is equal to the divergence of its flux. This equa- tion is not satisfied by the approximate wave function in the mole- cular expansion. However it is just the kind of equation we need to describe the fact that the transfer of electronic density from one nucleus to the other is accompanied by a transfer of momentum. Suppose that we now write :

$$\Psi_{ap} = \exp \{ ig (\underset{\sim}{r}, t) \} \sum_n a_n (t) \, \chi_n (\underset{\sim}{r}, R) \tag{7}$$

The "common translation factor" g can be expanded as :

$$g (\underset{\sim}{r}, t) = \sum_{i=1}^{3} g_i (t) \, X_i + \frac{1}{2} \sum_{i,j=1}^{3} h_{ij} (t) \, X_i \, X_j + \dots \tag{8}$$

where the x are the orthogonal components of $\underset{\sim}{r}$. Then, we obtain a sequence of Ehrenfest relations :

$$< \Psi_{ap} \mid p_i \mid \Psi_{ap} > = \frac{\partial}{\partial t} < \Psi_{ap} \mid x_i \mid \Psi_{ap} > \qquad (9)$$

$$\frac{1}{2} < \Psi_{ap} \mid x_k \, p_i + p_i \, x_k + x_i \, p_k + p_k \, x_i \mid \Psi_{ap} > = \frac{\partial}{\partial t} < \Psi_{ap} \mid x_i \, x_k \mid \Psi_{ap} >$$

etc...

These equations impose on the average value of the momentum $p_i$ and position moments the corresponding classical relations. Hopefully, a (truncated) expansion (7) will yield a better description of the collision problem by comparison with the usual molecular expansion. It can be shown, for example, that the inconsistancies of the molecular expansion are waved by retaining up to quadratic terms in (7) and (8).


IV - ATOMIC OR PSEUDO-STATE EXPANSIONS

We shall consider again the one-electron case for the sake of simplicity. The generalization of the following discussion to many electron cases involves some more subtil aspects connected with the application of the Pauli principle. We would like to make a few remarks on the connection between the molecular and atomic expansions.

In actual calculations, the molecular wave function $\chi_n$ is expressed as a combination of basis orbitals (atomic orbitals, Slater orbitals, Gaussian orbitals, etc...) centered on the nuclei or elsewhere. These techniques are well developped and documented in quantum chemistry

$$\chi_n \, (\underset{\sim}{r}, \, R) = \sum_{i=1}^{M} C_i^n \, (R) \, \phi_i \, (\underset{\sim}{r}, \, R) \qquad (10)$$

where $\{ \phi_i \}$ is the set of orbitals used in the calculations.

In the atomic expansion, one uses directly :

$$\Psi_{at} = \sum_{n=1}^{M} b_n \, (t) \, \phi_n \, (\underset{\sim}{r}, \, R) \, B_n \, (\underset{\sim}{r}, \, R) \qquad (11)$$

As in general the basis orbitals $\phi_n$ are localized at a definite center, $B_n$ is usually chosen as a plane wave factor describing the translation of this center. For example, if $\phi_n$ is an atomic state centered on the nucleus A, $B_n$ is so that the electron "travels" with A. Of course, this choice is questionable and sometimes no such simple prescription can be made (e.g. bicentric orbitals). Comparing (11) with (5) and (10), it is obvious that the main difference lies in the choice of the translation factors.

More precisely if we retain in the molecular expansion all the states that can be formed by diagonilizing $H_{e\ell}$ in the subspace spanned by $\{ \phi_i \}$ , then the only difference lies in the treatment of translation factors. Furthermore, if a "common translation factor" is used (as discussed earlier - see 7), the two expansions are completely equivalent. An other case where such an equivalence holds is by using the definition of ETF proposed by Riera and Salin and adopted by Delos[14] : they proposed that each orbital in (10) be alloted its translation factor, which is obviously equivalent to (11)[†].

†*This approach, however, meets with serious difficulties*[11]

Under this condition, the only advantage of the molecular approach
is that it could be a mean of reducing the number of orbitals $\phi_i$
in the expansion.

This close relation between atomic and molecular expansions has been
fully recognized for example by Fritsch et al.[15]. They compare a
two state atomic expansion (with plane wave factors for $B_n$) with a
two-state molecular expansion (with $D_n = 1$). If the molecular wavefunctions are calculated in a two-state LCAO approximation, then the
two theories differ only by the choice of ETF.

The recognition of this close relationship is very useful in order
to interpret the extention to atomic collision of the use of expansions first introduced in quantum chemistry. It has been recognized
for a long time that a three-center expansion is of great value for
the computation of molecular properties at small internuclear distances. Similarly Antal et al. and Lin et al.[16] have used a three
center atomic expansion for simple collision processes. They introduce in the expansion united-atom orbitals at the center of charge.
Obviously, with a suitable choice of ETF in these treatments, it is
possible to get a theory which joins smoothly the molecular expansion in the low energy range with the atomic expansion at medium
and high energies.


IV - CONCLUSION

Electron translation factors are essential for a proper definition
of close coupling expansions in ion-atom collisions. Though their
role has been emphasized mostly for the molecular expansion, it is
worthwhile to look into their relevance for the atomic (or pseudo
state) expansion. The molecular and atomic expansion differ through
the choice of translation factors. They are identical if a "common
translation factor" is used.

Unfortunately a proper determination of translation factors is made
difficult by the absence of an unambiguous criterium on the accuracy
of approximate solutions - difficulty due to the fact that the evolution operator is unbound when an infinite number of channels are
open.

Finally we should mention that we have not discussed the extension
of C.C. method to ionization processes. This is difficult since one
has to introduce continuum states as the final state of the collision. Pseudo state expansions do achieve this in some way but we
have the feeling that this is mostly by accident : the pseudo states have often a non zero overlap with the continuum. A more systematic approach has been proposed more recently by Micha and
Piacentini[17].


ACKNOWLEDGMENTS

The author is grateful to Dr. Ponce for useful discussions on variationnal methods and translation factors. Thanks are due to Pr. Riera
for his extensive criticisms on the manuscript and to Drs. Fritsch,
Lin and Rivarola for useful discussions on the subject.

REFERENCES

| 1| Shakeshaft, R., Phys. Rev. A $\underline{18}$ (1978) 1930-1934

| 2| Morrisson, H.G., and Öpik, U., J. Phys. B $\underline{11}$ (1978) 473-492 ; ibid. $\underline{12}$ (1979) L685-688

| 3| Errea, L.F., Méndez, L., and Riera, A., J. Phys. B $\underline{12}$ (1979) 69-82

| 4| Sil, N.C., Proc. Phys. Soc. $\underline{75}$ (1960) 194

| 5| Cheshire, I.M., J. Phys. B $\underline{1}$ (1968) 428-37

| 6| Storm, D., Phys. Rev. A $\underline{10}$ (1974) 1008-1009 ; Shakeshaft, R., and Spruch, L., Phys. Rev. A $\underline{10}$ (1974) 92-101 ; Weglein, A.B., Phys. Rev. A $\underline{17}$ (1978) 1810-1818

| 7| Storm, D., and Rapp, D., Phys. Rev. A $\underline{14}$ (1976) 193-203

| 8| Winter, T.G., and Lane, N.F., Phys. A $\underline{17}$ (1978) 66-79

| 9| Hatton, G.J. Lane, N.F., and Winter, T.G., J. Phys. B $\underline{12}$ (1979) L571-577

|10| Bates, D.R., and McCarroll, R., Proc. R. Soc., A 245 (1976) 175-183

|11| Errea, L.F., Méndez, L., and Riera. A., preprint, Universidad Autónoma de Madrid (May 1981)

|12| Salin, A., Comm. in At. and Mol. Physics $\underline{9}$ (1980) 165-171 Delos, J.D., Comm. in At. and Mol. Physics, in course of publication

|13| Ponce, V.H., J. Phys. B, in course of publication

|14| Riera, A., and Salin, A., J. Phys. B $\underline{9}$ 2877-2891 Delos, J.D., and Thorson, W.R., J. Chem. Phys. $\underline{70}$ (1980) 1774-1790

|15| Fritsch, W., Lin, C.D., and Tunnell, L.N., Phys. Rev. A, to be published

|16| Antal, M.J., McElroy, M.B., and Anderson, D.G.M., J. Phys. B 8 (1975) 1513-1521 Lin, C.D., Winter, T.G., and Fritsch, W., Phys. Rev. A, to be published

|17| Micha, D.A., and Piacentini, R.D., submitted to Phys. Rev. A

|18| Hahn, Y., O'Malley, T.F., and Spruch, L., Phys. Rev. $\underline{128}$ (1962) 932-942

PHYSICS OF ELECTRONIC AND ATOMIC COLLISIONS
S. Datz (editor)
© North-Holland Publishing Company, 1982

APPLICATION OF COMPLEX SCALING TO RESONANCES

B. R. Junker

ONR, Code 412
800 N. Quincy St.
Arlington, VA., 22217

The complex-coordinate theorem is described along with the
implied analytical properties of the wavefunctions for the
bound, scattering, and resonance states. These results and
previous numerical calculations imply the rotation angle is
merely playing the role of a nonlinear variational parameter.
This then suggests a technique for computing resonance parameters (position and width) using the unrotated Hamiltonian
with a square-integrable basis and without explicitly imposing a Siegert boundary condition on the resonance wavefunction.
This technique is illustrated with several examples.

INTRODUCTION

Many phenomena in atomic physics can be described or analyzed in terms of a
resonance structure. That is, a position and width can be associated with the
observed experimental structure. A number of techniques have been formulated to
compute these parameters either by fitting a complete scattering calculation in
the vicinity of the resonance to a Breit-Wigner form or by computing directly
the complex energy at which the S-matrix has a pole. Throughout this paper the
potentials will be assumed to have the necessary properties for the poles of the
S-matrix to be the same as the poles of the resolvent.

COMPLEX COORDINATE THEOREMS

Until the advent of the complex coordinate theorems[1,2] (or Balslev-Combes-
Simon theorems) the asymptotic form of the wavefunction played a significant
role in the calculation and/or the extraction of the resonance parameters. In
the complex-coordinate theorems the following transformation is defined.

$$r \longrightarrow r \exp{(i\theta)} \qquad\qquad (1)$$

When a Hamiltonian, $H(r)$, is transformed according to EQ.(1), the spectrum is
altered in the following manner:

(1) The bound state energies of the transformed Hamiltonian, $H(\theta)$, are
identical to those of $H(r)$.

(2) The real energy thesholds of the cuts of $H(\theta)$ are the same as those
of $H(r)$, but the cuts themselves are rotated down onto the lower sheets
in the complex energy plane by an angle $2\theta$. That is, the energies
associated with the cut of $H(\theta)$ are of the form $(E^T + E \exp(-2i\theta))$
where $E$ is the energy on the cut for $H(r)$ above the threshold energy,
$E^T$.

(3) Additional discrete complex poles of the form $E_R - iE_I$ $(E_I > 0)$ of the

resolvent appear when $\theta > \frac{1}{2} |Arg\ (E_R - iE_I)| = \beta$ where $E_R$ and $-E_I$ are the real and imaginary parts of the energy associated with these poles, respectively. These complex poles of the resolvent, as noted above, are assumed to be associated with resonances. In addition the wavefunctions associated with these poles are square-integrable when $\theta > \beta$.

(4) Finally there are a number of other elements of the spectrum which do not concern us here.

At first insight into the structure of the wavefunctions may appear to be a formidable task. However, several very useful analytical properties of the wavefunctions can be obtained quite easily. Balslev and Combes [1] showed that the domain of the bound states of $H(\theta)$ is obtained by transforming the domain of the bound states of $H(r)$ by EQ.(1). That is, the bound states of $H(\theta)$ are simply the analytical continuation of the bound states of $H(r)$. Similarly the resonance wavefunctions can be shown [3,4] to be an analytical continuation of the Siegert states of $H(r)$ by EQ.(1). The square-integrability can be seen to arise quite simply [4,5] in the following way. From the fact that

$$E = E_R - iE_I = \frac{1}{2}\ k^2 = \frac{1}{2}|k|^2 \quad \exp\ (-2i\beta) \qquad (2)$$

so that the Siegert function goes asymptotically as

$$\varphi^s \longrightarrow c\ \exp\ (ikr) = c\ \exp\ [i|k|r\ \exp\ (-i\beta)], \qquad (3)$$

analytical continuation of EQ. (3) by EQ. (1) yields

$$\varphi^s(\theta) \longrightarrow c \quad \exp\ [i|k|\ r\ \exp(i\ (\theta - \beta))] \qquad (4)$$

This is now square-integrable for $\theta > \beta$ as noted in the above complex-coordinate theorem.

Several comments are in order here. First the Siegert states are eigenfunctions of $H(r)$ corresponding to the complex poles regardless of any analytical continuation or complex-coordinate transformation. The effect of the cut rotating through the resonance pole, i.e., the resonance wavefunction being analytically continued from a value $\theta$ which is infinitismally smaller than $\beta$ to a value $\theta$ infinitismally larger than $\beta$ is to discontinuously convert the Siegert function from an asymptotically divergent function to an asymptotically convergent function. Secondly, the difference between the wavefunctions for the bound states and those for the resonance states is that the former are innately real functions of $r$ for $H(r)$ and thus real functions of $r$ $\exp\ (i\theta)$ for $H(\theta)$, while the resonance functions are innately complex functions of $r$ and $r$ $\exp\ (i\theta)$ for $H(r)$ and $H(\theta)$, respectively.

The wavefunctions for the energies along the cuts must be handled more carefully. For $E > 0$ and $\theta > 0$ the wavefunctions corresponding to energies along the cut are not just functions of $r$ $\exp\ (i\theta)$. The transformation (1) implies

$$P \longrightarrow P \quad \exp\ (-i\theta) \qquad (5)$$

The cuts can be chosen to lie in any direction one wishes, but the natural choice for the cut [6] is to use the momentum eigenvector basis. Thus

$$\psi_k \sim A \quad \exp(i k \cdot r) + B\ \exp(-i k \cdot r) \qquad (6)$$

and

$$E = \frac{1}{2} k \cdot k \quad \exp(-2i\theta) = |E| \quad \exp\ (-2i\theta) \qquad (7)$$

In conclusion the bound and resonance state wavefunctions for $H(\theta)$ are simply analytical continuations of those for $H(r)$. On the other hand, those corresponding to the cut are linear combinations of eigenfunctions of the analytically continued momentum operators.

## ANALYTICAL EXAMPLES

Previously the bound states for the analytically continued atomic hydrogenic Hamiltonian had been obtained by the direct solution of the analytically contin- ued Schrocdinger equation subject to square integrability.[3,7] More recently they have been obtained by identifying the poles of the hydrogenic S-matrix.[4] In the latter case the wavefunctions corresponding to the cut were also obtained. As expected from above the former states corresponded to analytical continuations of the bound states of H(r), while the latter were functions of $kr$ where $k$ and r are real.

Here we will consider instead the s-wave solutions for the short range poten- tial[8] $-V_0 \exp(-r/a)$.

The unrotated Hamiltonian is

$$H(r) = -(\hbar^2/2\mu)(d^2/dr^2) - V_0 \exp(-r/a) \qquad (8)$$

while the rotated Hamiltonian is

$$H(\theta) = -(\exp(-2i\theta)\hbar^2/2\mu)(d^2/dr^2) - V_0 \exp[-\exp(i\theta)r/a] \qquad (9)$$

After some algebra the Schroedinger equation for EQ.(9) can be put in the form

$$[(d^2/d^2\eta) + (1/\eta)(d/d\eta) + 4b^2((k^2/\eta^2) + \kappa'^2)]\chi = 0 \qquad (10)$$

where

$$\eta = \exp(-r/b) \qquad (11a)$$
$$b = a \exp(-i\theta) \qquad (11b)$$
$$k' = k \exp(i\theta) = (2\mu E/\hbar^2)^{\frac{1}{2}} \exp(i\theta) \qquad (11c)$$

and

$$\kappa' = \kappa \exp(i\theta) = (2\mu V_0/\hbar^2)^{\frac{1}{2}} \exp(i\theta) \qquad (11d)$$

The solutions are

$$\chi = J_{-2bk'i}(2b\kappa') J_{2bk'i}(2b\kappa'\eta)$$

$$- J_{2bk'i}(2b\kappa') J_{-2bk'i}(2b\kappa'\eta)$$

$$\sim J_{2bk'i}(2b\kappa') \exp(2bk'i \ln(b\kappa')) \exp(-ik'r) \\ *(\Gamma(2bk'i+1))^{-1}$$

$$- J_{2bk'i}(2b\kappa') \exp(-2bk'i\ln(b\kappa')) \exp(ik'r) \\ *(\Gamma(-2bk'i+1))^{-1}$$

$$= [-J_{-2bk'i}(2b\kappa') \exp(2bk'i\ln(b\kappa'))/\Gamma(2bk'i+1)]$$

$$*\left\{\left[\frac{J_{2bk'i}(2b\kappa')}{J_{-2bk'i}(2b\kappa')} \frac{\Gamma(2bk'i+1)}{\Gamma(-2bk'i+1)} \exp(-2bk'i\ln(b\kappa'))\right]\right.$$

$$\left. * \exp(ik'r) - \exp(-ik'r)\right\}$$

$$= C [S \exp(ik'r \exp(i\theta)) - \exp(-ik'r \exp(i\theta))] \qquad (12)$$

Since b is proportional to $\exp(-i\theta)$ and $\kappa'$ is proportional to $\exp(i\theta)$, $b\kappa'$ is independent of $\theta$ for both bound and scattering states. In addition, for bound states $k$ is proportional to $\exp(i\theta)$ so that $bk'$ is also independent of $\theta$. Then S is independent of $\theta$ and thus the wavefunctions for bound states depend on r and $\theta$ only in the exponential in the form $r \exp(i\theta)$. For the states along the cut E is proportional to $\exp(-2i\theta)$. Then $k'$ is independent of $\theta$ and k is proportional to $\exp(-i\theta)$. S now becomes a function of $\theta$ through the product $b \ell'$ which is proportional to exp $(-i\Theta)$ and the exponential functions become $\exp(\pm i \ell' r)$, i.e., $\Theta$-independent. Since the momentum operator is of the form exp $(-i\Theta)$ (d/dr), these functions correspond to momentum eigenfunctions with complex momentum, $\ell' \exp(-i\Theta)$. These results illustrate the functional independencies discussed in the previous section.

Analytical models which possess resonances are more difficult to develop. Doolen[9] has suggested one such model potential which illustrates the functional dependencies of resonance wavefunctions even though it has an undesirable divergence at the origin. The potential is of the form

$$V(r) = -\gamma r^{-2} + r^{-1} \tag{13}$$

Elsewhere[4] we have discussed the details for the generalization of the solutions of Doolen's potential to arbitrary partial waves. Here we will simply give the results for the solution of the Schroedinger equation

$$0 = [-(\exp(-2i\Theta)/2r^2) \ (d/dr) \ r^2(d/dr) + \exp(-i\Theta)/r$$
$$+ \ell'(\ell'+1) \exp(-2i\Theta)/(2r^2) \ -E] \chi \tag{14}$$

where $\ell'$ is defined as the solution of

$$\ell'(\ell'+1) = \ell(\ell+1) - 2 \tag{15}$$

which goes over to $\ell$ as $\gamma \longrightarrow 0$. One finds wavefunctions for the resonances (which arise when $8\gamma > (2\ell+1)^2$ are given by

$$F_\ell (\ell,r) = C_{\ell'} \ \exp (i\ell r \exp(i\Theta)) \ (\ell r \exp (i\Theta))^{\ell'}$$
$$*_1F_1 ( \ell' + 1 + \ell^{-1}; \ 2\ell' + 2; \ -2i\ell r \exp (i\Theta)) \tag{16}$$

where

$$\ell = 2 \{\mp[8\gamma - (2\ell+1)^2]^{\pm} -i [2n+1]\} / \{[2n+1]^2 + [8\gamma - (2\ell+1)^2]\} \tag{17}$$

and

$$n = 0,1,2,\ldots \tag{18}$$

From EQ. (17) one observes that $\ell$ is independent of $\Theta$ and, consequently, $F_\ell(\ell,r)$ is a function of only $r \exp(i\Theta)$.

NUMERICAL CALCULATIONS

A large number of complex-coordinate calculations[3,10,11] have been reported since the Balslev-Combes theorem was published. In order to put these calculations in the context of the previous discussion it will be useful to first note the assumptions made and the techniques used in performing these calculations.

First, consider calculations[10] such as those reported by Doolen, Nuttall, Ho, etc. where the Hamiltonian is rotated, but a set of real basis functions such as Slater type orbitals, hydrogenic functions, etc. are used. In these calculations approximate resonance energies, $\tilde{E}_r$ $(\Theta)$, are obtained as a function of $\Theta$ . Although it should be independent of $\Theta$, it is found to be at best a slowly varying function of $\Theta$ over a small range of $\Theta$- values and the assumption is made that the resonance energy is that for which $\tilde{E}_r(\Theta)$ is most stable with respect to variations in $\Theta$. However, above we have shown that the resonance wavefunctions depend on r and $\Theta$ only in the form $r \exp(i\Theta)$. In a basis set expansion this implies that the exponentials should have an oscillatory component dependent on $\Theta$. Since only real functions of r are used in these calculations, this dependence is forced into the linear coefficients. The reason for slow configuration interaction convergence even for two particle systems is thus evident. If each particle requires say ten basis functions for a reasonable approximation, a two particle system would require a hundred configurations, a three particle system would require a thousand, etc. The structure of the " trajectories" (the set of $\tilde{E}_r(\Theta)$-values) can be understood in the following manner. The resonance wavefunction for a given resonance of $H(\Theta)$ depends on $\Theta$ as discussed above. On the other hand, the basis sets used are independent of $\Theta$. $\widetilde{E}_r(\Theta)$ then stabilizes at a value of $\theta$ where the theta independent basis can best represent the $\Theta$-dependent wavefunctions. That is, $\Theta$ is really nothing more than a nonlinear variational parameter. The fact that theta is really nothing more than this is easily seen by viewing these calculations in a slightly different though entirely equivalent form, i.e.,

$$0 = \langle \psi_t \mid (H(\Theta)-E) \mid \psi_t \rangle = \langle \psi_t(\exp(-i\Theta)) \mid (H(r)-E) \mid \psi_t(\exp(-i\theta)) \rangle \quad (19)$$

The initial calculations[3,12] we performed for the $(1s2s^2)$ $^2S$ He$^-$ resonance properly employed basis functions of $r$ $\exp(i\theta)$ for bound particles, while using complex (though independent of $\Theta$) functions of $r$ for describing the unbound particle. Consequently the $\Theta$-induced oscillations are properly accounted for for most of the particles. This yielded significantly more rapid convergence for this three particle system than had been obtained for any two particle system. The role played by theta in these calculations is the same as discussed above. This choice of basis functions, however, effectively reduces the problem of representing the resonance wavefunctions to a level of difficulty equivalent to a one particle problem, while retaining the full (N+1) - particle correlations.

Rescigno, et al.[11] have constructed an effective one-electron Hamiltonian for shape resonances which is in the spirit of our He$^-$ calculations in that their technique effectively incorporates the $\Theta$-induced oscillations for the bound particles. They note that one could use a large CI wavefunction for the N-electron target state. However, to properly incorporate the (N+1) -electron correlations one must determine the linear coefficients in the solution of the effective one-electron Schoedinger equation and not in the solution of the N-electron problem. If the former approach is taken, one could conceivably also compute the resonance parameters for Feshbach resonances though the calculation would of necessity be larger than that required in the above mentioned He$^-$ calculations. As discussed above $\Theta$ is again playing the role of a non-linear variational parameter.

COMPLEX STABILIZATION METHOD

Based on the above theorems and the results of previous "complex-coordinate calculations" we have suggested a computational technique we call the complex stabilization method.[13]

Consider first a variational wavefunction of the form used by Doolen[10] for the H$^-$ resonance calculations. It is a linear combination of Hyleraas functions with $\Theta$-dependent coefficients, i.e.,

$$\psi_t(\theta) = \sum_{\ell,m,n} C_{\ell mn}(\theta) \, (r_1^\ell \, r_2^m + r_1^m \, r_2^\ell) \, r_{12}^n \, \exp[-a(r_1 + r_2)] \quad (20)$$

Let $\theta_1$ be the value of theta at which $\widetilde{E}_r(\Theta)$ stabilizes. On the other hand, since the resonance wavefunctions are analytic functions of $\Theta$, a completely equivalent wavefunction at $\Theta_2$ can be obtained from $\psi_t(\theta_1)$ by the transformation

$$r \longrightarrow r \, \exp[i(\Theta_2 - \Theta_t)] \quad (21)$$

If one applies this transformation to EQ. (20) for $\Theta_2 = 0$, one obtains

$$\psi_t(0) = \sum_{\ell,m,n} C_{\ell mn}(\theta_1)[(r_1 \, \exp(-i\Theta_1))^\ell \, (r_2 \, \exp(-i\Theta_1))^m$$
$$+ (r_1 \, \exp(-i\Theta_1))^m \, (r_2 \exp(-i\Theta_1))^\ell \, ] \, (r_{12} \, \exp(-i\Theta_1))^n$$
$$* \exp[-a(r_1 + r_2)\cos(\Theta_1)] \, \exp[ia(r_1 + r_2)\sin(\Theta_1)] \quad (22)$$

Equation (22) shows that the wavefunction for H(r) which is equivalent to that of H($\ominus_1$), in that it yields the <u>exact</u> same complex eigenvalue, is square-integrable even though the exact resonance wavefunction goes asymptotically as a divergent Siegert function. That is, none of the previous "complex-coordinate calculations" incorporate the proper asymptotic form of the wavefunction when rotated back to $\ominus = 0$.

First this implies that the exact boundary condition need not be explicitly imposed on the resonance wavefunction just as is the case with bound state wavefunctions. This arises because there exists a region in the complex coordinate plane in which the resonance wave function is square-integrable just as is the case for bound states. Siegert calculations are simply special cases of such calculations in which the exact boundary condition is explicitly imposed.

Secondly, the effect, discussed above, on the exact resonance wavefunction of the cut rotating through the resonance is not manifested in any of these calculations. That is, the basis sets used in all previous calculations are such that they are square-integrable under analytical continuation for values of $\ominus$ greater than and less than $\beta$. This implies the square-integrability property of the exact resonance wavefunctions in the complex-coordinate theorems plays no explicit role in these calculations, but only an implicit one in that it implies, as just stated, that the exact boundary condition does not have to be imposed.

This background then suggests the following technique for computing the complex resonance eigenvalues directly with the unrotated Hamiltonian with a square-integrable basis and without explicitly imposing any boundary condition. Basically one computes nonlinear parameter trajectories instead of theta trajectories using the real Hamiltonian and a square-integrable basis. $E_r$ is then stabilized with respect to variations (including complex, as well as real, scaling) of the various nonlinear parameters in the basis set. This stabilization can be carried out with each parameter individually, with groups of parameters, or with all parameters simultaneously. Calculations such as those of Doolen[10] correspond to a global complex scaling of all parameters simultaneously, while those of Junker[3,13] and Rescigno, et. al.[11] correspond to a complex scaling of a certain group of nonlinear parameters.

In addition to the fact that the basis set should be square-integrable, there must not exist a unitary transformation which converts it to a real basis. If such existed all eigenvalues would be real. While this may at first appear rather arbitrary, one should recall that unlike wavefunctions for scattering and bound states, those for the resonances are innately complex. Suggestions for structuring trial wavefunctions for various types of phenomena are discussed elsewhere[4].

Several recent numerical calculations lend additional support for this interpretation. Donnelly and Simmons[14] have recently reported a calculation of the position and width of the $(1s)^2 (2s)^2 (2p)^2 {}^2P^o$ Be$^-$ resonance. Their complex eigenvalue, $\overline{E}_r(\ominus)$, stabilized at a value of $\ominus$ <u>less than</u>[15] $\frac{1}{2} | Arg(E_r) |$. More recently McCurdy, et. al.[16] reported complex self-consistent field calculations for the $^2D$ resonance in Ca$^-$. Using the real Hamiltonian and real square-integrable basis functions, they were able to converge to a <u>complex</u> energy.

Since we have stated that previous "complex coordinate calculations" are just special cases of a more general complex stabilization method, one might ask what is to be gained from the latter method. If $\ominus$ is just a nonlinear scale factor, negative as well as positive values of theta are meaningful as opposed to just $\beta < \ominus < \pi/4$. Using H(r) chemical and physical insight can be incorporated straightforwardly into the construction of the variational wavefunction, i.e., the independent particle model is restored. Also one can use different $\Theta$'s for different contributions, e.g. different $\Theta$'s for the different (s,p,d, etc.) polarization contributions. Another situation in which more than one $\Theta$-value is useful is when there are several thresholds below a given resonance so that one must describe several continua. The value of $\Theta$ which stabilizes $\overline{E}_r$ with respect to functions describing one continuum is generally quite different from that which describes another continuum. Also molecules present no conceptual problems. That is, while the total molecular Hamiltonian including nuclear coordinates is dilation analytic, the Born-Oppenheimer Hamiltonian is not. Although there is no innate problems with using complex nuclear coordinates, the approximate physical picture offered by the Born-Oppenheimer Hamiltonian with real nuclear coordinates is appealing. Since the Hamiltonian is not scaled in the complex stabilization method, this problem is trivally averted. Finally, by using H(r) instead of H($\Theta$), one avoids the difficulty of describing the unnecessary kinematic $\Theta$-induced oscillations in the wavefunctions for bound and resonance states as well as for the bound particles in scattering states. As discussed above, these can lead to requiring very large CI wavefunctions. The complex SCF calculations of McCurdy, et al.[16] avoid the need of large CI wavefunctions by using basis functions which are linear combinations of orbitals from large basis sets. This substitutes integral transformations for large CI calculations. If instead they were to employ H(r) with a basis set of real and complex functions, they could probably use much smaller basis sets and thus save considerable effort in integral transformations. In the next sections we will give the results for the application of this method to several problems.

MODEL POTENTIAL

The model potential[17]

$$V(r) = 7.5r^2 \exp(-r) \tag{23}$$

has been used in a number of studies of resonances. From direct numerical integration the position and width ($\Gamma = 2|E_I|$) are known to be 3.42639a.u. and 0.025549 a.u., respectively. This implies $E_R - iE_I = 3.42639 - 0.012775i$a.u. We have used a variety of trial wavefunctions to study basis set and convergence problems in resonance calculations. The wavefunctions used had the forms

$$\Psi_1 = \sum_{i=1}^{M} C_i [r \exp(-i\Theta)]^{i-1} \exp[-\gamma r \exp(-i\Theta)] \tag{24a}$$

$$\psi_2 = \sum_{i=1}^{M} C_i [r \exp(-i\Theta)]^{i-1} \exp[-\gamma r \exp(-i\Theta)]$$
$$+ C_{M+1} r^{-1} [1-\exp(-r)] \exp(ik_R r) \exp(k_I r) \tag{24b}$$

$$\Psi_3 = \sum_{i=1}^{M} C_i [r \exp(-i\Theta)]^{i-1} \exp[-\gamma r \exp(i\Theta)]$$
$$+ C_{M+1} \exp(ik_R r) \exp(-k_I r) \tag{24c}$$

$$\Psi_4 = \sum_{i=1}^{L} C_i r^{n_i} \exp(-\gamma_i r)$$
$$+ \exp(ikr) [C_{L+1} \exp(-\alpha r) + C_{L+2} \exp(-(\alpha+\epsilon)r)] \tag{24d}$$

The Hamiltonian used in all of these calculations was H(r). The first function is equivalent to a Doglen basis calculation. The second is a Siegert calculation ($E_R - iE_I = \frac{1}{2}(k_R^2 - k_I^2 - ik_R k_I)$). The third function is similar to the second except that the last basis function is square-integrable and the exponent of r is 0 instead of -1. These changes eliminate the computational difficulty associated with Siegert functions as well as producing a basis function which no longer satisfies a Siegert boundary condition. For all three of these trial functions values of ten and fourteen were taken for M and $\overline{E}_r$ was stabilized with respect to variations in $\gamma$. The calculations in each case were performed at three values of theta - 0.05, 0.1, and 0.5. The optimized value for $\gamma$ and the complex energy for each case is given in Table I.

Table I:   Resonance Parameters for    (M),    (M), and    (M).

|          | 0.05 | | | 0.1 | | | 0.5 | | |
|----------|------|--------|--------|-----|--------|--------|-----|--------|--------|
|          |      | $E_R$ | $-E_I$ |     | $E_R$ | $-E_I$ |     | $E_R$ | $-E_I$ |
| $\Psi_1(10)$ | 2.8 | 3.43046 | .00688 | 2.8 | 3.42842 | .01063 | 4.0 | 3.42640 | .01278 |
| $\Psi_2(10)$ | 1.6 | 3.42652 | .01284 | 2.8 | 3.42649 | .01285 | 5.6 | 3.42639 | .01278 |
| $\Psi_3(10)$ | 1.6 | 3.42648 | .01294 | 1.6 | 3.42645 | .01290 | 5.4 | 3.42639 | .01278 |
| $\Psi_4(14)$ | 4.0 | 3.42672 | .00711 | 4.0 | 3.42662 | .01090 | 4.4 | 3.42639 | .01277 |
| $\Psi_2(14)$ | 2.6 | 3.42639 | .01276 | 2.6 | 3.42639 | .01277 | 5.2 | 3.42639 | .01277 |
| $\Psi_3(14)$ | 2.6 | 3.42640 | .01277 | 2.0 | 3.42639 | .01277 | 5.2 | 3.42639 | .01277 |

Although $\Psi_2$ does have improved convergence properties relative to $\Psi_1$, it is no better than $\Psi_3$. This is another vivid example of how unimportant the exact long range behavior of the wavefunction is. The smaller values for $\gamma$ for the smaller angles are simply a reflection of the need for the basis functions to extend outward a sufficient distance which depends on $\gamma \cos \theta$.

For a trial function of the form $\Psi_4$ with L=10, $E_r$ was stabilized with respect to each nonlinear parameter separately, except that $\alpha$ and $(\alpha + \epsilon)$ where varied together. The $\gamma_i$ with the largest structure projection, $P_i$,

$$P_i = \langle \chi_i | \Psi^r \rangle = \sum_j C_j \langle \chi_i | \chi_j \rangle, \qquad (25)$$

was varied first, followed by the $\gamma_i$ with the second largest structure projection, etc. This stabilizes the stabilization procedure. Two nonlinear parameter trajectories are shown in Tables II and III. $\Delta E_R$ and $\Delta E_I$ are the incremental changes in $E_R$ and $E_I$, respectively, for successive values of the nonlinear parameter.

TABLE II:  Nonlinear Parameter Trajectory For Term With Largest Structure Protection.

| $\gamma$ | $\Delta E_R$* 3.42640a.u. | $\Delta E_I$* -0.012753a.u. |
|------|------|------|
| 2.56 | 1.92 | 0.62 |
| 2.72 | -4.69 | 0.15 |
| 2.88 | 2.32 | -0.10 |
| 3.04 | 0.05 | 0.20 |
| 3.20 | -0.02 | -0.06 |
| 3.36 | 0.03 | -0.14 |
| 3.52 | 0.09 | -0.17 |
| 3.68 | 0.13 | -0.19 |
| 3.84 | | |

TABLE III:  Nonlinear Parameter Trajectory For k.

| k | $\Delta E_R$* 3.42640a.u. | $\Delta E_I$* -0.012757a.u. |
|-------|------|------|
| 1.75 | -1.20 | 0.83 |
| 1.785 | -0.89 | 0.60 |
| 1.82 | -0.60 | 0.38 |
| 1.855 | -0.32 | 0.18 |
| 1.89 | -0.05 | 0.00 |
| 1.925 | -0.22 | -0.16 |
| 1.96 | -0.51 | -0.29 |
| 1.995 | -0.82 | -0.40 |
| 2.03 | | |

* The numbers under $\Delta E_R$ and $\Delta E_I$ correspond to $E_R - iE_I$ for $\gamma$ = 3.36. The units of $\Delta E^R$ and $\Delta E_I$ are $10^{-4}$a.u.

* The numbers under $\Delta E_R$ and $\Delta E_I$ correspond to $E_R - iE_I$ for k=1.925. The units of $\Delta E_R$ and $\Delta E_I$ are $10^{-4}$a.u.

These calculations along with other studies of basis set properties will be discussed elsewhere[4].

2p° BERYLIUM MINUS SHAPE RESONANCE

A number of calculations have been reported for this resonance, [11,14,18,19] although no experimental results are available. We have used several trial wavefunctions of the following form

$$\Psi = \mathcal{A}\{\sum_{i=1}^{L} c_i \; \varphi\,(1s^2 2s^2)\; \chi P_i[r\; \exp(i\theta_1)]$$
$$+ \sum_{j=1}^{M} P\,[c_j\; \varphi\,(1s^2\; 2s'\; 2p;\; ^3P)\;\; \chi_{cj}^{S}\,[r\; \exp(i\theta_2)]$$
$$+ c_{M+j}\; \varphi(1s^2 2s'\; 2p;\; ^1P)\;\; \chi_{cj}^{S}\,[r\; \exp(i\theta_2)]]$$
$$+ \sum_{k=1}^{N} P\,[c_k\; \varphi\,(1s^2\; 2s'\; 3d;\; ^3D)\;\; \chi_k^{'P}[r\; \exp(i\theta_2)]$$
$$+ c_{N+k}\; \varphi(1s^2\; 2s'\; 3d;\; ^1D)\;\; \chi_k^{'P}[r\; \exp(i\theta_2)]] \tag{26}$$

where $\mathcal{A}$ is the antisymmetrizer and P symbolically denotes proper angular momentum and spin coupling. Calculations have been performed using the first term, the first two terms, and all three terms. In addition L, M, and N have been varied to insure convergence. Finally calculations were also performed in which $E_r$ was stabilized with respect to variations in the 2s nonlinear parameter. While the details will be discussed elsewhere, Table IV summarizes the results of these calculations and compares them where possible to other calculations.

TABLE IV: Position and Width of $^2P^o$ Be Resonance from Various Trial Wavefunctions

| Wavefunction | Position (ev) | Width (ev) |
|---|---|---|
| $(2s_A)^2\,kp$ | 0.69 | 1.19 |
| $(2s_A 2s_a\,)\,kp$ | 0.69 | 1.19 |
| $(2s_{ASCF})^2\,kp$ | 0.76 | 1.11 |
| Harris Method with $V_{HF}$ | 0.77 | 1.63 |
| $(2s_I)^2\,kp$ | 0.63 | 0.73 |
| $(2s_{ISCF})^2\,kp$ | 0.70 | 0.51 |
| Closed shell 2nd order MBPT | 0.57 | 0.99 |
| $(2s_A)^2\,kp + (2s2p)\,ks$ | 0.37 | 0.52 |
| $(2s_A)^2\,kp + (2s2p)\,ks + (2s3d)\,kp'$ | 0.19 | 0.31 |
| Harris Method with $V_{HF} + V_{pol}$ | 0.20 | 0.28 |
| $(2s_I)^2\,kp + (2s2p)\,ks$ | 0.44 | 0.42 |
| $(2s_I)^2\,kp + (2s2p)\,ks + (2s3d)\,kp'$ | 0.27 | 0.22 |

Note that in Table IV the $2s_A$ corresponds to optimization of the 2s nonlinear parameter in a neutral Be atom calculation, while $2s_I$ implies that $E_r$ was stabilized with respect to variations in the 2s orbital parameter. Note that this is an example were two distinct values for $\theta$ are useful.

STARK EFFECT

In the presence of an electric field, all atomic states become unstable with respect to ionization. Due to this additional decay channel, all become wider and more diffuse as the field strength is increased.

Damburg and Kolosov[20] have performed a separation of variables for the Schroedinger equation for the Stark Hamiltonian

$$H = H_0 + Fr\cos(\theta) \tag{27}$$

in terms of the coordinates

$$X = \mu\nu\;\cos(\phi) \tag{28a}$$
$$Y = \mu\nu\;\sin(\phi) \tag{28b}$$
$$Z = (1/2)\,(\mu^2 - \nu^2) \tag{28c}$$

In this case the $\nu$-function is the one which diverges asymptotically.

$$N(\nu) \sim (B/\nu)\, \exp[E_I\nu/F^{\frac{1}{2}}]\, \exp\left[i\,(F^{\frac{1}{2}}\nu^3/3 + E_R\nu/F + \lambda)\right]$$

$$= (B\,\exp(i\lambda)/\chi)\,\exp[E_I\chi/F^{\frac{1}{2}}]$$

$$*\exp\left[i\chi\,(F^{\frac{1}{2}}(r - r\cos(\Theta))/3 + E_R/F^{\frac{1}{2}})\right] \tag{29}$$

$$\chi = (-2Z)^{\frac{1}{2}}\left\{1 + \left[1 + \frac{x^2 + y^2}{(-Z)^2}\right]^{\frac{1}{2}}\right\}^{\frac{1}{2}} \tag{30}$$

Note that if a Siegert boundary condition were explicitly imposed, integrals would result which contain $r^{\frac{1}{2}}$, $r^{\frac{3}{2}}$, $\cos^{\frac{1}{2}}(\Theta)$, and $\cos^{\frac{3}{2}}(\Theta)$ in exponentials. Consequently avoidance of the explicit boundary condition is a considerable computational advantage.

Reinhart[21] first applied the "complex-coordinate method" to the Stark effect in hydrogen. Later Reinhart and Wendoloski performed similar calculations on the $^1P^o$ resonance of H$^-$. In both cases a Doolen type basis was used and for sufficiently large wavefunctions excellent results were obtained.

We have used the unrotated Hamiltonian and the following trial function

$$\Psi = \sum_{\ell=m_\ell}^{L} Y_\ell^{m_\ell}(\theta,\varphi)\left\{\sum_{i=1}^{M_L} c_i^\ell\, r^{(n_i - 1)}\exp(-\gamma_i^\ell r) + r^\ell\,\exp(i\ell_\ell r)(d_1^\ell\exp(-\alpha_\ell r) + d_2^\ell\exp(-(\alpha_\ell + \epsilon_\ell)r))\right\} \tag{31}$$

to compute the complex pole associated with field ionizing the ground state of hydrogen in an electric field of 0.08a.u. Two wavefunctions were used - one with L=3 and one with L=6. The details will be reported elsewhere. Here we note the $\gamma_i^\ell$'s for a given $\ell$ were kept equal and $\widehat{E}_r$ was stabilized with respect to each group of $\gamma_i^\ell$'s. Additional stabilization with respect to the $\ell_\ell$'s, $\alpha_\ell$'s, and $\epsilon_\ell$'s was also performed. The nonlinear parameters from the smaller calculation were kept constant in the larger wavefunction. No attempts were made to reduce the size of the CI wavefunction. The results are given in Table V and compared with the calculations of Damburg and Kolosov[20] and Hehenberger.[23]

TABLE V:   Field Ionization of Hydrogen Ground State

|                          | $E_R$(a.u.) | $\Gamma$(a.u.) |
|--------------------------|-------------|----------------|
| (L = 3)                  | -0.517598   | 0.004672       |
| (L = 6)                  | -0.517561   | 0.004540       |
| Hehenberger, et. al.     | -0.51756    | 0.00454        |
| Damburg and Kolosov      | -0.517495   | 0.004511       |

## DISCUSSION

We have discussed and illustrated with analytical examples the analytical properties of the wavefunctions of H($\Theta$) with respect to $\Theta$. This has enabled us to elucidate the role of theta in previous calculations. That is, theta is simply acting as a nonlinear variational parameter. As a result we have suggested a procedure (complex stabilization method) for computing the complex poles of the resolvent using the unrotated Hamiltonian with a square-integrable basis and without explicitly imposing any boundary condition or long range behavior on the trial wavefunction. This procedure has been illustrated with three examples.

## REFERENCES

1.   Balslev, E. B. and Combes, J. M., Spectral Properties of Many-Body Schroedinger Operators with Dilatation-Analytic Interactions, Commun. Math. Phys. 22 (1971) 280-294.
2.   Simon, B., Quadratic Form Techniques and the Balslev-Combes Theorem, Commun. Math. Phys. 27 (1972) 1-9 and Resonances in n-Body Quantum Systems with Dilatation Analytic Potentials and the Foundations of Time Dependent Perturbation Theory, Ann. Math. 97 (1973) 247-274.

3. Junker, B. R. and Huang, C. L., Computation of Resonances in Atoms Using the Complex Coordinate Method, in: Barat, M. and Reinhardt, J. (eds.), Abstracts of the Tenth International Conference on the Physics of Electronic and Atomic Collisions (Commisariat A L'Entergie Atomique, Paris, 1977) and Complex-Coordinate Method. Structure of the Wavefunction, Phys. Rev. A18 (1978) 313-323.

4. Junker, B. R., Complex Stabilization Method: General Theory for Application to Resonant Phenomena, in: Temkin, A. (ed.), Autoionization II: Developments in Theory, Calculation, and Application to Diagnostics of Solar and Other Plasmas. (Plenum, New York, to be published).

5. Junker, B. R. Complex Coordinate Calculations of Resonances in N-Electron Atoms, Int. J. of Quan. Chem. XIV (1978) 371-382.

6. Newton, R. G., Scattering Theory of Waves and Particles (McGraw-Hill, New York, 1966).

7. Nicolaides, C. A. and Beck, D. R., Time Dependence, Complex Scaling, and the Calculation of Resonances in Many-Electron Systems, Int. J. of Quan. Chem. XIV (1978) 457-513.

8. terHaar, D., Selected Problems in Quantum Mechanics (Academic Press, New York, 1964), p. 360.

9. Doolen, G. D., An Analytic Model and Some New Results for $e^{+}$-H Resonances, Int. J. of Quan. Chem. XIV (1978) 523-528.

10. For example, see Doolen, G. D., A procedure for Calculating Resonance Eigenvalues, J. Phys. B8 (1975) 525-528; Doolen, G. D., Nuttall, J., and Stagat, R. W., Electron-Hydrogen Resonance Calculation by the Coordinate-Rotation Method, Phys. Rev. A10 (1974) 1612-1615; and Ho, Y. K., A Resonant State and the Ground State of Positronium Hydride, Phys. Rev. A17 (1978) 1675-1678.

11. Rescigno, T. N., McCurdy, Jr., C. W. and Orel, A. E., Extensions of the Complex-Coordinate Method to the Study of Resonances in Many-Electron Systems, Phys. Rev. A17 (1978) 1931-1938.

12. Junker, B. R., Complex-Coordinate Method. II. Resonance Calculations with Correlated Target-State Wave Functions, Phys. Rev. A18 (1978) 2437-2442.

13. Junker, B. R., Complex Stabilization Method for Resonant Phenomena, Phys. Rev. Lett. 44 (1980) 1487-1490.

14. Donnely, R. A. and Simmons, J., Complex Coordinate Rotation of the Electron Propagator, J. Chem. Phys. 73 (1980) 2858-2866.

15. This was first noted by J. N. Bardsley.

16. McCurdy, C. W., Lauderdale, J. G., and Mowrey, R. C., Complex Self-Consistent-Field Calculations on Shape Resonances in Electron-Mg and Electron-Ca Scattering, to Appear in J. Chem Phys. (1981).

17. Bain, R. A., Bardsley, J. N., Junker, B. R., and Sukumar, C. V., Complex-Coordinate Studies of Resonant Electron-Atom Scattering, J. Phys. B: Atom. Molec. Phys. 7 (974) 2189-2202.

18. Kurtz, H. A. and Ohrn, Y., Calculation of $_2$P Shape Resonances in Be and Mg, Phys. Rev. A19 (1979) 43-48.

19. McCurdy, C. W., Jr., Rescigno, T. N., Davidson, E. R., and Lauderdale, J. G., Applicability of Self-Consistent Field Techniques Based on the Complex Coordinate Method to Metastable Electronic States, J. Chem. Phys. 73 (1980) 3268-3273.

20. Damburg, R. J. and Kolosov, V. V., A Hydrogen Atom in a Uniform Electric Field, J. Phys. B: Atom. Molec. Phys. 9 (1976) 3149-3157.

21. Reinhardt, W. P., Method of Complex Coordinates: Application to the Stark Effect in Hydrogen, Int. J. Quan. Chem. S10 (1976) 359-367.

22. Wendoloski, J. J. and Reinhardt, W. P., Effects of an External Electric Field on $^1p^0$ Resonances of H$^-$, Phys. Rev. A17 (1978) 195-200.

23. Hehenburger, M., McIntosh, H. V., and Brandas, E., Weyl's Theory Applied to the Stark Effect in The Hydrogen Atom, Phys. Rev. A10 (1974) 1494-1506.

PHYSICS OF ELECTRONIC AND ATOMIC COLLISIONS
S. Datz (editor)
© North-Holland Publishing Company, 1982

THE VARIATIONAL APPROACH TO ATOMIC COLLISIONS
AND ITS RELATION TO PERTURBATION THEORY

M. Kleber

Physik-Department
Technische Universität München
D-8046 Garching
West Germany

Unlike perturbation theory, the variational method does not
require a small expansion parameter. In spite of their
different nature the two methods can be related to and combined with each other. Some recent theoretical investigations
are reported. Numerical results are given for selected topics
in heavy particle collisions.

INTRODUCTION

This contribution aims to point out the use of variational methods in ion-atom
collisions. Special attention will be paid to the formation of collision-induced
molecules. This problem requires the use of rather flexible electron wave
functions which, as a function of the internuclear distance, should be able to
interpolate smoothly between the separated atom and the united atom configuration.
One possibility to meet this requirement is to parametrize the electron wave
function through time-dependent variational parameters. The use of variational
methods for heavy particle impact is, of course, not new [1-3]. A new aspect is
brought in [4-8] if the variational parameters are allowed to become complex because then electron excitation is automatically accounted for.

The reason for introducing complex parameters is quite simple: In classical mechanics the evolution in time of a coordinate $q(t)$ is known to follow from a differential equation which is of second order in time. This single equation is equivalent to two coupled first-order equations for the coordinate $q(t)$ and the
momentum $p(t)$. Hence, one could likewise introduce a complex variable
$\lambda(t) = q(t) + ip(t)$, but this would not simplify the problem in most cases. The
time-dependent Schrödinger equation, however, is a complex differential equation
of first order in time. One is therefore lead to introduce complex variational
parameters $\lambda_i(t)$ in the wave function.

The possibly collective variational parameters must be chosen on physical grounds.
For the variational space fixed, it is then straightforward to calculate the time
evolution of the parameters, to estimate the error and to determine transition
probabilities.

VARIATIONAL EQUATIONS

The time-dependent Schrödinger equation

$$(H - i\hbar\partial_t)|\psi\rangle = 0 \qquad (2.1)$$

can be cast into an eigenvalue problem by separating from the wave function $\langle\vec{r}|\psi\rangle$
a purely time-dependent factor $N(t)$:

$$|\psi\rangle = N(t)|\phi\rangle . \qquad (2.2)$$

We put (2.2) in (2.1) with the result that

$$N(t) = N(t_0) \exp\left(\frac{1}{i\hbar} \int_{t_0}^{t} dt' \frac{\langle\phi|H - i\hbar\partial_t|\phi\rangle}{\langle\phi|\phi\rangle}\right) \qquad (2.3)$$

and

$$(H - i\hbar\partial_t)|\phi\rangle = \left[\frac{\langle\phi|H - i\hbar\partial_t|\phi\rangle}{\langle\phi|\phi\rangle}\right] |\phi\rangle . \qquad (2.4)$$

In contrast to (2.1), the eigenvalue equation (2.4) is automatically invariant under the following change of the Hamiltonian

$$H \rightarrow H + F(t) , \qquad (2.5)$$

with $F(t)$ an arbitrary, real time-dependent function. For gauge transformations of type (2.5) the use of (2.3) and (2.4) guarantees the correct transformation properties for all wave functions including approximate ones.

Let us now parametrize the state $|\phi\rangle$ by a set of N complex time-dependent parameters $\underline{\lambda} = (\lambda_1, \lambda_2, \ldots \lambda_N)$

$$|\phi\rangle = (\langle\underline{\lambda}|\underline{\lambda}\rangle)^{-1/2} |\underline{\lambda}\rangle . \qquad (2.6)$$

With (2.4) we readily find the eigenvalue equation for $|\underline{\lambda}\rangle$

$$(H - i\hbar\partial_t) |\underline{\lambda}\rangle = \left[\frac{\langle\underline{\lambda}|H - i\hbar\partial_t|\underline{\lambda}\rangle}{\langle\underline{\lambda}|\underline{\lambda}\rangle}\right] |\underline{\lambda}\rangle \qquad (2.7)$$

which is equivalent to the variational principle

$$\delta I = \delta \int_{t_i}^{t_f} dt \frac{\langle\underline{\lambda}|H - i\hbar\partial_t|\underline{\lambda}\rangle}{\langle\underline{\lambda}|\underline{\lambda}\rangle} = 0 . \qquad (2.8)$$

Since $|\underline{\lambda}\rangle$ is supposed to depend on time only through the parameters $\lambda_i(t)$, we are allowed to write

$$\partial_t = \sum_{i=1}^{N} \dot{\lambda}_i \partial_{\lambda_i} = \dot{\underline{\lambda}} \, \underline{\nabla}_\lambda . \qquad (2.9)$$

(For a more general time dependence see Refs.[5,8]). The integrand in (2.8) is apart from a minus sign the Lagrangian $L(t)$ of the Schrödinger equation:

$$L(\underline{\lambda}^*, \underline{\lambda}, \dot{\underline{\lambda}}) = -\frac{\langle\underline{\lambda}|H - i\hbar\partial_t|\underline{\lambda}\rangle}{\langle\underline{\lambda}|\underline{\lambda}\rangle}$$

$$\qquad (2.10)$$

$$= -\langle H\rangle + i\hbar \, \dot{\underline{\lambda}} \, \underline{\nabla}_\lambda \ln \langle\underline{\lambda}|\underline{\lambda}\rangle .$$

In (2.10) we have used an obvious notation for the expectation value of the Hamiltonian. From the variational principle (2.8) we obtain as usual N Euler-Lagrange equations

$$\partial_{\lambda_i^*} L(\underline{\lambda}^*, \underline{\lambda}, \dot{\underline{\lambda}}) = \frac{d}{dt} \partial_{\dot{\lambda}_i^*} L(\underline{\lambda}^*, \underline{\lambda}, \dot{\underline{\lambda}}), \qquad (2.11)$$

which determine the time evolution of the variational parameters $\lambda_i(t)$. In our case the right-hand side of (2.11) is zero. We now put (2.10) in (2.11) with the result that

$$\partial_{\lambda_i^*}[\langle H\rangle - i\hbar \, \dot{\underline{\lambda}} \, \underline{\nabla}_\lambda \ln \langle\underline{\lambda}|\underline{\lambda}\rangle] = 0. \qquad (2.12)$$

In deriving (2.12) it is understood that the functions $\lambda_i(t)$ take on prescribed values at $t_i$ and $t_f$, but may be arbitrarily varied in between. In other words, we have omitted possible contributions from the boundaries at $t_i$ and $t_f$. Neglecting them is certainly correct if the number of variational parameters is sufficiently large so that $|\lambda\rangle$ will remain the exact solution for all times. In practice, however, we shall use only a small number of parameters. But even then there will be no contributions from the boundaries. This can be seen from looking at the error vector

$$|\varepsilon\rangle = (H-i\hbar\partial_t)\ |\psi\rangle\ . \tag{2.13}$$

Minimization of the corresponding norm

$$\partial_{\lambda_i^*} <\varepsilon|\varepsilon> = 0 \tag{2.14}$$

will again yield [9] the equations of motion (2.12). Note that the error $|\varepsilon\rangle$ is orthogonal to $|\psi\rangle$.

For time-independent problems the equations of motion (2.12) become the Rayleigh-Ritz principle

$$\partial_{\lambda_i^*} <H> = 0. \tag{2.15}$$

Then, assuming H to be real, the wave function will be a real function of the variational parameters $\lambda_i$. For this case we find from (2.15) that $\lambda_i = \lambda_i^*$ as it should be.

Time-dependent problems arise when H depends on time, or when the system is not in an eigenstate. As will be pointed out in the next section, both situations are realised in a heavy-ion collision.


EXAMPLE AND CLASSICAL CORRESPONDENCE

As an illustrative example we consider [6] monopole excitation of a target electron during the collision between a projectile nucleus and a target atom (Fig.1).

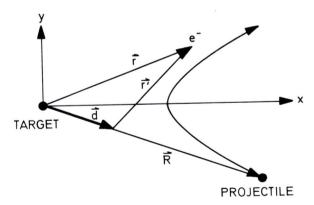

Figure 1
Coordinates used for the collision between a projectile
nucleus (charge $Z_p$) and a target atom (nuclear charge $Z_t$).
The electron is denoted by e⁻.

As an approximation, let us ignore all electrons except the K electron under consideration. Then, in the monopole approximation this K electron will feel an effective nuclear charge Z. In slow collisions it will vary between $Z = Z_p$, the value of the separated configuration ($R \to \infty$), and $Z = Z_p + Z_t$, the united atom limit $R \to 0$. The corresponding trial wave function becomes

$$\langle \vec{r} | \phi \rangle = \pi^{-1/2} \left(\frac{\lambda + \lambda^*}{2}\right)^{3/2} \exp(-\lambda r). \tag{3.1}$$

The real part $\lambda_1$ of the variational parameter $\lambda$ is the inverse of the K-shell Bohr radius

$$\lambda_1 = \langle \phi | \frac{1}{r} | \phi \rangle, \tag{3.2}$$

whereas the imaginary part $\lambda_2$ is proportional to the average radial velocity

$$\langle v_r \rangle = \frac{\hbar}{2mi} \left(\langle \phi | \frac{\partial \phi}{\partial r} \rangle - \langle \frac{\partial \phi}{\partial r} | \phi \rangle\right) = - \frac{\hbar \lambda_2}{m} \tag{3.3}$$

with m the electron mass. From (3.1) and the one-electron Hamiltonian

$$H = - \frac{\hbar^2}{2m} \vec{\nabla}^2 - \frac{Z_t e^2}{r} - \frac{Z_p e^2}{|\vec{r} - \vec{R}|} \tag{3.4}$$

we calculate the energy of the electron and then, by utilizing (2.12), the equations of motion for $\lambda_1$ and $\lambda_2$:

$$m \dot{\lambda}_1 = \frac{2}{3} \hbar \lambda_1^2 \lambda_2$$
$$m \dot{\lambda}_2 = \frac{2}{3} \hbar \lambda_1^2 (Z(t)/a_0 - \lambda_1) \tag{3.5}$$

with $Z(t) = Z_t + Z_p(1 + 2\lambda_1 R(t)) \exp(-2\lambda_1 R(t))$ and $a_0$ the hydrogen Bohr radius.
The dependence on time enters through $R(t)$ which is the time-dependent internuclear distance.

In strictly adiabatic collisions one has $\lambda_1 = Z(t)/a_0$ and $\lambda_2 = 0$. This means that the electron follows a path of minimum energy. In realistic collisions the electron cannot completely follow the changes of $Z(t)$. This leads to a non-vanishing radial acceleration $\lambda_2$ and, therefore, to a finite radial velocity. Hence, if one solves (3.5) for a given internuclear trajectory $R(t)$ and the initial conditions $\lambda_1(t \to -\infty) = Z_t/a_0$ and $\lambda_2(t \to -\infty) = 0$, one finds the electron in vibration around the equilibrium radius $a_0/Z_t$ when the collision is over. This 'breathing' signals monopole excitation.

It is useful to point out the close analogy to classical mechanics. To exhibit the classical correspondence we compare the phase in (2.3)

$$\int_{t_0}^{t} dt' \frac{\langle \phi | H - i\hbar \partial_{t'} | \phi \rangle}{\langle \phi | \phi \rangle} = \int_{t_0}^{t} dt' [\langle H \rangle + \mathrm{Im}(\hbar \underline{\dot{\lambda}} \cdot \nabla_\lambda \ln \langle \underline{\lambda} | \underline{\lambda} \rangle)] \tag{3.6}$$

with the classical action

$$S = S_0 - \int_{t_0}^{t} dt' [H + \sum_{i=1}^{N} \dot{p}_i q_i]. \tag{3.7}$$

From the last two equations it is obvious how one can introduce conjugate variables $p_i$ and $q_i$. For the example given above, a possible choice for the momentum is

$$\dot{p} = - \mathrm{Im}(\hbar \dot{\lambda}) = - \hbar \dot{\lambda}_2. \tag{3.8}$$

The corresponding coordinate then becomes

$$q = \frac{\partial}{\partial \lambda} \ln \langle \lambda | \lambda \rangle = \frac{3}{2\lambda_1} \quad . \tag{3.9}$$

By use of (2.12) we recover Hamilton's equation of motion

$$\dot{p} = - \frac{\partial \langle H \rangle}{\partial q} \quad ; \qquad \dot{q} = \frac{\partial \langle H \rangle}{\partial p} \quad . \tag{3.10}$$

Let us illustrate monopole breathing of the K electron in hydrogen. In Fig.2 we plot trajectories of p and q in phase space as a function of energy $E = \langle H \rangle$ [6].

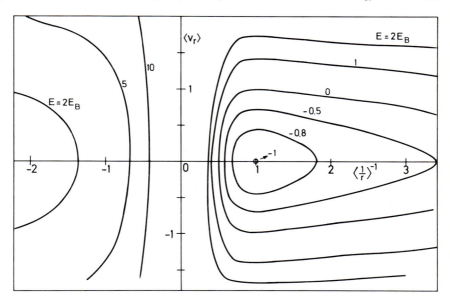

Figure 2
Phase space flow for monopole breathing at different energies
E (in units of K-shell binding $E_B = 1$). All quantities in
atomic units

For the ground state one has $E_0 = - 1$ (atomic units) and p = 0. Consider now a non-vanishing energy transfer $\Delta E$ to the electron due to a collision. If after the collision $E = E_0 + \Delta E < 0$, the electron motion is finite, then the trajectories of p and q are closed. For $E \geq 0$ the electron motion is no longer finite. As a result the trajectories are open. However, because of Hamilton's equation the flow in phase space is incompressible. Hence, those trajectories which extend to plus infinity must come back from minus infinity.


TRANSITION AMPLITUDES AND RELATION TO PERTURBATION THEORY

In order to calculate transition amplitudes and to control the accuracy of a variational wave function of type (2.6) it is convenient to establish the relation
between perturbation theory and variational method. For this purpose we follow
Demkov's investigation [1] of the non-diagonal action

$$I(\chi_f, \chi_i) = \int_{t_i}^{t_f} dt \, \langle \chi_f | H - i\hbar \partial_t | \chi_i \rangle \quad . \tag{4.1}$$

Here $|\chi_i\rangle$ and $|\chi_f\rangle$ are normalized trial states which correlate asymptotically to the unperturbed initial or final states of the target electron (s), for example.

Suppose that $|\chi_i\rangle$ and $|\chi_f\rangle$ are close to the true solutions $|\psi_i\rangle$ and $|\psi_f\rangle$; then we may write

$$|\chi_i(t)\rangle = |\psi_i(t)\rangle + |n_i(t)\rangle$$

$$|\chi_f(t)\rangle = |\psi_f(t)\rangle + |n_f(t)\rangle$$
(4.2)

with $\langle n|n\rangle$ a negligibly small quantity.

Furthermore, we assume $|\chi_i(t)\rangle$ exact prior to the collision and $|\chi_f(t)\rangle$ exact after the collision:

$$|n_i(t_i)\rangle = 0 ; \qquad |n_f(t_f)\rangle = 0.$$
(4.3)

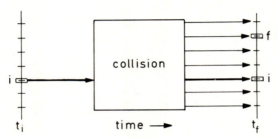

Figure 3
Collision-induced spreading of an initial state i
into various final states f

Now the collision takes place between $t=t_i$ and $t=t_f$ (see Fig.3) and, since we are unable to treat the collision correctly, we end up with non-vanishing errors

$$|n_i(t_f)\rangle \neq 0; \qquad |n_f(t_i)\rangle \neq 0.$$
(4.4)

Substituting from (4.2) into (4.1) we get

$$I(\chi_f,\chi_i) = \int_{t_i}^{t_f} dt \ \langle\psi_f + n_f|H - i\hbar\partial_t|\psi_i + n_i\rangle$$

$$= -i\hbar\langle\chi_f(t_f)|n_i(t_f)\rangle + \int_{t_i}^{t_f} dt \ \langle(H - i\hbar\partial_t) \ n_f|n_i\rangle .$$
(4.5)

The last equation is obtained by a partial integration where the boundary condition (4.3) is taken into account. We discard the second term in (4.5) because it is of the order $\langle n|n\rangle$. The first term is related to the exact transition amplitude $T_{if}$ in a simple way:

$$T_{if} = \langle\psi_f(t_f)|\psi_i(t_f)\rangle = \langle\chi_f(t_f)|\chi_i(t_f)\rangle - \langle\chi_f(t_f)|n_i(t_f)\rangle .$$
(4.6)

Combining (4.1), (4.5) and (4.6) we arrive at Demkov's result for the transition amplitude

$$T_{if}^{var} = \langle\chi_f(t_f)|\chi_i(t_f)\rangle + \frac{1}{i\hbar} \int_{t_i}^{t_f} dt \ \langle\chi_f H - i\hbar\partial_t|\chi_i\rangle$$
(4.7)

which is correct up to <u>second</u> order in the error of the wave function. Therefore, the transition amplitude is <u>stationary</u> with respect to a variation of the wave functions.

It should be noted that, in order to guarantee gauge invariance of the transition amplitude, one must choose $|\chi(t)>$ in analogy to equation (2.2):

$$|\chi(t)> = \exp\left[-\frac{i}{\hbar}\int^{t}dt'\frac{<\phi|H-i\hbar\partial_{t'}|\phi>}{<\phi|\phi>}\right] |\phi(t)> . \tag{4.8}$$

We observe that (4.7) becomes perturbation theory

$$T_{if}^{pert} = \delta_{if} + \frac{1}{i\hbar}\int_{t_i}^{t_f}dt <\phi_f^{as}|V(t)|\phi_i^{as}> \exp(\frac{i}{\hbar}(E_f-E_i)t) \tag{4.9}$$

if asymptotically unperturbed, time-independent and orthogonal wave functions $|\phi^{as}>$ are used. $V(t)$ is the perturbation and $E_f-E_i$ the asymptotic (separated atom) energy difference between initial and final state.

It can be readily verified that (4.7) becomes the perturbed united-atom approach [10] if molecular wave functions are used. Variational bounds [11-13] on the transition amplitude are obtained by inspection of the second-order correction to the exact transition amplitude. Writing

$$T_{if} = T_{if}^{var} + \Delta_{if} \tag{4.10}$$

introducing $|\varepsilon> = (H - i\hbar\partial_t)|\chi>$ and using the Schwarz inequality one finds [13]

$$|\Delta_{if}| \leq \frac{1}{\hbar^2} \int_{t_i}^{t_f}dt \|\varepsilon_f\| \int_{t_i}^{t} dt' \|\varepsilon_i\| \tag{4.11}$$

where $\|\varepsilon\| = (<\varepsilon|\varepsilon>)^{1/2}$. Since the variational method minimizes $\|\varepsilon\|$, the variational bound

$$|T_{if}^{var}| - |\Delta_{if}| \leq |T_{if}| \leq |T_{if}^{var}| + |\Delta_{if}| \tag{4.12}$$

on the amplitude is expected to be useful. Unfortunately, a calculation of $\Delta_{if}$ requires the computation of the fluctuation term $<(H-i\hbar\partial_t)^2>$. For a more practicable error control we refer to Ref.[14].


SELECTED RESULTS

It has been shown elsewhere [8] that the variational method is able to yield accurate results for ionization. Rather than compare experiment with theory, we want to point out some aspects which characterize the variational method described so far.

Dynamical Binding

Realistic collisions are neither adiabatic nor sudden. A useful measure for the extent of adiabaticity is

$$A = \frac{\Delta E_{dyn}}{\Delta E_{stat}} = \frac{E_{fast} - E_{dyn}}{E_{fast} - E_{stat}} , \tag{5.1}$$

a quantity which varies between one (strictly adiabatic collision) and zero (sudden collision). $E_{stat}$ and $E_{fast}$ denote electron energies in adiabatic and sudden heavy particle collision, respectively. $E_{dyn}$ is the actual electron energy as calculated

[14] from the equations of motion (2.12). Clearly, A will depend on the collision characteristics. An example is given in Figure 4.

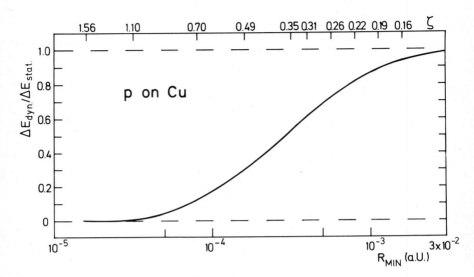

Figure 4

Ratio of dynamical to adiabatic increase of binding energy. Results are for a K electron in Cu as a function of $\zeta = v/v_k$ ($v$ = impact velocity, $v_k$ = K-shell velocity). All values for head-on collisions at the distance of closest approach ($R_{MIN}$)

## The 1sσ State in a Symmetric Collision

In slow symmetric collisions ($Z_0 = Z_t = Z_p$) the trial wave function (3.1) should be symmetrized

$$\langle \vec{r} | \phi \rangle = [2\pi(1+S)]^{-1/2} \left(\frac{\lambda+\lambda^*}{2}\right)^{3/2} [e^{-\lambda r_t} + e^{-\lambda r_p}] \tag{5.2}$$

where $S$ is the overlap of the two parts of the wave function, centered at target ($\vec{r}_t = \vec{r}$) and projectile ($\vec{r}_p = \vec{r} - \vec{R}$). Translational factors have been ignored in (5.2).

In Figure 5 we plot $\mathrm{Re}Z(t)/a_0 := \lambda_1$ for straight-line collisions, $\vec{R} = \vec{v}t$, at different velocities. In this approximation all charges scale with $Z_0$ and distances with $1/Z_0$.

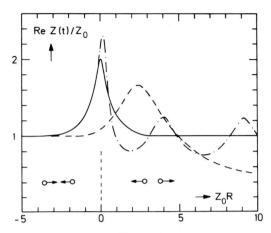

Figure 5
Fluctuations of the charge seen by a $1s\sigma$ electron in a symmetric
collision, as a function of the internuclear distance. All quantities in atomic units.
——— $v/v_k = 0$, $-\cdot-\cdot-$ $v/v_k = 0.5$, $-----$ $v/v_k = 2$

## Relativistic Trajectories [15]

For relativistic K electrons the Dirac Hamiltonian must be used. We note that
the equations of motion (2.12) remain valid if the Schrödinger Hamiltonian is
replaced by the Dirac Hamiltonian and, at the same time, the scalar trial wave
function by a spinor trial wave function [16]. An exact dynamical two-centre
calculation of K-shell excitation in heavy collision systems is not feasible at
present. The best we can do is to use monopole wave functions with variable charge
$\lambda$ (in analogy to (3.1)). In addition, we let the electron centre move a distance d
out of the target (see Fig.1). Hence d is a real variational parameter with
$d(t = -\infty) = 0$. Extending (3.2) and (3.3) to the relativistic case we find

$$< \frac{1}{r} > = \frac{1}{a_0} \frac{Z+Z^*}{s+s^*} = \frac{mc}{h} \frac{(Z+Z^*)\alpha}{s+s^*} \qquad (5.3)$$

and

$$< v_r > = i \ c \ \frac{(Z-Z^*)\alpha}{s+s^*} \qquad (5.4)$$

where $Z = a_0\lambda$ and $s = [1 -(Z\alpha)^2]^{1/2}$ . As usual, $\alpha$ denotes the fine structure
constant.

An example for a relativistic trajectory is shown in Figure 6. We observe that
electron excitation takes place in the second half of the collision. This is,
indeed, the finding of a coupled channel analysis [17]. If we compare the relativistic calculation with the corresponding non-relativistic one, then we find
relativistic effects responsible for increased non-adiabaticity.

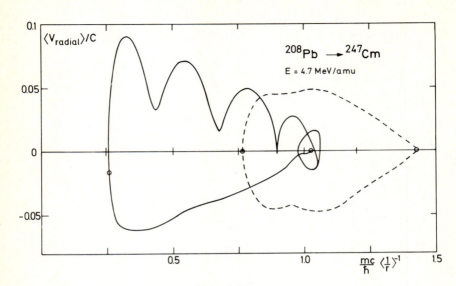

Figure 6
Average radial velocity and localization of a K electron
in Cm during a head-on collision between Pb and Cm. Starting point (t = -∞) and point of closest internuclear
distance are marked by small circles. Sense of circulation
is clockwise. Dashed line is the result of a non-relativistic
calculation.

Acknowledgment: This work was supported by the Bundesministerium für Forschung und
Technologie.

REFERENCES:

[ 1] Demkov, Yu.N., Sov.Phys. JETP 11 (1960) 1351.
[ 2] Sil, N.C., Proc. Phys. Soc. 75 (1960) 194.
[ 3] McCarroll, R., Piacentini, R.D. and Salin, A., J. Phys. B3 (1970) 137.
[ 4] Jancovici, B. and Schiff, H.D., Nucl. Phys. 58 (1964) 678.
[ 5] Kleber, M. and Zwiegel, J., Z. Physik A280 (1977) 137.
[ 6] Kleber, M., J. Phys. B11 (1978) 1069.
[ 7] Kleber, M. and Zwiegel, J., Phys. Rev. A19 (1979) 579.
[ 8] Kleber, M. and Unterseer, K., Z. Physik A292 (1979) 311.
[ 9] Reinhard, P.G., Rowley, N. and Brink, D.M., Z. Physik 266 (1974) 149.
[10] Briggs, J.S., J. Phys. B8 (1975) L485.
[11] Spruch L., in: Boulder Lectures in Theoretical Physics (Gordon and Breach,
      New York, 1969), vol. XI-C, p. 77.
[12] Shakeshaft, R. and Spruch, L., Phys. Rev. A10 (1974) 92.
[13] Storm, D., Phys. Rev. A10 (1974) 1008.
[14] Unterseer, K. and Kleber, M., to appear in Nucl. Instrum. Meth.
[15] Krause, J. and Kleber, M., in: NATO Advanced Course on 'Quantum Electrodyna-
      mics of Strong Fields', Lahnstein, Germany 1981 (Plenum Press, in prepar.).
[16] Sakurai, J.J., Advanced Quantum Mechanics (Addison-Wesley, Reading, 1978),
      p. 128.
[17] Soff, G., Reinhardt, J., Müller, B. and Greiner, W., Z. Physik A294 (1980) 137.

PHYSICS OF ELECTRONIC AND ATOMIC COLLISIONS
S. Datz (editor)
© North-Holland Publishing Company, 1982

STATE SELECTION WITH LASERS IN ATOMIC
AND MOLECULAR COLLISION PHYSICS

I.V. Hertel

Institut für Molekülphysik der
Freien Universität Berlin
Boltzmannstraße 20
1000 Berlin 33
Germany

INTRODUCTION

The rapid development and ready availability of tuneable laser
sources over a broad wavelength range (during the last years), has
made a tremendous impact on atomic and molecular collision and
reaction studies. Ever since the feasibility of scattering
experiments of laser excited Na-atoms was first demonstrated in
1973 [1] it was clear that lasers open up a completely new world of
possibilities for experimental physicists investigating collision
problems. State selective preparation and detection of colliding
particles are the most prominent aspect of laser application in this
field. The new techniques have, already now when we are still in the
beginning, produced a degree of detailed insight into various
problems of collision dynamics of which keen experimentalists
hitherto might only have dreamed. The collision system may be
prepared before and/or after the interaction process by lasers as
illustrated.

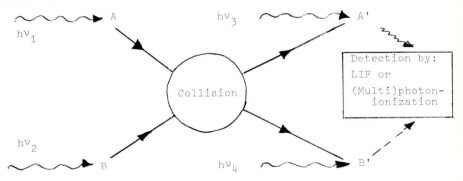

The detection may be performed by laserinduced fluorescence (LIF) or
by state selective (multi-) photonionisation, a new technique which
receives great attention recently [2]. Any combination of lasers
$\nu_1$ to $\nu_4$ may be used, alone or togethter with more conventional
preparation and selection techniques such as electric and magnetic
fields, mechanical velocity selectors, TOF-equipment, collision
induced fluorescence etc. It should be pointed out, that we are here
not discussing laser assisted collisions where the laser field acts
essentially only during the actual collision process [3,4]. We refer
here only to situations where the laser prepares or detects the

reactants long before or after collision, respectively. This holds
even if (as in many cases) the collision volume and excitation
(or detection) region overlap in space. Still, in the time domain
they are separated, since typical laser excitation times or atomic
life times may be of the order $10^{-8}$ to $10^{-9}$ sec while the actual
collision time is at least three orders of magnitudes shorter.

It cannot be the purpose of this brief article to review all the work
done so far. Many experiments have been concerned with atomic
collision problems, some with molecular collisions, but most state
selective work has been done in gas cell chemistry. It is an obvious
choice for a talk given at ICPEAC, to concentrate on the aspects of
collision <u>physics</u> which I feel is most clearly represented in
<u>experiments with</u> atomic and molecular <u>beams</u>. Even with this
<u>restriction</u> it will be impossible to discuss all the interesting
experiments which have been performed so far. I will try to give a
fair crossection of what are the possibilities and problems in the
field, what can be learned, where are the challenges, and where is
the field going. In particular, I will not elaborate on our own work
which can be found in the literature [5,6,7,8].

I give a brief survey of what states can be selected by lasers in

I. Atoms:

  A. Electronic excited states n,l
     Fine structure and hyperfine structure levels j,F
     (by virtue of the high resolution of laser-beam excitation)

  B. Magnetic substates $m_l$, $m_j$, $m_F$
     (by virtue of the laser polarization)

  C. Electronic ground states
     Fine structure levels j $m_j$ (i.e. electron spin polarization)
     and HFS-levels (i.e. nuclear spin polarization)

  D. Momentum states
     i.e. velocity and angular selected species
     (by virtue of the tuneability of the laser)

II. Molecules:

  A. to D. as above

  E. Vibrational, rotational levels (up until now only for electronic
     ground state molecules realized).

I will try to illustrate typical examples from the above list and
start with excited atoms, where most of the typical problems can be
illustrated best and I would like once again to offer my apologizes
to all those who have made important contributions to the field
which I do not mention due to lack of space and time.

## I. A. ELECTRONICALLY EXCITED ATOMS

The principle is simple: Take an atom
in its ground state (or a metastable
state) and expose it to a laser tuned
to an optically allowed transition.
In a stationary equilibrium between
excitation induced, and spontaneous
decay up to 50% of the atoms may be
excited.

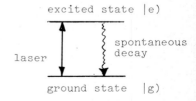

<u>Problem 1</u>: Find the right tuneable laser. A wide choice is now
available ranging from the IR (diodes ~5 μm, color center 2 μm) over
the whole visible (dye lasers) down to 250 nm (frequency doubled
dye lasers) if one takes only cw lasers into account as has been
done mainly up until now. This already allows us to excite a large
variety of species but of course powerful pulsed laser sources are
now available over an even broader range down to the Lyman α wavelengths suitable to study atomic hydrogen [9]. The true challenge to
atomic collision physics for the future may well be to overcome the
difficulties of the poor duty cycle of pulsed lasers. Pulsed lasers
are widely applied to gas cell experiments e.g. for state selective
detection of molecular reactions [10,11]. Detection levels as low as
some particles in the detection volume are quoted. As sensitive as
this may seem, when converted into currents (multiply essentially by
the particle velocity) this corresponds, however, to $10^5$ to $10^7$
particles/sec detection limit, which for a detector in a scattering
experiment does not compare well with state of the art particle
counting techniques.

<u>Problem 2</u>: The spontaneous life time of atoms is typically $10^{-8}$ to
$10^{-9}$ sec, the passage time through the collision region is of the
order $10^{-6}$sec. So collision region and excitation region have to be
at the same position, which sometimes may cause some inconvenience
(see also Problem 7).

<u>Problem 3</u>: The Doppler width of the atom beam limits the excitation
region. The Doppler detuning for a divergence angle α is given by
the average beam velocity $\bar{v}$ and the resonance wavelength λ as

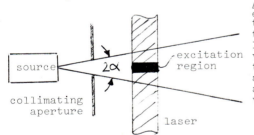

$\Delta\nu_D \approx (\bar{v} / \lambda) \cdot \alpha$. In order to
excite the atomic beam across
the whole profile one has thus
to keep $\Delta\nu_D < \Delta\nu_{nat}$, the natural
line width. For thermal beam
velocities and alkali atoms
this limits the beam divergence
angle to typically α < 1/100
and thus restricts the collision
volume.

One should be aware that this problem cannot be overcome by using a
"broad band" laser since a usual multimode laser is not broad band
in the sence needed here, but consists rather out of many narrow
modes, rapidly varying in frequency and thus shifting the excitation
zone in an uncontrolled way.

In favourable cases one may obtain some help from power broadening
when working in saturation. The power broadened line width is
∝ $\sqrt{\text{Laserpower}}$ and a high laser power may thus help to overcome the
Doppler broadening. Also, it seems possible [12] to use RF techniques
and acusto-optical modulation (at frequencies larger than the
reciprocal spontaneous life time) to artificially broaden the line
of a single mode laser. This scheme however can only be successful if
we do not encounter another problem, found in many cases:

<u>Problem 4</u>: Usually there are more than 2 levels involved in the
optical excitation problems. The situation for alkali atoms which
have a pronounced hyperfine structure (HFS) is typical. The optical

excitation of sodium [5] (nuclear spin I = 3/2) is only possible by
excitation of the $3^2S_{1/2}$ F = 2 ↔ $3^2$ $P_{3/2}$ F = 3 transition because
due to the ΔF = 0,+1 selection rule only
the F = 3 excited state can decay back
only to the F = 2 groundstate from where
it is originally excited. Power broadening
(which becomes important for laser powers
above 0.1 W/cm²) decreases the excited
state density since the HFS overlap
between the F = 3 and F = 2 excited
states leads to trapping of atoms in
the lower F = 1 ground state [13].
The situation is even worse in lithium,
where the HFS overlaps already within

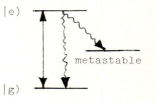

the natural line width. Circular polarized pumping will allow at
least some excited state population if very special pumping
conditions are chosen [14].

But even if there is no HFS, other complications may arise as in the
alkaline earth elements, where typically
one or more metastable states lay in
between the ground and excited state.
Nevertheless it is possible to perform
successfully scattering experiments
with these species, as first shown in
a pioneering electron scattering
experiment of excited Ba* [15] and
recently for the first reactive
system Ca ($^1$P) + HCl [16] using
laser excited species.

Metastable rare gas atoms may be laser excited state selectively.
A favourable case is $^{20}$Ne, where a stationary equilibrium between
the 3s $^3P_2$ and the 3p $^3D_3$ levels [17] may be achieved. In a pump
and probe technique other transitions between 3s and 3p Ne levels
may also be beautifully exploited [18] to selectively populate and
depopulate specific levels of the metastable 3s $^3P$ system in a
precision differential scattering experiment of Ne*+ Ne.

A very difficult situation is encountered
when the excited state of interest mainly
decays not back to the initial state.
Such situation is encountered in an
ambitious experiment [19] where He ($3^1$P)
atoms are excited by a dye laser from
the $2^1S_0$ metastable. Each excited atom
remains only for nsec in the excited
state and decays into the ground state.
Nevertheless it was possible to study
the associative ionisation process
He ($3^1$P) + Ne → HeNe$^+$ successfully.

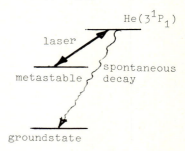

## I.B. MAGNETIC SUBSTATE EXCITATION

One of the nice possibilities using laser excitation is the
preparation of excited species in well defined magnetic substates
(without having to apply a magnetic field)by the well known optical
selection rules are ΔM = + 1 for circularly and ΔM = 0 for linearly
polarized light. When dealing with a pure $^1S_0$ → $^1P_1$ transition the
latter selection rule allows to prepare a pure M = 0 excited state

i.e. an atomic $p_z$ orbital with respect to a z - axis parallel to the electric field vector of the pumping light. Such a pure case is found e.g. in [136]Ba and Register et. al [20] where indeed able to exploit this in an electron scattering experiment to detect interesting spin orbit coupling effects.

Usually complications arize due to transitions other than S → P [17] or due to the presence of FS and HFS which in general couples several magnetic sublevels $M_L$ of the orbital electronic momentum. Nevertheless, a preferential excitation of certain magnetic levels can be achieved, the evaluation of the scattering data is merely a question of angular momentum algebra [21] and has been widely exploited in e + Na* scattering [22]. For the above discussed Na ($3^2P_{3/2}$ F = 3) excitation it is possible to prepare the atoms by linearly polarized light predominately in the $3p_z$ orbital, the ratio between the $3p_z$, $3p_x$, $3p_y$ population being up to 2.5 : 1 :1.

Problem 5: The above values are obtained for stationary pumping conditions and no disturbing environment. In practice one has to be aware of the fact that small magnetic stray fields (earth field) and radiation trapping in too dense atomic beams (above $10^{10}/cm^3$) tend to destroy the ideal polarization of the atoms. For quantitative collisional alignment and orientation experiments one has to measure first the actual atomic alignment and orientation prepared in the pumping process. This is easily done by studying the atomic fluorescence, some recipes and typical results for such studies are given in ref. [23].

Let us now turn to some application of $M_L$-state selective excitation.
a.) Elastic scattering from excited alkali atoms has first been investigated as early as 1975 [24,25,26]. But only recently the possibilities of atomic alignment have been exploited for suprathermal Na (3p) + Hg collisions: Elastic scattering from Na (3p) is determined by a Σ and a Π potential, correlating with the $p_z$ and $p_x$ (or $p_y$) orbital of the Na (3p).

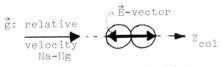

$\vec{g}$: relative velocity Na-Hg

$p_z$-orbital:  Σ - potential

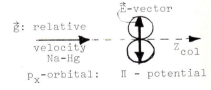

$\vec{g}$: relative velocity Na-Hg

$p_x$-orbital:  Π - potential

Depending on whether the electric field vector is parallel or perpendicular to the internuclear axis one state selects the Σ or the Π potential for the collision (predominately). The differential scattering experiment [27] shows clearly two rainbow structures (lab energy 3 eV) as a function of lab angle θ. The two experimental curves refer to the two directions of the laser polarization:

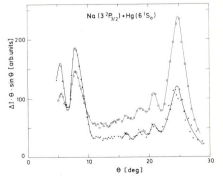

Na ($3^2P_{3/2}$) + Hg($6^1S_0$)

Crosses stand for $\vec{E}$ vector parallel to relative velocity $\vec{g}$
before collision ($\Sigma$) and open circles for $\vec{E}$ perpendicular $\vec{g}$ ($\Pi$).
The experiment illustrates beautifully how the $\Sigma$ rainbow at
around $8^\circ$ is predominant for preparation of the $p_z$ orbital of
Na*, while the $\Pi$-rainbow (deeper potential well) at $25^\circ$ is
strongly favoured for $p_x$-preparation.

b.) As a somewhat more complicated case Schmidt and collaborators
[28] have studied the superelastic process $Na^+ + Na(3p) \rightarrow$
$Na^+ + Na(3s)$ at kinetic energies up to 50 eV (c.m.). As well
known [29] a crossing between a $\Pi$-excited state potential and
the $\Sigma$-ground state at 4 a.u. is responsible for the process,
while the excited $\Sigma$-state (also converging to the Na(3p) state)
does not have such a crossing with the ground state and thus
cannot contribute to the deexcitation at the low energies under
consideration. Consequently the $p_x$-$\Pi$ preparation exclusively
should lead to the process while a minimum cross section is
expected for having the $\vec{E}$-vector parallel to the ion beam axis
(which is approximately equivalent to the relative velocity
before and after collision at the small scattering angles
observed). The experimental observation shows indeed that the
superelastic crossection varies drastically as the polarization

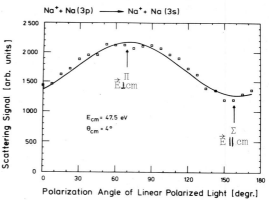

angle is changed
having a maximum for
$\Pi$-preparation (as
expected) with an
anisotropy ratio
$I_{max} : I_{min}$
approaching nearly
the maximum value of
2.5 : 1 for the $p_z$
and $p_x$ content in
the HFS-pumped excited
atom. It should be
noted that the above
data are obtained for
pumping light incident
perpendicular on to
the scattering plane
(the latter being
given by the relative
momentum of the particles before and after collision). A compli-
mentary experiment with light incident in the scattering plane
shows only neglibible variation of the scattering signal to a
level corresponding to the $\Sigma$ preparation in the above picture.
This experimental finding is a strict consequence of the
conservation of reflection symmetry in a $\Pi \rightarrow \Sigma$ transition: The
$\Pi$ state excited with an $\vec{E}$ vector perpendicular to the scattering
plane has negative symmetry ($\Pi^-$) and cannot be transfered into
a $\Sigma$-state with + symmetry with respect to reflection at the
scattering plane. In the previous case (perpendicular light
incidence) the $\Pi$ state has + symmetry also, and thus can couple
to the ground state.

The above experiments are two beautiful and clear cases for
state selective $\Pi$ and $\Sigma$-potential preparation. They illustrate
the full power of the magnetic substate selective laser excitation to investigate specific curve crossings and symmetry
properties of the collision process. However:

Problem 6: Can we always expect in a heavy particle collision that $p_z$, $p_x$, and $p_y$ lab preparation transform uniquely into a $\Sigma$, $\Pi^+$ or $\Pi^-$ potential with respect to the molecular frame? A number of examples show that this is not always the case [8] but we should be aware that this is not a methodological question but relates directly to the physics of the collision dynamics under study. In the completely equivalent time inverse studies of particle-photon coincidences which are now so elaborately used in many elegant experiments [30] one is faced with much the same problems. The key question is at which internuclear distance $R_M$ do the lab prepared atomic orbitals couple into the body fixed molecular frame. The potential energies $V(R)$ may typically behave as indicated.

For $R > R_M$ the atomic orbitals stay space fixed, for $R < R_M$ they are fixed to the internuclear axis (i.e. rotate with respect to the laboratory). A rough estimate for $R_M$ is obtained from the splitting between the $\Sigma$ and $\Pi$ curve $\Delta E$ which determines the angular velocity of the orbital rotation $\omega_{el} = \Delta E/\hbar$, and from the angular velocity of the collision which determines the torque on to the charge cloud due to nuclear motion $\omega_{col} = L/mR^2$ which depends on reduced nuclear mass $m$, angular momentum $L$ and instantaneous

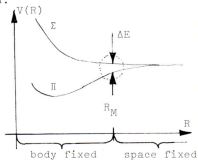

internuclear distance $R$. For $\omega_{el} \ll \omega_{col}$ we are in the space fixed situation, for $\omega_{el} \gg \omega_{col}$ in the body fixed frame. Thus roughly $R_M$ is given by $\omega_{el} \approx \omega_{col}$, i.e. the splitting of the potential curve at $R_M$ has to be $\Delta E(R_M) = \hbar L/mR_M^2 = \hbar b\sqrt{2 E_T/m} \cdot R_M^{-2}$, where $E_T$ is the initial kinetic energy and $b$ the impact parameter under observation. At not too short internulear distances the order of magnitude for $R_M$ may be obtained from setting $\Delta E = C_n R^{-n}$ and one obtains

$$R_M^{n-2} = \sqrt{m}\,(\hbar b C_n\sqrt{2 E_T})^{-1}.$$

Our original question wether a $p_z$ orbital is transfered into a $\Sigma$ potential (and $p_x$ into $\Pi$) is equivalent to find wether $b \ll R_M$. Take $b$ some a.u. we find e.g. for scattering of keV protons that $R_M$ becomes of the order of $b$ and one certainly cannot expect the one to one transfer from lab to body fixed frame, in the contrary under such conditions $p_z$ may become $\Pi$ and $p_x$ may turn into $\Sigma$. However, for the experimental examples discussed above, e.g. for low energy heavy particle scattering one estimates $R_M \sim 20$ to $50$ a.u. and in both experiments the impact parameter of observation is less than $10$ a.u. which explains the clean anisotropy effects observed.

There are complications, however, which may scramble the picture especially when working at very low (i.e. thermal) energies: Usually one has to expect additional $\Sigma$-$\Pi$ crossings at large inter-nuclear distances $R_L$ and also spin orbit interaction becomes noticeable in that region. For a clean case one therefore must have $b \ll R_M < R_L$ which sets conditions to the high and low energy limit. Furthermore, in collisions with molecules $\Sigma$-$\Pi$ transitions may be induced by molecular rotation. While the particles approach, the molecule performs one full rotation over a distance $R_{rot} = v \cdot t_{rot} = \sqrt{\mu/m} \cdot \sqrt{E_T/E_{rot}} \cdot r_o$ with the molecules reduced mass $\mu$ and equilibrium distance $r_o$. Thus we have to ensure that a rotation does not occur when the potentials are already body fixed i.e. $R_M < R_{rot}$

which places again a lower limit to the energy.

In fact, polarization effects have been observed even where these conditions do not apply, such as in thermal $Na^* + H_2$, $N_2$ quenching collisions [8], the results are interesting but difficult to interprete and reflect the dynamics at large internuclear distances.

In conclusion we can say that selection of magnetic substates with polarized laser light can be very informative in many collision problems. The clear selection of specific molecular interaction potentials in heavy particle collisions however is in general possible only for a limited range of intermediate energies.

I.C.   SPIN SELECTION IN ATOMIC GROUND STATES

Optical pumping into particular fine structure magnetic substates is equivalent to obtaining a spin polarization of the atomic electron. This possibility is very promising and many eventually replace the more conventional techniques using magnetic fields. The $^2S_{1/2}$ ground state of the alkali atoms is traditionally the main object in these studies and the basic idea of optically pumping all atoms with circularly polarized light into the $m = +1/2$ (i.e. spin up) level is very simple. However, HFS prevents such a simple schematic and more refined techniques have to be applied. By using a combination of magnetic hexapole fields and optical pumping [31] it was possible to polarize a atomic $^{23}Na$ beam to 92%, which certainly is the record so far obtained in this field. Another elegant scheme is to use two frequencies to pump simultaneously both hyperfine levels of the ground state. Using optoaccustical light modulation for the preparation of two frequencies Baum et al [32] were thus able to obtain 75% spin polarization in $^6Li$ by pumping predominately in the $2^2S_{1/2}$ F = 1 $M_F$ = 1 level. Why no higher polarization was achievable (one expects theoretically 100%) seems difficult to understand. Baum tends to explain this in terms of coherence effects in the dual frequency pumped atom [33], which leads us to

Problem 7: The coherent pumping laser field drives the atoms into a coherent mixture of ground and excited state (rather than simply into a statistical distribution) which leads to Rabi oscillations etc. The above example shows that this may effect the preparation process. Does it also effect the collision dynamics to be studied? There are a number of experiments (mainly in gas cells) studying these questions of collisional redistribution of coherent radiation, searching for "laser assisted" collisions etc. For the typical problems, however, studied with laser excited species in crossed beams the coherence should not be noticable: The Rabi oscillation time is typically of the order $10^{-9}$ sec while collision times are of the order $10^{-12}$ to $10^{-15}$ sec. Thus the colliding systems find its target either in the excited state or in the ground state. Unless interaction times of the order of $10^{-9}$ sec are probed, i.e. measurements are done with a resolution of $\hbar / 10^{-9}$ sec (i.e. $\mu V$) no coherence effects are to be expected [34].

## II. MOLECULES

Let us now turn to some applications of state selected molecules. Here we have all the above problems 1 through 7 plus some additional ones.

<u>Problem 8</u>: The main problems arizes from the fact that many rotational vibrational levels are involved. Even in a jet beam one starts off with a distribution of v,j states out of which according to all that was discussed above only one can be pumped with reasonable efficiency.

<u>Problem 9</u>: Electronic excitation is very difficult to maintain, since

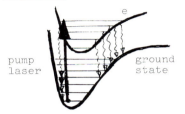

the molecules decay over the Franck-Condon region back into many ground state levels and get thus lost for further excitation. The problem is thus similair to the He($3^1$P) case discussed previously. $I_2^*$ may be a serious candidate, its life time beeing of the order μsec, but this implies of course also that the transition is difficult to saturate [35].

So far only collisions with selected ground state vib-rot levels have been carried out. Brooks et al [36] reported first beam studies of HCl$^{\#}$+ K → KCl + H with HCl-laser excited HCl (v=1) and were even able to state select specific rotational states [37].

In a very efficient excitation scheme using counter-propagating HF supersonic molecular and HF-laser beam the Göttingen group [38] was able to excite up to 4% of all HF-molecules into the HF (v=1, j=1) by using the 1P2 line focussed onto the jet beam nozzle. The excitation was detected in an indirect but unique way by monitoring the decrease in the $H_2F^+$ signal from $(HF)_2$ dimers, which are distroyed in collisions with HF$^{\#}$.

Multiphoton excitation is in principle another very efficient way to excite molecules [39] but state selectivity cannot be expected.

A much more powerful method for state selective collision studies with ground state molecules has been devised and very succesfully been used by Bergmann et al [40]:

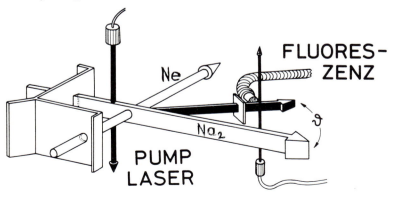

By optically pumping the molecule through an electronic excited
state one selectively depopulates one level, which is a nearly 100%
efficient process. After the collision the change in population of
individual rotational levels of the scattered molecules may be
monitored in the same way.

The experiment gives state to state differential cross sections for
$Na_2(v,j) + Ne \rightarrow Na_2 (v,j') + Ne$. There is no place to
discuss the exciting results of this pioneering work, such as
rotational rainbows but it is clear that such detailed information
on scattering processes can only be obtained by state selction with
lasers. Unfortunately, to date the number of molecules accessible
for such kind of studies is still very limited.

## MOMENTUM STATE SELECTION

Another very elegant way to detect scattered particles selected
according to their velocity and/or direction is to probe the
collision region in the scattering plane with a scanning laser and
thus to make use of the Doppler shift. This scheme was used in a

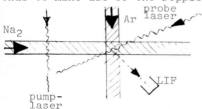

similair study as the one just
described by Pritchard and collaborators [41]. By directing the laser
in the direction of the relative
velocity of the colliding particles
one can by Doppler tuning measure
directly cm differential scattering
cross sections. Each scattering
angle Θ corresponds to a defined
Doppler shift and all azimuthal
angle are collected simultaneously. In this way a very high sensitivity can be obtained and it is even possible to measure   vibrational
inelastic $vj \rightarrow v'j'$ transitions. On the other hand the angular
resolution is not so good as in Bergmann's experiment. So each method
has its benifits. The momentum state detection seems now even
possible with pulsed lasers and has been applied to photofragmented
H-atoms [42], leading us into a promising future.

## FUTURE

It thus may be useful to conclude this article with some specula-
tions on the future.
It is clear that more laser sources ( eing more readily available)
will greatly widen the applicability of the new techniques.
Frequency doubling, mixing, and shifting techniques will be improved
in the near future and will facilitate the task of applying complicated laser systems to scattering apparatus which at any rate are
much more difficult to handle than usual spectroscopic tools. On the
detection side state selective multiphoton excitation looks very
promising. The great challenge seems to me the future use of pulsed
lasers. With the powers available there, one may well think of
coherent excitation techniques in
molecules. Coherent Raman Antistokes
excitaticn of molecules is possible by
such lasers with high efficiency. This
would make state selection accessible
to any mclecule and allow a world full
of new and fascinating studies...
if the duty cycle problem can be overcome.

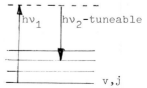

Even without these tempting possibilities one may expect a big wealth of new information from the use of lasers in state selective collision studies in the near future.

[1]    I.V. Hertel and W. Stoll (1973), VIII ICPEAC Beograd,
       Abstracts of papers, 321
[2]    Conf. on Resonant Ionization Spectroscopy (1981) Gatlingburg,
       Hurst ed. ORNL, Oak Ridge
[3]    Symposium on laser assisted collisions in "The Physics of
       Electronic and Atomic Collisions" (1980), K. Takayanagi and
       N. Oda eds.,
[4]    R.S. Berry (1981), What is really happening in laserinduced
       atomic collisions, this volume.
[5]    I.V. Hertel and W. Stoll (1978), Adv. Atom.Mol.Phys. 13,113
[5]    I.V. Hertel (1981) in,"The Excited State in Chemical Physics 2"
       McGowan ed, 341
[7]    I.V. Hertel, Adv.Atom.Mol.Phys.ed. K. Lawley, in press
[8]    W. Reiland, G. Jamieson, U. Tittes and I.V. Hertel (1981),
       to be published in Z.Phys.A
[9]    H . Zacharias, H. Rottke, J. Danon and K.H.Welge (1981),
       Opt. Commun. 37, 15
[10]   K.H. Gericke, F.J. Comes and R.D. Levine (1981),
       J.Chem.Phys. 74, 6106
[11]   H. Reisler, M. Mangir and C. Wittig (1980), Chem.Phys. 47, 49
[12]   S.J. Smith, private commun.
[13]   M. Fink, I.V. Hertel, G. Jamieson, to be published
[14]   A. Bähring, M. Fink and I.V. Hertel, to be published
[15]   D. Register, S. Trajmar, S. Jensen and R.T. Poe (1978),
       Phys.Rev.Let. 41,749
[16]   C.T. Rettner and N. Zare (1981), to be published
[17]   W. Bußert, J. Ganz, H. Hotop, M.-W. Ruf, A. Siegel and
       W. Waibel (1981), XII ICPEAC Book of Abstracts, 522
[18]   W. Beyer, H. Haberland, D. Hausmann (1981), XII ICPEAC,
       Book of Abstracts, 518
[19]   A. Pesnelle, S. Runge, D. Sevin, N. Wolffer and G. Watel
       (1981), XII ICPEAC, Book of Abstracts, 493
[20]   D. Register and S. Trajmar, to be published
[21]   J. Macek and I.V. Hertel (1974), J.Phys.B 7, 2173
[22]   H.W. Hermann, I.V. Hertel, M.H. Kelley (1980), J.Phys.B 13,3465
[23]   A. Fischer and I.V. Hertel (1981), Z.Phys.A submitted
[24]   G.M. Carter, D.E. Pritchard, M. Kaplan and T.W. Ducas (1975),
       Phys.Rev.Lett. 35, 1144
[25]   R. Düren, H.O. Hoppe and H. Pauly (1976), Phys.Rev.Lett.
       37, 743
[26]   R. Düren (1980), Adv.Atom.Mol.Phys. 16, 55
[27]   L. Hüwel, J. Maier, R.K.B. Helbing and W. Pauly (1980),
       Chem.Phys.Lett. 74, 459
[28]   H. Schmidt, E. Meyer and B. Miller (1981), XII ICPEAC
       Book of Abstracts, 752/754
[29]   P. Habitz and W.H.E. Schwarz (1975), Chem.Phys.Lett. 34, 248
[30]   D.H. Jaecks, Alignment and Orientation in Ion-Atom (Molecule)
       Collisions, this volume
[31]   D. Hils, W. Jitschin and H. Kleinpoppen (1981),
       Appl.Phys. 25, 1823
[32]   G. Baum, C. Caldwell and W. Schröder (1980),
       Appl.Phys. 21, 121
[33]   H.R. Gray, R.M. Whitley and C.R. Stroud jr. (1978),
       Opt.Lett. 3, 218
[34]   L. Hahn and I.V. Hertel (1972), J.Phys.B 5, 1995

[35]    R.B. Bernstein, private commun.
[36]    T.J. Odiorne, P.R. Brooks and J.V. Kasper (1971),
        J.Chem.Phys. $\underline{55}$,1980
[37]    H.H. Dispert, M.W. Geis and P.R. Brooks(1979)
        J.Chem.Phys. $\underline{70}$,5317
[38]    T. Ellenbroek, J.P. Toennies, M. Wilde and J. Wanner (1981),
        J.Chem.Phys. submitted
[39]    H.I. Lester, D.R. Coulter, L.M. Casson, G.W. Flynn and
        R.B. Bernstein (1981), J.Phys.Chem. $\underline{85}$,751
[40]    U. Hefter, P.L. Jones, A. Mattheus, J. Witt, K. Bergmann and
        R. Schinke (1981), Phys.Rev.Let. $\underline{46}$,915 and ref.quoted there
[41]    J.A. Serri, C.H. Becker, W.P. Moskowitz, M.B. Elbel,
        J.L. Kinsey and D.E. Pritchard (1981), J.Chem.Phys. submitted
[42]    R. Schmiedl, Ch.H. Dugan, W. Meier and K.H. Welge (1981),
        Z.Phys. A,to be published

PHYSICS OF ELECTRONIC AND ATOMIC COLLISIONS
S. Datz (editor)
© North-Holland Publishing Company, 1982

# MULTICOINCIDENCE TECHNIQUES IN ATOMIC PHYSICS

J.A. Fayeton, J.C. Brenot and J.C. Houver

Laboratoire des Collisions Atomiques et Moléculaires
Université Paris XI, Bât. 351, 91405 ORSAY Cedex,
FRANCE

The "multicoincidence technique" results from the combination  of two experimental methods which are  widely developped in atomic physics ; namely the "position sensitive
detection" and the coincidence technique. This new technique which actually consists in taking coincidence data
in several channels, *simultaneously*, combines the advantage of both methods reduction of data acquisition time
and more accurate identification of the various processes
occuring in a collision. The purpose of this paper is to
describe the electronic devices developped by various
authors and to compare their relative performances and
limitations. Applications to atomic collisions are exemplified.

Identification of the various processes occuring in atomic collision, requires
detailed information on the kinematics of the various collision  partners  which
can be obtained, for instance by differential scattering measurements. For this
purpose conventional methods consisting in measuring *sequentially* the scattered
particles with a single detector (say channel electron-multiplier) requires a
continuous monitoring of experimental parameters, for consistancy of data. Owing
to the development of the microchannel plates (MCP) detectors this method is, at
present, often replaced by "a multidetection system". The particles scattered at
various angles are detected *simultaneously on a position sensitive detector*,
thus reducing considerably the data acquisition time, and avoiding the continuous
monitoring of parameters.

Even more detailed information is provided by *coincidence* experiments establishing
a time correlation between two sets of data events stemming from the same collision
and hitting *two* detectors. For example, the coincidence between a scattered particle and the photon emitted by a particle excited in a collision associates the
information on the collision kinematics with the accurate identification of the
process through the optical spectroscopy. However the high level of comprehension
gained  by such accurate data is usually payed by a very time  consuming
experiment.

The combination of both techniques allows a *simultaneous* measurement of coincidence
data on several channels. More generally, this *multicoincidence*  method essentially
gives the *spatial* and *time* correlation between two sets of events characteristic
of a given type of collisions. Several groups have developed their own methods to
achieve this goal. However, a common general sheme can be used to describe  the
various methods. As shown on Fig. 1, the general sheme is composed of different
"blocks" each of them having a distinctive function. For each channel A and B corresponding to the two sets of events (A and B), Block I are related to the detection
of events, amplification and pulse shaping. These pulse are then treated in Block
II to get the *position informations*. The same data also enter  Block III the

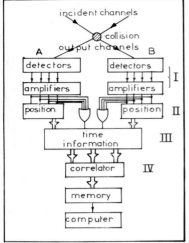

purpose of which is to get the *timing informa-tion*. Position and time information are then combined in Block IV ("correlator") before being stored in a *memory*. The stored data can be processed ("off or on line") in a computer and visualized on a CRT. The purpose of this paper is to describe the various *multicoinci-dence techniques*, within the frame work of the general diagram given above, and discuss the suitability for a given physical situation.

Figure 1.
Block diagramm of a multicoincidence experiment.

# I - BLOCK I - DETECTORS, AMPLIFIERS AND SHAPING

Although an array of individual electron multipliers can be used to get position information, the development of MCP detectors gives a simple way to get an accurate position information. The various methods to get this information, extensively reviewed by R.W. Van Resand and J. Los (1) will only be summarized here. They can be classified in two categories.

(i) the first category includes devices in which the position information is obtained by a charge division technique. Pulses issued from the MCP are received on a continuous anode collector made with a resistive material. For each pulse, the initial charge spreads out on the collector reaches two (or four) amplifiers (2). This method gives a very accurate position determination (50 µm or better) and requires only few amplifiers. However the poor rise time, and the broad out-put pulse of charge amplifiers (0.5-5µs) make this technique unsuitable for coincidence experiment since an accurate time pick up is necessary. Furthermore, such wide pulses drastically limit the maximum count rate (2000 c/s). The charge division technique can also be used with a set of discrete anodes connected to each other by capacitors (3). These methods which give a shorter rise time (10-100 ns) can be used for coincidence experiments but the maximum count rate capacity is about 50.000 c/s and only a poor time resolution can be reached.

(ii) the second category comprises devices where *individual* amplifiers are con-nected to *discrete* anodes. The accuracy in the position is limited by the minimum distance between two adjacent anodes ($\gtrsim$ 100 µm) but, as compared with the previous case, the better time resolution can be achieved and, in principle, is primarily determined by the time jitter of the amplifier. Development of such technique has been made practically feasible at low cost with hybrid circuit amplifiers. Such chips developed in connection with multi-wire chambers are now available from many companies (Le Croy, LMT, CIT Alcatel). They deliver 10-20 ns wide MECL signals with a time jitter better than 1 ns, and a reset time of 150 ns. With such device, the *maximum count rate* is only limited by the charac-teristics of the MCP up to 400.000 $c/s.cm^2$ (4).

## II - BLOCK II. POSITION INFORMATION

Since as discussed above, the charge division devices are not suitable for coincidence experiments, only devices using individual amplifiers will be considered. In all methods, the position information can be encoded both in an analog or logical way.

a) As an exemple of an *analog* coding of the position the pulse develivered by a given amplifier is *delayed* by an amount of time characteristic of this detector (5). Anticipating on the time information Block, it should be mentionned that this technique can only be used in association with a time to amplitude (TAC) method method and therefore can only be used for *low* count rate.

b) The *position information* can also be obtained in a fully logical device, by connecting the output of each detector to a *"priority encoder line"* (6) delivering a binary code. Such priority encoder line will soon be available commercially for systems up to 32 positions (Le Croy). In the experimental set up developed in our laboratory (7) a 64 "line priority encoder" made of MECL 10000 logic circuits delivers a position code within 20 ns. It should be noticed that a drawback in the use of discrete anode devices is the spreading ($\sim$ 300 µm) of the electron cloud on several adjacent collectors (Fig. 2). In a priority encoder, only the anode with the smaller code will respond to such an electron cloud, thus limiting the position accuracy. However improvement can be obtained using a *double* priority encoder line (8). As shematically shown on fig. 2, with such a device, the mean position of the electron cloud can be determined, as well as the size of the corresponding spot.

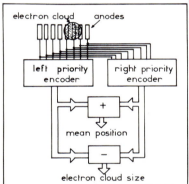

Fig.2. Block diagram of the double priority encoder line.

c) The logical device described above can in principle be used to correlate two sets of events A and B (fig. 1) by associating the codes of both sets of detectors. Although in most applications, correlation between events coming from several detectors A associated with only one detector B, are generally found, examples with several channels B can be found in the litterature. For example at the University of Maryland, a coincidence device involving two sets of 5 detectors (A=5, B=5) has been developed (9). A special feature of this device is provided by the association of two sets of detectors A and B in *pairs* using a set of 25 AND gates. The outputs of these gates are connected to an encoder generating a logical code identifying the detector-pair.

## III - BLOCK III. TIME INFORMATIONS

a) Devices using a time to amplitude converter (TAC). The most conventional technique to time correlate two events utilizes a time to amplitude converter. In the present case, it is first necessary to multiplex the outputs of detectors in each set A and B, using two OR-gates (Fig. 1). One of the outputs is conveniently delayed and used to start the TAC. The analog output is then converted in a logical word through an Analog to Digital Converter (ADC). The TAC and ADC are sometimes replaced by a TDC which directly converts time to a digital code. The advantage of this method lies in its simplicity since only commercially available devices are involved.

b) Multicorrelator method (Fig. 3). A more versatile method consists in deter-
ming the arrival times $T_A$ and $T_B$ of any events with respect to a common master clock, in both channels A and B. $T_A$ and $T_B$ are then converted in logical words. The whole time information treatment is processed using fast logical circuitry in a fully synchronous way with respect to the master clock. The interesting feature of this method is provided by the *fully logical* treatment of the time resulting in the high count rate ability of such a system. Actually the overall dead time is only determined by the period of the master clock. For instance, in the experiment developed in Orsay, the use of a 5 MHz master clock allows to process up to 500 000 c/s. On the other hand, the ultimate time accuracy is limited by the logical circuitry. In the present case, the rise time of MECL 10 000 circuits conveys an ultimate resolution of the order of the nanosecond.

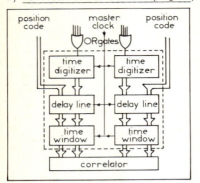

Fig.3.Detailed diagram of the time
information Block (dotted line)
used in the multicorrelator method.

This method, developed at LCAM (Orsay) will be summarized below. A detailed description has been published elsewhere (7).

(i) Time digitizer :
At first arrival time of each event is determined within the period of the master clock (200 ns) as a 6 bits code leading to an accuracy of 3.1 ns. This logical information is available at the end of the 200 ns period during which the event occurs. The next period is used to reset the signal for the position encoder and the time digitizer. This results in a maximum dead time Of 400 ns.

(ii) Delay line :
For each channel, the time information (6 bits) associated with the corresponding position code (6 bits) can be conveniently delayed. For this purpose this 12 bit word is written sequentially at every period of the master clock in a RAM (Random Access Memory) used as a ring buffer. Simultaneously to a writing in a given memory (a), the data already stored in a memory (m) is read. This results in an effective delay of (m-a)x 200 ns between the writing and the reading. This "delay line" operates between 0.2- and 51 μs by steps of 200 ns.

(iii) Time window :
The "time window" is defined as the time interval during which, two correlated events will be searched. In any coincidence experiments, one channel, say the A channel, is privileged. For instance, in devices using a TAC, one should better "start" the conversion with the lower count rate channel (A). In our method the time code of a given event from the A channel (Master channel) is latched during a preset "time window" in order to be correlated to *any* event from the B channel arriving during this time interval. This device allows to correlate a *single* low count rate event (A) with *several* high count rate events. This method presents two interesting features : first the loss of "correlated" B events is strongly reduced, then, increasing the time window does not result in an increasing dead time.


IV - CORRELATOR


The *time difference* $T_A$- $T_B$ of arrival time of A and B events, expressed is a 9 bit word is associated with the two corresponding *position* codes. This results in a 15 bit word which is temporarily stored in a memory waiting for an "off" line data processing using a computer.

ultimate time resolution of 3.1 ns per channel. The width of the peaks $\Delta T \simeq 9$ ns is mainly due to geometric parameters of the collision volume, showing that, for such experiment, it is useless to look for a better time resolution. Typical count rates of $2.10^5$ c/s (channel A) and $2.10^3$ c/s (channel B) result in a "true" coincidence rate of 0.02 to 0.2 coincidence per second according to the scattered angle. A typical accumulation time of about 8 hours allows to keep the statistical uncertainty below 0.07. The data corresponding to (a) and (b) processes have been processed using a PDP 11/34 computer to obtain the differential cross sections shown on figure 6 and compared to theory (12). Information about excitation of

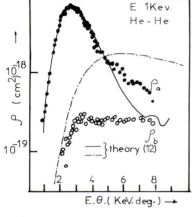

*magnetic* sublevels can be obtained by measuring the *angular* correlation between the emitted photon and the scattered particle (13). For this purpose the photon detector has been set at different position in the collision plane (Fig. 4). The data corresponding to a *same* scattering angle correspond to the so called "radiation pattern" for a given reaction (a) or (b) (Fig. 7). In this particular case the data are well fitted by a sine shaped curve demonstrating that a nearly pure He($2^1$P m = ± 1) is excited.

Fig. 6. Differential cross sections $\rho(\tau) = \sigma(\theta).\theta. \sin\theta$ for a and b reactions.

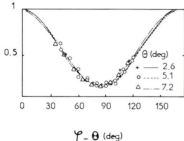

Fig. 7. Radiation pattern He($2^1$P) for reactions a represented with respect to the quantification (internuclear) axis. at E = 500 eV

## CONCLUSION

The above example describes a single 512 x 64 x 1 configuration (time x scattering angle x UV emission angle). The versality of this device allows different sharing of the core memory. The utilisation of such multi-configuration experiment can easily be extended to other cases. In Fig. 8a coincidences between particles scattered at various angles and energy analysed charged particles B, are looked for. In this experiment two sets of MCP are used, one determing the scattering angle, the other one, the energy of the particle B. An other application, requiring time of flight technique for *both* channels (Fig. 8b) uses a chopped incident beam. In both type of experiments, the data can be represented in a similar tridimentional surface. The size of the memory will be the ultimate limitation of the technique. As an example, in the above case, 8 channels for scattering angles, 64 channels for energy of B, 128 channels for TOF of B require a 8 x 64 x 128 configuration, twice the size of that of our device. The rapid increase of the memory size commercially available will open up for even more versatile devices. Increasing memory size makes shorter the working time.

## V - EXAMPLES OF APPLICATION

The multicorrelator method has been first applied to multicoincidences in atom-atom collisions (10). The He-He system selected because of its relative simplicity allowed a rather good understanding of collision mechanisms. As shown in Fig. 4, the experiment consists in a correlation between scattered Helium-atom (channel B) and any events received on a single electron multiplier (channel A). Actually two types of events are detected on channel B namely (i) UV photons issued from the He($2^1$P) decay and (ii) recoiled metastable particles which respectively correspond to the following reactions :

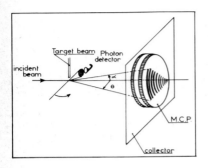

Fig. 4. Schematic diagram of the multicoincidence between photons and scattered particles

(i)      $\overline{\text{He}}$ + He → He + He*($2^1$P)        (a)
                    → He*($2^1$P) + He*($2^1$P)    (b)

(ii)     $\overline{\text{He}}$ + He → $\overline{\text{He}}$ + He*($2^1$S)    (c)
                    → He*($2^{1,3}$S) + He($2^{1,3}$S)   (d)

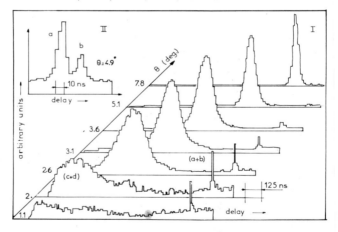

Fig. 5. Typical coincidence spectra at impact energy E = 1000 eV. For He-He collisions. I) Multicoincidence spectra with coarse time resolution (12.5 ns) displaying metastables (c+d) and photons (a+b) in coincidence with scattered particles for various scattering angles. II) Coincidence spectrum obtained with the ultimate time resolution (3.1 ns) resolving a and b reactions.

Figure 5.I shows some of the 64 simultaneous coincidence spectra. Actually such spectra display several peaks corresponding to different delay times $T_A$-$T_B$ characteristic of the various processes. The broad peak corresponds to (C+d) reactions whereas the narrow peaks are attributed to reactions a and b. Then such coincidence experiment results in an *energy loss spectroscopy* by time of flight (TOF). Spectrum shown in Fig. 5.II displaying the two a and b peaks is obtained with the

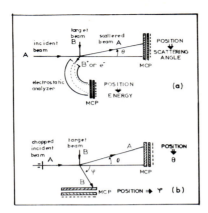

Fig. 8. Examples of multicoincidence experiments.
a) processes corresponding to a given scattering angle, energy loss of the
scattered particle and energy of another product of the collision are identified
(four-dimentional information)
b) processes corresponding to a given scattering and recoil angle, with
simultaneous identification of the energy of both particles are determined (
five dimentional information).

## ACKNOWLEDGEMENTS

The authors wish to thank M. Barat for his continuous support and encouragement in
the present work.

(1) R.W. Wijnaendts van Resandt and J. Los Invited Paper and progress reports
    XI ICPEAC, Kyoto (1979) edited by N. Oda and K. Takayanagi (North Holland,
    Amsterdam, 1979)

(2) M. Lampton and F. Paresce Rev.Sci.Inst. 45 (1974) 1098
    W. Parkes, K.D. Evans and E. Mathieson Nucl. Inst. Meth., 121 (1974) 151.

(3) R.W. Wijnaendts van Resandt, H.C. den Harink and J. Los J.Phys.E 9 (1976) 503

(4) J.L. Wiza Nucl. Inst. and Meth. 162 (1969) 587

(5) J. Herman B. Menner E. Reisacher L. Zehnle and V. Kempter J.Phys.B : Atom.
    Molec.Phys.B (1980) L165

(6) Semi conductor data library MECL integrated circuits 4 Motorola (1974) 139

(7) J.C. Brenot J.A. Fayeton and J.C. Houver Rev.Sci.Inst. 51 (1980) 1623

(8) J.C. Brenot and J.C. Houver LCAM, Orsay, work in progress

(9) T.L. Skillman Jr. E.D. Brooks III M.A. Coplan and J.H. Moore Nucl. Ins. and
    Meth. 155 (1978) 267

(10) J.A. Fayeton J.C. Brenot J.C. Houver and M. Barat J.Phys.B : Atom.Molec.Phys.
     (1981) in press

(11) J. Macek and D.H. Jaecks Phys.Rev. A4 (1971) 2288

(12) J.P. Gauyacq J.Phys.B : Atom.Molec.Phys. 9 (1976) 2289

PHYSICS OF ELECTRONIC AND ATOMIC COLLISIONS
S. Datz (editor)
© North-Holland Publishing Company, 1982

CHARGE TRANSFER TO MULTI-CHARGED IONS
IN PENNING TRAPS

D. A. Church

Physics Department
Texas A&M University
College Station, Texas 77843, U. S. A.

Measurements of rate constants for charge transfer from molecules and atoms to near-thermal multi-charged ions stored under
ultra-high vacuum conditions in Penning ion traps are described.
Advantageous features of the technique, and methods for producing the multi-charged ions in the trap by impact ionization are
also discussed.

INTRODUCTION

The Penning trap[1],[2] a device for confining ions in a vacuum environment, consists
of electrodes between which a dc potential is applied. The electrodes are basically figures of revolution about a symmetry axis, shaped to generate a nominally
quadrupole electrostatic potential in their interior. A charge moving within the
electrodes would be focussed axially, but de-focussing radially by the static potential. Confinement is obtained by the application of a strong uniform external
magnetic field B along the symmetry axis of the structure. The details of the
electrode construction and auxiliary circuits vary with intended trap use.
Figure 1 shows a diagram of the electrode structure and circuits used in multi-
charged ion charge transfer measurements.

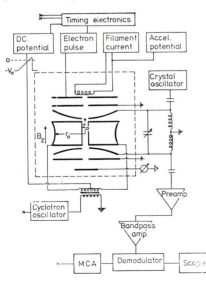

Fig. 1. Trap structure and
circuits to select and detect
ions (ref. 40).

Penning traps have seen extensive use recently for precision spectroscopy of free electrons[3-6] and ions,[7-13] which in some cases employed certain collision processes. Spin-dependent charge transfer and spin-exchange collisions of polarized atoms with singly-charged ions stored in traps were reviewed several years ago by Dehmelt.[14] Dunn and his collaborators have described the measurement of recombination cross-sections of stored molecular ions in their vibrational ground states in Penning traps.[15-18]

More recently, considerable interest has arisen in charge-transfer processes of low-energy multi-charged ions with atoms and molecules, due to the presence of impurity ions in fusion-type plasmas,[19-21] as well as to processes of astrophysical significance.[22,23] The highly-charged ions in a laboratory plasma may capture electrons into excited states, followed by radiation of useful energy from the plasma. Interaction with molecules near the boundary may modify plasma properties in significant ways.[24] Some important astrophysical processes are discussed by Dalgarno and his collaborators[22,23] who show that capture may be the dominant process for producing excited levels under certain conditions. This range of interest has led to theoretical calculations of charge transfer rate constants for near-thermal multi-charged ions with hydrogen[23,25,26] and helium[27-29] atoms, and to a number of measurements of doubly-charged ion rate constants with flow-drift tube methods by several groups.[30]

The Penning trap offers several advantageous features for collision studies of multi-charged ions:
- There is no inherent limitation on the charge of the ion to be studied, since the ions are confined in ultra-high vacuum.
- For a confining static potential $V_o$ in the Penning trap, the axial confining energy is $qV_o/2$ and the cyclotron radius $R=mv/qB$. Ions with large charge q are more effectively confined.
-The mean energies of the confined ions are the order of eV or less, and can be adjusted to some extent by varying the confining potential.
-Since the ions move harmonically with frequencies dependent on their mass to charge (m/q) ratio, specific m/q ratios can be selectively accumulated and stored by removing unwanted ions via resonant excitation.
-The ions are confined to a relatively small region in the trap, where they can interact effectively with photon, electron, or neutral atom beams.
-Production of multi-charged ions in the trap by impact ionization of gases with electrons or heavy ions is facilitated by the magnetic field, which guides the charged beams through the trap.
-The number and mean energy of the stored charges can be effectively measured by resonance methods or particle counting techniques.[31]

The recent implementation of multi-charged ion charge transfer measurements with these traps has taken two different routes, distinguished by the methods by which the ions are produced in the trap. Electron impact ionization has been found effective in creating doubly- or triply-charged ions,[32] but due to decreasing cross-sections, charges higher than four-times ionizated are expected to be difficult to attain by this method. However, x-ray spectroscopy[33] has demonstrated that fast beam stripped heavy ions can remove large numbers of electrons from a target atom in a single collision. Cocke[34] first showed clearly that the recoil energy of these highly ionized target atoms was less than 30 eV, sufficiently low to consider confining the recoils. The cross-sections for producing these higher charge to mass ratios are relatively favorable. The heavy-ion impact method has recently been successfully used to study charge transfer in an electro-static trap[35], and to generate the multi-charged ions stored in a Penning trap.[36] Since most progress has been made so far with electron impact ionization, the following discussion will emphasize this aspect of the work.

## ION MOTION

Assuming that the electrostatic potential is purely quadrupolar, and that the magnetic field is uniform, the potentials from which the ion motion can be derived are $\Phi = - ((V_0/4z_0^2)(r^2-2z^2) + V_0/2)$ and $\vec{A} = B/2(-y,x,0)$ where $2z_0$ is the end electrode separation, the end electrodes of the trap are at zero potential, and the ring electrode is at $-V_0$. The equations of motion of an ion with mass m, charge q are

$$\ddot{Z} + \omega_z^2 Z = 0 \qquad\qquad \omega_z^2 = qV_0/mz_0^2 \qquad (1)$$

$$\ddot{X} - \omega_c\dot{Y} - (\omega_z^2/2)X = 0 \qquad \omega_c = qB/m \qquad (2)$$

$$\ddot{Y} + \omega_c\dot{X} - (\omega_z^2/2)Y = 0 \qquad\qquad\qquad (3)$$

The harmonic motion given by equation (1) has a frequency $\omega_z$ dependent only on the electric potential $V_0$. Equations (2) and (3) can be solved jointly to produce a solution of the form $X + iY = c e^{-i\omega_+ t} + d e^{-i\omega_- t}$ where c and d are oscillation amplitudes and $\omega_+ = \omega_c/2 + \omega_0$, $\omega_- = \omega_c/2 - \omega_0$, and $\omega_0 = (\frac{\omega_c^2}{4} - \frac{\omega_z^2}{2})^{1/2}$. For $\omega_+ \gg \omega_z$, the usual condition, $\omega_+ \simeq \omega_c - \omega_-$ and $\omega_- \simeq V_0/2Bz_0^2 + (m/q)V_0^2/4B^3z_0^4$. Since the frequencies $\omega_+ \gg \omega_z \gg \omega_-$ depend on m/q, they may all be used to select certain m/q ratios by applying a driving radio-frequency field at resonances of the undesired ions. The motion is then rapidly excited to large amplitudes, and the ions leave the trap, leaving the desired ions undisturbed.

## ION DETECTION

Although several methods of ion detection exist,[14] a resonant detection method is quite sensitive and convenient for most purposes. An inductor is connected across the end electrodes of the trap, forming a tuned circuit as in Fig. 1. The tuned circuit is excited by an external oscillator. The ion axial oscillation frequencies $\omega_z$ are successively brought to resonance with the tuned circuit by sweeping the dc potential linearly to zero. Upon passing through resonance, the ions at each m/q ratio sequentially absorb energy from the exciting rf, modulating its amplitude. The basic time cycle is shown in Fig. 2. The amplitude modulation is detected, producing a succession of absorption signals for the different ion m/q ratios as shown in Fig. 3. Each signal is proportional to ion number N through the relation S= qNPA, where P is the average rf power dissipated in the tuned circuit and A is a constant which depends on the parameters of the dc voltage sweep and of the tuned circuit. The thermal noise energy in the same bandwidth is just $kT_0$ where $T_0$ is the effective temperature of the tuned circuit and input amplifier. For the typical parameters used, one calculates unity signal to noise for as few as 50 singly-charged ions detected in a single sweep. Integration by storing successive signals improves the effective sensitivity greatly. Such signals for light singly-charged ions are shown in Fig. 4(a). Similar signals can be produced by cyclotron excitation and radial detection with the split ring electrodes.

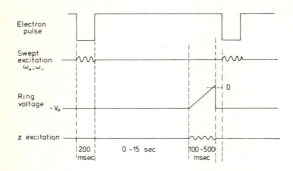

Timing  sequence  for  single  sweep

Electron
pulse

Swept
excitation
$\omega_+,\omega_-$

Ring
voltage  $-V_0$

z excitation

200        0 -15 sec     100 -500
msec                         msec

Fig. 2.  Timing sequence
used to produce, select,
store and detect stored
ions.

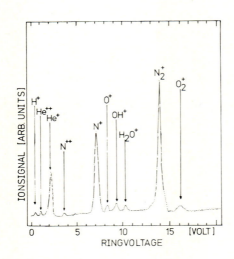

IONSIGNAL [ARB. UNITS]

$H^+$
$He^{++}$
$He^+$
$N^{++}$
$O^+$
$N^+$
$OH^+$
$H_2O^+$
$N_2^+$
$O_2^+$

0          5          10          15   [VOLT]
RINGVOLTAGE

Fig. 3.  Residual gas ion
spectrum made in the $10^{-9}$
Torr range in "ion-gauge"
mode of the trap.

A second method of resonantly detecting ions is to sweep them through the tuned
circuit resonance in the manner described above, but with no external rf excitation applied to the tuned circuit.[37,38]  The random motion of the charges in the
trap induces an additional fluctuation current across the effective tuned circuit
resistance R with a mean-squared value proportional to the number and to the mean
energy per ion $kT_i$: $I^2 = NkT_i/R\tau_d$.  Here $\tau_d = 4mz_0^2/q^2R$ is a damping constant.
The signal now is the additional energy dissipated by the ions in the tuned circuit as they pass through resonance.  The signal to noise ratio for signals of
this type in the tuned circuit bandwidth $t_0^{-1} = \omega_0/Q$ can be written as S/N =
$N(t_0/\tau_d)(\frac{T_i - T_0}{T_0})$.  When the ion number is independently measured and the apparatus
parameters are known, a mean ion temperature relative to the tuned circuit temperature can be determined.  Ion temperature signals of this type are shown in
Fig. 4(b).  Mean energies near 1 eV are observed in 10 to 15 V potential wells.

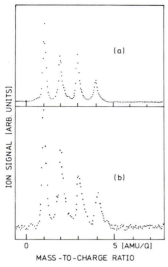

Fig. 4(a)  Ion number signals for
$H^+$, $H_2^+$, $H_3^+$, and $He^+$ detected by rf
absorption, integrated for 100
sweeps.  (b) Corresponding temperature signals integrated 1000 sweeps.

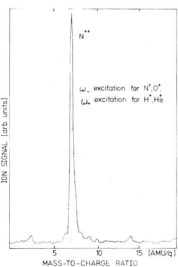

Fig. 5.  Selected $N^{+2}$ ion number
signal with other major peaks
removed by swept $\omega_+$ and $\omega_-$ excitation.

## ION SELECTION

As shown in Fig. 2, unwanted m/q ratios are selectively driven from the trap by
resonant excitation at the frequencies $\omega_+$ and $\omega_-$ applied across the split ring
electrode of the trap during the ion creation interval.  The driving frequencies
are repetitively swept.  It is possible in this way to accumulate desired ions
with relatively low cross-sections for production, as Fig. 5 shows for $N^{+2}$.

ION STORAGE

After the ions are formed in the trap, they are held for a variable storage time t and then detected. Due to charge transfer collisions, the ion number with a given m/q changes with t. It is frequently observed that the ion number decreases exponentially with time, at least for times the order of seconds, so that the storage time can be characterized by a decay constant $\tau$: $N(t) = N(0)e^{-t/\tau}$. On the other hand, other ion m/q ratios are observed to increase and then decay due to the formation of a lower charge state from neutrals by capture. Fig. 6 shows examples of both occurances for the ions $N^{+2}$ and $N^{+}$ respectively.[39] The primary ion number decay may not be a single exponential due to the presence of metastable ion levels which react at different rates than do the ground state levels.[40] In this case the decay may be represented by a sum of exponentials, e.g. $N(t) = N_{10}e^{-t/\tau_1} + N_{20}e^{-t/\tau_2}$ where $N = N_1 + N_2$.

In cases when the reaction rate is very small, ions may be lost by radial diffusion across the magnetic field lines, induced by collisions. This has not been observed for ions with typical measurement parameters used to far, but is found at reduced magnetic field strengths.

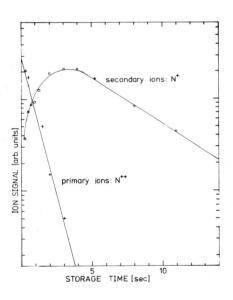

Fig. 6. Primary and secondary ion number changes with storage time. Fast secondaries may not be observed.

PRESSURE MEASUREMENT

To compute a charge transfer rate constant, it is necessary to obtain an absolute number density $n_g$ of the reactant gas, often at pressures in the low $10^{-9}$ Torr range. A nude ionization gauge was calibrated against several gases in the $10^{-4}$ to $10^{-5}$ Torr range using a capacitance manometer, and linearity was assumed in extrapolating to $10^{-9}$ Torr and below. The partial pressures of individual gases were obtained with a quadrupole residual gas analyzer in the $10^{-8}$ Torr range, and correlated both with the total pressure measurement of the ionization gauge, and with the singly-charged ion signals observed with the ion trap, similar to those shown in Fig. 3. At the lowest partial pressures, the trap alone was used in this "ion gauge" mode to obtain the partial pressures. Comparison of rate constants for charge transfer to singly-charged ions measured in the $10^{-8}$ to $10^{-9}$ Torr ranges with data from other techniques taken at pressures 100 to $10^5$ times higher indicate that this calibration procedure is adequate.[40]

## CHARGE TRANSFER RATE CONSTANTS

The rate constant $k = \int_0^\infty f(v)\sigma(v)v\,dv$ can be related to the measured value of the decay constant $\tau$ and number density $n_g$ by $k = (n_g\tau)^{-1}$. $\tau$ is obtained from a least-squares fit to the ion number signals plotted vs. storage time. The rate constant $k$ is often computed from calculated cross-sections and mean relative motion temperatures assuming a Maxwellian distribution of velocities $f(v)$. The harmonic motion of the ions in the trap is likely to approach such a distribution if the relaxation time constant for energy transfer between the degrees of freedom[41] of the motion is shorter than the typical measurement times used. At low pressures, ion-ion collisions will provide the dominant collisional relaxation mechanism. The Spitzer[42] time constant $t_s$ provides a good approximation to the observed relaxation in similar traps[4,43] and estimates with typical parameters indicate relaxation times less than 0.5 seconds for doubly-charged ions. Considerably faster relaxation should occur for higher charges, since $t_s$ varies as $q^{-4}$.

## CHARGE TRANSFER TO DOUBLY- AND TRIPLY-CHARGED IONS

Several rate constants for charge transfer to multi-charged ions have been measured to demonstrate the method and to provide information for use in other measurements.[32,39,40] Results are shown in Table 1. The double charge transfer process $Ar^{+2} + N_2 \rightarrow Ar + N^+ + N^+$ was studied since flow-drift tube measurements had shown that the metastable $^1S$ and $^1D$ levels have rate constants that differ from that of the $^3P$ ground state. The ion storage method is not ideally suited to such multi-exponential measurements, unless certain levels can be selectively quenched by introducing a gas which rapidly charge-exchanges with one level but not others, or unless favorable decay rates rapidly deplete metastable populations. The reaction $O^{+2} + O_2 \rightarrow$ products did not show any significant departure from a single exponential despite the presumed presence of metastable level populations. It is interesting that the reaction $N^{+2} + N_2 \rightarrow N^+ + N^+ + N$ appears to proceed by single charge transfer despite level energies similar to those in the $Ar^{+2}$, $N_2$ collision. The $O^{+3} + O_2 \rightarrow$ products rate constant is essentially that predicted by Langevin theory. The mean energies of the $O^{+2}$ and $O^{+3}$ ions were measured to be $\simeq 1$ eV.

TABLE 1

| REACTION | MEASURED RATE CONSTANT ($cm^3/sec$) | |
|---|---|---|
| $N^{+2} + N_2 \rightarrow N^+ + N^+ + N$ | $2.8(0.6) \times 10^{-9}$ | ref. 39 |
| $Ar^{+2}(^1D) + N_2 \rightarrow Ar + N^+ + N^+$ | $1.4(0.2) \times 10^{-9}$ | ref. 40 |
| $Ar^{+2}(^3P) + N_2 \rightarrow Ar + N^+ + N^+$ | $9(1.0) \times 10^{-10}$ | ref. 40 |
| $O^{+2} + O_2 \rightarrow products$ | $1.0(0.3) \times 10^{-9}$ | ref. 32 |
| $O^{+3} + O_2 \rightarrow products$ | $2.5(0.3) \times 10^{-9}$ | ref. 32 |

## MULTI-CHARGED ION CAPTURE FROM H ATOMS

Charge transfer from reactive atoms requires an atomic beam source of the type shown in Fig. 7, with the concomitant low atom density in the trap. The residual $H_2$ pressure in the trap region is increased by the atom beam flux, and it is necessary to introduce the parent gas of the ions being studied. Since complete dissociation is not feasible, the atom beam contains a fraction of $H_2$. Fortunately discharge sources do not produce significant metastable populations. A method has been developed using a valve to isolate the source or to block the beam from passing through the trap in certain measurements. The reciprocals of storage time constants measured under the different gas and atom beam partial density conditions are added or subtracted to arrive at the rate constant for capture from H. Of course, this multiple-measurement procedure reduces the measurement precision. Preliminary data[44] for $O^{+2}$ and $O^{+3}$ capturing an electron from H are in reasonable agreement with the calculations of Dalgarno and his collaborators.[23]

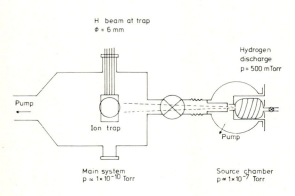

Fig. 7. Beam and trap configuration for measurements of charge transfer from atomic hydrogen.

## STORAGE OF HIGHTY-CHARGED IONS PRODUCED BY HEAVY-ION IMPACT

The Penning trap configuration was modified for use with a stripped fast heavy-ion beam accelerated by the Oak Ridge EN Tandem Van de Graaff as the means of impact ionization of the target gas. An auxilliary hot filament electron source was mounted off-axis to produce lower charge states for trap operation tests and partial pressure calibrations. Initial tests were made with a 35 MeV Cl$^{+5}$ beam post-stripped to a mean charge state near 10, incident on a static neon gas target. The neon pressure was varied between $10^{-8}$ and $10^{-6}$ Torr in the cryo-pumped uhv system. The uhv beam line was effectively isolated from the target chamber by small diameter tubes which transported the beam through the magnet yoke and pole tips. Fig. 8 shows a spectrum of stored neon charge states from +2 to +5, made with $B \leq 0.7T$ and $V_o \leq 50$ V. The +1 charge state was not confined under these conditions. The number of stored +4 and +5 ions was reduced when the post-stripping foil was removed. Preliminary analysis indicates rapid reaction rates for Ne$^{+2}$ and Ne$^{+3}$, and that capture from the residual gas near $10^{-8}$ Torr is not negligible.[36]

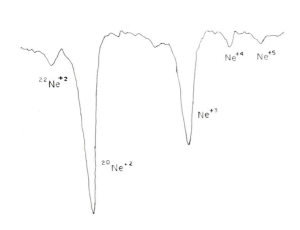

$^{22}$Ne$^{+2}$

$^{20}$Ne$^{+2}$

Ne$^{+3}$

Ne$^{+4}$   Ne$^{+5}$

Fig. 8. Spectrum of stored neon charge states produced by heavy ion impact. Signals are inverted by detector.

Assuming that the ions are produced from, and react with, the same static parent gas with density $n_g$, and that all ions produced are trapped, then an equilibrium number of each charge will eventually be established in the trap, given by $N_e = \bar{\sigma}lk^{-1}(I/e)$ where $I/e$ is the particle current, $\bar{\sigma}$ is the mean cross-section for the production of each m/q ratio, k is the mean rate constant for loss by capture, and l is the interaction length in the trap. Measured parameters may be used to set a lower bound on the production cross-sections by this means. First estimates of the mean energy of the stored ions indicates values of a few eV for Ne$^{+2}$.

CONCLUSION

The techniques and results discussed here support the claim that the Penning
trap will be an increasingly useful tool for the study of collision processes
of highly-charged ions at low energies. Other types of ion traps can of course
be used, and very interesting measurements have already been published with an
electrostatic trap.[35] The possibilities of the method for precision spectroscopy
on multi-charged ions[43] have not yet been exploited.[45]

ACKNOWLEDGMENT

This research would not have reached its current level of development without
the important contributions of my co-workers Dr. H. M. Holzscheiter and Prof. R.A.
Kenefick at Texas A&M, Professors I. A. Sellin and S. Elstron and Dr. S. Huldt
of the University of Tennessee and ORNL, and my students W. S. Burns, R. Doerner,
and J. S. Fant. Research supported by U. S. DOE, Div. of Chem. Sci., Contract No.
DE-AS05-78ER06043, ORNL, and by TAMU Center for Energy and Mineral Resources.

REFERENCES

[1]  Byrne, J. and Farago, P. S., Proc. Phys. Soc. (London) 88, (1965) 801.

[2]  Dehmelt, H. G., Adv. At. Mol. Phy., Bates, D. R. and Esterman, I., Eds.,
     (Academic, New York) Vol. 3, (1967) 53 and Vol. 5, (1969) 109.

[3]  Graff, G., Klempt, E., and Werth, G., Z. Phys. 222 (1969) 201.

[4]  Church, D. A. and Mokri, B., Z. Phys. 244 (1971) 6.

[5]  Walls, F. L. and Stein, T. S., Phys. Rev. Letters 31 (1973) 975.

[6]  Van Dyck, R. S., Schwinberg, P. B. and Dehmelt, H. G., Phys. Rev. Letters
     38 (1977) 310.

[7]  Prior, M. H. and Shugart, H. A., Phys. Rev. Letters 27 (1971) 902.

[8]  Prior, M. H. Phys. Rev. Letters 29 (1972) 611.

[9]  Blumberg, W. A. M., Jopson, R. M., and Larson, D. J., Phys. Rev. Letters
     40 (1978) 1320.

[10] Jopson, R. M. and Larson, D. J., Opt. Letters 5 (1980) 531.

[11] Drullinger, R. E., Wineland, D. J., and Bergquist, J. C., Appl. Phys.
     Letters 22 (1980) 365.

[12] Wineland, D. J., Bergquist, J. C., Itano, W. M., and Drullinger, R. E.,
     Opt. Letters 5 (1980) 245.

[13] Wineland, D. J. and Itano, W. M., Phys. Letters 82A (1980) 75.

[14] Dehmelt, H. G., Invited Papers of the IX ICPEAC, Risley, J. S. and
     Geballe, R., Eds, (University of Washington, Seattle and London, (1976) 857.

[15] Walls, F. L. and Dunn, G. H., J. Geophys. Res. 79 (1974) 1911.

[16] Walls, F. L. and Dunn, G. H., Physics Today 27 (1974) 31.

[17] Heppner, R. A., Walls, F. L., Armstrong, W. T., and Dunn, G. H. Phys. Rev.
     A13 (1976) 1000.

[18] Dubois, R. D., Jeffries, J. B., and Dunn, G. H., Phys. Rev. A17, (1978) 1314.

[19] Barnet, C. F., Invited Papers of the IX ICPEAC, Risley, J. S. and Geballe, R.
     Eds., (University of Washington, Seattle and London (1976) 846.

[20] Barnet, C. F., Atomic Physics 5, Marrus, R., Prior, M. H., and Shugart, H. A.,
     Eds. (Plenum, New York 1977) 375.

[21]  Drawin, H. W., Journal de Physique C1 (1979) 73.

[22]  Dalgarno, A. and Butler, S. E., Comments At. and Mol. Phys. 7 (1978) 129.

[23]  Butler, S. E., Heil, T. G., and Dalgarno, A., Astrophys. J. 241 (1980) 442.

[24]  Post, D., private communication.

[25]  Butler, S. E., Guberman, S. L., and Dalgarno, A., Phys. Rev. A16 (1977) 500.

[26]  Butler, S. E., Bender, C. F., and Dalgarno, A., Astrophys. J. 230 (1979) L59.

[27]  Cohen, J. S. and Bardsley, J. N., Phys. Rev. A18 (1978) 1004.

[28]  Bardsley, J. N., Cohen, J. S., and Wadehra, J. M., Phys. Rev. A19 (1979) 2129.

[29]  Dalgarno, A., Butler, S. E., and Heil, T. G., J. Geophys. Res. 85 (1980) 6047.

[30]  Although the literature is too extensive for complete citation here, publications by several groups include: Okuno, K., Koizumi, T., and Kaneko, Y., Phys. Rev. Letters 40 (1978) 1708; Adams, N. G., Smith, D., and Grief, J., Phys. B12 (1979) 791; Stori, H., Alge, E., Villinger, H., Egger, F., and Lindinger, W., Int. J. Mass Spectrom. Ion Phys. 30 (1979) 263; Johnsen, R. and Biondi, M. A., Phys. Rev. A20 (1979) 87; Howorka, F., Viggiano, A. A., Albritton, L., Ferguson, E. E., and Fehsenfeld, F. C., J. Geophys. Res. 84 (1979) 5941.

[31]  Kienow, E., Klempt, E., Lange, F., and Neubecker, K., Phys. Letters 46A (1974) 441.

[32]  Holzscheiter, H. M. and Church, D. A., (to be published)

[33]  see e.g. Brown, M. D., MacDonald, J. R., Richard, P., Mowat, J. R., and Sellin, I. A., Phys. Rev. A9 (1974) 1470 and references cited.

[34]  Cocke, C. L., Phys. Rev. A20 (1979) 749.

[35]  Vane, C. R., Prior, M. H., and Marrus, R., Phys. Rev. Letters 46 (1981) 107.

[36]  Burns, W. S., Church, D. A., Elston, S., Huldt, S., Kenefick, R. A., and Sellin, I. A., (unpublished work).

[37]  Dehmelt, H. G., and Walls, F. L., Phys. Rev. Letters 21 (1968) 127.

[38]  Church, D. A. and Dehmelt, H. G., J. Appl. Phys. 40 (1969) 3127.

[39]  Church, D. A. and Holzscheiter, H. M., Chem. Phys. Letters 76 (1980) 109.

[40]  Holzscheiter, H. M. and Church, D. A., J. Chem. Phys. 74 (1981) 2313.

[41]  Wineland, D. J. and Dehmelt, H. G., J. Appl. Phys. 46 (1975) 919.

[42]  Spitzer, Lyman, Jr., "Physics of Fully Ionized Gases", (Interscience, New York, 1956).

[43]  Church, D. A., unpublished work.

[44]  Holzscheiter, H. M. and Church, D. A., Abstracts, this conference.

[45]  For spectroscopy used in collision studies, see Mann, R., Beyer, H. F., and Folkmann, F., Phys. Rev. Letters $\underline{46}$ (1981) 646; and J. Phys. B. $\underline{14}$ (1981) 1161.

PHYSICS OF ELECTRONIC AND ATOMIC COLLISIONS
S. Datz (editor)
© North-Holland Publishing Company, 1982

POLARIZED ELECTRONS*

R.J. Celotta, D.T. Pierce, M.H. Kelley[+] and W.T. Rogers[+]
National Bureau of Standards, Washington, D.C. 20234

INTRODUCTION

Previous papers on this topic at ICPEAC and ICAP have described clever techniques
for producing polarized electron beams, novel experiments that could be performed
if sufficient beam intensity were available, and a number of promising experiments
skillfully completed with the huge effort necessary to overcome the inefficient
processes of producing and detecting electron spin polarization.  The spirit of
these papers is well summarized in the concluding remarks of Ob'edkov in his
paper[1] on this subject, presented at the last ICPEAC: "It is easy to continue the
list of problems where the application of polarized electrons could give unique
information, which is unthinkable to get by any other way... further progress
first of all depends on the success of experimental research which will stimulate
theoretical work in the different fields."  We report here on recent experimental
advances, coming primarily from the rapidly developing area of surface physics,
that have given us the technology necessary to obtain this long sought information.
After decribing the two main spin dependent effects,[2] the spin-orbit and exchange
interactions, we describe the GaAs polarized electron source.  We then illustrate
its usefulness with studies of the spin-orbit and exchange interactions on solid
surfaces and, in both cases, describe new, and promising ways of measuring electron polarization.  Finally, we offer our view of the prototype future electron-
atom collision experiment, including full quantum state selection.

Appearing naturally in the Dirac formalism, and giving rise to atomic fine-
structure, the spin-orbit interaction can be simply understood and added in an
ad hoc manner to the non-relativisitic scattering Hamiltonian.  As the incident
electron approaches the nucleus, it experiences a magnetic field due to the
rapidly changing electric field.  The orientation of the electron spin with
respect to this field gives rise to a $-\mu \cdot B$ term in the scattering potential.  This
is usually rewritten in terms of $\ell \cdot s$, $\ell$ being the angular momentum of the collision. The overall strength of the interaction is dependent on the rapid change of
the electric field, and hence, this interaction is most important for high-Z

546                            *R.J. Celotta et al.*

materials.

The exchange interaction can be illustrated in a conceptually straightforward way
if a polarized electron beam and polarized target atom are considered.  The
following processes are possible:

$$e\downarrow + A\uparrow \rightarrow e\downarrow + A\uparrow \qquad\qquad f(\theta) \qquad\qquad (1)$$
$$e\downarrow + A\uparrow \rightarrow e\uparrow + A\downarrow \qquad\qquad g(\theta) \qquad\qquad (2)$$
$$e\uparrow + A\uparrow \rightarrow e\uparrow + A\uparrow \qquad\qquad f(\theta) - g(\theta) \qquad\qquad (3)$$

In the first, no exchange has occurred and the scattering process is characterized
by the direct scattering amplitude, f.  In the second, exchange has occurred, so
the exchange scattering amplitude applies.  The third does not allow a separation
of the two amplitudes, resulting in f-g as the effective amplitude.[2]  By using an
oriented target, and with proper choice of the electron scattering plane, the
effects of the spin-orbit and exchange interactions may be separately discerned.[3]

POLARIZED SOURCES

The quest for an "ideal" polarized electron gun has proven to be a stimulating
intellectual pursuit for many years.[4]  The Fano effect source, described by
Kessler in this volume, is notable for having achieved 100% polarization.  Two
other successful devices were based upon chemi-ionization of optically oriented
metastable helium[5] and field emission[6].  The former has produced high currents
from an atom beam, while the later has exceptional electron optical properties,
akin to ordinary field emission electron guns.

In general, because of beam density and space charge limitations, the trend has
been toward solid state sources.  The most widely accepted device is the GaAs
polarized electron gun;[7] it is being used or constructed by practically every
group involved in polarized electron scattering.  Figure 1 shows the basis for
its operation.  On the left we see the GaAs band structure, with the spin-orbit
split valance band below and the conduction band above, requiring a minimum
photon energy of 1.52 eV (815 nm) for excitation.  Shown to the right are the
degenerate $m_j$ levels and the transitions ($\Delta m_j = \pm 1$) possible with circularly
polarized light.  For $\sigma^+$ ($\sigma^-$) light only the solid (dashed) transitions are
possible, with the relative transition probabilities shown inside of the circles.
If photons of energy near 1.52 eV are used, then no excitations occur from the
lower states.  For $\sigma^+$ ($\sigma^-$) incident radiation we would then find three electrons
in the $m_j = -1/2$ ($m_j = +1/2$) state for every one electron in the $m_j = +1/2$ ($m_j =
-1/2$) state, resulting in a 50% polarization in the excited state.  If nothing

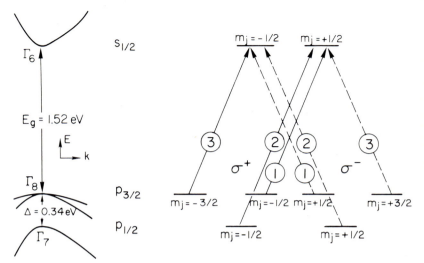

Figure 1

more were done, this state would decay and release its angular momentum through polarized fluorescence. However, when the surface of GaAs is coated with cesium and oxygen, it acquires a negative electron affinity, meaning that the conduction band lies ~0.25 eV *above* the vacuum level. Hence, the polarized excited state can decay via electron emission. The electrons emerge with a mean energy of ~0.25 eV, an energy width of ~0.13 eV (FWHM), and a measured polarization of 43%. Since GaAs is a highly efficient photoemitter, a yield of 20μA for a milliwatt of incident light is readily obtainable. The electron polarization, which lies along the photon direction, is easily reversed optically by changing the circular polarization of the light. This is an important characteristic, because frequently a small spin dependent effect must be observed by phase sensitive detection to discriminate against an overwhelming spin-independent signal. The attractive features of this device have stimulated its use in fields as diverse as tests of the unified field theory[8] and studies of solid surfaces.[3]

## OBSERVATION OF THE SPIN-ORBIT EFFECT

When a collimated electron beam is scattered from a metal single crystal, a specularly diffracted beam (i.e. angle of incidence equals angle of reflection) results. If, in addition, the crystal is a high-Z material and analysis of spin dependent scattering is possible, then data such as seen in Fig. 2 may be obtained. The lower curves show the specularly scattered elastic intensity for spin-up or spin-down incident electrons scattered at an angle of incidence of 16° over the

energy range of 50 to 150 eV. Conventional
measurements would yield the average of these
two curves. It is apparent that the cross
sections for spins up and down differ by a
factor of two at many energies and at some
energies (e.g. 79 eV), near intensity minima,
the factor can be as large as six! The
magnitude of this spin dependence is described
by the top curve, S, where $S = (I_\uparrow - I_\downarrow)/(I_\uparrow + I_\downarrow)$
and $I_\uparrow$ ($I_\downarrow$) is the intensity scattered when
the incident spin is oriented up, i.e.
parallel, (down, i.e. antiparallel) relative
to the scattering plane normal.

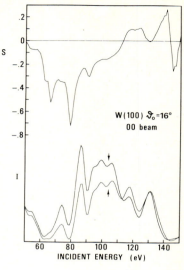

Figure 2

It is straightforward to see from Fig. 2 how
such a scattering process can be used as
either a polarizer or a polarization detector.
An incident, unpolarized beam, composed of
equal numbers of spin up and down electrons,
will become partially polarized because of the cross section difference for spin
up or down. Alternatively, if the polarization direction of a partially polarized beam is reversed then the scattered intensity will change. These curves were
obtained[9] using a GaAs source and modulated incident polarization; the modulated
part of the scattered signal measured the spin dependent part of the interaction
while the dc signal component measured the spin averaged part normally obtained.


RESONANCE EFFECTS


Since strong polarization features have been observed in electron-atom resonance
scattering,[10] the GaAs source was used to study surface resonances.[11,12] The
observation of these resonances was only possible because of the relatively narrow
energy spread (0.13 eV FWHM) produced with the GaAs source. The formation of the
"Rydberg" type resonances we observe can best be understood in the normal incidence geometry. Unlike electron-atom scattering, the periodicity of the crystal
causes the scattering to be collected in a few, very well collimated beams. Here
incident electrons can be specularly scattered back into the incident direction or
diffracted into beams symmetrically located to each side of the incident beam.
Because the momentum transfer from the crystal parallel to the surface is quantized, as the incident energy is reduced, the symmetric diffracted beams move away
from the backscattering direction and eventually an energy is reached which
corresponds to these non-specularly diffracted beams skimming along the crystal

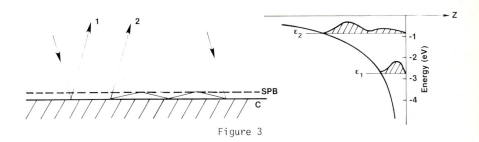

Figure 3

surface. In a metal, such an electron trajectory gives rise to an image charge moving just below and parallel to the surface, and the diffracted electron can find itself trapped in the image charge potential. This potential, shown in the right half of Fig. 3, supports bound "Rydberg" states because of its 1/z character, where z is the coordinate perpendicular to the surface. In order to observe these states, we use a non-normal incidence geometry. Then, as diagrammed in the left half of Fig. 3, we observe the interference between the specularly diffracted beam (1) and one of the non-specular beams (2) which has been diffracted into a quasi-bound surface Rydberg state and then diffracted again to emerge in the specular direction. Observation of elastically scattered electrons discloses an interference between the indistinguishable processes (1) and (2). As the incident energy is scanned below the beam emergence threshold a series of peaks, shown in Fig. 4, is observed,[13] which corresponds to the Rydberg-like energy levels of Fig. 3, right. Figure 4 shows the resonance spectrum as observed with both spin-up and spin-down electrons. The lower energy (n=1) peak is the broadest because of the wave functions larger overlap with the solid and the consequential higher inelastic decay probability. The dashed and solid curves show a spin-orbit splitting which is also dependent upon Rydberg state, with the smallest splittings for the higher, mostly external states which interact weakly with the high-Z tungsten ion cores of the solid. The splitting-to-width ratio is approximately the same for all of the observed peaks at angles of incidence above 23°, but for smaller angles significant variations occur, which can be used to better understand the surface resonance phenomena.[14]

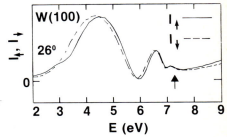

Figure 4

## PLEED DETECTORS

While polarization analysis is possible using specular reflection from single

crystals, the use of non-
specularly diffracted beams
is preferred.  As shown in
Fig. 5, the beam to be analyzed
is put at normal incidence to
the crystal and beams diffracted symmetrically to both
sides are observed.  These
beams show spin dependences
which vary with energy.  The
two detectors will register
different intensities for a

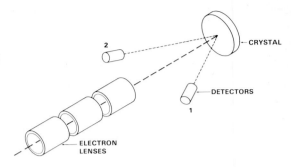

Figure 5

partially polarized beam since an electron with spin "up" (i.e. parallel to the
scattering plane normal, $\hat{n} = (k_i \times k_f / |k_i||k_f|)$) for one detector, will be "down"
(antiparallel) for the other.  This is exactly the same principle as is used in
the traditional Mott scattering method, with a few important differences.  The
energy used is about 100 eV, rather than 100 KeV and consequently the device can
be quite small, and convenient to move about.  Collimated input beams are required
and beam collimation is preserved, so that small solid angle detectors can be
used.  Unfortunately, while high energy Mott scattering can be calculated to the
required accuracy, a previous measurement must have been made of any crystal face
chosen for use as a LEED detector.  So far, such measurements have been made on
tungsten[9,15,16] and gold.[17]  Detectors based on this principle have proven to
have a figure of merit[2] of $S^2 I/I_0 \sim 10^{-4}$, comparable to that of the best Mott
detectors available.  It must be remembered, however, that because of its higher
scattering energy the Mott detector will be able to accept electron beams with
much larger area-solid angle products, so the LEED detector is best suited to the
measurement of low energy, collimated beams.

## OBSERVATION OF THE EXCHANGE INTERACTION

The surface analogue to scattering electrons from spin oriented atomic beams
employs magnetically saturated ferromagnetic materials.  This is accomplished[18] by
using a ferromagnetic Ni target crystal as the "keeper" of an electromagnet as
shown in Fig. 6.  In this way, the stray field is minimized and even low energy
electron scattering is practical.  The exchange interaction produces an intensity
modulation in the scattered signal when the incident spin direction is alternated
between parallel and anti-parallel to the net spin orientation in the sample.  In
contrast to the electron-atom case, the first objective of these measurements is
not to learn about the strength of the collisional exchange interaction, but
instead to use it as a means of studying the surface properties of ferromagnets,

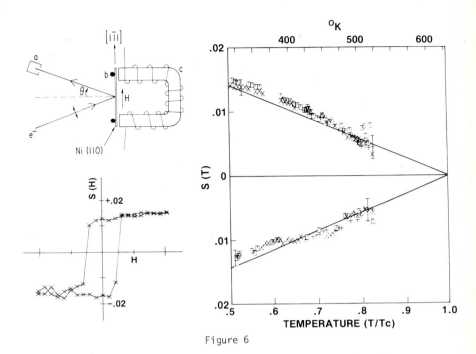

Figure 6

i.e. the surface magnetization. Magnetic studies are usually done by neutron scattering where the weakly interacting neutron provides a wealth of information about *bulk* magnetism. The strongly interacting electrons we observe have mean free paths of ~7Å, so this type of measurement tells us about the magnetic properties of the top one or two atomic layers.

The most basic surface magnetic property to be observed is the hysteresis curve. In Fig. 6 we show the measured parameter $S=(I_\uparrow-I_\downarrow)/(I_\uparrow+I_\downarrow)$, as a function of applied magnetic field. Here $I_\uparrow(I_\downarrow)$ is the scattered intensity when the incident spin is parallel (anti-parallel) to the net substrate spin. The observation of such curves is crucially dependent upon surface cleanliness, with a fraction of a monolayer of impurity totally destroying the effect. Next, the temperature dependence of the surface magnetization was measured[18] over the temperature range of 0.5-0.8 $T_c$ where $T_c$ is the Curie temperature of nickel. The results are shown in Fig. 6 for the two possible externally applied magnetic field directions. The linear dependence discovered is quite unlike the normally observed bulk magneti- zation curve, which rises rapidly below the Curie temperature and is relatively flat over the measured temperature range.

An extension of these surface magnetism measurements was made to amorphous mater-

ials, e.g. ferromagnetic glasses.  In spite of the large signal loss due to the
continuous distribution of the scattered signal throughout space, these measurements[19] proved to be very practical because of the high sensitivity possible with
a modulated GaAs source.  We found that an alternative to detecting a small
fraction of the scattered current was to observe the spin dependent modulation of
the current *collected* by the ferromagnetic target.  An electrometer and a phase
sensitive detector were connected to the surface of the magnetic glass and the
spin dependent asymmetry A in the collected current was measured.  We use the same
definition for A as was used previously for S, $A = (i_\uparrow - i_\downarrow)/(i_\uparrow + i_\downarrow)$, except we now
speak of absorbed rather than scattered currents.  For low incident energies, A
and S have[19] a magnitude of order of 1% and are of opposite sign.  That is, when
up-spins are preferentially backscattered, down-spins give rise to larger net

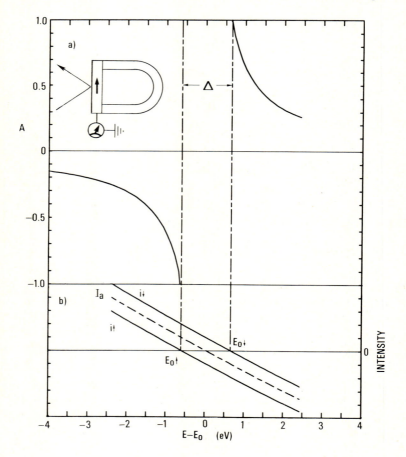

Figure 7

absorbed current.    However, as the incident energy approaches 150 eV, as shown in Fig. 7a, the value of A diverges to -∞ and then returns from +∞.   The reason for this spectacular variation in the spin dependence comes from the intensity normalizing denominator of A going to zero as a result of the energy dependence of the secondary emission coefficient of the target material.   For an incident unpolarized beam, as the energy is increased, more secondaries are produced and near 150 eV the point is reached where the current collected by the sample is zero, as indicated by the dashed curve in Fig. 7b.   This zero crossing energy is spin dependent[19] because of the exchange interaction and a fully polarized incident beam will give rise to an absorbed current of $i_\uparrow$ ($i_\downarrow$) that will have a zero crossing at energy $E_{0\uparrow}$ ($E_{0\downarrow}$) if its spin is oriented parallel (anti-parallel) to the substrate spin orientation.   Hence, at the energy $E_{0\uparrow}$ the spin up component of an incident beam will give rise to no net absorbed current and any such current is due only to spin down incident electrons, corresponding to a spin asymmetry of -1. Likewise, at $E_{0\downarrow}$ only spin up incident electrons give rise to a signal and A=+1.

This effect can be used for polarization detection[20] in many ways, e.g. measure currents at $E_{0\uparrow}$ and $E_{0\downarrow}$, measure the energy intercept, etc.   Also, although discovered in experiments studying the exchange interaction, the source of the spin dependence is immaterial; scattering at an oblique angle from a high-Z target also gives rise to spin dependent absorbtion.[20,21,22]   So far, detectors using a metallic glass,[19,21] tungsten[21], gold[20,22], and even uranium[23] have been tested. They are characterized[20] by their simplicity and their efficiency, which is comparable to the best Mott detectors.   The disadvantage of not being able to count single pulses is mitigated in part by the high intensities available in many experiments with the new source technology.

FUTURE PROSPECTS

The usefulness and capabilities of the GaAs source, as illustrated in the examples given of its application to surface physics, and the promise of the new, compact solid state polarization detectors suggest the possibility of many new experiments in atomic physics.   One avenue that is sure to be explored with the new tools available, is fully state selected collisions between electrons and atoms.   As Hertel has told us in this volume, laser state selection of atomic beams is an exceptionally powerful and practical tool for collision studies.   The optical state selection of electron polarization in the GaAs source can be viewed as an extension of this philosophy, and crossed electron-atom beam experiments with optical state selection of both collision partners is being investigated by a number of groups as the next generation of scattering experiments.   The NBS

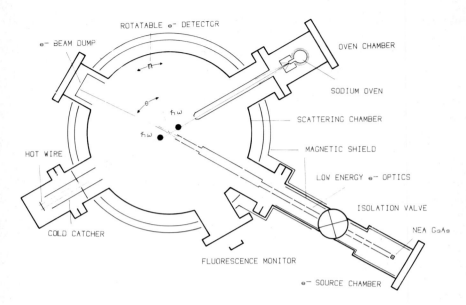

Figure 8

version of such an apparatus is shown in Fig. 8.  Polarized electrons scatter from
an optically prepared Na beam at the center of an ultra-high vacuum, magnetically
shielded chamber.  Single frequency laser pump light, directed into the plane of
the figure, selects a single hyperfine transition to polarize the atom beam.  A
second laser interaction region is used to monitor the atom beam polarization by
fluorescence measurements.  A rotatable scattered electron collector provides
energy analysis and, if required, polarization analysis.

Such an apparatus can be used to prepare the collision partners in completely
defined states, including even the nuclear spin direction, and allows the direct
measurement of the quantum amplitudes and phases of Eq. 1-3.  By moving the first
laser pumping region into the scattering volume, "complete" scattering experiments
can be performed on excited state targets.  The role of exchange in elastic,
inelastic or superelastic scattering can be probed; our knowledge of the spin-
orbit interaction can be extended to low energies.

Surely there will continue to be advances made in polarization sources and
detectors in the future.  But the current state-of-the-art has greatly expanded
the catalogue of possible experiments.  The wide range of application of electron
polarization measurements described at this conference is indicative of the
future vitality of the field.  Now that polarized electron guns rival conventional

electron sources in their performance the appearence of spin resolved experimental
data promises to be a regular occurrence.

REFERENCES

\* This work was supported in part by the Office of Naval Research.

\+ NBS-NRC Postdoctoral Fellows.

1. Ob'edkov, V.D., Electronic and Atomic Collisions, Proc. XIth ICPEAC, Invited
   Papers and Progress Reports, (North-Holland, 1980), pp. 219-223.

2. Kessler, J., Polarized Electrons (Springer-Verlag, 1976).

3. Pierce, D.T., and Celotta, R.J., Adv. in Elec. and Elec. Phys. $\underline{56}$ (1981) 219.

4. Celotta, R.J., and Pierce, D.T., Adv. Atomic & Molecular Phys. $\underline{16}$ (1980) 101.

5. Hodge, L.A., Dunning, F.B., and Walters, G.K., Rev. Sci. Instrum. $\underline{50}$ (1979) 1.

6. Kisker, E., Baum, G., Mahan, A.H., Raith, B., and Reihl, B., Phys. Rev. B $\underline{18}$
   (1976) 2256.

7. Pierce, D.T., Celotta, R.J., Wang, G.-C., Unertl, W.N., Galejs, A., Kuyatt,
   C.E., and Mielczarek, S.R., Rev. Sci. Instrum. $\underline{51}$ (1980) 478.

8. Prescott et al, Phys. Lett. B $\underline{77}$ (1978) 347.

9. Wang, G.-C., Celotta, R.J. and Pierce, Phys. Rev. B $\underline{23}$ (1981) 1761.

10. Heindorff, T., Höfft, J., Reichert, E., J. Phys. B $\underline{6}$ (1973) 477.

11. McRae, E.G., Rev. Mod. Phys. $\underline{51}$ (1979) 541.

12. Adnot, A., and Carette, J.D., Phys. Rev. Lett. $\underline{38}$ (1977) 1084.

13. Pierce, D.T., Celotta, R.J., Wang, G.-C., and McRae, E.G., Solid State Comm.
    $\underline{39}$ (1981) 1053.

14. McRae, E.G., Pierce, D.T., Wang, G.-C., and Celotta, R.J., Phys. Rev. B
    (1981).

15. Kirschner, J. and Feder, R., Phys. Rev. Lett. $\underline{42}$ (1979) 1008.

16. Feder, R. and Kirschner, J., Surf. Sci. $\underline{103}$ (1981) 75.

17. Müller, N., Wolf, D., and Feder, R., In "Electron Diffraction 1927-1977,"
    (P.J. Dobson, J.B. Pendry and A.J. Humphreys, Eds.), Inst. of Physics Conf.
    Series No. 41, (1978) 281.

18. Celotta, R.J., Pierce, D.T., Wang, G.-C., Bader, S.D., and Felcher, G.P.,
    Phys. Rev. Lett. $\underline{43}$ (1979) 728.

19. Siegmann, H.C., Pierce, D.T., and Celotta, R.J., Phys. Rev. Lett. $\underline{46}$ (1981)
    452.

20. Pierce, D.T., Girvin, S.M., Unguris, J., and Celotta, R.J., Rev. Sci. Instrum.
    (1981).

21. Celotta, R.J., Pierce, D.T., Siegmann, H.C., and Unguris, J., Appl. Phys.
    Lett. $\underline{38}$ (1981) 577.

22. Erbudak, M. and Müller, N., Appl. Phys. Lett. $\underline{38}$ (1981) 575.

23. Erbudak, M., private communication.

PHYSICS OF ELECTRONIC AND ATOMIC COLLISIONS
S. Datz (editor)
© North-Holland Publishing Company, 1982

ELECTRON MOLECULE COLLISION AT LOW ENERGY
RESONANCE AND THRESHOLD PHENOMENA

Roger Azria

Laboratoire de Collisions Atomiques et Moléculaires
(associé au CNRS)
Université Paris-Sud, Bât. 351, 91405 ORSAY CEDEX
(France)

INTRODUCTION

During the last five years electron molecule scattering experiments at low energy
have been dominated by the study of HX molecules (X = F, Cl, Br, I), because of
their unusual scattering properties at low energy.

In vibrational excitation functions of HF, HCl, HBr, threshold resonances charac-
teristised by isotropic angular distributions have been observed indicating that
short-range interactions are involved in the process. The same phenomenon have
been reported for the vibrational excitation functions of others polar ($H_2O$, $H_2S$,
$CCl_3F$) and non polar ($CH_4$, $SF_6$) molecules (for a review see Rohr 1979).

In dissociative attachment the cross sections for $X^-/HX$ (F, Cl, Br) were obtained
showing
i) a vertical onset for these processes and an unusual step structure at incident
energies corresponding to vibrational thresholds of HX (Abouaf et al. 1977, Ziesel
et al. 1975, Azria et al. 1980)

ii) a small structure located close to the maximum of $X^-/HX$ cross section, appearing at any scattering angle (fig. 1 and 2). This structure which is not understood at the present time seems characteristic of $X^-/HX$ cross section as the step
structure previously mentioned

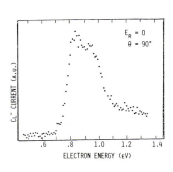

Fig.1 - Constant ion energy
spectrum for $Cl^-/HCl$ taken
at $E_R = 0$

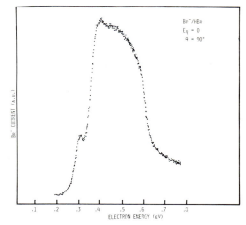

Fig.2 - Constant ion energy spectrum for $Br^-/HBr$
taken at $E_R = 0$

iii) a strong increase of cross sections for $Cl^-/HCl$ and $F^-/HF$ with initial nuclear excitation of the HCl and HF targets. (Allan et al. 1981).

Extensive theoretical work was originated from these experimental results (see for instance Taylor et al. 1977, Herzenberg 1979, Domcke et al. 1981). However in spite of this work, the physical mecanisms invoked in order to account for these observations are still controversial. The main question being whether resonant electron-molecule scattering in a presence of a dipole scattering potential (Domcke et al. 1981) or virtual states (Herzenberg 1979) dominate electron scattering at low energy.

In this report, we will first discuss some experimental aspects of the threshold resonances observed in vibrational excitation functions in the light of recent results concerning vibrational excitation of HBr near threshold. Secondly we will present new results on $H^-$ formation by dissociative attachment in HX molecules (X = Cl, Br, I). The case of HF will be treated separatly because of the special features observed in $F^-$ and $H^-$ formations.

EXPERIMENTAL

Two different apparatus have been used for this studies. The fist one, used to study $F^-$ ions from HF, already described in details (Abouaf et al. 1976), consists of a magnetic mass spectrometer using a trochoïdal monochromator as electron gun.

The second apparatus (fig. 3) consists of a cross beam electron impact spectrometer with two 127° electrostatic energy filters to produce the incident electron beam and analyse in energy all the scattered negative particles. Because an electrostatic filter does not distinguish between electrons and negative ions with the same

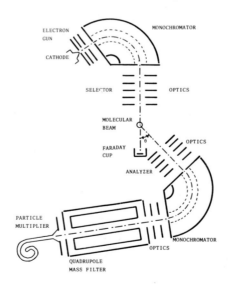

Fig.3 - Electron impact spectrometer with
quadrupole mass filter

energy, negative ions are separated from electrons and mass analysed through a quadrupole mass filter. When the quadrupole radio frequency is off, electrons plus negative ions are detected and this set up is used as a classical electron impact spectrometer. The whole analyser can rotate from 10° to 120° with respect to the incident electron beam.

## Negative ion contribution to threshold peaks

As mentioned in the introduction threshold resonances in vibrational excitation functions of a number of molecules have been reported.

Experimentally, these excitation functions have been obtained by means of an electron scattering apparatus which consists mainly of an electron gun and a rotatable electron detector, each system having a 127° energy selector. Practically these excitation functions are obtained by setting the incident electron energy $E_i$ and the analysis energy of the scattered electrons $E_r$ to the energy $E_v$ of the vibrational level under consideration and zero respectively and then by sweeping these energies. The threshold peaks in the excitation functions are then due to scattered electrons having a residual energy close to zero.

Because an electrostatic energy filter does not distinguish between negative particles with the same charge and the same kinetic energy, if negative ions are formed by dissociative attachment with no kinetic energy and a large cross section in the energy range of the threshold peaks, they must contribute to these peaks.

It follows then that when a excitation threshold $E_K$ of a molecule is in the energy range of negative ions formed with no kinetic energy and a large cross section, its corresponding excitation function will, in any case, exhibit a threshold peak due to these negative ions since these particles which are detected when $E_i = E_K$ and $E_r = 0$ are no longer collected when the residual energy becomes greater than the maximum kinetic energy of these ions. Such situation is illustrated on figure 4.

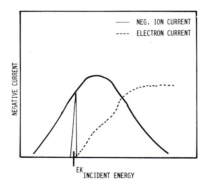

Fig. 4 - Illustration a threshold peak simulation, by negative
ions, in the excitation function of an hypothetical
level $E_K$.

Using the electron impact spectrometer described in section 2, we have reinvestigated vibrational excitation close to threshold in HBr (Azria et al. 1980) as for this molecule (i) strong threshold peaks are observed in vibrational excitation functions up to v = 5, (ii) the dissociative attachment cross section for $B_r^-$ formation has been obtained in the energy range 0-2 eV (Ziesel et al. 1975, Abouaf et al. 1980), the absolute value of the cross section being $2.7 \times 10^{-16} cm^2$ at .28 eV (Christophorou et al. 1978) and the kinetic energy of $Br^-$ ions being close to zero in the whole energy range of formation of these ions because the ratio m/M is ~ 1 m and M being the masses of $Br^-$ ions and HBr molecule respectively. We have shown that the narrow peaks close to threshold reported by Rohr in the vibrational cross section of HBr for v = 3 up to v = 5 are due to $Br^-$ ions and that for v = 1 and v = 2 the threshold peaks observed contain about 5 % and 40 % of $Br^-$ ions respectively.

The above considerations and results indicate that threshold peaks in vibrational excitation functions must be regarded with some caution every time that negative ions with kinetic energy close to zero are formed with large cross section in the energy range of these peaks. In particular in $SF_6$, which is a very important case because of the non polar nature of this molecule, vibrational excitation of the $\nu_1$ mode (Rohr 1977a, b) must be reinvestigated carefully close to threshold as stable $SF_6^-$ ions are formed by electron impact on $SF_6$ with a very large cross section (Melton 1970).

## $H^-$ formation in HX molecules (X = Cl, Br, I)

$H^-$ formation in these molecules proceeds through the reactions

$$e + HX \rightarrow HX^{-*} \rightarrow H^- (^1So) + X(^2P_{3/2})$$
$$\rightarrow H^- (^1So) + X(^2P_{1/2}),$$

the lower limit for the dissociating products being $H^-(^1So) + X(^2P_{3/2})$ and the spin orbit splitting $X(^2P_{3/2} \leftrightarrow {}^2P_{1/2})$ being 109, 457 and 987 meV respectively for Cl, Br and I atoms. According to Wigner and Witmer correlation rules, three states $^2\Sigma_{1/2}$, $^2\Pi_{3/2}$ and $^2\Pi_{1/2}$ are correlated to these limits the two first to the limit $H^-(^1So) + X(^2P_{3/2})$ and the third one to the limit $H^-(^1So) + X(^2P_{1/2})$.

In a previous study of $H^-$/HCl (Azria et al.1980), effects due to the spin orbit splitting of Cl atoms were missed as the electron energy resolution used for this work was typically 120 meV. It has then been shown that $H^-$ ion yield associated with $Cl(^2P)$ fragments exhibits two peaks at 7.1 and 9.3 eV. The first one has a non-vertical onset and the angular distribution of $H^-$ ions are charcteristic of a $d\sigma$ wave indicating that $H^-$ ion formation in the energy range of this peak proceeds via a $^2\Sigma^+$ $HCl^-$ state. The second peak with angular distribution of $H^-$ ions characteristic of a $d\Pi$ wave proceeds via a $^2\Pi$ $HCl^-$ state.

Figure 5 shows a kinetic energy distribution of $H^-$/HCl at 6.8 eV i.e. in the energy range of the $^2\Sigma^+$ $HCl^-$ state. This curve exhibits two peaks not completely resolved

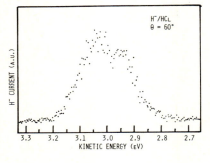

Fig. 5 -Kinetic energy distribution of $H^-$/HCl at 6.8 eV electron incident energy.

at 3.04 and 2.940 which correspond respectively to H$^-$ ions associated to Cl($^2P_{3/2}$) and Cl($^2P_{1/2}$) atoms in agreement with the balance energy. Such a result is not expected from the adiabatic correlation rules.

In fact this effect was first observed in H$^-$ formation in HBr. Figure 6 shows H$^-$ ion yield associated with Br($^2P_{3/2}$) (upper curve) and with Br($^2P_{1/2}$) atoms (lower curve) for an angle of observation of 40°. Both curves exhibit two peaks in the same energy range, 5-7 eV for the first one, 7.5-10 eV for the second one. Such

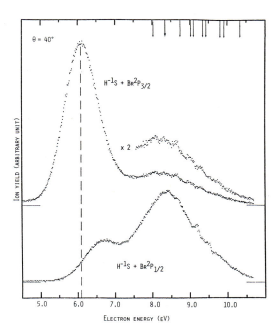

Fig. 6 - Energy dependence of H$^-$ formation from HBr taken at $\theta$ = 40°. The upper curve corresponds to H$^-$ associated with Br($^2P_{3/2}$) atoms and the lower are to H$^-$ associated with Br($^2P_{1/2}$) atoms.
The structure observed on the high energy tail of the second peak of these curves are due to the interaction of the $^2\Pi_{3/2, 1/2}$ HBr$^-$ states with Feshbach resonance.

result concerning H$^-$ ions associated with Br($^2P_{1/2}$) atoms is not expected. The angular distributions of H$^-$ ions associated with Br($^2P_{3/2}$) and Br($^2P_{1/2}$) atoms at 6.1 and 8.3 eV are shown on figure 7.

At 6.1 eV, the angular distributions observed which are quite similar for H$^-$(Br$^2P_{3/2}$) or H$^-$(Br$^2P_{1/2}$) ions are characteristic of a d$\sigma$ wave (Azria et al. 1980) indicating that H$^-$ ions associated with Br($^2P_{3/2}$ or $^2P_{1/2}$) are formed via a $^2\Sigma^+$HBr$^-$ state. At 8.3 eV tha angular distributions observed are also similar and characteristic of a d$\Pi$ wave in agreement with a $^2\Pi$ HBr$^-$ resonant state.

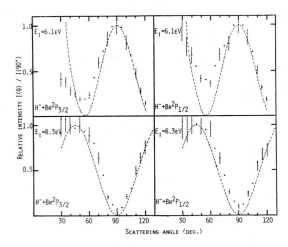

Fig. 7 - Measured angular distribution of H⁻ ions from HBr at
6.1 and 8.3 eV.
For the upper cures, the broken curves represent a
$(1-3\cos^2\theta)^2$ distribution characteristic of a pure dσ
wave.For the lower curves, the broken curves represent
a $(\cos^2\theta\sin^2\theta)$ distribution characteristic of a pure
dΠ wave.

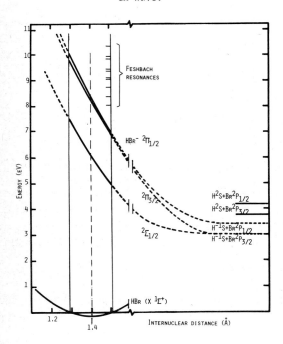

Fig. 8 - Possible set of
potential energy curve of
HBr⁻ deduced from experimental
observation.

If we look to a plausible set of adiabatic potential energy curves for HBr⁻ states correlated to the limits H⁻ + Br($^2P_{3/2}$) and H⁻ + Br($^2P_{1/2}$) (fig.8), this results indicate that in energy range of $^2\Sigma_{1/2}$ HBr⁻ state, H⁻ ions associated with Br($^2P_{1/2}$) atoms are formed via a non adiabatic transition : at large internuclear separation H-Br, HBr⁻($^2\Sigma_{1/2}$ and $^2\Pi_{1/2}$) resonant states are strongly coupled. It is the first observation of such transitions in dissociative attachment processes.

The ratio of H⁻ ion current (Br$^2P_{1/2}$) to H⁻ ion current (Br$^2P_{1/2}$) have been measured as a function of incident energy in the energy range of the $^2\Sigma_{1/2}$ HBr⁻ state (table 1)

| $E_i$ (eV) Incident electron energy | 5. | 5.25 | 5.5 | 5.75 | 6. | 6.25 | 6.5 | 6.75 | 7. | 7.25 | 7.5 |
|---|---|---|---|---|---|---|---|---|---|---|---|
| $\dfrac{I(H^- + Br^2P_{1/2})}{I(H^- + Br^2P_{3/2})}$ | .00 | .03 | .05 | .09 | .10 | .20 | .28 | .36 | .44 | .53 | .70 |

TABLE 1

This ratio increases very fast with the incident electron energy indicating a strong energy dependant coupling at large internuclear distance, between the two resonant HBr⁻ states.

Concerning H⁻ formation in the energy range 7.5-10 eV, the ratio H⁻(Br$^2P_{3/2}$ / H⁻(Br$^2P_{1/2}$) cannot be easely related to a coupling between $^2\Sigma_{1/2}$ and $^2\Pi_{1/2}$ resonant HBr⁻ states, since the $^2\Pi_{3/2}$ state which is in the same energy range as the $^2\Pi_{1/2}$ state is involved in H⁻ formation associated with Br$^2P_{3/2}$ atoms.

These different observations are not understood at the present time.

We have studied H⁻ formation in HI. Figure 9 shows H⁻ ion yield associated with I($^2P_{3/2}$) (upper curve) and I($^2P_{1/2}$) (lower curve) atoms.

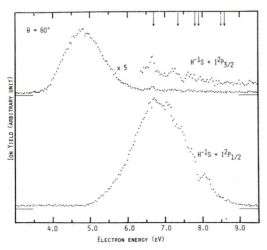

Fig. 9 - Energy dependence of H⁻ formation from HI.

The H⁻(I²P₃/₂) ion yield exhibits two peaks at 4.8 and around 6.5 eV the latter being very small and the H⁻(I²P₁/₂) yield exhibits a single peak at 6.7 eV.

The peaks of the upper curve are attributed to the $^2\Sigma_{1/2}$ and $^2\Pi_{3/2}$ HI⁻ resonant states and the peak of the H-(I²P₁/₂) to the $^2\Pi_{1/2}$ state in agreement with the Wigner and Witner correlation rules. Then in the case of HI, because of the large spin-orbit splitting of I atom (987 meV) there is no coupling between the $^2\Sigma_{1/2}$ and $^2\Pi_{1/2}$ HI⁻ states. The dissociation of HI⁻ states leading to H⁻ formation occurs adiabatically.

In conclusion, strong dynamical coupling between $^2\Sigma_{1/2}$ and $^2\Pi_{1/2}$ HX⁻(X = Br, Cl) states is observed in H⁻ formation by dissociative attachment in these molecules. This effect has been studied in details in the case of HBr and evidenced in HCl. The details studies in HCl which are in progress seem to indicate a stronger effect than in HBr due to the smaller spin-orbit splitting in Cl atom (109 meV). These effects are not understood at the present time.

## NEGATIVE ION FORMATION IN HF

### F⁻ IONS

The variation of F⁻/HF cross section versus electron energy is shown in figure 10 and 11 from the onset up to 16 eV (Abouaf et al. 1980 a, b).

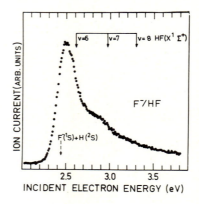

Fig. 10 - F⁻/HF ion current versus electron energy. The arrows indicate the energy position of the vibrational levels of HF.
The dashed arrow shows the location of the thermodynamical threshold.

The breaks in the cross section shown in figure 10 occur at the energy position of the v = 6 and v = 7 levels of the HF molecule and attributed to the same phenomenon as observed in Cl⁻/HCl and Br⁻/HBr cross section i.e. cusps due to the opening of new decay channels for the HF⁻($^2\Sigma^+$) state.

In figure 11 the cross section appears to consist of one strong series of peaks between 12.5 and 14 eV and two weaks series between 14.5 and 16 eV. The strong

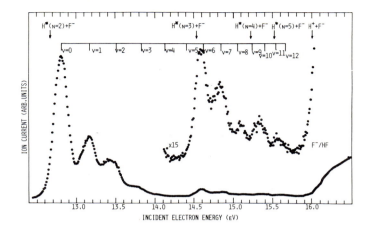

Fig. 11 - F⁻/HF ion current versus electron energy.
The arrows indicate the position of the various limits
F⁻ + H* (n = 2 to 5) and the ion pair threshold (F⁻ + H⁺).
Also are represented the possible location of the HF⁻($^2\Pi$)
vibrational levels.

rise observed around 16 eV is due to the ion pair formation F⁻ + H⁺ (16.07 eV).
These features have been attributed to production of F⁻ ions via the predissociation of the $^2\Pi$ Feshbach resonance observed in transmission experiments (Spence
et al. 1975, Abouaf et al. 1980 a). This $^2\Pi$ state would be predissociated by
negative ion states leading to F⁻ + H*(n).

The threshold for the process leading to F⁻ + H*(n = 2) is 12.67 eV and the first
F⁻ peak observed corresponds to the v = 0 vibrational level of the $^2\Pi$ HF⁻ state at
12.8 eV, the smaller peaks corresponding to v = 1, 2, 3 and 4.

The second series of peaks around 14.6 eV are sttributed to the predissociation of
the same $^2\Pi$ HF⁻ state ( v > 6) by the dissociative state correlated to F⁻ + H*(n=3)
(limit 14.56 eV). The third and the fourth series are interpreted in the same
way the dissociative HF⁻ states being correlated to F⁻ + H* (n = 4) and F⁻ + H*
(n = 5) respectively.

Processes of this type were looked for and not observed in Cl⁻/HCl and Br⁻/HBr
cross section.

## H⁻/HF

The yield of H⁻ ions associated with F($^2$P) fragments is shown on figure 12 for an
angle of observation of 60°.

This curve exhibits two peaks at 9.5 and 12.5 eV in agreement with the Wigner-
Witmer correslation rules applied to the limit H⁻($^1$So- + F($^2$P). The first process
does not show a vertical onset indicating that the potential energy curve corresponding to this process is repulsive in the Franck-Condon region.

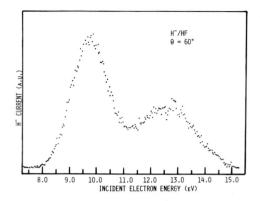

Fig. 12 - Energy dependence of H⁻ formation from HF.

The angular distribution of H⁻ ions in the energy range of these peaks is shown on figure 13. At 9.5 eV the angular distribution is maximum around 120° and at 12.5eV the angular distribution presents a maximum around 60° and then fall off rapidly.

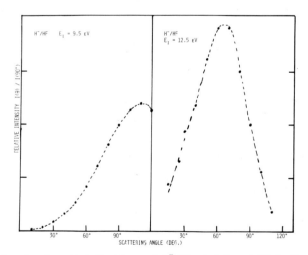

Fig. 13 - Angular distributions of H⁻/HF at 9.5 and 12.5 eV.

These angular distributions look totally different from those obtained for H⁻/HCl or H⁻/HBr. It seems that the angular distribution at 9.5 eV may be understood if we assume a $\Pi$ symmetry for the HF⁻ resonant state involved, the partial waves being $Y_1^1$ and $Y_2^1$.

Concerning the angular distribution at 12.5 eV, we must assume a $\Sigma$ HF⁻ state with two partial waves $Y_1^0$ and $Y_3^0$. It turns then out that for the HF⁻ states leading to H⁻ ions formations the $^2\Pi$ state is lower than the $^2\Sigma$ state in contradiction with the situation in the corresponding HX$^-$ states (X = Cl, Br, I).


ACKNOWLEDGMENT

These experimental results were obtained with the collaboration of Y. Le Coat and M. Tronc for the H$^-$/HX (X = Cl, Br, I) formation and R. Abouaf and D. Teillet-Billy for HF. The author wishes to express the pleasure he experiences working with them and is grateful to J.P. Ziesel and J.P. Gauyacq and V. Sidis for many interesting discussions.


REFERENCES

Abouaf R., Paineau R. and Fiquet-Fayard F., 1976, J. Phys. B : Atom. Mol. Phys. 9 303

Abouaf R. and Teillet-Billy D., 1977, J. Phys. B : Atom. Molec. Phys. 10 2261-8

Abouaf R. and Teillet-Billy D., 1980 a, J. Phys. B. Atom. Molec. Phys. 13L 275
                                    1980 b, Chem. Phys. Lett. 73 106

Allan M. and Wong S.F., 1981, J. Chem. Phys. 74 (3) 1687

Azria R., Le Coat Y., Simon D. and Tronc M., 1980 J. Phys. B : Atom. Molec. Phys. 13 1909

Azria R., Le Coat Y. and Guillotin J.P. 1980 J. Phys. B : Atom. Molec. Phys. 13L 505

Christophorou L.G., Compton R.N. and Dickson H.N. 1968 J. Chem. Phys. 48 1949

Domcke W. and Gederbaum L.S. 1981 J. Phys. B : Atom. Molec. Phys. 14 149

Herzenberg A. 1979 Symposium on Electron Molecule Collisions, Tokyo, Ed I. Shimamura and M. Matsuzawa.

Melton C.E. 1970 Principles of Mass Spectrometry and Negative Ions (New York : Maral Dekker)

Rohr K. 1977 a J. Phys. B : Atom. Molec. Phys. 10L 399
        1977 b J. Phys. B : Atom. Molec. Phys. 10  1175
        1979 Symposium on Electron Molecule Collisions, Tokyo, Ed. L. Shimamura and M. Matsuzawa

Spence D. and Noguchi T. 1975 J. Chem. Phys. 63 505-13

Taylor H.S., Goldstein E and Segal G.A. 1977 J. Phys. B : Atom Molec. Phys. 10 2253-9

Ziesel J.P., Nenner I. and Schulz G.J. 1975 J. Chem. Phys. 63 1943-9

PHYSICS OF ELECTRONIC AND ATOMIC COLLISIONS
S. Datz (editor)
© North-Holland Publishing Company, 1982

# THEORETICAL METHODS FOR LOW-ENERGY
# ELECTRON-MOLECULE SCATTERING

W. Domcke*

Arthur Amos Noyes Laboratory of Chemical Physics
California Institute of Technology
Pasadena, California 91125, U. S. A.

and

L. S. Cederbaum

Phys.-Chem. Institut, Universität Heidelberg
D-69 Heidelberg, West Germany

Some basic concepts of the theory of resonant electron-
molecule scattering, dissociative attachment and related
processes involving a transfer of energy from low-energy
electrons to the internal nuclear motion of a molecule are
discussed. The complex non-local potential theory of nuclear motion in resonance states is reviewed. It is shown
that the analytic structure of the fixed-nuclei S-matrix for
electron scattering near threshold is completely determined
by the threshold law for the width function. A unified description of resonances, virtual states and bound states is
obtained and specifically applied to polar molecules. Exact
multi-channel calculations of vibrational excitation functions are performed for a simple model system, yielding a
qualitative explanation of the threshold peaks observed in
electron-polar molecule scattering.

## INTRODUCTION

It has long since been recognized that scattering or reaction processes involving
a transfer of energy from low-energy electrons to the internal nuclear motion
of a molecule or vice versa proceed most effectively via short-lived negative
ion states. Although much work has been done on the theory of electron impact
induced vibrational excitation, dissociative attachment and recombination, associative detachment and so forth, a comprehensive understanding of these important reaction processes is still missing. This fact is highlighted, for
example, by the controversial discussion on the origin of the threshold peaks in
the vibrational excitation functions of polar molecules [1, 2]. Several theoretical explanations have been put forward [3-6], but there is still no consensus
on the basic nature of the phenomenon. In this article we present some simple
theoretical concepts which might form the basis for a unified description of the
above-mentioned processes.

## DYNAMICAL THEORY OF ELECTRON-MOLECULE SCATTERING

In the present context, "dynamical" stands for "full inclusion of nuclear motion
effects", well beyond the adiabatic nuclei and Born-Oppenheimer approximations. This section is actually a brief and slightly modified review of theories
developed independently by Chen [7], O'Malley [8], Bardsley [9] and others
for electron scattering and dissociative attachment, starting from Feshbach's
projection operator approach [10] or Fano's theory of configuration interaction

in the continuum [11]. To understand the essentials, it suffices to consider just one discrete electronic state $|d\rangle$ interacting with one electronic continuum $|k\rangle$. Including one vibrational degree of freedom and omitting the rotations, we have the Hamiltonian

$$H = H_0 + V \tag{1a}$$

$$H_0 = |d\rangle(\tilde{H}_0 + \epsilon_d)\langle d| + \sum_k |k\rangle(\tilde{H}_0 + \epsilon_k)\langle k| \tag{1b}$$

$$\tilde{H}_0 = T_N + V_0(R) \tag{1c}$$

$$V = \sum_k |d\rangle V_{dk}\langle k| + h.c. \tag{1d}$$

$T_N = -(2\mu)^{-1} d^2/dR^2$ is the kinetic energy of vibrational motion ($\mu$ is the reduced mass corresponding to the vibrational coordinate R) and $V_0(R)$ is the electronic potential energy of the target. $\epsilon_d$ is the energy of the discrete state relative to the electronic energy of the target and is assumed to depend on R. $\epsilon_k$ is the energy of a continuum electron and is thus independent of R. The interaction V mixes $|d\rangle$ with $|k\rangle$ and converts the discrete state into a resonance. The R-dependence of $\epsilon_d$ is the essential ingredient which leads to the coupling between electronic and vibrational motions.

Considering $H_0$ as the unperturbed Hamiltonian and V as the interaction, the electron scattering problem defined by eq. (1) may be solved by the standard T-matrix formalism of scattering theory [12]. The only additional approximation to be introduced is the Born-Oppenheimer approximation for the basis states $|d\rangle$ and $|k\rangle$, implying that the nuclear kinetic energy operator $T_N$ commutes with the electronic wave functions of the discrete and continuum states. The resulting expression for the integral multi-channel vibrational excitation cross sections is [13]

$$\sigma_{v0} = v\pi/(2E_i) \Gamma(E_i) \Gamma(E_f) |\langle v|(E_t - \mathcal{K})^{-1}|0\rangle|^2 \tag{2a}$$

$$\mathcal{K} = T_N + V_{opt} \tag{2b}$$

$$V_{opt} = V_1(R) + \Delta(E_t - \tilde{H}_0) - \tfrac{1}{2}i\Gamma(E_t - \tilde{H}_0) \tag{2c}$$

$$V_1(R) = V_0(R) + \epsilon_d(R) \tag{2d}$$

$$\Gamma(E) = 2\pi \sum_k V_{dk}\,\delta(E - \epsilon_k) V_{kd} \tag{2e}$$

$$\Delta(E) = \frac{1}{2\pi} P \int dE'\, \Gamma(E')/(E - E'). \tag{2f}$$

$E_i$ and $E_f$ are the initial and final energies of the scattered electron and $E_t = E_i + \langle 0|\tilde{H}_0|0\rangle$ is the total energy which is conserved in the scattering process. $^1v$ counts the electronic degeneracy of the resonance state. $|0\rangle$ and $|v\rangle$ are the ground and the v-th excited vibrational state of the target molecule. $\Gamma$ and $\Delta$ are the width and the level shift of the discrete state acquired by the interaction with the continuum. $V_{opt}$ is the well-known optical potential [10] governing the nuclear motion in the resonance state. It is complex, energy-dependent and, moreover, dependent on the kinetic energy of nuclear motion. As a consequence of the latter fact, $V_{opt}$ is non-diagonal in the nuclear coordinate representation, i.e., it is a non-local potential. Equation (2) is essentially equivalent to the expressions derived by Chen [7] and O'Malley [8] (for dissociative attachment) and the integro-differential equation for the nuclear wave function given by Bardsley [9]. These expressions are extremely difficult to evaluate, in general, due to the non-locality of the optical potential. Usually one resorts very quickly to the local approximation, replacing the energy-dependent operators $\Gamma$ and $\Delta$ by functions of R, though this is known to be inadequate near threshold [9]. It is very useful to notice that there exists a representation of the effective Hamiltonian $\mathcal{H}$, where $\Gamma$ and $\Delta$ become exactly local, namely the representation in the target vibrational states. This will be the clue to the rigorous evaluation of eq. (2) to be discussed below.

## ANALYTIC S-MATRIX THEORY IN THE FIXED-NUCLEI LIMIT

Analytic continuation of the scattering matrix into the complex momentum or energy planes has proven to be a useful tool to rationalize scattering phenomena in simple terms, especially in nuclear and particle physics. It is usually assumed that any pronounced feature in the cross section such as a resonance peak is associated with a singularity of the analytically continued S-matrix. Equation (2) provides the basis for an analytic theory of resonant electron-molecule scattering. As we are dealing with a multi-channel problem, analytic continuation leads to a multi-sheeted Riemann surface [19]. We will not enter into this rather specialized topic here. Rather we confine ourselves to the fixed-nuclei limit, where we have a single-channel problem with the inter-nuclear distance R as a variable parameter.

In the fixed-nuclei limit ($\mu \to \infty$, $T_N \to 0$) eq. (2) reduces to the standard Breit-Wigner type expression for a resonance

$$\sigma = \frac{2\pi\nu}{E} \frac{[\Gamma(E)/2]^2}{[E - \epsilon_d(R) - \Delta(E)]^2 + [\Gamma(E)/2]^2} . \tag{3}$$

The corresponding S-"matrix" reads

$$S(E) = [1 + iK(E)][1 - iK(E)]^{-1} \tag{4}$$

with the real K-"matrix"

$$K(E) = \tan \delta(E) = -\tfrac{1}{2}\Gamma(E)/[E - \epsilon_d(R) - \Delta(E)] . \tag{5}$$

$\delta(E)$ denotes the fixed-nuclei phase shift. These are the standard expressions for resonances in potential scattering [12]. In contrast to the usual treatments, however, we will take full account of the energy dependence of the width and the level shift.

To find the analytic continuation of $S(E)$ we realize that $\Delta$ and $\Gamma/2$ are the real and imaginary parts of the complex function

$$F(z) = \frac{1}{2\pi} \int dE \, \Gamma(E)/(z - E), \tag{6}$$

for z on the real axis. Considering F as a function of the complex momentum variable k, we define the Jost function [12] as

$$f(k) = 1 - F(k)/[k^2/2 - \epsilon_d(R)] . \tag{7}$$

The analytic S-matrix is given in terms of the Jost function as [12]

$$S(k) = f(-k)/f(k) .$$ (8)

The poles of $S(k)$ are thus given by the zeros of $f(k)$, i.e., by the solutions of the complex equation

$$k^2/2 - \epsilon_d(R) - F(k) = 0.$$ (9)

The pole positions depend on the internuclear distance R via the discrete level energy $\epsilon_d(R)$.

Inspection of eqs. (2-9) shows that the width function $\Gamma(E)$ is the key quantity of the formalism. First of all, it determines the level shift function $\Delta(E)$ via the Hilbert transform (2f). The energy dependence of $\Gamma$ and $\Delta$ in turn governs the importance of non-adiabatic or non-local effects, as is seen from eq. (2c). $\Gamma(E)$ determines also the analytic structure of the S-matrix as shown by eq. (6). We emphasize that the discrete level energy $\epsilon_d$ and the width function $\Gamma(E)$ are rigorously defined quantities within the Feshbach projection operator approach [10]. Methods for the explicit calculation of $\epsilon_d$ and $\Gamma(E)$ have been developed recently [14].

We can proceed without explicit calculation of $\Gamma(E)$ by noting that the width function is subject to Wigner's threshold laws [15] for $E \to 0$. According to these laws, the threshold onset of $\Gamma(E)$ depends only on the angular momentum of the scattered electron and the leading multi-pole moments of the molecular charge distribution. This allows us to derive universal results on the analytic structure of the S-matrix close to threshold.

As an example for the application of these ideas we consider electron scattering from polar molecules, where the long-range dipole potential poses non-trivial problems at low energies. As is well-known from work on photodetachment cross sections [16], the usual threshold onset for s-wave scattering, $\Gamma \sim E^{1/2}$, is replaced by $\Gamma \sim E^{\alpha}$ in the presence of a dipole potential, where $\alpha$ approaches zero as the dipole moment approaches its critical value [17]. A simple parametrization of the width function which is in accord with this threshold law is

$$\Gamma(E) = A(E/B)^{\alpha}(2-E/B)^{\alpha}$$ (10)

where A and B are free parameters. With this simple form for $\Gamma(E)$, the integral (6) can be expressed in terms of hypergeometric functions [18]. The poles of the S-matrix in the complex momentum plane are obtained by the solution of eq. (9). Taking A = 3 eV, B = 7 eV, we obtain the pole trajectories shown in fig. 1 for four values of the threshold exponent $\alpha$. Note that the form of these trajectories depends only on the width function $\Gamma(E)$. The dependence of $\epsilon_d$ on the internuclear distance R, on the other hand, determines the "velocity" of movement of the poles as a function of internuclear distance.

For $\alpha = 0.5$ (pure s-wave scattering) we obtain two complex resonance poles in the lower half plane, situated symmetrically to the imaginary axis. Assuming $\epsilon_d(R)$ to decrease with increasing R, the poles move towards the imaginary axis as R increases. Having met the imaginary axis, one pole moves up and one down on the imaginary axis. These poles are called virtual state poles [12,19]. Once the upward moving pole has crossed the origin, it represents a bound state. This pattern of pole motion is in full accord with general analytic S-matrix theory for s-wave scattering from short-range potentials [19]. This confirms that the Breit-Wigner type expression (eq. (3)) combined with the threshold law for $\Gamma(E)$ and analytic continuation yields the exact analytic structure of the S-matrix near threshold.

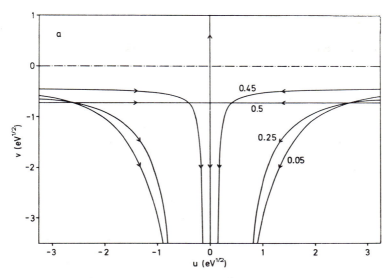

Figure 1

Trajectories of the poles of the fixed-nuclei S-matrix for electron-polar molecule scattering in the complex momentum plane $k = u + iv$ for various values of the threshold exponent $\alpha$. $\alpha = 0.5$ represents pure s-wave scattering, $\alpha = 0$ corresponds to the critical dipole moment.

The remaining trajectories in fig. 1 illustrate the impact of a dipole potential on the analytic structure of the S-matrix. Even for a very weak dipole potential ($\alpha = 0.45$) the trajectories change profoundly. The resonance poles no longer reach the imaginary axis. They are deflected downwards and disappear into the lower half-plane, while a virtual state pole moves up the axis and crosses the origin to become a bound state pole. Note that the resonance and bound state pole trajectories have become disconnected.

The most convenient representation of the results for the understanding of nuclear motion effects is in the form of potential energy curves of resonances, virtual states and bound states. These are shown in fig. 2 for s-wave scattering ($\alpha = 0.5$) and for s-wave scattering in the presence of a dipole potential of intermediate strength ($\alpha = 0.25$). For simplicity, we have chosen a harmonic target potential energy curve $V_0(R)$ and a linear function for $\epsilon_d(R)$, which gives $V_1(R) = V_0(R) + \epsilon_d(R)$ as a shifted harmonic potential. The full curve, denoted by RES in fig. 2a, represents the real part of the complex resonance pole, VS denotes the virtual state pole and BS the bound state pole of the S-matrix. The dotted curve, denoted by RES′, represents the resonance position as determined from the phase shift in eq. (5), i.e., $\delta(E, R) = \pi/2$. RES and RES′ are not two resonances, but just two different definitions of the resonance position. Figure 2a gives a unified description of s-wave resonances, virtual states and bound states and their interrelation at threshold.

When a dipole potential is included, the complex resonance pole decouples from the poles on the imaginary k-axis as discussed above. Figure 2b shows that the complex resonance pole (RES) moves smoothly through threshold. The resonance potential curve determined from the phase shift (RES′) and the

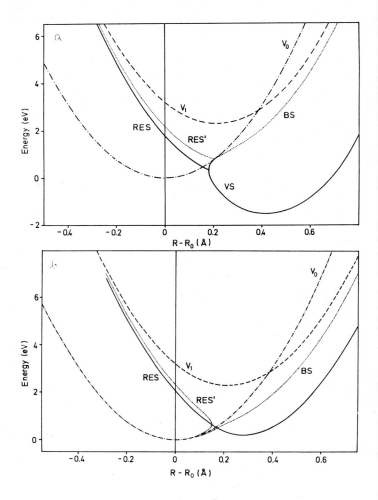

Figure 2
Fixed-nuclei potential energy curves of resonances, virtual
states and bound states for (a) $\alpha$ = 0. 5 (pure s-wave scattering)
and (b) $\alpha$ = 0. 25 (intermediate dipole moment).

bound state (BS) form a connected potential curve which exhibits a kind of
avoided crossing behaviour with the target potential curve $V_0$.   This unusual
behaviour of resonances and bound states close to threshold is clearly of crucial importance for the understanding of electron scattering and dissociative
attachment in polar molecules.

## CALCULATION OF MULTI-CHANNEL VIBRATIONAL EXCITATION CROSS SECTIONS

Despite the complexity of the fixed-nuclei potential energy curves shown in fig. 2, the corresponding electron scattering cross sections can be calculated exactly starting from eq. (2). Choosing the target vibrational states $|v\rangle$ as the basis for a representation of the effective Hamiltonian $\mathcal{H}$, the width and level shift operators become diagonal. It is only the R-dependence of $\epsilon_d$ which gives rise to non-diagonal elements of $\mathcal{H}$. In the simple shifted harmonic oscillator model introduced above, $\langle v|\mathcal{H}|v'\rangle$ becomes tri-diagonal and the matrix elements of the resolvent $(E-\mathcal{H})^{-1}$ are given by continued fractions [5]. This renders the evaluation of the multi-channel cross sections a trivial numerical task.

Figure 3 and fig. 4 show the exact $v = 0 \rightarrow 1$ and $v = 0 \rightarrow 2$ vibrational excitation cross sections corresponding to the potential curves in fig. 2a and fig. 2b,

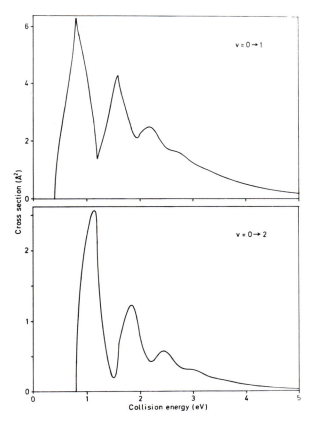

### Figure 3
Exact dynamical $v = 0 \rightarrow 1$ and $v = 0 \rightarrow 2$ vibrational excitation cross sections for the potential energy curves of fig. 2a.

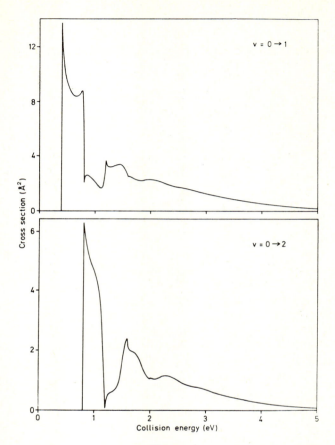

Figure 4
Exact dynamical v = 0 → 1 and v = 0 → 2 vibrational
excitation cross section for the potential energy
curves of fig. 2b.

respectively.  For pure s-wave scattering (fig. 3) one observes "boomerang"
oscillations [20] distorted by Wigner cusps at the opening of higher inelastic
channels.  Figure 4 shows the development of threshold peaks upon inclusion of
the dipole potential.  Although the harmonic model considered here is too simplified to be directly applicable for the description of real molecules, one may
gain from its solution deeper insight into the nature of non-adiabatic effects in
low-energy electron molecule scattering.

Acknowledgment

One of the authors (W. D.) would like to thank the Deutsche Forschungsgemeinschaft for a research grant and Vincent McKoy for his kind hospitality at the
California Institute of Technology.

# REFERENCES

*Fellow of the Deutsche Forschungsgemeinschaft (DFG). Permanent address: Phys. -Chem. Institut, Universität Heidelberg, D-69 Heidelberg, West Germany.

[1] Rohr, K. and Linder, F., J. Phys. B 9, (1976) 2521.

[2] Rohr, K., J. Phys. B 11 (1978) 1849.

[3] Taylor, H. S., Goldstein, E. and Segal, G. A., J. Phys. B 10 (1977) 2253.

[4] Dubé, L. and Herzenberg, A., Phys. Rev. Lett. 38 (1977) 820.

[5] Domcke, W., Cederbaum, L. S. and Kaspar, F., J. Phys. B 12 (1979) L359.

[6] Rudge, M. R. H., J. Phys. B 13 (1980) 1269.

[7] Chen, J. C. Y., Phys. Rev. 148 (1966) 66; 156 (1967) 12.

[8] O'Malley, T. F., Phys. Rev. 150 (1966) 14.

[9] Bardsley, J. N., J. Phys. B 1 (1968) 349.

[10] Feshbach, H., Ann. Phys. (N. Y.) 5 (1958) 357; 19 (1962) 287.

[11] Fano, U., Phys. Rev. 124 (1961) 1866.

[12] Taylor, J. R., Scattering Theory (Wiley, New York, 1972).

[13] Domcke, W. and Cederbaum, L. S., Phys. Rev. A 16 (1977) 1465.

[14] Hazi, A. U., J. Phys. B 11 (1978) L259; Hazi, A. U., Rescigno, T. and Kurilla, M., Phys. Rev. A 23 (1981) 1089.

[15] Wigner, E. P., Phys. Rev. 73 (1948) 1002.

[16] O'Malley, T. F., Phys. Rev. 137 (1965) A1668.

[17] Crawford, O. H., Proc. Phys. Soc. 91 (1967) 279.

[18] Domcke, W., J. Phys. B, submitted for publication.

[19] Newton, R. G., Scattering Theory of Waves and Particles (McGraw-Hill, New York, 1966).

[20] Herzenberg, A., J. Phys. B 1 (1968) 548.

PHYSICS OF ELECTRONIC AND ATOMIC COLLISIONS
S. Datz (editor)
© North-Holland Publishing Company, 1982

ELECTRON SCATTERING AT HIGH AND INTERMEDIATE ENERGIES
-QUANTITATIVE MEASUREMENTS IN MOLECULAR SPECTROSCOPY

C.E. Brion

Department of Chemistry
University of British Columbia
Vancouver, B.C.
CANADA

This work shows how absolute dipole oscillator strengths are
measured for photoabsorption, photoionization and photofrag-
mentation using the inelastic scattering of fast electrons
at negligible momentum transfer.

INTRODUCTION

This is a broad topic,therefore this discussion is limited to a consideration of
some aspects of the application of the techniques of inelastic electron scattering
to studies in molecular spectroscopy and molecular electronic structure of gaseous
free molecules.

Spectroscopy has been traditionally carried out mainly by studies of the absorption and emission of electromagnetic radiation over wide ranges of the electromagnetic spectrum.  However, a detailed examination of the literature shows that
rather limited information is available for molecules in the far UV and X-ray regions of the spectrum, i.e. at energies above about 10 eV - that is $\lambda$ shorter
than about 1200 Å.  These higher energy regions are important as they cover the
range of many highly excited electronic states of valence electrons as well as all
inner shell transitions and most ionization and fragmentation processes.  Although
a large body of information on ionization energies is available from photoelectron
spectroscopy using resonance line sources at HeI (21.22 eV), HeII (40.8 eV), and
a few much higher energy X-ray lines such as MgK$\alpha$ (1254 eV) and AlK$\alpha$ (1487 eV)
there is very little absolute cross-section information on the photoabsorption or
total and partial photoionization and fragmentation of molecules at continuous
energies beyond 20 eV.

The reason for this paucity of data, both spectral and particularly cross-sectional
is the limited availability of continuum light sources at energies above 20 eV
($\lambda$ < 600 A).  Conventional light sources using hydrogen and noble gas continua
have provided useful but relatively weak structured continua up to $\sim$ 20 eV [1].
Until the relatively recent advent of tuneable synchrotron radiation, which
extends right through to X-ray energies, there has been no effective source of
far UV and X-ray continuum radiation with the exception of few weak bremmstrahlung
continua in restricted regions.  Earlier cross-section measurements were often
obtained using line spectra at randomly and frequently widely spaced energies
while detailed spectroscopy has not been possible until quite recently due to the
lack of sufficiently intense continuum sources.  Even where synchrotron radiation
has been available optical studies have not always been easy to carry out due to
the problems involved in the dispersion of short wavelength radiation.  For example
the very low reflectivity of mirrors and gratings at short wavelengths seriously
attenuates the useable photon intensity despite the high flux from the synchrotron
itself.  Also such sources are of enormous expense and usually remotely located
from most users laboratories.  Other difficulties, particularly at short wavelengths, include order overlapping, stray light corrections and absorption by
optical components, particularly in the carbon K region due to surface contamina-

tion.  Despite these difficulties great progress [2] is being made in the utiliza-
tion of synchrotron radiation and some elegant experiments are being done as can
be seen from contributed papers at this conference [3].  However, the scope of
many types of molecular measurements is still rather limited.  In the meantime
there is the hope that tuneable short wavelength lasers (which would avoid the
problems of monochromation) will provide a convenient light source in the future.

In view of this situation what alternative methods exist for exciting and ionizing
molecules?  It has long been known that inelastic collisions of particles, in
particular electrons, can also be used to achieve energy transfer to molecules,
often with larger cross-sections than with photons.  Indeed the early experiments
of Franck and Hertz [4] and others showed that in many ways an electron interacted
with an atom or molecule in a similar fashion to a photon in that excitation,
ionization and dissociation were observed.  With electrons continuous energy
transfer is possible over the whole electromagnetic spectrum.  In 1935 Bethe [5],
using the Born approximation showed that for fast electrons there was a quantitative relationship between the differential electron scattering cross-section and
the generalized oscillator strength df/dE(K).

$$\frac{d\sigma}{dE} = \frac{2}{E} \frac{\overline{k}_n}{\overline{k}_0} \frac{1}{|K|^2} \frac{df(K)}{dE} \tag{1}$$

where $\overline{k}_0$, $\overline{k}_n$ and K are the incident, scattered and transferred momenta respectively and E is the energy loss.

Bethe also showed that the generalized oscillator strength could be expanded in a
series of terms in which the first term is the optical or dipole oscillator
strength with higher terms involving even powers of K, the momentum transfer.

$$\frac{df}{dE}(K) = \frac{df_0}{dE} + AK^2 + BK^4 + \ldots \tag{2}$$

In the limit of zero momentum transfer it can be seen that the generalized oscillator strength becomes equal to the optical oscillator strength (OOS) $df_0/dE$.
The OOS is simply related to the cross-section for photoabsorption or photoionization ($\sigma ph$)

$$\frac{df_0}{dE} = \frac{mc}{\pi e^2 h} \sigma ph = \frac{E}{2} \frac{\overline{k}_0}{\overline{k}_n} K^2 \frac{d\sigma}{dE} \tag{3}$$

This important relationship shows that not only are dipole transitions induced
in atomic or molecular targets by fast electrons but that their absolute intensity
can be obtained by kinematic conversion of the differential electron scattering
intensities.  Thus it is possible to carryout quantitative optical spectroscopy
without any photons! - i.e. "spectroscopy in the dark" - and at any energy
transfer (or equivalent wavelength).  In 1961 Bethe's theoretical work was
discussed and clearly explained in Inokuti's noteable review article "The Bethe
Theory Revisited" [6].  Such theoretical ideas were also discussed by Lassettre
et al. [7] who then, in an elegant series of electron impact studies, obtained
optical oscillator strengths for a number of discrete atomic and molecular
transitions by extrapolating scattering intensity measurements at varying scattering angle for each transition to zero momentum transfer [8,9].  Hertel and Ross
[10,11] made similar measurements for discrete valence transitions in alkali
metals by variation of impact energy together with extrapolation to $K^2 = 0$.
Meanwhile Van der Wiel, at the FOM Institute in Amsterdam had shown that it was
possible to avoid the tedious procedure of extrapolation by making direct electron
scattering measurements sufficiently close to the optical limit [12].  In practical
terms this means using fast electrons (> 3 keV for valence electrons) at zero
scattering angle (typically with a solid angle of $10^{-4}$ steradians, about $\theta = 0^\circ$).

Van der Wiel and his co-workers made a series of absolute measurements on the multiple photoionization of atoms and also the photoionization and fragmentation of molecules using this so called "poor man's synchrotron" over an equivalent photon energy range (electron energy loss) up to 400 eV (i.e. $\lambda$ down to 30Å). These and later experiments [13] have shown that the use of fast electrons and coincidence techniques together with the Bethe-Born kinematic conversion factors produces <u>quantitative</u> results equivalent to those that would be obtained optically for a wide range of photoabsorption and photoionization phenomena.

The ability of fast electrons to induce dipole allowed (optical) transitions can be qualitatively understood in terms of the "virtual photon field". Figure 1 illustrates the principal effects occurring [13] when a fast electron interacts with a target molecule via a distant collision (large impact parameter and thus small scattering angle). As the electron passes by, the target experiences a sharply pulsed electric field of which the perpendicular component is significant. Ideally, in the limit the E field will approach a delta function that, if Fourier transformed into the frequency domain, would afford the perfect spectroscopic "light" source consisting of a continuum composed of all frequencies at equal intensities. In practice, the pulse will have a narrow but finite width, and there will be a falloff in intensity of "pseudophotons" at high frequencies. (It should be stressed that this method does not simulate photons; rather, under the appropriate conditions, the electron-impact differential cross section is related to the optical cross section by kinematic factors alone - see equation 3). Nevertheless, a sufficiently wide spectral range can readily be achieved in the laboratory, and the effective high-energy-transfer limit in electron-scattering experiments is usually determined by other factors[13].

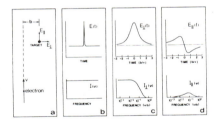

Figure 1   Virtual photon field of a fast electron

Compare the processes of resonant photoabsorption and electron impact excitation

$$h\nu + AB \rightarrow AB^* \qquad\qquad (4)$$
$$E$$

$$e + AB \rightarrow AB^* + e \qquad\qquad (5)$$
$$E_0 \qquad\qquad (E_0 - E)$$

It is clear that the energy loss (E) of the incident electron of energy $E_0$ is analogous to the resonant photon energy (E) required to produce the excited state of quantum energy $E^*$. It is therefore possible to study the process of energy transfer [13] by electron energy loss spectroscopy (EELS). In effect, a fast electron ($E_0$) offers the target a "white-light" continuum of "pseudophotons" that are absorbed with a frequency-dependent probability that can be quantitatively related to the optical oscillator strength via the Bethe-Born relation.   The

<u>net</u> result is that under the appropriate experimental conditions we may perform quantitative measurements equivalent to photoabsorption using techniques of fast electron-impact and electron-energy-loss spectroscopy.  It is of particular advantage to exploit this relationship in the UV and soft-X-ray regions of the spectrum where continuum light-source availability is restricted.

Using this simulation technique photoabsorption studies can be carried out at a wide range of energies both for the valence shell and core electrons.  Inner shell absorption spectroscopy is particularly advantageous for study with electron impact due to the energy resolution advantages in electron impact spectroscopy at high energy losses [13].  This advantage arises due to the inverse relationship between energy (loss) and wavelength and has resulted in much higher acheivable resolution in electron energy loss than in photoabsorption [14,15].  A further advantage of electron energy loss spectroscopy is that a wide spectral range (IR → X-ray) can be covered in a single scan with a single spectrometer [13,16].

Since ionization is only a special case of excitation it is also possible to simulate the photoionization process using fast electrons.  We may compare the processes of photoionization and electron impact ionization

$$\underset{E}{h\nu} + AB \rightarrow [AB^+ + e_{ej}] \tag{6}$$

$$\underset{E_0}{e} + AB \rightarrow [AB^+ + e_{ej}] + \underset{(E_0 - E)}{e_{sc}} \tag{7}$$

where $e_{ej}$ is the electron ejected from AB on ionization and $e_{sc}$ is the fast-scattered electron.

It is apparent that in both cases energy E is deposited in $[AB^+ + e_{ej}]$ and that, as in the case of excitation, the photon energy is analogous to the electron energy loss.  However, since there are now two electrons sharing the excess energy in electron-impact ionization, it is necessary to use time correlation (coincidence techniques) for the simulation of photoionization experiments [13].  A variety of photoabsorption and photoionization experiments have been simulated effectively to give quantitative measurements [13].  These are shown in Table 1.

Table 1

PHOTON-SIMULATION EXPERIMENTS

| Photon Experiment | Electron-Impact Equivalent |
|---|---|
| Photoabsorption | Electron-energy-loss spectroscopy (e,e) |
| Photoionization mass spectrometry (fragmentation) | Electron-ion coincidence (e,e + ion) |
| Total photoionization | Electron energy loss - total ejected electron coincidence (e,2e) |
| Photoelectron spectroscopy | Electron energy loss - selected ejected electron coincidence (e,2e) |
| Photofluorescence (of ionic states) | Electron energy loss - ion - photon (triple) coincidence (e, e + ion + hν) |

A detailed discussion of the background theory, technique, applications and results of using electron impact "photon" simulation experiments for optical oscillator strength measurements has been given in a recent review article [19]. Much of the earlier work was carried out in a series of collaborative studies at the FOM Institute, Amsterdam. A large number of measurements of dipole oscillator strengths for molecular photoabsorption as well as total and partial photoionization and fragmentation have been obtained. More recently the electron-ion coincidence spectrometer [17] has been moved from Amsterdam to UBC where the innershell electron energy loss [18] and dipole electron-electron (e,2e) coincidence [19,20] spectrometers are also located. A comprehensive programme of oscillator strength measurements for photoabsorption, photoionization and fragmentation of molecules in both valence and core regions is continuing over the energy loss (photon-energy) range up to 1000 eV. As well as the basic spectral and oscillator strength (cross-section) information the data is being used to investigate the detailed dipole breakdown pattern of molecules. This work finds application in areas including radiation induced decomposition, radiation chemistry, biology and physics, dosimetry, aeronomy and space physics, fusion, high temperature chemistry, electron microscopy and the development and evaluation of quantum mechanical procedures. As the body of data grows we are now able to begin the systematic study of trends of total and partial oscillator strength distributions in molecular systems.

EXPERIMENTAL

Typical experimental arrangements are shown in Figures 2, 3 and 4 for the tuneable energy simulations of (a) Absorption Spectra—Electron Energy Loss Spectrometer (EELS) [16] and (b) Photoionization and Photoabsorption (i) Photoelectron Spectroscopy—dipole (e,2e) spectrometer [18] and (ii) Photoionization mass spectrometry—electron-ion (e, e + ion) spectrometer [17].

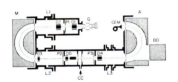

Figure 2   Electron energy loss
           spectrometer

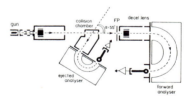

Figure 3   (e,2e) spectrometer

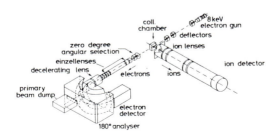

Figure 4   (e, e + ion) spectrometer

RESULTS AND DISCUSSION

In order to illustrate the use and application of various techniques the results
of a number of recent studies are shown below together with references to other
work already completed.

a.   Valence and Inner Shell Electron Energy Loss Spectra

The high resolution electron energy loss valence shell spectrum of HF obtained at
2.5 keV and 0° is shown in figure 5.  This highly corrosive and reactive gas is
difficult to handle in optical instruments and only limited portions of the
optical spectrum have been reported.  Due to the highly polar character of HF
great care is needed to avoid dominant spectral contamination due to desorbed
species [20].  A detailed Rydberg and valence transition analysis of this spectrum
over the energy range up to 50 eV is currently in press [21].  Continuum valence
shell photoabsorption measurements are discussed in Section (b) below.

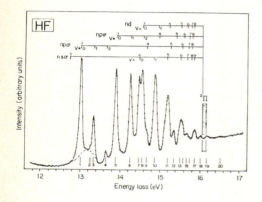

Figure 5   Valence shell electron
           energy loss spectrum of HF

The F 1s K-shell spectra of HF and $F_2$ have recently been measured [22] and are
shown in figures 6 and 7.  These spectra, at energy losses in the region of 700

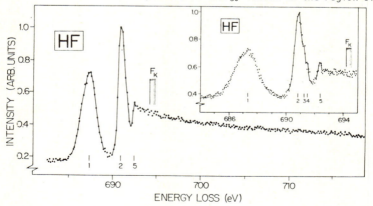

Figure 6  Fluorine K-shell spectrum of HF

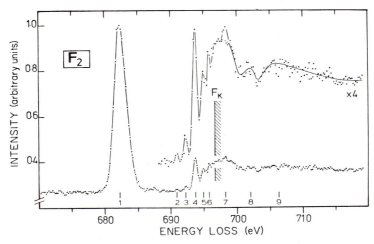

Figure 7   Fluorine K-shell spectrum of $F_2$

eV, correspond to an equivalent wavelength region of $\sim$ 17Å.  While the HF spectrum
is essentially atomic-like the F  core spectrum shows the strongly resonant behaviour (the transition at $\sim$ 682 eV is due to a $\sigma^*$ resonance) caused by the
anisotropic molecular field [23,24].  Similar resonances have been observed in
other diatomic and related molecules [25-30].  The core spectra of HF and $F_2$ and
also some resonances in a series of small molecules have been discussed in detail
both at this conference [31] and in a forthcoming publication [22].  Using the
equivalent core model [32,33] information can be obtained on the excitation
energies of the species NeH (from HF) and NeF (from $F_2$).  In particular NeF is of
interest since it is an attractive candidate for a short wavelength rare gas-
halide laser [34].  Similar EELS studies on $C\ell_2$ by Shaw et al. [35] have provided
information on $ArC\ell$.

In recent years very high resolution electron energy loss studies have been made
for a number of small molecules at the University of Manchester by King, Read &
Co-workers [36-39] as well as in the authors laboratory at UBC [26,40].   The
resolution far exceeds that attainable by existing optical techniques for inner
shell transitions and has enabled the study of lifetimes [14,26], vibrational
structure [14,26,36-39], and isotopic substitution [40].  Other details of ISEELS
are discussed in recent review articles [41,42,43].

Inner shell electron energy loss spectroscopy has now been used to study a large
number of molecules in this laboratory.  These include $N_2$ and CO [25,26], $CO_2$ and
$N_2O$ [44], $CH_4$, $NH_3$, $H_2O$, $CH_3OH$, $CH_3OCH_3$ and $CH_3NH_2$ [45,50], NO and $O_2$ [26,46],
$CF_4$ [47], $CS_2$ and COS [48], $CH_3COCH_3$ [49], $C_2H_6$, $C_2H_4$, $C_2H_2$ and $C_6H_6$ [50], $CH_4$ and
$CD_4$ [51], Methyl halides [52], monohalobenzenes [54], $SF_6$ [55], chlorosubstituted
methanes [56], HCN and $C_2N_2$ [57,58], HCHO, $CH_3CHO$ and $CH_3COCH_3$ [59], HF and $F_2$
[22] and $C_2H_6$ [60].

Absolute oscillator strength measurements have been made for K-shell excitation
of $N_2$ and CO by Kay et al  [61].  Hitchcock et al [62] have also made oscillator
strength measurements on the sulphur L-shell of $SF_6$.  More recently Hitchcock
[63] has prepared a comprehensive bibliography of innershell spectroscopy including a variety of other measurements in addition to those made by ISEELS.

## b.  Photoionization

### (i) Dipole (e,2e) Spectroscopy

This method has provided an accurate and versatile simulation of tuneable energy
photoelectron spectroscopy for the measurement of oscillator strengths for photoabsorption (noncoincident mode) and total as well as partial photoionization
(coincident mode) for production of electronic states of ions in the (photon)
energy range below 100 eV [13].  In those few cases where conventional optical
or synchrotron radiation studies have been made excellent agreement with the
dipole (e,2e) results has been observed - see for example $N_2$ [13,64], $O_2$ [65] and
$CO_2$ [66].  The absolute photoabsorption cross-section of COS [67] is shown in
figure 8 in excellent agreement with recent synchrotron studies by Wu and Judge
[58].  It should be noted that our electron impact data has been made absolute by
sum-rule normalization [13] which is a "free bonus" of the method permitting
absolute results without absolute measurements.

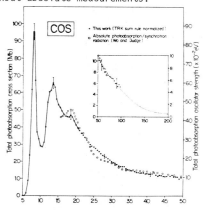

Figure 8   Photoabsorption of COS

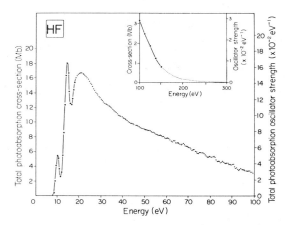

Figure 9   Photoabsorption of HF

Recent results for HF [59] are shown in figure 9 (photoabsorption), figure 10 (photoelectron spectra) figure 11 (branching ratios), figure 12 (partial photo-ionization cross-sections). In figure 13 are shown recent theoretical calculations of HF partial photoionization cross-sections by Hush using the GIPM method [70] and by Faegri and Kelly using the single centre and Tchebychev methods [71]. In figure 14 are shown the total photoabsorption oscillator strengths for the isoelectronic series Ne [72], HF [69], $H_2O$ [73], $NH_3$ [74] and $CH_4$ [75].

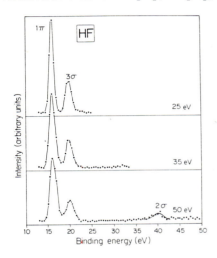

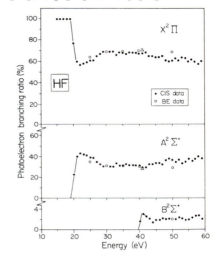

Figure 10  Binding energy spectra

Figure 11  Photoelectron branching ratios

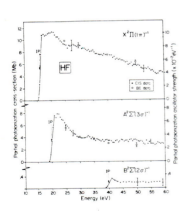

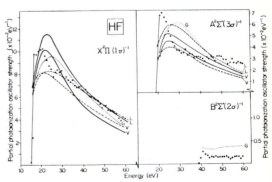

Figure 12

Figure 13

Partial oscillator strengths
for photoionization of HF

Partial oscillator strengths of HF
comparison with theory.  Solid lines;
Tchebychev; Dashed lines, single
centre [71]; Dotted line, GIPM method [70]

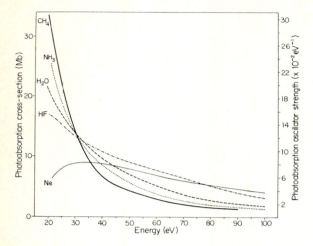

Figure 14

Photoabsorption of
isoelectronic species

The total photoabsorption oscillator strengths of CO [66,76], COS [67] and $CS_2$ [77] show the dramatic effects (figure 15) of the sulphur 3p orbitals in concentrating oscillator strength at low energies as has been predicted in atomic calculations by Manson et al [78].

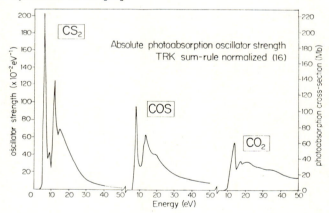

Figure 15   Photoabsorption of $CS_2$, COS and $CO_2$

## (ii)   Electron-ion (e, e + ion) Spectroscopy

This method, an accurate and facile simulation of tuneable energy photoionization mass spectrometry in the (photon) energy range up to at least 550 eV, produces dipole oscillator strengths (cross-sections) for photoabsorption (non-coincident mode) as well as for photoionization and fragmentation (coincident mode).   In this method [13] ion fragments are collected with 100% efficiency regardless of kinetic energy of fragmentation (up to 20 eV excess energy).   A typical series of results are those obtained for $H_2O$ [73] and $CO_2$ [76].   Recent (e,e+ion) results for HF [69] are shown in figure 16 (TOF mass spectrum), figure 17 (ion branching ratios) and figure 18 (molecular and fragment ion oscillator strengths).

In figure 18 linear combinations of the (e,2e) oscillator strengths for electronic states of HF⁺ are shown in comparison with the (e, e + ion) data. This analysis suggests that HF⁺ is formed from the X̃ state and 20% of the Ã state whereas H⁺ is formed from the remaining 80% of the Ã state in accord with expectations based on the appearance potential of H⁺. It is evident that some F⁺ is formed at energies below the B̃ state threshold but above the upper limit of the Ã state Franck-Condon region.

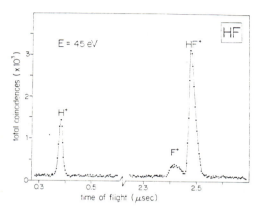

Figure 16   TOF Mass Spectrum of HF

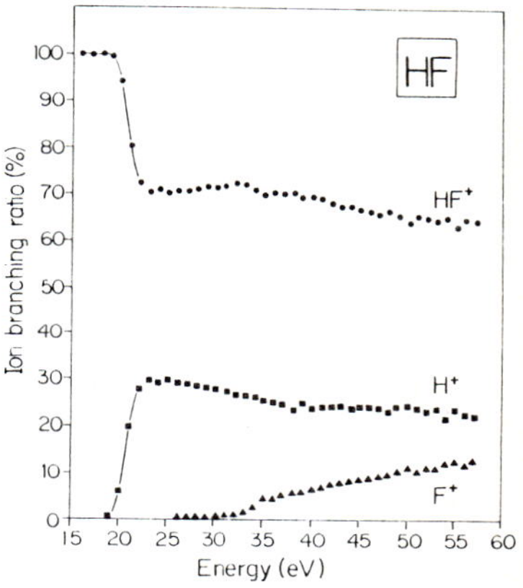

Figure 17   Ion branching ratios

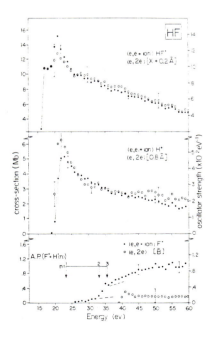

Figure 18

Oscillator strengths for molecular and fragment ion formation in HF

Recent results for COS and CS  [80] have been combined together with results
obtained using the dipole (e,2e) technique [67,77] to measure the partial photoionization cross sections and breakdown pathways.  The spectral decomposition
indicates that the molecular ion, $COS^+$, is formed from the $\tilde{X}$ state of $COS^+$ while
$CS_2^+$ is formed from the $\tilde{X}$, $\tilde{A}$ and $\tilde{B}$ states of $CS_2^+$.  Fragment ion oscillator
strengths and spectral oscillator strength sums are in accord with known ioniza-
tion and appearance potentials.  These results are in agreement with photoelectron
- photoion coincidence measurements [81,82] at a single photon energy of 21 eV.
The excellent agreement in shape and in absolute cross-sections between the
molecular (fragment) ion oscillator strengths and the linear combination of ion
electronic state oscillator strengths is strong confirmation of the quantitative
reliability of both the dipole (e,2e) and (e, e + ion) methods.  These types
of considerations lead to a detailed dipole breakdown picture [80] as has been
discussed for example in the case of $H_2O$ [73], $NH_3$ [74] and $CO_2$ [76].

Using the dipole (e,2e) and/or (e, e + ion) techniques photoionization and photoabsorption oscillator strengths have been obtained for the following molecules.
$N_2$ and CO [64,84,85], $H_2$, HD and $D_2$ [86,87], $O_2$ [65], NO [88,89], HF [69,79] $NH_3$
[74,18], $H_2O$ [73], $H_2S$ [90], $SF_6$ [62,91], COS [67,79], $CS_2$ [77,80], HCℓ and HBr
[92] $CH_4$ [17,75,93], $N_2O$ and $CO_2$ [66,76].

The (e, e + ion) method has also been used to study fragmentation arising from
innershell electronic process [61,62].  A recent example is that of $N_2O$ where
fragmentation was studied following resonant·K shell excitation of terminal and
central N atoms of $N_2O$ respectively [83].  Typical results are shown in figure 19.

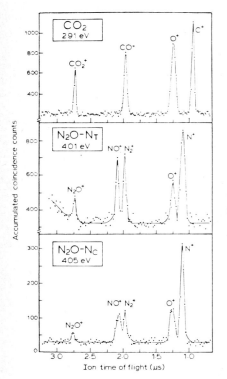

Figure 19

Time of flight mass spectra at innershell
excitation energies in $CO_2$ and $N_2O$

In summary then fast electron impact methods, using techniques of electron energy loss spectroscopy and coincidence counting, are providing accurate and versatile laboratory based methods of studying molecular electronic spectroscopy of inner and valence electrons in the 10-1000 eV range. Furthermore, application of the Bethe-Born factors permits measurement of accurate absolute dipole oscillator strengths (cross-sections) for photoabsorption, partial and total photoionization and also fragmentation of molecules. Detailed information can be deduced concerning the dipole breakdown pattern and radiation induced decomposition of molecules. The results are considerably more numerous and cover a wider energy range than those reported to date using synchrotron radiation. The results have also been useful for evaluation of theoretical calculation methods [70,71,94].

ACKNOWLEDGMENT

I would like to thank numerous co-workers at the University of British Columbia and the FOM Institute, Amsterdam, who have collaborated in the work discussed in this article. Financial support for this work has been provided by the Natural Sciences and Engineering Research Council of Canada and the Petroleum Research Fund administered by the American Chemical Society.

REFERENCES

[1] Samson, J.A.R., "Techniques of Vacuum Ultraviolet Spectroscopy" John Wiley (New York, 1967).
[2] Winick, H. and Doniach, S., "Synchrotron Radiation Research" Plenum Publishing Corporation (New York, 1980).
[3] Abstracts of Papers XII ICPEAC, Gatlinburg 1981. Ed. S. Datz.
[4] Franck, J. and Hertz, G., Verh. Deutsch Phys. Ges. 16 (1914) 10.
[5] Bethe, H., Ann. Phy. (Leipzig) 5 (5) (1930) 325.
[6] Inokuti, M., Rev. Mod. Phys. 43 (1971) 297.
[7] Lassettre, E.N., Rad. Research (Supp) 1 (1959) 530.
[8] Lassettre, E.N. and Skerbele, A., Meth. Exp. Phys. 3B (1974) 868.
[9] Lassettre, E.N. in Sandorfy, C., Ausloos, P.J. and Robin, M.B., Eds. Chemical Spectroscopy and Photochemistry in the Vaccuum Ultraviolet, Reidel (Boston, 1974).
[10] Hertel, I.V. and Ross, K.J., J. Phys. B. 2 (1969) 285.
[11] Hertel, I.V. and Ross, K.J., J. Phys. B. 1 (1968) 697.
[12] Van der Wiel, M.J., Physica 49 (1970) 411.
[13] Brion, C.E. and Hamnett, A., Continuum Optical Oscillator Strength Measurements by Electron Spectroscopy in the Gas Phase, in "The Excited State in Chemical Physics, Part 2", Adv. Chem. Physics, volume 45 (Ed. J.W. McGowan) John Wiley New York (1981).
[14] King, G.C., Read, F.H. and Tronc, M., Chem. Phys. Letters 52 (1977) 50.
[15] Hitchcock, A.P. and Brion, C.E., J. Electron Spectrosc. 18 (1980) 1.
[16] Hitchcock, A.P. and Brion, C.E., J. Electron Spectrosc., 13 (1978) 193.
[17] Backx, C. and Van der Wiel, M.J., J. Phys. B. 8 (1975) 3020; 9 (1975) 315.
[18] Brion, C.E., Hamnett, A., Wight, G.R. and Van der Wiel, M.J., J. Electron Spectrosc. 12, (1977) 323.
[19] Hamnett, A., Stoll, W., Branton, G., Van der Wiel, M.J. and Brion, C.E., J. Phys. B, 9, (1976) 945.
[20] Brion, C.E. and Hitchcock, A.P., J. Physics B., 13, (1980) L677.
[21] Hitchcock, A.P. and Brion, C.E., Chem. Phys. in press.
[22] Hitchcock, A.P. and Brion, C.E., J. Phys. B. in press.
[23] Dehmer, J.L. and Dill, D., Phys. Rev. Letters 35 (1975) 213.
[24] Dehmer, J.L. and Dill, D., J. Chem. Phys. 65 (1976) 5327.
[25] Wight, G.R., Brion, C.E. and Van der Wiel, M.J., J. Electron Spectrosc. 1 (1972/73) 457.
[26] Hitchcock, A.P. and Brion, C.E., J. Electron Spectrosc. 18 (1980) 1.
[27] Wight, G.R. and Brion, C.E., J. Electron Spectrosc. 4 (1974) 313.

[28] Hitchcock, A.P. and Brion, C.E., J. Electron Spectrosc. 19 (1980) 231.
[29] Hitchcock, A.P. and Brion, C.E., Chem. Phys. 33 (1979) 319.
[30] Hitchcock, A.P. and Brion, C.E., J. Electron Spectrosc. 10 (1977) 317.
[31] Abstracts of Papers, XII ICPEAC, Gatlinburg 1981, Ed. S. Datz, page 355.
[32] Wight, G.R. and Brion C.E., J. Electron Spectrosc. 3 (1974) 191.
[33] Schwarz, W.H.E., Chem. Phys. 11 (1975) 217.
[34] Ewing, J.J., Physics Today (1978) page 32.
[35] Shaw, D.A., King, G.C. and Read, F.H., J. Phys. B. 13 (1980) L723.
[36] Tronc, M., King, G.C. and Read, F.H., J. Phys. B. 13 (1980) 999.
[37] King, G.C., Tronc, M., Read, F.H. and Bradford, R.C., J. Phys. B. 10 (1977) 2479.
[38] Tronc, M., King, G.C., Bradford, R.C. and Read, F.H., J. Phys. B. 9 (1976) L555.
[39] Tronc, M., King, G.C. and Read, F.H., J. Phys. B. 12 (1979) 137.
[40] Hitchcock, A.P. and Brion, C.E., J. Electron Spectrosc. 17 (1979) 139.
[41] Read, F.H. and King, G.C. in Symposium on Electron-Molecule Collisions (University of Tokyo, Sept. 1979), Invited Papers, eds. I. Shimamura and M. Matsuzawa, page 155.
[42] Van der Wiel, M.J., in Electronic and Atomic Collisions, Invited papers and progress reports, XI ICPEAC, Kyoto eds. N. Oda and K. Takayanagi (North Holland, 1980) page 209.
[43] Read, F.H., J. de Physique, Colloque C1, supp. 5, 39 (1978) 82.
[44] Wight, G.R. and Brion, C.E., J. Electron Spectrosc. 3 (1974) 191.
[45] Wight, G.R. and Brion, C.E., J. Electron Spectrosc. 4 (1974) 25.
[46] Wight, G.R. and Brion, C.E., J. Electron Spectrosc. 4 (1974) 313.
[47] Wight, G.R. and Brion, C.E., J. Electron Spectrosc. 4 (1974) 327.
[48] Wight, G.R. and Brion, C.E., J. Electron Spectrosc. 4 (1974) 335.
[49] Wight, G.R. and Brion, C.E., J. Electron Spectrosc. 4 (1974) 347.
[50] Hitchcock, A.P. and Brion, C.E., J. Electron Spectrosc. 10 (1977) 317.
[51] Hitchcock, A.P., Pocock, M. and Brion, C.E., Chem. Phys. Letters 49 (1977) 125.
[52] Hitchcock, A.P. and Brion, C.E., J. Electron Spectrosc. 13 (1978) 193.
[53] Hitchcock, A.P. and Brion, C.E., J. Electron Spectrosc. 17 (1979) 139.
[54] Hitchcock, A.P., Pocock, M., Brion, C.E., Banna, M.S., Frost, D.C., McDowell, C.A. and Wallbank, B., J. Electron Spectrosc. 13 (1978) 345.
[55] Hitchcock, A.P. and Brion, C.E., Chem. Phys. 33 (1978) 55.
[56] Hitchcock, A.P. and Brion, C.E., J. Electron Spectrosc. 14 (1978) 417.
[57] Hitchcock, A.P. and Brion, C.E., J. Electron Spectrosc. 15 (1979) 401.
[58] Hitchcock, A.P. and Brion, C.E., Chem. Phys. 33 (1979) 319.
[59] Hitchcock, A.P. and Brion, C.E., J. Electron Spectrosc. 19 (1980) 231.
[60] Hitchcock, A.P. and Brion, C.E., J. Electron Spectrosc. 22 (1980) 283.
[61] Kay, R.B., Van der Leeuw, Ph. E. and Van der Wiel, M.J., J. Phys. B. 10 (1977) 2153.
[62] Hitchcock, A.P. Brion, C.E. and Van der Wiel, M.J., J. Phys. B. 11 (1978) 3245.
[63] Hitchcock, A.P. - to be published.
[64] Hamnett, A., Stoll, W. and Brion C.E., J. Electron Spectrosc. 8 (1976) 367.
[65] Brion, C.E., Tan, K.H., Van der Wiel, M.J. and Van der Leeuw, Ph. E., J. Electron Spectrosc. 17 (1979) 101.
[66] Brion, C.E. and Tan, K.H., Chem. Phys. 34 (1978) 141.
[67] White, M.G., Leung, K.T. and Brion, C.E., J. Electron Spectrosc. 23 (1981) 127.
[68] Wu, K. and Judge, D.L., private communication.
[69] Carnovale, F., Tseng, R. and Brion, C.E., J. Phys. B, in press.
[70] Hush, N.S., private communication.
[71] Faegri, K.F. and Kelly, H.P., Phys. Rev. A23 (1981) 52.
[72] West, J.B. and Marr, G.V., Proc. Roy. Soc. A349 (1976) 397.
[73] Tan, K.H., Brion, C.E., Van der Leeuw, Ph. E. and Van der Wiel, M.J., Chem. Phys. 29 (1978) 299.
[74] Wight, G.R., Van der Wiel, M.J. and Brion, C.E., J. Phys. B. 10 (1977) 1863.
[75] Backx, C., Wight, G.R., Tol, R.R. and Van der Wiel, M.J., J. Phys. B. 8

(1975) 3007.

[76] Hitchcock, A.P. Brion, C.E. and Van der Wiel, M.J., Chem. Phys. 45 (1980) 461.

[77] Carnovale, F., White, M.G. and Brion, C.E., J. Electron Spectrosc. 24 (1981) 63.

[78] Manson, S.T., Msezane, A., Starace, A.F. and Shahabi, S., Phys. Rev. A 20 (1979) 1005.

[79] Carnovale, F. and Brion, C.E., to be published.

[80] Carnovale, F., Hitchcock, A.P., Cook, J.P.D. and Brion, C.E., to be published.

[81] Eland, J.H.D., Int. J. Mass Spectrom. Ion Phys. 12 (1973) 389.

[82] Brehm, B., Eland, J.H.D., Frey, R. and Kustler, A., Int. J. Mass Spectrom. Ion Phys. 12 (1973) 213.

[83] Hitchcock, A.P., Brion, C.E. and Van der Wiel, M.J., Chem. Phys. Letters 66 (1979) 213.

[84] Wight, G.R., Van der Wiel, M.J. and Brion, C.E., J. Phys. B. 9 (1976) 675.

[85] Backx, C., Klewer, M. and Van der Wiel, M.J., Chem. Phys. Letters 20 (1973) 100.

[86] Backx, C., Wight, G.R. and Van der Wiel, M.J., J. Phys. B. 9 (1976) 315.

[87] Van Wingerden, B., Van der Leeuw, Ph. E., de Heer, F.J. and Van der Wiel, M.J., J. Phys. B. 12 (1979) 1559.

[88] Tan, K.H. and Brion, C.E., J. Electron Spectrosc. 23 (1981) 1.

[89] Carnovale, F. and Brion, C.E., to be published.

[90] Brion, C.E., Cook, J.P.D. and Tan, K.H., Chem. Phys. Letters 59 (1978) 241.

[91] Hitchcock, A.P. and Van der Wiel, M.J., J. Phys. B. 12 (1979) 2153.

[92] Carnovale, F. and Brion, C.E., to be published.

[93] Van der Wiel, M.J., Stoll, W., Hamnett, A. and Brion, C.E., Chem. Phys. Letters 37 (1976) 240.

[94] Padial, N., Csanak, G., McKoy, B.V. and Langhoff, P.W., Phys. Rev. A 23 (1981) 218.

**PHYSICS OF ELECTRONIC AND ATOMIC COLLISIONS**
S. Datz (editor)
© North-Holland Publishing Company, 1982

EXPERIMENTS ON COLLISIONS OF ELECTRONS
AND MULTICHARGED IONS

D. H. Crandall

Oak Ridge National Laboratory
Oak Ridge, Tennessee
U.S.A.

Controlled experiments with single collision conditions
between multiply charged ions and electrons are few in
number. Most of the results are for ionization, but many
of these have exhibited structure in the measured cross
sections which is attributed to excitation of autoionizing
resonances. This coupling of direct and indirect processes
in the ionization experiments together with the anticipated
importance of dielectronic recombination for multicharged
ions suggests that some unique physics characterizes certain
aspects of these collisions. The experimental approach and
difficulties and the insights obtained thus far from the
experiments are discussed.

INTRODUCTION

From an experimental viewpoint, the study of collisions of electrons and multicharged
ions is still in its infancy. Few experiments have been performed - no detailed
measurements for ions of charge greater than +5 and no definitive measurements at
all for recombination. The simple reason for the paucity of results is that the
experiments are difficult. For the most part, the present paper will be restricted
to experiments in which the measured signal is attributable to a single electron-ion
collision for ions of initial charge of +3 or greater. Some details of experimental
difficulties will be discussed, primarily for colliding-beams experiments.

Both opportunity and motivation currently exist and should result in considerable
future experimental study of electron-ion collisions. The motivation derives
from applied needs, significant (but untested) theoretical predictions, and basic
physics questions arising from the few experiments accomplished. In the opening
paper of this meeting, Dalgarno demonstrated the interest of astrophysics, and
another symposium will be devoted to "Atomic Collision Processes in Fusion";
these are the two "applied" areas with most pressing interest in the present
collisions studies. However, in this symposium we will concentrate on the basic
physics issues.

The opportunity for experimental studies rests on two separate technology developments. Colliding-beams experiments were developed in the 1960s and 1970s and
have been variously reviewed.[1-5] The techniques are now reasonably well tested
and, while the experiments are demanding, the ability to control and systematically vary nearly all the collision parameters is established. In order for these
techniques to be applied to studies of multiply charged ions it is critical that
ion beams also be available, and for most experiments the ions must be slow compared to beams at major accelerators.

Significant ion source development has been ongoing,[6] and the new electron beam-
driven(EBIS)[7,8] and the electron cyclotron-driven(ECRIS) sources[8] are suitable for
future colliding-beams experiments. Problems do remain in adapting beams techniques to multicharged ion studies, but we have the basic tools.

## EXPERIMENTAL TECHNIQUES

### Plasmas

Careful observations on laboratory plasmas have attempted to provide atomic collisions data for multicharged ions.[8]  Ionization rate measurements are ongoing.[9,10]
Plasma spectroscopic observations have provided the only convincing experimental
evidence for dielectronic recombination.[11-13]  However, these optical measurements
are rarely ascribable to a single collision event and are unable to provide enough
detail to rigorously test our basic understanding of collisions.

### Trapped Ions

Several approaches have been employed to create and hold ions by a combination of
electric and magnetic fields.  Attempts to extract ionization cross sections by
bombarding such trapped ions with a monoenergetic electron beam have been carried
out by two groups.[14,15]  The work of Donets and Ovsyannikov (Fig. 1) is based on
time evolution of the charge state distribution of ions in the EBIS source.

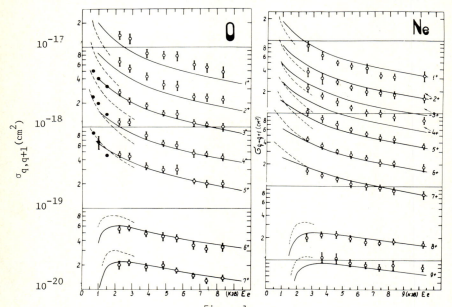

Figure 1
Ionization cross sections for oxygen ions and neon ions,
as determined from analysis of EBIS ions (Ref. 15) - open
circles; from crossed-beams measurements (Ref. 31) - solid
points.  Solid curves are Lotz formula (Ref. 33), and dashed
curves are classical calculations (Ref. 46).

Low-charge state ions are created and trapped by injection of electrons into a low
density gas in a magnetic solenoid with appropriate potentials at the ends of the
solenoid for electrostatic trapping of ions.  The electron beam is then switched
to the desired bombarding energy (usually a few keV above the initial trapping
energy).  After a predetermined time of bombardment the ion trap is dumped and the
charge state distribution of the ejected ion population is determined by time-of-
flight mass spectroscopy.  By repeated samplings at different times, the evolution

of charge for given electron energy (and current distribution) is measured. In a manner similar to plasma-based rate measurements, this evolution of charge is represented by a model in which the impact ionization cross sections are adjusted to provide a best fit of the model to the measured evolution of charge states. The determination of cross sections by this technique does not meet our criterion of being attributable to a single collision of an electron and ion. In addition, assumptions about the types of processes included (e.g., single vs multiple ionization in a given collision), about recombination, and about losses from the trap are all implicit in the technique. Cross sections can only be determined down to a few times threshold energy. In spite of these difficulties, carefully conducted studies have been ongoing since the early 1970s, and results agree with crossed-beams data where overlap occurs. The only cross-section data available for ions of charge greater than +5 have been produced by these experiments. Ionization data for hydrogen-like ions up to $Ar^{+17}$ have been produced and agree reasonably with Coulomb-Born theory for this case.[15]

Colliding Beams

Figure 2 can be used to illustrate most colliding-beams experiments. Ions from any source must often be pre-analyzed (or purified) at the experiment to remove unwanted charge states. The angle $\theta$ between the electron and ion beams is selected to optimize different conditions. The photomultiplier, viewing the interaction region from above, can be used to study excitation of the ions via light produced. Ion-charge-state analysis can be employed to study both recombination and ionization cross sections (the channel plate detector is not a general feature).

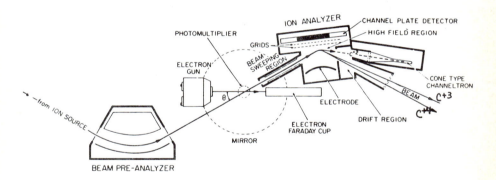

Figure 2
Schematic of a colliding-beams experiment (from Ref. 21).

In any of these experiments, cross sections would be obtained from an expression like

$$\sigma(E) = \frac{R}{D} \frac{F}{I_i I_e} K \tag{1}$$

where R is a measured signal rate, D the detection efficiency of the apparatus for signal produced by beams interaction, $I_i$ and $I_e$ are beam currents at the beams intersection, F is the beam overlap factor, and K is given by

$$K = \frac{q_i e^2 \, v_i v_e \, \sin\theta}{\left(v_i^2 + v_e^2 - 2v_i v_e \cos\theta\right)^{1/2}} \tag{2}$$

where $q_i$ is the ionic charge, $v_i$ and $v_e$ are velocities, and $\theta$ is the beams intersection angle.  For crossing beams the overlap factor is accurately represented
by only the vertical beam distributions as

$$F = \frac{\int i_i(z)dz \int i_e(z)dz}{\int i_i(z)\ i_e(z)dz} \qquad (3)$$

(assuming no variation of efficiency of detection of signal produced at different
z).  For merged beams the overlap in two dimensions is required.  It is instructive to note that for $\theta = 90^0$ and $v_e \gg v_i$ as for most of the experiments, Expression 2 reduces to $K = q_i e^2 v_i$, and the attainable signal is directly proportional to
$v_i$ making slow ions most desirable.

Excitation Experiments

The only excitation experiments on multicharged ions have been for resonance excitation, 2s-2p, of Li-like ions[16,17] $C^{+3}$(155nm) and $N^{+4}$(124nm).  The apparatus was
similar to that of Fig. 2 with $\theta = 90^0$.  The mirror illustrated in Fig. 2 to increase photon collection efficiency was not present in these experiments, but a
similar "trick" was employed for relative measurements.  For absolute measurements
the solar-blind photomultiplier with $MgF_2$ window (band pass about 30 nm) viewed the
interaction region through defined geometry and with all surfaces coated with gold
black to absorb (rather than reflect) photons.  However, insertion of an aluminized
reflector tube between the interaction region and the photomultiplier increased
signal to noise by about 3 times for relative measurements.

Table 1 gives detailed parameters for the absolute $N^{+4}$ experiment and compares
with a rather easier $Ba^+$ excitation measurement[18] pursued with the same techniques.
In spite of increases of about 10-fold in ion and electron currents, the $N^{+4}$
signal is about 5 times smaller with a background 30 times greater.

TABLE 1.  Parameters from electron impact excitation experiments

| Parameter | $Ba^+$ (Ref. 18) | $N^{+4}$ (Ref. 17) |
|---|---|---|
| Electron Energy | 4 eV | 15.5 eV |
| Electron Current | 10 μA | 90 μA |
| Ion Energy | 750 eV | 40 keV |
| Ion Current | 0.1 μA | 1.0 μA |
| Pressure (both beams on) | $1 \times 10^{-9}$ torr | $1.2 \times 10^{-9}$ torr |
| Wavelength | 455 nm | 124 nm |
| Band Pass | 10 nm | 30 nm |
| $D(z_0,\lambda)$ | $7.4 \times 10^{-4}$ | $4 \times 10^{-4}$ |
| Signal (S) (both beams) | 10 Hz | 2 Hz |
| Background (B) (both beams) | 3 Hz | 90 Hz |
| Emission Cross Section | $17.4 \times 10^{-16}$ $cm^2$ | $2.7 \times 10^{-16}$ $cm^2$ |

A principal difficulty with the $N^{+4}$ case is the detection of the photons.  An
interference filter obtained for the experiment would have reduced the band pass
and background but had low peak transmission and thus reduced signal to an unten-
able level.  At energies above 20 eV (specifically tested at 52 eV) a spurious
signal arose.  Presumably the electron beam created and trapped an ion (perhaps
$H_2^+$ or $N_2^+$) which could produce photons within the band pass when excited by the
ion beam).  This spurious signal was detected because of its dependence on electron
current which was systematically varied.  The final quoted cross section was
obtained by extrapolation of the apparent cross section to zero electron current.

These details are presented to illustrate the fact that the excitation experiments
for highly charged ions are essentially stalled against such difficulties.  New
techniques are required.  To overcome low signals and eliminate photon detection,
it is tempting to suggest merged electron-ion beams with detection of energy loss

of the incident electrons to identify specific excitation events. Establishing the detailed techniques and "calibrating" such an approach could not be accomplished quickly, however, so new direct measurement of excitation cross sections is not likely in the near future.

A current and rather complete overview of all ion excitation for both theory and experiment can be obtained from combining References 5, 19, and 20.

Recombination

Figure 2 is borrowed from Kohl and Lafyatis[21] who designed the experiment to measure dielectronic recombination of $C^{+3}$. The $C^{+3}$ ions that undergo recombination via excitation of $C^{+3}(2s-2p)$ accompanied by capture of the incident electron to a high $n\ell$ orbital are to be detected. Such a recombination event would produce doubly excited $C^{+2}(1s^22p\ n\ell)$ which can stabilize to $C^{+2}(1s^22s\ n\ell)$ by emission of a photon slightly shifted from the resonance lines of $C^{+3}(2s-2p)$ at 155 nm. Photons near 155 nm are to be collected by the mirror plus photomultiplier and could be counted in coincidence with $C^{+2}$ ions detected to separate out the signal for the specific recombination process.

This experiment has all the difficulties associated with an excitation experiment and a few others as well. Even at chamber pressures near $10^{-11}$ torr, electron capture provides a large background of $C^{+2}$ while many of the photons are not due to recombination. The stabilized $C^{+2}(1s^22s\ n\ell)$ ions are still fragile and may be field ionized during charge state analysis. The experiment is designed to accomodate this latter problem by accelerating the electrons from the field ionization into the channel plate detector (Fig. 2) and then adding that signal to the $C^{+2}$ counted ions to represent total recombination. The experiment is appealing in that all problems and parameters are (in principle) accounted for and could provide the first definitive recombination cross-section measurement.

A number of other researchers are attempting dielectronic recombination experiments. McGowan et al. are adapting their merged-beams apparatus[22] to measurements now in progress with singly charged $C^+$ and might eventually pursue multicharged ion cases. A group at Oak Ridge[23] is using a merged-beams approach which couples accelerator-produced ions with an electron source similar to the EBIS, without ion trapping but with the same electron beam geometry. Fast ions are passed through a small hole in the cathode and along the solenoid colinear with the electrons. Outside the magnetic field the electrons diverge and are collected, and the ions are charge state analyzed to detect recombination.

Signal rates should be best from these merged-beams approaches which can in principle also achieve the required small relative energy spread and charge exchange with background gas is small for fast ions. However, calibration of the beams overlap factor and effects of field ionization on the signal together with non-trivial construction and operation of the electron sources complicates these experiments significantly; results will be slow to appear.

Ionization

The most numerous and accurate experimental results for electron-multicharged ion collisions will likely continue to be for ionization measurements. Such experiments can be pursued with crossed-beams approach with accurately determined overlap factors and the detection efficiency (D in Eq. 2) near unity. The principal new difficulty facing these experiments is adequate separation of the primary ion beam from the signal ions at the charge state analyzer.

Though a number of groups have produced colliding-beams results for singly and doubly ionized ions (see Refs. 2 and 24 for examples), only the ORNL-JILA collaboration in the United States and the Giessen-Frankfurt collaboration in West Germany have reported measurements for multicharged ions.

RESULTS

Excitation

For the Li isoelectronic studied, the excitation of singly charged $Be^+(2s\text{-}2p)$[25] of Fig. 3 has been treated as something of a benchmark study. The results are appealing in that the theory[26,27] converges toward experiment with increasing sophistication of the approximations in the way one might anticipate. However, agreement is never attained even for 98% confidence level uncertainty in the experiment. However, for the $C^{+3}(2s\text{-}2p)$ case[16] of Fig. 4, and equally the $N^{+4}$ case,[17] any of the Coulomb-Born, distorted-wave, or 2-state close-coupling calculations[19,28,29] (exchange included in all) are found to be in excellent agreement with the independently absolute experiments. This result might be partly attributed to the nature of the energy scaling peculiar to this Li-like sequence - Fig. 5. However, the increasing dominance of the ionic charge term of the interaction potential compared to the electron-electron interaction should lead to improving reliability of approximations (like the distorted-wave) as ionic charge increases. Since many more cases can be calculated with distorted-wave than with coupled-state approximation it remains important to experimentally confirm this improvement of theoretical predictions of optically allowed excitation with increasing ionic charge in other sequences.

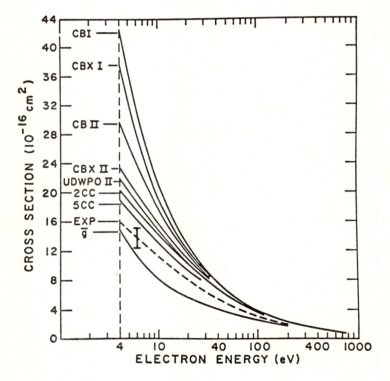

Figure 3
Excitation of $Be^+(2s\text{-}2p)$ by electron impact - from Ref. 25.
Theories are from Ref. 26, except 5cc from Ref. 27. Error bar
represents 98% confidence level absolute uncertainty.

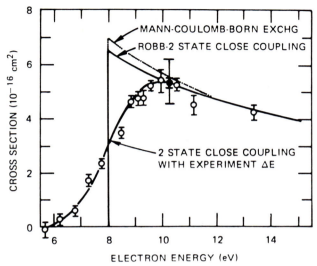

Figure 4
Excitation of $C^{+3}$(2s-2p) by electron impact - from Ref. 16.
Theories are from Ref. 28. Error bars are 90% confidence
level statistics with total absolute uncertainty at 10.2 eV only.

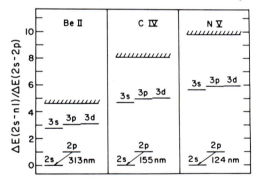

Figure 5
Energy levels of $Be^+$, $C^{+3}$, and $N^{+4}$ scaled to the 2s-2p
transition energy - from Ref. 17.

## Ionization

Accurate ionization cross sections for at least 20 ions of charge +3, +4, or +5
have now been measured over fairly broad energy ranges beginning at threshold in
each case (not all are published). While this is a small fraction of possible
cases, it is too much to cover in detail here.

Figure 6 shows comparison of distorted wave theory[30] and all of the beams experiments for He-like multicharged ions.[31] The cross sections scale conveniently and
follow the theory reasonably well. However, all theories (even Thomson,[32] 1912,
or the Lotz formula[33]) give comparable cross sections for He-like ions so that

Fig. 6 only demonstrates that the experiments are not grossly incorrect.

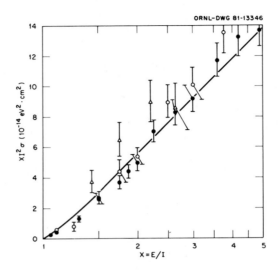

Figure 6
Electron impact ionization of He-like multicharged ions
scaled by the square of the ionization potential, I, as
a function of energy in threshold units.  Solid curve from
Ref. 30 is for $N^{+5}$ but, for this scaling, represents any
multicharged He-like.  Experimental values for $B^{+3}$ - solid
points;   $C^{+4}$ - open circles; and $N^{+5}$ - triangles (all from
Ref. 31).  Bars are absolute total uncertainty at 67%
confidence level.

For the next simplest case, Li-like ions, basic physics complications arise which
are not presently resolved satisfactorily.  The ionization cross sections for $Be^+$,
$C^{+3}$, $N^{+4}$, and $O^{+5}$ have been measured in detail.[34,35]  For direct ionization of the
outer electron, $N^{+4}$ and $O^{+5}$ cases are in good agreement with Coulomb-Born[36] or
scaled-hydrogenic-Coulomb-Born[37] ionization calculations, but the $C^{+3}$ data are
about 30% lower than the best calculation[36] at the cross-section peak (200 eV).
In all cases inner-shell excitation ($1s^2 2s \rightarrow 1s2sn\ell$) followed by autoionization
is observed.

Figure 7 shows the data for $O^{+5}$ which are the least precise of all cases but show
the largest excitation autoionization contribution.  By subtracting the direct
ionization component (assuming the theoretical shape of the direct ionization cross
section), an experimental excitation cross section is obtained and can be compared
with excitation calculations.[38,39]  Such a subtraction assumes that excitation and
ionization simply add (no interference) and for the present cases assumes that all
of the inner-shell excitation decays via autoionization.  Table 2 shows results
and comparisons after such subtraction for Li-like ions.  Except for the $O^{+5}$ case,
the six-state, close-coupling cross sections are consistently about 40% lower than
experiment, but the assumptions and uncertainties of the subtraction process are
probably about equal to this discrepancy.  For $O^{+5}$ the statistical uncertainties
are large but even taking the extremes of the 90% confidence level relative uncertainties gives minimum deduced excitation cross sections about twice the 6-cc
theory.  The discrepancy could be failure of the experiment, of excitation theory,
or of the assumption of simple addition of excitation and ionization.  The

situation provides strong motivation for improving the $O^{+5}$ data and extending studies along this sequence to higher charge states.

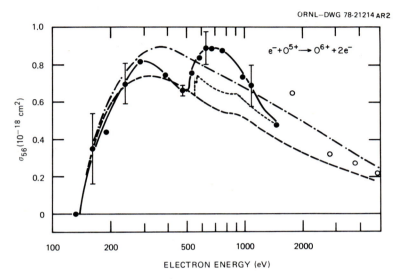

Figure 7

Electron impact ionization of $O^{+5}$ with excitation-autoionization beginning at 520 eV. Solid points - Ref. 35 with 90% C.L. relative uncertainties; open points - Ref. 15; dot-dashed curve is Lotz - Ref. 33; long-dashed curve is scaled-Coulomb-Born - Ref. 37; and short-dashed curve is 6-cc excitation theory by Henry - Ref. 39 - added to scaled Coulomb-Born.

TABLE 2. Comparison of theoretical and experimental cross sections for inner-shell excitation of Li-like ions, $1s^2 2s \rightarrow 1s 2s 2\ell$ at energies about 1.1 times excitation threshold.

Cross Sections in $10^{-19}$ cm$^2$

| Ion | Energy (eV) | Experiment[a] | Scaled Coulomb-Born[b] | Six-state close-coupling[c] | Ratio Exp/6cc Theory |
|-----|-------------|---------------|------------------------|-----------------------------|----------------------|
| $Be^+$ | 130 | 17 | 23 | 12.2 | 1.4 |
| $C^{+3}$ | 340 | 3.2 | 3.7 | 2.15 | 1.5 |
| $N^{+4}$ | 460 | 1.8 | 2.0 | 1.27 | 1.4 |
| $O^{+5}$ | 612 | 2.8 (1.4) | 1.1 | 0.74 | 3.8 (1.9) |

a. Cross sections determined by estimating the increase in total ionization cross section at energy 1.1 times excitation threshold. The $Be^+$ case is from Ref. 34 while $C^{+3}$, $N^{+4}$, and $O^{+5}$ are from Ref. 35. For $O^{+5}$ the uncertainties are large and the value in parenthesis is the smallest excitation cross section derived from the ionization data allowing for the 90% confidence level error bars.

b. From Ref. 38 which cautions that the technique is not appropriate to low ionic charge cases such as $Be^+$.

c. From Ref. 39 with $Be^+$ case extrapolated according to $(z - 1.4)^3 \sigma = $ constant for a given energy in threshold units.

Similar experimental and theoretical investigation along the Na-like sequence has been initiated. The only multicharged case studied experimentally is $Si^{+3}$ which is presented in Fig. 8. In all cases ($Mg^+$, $Al^{+2}$, $Si^{+3}$), the excitation $2p^63s \rightarrow 2p^53s3p$, indicated as 2p-3p on the figure, is the strongest transition in distorted-wave theory,[40] or, for $Mg^+$, in Coulomb-Born.[41] This transition is missing in the experiments[42],[43] suggesting a specific problem in the theory.

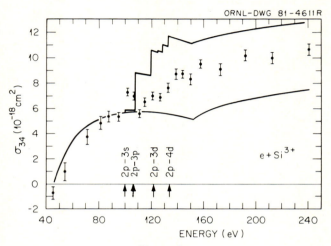

Figure 8
Electron impact ionization of $Si^{+3}$ from Ref. 43. Solid curve is distorted wave theory of Ref. 44 with distorted wave excitations of a 2p electron (Ref. 40) added at energies indicated by arrows.

The effects of excitation-autoionization appear in many of the ionization experiments. Figure 9 shows the $Ar^{+4}$ ionization which has been measured by both groups.[31],[45] Excitation of a 2p-inner electron followed by autoionization gives the feature near 250 eV (which was specifically included in the classical calculation).[46] Excitation of a 3s electron might lead to autoionizing levels near threshold which could be responsible for the fast rise of the ionization cross section and the discrepancy with the scaled-Coulomb-Born result[37] which is guessed to be the best of the available predictions. Of course, metastables in the incident beams could also be responsible for some of the pecularities near threshold.

The original ionization experiments showing strong excitation-autoionization were for alkali-like ions[47] for which $\Delta n = 0$ transitions np-nd lead to autoionizing levels. Recent work[48] on alkali-like $Ti^{+3}$, $Zr^{+3}$, and $Hf^{+3}$ (represented by Fig. 10 for $Ti^{+3}$) shows the most dramatic excitation-autoionization yet observed with excitation dominating the ionization cross section by factors of 10 for $Ti^{+3}$ and $Hf^{+3}$ and 20 for $Zr^{+3}$ beginning very near threshold. Calculations of the atomic structure (energies) together with distorted-wave excitation cross sections reproduce the observed shape of the cross section (Fig. 10) for these cases but overestimate the excitation by a uniform factor of 2.5. An additional interesting feature of these alkali-like cases is that for higher charge states the $\Delta n = 0$ transitions are predicted to become bound so that the anticipated excitation-autoionization should decrease dramatically. These cases hold basic physics interest for atomic structure (spectroscopy), excitation theory, and our understanding of the coupling of direct and indirect ionization processes. Current theoretical work including the interference between direct and indirect processes[49] will provide interesting comparisons with experiments.

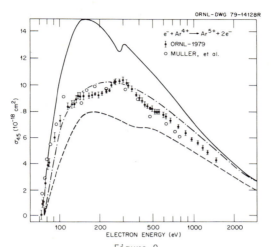

Figure 9
Electron impact ionization of $Ar^{+4}$. Solid points with 90% C.L.
relative bars from Ref. 31; open points from Ref. 45. Solid curve
is classical theory of Ref. 46; dashed curve is scaled-Coulomb-
Born of Ref. 37, and dot-dashed curve is Lotz - Ref. 33.

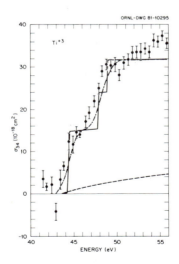

Figure 10
Electron impact ionization of $Ti^{+3}$ - from Ref. 48. Experimental data have 1 std. dev. relative error bars. Dashed
curve is Lotz formula (Ref. 33) taken as representative of
direct ionization. Solid curve is distorted wave calculation
of excitation of levels of $3p^6 3d \rightarrow 3p^5 3d^2$ divided by 2.5 and
added to Lotz. Chain curve is solid curve convoluted with
2-eV energy spread appropriate to the experiment.

Another dramatic example of the importance of autoionization for multicharged ions
has been provided by studies of multiple ionization in a single collision.[50]
Figure 11 shows cross-section results for ionizations +2 to +4, +2 to +5, and
+3 to +5 for Argon ions.  At the threshold for direct ionization of an L-shell
electron, the multiple ionization cross sections increase by up to an order of
magnitude attributable to inner-shell ionization-autoionization.

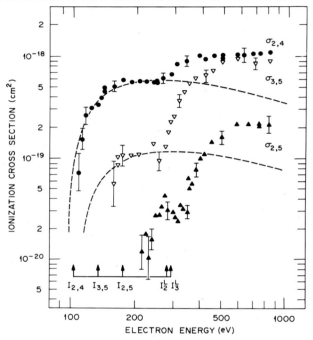

Figure 11
Multiple ionization of $Ar^{+2}$ and $Ar^{+3}$ in a single collision
by electron impact - from Ref. 45.  Arrows indicate threshold
energies for +2 → +4, +3 → +5, +2 → +5, and L-shell ionization.

CONCLUSION

While only ionization experiments have provided much data on collisions of electrons
and multicharged ions, these data have contained a wealth of detailed information
which is challenging our understanding of collisions involving these ions.  It
seems reasonable to guess that high quality experiments on dielectronic recombination would provide just as much challenge to our understanding.  It is clear that
excitation is important in all of the collision processes involving these ions.
Thus, while the experiments are difficult, the physics is exciting.

This work was supported through the Office of Fusion Energy, U.S. Department of
Energy, under contract W-7405-eng-26 with the Union Carbide Corporation.

BIBLIOGRAPHY


1.  M.F.A. Harrison in Methods of Experimental Physics Vol. 7a, B. Bederson and
    W. L. Fite (eds.) (Academic Press, New York and London, 1968) p. 95.
2.  K. T. Dolder and B. Peart, Rep. Prog. Phys. 39, 693 (1976).
3.  K. T. Dolder in Electronic and Atomic Collisions, Invited paper for X ICPEAC,
    G. Wate (ed.) (North-Holland, 1978) p. 281.
4.  J. Wm. McGowan in Electronic and Atomic Collisions, Invited papers for
    XI ICPEAC, N. Oda and K. Takayanagi (eds.) (North-Holland, 1980) p. 237.
5.  G. H. Dunn, "Electron Ion Collisions" to be published in Physics of Ionized
    Gases-1980 (proceedings of 1980 Symposium in Yugoslavia).
6.  International Conference on Heavy Ion Sources, IEEE Trans. Nucl. Sci. NS-23,
    2 (1976).
7.  E. D. Donets and V. P. Ovsyannikov, "The Cryogenic Electron Beam Ionizer,
    KRION-2," Report P7-80-515 of J. Inst. Nucl. Res., Dubna, USSR (1980).
    [Translation ORNL-tr-4703, available from Tech. Info. Ctr., P.O. Box 62,
    Oak Ridge, TN  37830, U.S.A.].
8.  J. Arianer and R. Geller, "The Advanced Positive Heavy Ion Sources" (review
    paper on EBIS and ECRIS to be published in 1981).
9.  H.-J. Kunze, Space Sci. Rev. 13, 565 (1972).
10. P. Greve, M. Kato, H.-J. Kunze, and R. S. Hornady, Phys. Rev. A (scheduled
    for July 1981).
11. R. L. Brooks, R. U. Datla, and H. R. Griem, Phys. Rev. Lett. 41, 107 (1978).
12. C. Breton, C. de Michelis, M. Finkenthal, and M. Mattioli, Phys. Rev. Lett. 41,
    110 (1978).
13. M. Bitter, K. W. Hill, N. R. Sautholf, P. C. Efthimion, E. Meservey, W. Roney,
    S. von Goeler, R. Horton, M. Goldman, and W. Stodick, Phys. Rev. Lett. 43, 129
    (1979) and additional measurements reported at the Third Conference on Atomic
    Processes in High Temperature Plasma, Baton Rouge, LA. February 1981 (in
    preparation for publication).
14. M. Hamdam, K. Birkinshaw, and J. B. Hasted, J. Phys. B 11, 331 (1978).
15. E. D. Donets and V. P. Ovsyannikov, Reports P7-10780 (1977) and P7-80-404
    (1980) of J. Inst. Nucl. Res., Dubna, USSR. [Translations ORNL-tr-4616 and
    ORNL-tr-4702, available from Tech. Info. Ctr., P.O. Box 62, Oak Ridge, TN
    37830, U.S.A.].  See also p. 897 of Ref. 6.
16. P. O. Taylor, D. Gregory, G. H. Dunn. R. A. Phaneuf, and D. H. Crandall,
    Phys. Rev. Lett. 39, 1256 (1977).
17. D. Gregory, G. H. Dunn, R. A. Phaneuf, and D. H. Crandall, Phys. Rev. A 20,
    410 (1979).
18. D. H. Crandall, P. O. Taylor, and G. H. Dunn, Phys. Rev. A 10, 141 (1974).
19. R.J.W. Henry, Phys. Rep. 68, 1 (1981).
20. D. H. Crandall, "Electron Impact Excitation of Ions" prepared for NATO
    Institute 1981, and to be published in Physics of Ion-Ion and Electron-Ion
    Collisions, F. Brouillard and J. Wm. McGowan (eds.) (Plenum Press).
21. J. L. Kohl and G. P. Lafyatis, Center for Astrophysics, Cambridge, Mass.
    (private communication, 1981).
22. D. Auerbach, R. Cacak, R. Caudano, T. D. Gaily, C. J. Keyser, J. Wm. McGowan,
    J.B.A. Mitchell, and S.F.J. Wik, J. Phys. B 10, 3797 (1977).
23. P. F. Dittner, G. D. Alton, W. B. Dress, C. D. Moak, P. D. Miller, and
    S. Datz, Oak Ridge National Laboratory, Oak Ridge, TN  (private communication,
    1980).
24. D. H. Crandall, Phys. Scr. 23, 153 (1981).
25. P. O. Taylor, R. A. Phaneuf, and G. H. Dunn, Phys. Rev. A 22, 435 (1980).
26. M. A. Hayes, D. W. Norcross, J. B. Mann, and W. D. Robb, J. Phys. B 10,
    L429 (1977).
27. R.J.W. Henry and W. L. van Wyngaarden, Phys. Rev. A 17, 798 (1978).
28. Coulomb-Born calculations by J. B. Mann and distorted wave calculations by
    W. D. Robb are reported by N. H. Magee, J. B. Mann, A. L. Merts, and
    W. D. Robb in Los Alamos Sci. Lab. Report LA-6691-M5 (April 1977).
29. W. L. van Wyngaarden and R.J.W. Henry, J. Phys. B 9, 146 (1976).

30. S. M. Younger, Phys. Rev. A $\underline{22}$, 1425 (1980).
31. D. H. Crandall, R. A. Phaneuf, and D. C. Gregory, Oak Ridge National Laboratory Report ORNL/TM-7020 (1979).
32. J. J. Thomson, Philos. Mag. $\underline{23}$, 449 (1912).
33. W. Lotz, Z. Phys. $\underline{216}$, 241 (1968); ibid, Z. Phys. $\underline{220}$, 466 (1969).
34. R. A. Falk and G. H. Dunn, "Electron-Impact Ionization of $Be^+$" (submitted to Phys. Rev. A 1981).
35. D. H. Crandall, R. A. Phaneuf, B. E. Hasselquist, and D. C. Gregory, J. Phys. B $\underline{12}$, L249 (1979).
36. D. L. Moores, J. Phys. B $\underline{11}$, L403 (1978).
37. L. B. Golden and D. H. Sampson, J. Phys. B $\underline{10}$, 2229 (1977). See also D. L. Moores, L. B. Golden, and D. H. Sampson, J. Phys. B $\underline{13}$, 385 (1980) and L. B. Golden and D. H. Sampson, J. Phys. B $\underline{13}$, 2645 (1980).
38. D. H. Sampson and L. B. Golden, J. Phys. B $\underline{12}$, L785 (1979).
39. R.J.W. Henry, J. Phys. B $\underline{12}$, L309 (1979).
40. D. C. Griffin, C. Bottcher, and M. S. Pindzola, "The Contribution of Excitation-Autoionization to the Electron Impact Ionization of $Mg^+$, $Al^{+2}$, $Si^{+3}$" (submitted to Phys. Rev. A 1981).
41. D. L. Moores and H. Nussbaumer, J. Phys. B $\underline{3}$, 161 (1970).
42. S. O. Martin, B. Peart, and K. T. Dolder, J. Phys. B $\underline{1}$, 537 (1968).
43. D. H. Crandall, R. A. Phaneuf, R. A. Falk, D. S. Belić, and G. H. Dunn, "Electron Impact Ionization of Na-like Ions - $Mg^+$, $Al^{+2}$, $Si^{+3}$" (submitted to Phys. Rev. A 1981).
44. S. M. Younger, "Cross Sections and Rates to Direct Electron Impact Ionization of Sodium-like Ions" (submitted to Phys. Rev. A).
45. A. Müller, E. Salzborn, R. Frodl, R. Becker, H. Klein, and H. Winter, J. Phys. B $\underline{13}$, 1877 (1980).
46. A. Salop, Phys. Rev. A $\underline{14}$, 2095 (1976).
47. B. Peart and K. T. Dolder, J. Phys. B $\underline{8}$, 56 (1975).
48. R. A. Falk, G. H. Dunn, D. C. Griffin, C. Bottcher, D. C. Gregory, D. H. Crandall, and M. S. Pindzola, "Excitation-Autoionization Contributions to Electron Impact Ionization" (submitted to Phys. Rev. Lett. 1981).
49. H. Jakubowicz and D. L. Moores, "Electron Impact Ionization of Li-like and Be-like Ions" (to be published in 1981). See also thesis of Jakubowicz, University College London, July 1980, unpublished.
50. A. Müller and R. Frodl, Phys. Rev. Lett. $\underline{44}$, 29 (1980).

PHYSICS OF ELECTRONIC AND ATOMIC COLLISIONS
S. Datz (editor)
© North-Holland Publishing Company, 1982

EXCITATION PROCESSES IN COLLISIONS OF ELECTRONS WITH MULTICHARGED IONS

Ronald J. W. Henry

Department of Physics and Astronomy
Louisiana State University
Baton Rouge, Louisiana   70803
U.S.A.

The current state of theory and its application to electron
impact excitation of atomic ions of charge +3 or greater is
reviewed.  Results of calculations are compared with the
two existing experimental measurements.  Essential physics
to be considered for all calculations is the role of target
state correlations, distortion, exchange, resonances, and
intermediate coupling of target states.

INTRODUCTION

Electron impact excitation of positive ions has been studied for many years by
astrophysicists who wish to interpret spectroscopic observations and to construct
models of stellar atmospheres.  The quest for controlled thermonuclear fusion by
use of magnetically confined plasmas such as a tokamak has given further impetus
to the field of electron-ion scattering.

Due to experimental difficulties encountered especially for multiply-ionized
species, determination of collision strengths must rely primarily on calculations.
Crandall [1] has reviewed the crossed-beam experimental measurements in the
previous paper.  The present review is specifically directed to calculations of
electron impact excitation of ions of charge +3 or greater.  Previous reviews of
calculations include Bely and van Regemorter [2], Seaton [3], and Henry [4].

The collision strength $\Omega(i,f)$ is related to the excitation cross section $\sigma(i \to f)$
(measured in units of $\pi a_0^2$) by:

$$\Omega(i,f) = \omega_i \, k_i^2 \, \sigma(i \to f) \tag{1}$$

where $k_i^2$ is the energy (in Ry.) of the incident electron relative to the lower
state i, and $\omega_i$ is the statistical weight of the lower atomic state (either
$(2S_i+1)(2L_i+1)$ for LS-coupling or $(2J_i+1)$ if fine structure levels are being
taken into account).

We introduce the parameter x, the energy in threshold units, defined by

$$x = k_i^2/\Delta E_{if} = k_i^2/(E_f - E_i) \tag{2}$$

where $\Delta E_{if}$ is the excitation energy (in Ry.) for the transition from level with
energy $E_i$ to level with energy $E_f$.

The excitation rate coefficient $q_{if}$ is given by

$$q_{if} \ (T) = \frac{8.63 \times 10^{-6}}{\omega_i T^{\frac{1}{2}}} \ \exp \ [-\Delta E_{if}/(kT)] \ \gamma_{if}(T) \ cm^3 \ s^{-1} \tag{3}$$

where the rate parameter $\gamma$ is defined as

$$\gamma_{if} \ (T) = \int_0^{\infty} \Omega \ (i,f) \ \exp \ [-\Delta E_{if} \ x/(kT)] \ d[\Delta E_{if} \ x/(kT)] \tag{4}$$

and kT is in Ry and T is the electron temperature.

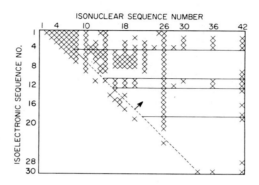

Figure 1
Electron-ion Calculations

An extensive bibliography of calculations made through early 1980 is included in
the review article by Henry [4]. Figure 1 summarizes the various ions on which
calculations have been made. In addition to the light ions, the impetus for the
field of electron-ion scattering has come from astrophysics as evidenced by the
work on cosmically-abundant ions Ne, Fe, Si, Mg, S, and Ar. Also, tokamak
plasmas have introduced data needs for Ti, Fe, and Mo. In contrast to the number
of calculations, there are only two experiments to date on multi-charged ions
which have yielded directly values for cross sections or collision strengths.
They are crossed-beam experiments on Li-like CIV by Taylor et al. [5] and NV by
Gregory et al. [6]. I will not discuss rate coefficient measurements which may
be deduced from plasma experiments. Such measurements have been reviewed by
Kunze [7].


COMPARISON WITH EXPERIMENT

Figure 2 gives $\Omega(2s,2p)$ for CIV. Absolute collision strengths to a good confidence level of ±17% for the energy region from threshold to 530 eV (x ~ 66) were
obtained by Taylor et al. [5]. This represented the first definite experimental
results for a multiply-charged ion. Curves A and B give Coulomb-Born calculations
CBI of Callaway et al. [7] and CBXII of Mann [8]. Distorted wave results of
Blaha and Davis [10] are given by curve C and two-state close-coupling

calculations 2CC of Robb [11] and Gau and Henry [13] by curve D.

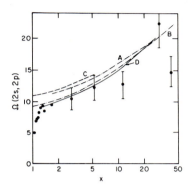

Figure 2
$\Omega(2s,2p)$ for CIV

Exchange effects are important. CBI and nonexchange distorted wave calculations (curves A and C) lie at least 15% above experiment. Both CBXII and 2CC calculations are in excellent agreement with experiment. Thus, distortion is not important for this case.

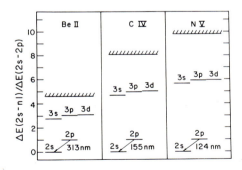

Figure 3
Energy Levels for Li Isoelectronic Sequence

The effect of coupling to higher states is also unimportant. Figure 3 shows the relative position of various states as the nuclear charge is increased, with 2s-2p energy splitting taken as one unit. Since the energy difference between 2s and 2p states increases linearly with Z and the energy difference between 2s and higher n states increases approximately as $Z^2$, the coupling becomes less important as Z increases.

Another important reason that there is excellent agreement between theory and experiment is that correlation between the scattered and target electrons decreases with increasing Z [3]. This is probably true for multiply-ionized systems considered here. However one must note that as the system increases initially in Z from the neutral end of an isoelectronic sequence, short range correlation effects first become more important since the increased nuclear charge pulls in the electrons and increases the probability of finding them close together.

For NV, very satisfactory agreement exists between theory and experiment for $\Omega(2s,2p)$. Gregory et al. [6] obtained absolute cross sections to $\pm 18\%$ in a crossed-beam experiment for the energy region from threshold to 52 eV (x ~ 5.2). When two-state and five-state close coupling calculations of van Wyngaarden and Henry [14] are folded in with the experimental energy beam profile, very good agreement is obtained between theory and experiment over the entire energy range. The five-state calculation includes the 2s, 2p, 3s, 3p, and 3d states.


ESSENTIAL PHYSICS

The essential physics which should be considered for all calculations of electrons scattering from multiply-charged ions are exchange, coupling, distortion, target state correlations, resonances, and intermediate coupling of target states.

Let us consider the collision of an electron with an N-electron ion of nuclear charge Z and net ionic charge z = Z-N whose target wave functions are known. The most exact quantum mechanical description for the solution of the collision problem is a converged close-coupling method. A discussion of the equations may be found in the review articles [3,4] and in Burke and Seaton [15]. Other approximate methods are also discussed in the various reviews. We will consider only the physical effects in this review and so we will not repeat the mathematical equations.


TARGET STATE CORRELATIONS

Target wave functions for lithiumlike ions may be represented to sufficient accuracy by single configuration, Hartree-Fock functions. Thus, target state correlations did not play a role in the above discussion on CIV and NV. However, for many systems, consideration of configuration mixing in the description of the target ion must be given. It follows from the variational principle used in the formulation of the scattering problem, that the error in the collision strengths is directly related to the first order error in the target wave functions. Thus, target wave functions must be chosen carefully.

A figure of merit for collision strengths is probably provided by the accuracy of oscillator strengths obtained with the same target wave functions. This is correct at least at very high energies where the collision strength is directly proportional to the oscillator strength.

The effect of correlations due to the presence of an open 3d shell is shown in Table I, which gives oscillator strengths calculated in the dipole length ($f_L$)

and dipole velocity ($f_v$) approximations for SIV. Two sets of configurations A and B were used by Bhadra and Henry [16] to describe the target functions. They retained configurations $3s^2 3p$, $3s 3p^2$, and $3s^2 3d$ in set A. Set B included set A plus terms with the largest coefficients for configurations $3s 3p 3d$, $3p 3d^2$, $3s 3d^2$, and $3p^2 3d$. Glass [17] used set C, a large number of configurations which included set B as a subset. Also given is $f_L$ obtained by Bhatia et al. [18] using set A with a different description of the orbitals. The disagreement between $f_L$ and $f_v$ for configuration set A for a given oscillator strength indicates that the representation of the target functions is not very accurate. In contrast, agreement between $f_L$ and $f_v$ for set B and agreement between sets B and C indicates fair convergence of the configuration interaction expansion.

Table I
Oscillator Strengths for SIV

| Configurations | A[16] $f_L$ | A[16] $f_v$ | A[18] $f_L$ | B[16] $f_L$ | B[16] $f_v$ | C[17] $f_L$ | C[17] $f_v$ |
|---|---|---|---|---|---|---|---|
| $^2P^o, {}^2D$ | 0.057 | 0.019 | 0.130 | 0.040 | 0.035 | 0.040 | 0.037 |
| $^2P^o, {}^2S$ | 0.103 | 0.066 | 0.105 | 0.090 | 0.092 | 0.092 | 0.090 |
| $^2P^o, {}^2P$ | 1.068 | 0.511 | 1.089 | 0.771 | 0.705 | 0.759 | 0.716 |

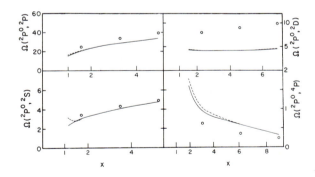

Figure 4
Collision Strengths for SIV

Figure 4 gives some collision strengths versus x for SIV. Circles represent distorted wave calculations of Bhatia et al. [18] with configurations of set A. Solid and dashed lines give five-state and two-state close coupling results of Bhadra and Henry [16] who used set B wave functions. The five-state calculation includes the $^2P^o$, $^4P$, $^2D$, $^2P$, and $^2S$ states, whereas the two-state calculation

includes the initial $^2P^o$ ground state and one of the excited states.  Comparison
of 2CC and 5CC indicates that coupling to intermediate states is unimportant for
energies greater than 1.5 times threshold.  We can understand qualitatively why
collision strengths of Bhatia et al. are larger than those of Bhadra and Henry for
optically allowed transitions.  For $\Omega(^2P^o, ^2D)$, where the dipole length oscillator
strengths differ by a factor of 3, the collision strengths differ by a factor of
2.  For the other optically allowed transitions, differences in $f_L$ values are less
and so the agreement in the collision strengths is better.  Differences between
the calculations are due to the choice of wave function description rather than to
the choice of scattering approximation.


RESONANCES

The main type of resonance which dominates electron-ion collisions is the Feshbach
or closed-channel resonance.  An infinite series of resonances converges on to
each of the states of the target ion due to the attractive Coulomb potential.
Consider an idealized resonance system in Figure 5 which has three levels of the
target ion labelled initial, final and upper (Robb [19]).  The upper level u may
capture the incident electron into an orbital of negative energy to form a compound state j of the (N+1)-electron system.  This resonance state j will either
undergo dielectronic recombination by radiatively decaying to a bound state b of
the (N+1)-electron system, or it will autoionize into the continua associated with
the N-electron ion states i or f.

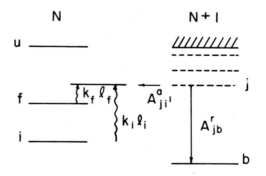

Figure 5
Idealized Resonance System

Large enhancements in the collision strengths may result due to resonance effects
particularly if the autoionization states lie close in energy to the excitation
thresholds.  Calculations have included resonance effects in a number of different
methods, some of which are given below.

A close-coupling calculation which involves the three target states, i, f, and u
will automatically include all of the resonance states j.  An example is given in

Figure 6 for $\Omega(2^1S, 2^3P^o)$ in OV by Berrington et al. [20]. The dashed line represents a 6CC calculation in which the six n=2 states $2s^2\ ^1S$, $2s2p\ ^3P^o$, $^1P^o$ and $2p^2\ ^3P$, $^1D$ and $^1S$ are retained. The solid line represents a 12CC calculation which includes the six n=2 states and six n=3 states $2s3s\ ^3S$, $^1S$, $2s3p\ ^3P^o$, $^1P^o$, and $2s3d\ ^3D$, $^1D$. Below the resonances converging to the n=3 states the 6CC and 12CC results are in good agreement. However, in the resonance region, $\Omega(2^1S, 2^3P^o)$ is increased on average by a factor of two. This enhancement is probably due to the strong coupling between the initial $2s^2\ ^1S$ state and the closed channel $2s2p\ ^1P^o$ for energies below the n=2 threshold. Below the n=3 threshold, the effect may be due to coupling to the $2s3s\ ^1S$ and $2s3d\ ^3D$ levels. Collision strengths to these states from the ground $2s^2\ ^1S$ state are of comparable magnitude to $\Omega(2^1S, 2^3P^o)$ in the above threshold region. For the allowed transition, $\Omega(2^1S, 2^1P^o)$ was increased by 10% on average by the inclusion of resonance effects.

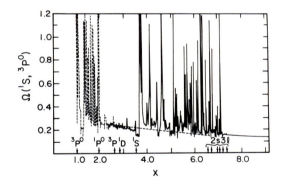

Figure 6

$\Omega(2^1S, 2^3P^o)$ for OV

The 12CC calculations were performed for several hundred energies below the n=3 thresholds and for a few energies above these thresholds. Such direct computations are relatively expensive. A program based on multichannel quantum defect theory (Seaton [21]) has been developed by Pradhan and Seaton [22] to yield both detailed and averaged collision strengths. In it, reactance matrices obtained at sacttering energies where all states are energetically accessible, are extrapolated to the resonance region.

An example is given in Figure 7 for $\Omega(1^1S, 2^3S)$ in CV by Pradhan et al. [23]. The effective average collision strength just above the $2^3S$ threshold, represented by the dashed line, is more than six times larger than the background just above the $2^1P^o$ threshold. A Gailitis-averaged value between the $2^1S$ and $2^3P^o$ thresholds is given by $\overline{\Omega}$. At the $2^3P^O$ threshold, a large Gailitis jump occurs due to the re-distribution of flux into the newly opened $2^3P^O$ channels. The resonance enhance-ment below the $2^3P^O$ is large due to the strong coupling between the initial and final states $1^1S$ and $2^3S$, and the $2^3P^O$. The magnitude of the jump when a thres-hold is crossed reflects the strength of coupling to that state.

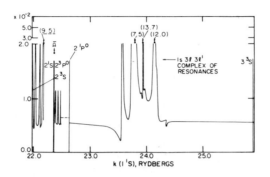

Figure 7

$\Omega(2^1S, 2^3S)$ for CV

In addition to the closed-channel resonances below the $2^1P^O$ state calculated indirectly by Pradhan et al. [23], they calculated resonance effects between the n=2 and 3 complexes. These were obtained by including bound channels of the form $1s3\ell3\ell'$ in the expansion of the total wave function for the system. Note that the n=3 states were not included explicitly in the close-coupling expansion, and so the resonances are due to the (N+1)-electron bound correlation states and not to closed channel functions belonging to the N-electron target plus scattered elec-tron. This approach was used successfully by Hayes and Seaton [24] for hydrogen-like ions.

Resonances do not in general affect the optically allowed transitions by a large factor. However, even for allowed transitions there could be a large resonance effect if the initial or final states are more strongly coupled to a closed channel state than to each other. Presnyakov and Urnov [25] used analytic pro-perties of the Coulomb Green's function to obtain $\Omega(2s,3s)$ and $\Omega(2p,3s)$ averaged over resonance structures for OVI. Below the 3p level, they found large enhance-ments of ~2 and ~30, respectively, due to resonances converging to the 3p and 3d levels. In this case, the forbidden transition was affected less than the allowed one. Figure 8 gives $\Omega(2p,3s)$ for OVI calculated by Bhadra and Henry [26], who used program RANAL [22] to yield detailed and averaged collision strengths. Resonances due to Rydberg series converging on the 3d state on average enhance $\Omega(2p,3s)$ by a factor of 2.8 in the energy region between the 3p and 3d states. Below the 3p state, the Rydberg series, which converge on both the 3p and 3d states, enhance $\Omega(2p,3s)$ by a factor of 6.8.

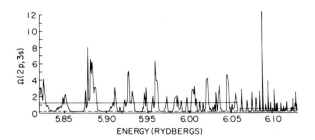

Figure 8

$\Omega(2p,3s)$ for OVI

Following the approach of Gailitis [27], collision strengths may be averaged over the resonances [21]. If the separation of the resonances is large compared with the widths of the resonances then:

$$\Omega(i,f) = \Omega^>(i,f) + \sum_{i'} \frac{\Omega^>(i,i') \; \Omega^>(i',f)}{\sum_{i''} \Omega^>(i',i'')} \qquad (5)$$

where $\Omega^>$ are collision strengths calculated above the new threshold and extra-polated to energies below this threshold. In equation (5), $i'$ is summed over degenerate closed channels of the new threshold and $i''$ is summed over all open channels. Table II gives the averaged resonance enhancement factors for colli-sion strengths for OVI in the energy region below the 3p threshold as calculated by the equation and by program RANAL. Thus, the relative values of $\Omega^>$ may be used to gauge the strength of coupling and the probable resonance enhancement.

Table II
Averaged Resonance Enhancements for OVI

|          | 2s3s | 2p3s | 2s3p | 2p3p |
|----------|------|------|------|------|
| Eq. (5)  | 1.7  | 5.1  | 2.9  | 3.8  |
| RANAL    | 2.0  | 6.8  | 2.1  | 2.8  |

For many applications, the quantity of interest is the rate coefficient. Cowan [28] has outlined a novel procedure in which the resonance contribution to the excitation rate coefficient can be calculated separately from the non-resonant contribution. Figure 9 gives rate coefficients $q(2s^2 \; {}^1S, \; 2s2p \; {}^3P^o)$ for OV. Curves A and B represent close coupling calculations 6CC [29] and 12CC [20], respectively. Distorted wave calculations of Mann [8] are given by curve C and represent the background contribution. To obtain curve D, resonance effects of

the n=2 and n=3 states are obtained by Cowan's method and added to curve C.   These
calculations agree within 10% with the 12CC ones of Berrington et al. [20] in
which the n=2 and n=3 states are included in the close coupling expansion.

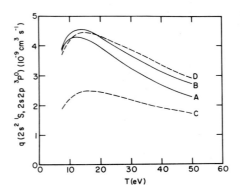

Figure 9
$q(2^1S, 2^3P^o)$ for OV

INTERMEDIATE COUPLING OF TARGET STATES

As the nuclear charge of the target ion increases, relativistic effects become
important, especially for Z > 20.  Figure 10 shows the separate effects of
resonances and intermediate coupling on the excitation rate $q(1s2s\ ^3S-1s2p\ ^1P^o)$
for FeXXV.  Curves A and B represent calculations by Clark et al. [30] and
Pradhan et al. [23].  The solid curves represent results without resonance effects
or intermediate coupling effects present.  The dashed curve shows the results [23]
with resonance effects only while the dash-dot curve shows the results [30]
including intermediate coupling but no resonance effects.  Both effects are seen
to be important for elements as heavy as iron.  Resonance effects may be smaller
than shown when allowance is made for radiative decay of the autoionizing state,
a physical process which increases as $Z^4$ with increasing Z.  The spin-orbit term
mixes the $1s2p\ ^1P_1^o$ and $1s2p\ ^3P_1^o$ configurations and so the dipole allowed transition $1s2s\ ^3S_1 - 1s2p\ ^3P_1^o$ dominates.  Provided Z is not too large, the additional
Breit-Pauli terms in the Hamiltonian may be treated as perturbations, and calculations in LS-coupling may be recoupled to yield collision strengths between fine-
structure levels using the procedures outlined by Saraph [31].

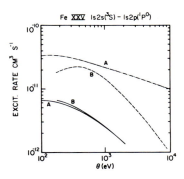

Figure 10

$q(1s2s^3S, 1s2p^1P^o)$ for FeXXV

INNER SHELL EXCITATION

An exciting area which is rapidly being explored both experimentally and theoretically is the contribution of inner-shell excitation to ionization of multicharged ions by electron impact. Crandall [1] has surveyed the recent experimental results and Moores [32] will discuss the theoretical results in the next review paper.

FUTURE DIRECTIONS

Since data needs are so vast for the fusion and astrophysics community, it is important to study trends along isoelectronic and/or isonuclear sequences.

Figure 11 gives the ratio of resonance enhancement as calculated by Eq. (5) for lithiumlike ions [26] as a function of $Z^{-1}$. For the 2s-3s, 2s-3p, and 2p-3p transitions the ratios are relatively independent of Z. However, for the 2p-3s transition, the enhancement increases approximately linearly with Z. Thus, resonances remain important for the heavier ions.

Figure 12 gives the energy levels for carbonlike ions relative to $2s^2 2p^2$ $^3P$ as a function of nuclear charge Z [33]. The energy differences for the energy levels in the $2s^2 2p^2$ and $2s2p^3$ configurations vary essentially as linear functions of Z. However, since the binding energy of the 3s electron increases as $Z^2$, the $2s2p^3$ 3s which is a resonant state for Z=7 becomes a bound state as Z increases. Thus, higher resonance states will sweep through the energy region of interest for $\Delta n=0$ transitions. [For $\Delta n\neq 0$ transitions, the difference in energy levels varies as $Z^2$ and so resonances will not sweep through the energy region of interest]. These

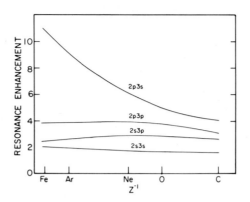

Figure 11
Resonance Enhancement for Li Isoelectronic Sequence

resonance states will probably be narrower since they belong to higher n states
and so their effect may be smaller.  In addition, photon emission will stabilize
the autoionizing levels for high Z and hence weaken the resonance contribution.
As the effect of resonances is weakened, intermediate coupling effects will be
increasing in importance.  The trends for these two processes should be investigated carefully.  It may be shown that the background collision strengths vary as
$Z^{-2}$ along an isoelectronic sequence.  However, resonance and/or intermediate
coupling effects may lead to dependence of rate coefficients on Z that is considerably less than $Z^{-2}$ [34].  In addition, an accurate representation of the
target state wave functions must first be given and exchange and distortion
effects should be included for energies up to at least three times threshold.

ACKNOWLEDGEMENTS

The work was supported in part by the U. S. Department of Energy (Office of Basic
Energy Sciences).  The author is grateful to A. L. Merts and A. K. Pradhan for
communicating some of their manuscripts prior to publication.

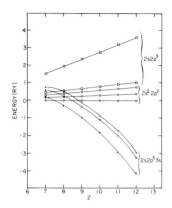

Figure 12
Energy Levels for C Isoelectronic Sequence

REFERENCES

[1]   Crandall, D. H., in: Datz, S. (ed.), XII ICPEAC 1981 (North-Holland, Amsterdam, 1982).
[2]   Bely, O. and van Regemorter, H., Ann. Rev. Astron. Astrophys. 8 (1970) 329.
[3]   Seaton, M. J., Adv. Atom. Molec. Phys. 11 (1975) 83.
[4]   Henry, R. J. W., Phys. Reports 68 (1981) 1.
[5]   Taylor, P. O., Gregory, D., Dunn, G. H., Phaneuf, R. A., and Crandall, D. H., Phys. Rev. Lett. 39 (1977) 1256.
[6]   Gregory, D., Dunn, G. H., Phaneuf, R. A., and Crandall, D. H., Phys. Rev. A20 (1979) 410.
[7]   Callaway, J., Msezane, A. Z., and Henry, R. J. W., Phys. Rev. A19 (1979) 1416.
[8]   Mann, J., quoted in [9].
[9]   Merts, A. L., Mann, J. B., Robb, W. D., Magee, N. H., and Argo, M. F., Los Alamos Scientific Laboratory Report No. LA-8267-MS, Los Alamos, New Mexico (1980).
[10]  Blaha, M. and Davis, J., quoted in [9].
[11]  Robb, W. D., quoted in [12].
[12]  Magee, N. H., Mann, J., Merts, A. L., and Robb, W. D., Los Alamos Scientific Laboratory Report No. LA-6691-MS, Los Alamos, New Mexico (1977).
[13]  Gau, J. N., and Henry, R. J. W., Phys. Rev. A 16 (1977) 986.
[14]  van Wyngaarden, W. L., and Henry, R. J. W., Can. J. Phys. 54 (1976) 2019.
[15]  Burke, P. G., and Seaton, M. J., Methods in Computational Phys. 10 (1971) 1.
[16]  Bhadra, K., and Henry, R. J. W., Astrophys. J. 240 (1980) 368.
[17]  Glass, R., J. Phys. B 12 (1979) 2953.
[18]  Bhatia, A. K., Doschek, G. A., and Feldman, U., Astron. Astrophys. 86 (1980) 32.
[19]  Robb, W. D., in: McDowell, M. R. C. and Ferendeci, A. M. (eds.), Atomic and Molecular Physics in Controlled Thermonuclear Fusion (Plenum Publ. Corp., New York, 1980).

[20]   Berrington, K. A., Burke, P. G., Dufton, P. L., Kingston, A. E., and
       Sinfailam, A. L., J. Phys B 12 (1979) L275.
[21]   Seaton, M. J., J. Phys. B 2 (1969) 5.
[22]   Pradhan, A. K., and Seaton M. J., private communication (1980).
[23]   Pradhan, A. K., Norcross, D. W., and Hummer, D. G., Phys. Rev. A23 (1981)
       619.
[24]   Hayes, M. A., and Seaton, M. J., J. Phys. B 11 (1978) L79.
[25]   Presnyakov, L. P., and Urnov, A. M., J. Phys. B 8 (1975) 1280.
[26]   Bhadra, K., and Henry, R. J. W., submitted to Phys. Rev. A.
[27]   Gailitis, M., Sov. Phys. JETP 17 (1963) 1328.
[28]   Cowan, R. D., J. Phys. B 13 (1980) 1471.
[29]   Dufton, P. L., Berrington, K. A., Burke, P. G., and Kingston, A. E., Astron.
       Astrophys. 62 (1978) 111.
[30]   Clark, R. E. H., Magee, N. H., Mann, J. B., and Merts, A. L., submitted to
       Astrophys. J.
[31]   Saraph, H. E., Comput. Phys. Commun. 15 (1978) 247; ibid. 3 (1972) 256.
[32]   Moores, D. L., in: Datz, S. (ed.), XII ICPEAC 1981 (North–Holland, Amsterdam,
       1982).
[33]   Eissner, W., Nussbaumer, H., Saraph, H. E., and Seaton, M. J., J. Phys.
       B 2 (1969) 341.
[34]   Pradhan, A. K., Norcross, D. W., and Hummer, D. G., Astrophys. J. 246 (1981)
       1031.

PHYSICS OF ELECTRONIC AND ATOMIC COLLISIONS
S. Datz (editor)
© North-Holland Publishing Company, 1982

IONIZATION PROCESSES IN COLLISIONS OF ELECTRONS
WITH MULTICHARGED IONS

D.L. Moores

Department of Physics and Astronomy
University College London
Gower Street
London WC1E 6BT, England

## INTRODUCTION

The construction of a general quantum-mechanical theory of the electron impact
ionization of atomic systems, even the simplest ones, presents theoretical
difficulties which do not arise in for example the treatment of the excitation
process. These difficulties are associated with the long-range nature of the
Coulomb interaction between the ejected and scattered electrons, which continue
to interact with the ion and with each other even out to infinity, so that a
proper treatment of the process would demand the full solution of a many-body
problem in the asymptotic region. The present state of the basic theory has been
summarized by Jakubowicz and Moores, [1]. In the formulations of Peterkop [2,3] and
Rudge and Seaton [4] (we take the case of ionization of a one-electron ion, for
convenience) the outgoing electrons move in potentials which behave asymptotically
as $z_1/r$, $z_2/r$ where $z_1$ and $z_2$ satisfy the equation

$$z_1/k + z_2/\varkappa = Z/k + Z/\varkappa - \frac{1}{|k - \varkappa|} \tag{1}$$

$\varkappa$ and $k$ being the wave vectors of the ejected and scattered electrons respectively,
and $Z$ the nuclear charge. The exchange amplitude $g(\varkappa, k)$ and the direct amplitude
$f(\varkappa, k)$ satisfy the equation

$$g(\varkappa, k) = e^{i\tau(\varkappa, k)} f(k, \varkappa) \tag{2}$$

In an exact theory (1) would be satisfied and $\tau(\varkappa, k)$ would be uniquely defined.
If (1) is not satisfied (and this is usually the case in approximate theories)
then $\tau(\varkappa, k)$ is essentially arbitrary and some choice for it must be made. This
affects the magnitude and sign of the interference term in the cross section,

$$Q(\varepsilon) = \int_0^{\varepsilon/2} \sigma(\varepsilon, \varkappa^2) \, d\varkappa^2 \tag{3}$$

$$\sigma(\varepsilon, \varkappa^2) = \frac{k\varkappa}{8\pi k_0 g_i} \sum_m \int \left( |f(\varkappa, k)|^2 + |g(\varkappa, k)|^2 - \text{Re} \, f^*(\varkappa, k) g(\varkappa, k) \right) d\hat{k}_0 d\hat{k} d\hat{\varkappa} \tag{4}$$

(In equation (4), $k_0$ is the incident wave vector, $g_i$ the statistical weight of the
initial state of the ion and

$$E = \frac{1}{2} k^2 + \frac{1}{2} \varkappa^2 = \frac{1}{2} k_0^2 - I \tag{5}$$

is the total energy of the continuum electrons and I the ionization energy, both
in atomic units. The sum in (4) is over degenerate magnetic quantum numbers.
The term Re f*g is referred to as the interference term.) In any approximate
calculation two distinct approximations have to be introduced, one for the
magnitude of the ionization amplitudes (which amounts to choice of potentials or
wave functions) and another for their relative phase.

The methods that have been used in practical calculations are chiefly variants of
the Coulomb-Born approximation, in which the wave functions representing the
incident and scattered electrons are taken to be Coulomb waves. The Coulomb-Born

amplitude for ionization of an (N+1)- electron ion is given by

$$f_{CB}\left(\chi,\underline{k}\right) = -\left(2\pi\right)^{-5/2}\int \Psi_o(N+1)\left(\phi^*(2-N-1,-\underline{k}_0,r_{N+2})\sum_{i=1}^{N+1}\frac{1}{r_{i,N+2}}\ \phi\left(2-N-1,-\underline{k},r_{N+2}\right)d\underline{s}_{N+2}\ \Psi_f(N+1,\underline{r})d\tau\right)$$ (6)

where $\Psi_o$ and $\Psi_f$ are wave functions of the (N+1)- electron target in its initial
state and of the system of N-electron ion plus ejected electron, respectively.
The functions $\phi$(Z-N-1, -$\underline{k}$, $\underline{r}$) are Coulomb wave functions satisfying

$$\left(\frac{1}{2}\nabla^2 + \frac{2-N-1}{r} + \frac{1}{2}k^2\right)\ \phi\left(2-N-1,-\underline{k},\underline{r}\right) = 0$$ (7)

Sometimes (7) is modified by introducing a spherically symmetric distortion
potential V(r) and the approximation is then referred to as the Distorted Wave
(DW) method. By writing the amplitude in the form (6) the ionization event is
seen as a bound-free transition of the target ion, provoked by electron impact.
The two continuum electrons are thus treated on a different basis, and equation (1)
is not satisfied unless an angle-dependent form is chosen for . This is not
a good description of ionization very close to threshold, but should improve with
increasing energy. Exchange may be included in a number of ways. In the
straight Coulomb-Born method (CB) the interference term is dropped in (4) and the
cross section becomes, using (2)

$$Q_{CB} = \int_o^E \frac{k\chi\,d(\chi^2)}{8\pi k_o g_i}\sum_m\int \left|f(\chi,\underline{k})\right|^2 d\hat{k}_o\,d\hat{k}\,d\hat{\chi}$$ (8)

This method does not really ignore exchange since it includes events which can
only take place by electron exchange. Because interference is not included, the
CB method often overestimates cross sections. In the modified or half-range
Coulomb-Born (CBNX) the amplitude g is set equal to zero in (4). Only events in
which the ejected electron is the slower are included. This corresponds to the
no-exchange method used for excitation cross section calculations. Calculations
show that at low energies g is comparable in magnitude to f ; however, the fact
that the exchange and interference terms in (4) tend to cancel makes the results
of this method quite good. In the Coulomb-Born Exchange (CBX) method, one sets
$|g(\chi,\underline{k})| = |f(\underline{k},\chi)|$ and $\tau(\chi,\underline{k})$ is chosen in such a way that when the algebraic
reduction is carried out, all Coulomb phase factors disappear from (4). It can be
shown that this approximation gives maximum interference in the case of highly
charged ions. In the Coulomb-Born-Oppenheimer (CBO) method a final state wave
function is chosen which is antisymmetric in all electrons. Events in which the
total spin of the target changes are thus included, an advantage of the method
over the CBX which does not include such events. The CBO method is however
algebraically more cumbersome and suffers from the disadvantage that in the
exchange terms the initial and final wave functions are not orthogonal.

In the case of large ionic charge the CBX and CBO methods tend to the same limit
and phase problems and lack of orthogonality both become less severe. At the same
time the basic assumption of all forms of the Coulomb-Born approximation, that
the central Coulomb potential, which is fully included in the theory, dominates
all other contributions to the interaction is more justifiable and one might
expect the method to yield accurate results for highly-charged hydrogenic ions.
All the Coulomb-Born approximations discussed above of course have their distorted
wave equivalents.

In addition to the problems associated with the collision dynamics of the simple
knock-out ionization process, for complex ions the description of the target
system (and by target system we include the state of the ejected electron) must
be considered. In photoionization work, it is well established that in many cases

accurate cross sections cannot be obtained without a fairly elaborate treatment of the target system; configuration interaction, and final state resonance effects can be very important. The same considerations will also apply to electron impact ionization. For complex systems, no statement about the applicability of the Coulomb-Born method can be made until the accuracy of the target functions $\Psi_0$ and $\Psi_f$ used in a given calculation has been firmly established. The inclusion of resonance effects in the final state leads naturally to the next consideration – the processes of inner shell ionization and inner shell excitation to a state higher in energy than the ionization energy, followed by autoionization. Recent experimental work (reviewed by Crandall [5] has confirmed the importance of this latter effect, particularly in more complex systems, but even in the relatively simple Li-like ions the effect can be quite important. Autoionization may manifest itself in two rather different forms, depending upon whether the original excitation is of a closed inner-shell electron or of an electron in a less tightly bound shell. In Li-like ions, the effect is caused by the excitation of a 1s electron as follows:

$$ 1s^2 2s + e^- \rightarrow 1s2s2\ell + e^- \rightarrow 1s^2 + e^- + e^- $$

In Na-like ions the effect is larger, involving the excitation of an electron from the closed 2p shell which contains six electrons.

$$ 1s^2 2s^2 2p^6 3s + e^- \rightarrow 1s^2 2s^2 2p^5 3s\, n\ell + e^- \rightarrow 1s^2 2s^2 2p^6 + e^- + e^- $$

In both these cases the result is an abrupt jump in the ionization cross section at the autoionization threshold, a good example of which is provided by the ionization cross section for $Ca^+$ measured by Peart and Dolder [6]. In the case of B-like ions, for example, the process is

$$ 1s^2 2s^2 2p + e^- \rightarrow 1s^2 2s 2p\ (^{1,3}P)n\ell + e^- \rightarrow 1s 2s^2 + e^- + e^- $$

The lower members of the rydberg series $1s^2 2s2p n\ell$ are pure bound states. As Z increases, the principal quantum number of the first autoionizing level increases and for large Z the autoionizing states form a quasi-continuum. The autoionization thus contributes right from the ionization threshold and manifests itself as an overall enhancement of the ionization cross section rather than as a discontinuous jump (Moores [7]). The contribution from autoionization may be taken into account by adding to the direct ionization cross section the sum of cross sections for excitation of the autoionizing levels. This was first done by Bely [8] for Na-like ions, and the method has been used by a number of authors since then. Alternatively, the effect may be taken into account by employing a close-coupled wave function for the final state of ionized ion plus ejected electron (Jakubowicz and Moores [9]). If the close coupling expansion is chosen in the appropriate way then resonance effects will arise naturally in the wave function and thus autoionization will automatically be included if this wave function is substituted in the matrix elements appearing in the definition of the ionization cross section.

For all ions the contribution from autoionization may be reduced if the auto-ionizing states can also decay radiatively to a bound state of the system. For a given rydberg series, the radiative process competes with autoionization for large n, and for a given state, as Z increases along an isoelectronic sequence, the autoionization rate $A_a$ remain approximately constant while the radiative rate $A_r$ varies like $Z^4$. Calculations of $A_a$ and $A_r$ for configurations 1s2s2l of li-like ions (Gabriel [44] and Bhalla et al [10]) show that for ions heavier than oxygen $A_r$ becomes comparable with $A_a$ for some states to that loss of flux into radiative channels should be taken into account in the calculation of the autoionization contribution to ionization. This may be achieved by multiplying the excitation cross section by the branching ratio $A_a / (A_a + A_r)$. Cowan and Mann [11] have shown that in the Na-like ion Fe XVI even when account is taken of the fact that a large fraction of the excited states decay radiatively, autoionization still gives the dominant contribution to the total ionization rate for kT between 100 and 1500 ev. For more highly charged Na-like ions they conclude

that this contribution will be approximately the same since the main mechanism
involves the excitation of levels with $A_r \ll A_a$, whose excitation rates scale
smoothly with Z.  It is worthy of note that even if radiative decay is ignored,
Cowan and Mann predict a much smaller autoionization contribution than that
predicted by Kim and Cheng [12] or by Bely [8].

THE SCALED HYDROGENIC METHOD

For application to plasmas and astrophysics it is often necessary to know large
numbers of ionization cross sections both for the ground and for excited states
of a number of ions in all stages of ionization.  Since elaborate calculations
for such a large quantity of data are at present impracticable, it is important
to have a fairly simple formula of known accuracy, such as the semi-empirical one
of Seaton [13] or that of Lötz [14, 15] capable of generating all the information
required.  It is of course important also that any such formula should be of a
form soundly based on quantum-mechanical theory.  The method of Golden and
Sampson [16] represents an improvement upon the Seaton formula, which was only
guaranteed by its proposer to be valid up to about twice the ionization energy
and to within a factor of two in magnitude.  It also stands on a firmer theoretical
base than the Lötz formula.  The work of Golden and Sampson is based on the fact
that for hydrogenic ions, if one neglects screening completely and assumes that
the incident, bound, ejected and scattered electrons all move in the potential
$V(r) = Z/r$ due to the nucleus, then the reduced CBX cross section for ionization
of state nl, defined by

$$Q_R(nl, X) = \frac{1}{\pi a_o^2} \frac{z^4}{n^4} Q_{CBX}(nl, X)$$                (9)

where X is the incident energy in units of the ionization energy $z^2/n^2$ , is
independent of Z and tends to a finite limit as $Z \to \infty$.  In this approximation
the CBX and CBO methods give identical results, and cross sections calculated
by this method for hydrogenic ions are said to 'scale exactly'.  The $Q_R(nl, X)$
are calculated numerically and then fitted to a parametric form which has the
correct Bethe logarithmic behaviour at high energies:

$$X Q_R(nl, x) = A_{nl} \ln X + D_{nl} \left(1 - \frac{1}{x}\right)^2 + \left(\frac{C_{nl}}{x} + \frac{d_{nl}}{x^2}\right)\left(1 - \frac{1}{x}\right)$$                (10)

The Bethe coefficients $A_{nl}$ and $D_{nl}$ may be calculated independently while $c_{nl}$ and
$d_{nl}$, and in some cases $D_{nl}$, are obtained by fitting to the calculated points.
For any ion in the cross section is then written

$$Q(nl, x) = \pi a_o^2 \left(\frac{n}{Z_{eff}^{(nl)}}\right)^2 \frac{I_H}{I_{nl}} Q_R(nl, x)$$                (11)

Where $r_{nl}$ is the number of electrons in the shell being ionized, $I_{nl}$ is the
ionization energy of the shell nl and $I_H$ that of hydrogen and where $Z_{eff}(nl)$
are obtained either from screening constant theory (Mayer, [17]) or from a
prescription given by Golden and Sampson [16].  Reduced cross sections have been
computed and parameterized and the screening factors $Z_{eff}$ tabulated for n = 1 to 4
and l = 0 to n-1.  From the nature of the approximation one might expect the
method to give best results for highly charged ions.  Comparisons that have been
made with more elaborate calculations (Jakubowicz and Moores [1]) confirm that
the Golden and Sampson method gives a close approximation to a full CBX cross
section for Li-like and Be-like ions.  This is illustrated in Table I where it
can be seen that the two sets of results agree to within a few percent.

Table I. Electron Impact Ionization cross sections for three positive ions.
JM; calculated by Jakubowicz and Moores [1] in Coulomb-Born Exchange approximation.
GS; calculated by the scaled hydrogenic method of Golden and Sampson [16].

| X | $C^{3+}$ $(10^{-22}m^2)$ | | $Ne^{7+}$ $(10^{-23}m^2)$ | | $Fe^{22+}$ $(10^{-25}m^2)$ | |
|---|---|---|---|---|---|---|
| | JM | GS | JM | GS | JM | GS |
| 1.125 | 1.01 | 1.08 | 0.77 | 0.85 | 2.63 | 2.79 |
| 1.25 | 1.69 | 1.77 | 1.29 | 1.40 | 4.38 | 4.52 |
| 2 | 3.11 | 3.15 | 2.38 | 2.48 | 7.88 | 8.02 |
| 3 | 3.20 | 3.15 | 2.44 | 2.48 | 7.96 | 8.04 |
| 4.5 | 3.83 | 2.72 | 2.67 | 2.68 | - | - |

## COULOMB-BORN AND DISTORTED-WAVE CALCULATIONS

Full Coulomb-Born-Exchange and/or Distorted Wave exchange calculations have
been carried out recently by Jakubowicz and Moores [1, 9] and by Younger [18-22].
These calculations follow previous work using the methods by a number of authors
[23-31].

In Younger's calculations, for ionization of an (N+1)- electron ion the incident
and scattered electron wave functions are generated using a spherically
symmetric potential $V^{N+1}(r)$ composed of the sum of the frozen core Hartree-Fock
potential of the target ground state, plus, in the case of ions with more than
three electrons, an energy-dependent semi-classical exchange potential $V_{SCE}$.

$$V^{N+1}(r) = V_D^{N+1}(r) + V_{SCE}^{N+1}(r) \tag{12}$$

where

$$V_D^{N+1}(r) = \frac{Z}{r} + \sum_{n\ell} r_{n\ell} J_{n\ell}(v) \tag{13}$$

where

$$J_{n\ell}(v) = \frac{1}{r} \int_0^r [P_{n\ell}(\rho)]^2 d\rho + \int_r^\infty \frac{[P_{n\ell}(\rho)]^2}{\rho} d\rho \tag{14}$$

and $P_{n\ell}(\rho)$ are the target radial wave functions.

$$V_{SCE}^{N+1}(r) = \frac{V_D^{N+1} - E}{2} + \frac{1}{2}\left[(E - V_D^{N+1})^2 + \frac{4}{r^2}\sum_{i=1}^{N+1}(P_i(r))^2\right]^{\frac{1}{2}} \tag{15}$$

The ejected electron wave function is computed from a potential $V^N(r)$ of the
same form. Younger's results are discussed below.

In the work of Jakubowicz and Moores [1] the wave functions $\phi$ are either Coulomb
Wave Functions or distorted waves generated by a Thomas-Fermi-Dirac-Amaldi
statistical model potential. The functions $\Psi_o$ and $\Psi_f$ in equation (6) are taken
to be solutions of the set of coupled integro-differential equations describing
the interaction of an electron with the N-electron system, and have the form
(Burke and Seaton [32])

$$\Psi_o = \sum_{i=1}^{N_c} \theta_i + \sum_{j=1}^{N_b} c_j \Phi_j \tag{16}$$

$$\Psi_o \underset{r \to 0}{\sim} 0 \tag{17}$$

$$\Psi_f(x_1 \cdots x_{N+1}) = \sum_{LSM_LM_S\ell_\gamma m_\gamma m_{s\gamma}} (-1)^{\ell_\gamma - L_\gamma - M_L + \frac{1}{2} - S_\gamma - M_{S\gamma}} \frac{2\eta}{\chi^{1/2}} i^{\ell_\gamma + 1} Y_{\ell_\gamma m_\gamma}^*(\hat\chi) (2L+1)^{\frac{1}{2}}(2S+1)^{\frac{1}{2}} \begin{pmatrix} L_\gamma \ell_\gamma L \\ M_\gamma m_\gamma - M \end{pmatrix} \begin{pmatrix} S_\gamma \frac{1}{2} S \\ M_{S\gamma} m_{s\gamma} - M_S \end{pmatrix} \Psi_\gamma(\alpha LSM_LM_S) \tag{18}$$

Where the functions $\mathcal{I}_\gamma$ have the same form as $\mathcal{I}_o$ but satisfy continuum S-matrix boundary conditions asymptotically. These functions are computed by the program IMPACT (Crees et al, [33]. For a detailed description of these functions, the system of equations they satisfy and the method of solution of this system, the reader is referred to the detailed treatments by Burke and Seaton [32] Seaton [34] and Crees et al [33].

In (16) the sum is over $N_c$ open or closed free channels $\Theta_i$ and $N_b$ bound channels $\Phi_j$. $\mathcal{I}_o$ is obtained form a solution with all channels closed. The inclusion of bound channels allows for some final state correlation and also enables resonance effects to be included even in the absence of closed channels in the first surmation. Since $\mathcal{I}_o$ and $\mathcal{I}_f$ are obtained from solution of the same problem but at different energies, they will be orthogonal.

The use of close-coupled wave functions in (6) gives the method a great deal of flexibility. $\mathcal{I}_o$ can represent the ground state or any excited state of the N+1-electron system and its chosen form means that in theory an arbitrary number of configurations may be included. The form of the final state enables inner shell ionization, simultaneous excitation and ionization and autoionization, either from bound channels $\Phi$ or from closed channels $\Theta$ , or both, to be included directly in the wave function $\mathcal{I}_f$ .

When the energy of either the ejected electron or the scattered electron passes through a resonance this gives rise to a sharp peak in the differential cross section $\sigma(E,\chi^2)$, which has the form shown in figure 1. When the integration over $\chi^2$ is carried out this leads to an increase in the total cross section Q(E). It should be noticed that since $k^2 > \chi^2$ , the resonant behaviour first appears in the exchange amplitude g, and does not appear in f until $\chi^2$ attains the resonance value. This means that the CBNX approximation gives the autoionization threshold in the wrong place, at too high an energy.

In order to deal with the integral over $\chi^2$ in cases where $\sigma(E,\chi^2)$ is dominated by resonances, techniques based on many-channel quantum defect theory have been introduced [41] . This avoids having to calculate the cross section at a large number of energy points, a procedure that clearly becomes intractible when rydberg series of closed channel resonances are involved.

The possibility of radiative decay of the resonant states may also be taken into account by incorporating the process into the final state wave function, using time-dependent perturbation theory. One writes the time-dependent wave function for the N+1- electron system in the form

$$\Psi(t,E) = \sum_\gamma \int dE\, \mathcal{I}_\gamma(E) f_\gamma(E,t) e^{-iEt}\, \psi(0) + \sum_{\beta,\mu} \int \rho\, d\omega\, \Phi_\beta(E_\beta) g_\beta(\mu,\omega,t_\beta,t) e^{i(E_\beta+\omega)t}\, \psi(\mu,\omega) \quad (19)$$

Where $f_\gamma$ (E,t) is the amplitude for the state in which the (N+1)- electron system is in continuum state $\gamma$ with energy E at time t with no photon present; $g_\beta$ ($\mu$, w, $E_\beta$, t) is the amplitude for the state in which it is in a bound state $\rho$ with energy $E_\beta$ and a photon of angular frequency w and polarization $\mu$ is present. The functions $\psi$ represent photon states and $\rho$ their density. Time dependent perturbation theory leads to a set of time dependent coupled first order differential equation for f and $g_\gamma$. The approximate solution of these equations gives f (E, $\infty$) and g ($\mu$, w, E , $\infty$) in terms of $f_\gamma$ (E, 0), which is taken to be such that the wave function at t = 0 is a wave packet with a Gaussian profile. Hence, the form of the final state wave function, modified by the presence of the radiation field, is determined. In the case of a single channel $\gamma$ and one bound state $\beta$ , it is found that the effect of radiative decay is to modify the scattering amplitude by a multiplicative factor which depends on the radiative dipole matrix element between the states $\gamma$ and $\beta$ [55] .

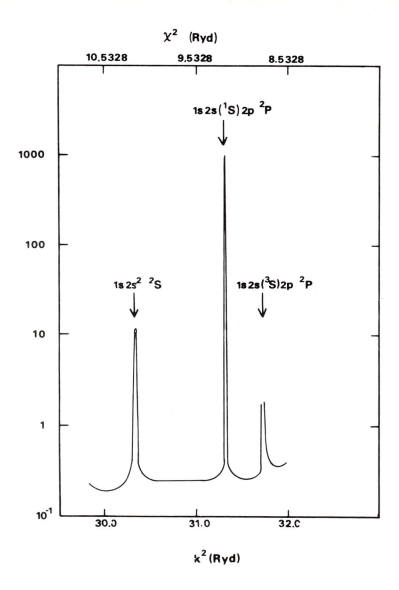

Figure 1
Differential Cross Section $\sigma(E,\chi^2)$ for OVII at Incident Electron
Energy E = 51 Showing Autoionizing Peaks

RESULTS

In this section we discuss the results of all the recent distorted-wave and
Coulomb-Born work that has been carried out, and compare with experimental results.

H-like ions

The recent calculations by Younger [18] for H-like ions and X ⩽ 2.25 confirm
previously-held ideas about the values of these cross sections.  The CBNX results
are in good agreement with the calculations of Rudge and Schwartz [25] and the
CBX results with those of Burgess and Rudge [24].  The difference obtained between
CBX and CBNX varies from below 5% at X = 1.125 to at most 20% at X = 2.25.  (It
should be recalled that the CBNX cross section, for a given choice of wave function
and if autoionization is unimportant, lies below the CBX cross section near
threshold but exceeds it above about X = 1.5 (Stingl 1972, [31] )).  The difference
between DW and CB calculations is less than 1% for C VI and Ne X, indicating that
the effects of the distortion potential, the second term on the right in equation
(12) are small, but up to 10%, depending on X, for He II.  For this ion good
agreement with the crossed beam experiment of Peart et al [35] is obtained.

He-like ions

The recent calculations by Younger [19] for He-like ions also show that the effects
of exchange are more important than those of distortion.  Younger's DWX results
differ by up to 30% from the corresponding DWNX results for 1.125 ⩽ X ⩽ 5.  Good
agreement is obtained between the DWX results for BIV, CV and NVI and the crossed-
beam experiments of Crandall et al [36].  In the case of Li II improved agreement
with the experimental results of Wareing and Dolder [37] and Lineberger et al [38]
compared with the CBNX calculations of Moores and Nussbaumer [27] is obtained.
This is due much more to the effects of exchange than to the use of distorted
waves.

Li-like ions

The recent experimental studies [36] reveal structure in the cross sections at
energies of about three times ionization energy due to inner shell excitation-
autoionization of configurations 1s2121'.

Below the autoionization threshold CBX and CBNX calculations of Jakubowicz and
Moores [1] give results within the experimental error limits for NV and OVI but
outside and above them for CIV.  The effects of distorted waves are found to be
small because the main contribution to the calculated cross section arise  from
high partial waves which are little affected by the distortion potential.  The
above conclusions are confirmed by the work of Younger [18].

The calculations of Jakubowicz and Moores use a final state wave function of the
form (16) with a single open channel together with bound channels of configuration
1s2s21.  Above the autoionization threshold the presence of the bound channels
$1s2s^2$ $^2$S and 1s2s ( $^{1,3}$S) 2p $^2$P gives rise to peaks in the differential ionization
cross section $\sigma(E, \chi^2)$, leading to jumps in the integrated cross section.  The
results obtained are similar to those by Sampson and Golden [39], with the positions
and magnitudes of the jumps agreeing with observation for CIV and NV but predicting
a jump of smaller magnitude than that observed in OVI.  The ratio of the second to
the first maxima in the cross section is found to increase slowly along the isoelectronic sequence in a similar manner to that predicted by Sampson and Golden.
The value for Fe XXIV is about 1.17.  The effects of radiative decay of the 1s2s21
states reduce the size of the jump due to autoionization.  This is illustrated by
calculations carried out by two methods.  First, the direct ionization cross
section was calculated by the method of Golden and Sampson [16].  The contribution
from autoionization but without radiative decay was then obtained by adding to this

cross section the cross section for excitation of states 1s2s2l calculated by
Sampson, Clark and Parks [43] with allowance for configuration mixing and in
intermediate coupling. The effects of radiative decay were subsequently taken
into account by multiplying each autoionization contribution by the appropriate
branching ratio $A_a/(A_a+A_r)$ where $A_a$ and $A_r$ where taken from the work of Gabriel
[44] and Bhalla et al [10] . The results are shown in figure 2, where it is
seen that radiative decay reduces the effect of autoionization by a factor of
about six.

Alternatively, calculations may also be made by the methods of Jakubowicz and
Moores [1] . Results obtained by Butler and Moores (to be published) are
also displayed in figure 2 where it is seen that the reduction produced by
radiative decay is of a similar magnitude to that obtained using the first
method.

Results for CIV, NV AND OVII are shown in figures 3, 4 and 5.

Be-like ions    The results obtained for Be-like ions by Younger [21] and by
Jakubowicz and Moores [1] are in good agreement. The different approximations
display a similar pattern to the results for Li-like ions, with the distorted-
wave results giving the lowest cross sections. Comparison with experiments for
CIII, NIV and OV is complicated by the fact that the measured cross sections
contain a contribution of uncertain magnitude from ionization out of the
metastable $1s^2 2s2p^3P$ level, resulting in a non-zero value at the threshold for
ionization of $1s^2 2s^2$ 1S. The distorted-wave results are a few percent lower
than the unmodified experimental results for CIII [30] and NIV [36] but well
below experiment [36] for OV, the difference being 25% at the peak.

B-like ions    As discussed above, for these ions autoionization is important right
from threshold but is unlikely to cause much structure in the cross sections.
CBX calculations have been carried out by Stingl [31] and CBNX by Moores [28, 30].
The latter calculations include an estimate of the autoionization contribution
but appear to give an overestimate of the observed cross sections for CII and
NIII [49] at low energies, even though the observations are contaminated by a
metastable contribution. The calculations clearly need to be repeated and more
accurate calculations are currently in progress at University College London
using the techniques of Jakubowicz and Moores [1] .

Ions with configurations $2p^q$, $1 < q < 6$    CBNX calculations for these ions were
made by Moores [28] and additional results may be obtained from these by
interpolation along isoelectronic sequences or between different values of q.
No allowance for autoionization was made. These should also be the
subject of future theoretical study.

Ne-Like ions    DWX calculations have been carried out for several ions by
Younger [20] and CBNX calculations by Moores [28] . Experimental results exist
for Na II [46, 47] and for Mg III [48] but agreement between theory and experiment is only fair. Younger found great sensitivity of the results to the
potential used for the ejected electron but insensitivity to the form of the
incident and scattered waves. None of the calculations gave good agreement
with experiment for Mg III or for Na II, for which the two existing experimental
determinations are mutually consistent. It should also be noted that for these
ions the Lotz formula considerably overestimates the measured cross sections.
It is clear that a more detailed treatment of the final state of ionized ion plus
ejected electron is required, using close-coupled wave functions of the form (16)
and the program of Jakubowicz and Moores.

Na-like ions    For the Na isoelectronic sequence, the calculations of Younger [22]
confirm the importance of inner shell ionization predicted in Mg II by Moores and
Nussbaumer [27] . If the autoionization contribution, which Moores and Nussbaumer
appear to overestimate, is disregarded then very good agreement is found between
these CBNX calculations, the DWX results of Younger and the experimental results
of Martin et al [45] . The effects of autoionization in Fe XVI are large and
have been  discussed above.

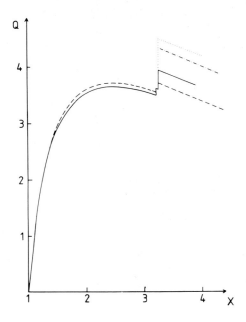

Figure 2
Electron Impact Ionization Cross Section for Fe XXIV. Continuous curve,
calculation by Butler and Moores, including radiative decay: dotted
curve same authors but without inclusion of radiative decay. Dashed
curve alternative method (see test) with radiative decay included;
dot-dash curve, without it.

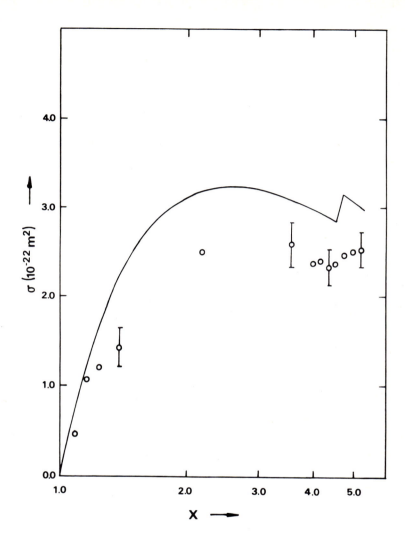

Figure 3
Electron Impact Ionization Cross Section of CIV. Continuous curve, DWX
calculation of Jakubowicz and Moores [1], experimental points,
Crandall, et al. [36].

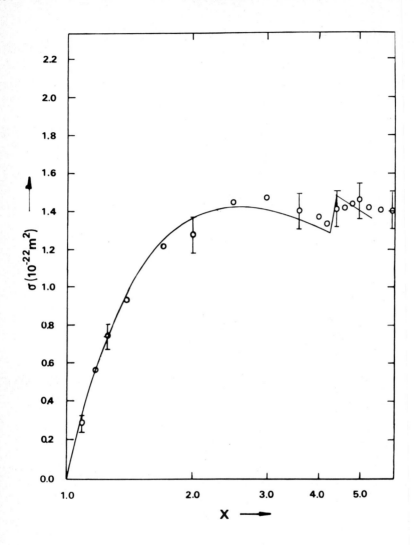

Figure 4
As Figure 3, for NV

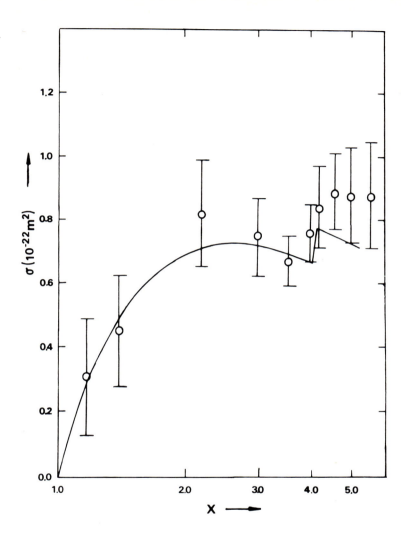

Figure 5
As Figure 3, for OVI

Ions with 12 or more electrons

Although a number of recent experimental determinations of cross sections for more
complex ions have been carried out, little theoretical work has yet been done on
them. Calculated values may be obtained from the Lötz formula or by the method of
Golden and Sampson. Because of the increasingly complex shell structure, autoionization and inner shell ionization can dominate at low energies and any calculation should pay due regard to this fact.

Conclusions

The results of the recent calculations and their comparison with crossed beam
experimental work enable a number of general conclusions to be drawn. The
difference between distorted-wave and Coulomb-Born calculations is small, in many
cases less than the experimental error limits. This indicates that the spherically
symmetric static potential of the target plays an unimportant role in determining
the ionization cross sections. It does not necessarily follow however that the
Coulomb Born approximation is adequate, since the effects of other off-diagonal
coupling potentials, especially dipolar ones, could be strong, particularly at
low energies and for the less highly charged systems. Such effects could possibly
be responsible for the apparent breakdown of the scaling law observed for the
lowest member of the Li sequence, CIV. More important than the distortion effect
is the effect of exchange, the difference between CBX and CBNX (or DWX and DWNX),
near the peak in the cross section, the energy at which they differ the most, being
of the order of 20-30%. Below X = 2 however the difference between them is much
less than this. The CB method gives cross sections which are 50-100% larger than
the other two over the whole energy range, and its use is not recommended. The
quantal approximation giving the best overall agreement with experiment is the
DWX, which gives results within the error bars in many cases. In those cases for
which comparison has been made, the scaled hydrogenic method seems to be a good
approximation to the complete CBX method. However, this may not be the case for
ions with more complex structure. The Lötz formula gives results which are often
very good and is usually within 40% of the DWX or CBX result.

Effects of configuration interaction and resonances in the target can be very
important, depending on the structure and it is anticipated that they will be of
fundamental importance for configurations $2p^q$, $1 < q < 6$.

Calculations of the autoionization structure observed in Li-like ions give good
agreement with experiment for CIV and NV, less good for OVI, for which theory gives
a smaller cross section than experiment. In Fe XXIV, radiative decay is found to
reduce the magnitude of the enhancement due to autoionization by a factor of about
six.

Finally, a word about the scaling of ionization cross sections along isoelectronic
sequences. Even though in many cases this is possible, care should be exercised
before blindly applying scaling laws. Even though many theoretical treatments
yield cross sections which scale very well, this is often inherent in the approximations made in the theory. One factor which can affect scaling laws is the
fact that distortion and strong coupling effects tend to be more important relative
to the Coulomb potential for ions of low charge, leading to deviations from scaling
for lower members of isoelectronic sequences. Another is the different relative
importance of autoionization and direct ionisation along a sequence. If possible,
the different components of a cross section could be scaled independently before
adding them together. Also, the shifting of autoionization thresholds to lower
energies will certainly affect the scaling along an isoelectronic sequence.

Ionization Balance Curves

Apart from their intrinsic interest the electron impact ionization cross section
for positive ions are required for the compilation of ionization balance curves,
giving the relative abundance of the different ionization stages of each element
as a function of temperature and density. These curves are required in turn for
the interpretation of the properties of laboratory plasmas such as occur in fusion

research and astrophysical plasmas like the solar corona. A number of different sets of calculations exist which employ different approximations for the ionisation and recombination rates. These are show in table 2 .

In the past the main effort has been in improving the recombination rates used, in particular the contribution from dielectronic recombination. The calculations of Jacobs et al [53] are the most up to date in this respect. The main contribution to the ionization rate coefficients below about $10^7$K comes from energies between threshold and about X = 2.5. In general, the Coulomb-Born calculations carried out before 1980 and the results of crossed-beam experiments tend to confirm the adequacy of the semi-empirical formulae in this region. However, the exchange classical impact parameter method, which tends to give lower ionization rates, is found to be in many cases in good agreement with plasma-derived experimental results. Since the use of ionization balance curves based on different rates can lead to conflicting or puzzling conclusions about the properties of plasmas, it is essential that the discrepancy be explained.

The results presented in this paper provide evidence in favour of the Coulomb-Born (or distorted wave) calculations and the crossed-beam experimental work. For a range of ions the difference between theory and experiment is good and never more than at the most, 25%, in contrast with the results of the ECIP method. A good example is provided by NV, where there are six mutually consistent sets of results, three theoretical and three experimental all of which are inconsistent with the plasma-derived point of Kunze 56 and the ECIP calculation.

This is illustrated in figure 6.

It is difficult to see especially in the case of highly ionized iron ions how any improvements in our theory, while still possible could bring the results down to the values required to make them agree with the ECIP values. If it is true that the CBX results are better than the ECIP then the ionization balance calculations of Jordan [52] or Jacobs et al [53] should be taken in preference to those of Summers [54].

Further evidence in favour of this viewpoint comes from the analysis of solar flare spectra of highly ionized iron ions by Cheng et al [50] and by Feldman et al [51] who predict temperatures of maximum fractional abundance which are a factor of two lower than those predicted by Summers.

Table 2.                          Ionization Balance Calculations

| Authors | Ionization Rates Used | Recombination Rates Used |
|---|---|---|
| Jordan [52] | Seaton [13] | Burgess and Seaton, Elwert |
| Jacobs et al [53] | Lütz [14, 15] | Burgess and Summers, with alternative autoionization channels |
| Summers [54] | ECIP (Summers) | Burgess and Summers |

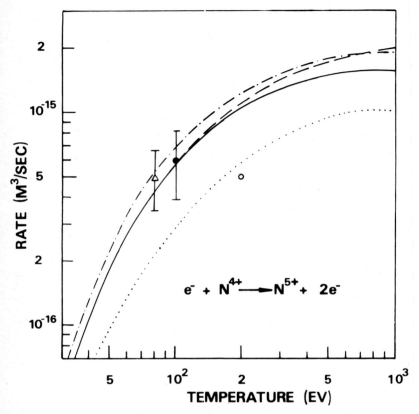

**Ionization rates for ground state N⁴⁺ as
a function of plasma electron temperature.**

Figure 6
Ionization Rate Coefficients for NV as a Junction of Temperature.  Solid curve,
DWX calculation by Jakubowicz and Moores [1].  The results of Golden and
Sampson [16] and Younger [20] lie close to this curve.  Dashed curve
Crandall et al. [26]; dot-dash, Lotz formula.  Open triangle and solid
circle, plasma measurements of Rowan and Roberts [57] and Källne and
Jones respectively.  The dotted curve is the ECIP result and the open
circle is from Kunze [56].

REFERENCES

1   Jakubowicz, H. and Moores, D.L.  J. Phys. B. 1981, in the press.
2   Peterkop, R.K., Proc. Phys. Soc. 77, 1961, 1220.
3   Peterkop, R.K., "Theory of Ionization of Atoms by Electron Impact", Colorado Associated University Press, 1977.
4   Rudge, M.R.H. and Seaton, M.J., Proc. Roy. Soc. A. 283, 1965, 262-90.
5   Crandall, D.H. Physica Scripta 23, 1981, 153-62.
6   Peart, B. and Dolder, K.T., J. Phys. B. 8, 1975, 56-62.
7   Moores, D.L. J. Phys. B. 12, 1979, 4171-8.
8   Bely, O. J. Phys. B. 1, 1968, 23-7.
9   Jakubowicz, H. and Moores, D.L., Comments on Atomic and Molecular Physics 9, 1980, 55-70.
10  Bhalla, C.P., Gabriel, A.H. and Presnyakov, L.P., MNRAS, 172, 1975, 359-75.
11  Cowan R.D. and Mann, J.B., Ap. J. 232, 1979, 736-40.
12  Kim, Y-K and Cheng K-T, Phys. Rev. A. 18, 1978, 36-
13  Seaton, M.J., Planet. Space. Sci. 12, 1964, 55-74.
14  Lütz, W., Ap. J. Suppl. 14, 1967, 207-38.
15  Lütz, W., Z. Phys. 216, 1968, 241-7.
16  Golden, L.A. and Sampson, D.H., J. Phys. B. 10, 1977, 2229-37.
17  Mayer, H. 1947, LASL Report LA-647.
18  Younger, S.M. Phys. Rev. A. 22, 1980, 111-7.
19  Younger, S.M. Phys. Rev. A. 22, 1980, 1425-31.
20  Younger, S.M. Phys. Rev. A. 23, 1981, 1138-46.
21  Younger, S.M. To be published.
22  Younger, S.M. To be published.
23  Malik, F.B. and Trefftz, E. Z. Naturf., 16, 1961, 583-98.
24  Burgess, A. and Rudge M.R.H. Proc. Roy Soc. A273, 1963, 372-86.
25  Rudge M.R.H. and Schwartz, S.B., Proc. Phys. Soc. 88, 1966, 563-78 and 579-85.
26  Bely, O. and Schwartz, S.B., Astron. Astrophys. 1, 1969, 281-5.
27  Moores, D.L. and Nussbaumer, H., J. Phys. B. 3, 1970, 161-72.
28  Moores, D.L., J. Phys. B., 5, 1972, 286-98.
29  Moores, D.L., J. Phys. B. 11, 1978, L403-5.
30  Moores, D.L., J. Phys. B. 12, 1979, 4171-8.
31  Stingl, E. J. Phys. B. 5, 1972, 1160-74.
32  Burke, P.G. and Seaton M.J., Meth. Comp. Phys. 10, 1971, 1-56.
33  Crees, M.A., Seaton, M.J. and Wilson, P.M.H., C.P.C. 15, 1978, 23-83.
34  Seaton, M.J. J. Phys. B. 7, 1974, 1817-40.
35  Peart, B., Walton, D.S. and Dolder K.T. J. Phys. B. 2, 1969, 347-
36  Crandall, D.H., Phaneuf, R.A. and Gregory, D.C.  Report No. ORNL/TM-7020, Oak Ridge National Laboratory, Oak Ridge TN 3 7830, USA, 1979, (available from NTIS, 5285 Port Royal Road, Springfield Va 22161, USA).
37  Wareing, J.B. and Dolder K.T., Proc. Phys. Soc. 91, 1967, 887-93.
38  Lineberger, W.C., Hooper, J.W. and McDaniel, E.W., Phys. Rev. 141, 1966, 151-64.
39  Sampson, D.H. and Golden, L.B. J. Phys. B. 12, 1979, L785-91.
40  Woodruff, P.R., Hublet, M.C. and Harrison, M.F.A. J. Phys. B. 11, 1978, L305-8.
41  Dubau, J. and Wells J. J. Phys. B. 6, 1973, 1452-60.
42  Bell, R.H., 1979, Thesis, University College London.
43  Sampson, D.H., Clark, R.E.H. and Parks, A.D., J. Phys. B. 12, 1979, 3257-72.
44  Gabriel, A.H., MNRAS, 160, 1972, 99-119.
45  Martin, S.O., Peart, B. and Dolder, K.T., J. Phys. B. 1, 1968, 537-42.
46  Peart, B. and Dolder, K.T. J. Phys. B. 1, 1968, 240-4.
47  Hooper, J.W., Lineberger, W.C. and Bacon, F.M., Phys. Rev. 141, 1966, 165-72.
48  Peart, B. Martin, S.O. and Dolder, K.T., J. Phys. B. 2, 1969, 1176-9.
49  Aitken, K.L., Harrison, M.F.A. and Rundell, R.D., J. Phys. B. 4, 1971, 1189-99.
50  Cheng, C-C, Feldman, U. and Doschek, G.A. Ap. J. 233, 1979, 736-40.
51  Feldman, U., Doschek, G.A., and Cowan, R.D., 1981, submitted to MNRAS.
52  Jordan, C. MNRAS 142, 1969, 499-512 ; 148, 1970, 17-23.
53  Jacobs, V.L., Davis, J., Kepple P.C. and Blaha, M., Ap. J., 211, 1977, 605-516; 215, 1977, 106-

54    Summers, H. MNRAS <u>169</u>, 1974, 663-80; Appleton Laboratory Report, AL-R-5, 1979.
55    Butler, K. and Moores, D.L., XII ICPEAC Abstracts 473-4, 1981.
56    Kunze, H-J., Phys. Rev. A <u>3</u>  655-78, 1971.
57    Rowan, W.L. and Roberts, J.R., Phys. Rev. A <u>19</u>, 90-8, 1979.
58    Kallne, E. and Jones, L.A.J., Phys. B <u>10</u>, 3637-48, 1977.

PHYSICS OF ELECTRONIC AND ATOMIC COLLISIONS
S. Datz (editor)
© North-Holland Publishing Company, 1982

# DIELECTRONIC RECOMBINATION IN COLLISION OF
# ELECTRONS WITH MULTICHARGED IONS

Larry J. Roszman

National Bureau of Standards, Washington, D.C. 20234

## INTRODUCTION

The process of dielectronic recombination can be thought of as a two-step process
[1]. An ion in state $|i>$ non-radiatively captures a free electron which has
energy $\varepsilon$. The free electron is captured into some excited orbital and the excess
energy is taken up by the excitation of one of the core electrons. The resulting
state of the recombining ion $|j>$ has a total energy $\varepsilon_j$ which is larger than the
energy of the first ionization limit of the recombining ion. Since energetically
the state $|j>$ lies in the first continuum of the initial ion, the excited state
$|j>$ can decay by autoionization, which produces a free electron and the initial
ion (not necessarily the initial state) or it can decay by photon emission. The
final state of the radiative transition can be a state $|k>$ which is energetically
below the first ionization limit and is stable to autoionization (radiative
stabilization) or it can be another autoionizing state [2] and a cascade through
autoionizing states can result [3]. Generally, the first radiative process is the
most likely with the consideration of cascades a fine tuning of results.

This process is the dominant recombination process for ionic structures more
complex than helium-like occuring in low density plasmas ($<10^{14}$ electrons/cm$^3$)
[4,5,6]. Rates of dielectronic recombination must be known accurately in order to
model the ionization and energy balances of many plasmas of astrophysical interest
and of interest in magnetically confined fusion research. Additionally, when the
excited orbital states of the core excited and the capture electrons lie close
together in energy, the stabilizing radiative transition appears satellitic to a
resonance transition of the initial ion. The ratio of intensities of the satellite
and parent lines is quite sensitive to the local density and temperature of the
plasma, providing a useful diagnostic probe [7,8]. When the energies of the two
orbital states are far apart, the wavelength of the stabilizing radiative transition nearly coincides with the resonant transition of the initial ion.

The history of the calculation of dielectronic recombination rates contains many

attempts to use very simple, but computationally rapid, approximations for the
large amounts of atomic data required [4,9]. In the absence of any successful
measurement of a rate of dielectronic recombination, these simple approximations
can be checked only by carrying out relatively sophisticated calculations and by
comparing the two sets of results. While the full set of advanced calculations is
not yet complete, enough calculations have been done to permit some conclusions to
be drawn concerning the sophistication necessary for the computation of accurate
rates. These considerations and conclusions will be discussed below.

Accuracy in the calculation of the positions and intensities of the spectral lines
produced by dielectronic recombination requires the full sophistication of atomic
structure theory for multicharged ions and a relatively sophisticated brand of
collision theory [8]. As interest in the cross sections of dielectronic recombination increases with the development of the merged beam experiments [10],
refinements in the basic theory of dielectronic recombination as well as the
atomic structure and collision theories must be made. A portion of these refinements is discussed next.

THE BASIC THEORY

Much of the theoretical work carried out on dielectronic recombination [3,11,12]
has attempted to establish that the original expression for dielectronic recombination through a particular autoionizing state into a particular bound state is
well approximated by the following expression:

$$\alpha(i;j;k) = \frac{4\pi^{3/2}a_0^3}{T^{3/2}} \frac{A_a(j;i\varepsilon_j)A_r(j;k)e^{-\varepsilon_j/kT}}{\sum_i A_a(j;i'\varepsilon_j)+\sum_k A_r(k;j)} \qquad (1)$$

where $A_a(j;i\varepsilon_j)$ is the autoionization rate from the state $|j>$ to the continuum
state $|i\varepsilon_j>$ and $A_r(j;k)$ is the radiative transition rate from the autoionizing
state $|j>$ to the bound state $|k>$. From the existing body of theoretical work,
one can conclude that the expression in eq. (1) should accurately represent the
rate of dielectronic recombination when (1) neither the autoionizing or radiative
widths of different autoionizing states interacting with the same continuum
overlap, (2) the interference between direct radiative recombination and dielectronic recombination is negligible, (3) the shift of the apparent center of the
autoionizing resonance due to second order interactions with the continuum and the
radiation field is negligible, (4) the atomic parameters do not vary significantly
over the width of the resonance, and (5) the resonance is very narrow (sharp).

A few calculations convince one that the variation of the atomic parameters across the width of a typical resonance is negligible and that most resonances are "narrow" for a multicharged ion, though the narrowness of a resonance depends somewhat upon the final quantity which is being examined and the energy resolution used. The neglect of the energy shift is most likely a reasonable approximation for multicharged ions though apparently it has never been examined. The remaining conditions for the applicability of the expression in eq. (1) should be examined for each case considered since their use when not applicable could lead to serious error. The first area of the problem discussed here is the development of a formalism which can be used to examine the five conditions listed above more closely.

Several derivations of dielectronic recombination exist [3,11,12], each emphasizing a different aspect of the problem, and each having different types and degrees of approximation. None of these derivations are well connected to the large body of literature on autoionizing states inspired by Fano's original treatment of autoionization [13,14]. By utilizing a generalization of Heitler's radiation damping theory [15] (somewhat different than one used by Davis and Seaton [16]) in combination with Fano's treatment of autoionization as configuration interaction in the continuum, one can develop a theoretical treatment which is connected to the existing atomic physics literature and provides a method for examining and handling the points discussed above.

The state of the recombining ion at time t can be expanded in the following manner [15,17]:

$$|\Psi(t,t_0)> = \sum_{\alpha} \int dE c_{\alpha}(E,t,t_0)|\alpha E> + \sum_{k} \int d\omega d_k(\omega,t,t_0)|k\omega>, \qquad (2)$$

where $|\alpha E>$ is a continuum state (including resonances--if any) resulting from Fano's theory, $|k>$ is a bound state, and $|\omega>$ is the single photon state of energy $\omega$. The initial condition occured at time $t_0$. The probability amplitude that at some time t the recombined ion is in the bound state $|k>$ and a photon is in state $|\omega>$ is just,

$$<k\omega|\Psi(t,t_0)> = d_k(\omega,t,t_0). \qquad (3)$$

The far future limit of this amplitude yields the probability of interest in dielectronic recombination,

$$P_k(\omega) = \lim_{t\to\infty} |d_k(\omega,t,t_0)|^2. \qquad (4)$$

The cross section for dielectronic recombination is given by Eq. (5),

$$\sigma_k(\omega)d\omega \;=\; \frac{\pi}{2g_i}\,\left(\frac{\hbar}{mv}\right)^2\, g_j P_k(\omega)d\omega, \qquad (5)$$

where $g_i$ is the statistical weight of the initial ion state $|i\rangle$, and $g_j$ is the statistical weight of the state $|j\rangle$. The initial free electron velocity is v.

By straightforward extension of Heitler's radiation damping theory, one can show that the probability amplitude of interest can be computed from the following equation.

$$\lim_{t\to\infty} d_k(\omega,t,t_o) \;=\; \sum_\alpha \sum_{k'} \{1 + D\}^{-1}_{kk'}\, G_{k'\alpha}. \qquad (6)$$

D is a matrix defined by its elements,

$$D_{k'k''} \;=\; i\pi \lim_{\eta\to 0} \sum_\alpha \int dE\; \frac{\langle k'; \omega+\varepsilon_k-\varepsilon_{k'}, |H|\alpha E\rangle\langle\alpha E|H|k''; \omega+\varepsilon_k-\varepsilon_{k''}\rangle}{\varepsilon_k + \omega - E + i\eta} \qquad (7)$$

and $G_{k'\alpha}$ which contains the initial conditions is defined by

$$G_{k'\alpha} \;=\; \lim_{\eta\to 0}\int dE\; \frac{\langle k', \omega+\varepsilon_k-\varepsilon_{k'},|H|\alpha E\rangle\langle\alpha E|\Psi(t_o,t_o)\rangle}{\varepsilon_k + \omega - E + i\eta} \qquad (8)$$

$\varepsilon_k$ is the eigenenergy of the state $|k\rangle$, and H represents the interaction between the radiation field and the recombining ion. It has been assumed that initially the system is composed of continuum states, i.e.,

$$d_k(\omega,t_o,t_o) \;=\; 0. \qquad (9)$$

In the case of a single autoionizing resonance and a single continuum, Fano has shown that the continuum state $|\alpha E\rangle$ can be written as [13]

$$|\alpha E\rangle \;=\; a_{\alpha j}(E)|\,j\rangle + \int dE' b_{\alpha i}(E,E')|i,E'\rangle, \qquad (10)$$

where

$$a_{\alpha j}(E) \;=\; -\frac{1}{\sqrt{2\pi}}\;\frac{\Gamma_{ji}^{1/2}(E)}{\sqrt{(E-\varepsilon_j-\Delta_j(E))^2+\tfrac{1}{4}\Gamma_{ji}^2}} \qquad (11)$$

and

$$b_{\alpha i}(E,E') \;=\; -\frac{1}{2\pi}\,\frac{P}{E-E'}\;\frac{\Gamma_{ji}^{1/2}(E')\,\Gamma_{ji}^{1/2}(E)}{\sqrt{(E-\varepsilon_j-\Delta_j(E))^2 + \tfrac{1}{4}\Gamma_{ji}^2}}$$

$$\quad -\frac{(E-\varepsilon_j-\Delta_j(E))}{\sqrt{(E-\varepsilon_j-\Delta_j(E))^2 + \tfrac{1}{4}\Gamma_{ji}^2}}\,\delta(E-E'). \qquad (12)$$

The autoionizing width $\Gamma_{ji}(E)$ is defined by

$$\Gamma_{ji}(E) = 2\pi|<j|V|iE>|^2 \tag{13}$$

where V is the configuration interaction potential between the bound-like auto-ionizing state $|j>$ and the purely continuum state $|iE>$. $P/(E-E')$ denotes the principal part distribution. The apparent shift in the autoionizing resonance energy is given by $\Delta_j(E)$,

$$\Delta_j(E) = \frac{1}{2\pi} P \int dE' \frac{\Gamma_{ji}(E')}{E-E'} . \tag{14}$$

Generally, both $\Gamma_{ji}(E)$ and $\Delta_j(E)$ are constant in the neighborhood of the resonance and need be evaluated at $E = \varepsilon_j$ only.

In the following analysis the contribution made by the continuous state $|iE>$ to the radiative transition will be neglected. This is a simplifying approximation made here in order to highlight the effects of overlapping resonances, which need not be made in general [17]. Since the retention of the purely continuum contribution is necessary to reproduce the characteristic line shape of autoionizing resonances, and can lead to quite interesting conclusions when the autoionizing and radiative widths are comparable [18], it should be retained in any reasonable theory of dielectronic recombination cross sections of satellites. Its neglect in a theory of the total dielectronic recombination rate for a multi-charged ion is reasonable since the rate of recombination through the autoionizing resonances is generally several order of magnitude greater than the rate of direct radiative recombination to the same final state.

With the neglect of the continuum, the radiative matrix element is

$$<k',\omega + \varepsilon_k - \varepsilon_{k'} |H|\alpha E> = - \frac{1}{2\pi} \sqrt{\frac{\Gamma_{ji}^{1/2}(E) \Gamma_{jk'}^{1/2}}{(E-\varepsilon_j-\Delta_j(E))^2 + \frac{1}{4}\Gamma_{ji}^2}} \tag{15}$$

where the radiative width of the transition from the autoionizing state $|j>$ to the bound state $|k>$ is $\Gamma_{jk'}^{1/2}$ defined by

$$\Gamma_{jk'}^{1/2} = 2\pi|<k',\omega + \varepsilon_k -\varepsilon_{k'} |A|j>|^2. \tag{16}$$

The use of the radiative matrix element expressed in eq. (15) in the definition of D (eq. (7)), and the evaluation of the energy integral by contour integration (assuming all atomic parameters are constant in the neighborhood of the resonance) results in

$$D_{k'k''} = -\frac{i}{2} \frac{\Gamma_{jk'}^{1/2} \Gamma_{jk''}^{1/2}}{\varepsilon_k + \omega - \varepsilon_j - \Delta_j + \frac{i}{2}\Gamma_{ji}} \tag{17}$$

The initial state of the system $|\Psi(t_0,t_0)>$ will be taken to be a continuum state with a gaussian distribution of energy of width $\delta$ centered about a central energy $E_0$, i.e.,

$$|\Psi(t_0,r_0)> = (2\pi)^{-1/4}\delta^{-1/2} \int dE e^{-1/4(E-E_0)^2/\delta^2} |iE>. \tag{18}$$

The use of this initial condition in the expression for $G_{k'\alpha}$ (eq. 8), and the evaluation of the energy integral by contour integration results in the following,

$$G_{k'\alpha} = -i(2\pi)^{-1/4}\delta^{-1/2} \frac{\Gamma_{ji}^{1/2} \Gamma_{jk}^{1/2}}{\varepsilon_k + \omega - \varepsilon_j - \Delta_j + \frac{i}{2}\Gamma_{ji}} e^{-1/4(\varepsilon_j+\Delta_j-E_0)^2/\delta^2}. \tag{19}$$

By using the expression for $D_{k'k''}$ and $G_{k'\alpha}$ of equations (17) and (18), the probability $P_k(\omega)$ defined in eq. (4) can be evaluated from

$$P_k(\omega) = \frac{1}{\sqrt{2\pi}} \frac{1}{\delta} \frac{\Gamma_{jk} \Gamma_{ji}}{(\varepsilon_k+\omega-\varepsilon_j-\Delta_j)^2 + \frac{1}{4}(\Gamma_{ji}+\sum_k \Gamma_{jk})^2} e^{-1/2(\varepsilon_j+\Delta_j-E_0)^2/\delta^2}. \tag{20}$$

The recombination rate associated with this probability is

$$\alpha(i;j;k) = \frac{\pi g_j}{4g_i} \frac{\hbar^2}{(mkT)^{3/2}} \frac{\Gamma_{jk} \Gamma_{ji} e^{-(\varepsilon_j+\Delta_j)/kT}}{(\varepsilon_k+\omega-\varepsilon_j-\Delta_j)^2 + \frac{1}{4}(\Gamma_{ji}+\sum_{k'} \Gamma_{jk'})^2} \tag{21}$$

where a Maxwellian distribution for the initial velocity $E_0$ has been assumed and the limit $\delta\to0$ has been taken. The final rate of dielectronic recombination is obtained by integrating the rate of eq. (21) over the photon energy $\omega$. By using the contour integration, one finds

$$\alpha(i;j;k) = \frac{\pi^2 g_j}{2g_i} \frac{\hbar^2}{(mkT)^{3/2}} \frac{\Gamma_{jk} \Gamma_{ji}}{\Gamma_{ji} + \sum_{k'} \Gamma_{jk'}} e^{-(\varepsilon_j+\Delta_j)/kT}. \tag{22}$$

It has not been necessary to make a narrow resonance approximation. The total rate for the dielectronic recombination through a set of isolated autoionizing resonances is obtained by summing $\alpha(i;j;k)$ as defined by eq. (22) over $j$ and $k$.

OVERLAPPING RESONANCES

The potentially difficult problem of overlapping resonances [11,12,19] can be analyzed analogous to the isolated resonance case by employing Fano's expressions for a single continuum state and several autoionizing resonances [13]. Extension of the theory to several continuum is straightforward but tedious [14,17]. For this case, the state $|\alpha E>$ is expanded in the following manner:

$$|\alpha E> = \sum_{\nu} a_{\alpha\nu}(E)|\nu> + \int dE' \, b_{\alpha i}(E,E')|iE'>. \qquad (23)$$

The autoionizing state $|\nu>$ is related to the original expansion states $|j>$ by [13]

$$|\nu> = \sum_{j} A_{\nu j}|j> \qquad (24)$$

where the coefficients $A_{\nu j}$ are the elements of the eigenvectors of the eigenvalue problem for $\varepsilon_{\nu}$,

$$\varepsilon_j A_{\nu j} + \sum_{j'} \Delta_{j'j} A_{\nu j'} = A_{\nu j}\varepsilon_{\nu}, \qquad (25)$$

with

$$\Delta_{j'j} \equiv \frac{1}{2\pi} P \int dE' \, \frac{\Gamma_{j'i}^{1/2}(E') \, \Gamma_{ji}^{1/2}(E')}{E-E'}. \qquad (26)$$

The configuration interaction coefficients in eq. (23) are expressed by the following equations [13],

$$a_{\alpha\nu}(E) = -\frac{1}{\sqrt{2\pi}} \, \frac{\Gamma_{\nu i}^{1/2}(E)}{E-\varepsilon_{\nu}} \, [1 + 1/4(\sum_{\nu'} \frac{\Gamma_{\nu'i}(E)}{E-\varepsilon_{\nu'}})^2]^{-1/2}, \qquad (27)$$

and

$$b_{\alpha i}(E,E') = [1 + 1/4(\sum_{\nu} \frac{\Gamma_{\nu i}(E)}{E-\varepsilon_{\nu}})^2]^{-1/2} \{\frac{1}{2\pi} \sum_{\nu} P \frac{\Gamma_{\nu,i}(E')}{E-E'} \, \frac{\Gamma_{\nu,i}(E)}{E-\varepsilon_{\nu}} + \delta(E-E')\}. \qquad (28)$$

When the continuum radiation is neglected, the radiative matrix element becomes,

$$<k',\omega + \varepsilon_k - \varepsilon_{k'}|H|\alpha E> = -\frac{1}{2\pi}(\sum_{\nu} \frac{\Gamma_{\nu i}^{1/2}(E)\Gamma_{\nu k}^{1/2}}{E-\varepsilon_{\nu}} \, [1 + \frac{1}{4}(\sum_{\nu} \frac{\Gamma_{\nu i}(E)}{E-\varepsilon_{\nu}})^2]^{-1/2}. \qquad (29)$$

Evaluation of the energy integral contained in the definitions of $D_{k'k''}$ and $G_{k'\alpha}$ (eqs. (7) and (8)), when the expression for the radiative matrix element contained in eq. (29) is used, requires that the zeros of the equation

$$1 + 1/4(\sum_{\nu} \frac{\Gamma_{\nu i}(E)}{E-\varepsilon_{\nu}})^2 = 0 \qquad (30)$$

be known. If a resonance is "isolated", i.e., when the energy variable E is close to a particular resonance energy $\varepsilon_{\nu}$, the contribution by all the other terms in the summation over $\nu$ in eq. (30) are negligible, the solution is

$$E = \varepsilon_\nu \pm i/2 \; \Gamma_{\nu i}.\tag{31}$$

If several resonances "overlap", i.e., the condition which leads to the solution of eq. (31) is not satisfied, a quadratic or higher order equation must be solved to find the solutions to eq. (30). For example, if two resonances overlap [20], the equation to be solved is

$$1 + 1/4 \left( \frac{\Gamma_1}{E-\varepsilon_1} + \frac{\Gamma_2}{E-\varepsilon_2} \right)^2 = 0,\tag{32}$$

which has the solutions,

$$E_{1\pm} = 1/2(\varepsilon_1 + \varepsilon_2 \pm i/2(\Gamma_1 + \Gamma_2)) + 1/2[((\varepsilon_1 - \varepsilon_2) \pm i/2(\Gamma_1 - \Gamma_2))^2 - \Gamma_1\Gamma_2]^{1/2}\tag{33}$$

and

$$E_{2\pm} = 1/2(\varepsilon_1 + \varepsilon_2 \pm i/2(\Gamma_1 + \Gamma_2)) - 1/2[((\varepsilon_1 - \varepsilon_2) \pm i/2(\Gamma_1 - \Gamma_2))^2 - \Gamma_1\Gamma_2]^{1/2}.\tag{34}$$

In the limit

$$\Gamma_1\Gamma_2 \ll \varepsilon_1 - \varepsilon_2,\tag{35}$$

The resonances become "isolated", i.e.,

$$E_{1\pm} \to \varepsilon_1 \pm i/2 \; \Gamma_1,$$

and

$$E_{2\pm} \to \varepsilon_2 \pm i/2 \; \Gamma_2.$$

In summary, to the order of approximation necessary in most calculations, the zeroes of eq. (30) are approximated by "isolated" resonances and are of the form expressed in eq. (31). If the zeros result from overlapping resonances, the form of the zeros is more complex, and, for the case of two overlapping resonances, is expressed in eqs. (33) and (34).

The matrix elements of D become

$$D_{k'k''} = -\frac{i}{2} \sum_{\nu \neq 1,2} \frac{\Gamma_{\nu k'}^{1/2} \Gamma_{\nu k''}^{1/2}}{\varepsilon_k + \omega - \varepsilon_\nu + \frac{i}{2}\Gamma_{\nu i}}$$

$$-\frac{i}{2} \sum_{\nu=1,2} \left[ \frac{\Gamma_1^{1/2}\Gamma_{1k'}^{1/2}}{E_{\nu -} - \varepsilon_1} + \frac{\Gamma_2^{1/2}\Gamma_{2k'}^{1/2}}{E_{\nu -} - \varepsilon_2} \right] \frac{(E_{\nu -} - \varepsilon_1)^2 (E_{\nu -} - \varepsilon_2)^2 (-i)(-1)^{\nu+1}}{(E_{\nu -} - E_{1+})(E_{\nu -} - E_{2+})(E_{1-} - E_{2-})}$$

$$\times \ [\frac{\Gamma_1^{1/2}\Gamma_{1k''}^{1/2}}{E_{\nu-} - \varepsilon_1} + \frac{\Gamma_2^{1/2}\Gamma_{2k''}^{1/2}}{E_{\nu-} - \varepsilon_2}] \ \frac{1}{\varepsilon_k + \omega - E_{\nu-}} \quad , \tag{36}$$

and those of $G_{k'\alpha}$ with use of the initial condition in eq. (18) are,

$$G_{k'\alpha} = -i \ \frac{1}{(2\pi)^{1/4}} \ \frac{1}{\delta^{1/2}} \ \underset{\nu\neq 1,2}{\Sigma} \ \frac{\Gamma_{\nu i}^{1/2} \ \Gamma_{\nu k'}^{1/2}}{\varepsilon_k + \omega - \varepsilon_\nu + \frac{i}{2}\Gamma_{\nu i}} \ e^{-1/4(\varepsilon_\nu - E_0)^2/\delta^2}$$

$$-i \ \frac{1}{(2\pi)^{1/4}} \ \frac{1}{1/2} \ \underset{\nu=1,2}{\Sigma}[\frac{\Gamma_1^{1/2}\Gamma_{1k'}^{1/2}}{E_{\nu-} - \varepsilon_1} + \frac{\Gamma_2^{1/2}\Gamma_{2k'}^{1/2}}{E_{\nu-} - \varepsilon_2}] \ \frac{2(E_{\nu-}-\varepsilon_1)^2(E_{\nu-}-\varepsilon_2)^2(-1)^{\nu+1}}{(E_{\nu-}-E_{1+})(E_{\nu-}-E_{2+})(E_{1-}-E_{2-})}$$

$$\times \frac{1}{\varepsilon_k + \omega - E_{\nu-}} \ e^{-1/4(R_e(E_{\nu-}) - E_0)^2/\delta^2} \tag{37}$$

Results similar to those for the single resonance case can be derived [17]. There are two significant differences, however. First, even if the resonances are isolated, a matrix $\theta_{\nu,\nu'}$ must be inverted. Its elements are

$$\theta_{\nu\nu} = \varepsilon_k + \omega - \varepsilon_\nu + \frac{i}{2}\Gamma_{\nu i} + \frac{i}{2}\underset{k'}{\Sigma}\Gamma_{\nu k'}, \tag{38}$$

and

$$\theta_{\nu\nu'} = -\frac{i}{2}\underset{k'}{\Sigma} \ \Gamma_{\nu k'}^{1/2} \ \Gamma_{\nu' k'}^{1/2}, \quad (\nu\neq\nu'). \tag{39}$$

If the off diagonal elements are negligible compared to the diagonal, the previous results can be reproduced [21]. If they are not negligible, the matrix must be diagonalized. These off-diagonal elements represent second order interactions among the resonances arising from their coupling to the radiation field, analogous to the second order interactions with the continuum [13] noted in eqs. (22) through (26). Essentially, they are radiative "shift and width" corrections to the resonance [15].

The second difference occurs when resonances overlap. The off-diagonal elements associated with the different solutions to eq. (30) for a set of overlapping resonances, e.g. $E_{1-}$ and $E_{2-}$, should not be neglected even if those associated with the "isolated" resonances are. The resulting submatrix in $\theta$ should be inverted.

Extension of the above theory to include radiative cascades through the auto-ionizing states is easily carried out and will not be discussed here.

## ATOMIC STATE CONSIDERATIONS

The remaining atomic physics problems in the theory and computation of dielectronic recombination parameters are the approximation and practical calculation of
the continuum states $|iE>$, the autoionizing states $|j>$, and, to a lesser extent,
the bound states $|k>$ [22]. The simplest approximation of the continuum state $|iE>$
for a multicharged ion is a distorted wave approximation. The most useful distorted wave method has the continuum orbital computed in the Hartree-Fock potential
produced by a frozen ion core. Various versions of this theory have the field
spherically or configuration averaged or term dependent. For multicharged ions
replacement of the global exchange potential by a local potential, such as the
semi-classical exchange potential [23], is usually adequate. Coulomb waves are
often a good approximation to the continuum orbital for large values of the
orbital angular momentum even for quite light, not highly ionized ions. Which
refinments must be included in order to calculate accurate continuum orbitals
depends upon the particular ionic structure being investigated and where it lies
in its isoelectronic sequence. One of the intriguing but uninvestigated problems
in computing continuum states is whether the continuum states should be represented as a superposition of distorted wave states, i.e., is there configuration
interaction among the continuum states? If there is, how is such a superposition
carried out?

The single configuration Hartree-Fock method is the simplest theory which can be
used for computing the autoionizing states. For the lowest members of a Rydberg
series, where the Rydberg orbital significantly penetrates the ion core either
because of a low value of the orbital angular momentum or because of the particular atomic structure being considered, a Hartree-Fock calculation of the entire
state $|j>$ must be carried out [24]. For less penetrating high series members or
orbitals with large angular momenta, a frozen core Hartree-Fock calculation will
suffice. This approach is analogous to the distorted wave technique discussed
above. Hydrogenic orbitals can be used for the Rydberg orbitals at sufficiently
high angular momentum. In the frozen core, Hartree-Fock approximation for a
multicharged ion a local approximation to the global exchange potential is quite
accurate and computationally more rapid. Cowan's HX approach [25] or a modification of this method is often very accurate for multicharged ions.

Since so many autoionizing configurations of the same symmetry often have nearly
the same energy, configuration interaction among the autoionizing states can have
a significant effect on the position of satellite lines and intensities [7,8,24].
Similar effects should appear in the cross sections of dielectronic recombination.

The influence of configuration interaction on total rates of dielectronic recombination is probably lessened due to the "averaging" introduced by the summations which are taken over all possible states. However, the critical investigation of these possible contributions is not complete [24].

The calculation of the bound states is not as critical a problem. The stabilizing radiative transition is most often undergone by the excited core electron, while the highly excited Rydberg electron is just a spectator which perturbs the core negligibly for Rydberg states higher than the first or second.

The largest practical problem remaining is the choice of a coupling scheme. Ideally, the Hamiltonian should contain all relevant interaction terms, including the spin-orbit and Breit terms, an "appropriate" zeroth order representation chosen, and the resulting energy matrix diagonalized. Such an approach is practical and necessary for computation of satellite properties [7,8] and of cross sections within a limited energy range, but it is currently quite impractical for the calculation of the large number of states involved in dielectronic recombination [24]. It may be that due to the "averaging" tendency noted above, ordinary LS coupling or some other simple variant is sufficient [11]. A detailed analysis of this problem has not been carried out.

PLASMA EFFECTS

In the previous discussion the recombining ion and electron were isolated. The actual recombination process is modified by the plasma environment in which it occurs [5,6,26,27]. At low to moderate electron and ion densities ($<10^{18}$ electrons/$cm^3$) the principal plasma effects are modification of the population densities (as compared to the "isolated atom," corona model densities) of the initial, autoionizing, and final states by inelastic electron and ion collision process [27], and some perturbation of energy levels and atomic states by the plasma produced microfields [28]. At higher densities the distinction between the recombining ion and the remainder of the plasma is lost. While dielectronic recombination is probably not an important process in the ionization and energy balances for most of the ions in a high density plasma [26], the shapes and relative intensities of the satellites produced by the process are quite sensitive to the plasma environment and can serve as diagnostic tools. Further discussion concerning dense plasmas is outside the scope of this review.

The principal effect in the low density plasmas ($<10^{14}$ electrons/$cm^3$) which are of interest in astrophysics and magnetic confinement fusion research arises from collisions between the recombining ion in one of its autoionizing states and

plasma ions [9,27,29]. In such collisions, the recombining ion undergoes a
transition from one angular momentum state of a configuration to another, nearly
degenerate, angular momentum state of the same configuration. Since the statistical weight of states with large values of angular momentum is large and their
autoionizing rate decreases rapidly (for $\ell > 7$), these collisions generally increase
the rate of dielectronic recombination. Similar effects appear in the satellite
intensities at somewhat higher densities.

In moderate density plasmas ($\geq 10^{14}$, $\leq 10^{18}$ electrons/cm$^3$), the plasma processes of
excitation, de-excitation, and ionization compete with resonant capture into an
autoionizing state, autoionization and radiative stabilization [26,29]. Each
individual process must be built into the plasma modeling code. In order to carry
out such a program the autoionizing resonances must be non-overlapping and the
higher order radiative effects must be negligible. Such approximations are
reasonable for overall ionization and energy balance calculations, but are not
reasonable for detailed diagnostic modeling.

REFERENCES

1.   Bates, D.R. and Dalgarno, A., in Atomic and Molecular Processes, ed. D.R.
     Bates, (Academic Press, 1962), p. 259.
2.   Blaha, M., Astro. Letters 10, (1972) 179.
3.   Gau, J.N. and Hahn, Y, JQSRT 23, (1980) 121.
4.   Burgess, A., Astrophys. J. 141, (1965) 1588.
5.   Merts, A.S., Cowan, R.D., and Magee, N.H. Jr., Los Alamor Scientific Lab-
     oratory Report No. LA-6220-MS (1976).
6.   Post, D.E., et al, Princeton Plasma Physic Laboratory Report No. PPPL-1352
     (1977).
7.   Gabriel, A.H., Mon. Not. 12 Astr. Soc. 160, (1972) 99.
8.   Bhalla, C.P., Gabriel, Presnyakov, L.P., Mon. Not. R. Astr. Soc. 172, (1975)
     359.
9.   Seaton, M.J. and Storey, P.J., in Atomic Processes and Applications, ed. P.G.
     Burke and B.S. Moiseiwitsch, (N. Holland, 1976), p. 133.
10.  Lafyatis, G.P. and Kohl, J.L., Bull. Amer. Phys. Soc. 26, (1981) 810;
     Mitchell, J.B.A. et al., Bull. Amer. Phys. Soc. 26 (1981) 810.
11.  Trefftz, E, J. Astro. 65 (1967) 299; in Physics of the One- and Two-Electron
     Atoms, ed. F. Boppand, H. Kleinpoppen (N. Holland, 1969), p. 839; J. Phys. B
     3, (1970) 763.
12   Shore, B., Rev. Mod. Phys. 39, (1967), 439; Astro. J. 158 (1969) 1205.
13.  Fano, U, Phys. Rev. 124 (1961) 1866; Fano, U, and Prats, F., Proc. Nat. Acad.
     Sci. India A33, (1963) 553.
14.  Mies, F.H., Phys. Rev. 175 (1968) 164.
15.  Heitler, W., The Quantum Theory of Radiation, 163 (Oxford Univ. Press, 1960).
16.  Davies, P.C.W., and Seaton, M.J., J. Phys. B 2, (1969) 757.
17.  Roszman, L.J., unpublished.
18.  Armstrong, L. et al., Phys. Rev. A 18, (1979) 2538.
19.  Feshbach, H., Annals of Physics 43 (1967) 410.
20.  Mower, L., Phys. Rev. 142 (1966) 799.
21.  Van Santen, R.A., Physica 62, (1972) 51.
22.  Roszman, L.J., Phys. Rev. A 20, (1979) 673.
23.  Riley, M.E., and Truhlar, D.G., J. Chem. Phys. 63 (1975) 2182; 65 (1976) 792.
24.  Roszman, L.J. and Wiess, A, unpublished.
25.  Cowan, R.D., Phys. Rev. 163 (1967) 54.

26. Weisheit, J.C., J. Phys. B $\underline{8}$, (1975) 2556.
27. Burgess, A., and Summers, H.P., Astro. J. $\underline{157}$ (1972) 255; $\underline{169}$ (1974) 663.
28. Davis, J. and Jacobs, V.L., Phys. Rev. A $\underline{12}$, (1975) 2017; Jacobs, V.L., and Davis, J., Phys. Rev. A $\underline{19}$, (1979) 776.
29. Jacobs, V.L., and Davis, J., Phys. Rev. A $\underline{18}$ (1978) 697.

PHYSICS OF ELECTRONIC AND ATOMIC COLLISIONS
S. Datz (editor)
© North-Holland Publishing Company, 1982

INTRODUCTION TO THE SYMPOSIUM ON COLLISIONS OF

MULTICHARGED IONS WITH ATOMS

P. Hvelplund

Institute of Physics, University of Aarhus
DK-8000 Aarhus C, Denmark

During the last decade, a great deal of information on electron-capture, ionization, and electron-loss processes involving multiply
charged ions has been accumulated. The situation as it appeared two
years ago, both theoretical and experimental, was reviewed at a similar symposium at the XI ICPEAC, and excellent reviews of the field
can be found in the proceedings of that conference. An extensive up-
to-date review has recently appeared (Janev and Presnyakov[1]) and is
recommended to those who want to learn about the present situation in
the field of highly charged ions interacting with atoms.

Research within this field was first initiated in connection with the
interaction of fission products with matter. Bohr[2] and later Bohr and
Lindhard[3] derived simple expressions for capture and ionization cross
sections, which have proved to be a good first-order estimate of the
dependences on various parameters.

In the late sixties, experimental investigations of interaction of
highly charged accelerator-produced ions with atoms started, and in
this connection, the work of Macdonald and Martin[4] should be mentioned. During the seventies, the late Jim Macdonald, to whom this
symposium is dedicated, initiated and completed a great variety of
ingenious collision experiments involving highly charged ions, a work
which was stopped so abruptly by his untimely death. In the experiment by Macdonald and Martin, the highly charged ions were produced
by stripping of heavy ions accelerated by a tandem accelerator. This
technique is still used for high- and medium-energy experiments. At
lower energies, a variety of methods for production of highly charged
ions are now in use, and at this symposium, we shall see examples of
the application of some of these techniques.

The papers presented give a short survey of selected topics within
the field. Completeness is not attempted since, as already mentioned,
the same field was treated at a symposium at the last ICPEAC conference. As a basis for the present selection of contributions, three
keywords may be listed: (1) Scaling rules. (2) Capture to excited
states, and (3) Low-energy, high-charge state. Scaling rules, which
are valid for a broad range of parameter variation, are extremely important for the users of these cross sections since not all cross
sections, which are important in, e.g., the field of controlled thermonuclear fusion, can be expected to be available in the near future.
Capture to excited states is also of great practical interest besides
being a crucial test of existing theoretical estimates.

The "low-energy, high-charge state" keyword points to a direction,
which future research within this field is most likely to take. This
is experimentally as well as theoretically the most difficult region,
but good results are extremely important in connection with fusion

*P. Hvelplund*

research and for a better understanding of atomic-collision processes
in general.

REFERENCES

[1]  Janev, R.K., and Presnyakov, L.P., Phys.Rep. 70, 1 (1981)
[2]  Bohr, N., K.Dan.Vidensk.Selsk.Mat.Fys.Medd. 18, No 8 (1948)
[3]  Bohr, N., and Lindhard, J., K.Dan.Vidensk.Selsk.Mat.Fys.Medd.
     28, No 7 (1954)
[4]  Macdonald, J.R., and Martin, F.W., Phys.Rev. A 4, 1965 (1971)

PHYSICS OF ELECTRONIC AND ATOMIC COLLISIONS
S. Datz (editor)
© North-Holland Publishing Company, 1982

# ELECTRON CAPTURE AND TARGET IONIZATION BY MEDIUM- AND HIGH-VELOCITY MULTIPLY CHARGED IONS

H. Knudsen
Institute of Physics, University of Aarhus
DK-8000 Aarhus C, Denmark

Electron capture and target ionization in collisions between medium- and high-velocity, multiply
charged ions and light atoms are discussed. For
ions of high charge, contrary to what is the case
for ions of low charge, these processes can be described within semiclassical theoretical models.
One of these models, which is due to Bohr and
Lindhard, is valid over broad regions of the relevant experimental parameters and provides a general framework for the understanding of experimental results. The validity of scaling laws and
quantitative results extracted from the semiclassical model as well as from other theoretical approaches is investigated on the basis of experimental data obtained in our laboratory  and by
other experimental groups.

## INTRODUCTION

When an ion collides with an atom, a large number of processes may
take place. The atom may be excited and later deexcite  by photon or
electron emission. It may be directly ionized or lose electrons in a
charge-transfer process. The ion may capture or lose electrons, and
after the collision, it may end up in an excited state and, in turn,
deexcite by the emittance of electrons or photons.

We shall discuss two of these processes, each one taking place with
a high probability under certain conditions, and each one leading to
the  removal  of  an  electron  from  the  target  atom.  One  is
the electron-capture process, where the ion picks up one of the target electrons, and the other the ionization process, where a target
electron is released due to the ion impact. Formally, these two processes can be written as

$$X^{q+} + A \rightarrow \begin{cases} X^{(q-1)+} + A & \text{(capture)} \\ X^{q+} + A^{+} + e & \text{(ionization)} \end{cases}$$

where the ion X of charge q collides with the target atom A.

We shall be concerned with collisions between an ion and an electron
bound in a target atom. In the theoretical treatment of such interactions, we may, as discussed by Bohr [1], divide up the collisions
into groups, which are defined by the values of two decisive parameters $\kappa$ and $\eta$.

The parameter $\kappa$ is the ratio between a typical distance in the ion-electron collision (collision diameter b) and the de Broglie wave-length $\lambda$ of the electron in the ion frame. We obtain

$$\kappa \equiv b/\lambda = 2q \frac{v_0}{V} \tag{1}$$

where V is the ion velocity and $v_0$ the Bohr velocity. When $\kappa \gg 1$, the target electron is well localized, and we may therefore use a classical description of the relevant trajectories. On the other hand, when $\kappa \ll 1$, a quantum-mechanical treatment should be used. Furthermore, in this case, the interaction is weak and brief, and it can be described by perturbation models.

The other decisive parameter, $\eta$, is essentially the ion velocity measured relative to the target-electron velocity, viz.,

$$\eta \equiv 2 \frac{V}{v} . \tag{2}$$

When $\eta \gg 1$, the electron under consideration may be regarded as quasi-stationary with respect to the ion impact, whereas when $\eta \ll 1$, the electron moves with a high velocity so that it can adjust to the presence of the ion. In that case, a quasi-molecule will be formed, and cross sections are determined by the formation of the quasi-molecular orbitals during the collision.

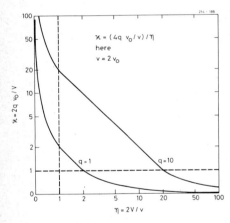

Figure 1. The $\kappa$–$\eta$ plane. Note the pseudo-logarithmic scales.

Each collision system can be characterized by a combination of $\kappa$ and $\eta$, which corresponds to a point in Fig. 1.

For a given target-electron velocity v, there is a curve showing the connection between $\kappa$ and $\eta$ for each value of the ion charge q. In Fig. 1, the solid curves show $\kappa$ as a function of $\eta$ for q = 1 and 10, respectively, for a typical light-atom electron velocity v = $2v_0$.

For ions with q = 1, it can be concluded from the figure that the regions of validity of molecular orbital models ($\eta < 1$) and of perturbation models ($\kappa < 1$) cover nearly all possible values of the ion velocity. On the other hand, for ions of high charges, this is not the case. Here there is a broad velocity region, where none of the theoretical approaches mentioned before applies.

In the following, we shall concentrate on such cases where $\kappa > 1$, and, accordingly, an impact-parameter description applies: We shall be concerned with ion velocities from $\sim v_0$ to $\sim 10v_0$, ion charges from 1 to $\sim 30$, and light targets (H, $H_2$, He). In this regime, semiclassical models as developed by Bohr [1] and Bohr and Lindhard [2] give a general framework for the discussion of experimental results.

We shall not try to review the field (excellent review papers have recently been published [3-4] but instead discuss the validity of the relevant theoretical results on the basis of new experimental data obtained in our laboratory as well as those obtained by other experimental groups.

EXPERIMENTAL TECHNIQUE

As an example of the experimental technique involved in measuring capture and ionization cross sections for collision systems charac- terized by the parameters mentioned in the previous section, we shall briefly describe the system used in our laboratory [5-7]. Here, a beam of ions of the proper velocity is extracted from an EN tandem accelerator. A post-stripper facility (carbon foil or gas) produces ions of the desired charge state, which are selected by magnetic de- flection. Immediately before entering the target cell, the beam is electrostatically deflected to remove ions experiencing charge- changing collisions in the beam line. The target cell may be a tung- sten furnace, with which atomic-hydrogen targets can be used, or a gas cell as that shown in figure 2.

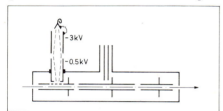

Figure 2. Target-gas cell.

The cell is equipped with a set of condenser plates, with which the total cross section for charge pro- duction can be measured. Further- more, a time-of-flight spectrometer can be used to obtain the distribu- tion of produced target charge states. The combination of these two measurements gives the ioniza- tion cross sections. Having passed through the target cell, the beam is electrostatically analyzed, and from the ion charge-state distribution, we have the charge-exchange cross sections. The signals from the target and beam ions can be de- tected in coincidence so that partial cross sections can be obtained (for example, the cross section for one electron capture when, in the same collision, two electrons are removed from the target atom).

ELECTRON CAPTURE, THEORY

The electron-capture process for highly charged ions has been treated by Bohr and Lindhard [2]. They used a semiclassical model, which is justified for the combination of experimental parameters considered here because $\kappa > 1$, but also because for highly charged ions, there is a high density of final states, to which the target electron can be transferred. Bohr and Lindhard introduces two interaction dis- tances: The release distance $R_r$ is the distance between the ion and the atom, where the force from the ion on the electron is equal to the binding force of the relevant electron in the target atom, and the capture distance $R_c$ is the corresponding distance, where the po- tential energy of the electron with respect to the ion is equal to the kinetic energy of the electron in the ion frame. Bohr and Lind- hard then finds the capture cross section for ions of velocity $V \lesssim v$ and $V \gtrsim v$, respectively, to be

$$\sigma_1 = \pi R_r^2 \qquad (V \lesssim v)$$
and
$$\sigma_2 = \pi R_c^2 (vR_c/aV) \qquad (V \gtrsim v)$$

, (3)

where a is the orbital radius of the target electron. In Ref. [6], we integrated these basic cross sections, using a rather crude

atomic-electron-density distribution, which is characterized by the
target-ionization potential, the target atomic number, and an adjustable parameter α. Inherent in the result is an important  scaling
law: For a given target atom, the capture cross section divided by
the ion charge depends only on the parameter

$$X \equiv E[keV/amu]/q^{4/7} \qquad\qquad\qquad (4)$$

where  E  is the ion energy. In Fig. 3, the integrated Bohr-Lindhard
(B-L) cross section for an atomic-hydrogen target has been plotted
(solid curve), using reduced variables as suggested by the scaling law.

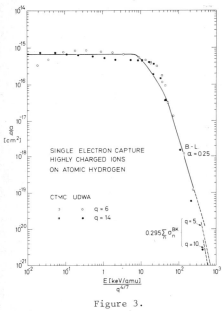

Figure 3.

For the combinations of experimental parameters we are concerned
with here, for atomic-hydrogen targets, there exists a number of more
detailed theoretical calculations.
Most notable are the classical-
trajectory Monte-Carlo calculations
(CTMC) of Olson and coworkers [8]
and the unitarized distorted wave
approximation (UDWA) results of
Ryufuku and Watanabe [9]. Results
for ions of charge q = 6 and 14
from these theoretical calculations
are included in Fig. 3. Within a
factor of two, they  agree with
the B-L result and support the scaling inherent in the B-L theory.
Ruyfuku and Watanabe [9] also obtained a scaling with q and V. They
used an empirical fitting procedure
and found $\sigma/q^{1.07}$ to depend on
$E[keV/amu]/q^{0.35}$.

For large ion velocities, κ becomes
less than one, and classical theories are not valid. As can be seen
from Fig. 3, the UDWA results tend
towards the often used perturbation
calculations of Brinkman and Kramers [10], which decrease more strongly with ion velocity than does the classical B-L result.

ELECTRON CAPTURE, EXPERIMENT

In Fig. 4 on the following page is shown a comparison between the B-L
theoretical cross section for electron capture for highly charged
ions colliding with atomic hydrogen and a large number of experimental data for ions of charge q ≥ 5.

The data support both the absolute magnitude of the B-L result and
its inherent scaling law for atomic-hydrogen targets. For other target atoms, the semiclassical result has been found to work equally
well. For a comparison between the integrated B-L cross section and
experimental data for helium, argon, and krypton targets, see Ref. [6].

The scaling with the reduced parameter X stems from the basic cross
sections, Eq. (3), and, consequently, is independent of the target-
electron distribution. It might therefore be expected that the scaling should be valid also for molecular targets. Of special interest

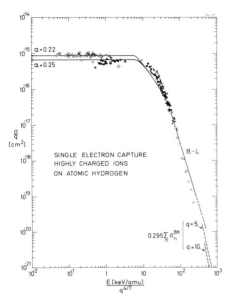

Figure 4. For a reference to the experimental data, see Ref.[7].

is the molecular-hydrogen ($H_2$) target. There are two reasons for this. The first reason is that for many years, it was not possible to obtain experimental data on atomic hydrogen. On the other hand, much theoretical work existed for that target. In comparisons, it was then arbitrarily assumed that the molecule acts like two separate atoms so that the ratio $\sigma(H_2)/\sigma(H)$ was set equal to two. This assumption is still used [9,11] where no experimental data for the atomic target exist. The other reason is that it is the simplest molecule of all; hence, any future theory on capture for molecular targets should be able to provide values of $\sigma(H_2)$.

In Fig. 5 is shown the ratio $\sigma(H_2)/\sigma(H)$ plotted as a function of the scaling parameter X. The points are experimental results. As can be seen, the ratio is <u>not</u> equal to two. On the other hand, it does scale with X, thus lending credit to the B-L scaling law, also for molecular targets.

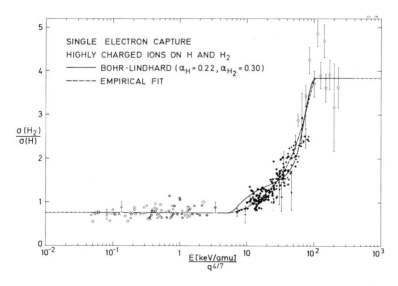

Figure 5. For a reference to the experimental data, see Ref. [7].

## TARGET IONIZATION, THEORY

The target-ionization process has been treated theoretically by Bohr
[1]. He obtained an analytical expression for the cross section
valid for $\eta > 1$ for all values of $\kappa$. The Bohr theory is based on the
Rutherford cross section for energy transfer T to a free, stationary
electron. The basic idea is that if an energy greater than the elec-
tron-ionization potential I is transferred, then the electron is ionized. The ionization cross section is therefore obtained by integrating the Rutherford cross section from I to the maximum transferable
energy $2mV^2$,

$$\sigma = 4\pi a_0^2 q^2 (v_0/V)^2 \{I_0/I - (v_0/2V)^2\} \quad . \tag{5}$$

Here, $a_0$ is the Bohr radius and $I_0 = \frac{1}{2}mv_0^2$, where m is the electron
mass.

However, real target electrons are not free and stationary. This
Bohr took into account by introducing two kinds of collisions: The
'free collisions' are characterized by their collision time being
shorter than the electron-revolution time, and their energy transfer being larger than the ionization potential. The 'resonance collisions' are characterized by the field from the ion being approximately uniform across the electron orbit and the interaction being
so weak that the probability for ionization is small.

If we regard the cases where $1 < \kappa < \eta$, it is valid to use an impact
-parameter description of the interaction ($\kappa > 1$). Bohr introduced
two decisive impact parameters: The first one i is defined as the
impact parameter where an amount of energy equal to the ionization
potential can be transferred to the electron if it were free and
stationary. It is easy to see that

$$i = \kappa a \quad . \tag{6}$$

The other decisive impact parameter d is the well known adiabatic
distance, where the collision time is equal to the electron-revolution time. This can be written as

$$d = \eta a \quad . \tag{7}$$

When $1 < \kappa < \eta$, it thus follows that $a < i < d$. The 'free' collisions
are those with impact parameters smaller than i. For those, the ionization cross section, Eq. (5), is obtained. The 'resonant' collisions are those with impact parameters between i and d. Using the
correspondence principle, Bohr argued that the ionization cross section for these collisions can be found from the <u>mean</u> energy transferred to a corresponding, classically bound <u>electron</u>. Integrating
the cross section for energy transfer ($\propto T^{-2}$) multiplied by the energy
transfer, we consequently obtain a cross section containing a logarithmic dependence on I, V, and q.

Bohr also treated the collisions, where $\eta > \kappa$ (very strong interaction) and collisions, for which $\kappa < 1$ (the perturbation region). He
found that in all cases where $\eta > 1$, the ionization cross section
for one electron can be written as

$$\sigma_{ion} = 4\pi a_0^2 q^2 (v_0/V)^2 (I_0/I)\{[\kappa/\eta]^{-1} + \delta\ln[\eta^2[\kappa]^{-2}] - \eta^{-2}\} \quad , \tag{8}$$

where the terms in $\{\}$ give the contribution from the 'free' collisions, the 'resonant' collisions, and the upper integration limit, respectively. The square brackets denote a function, which is equal to its argument if the argument is greater than one, and otherwise equal to one. In Eq. (8), $\delta$ is the fraction of resonant collisions leading to ionization. For a given target atom, it is a constant ($\delta = 0.28$ for hydrogen [12], $\delta = 0.45$ for helium [13]). For target atoms with more than one electron, Bohr suggested that the ionization cross section be obtained by summation of Eq. (8) for each electron.

From a closer examination of Eq. (8), it can be seen that for ions of high charge, $\sigma/q$ depends only on the scaling parameter,

$$Y = E[\text{keV/amu}]/q \qquad (9)$$

for a broad region of Y values. This scaling breaks down only for small ion velocities, where the term $\eta^{-2}$ becomes large, and for large ion velocities (in the perturbation region), where $\kappa < 1$. In Fig. 6 is shown Bohr's ionization cross section for ions of charges 1, 10, and 20, plotted as suggested by this inherent scaling law.

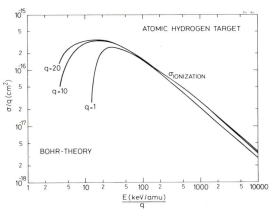

Figure 6. Single-ionization cross sections.

Also for target ionization by highly charged ions, a number of more elaborate theoretical calculations exists. The UDWA calculations of Ruyfuku [9] for atomic-hydrogen targets are valid over a broad range of ion parameters. For the 'classical' strong-interaction region ($\kappa > 1$, $\eta > 1$), Olson [14] has used his CTMC code to calculate a number of theoretical cross sections for hydrogen and helium targets. Janev and Presnyakov [15] have performed three-state, close-coupling calculations for hydrogen and helium targets for the same region of experimental parameters. They find that $\sigma/q$ is a function of parameter Y only for all Y values. For target ionization in the perturbation region ($\kappa < 1$), the Bethe-Born theory [16] has proven very successful.

TARGET IONIZATION, EXPERIMENT

Cross sections for target ionization by highly charged ions have been measured recently for an atomic-hydrogen target by Shah and Gilbody [17] and for a helium target by our group [5,18]. In both cases, coincidence techniques were used to obtain cross sections for 'pure' ionization, i.e., the process, in which the ion retains its charge state during the collision while a target electron is released. The experimental data are shown in Fig. 7, plotted as $\sigma/q$ as as a function of Y.

As can be seen from the figure, the data for the helium target scale very accurately around and above the cross-section maximum, as sug-

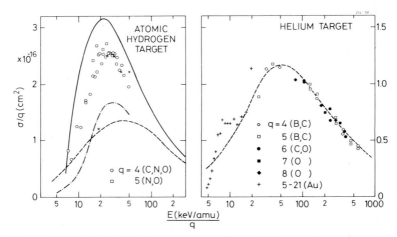

Figure 7. Single-ionization cross sections. Points are experimental
data. The theoretical results are: —— Bohr [1] (q = 4); --- close
coupling [15] (all q);—·—·— UDWA [9] (q = 5); × CTMC [14] (q = 4).

gested by the Bohr theory. Furthermore, for low Y values, the data
indicate a break-down of the scaling, which is also inherent in
Bohr's theory. For the atomic-hydrogen target, the available data
for highly charged ions support the scaling, but further data are
needed before a firmer statement can be made.

In Fig. 7, the available theoretical results have been compared to
the atomic-hydrogen data. These data make possible the most crucial
test of the theories as they are all (Bohr, CTMC, close coupling,
UDWA) basically one-electron calculations. From the figure, it can
be concluded that none of the theoretical results shows overall
agreement with the experimental data. The close-coupling result is
too low, does not have the right Y dependence, and its maximum fails
at too high a Y value. The Bohr and the UDWA theories lead to better
cross-section maximum positions, but the first gives too large and
the second too small absolute values. The CTMC calculations agree
very well with the experimental data above the maximum but decreases
much too fast with decreasing Y below the maximum.

The reason for this rather poor agreement between experiment and
theory may, as noted by Shah and Gilbody [17], stem from the fact
that there is actually two rather different mechanisms leading to
the observed ionization process. The first is the impact ionization,
with which the theories deal. The other is the socalled capture-to-
continuum process, which is closely related to capture to bound
states [19]. This second process has been shown [20] to give a large
contribution to the ionization of atomic hydrogen by  protons around
and below the cross-section maximum. It is not clear whether capture
to continuum has been taken properly into account in the theoretical
calculations discussed here.

From Fig. 7, the close-coupling calculations of Janev [15] are seen
to agree very accurately with the helium-target data. In the light
of the poor agreement between these calculations and the data for
atomic hydrogen, this is rather surprising. The gold-ion data seem
to suggest that the exact scaling of σ/q with Y for the lowest Y

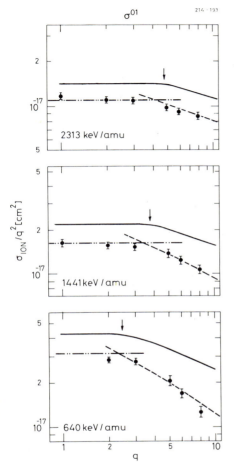

$\sigma^{01}$

214 - 193

2313 keV/amu

1441keV/amu

640 keV/amu

Figure 8. Single ionization of He. The arrows show where $\kappa = 1$. The theoretical curves are: —— Bohr [1]. —··—·· Bethe-Born [21]. --- close coupling [15].

values inherent in the close-coupling calculations is not correct.

For higher ion velocities, we reach the limit where $\kappa$ becomes equal to one, and above that limit, perturbation calculations become valid. The ionization in this transition region has been investigated experimentally by our group [18]. We measured the single-ionization cross section of helium for fully stripped H, He, Li, B, C, and O ions at three different ion velocities. The data are shown in Fig. 8.

In the perturbation region, the ionization cross section should be proportional to $q^2$ [12,13]. This can be seen from Eq. (8), which is proportional to $q^2$ when $\kappa < 1$ and $\kappa < \eta$ (see also Fig. 1). The dominating term inside the parenthesis is $\delta \ln(\eta^2)$, which is the socalled Bethe logarithm, and which is independent of $q$. However, when $\kappa$ becomes larger than one, this term becomes $\delta \ln(\eta^2 \kappa^{-2})$, and its $q$ dependence (via $\kappa^{-2}$) will weaken the $q^2$ proportionality valid in the perturbation region when the ion charge becomes large.

In Fig. 8, the data are plotted as $\sigma/q^2$ as a function of $q$. For low $q$ ($\kappa < 1$), we find, as expected, $\sigma/q^2$ to be constant. For high $q$ ($\kappa > 1$), however, the cross section increases less rapidly with $q$. The Bohr theoretical result is seen to describe very well the overall behaviour of the data in this transition region although quantitatively, it is some 20% too large. In the perturbation region, Bethe-Born calculations of Gillespie [21] are in excellent agreement with the data. In the strong-interaction region, the close-coupling calculations of Janev [15] agree with the experimental results.

## TARGET ELECTRON LOSS

The cross section for release of an electron from the target atom, irrespective of the fate of the ion, is the socalled target-electron-loss cross section. This can be obtained as the sum of the cross sections for electron capture and target ionization. This sum can be easily obtained from the Bohr-Lindhard [2] and the Bohr [1] calculations. Using the CTMC method, Olson [22] has calculated the loss cross section for hydrogen and helium targets. For highly charged

ions, he found that when divided by the ion charge, this cross section depends only on the scaling parameter Y. Duman and coworkers
[23] obtained the same scaling law using asymptotic and semiclassical methods.

Olson's theoretical results for atomic hydrogen have been compared
with experimental data obtained on $H_2$ by Berkner et al. [11] and
recently with atomic-hydrogen data obtained by Shah and Gilbody [17].
These comparisons show the CTMC result to agree quantitatively with
the data, except at high energies, where the CTMC result decreases
too rapidly with increasing Y. The scaling was found to be approximately obeyed. For helium targets, the target-electron-loss cross
section has been measured for carbon ions of charge 4-6 by Schlachter et al. [14] and recently by our group [18] for many different
ions and ion charges. Our results are shown in Fig. 9, plotted as
$\sigma/q$ as a function of Y.

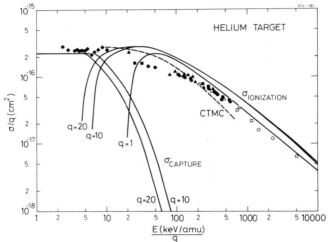

Figure 9. Single-target electron-loss cross sections. Solid
curves: Bohr-Lindhard and Bohr theoretical results. Dashed
curve: CTMC result [22]. The experimental data are: o: $H^+$,
□: $H^{2+}$, △: $Li^{3+}$, •: $B^{4,5+}$, ■: $C^{4,5,6+}$, ▲: $O^{6,7,8+}$, and ◆:
$Au^{6-17+}$.

The experimental data scale  as suggested by Olson. His universal
curve also agrees within a factor of two with the data. However, for
high values of Y, it decreases too rapidly with increasing Y. The
theoretical loss cross section obtained as the sum of the Bohr and
Bohr-Lindhard cross sections also agrees well with the data although
the ionization cross section is somewhat too high. The semiclassical
theoretical result quite adequatly describes the transition to the
perturbation region.


## PARTIAL CROSS SECTION

If we wish to learn more about the processes discussed in the previous sections, one possibility is to investigate partial cross sec-

tions: For a helium target, there are five dominating charge-exchange cross sections,

$$\sigma^{01} = \sigma^{01}_{qq} + \sigma^{01}_{qq-1}$$
$$\sigma^{02} = \sigma^{02}_{qq} + \sigma^{02}_{qq-1} + \sigma^{02}_{qq-2}$$

Here, superscripts indicate the target charge before and after the collision, and subscripts indicate the charge before and after the collision of the ion.

Such partial cross sections have been measured for helium targets by Cocke et al. [24], who used argon ions of low velocity, and by our group [25] for 20-MeV gold ions. In both experiments, coincidence techniques and 'LIST-MODE' data acquisition were applied. In Fig. 10 is shown our results plotted as a function of the ion charge.

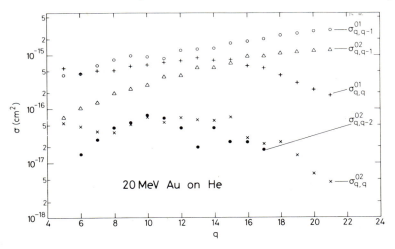

Figure 10.

These data are of special interest because they give information about the region, where the electron-capture and the ionization cross sections are of equal magnitude. It can be seen from the figure that for low q, the single-electron 'pure' ionization cross section is the larger. However, for increasing charge, this cross section decreases, and the single-capture cross section becomes dominant. For all ion charges, cross sections for processes involving the loss of two electrons from the target are small except for the socalled 'transfer-ionization' process, which involves the loss of two target electrons, accompanied by capture of one of these electrons. Next to single capture, this process is the dominating one at high q values.

It is interesting to compare these results with similar data obtained by Cocke et al. [24]. They used ions of much lower velocity, and they are well within the molecular-orbital region. Also in their case, the single-electron-capture cross section is the dominating one, and, not surprisingly, the direct ionization probability is very small. For the highest charges, they also observe a large contribution from the 'transfer-ionization' channel.

CONCLUDING REMARKS

It has been demonstrated that semiclassical models give a very convenient framework for the presentation of experimental and theoretical results for electron-capture and target-ionization processes involving multiply charged, medium- and high-energy ions and light targets.

In general, total cross sections obey simple scaling laws. Quantitatively, the Bohr-Lindhard, UDWA, and CTMC calculations agree well with experimental data for electron capture. For target ionization, the theoretical results agree less well with the data, and here, more work clearly remains to be done.

Future experimental and theoretical investigations will probably tend towards a better understanding of partial cross sections, both such as those presented in the previous section and (for electron capture) cross sections populating specific ionic energy levels.

ACKNOWLEDGEMENTS

The close cooperation with L. Andersen, H. Damsgaard, H.K. Haugen, and P. Hvelplund is gratefully acknowledged.

REFERENCES

[1] Bohr, N., K.Dan.Vidensk.Selsk.Mat.Fys.Medd. 18 (1948) No 8
[2] Bohr, N., and Lindhard, J., K.Dan.Vidensk.Selsk.Mat.Fys.Medd. 28 (1954) No 7
[3] Janev, R.K., and Presnyakov, L.P., Phys.Rep. 70 (1981) 1
[4] See Book of Invited Papers, XI Int.Conf. on the Physics of Electronic and Atomic Collisions, eds. N. Oda and K. Takayanagi, Kyoto, Japan (1979) p. 387-456
[5] Hvelplund, P., Haugen, H.K., and Knudsen, H., Phys.Rev. A 22 (1980) 1930
[6] Knudsen, H., Haugen, H.K., and Hvelplund, P., Phys.Rev. A 23 (1981) 597
[7] Knudsen, H., Haugen, H.K., and Hvelplund, P., Phys.Rev.A, Rapid Communications, Sept. 1981
[8] See, e.g., R.E. Olson in: Book of Invited Papers, XI Int.Conf. on the Physics of Electronic and Atomic Collisions, eds. N. Oda and K. Takayanagi, Kyoto, Japan (1979) p. 391
[9] Ruyfuku, K., and Watanabe, T., Phys.Rev. A 20 (1979) 1828 Ryufuku, H., JAERI-memo 9454 (to be published in Phys.Rev.A)
[10] Brinkman, H.C., and Kramers, H.A., Proc.Acad.Sci. (Amsterdam) 33 (1930) 973
[11] Berkner, K.H., Graham, W.G., Pyle, R.V., Schlachter, A.S., and Stearns, J.W., Phys.Rev. A 23 (1981) 2891
[12] Bethe, H.A., Ann.Phys. (5) 5 (1930) 325
[13] Kim, Y.K., and Inokuti, M., Phys.Rev. A 3 (1971) 665
[14] For references, see A.S. Schlachter, K.H. Berkner, W.G. Graham, R.V. Pyle, P.J. Schneider, K.R. Stalder, J.W. Stearns, J.A. Tanis, and R.E. Olson, Phys.Rev. A 23 (1981) 2331; R.E. Olson and A. Salop, Phys.Rev. A 16 (1977) 531

[15] Janev, R.K., and Presnyakov, L.P., J.Phys.B:Atom.Molec.Phys. 13 (1980) 4233; Janev, R.K., Phys.Lett. 83A (1981) 5
[16] See, e.g., M. Inokuti, Rev.Mod.Phys. 43 (1971) 297; ibid. 50 (1978) 23
[17] Shah, M.B., and Gilbody, H.B., to be published in J.Phys.B:Atom. Molec.Phys.
[18] Haugen, H.K., Andersen, L., Hvelplund, P., and Knudsen, H., to be published in Phys.Rev. A
[19] Rødbro, M., and Andersen, F.D., J.Phys.B:Atom.Molec.Phys. 12 (1979) 2883
[20] Shakeshaft, R., Phys.Rev. A 18 (1978) 1930
[21] Gillespie, H., Phys.Lett. 72A (1979) 329
[22] Olson, R.E., Berkner, K.H., Graham, W.G., Pyle, R.V., Schlachter, A.S., and Stearns, J.W., Phys.Rev.Lett. 41 (1978) 163; Olson, R.E., Phys.Rev. A 18 (1978) 2464
[23] Duman, E.L., Men'shikov, L.I., and Smirnov, B.M., Sov.Phys. JETP 49 (1979) 260 [Zh.Eksp.Teor.Fiz. 76 (1979) 516]
[24] Cocke, C.L., Dubois, R., Gray, T.J., Justiniano, E., and Can, C., submitted to Phys.Rev.Lett. (1981)
[25] Damsgaard, H., Knudsen, H., Haugen, H.K., and Hvelplund, P., to be published

PHYSICS OF ELECTRONIC AND ATOMIC COLLISIONS
S. Datz (editor)
© North-Holland Publishing Company, 1982

ELECTRON CAPTURE BY HIGHLY CHARGED
LOW-VELOCITY IONS

C. L. Cocke, R. Dubois, E. Justiniano, T. J. Gray and C. Can

J. R. Macdonald Laboratory
Physics Department
Kansas State University
Manhattan, Kansas 66506

Introduction

There may be no two expressions which more frequently characterize the many contributions of Jim Macdonald to physics than those of "electron capture" and "highly-charged ions". Although Jim did not live to see the KSU work discussed here on low-velocity capture come to fruition, he was very influential in the conception of the program. Indeed, his influence in all aspects of capture by highly-charged projectiles continues to be felt strongly by all of us at KSU.

This paper describes the use of a fast heavy ion beam to produce, by bombardment of gaseous targets, highly-charged low-velocity recoil ions, and the use of these secondary ions in turn as projectiles in studies of electron capture and ionization in low-energy collision systems. The interest in collisions involving low-energy highly-charged (LEHQ) projectiles comes both from the somewhat simplifing aspects of the physics which attend the long-range capture[1] and from applications to fusion plasmas,[2] astrophysics[3] and more speculative technology such as the production of X-ray lasers.[4] The ions of interest in such applications should have both electronic excitation and center-of-mass energies in the keV range and cannot be produced by simply stripping fast heavy ion beams. Several novel types of ion source have been developed to produce LEHQ ions, of which the secondary ion recoil source discussed here is one.

The Source

It has been known for some time that a fast highly charged heavy-ion projectile may remove, in a single collision, many electrons from a neutral target.[5-7] If the velocity of the projectile greatly exceeds that of the target electrons, the energy transferred to each electron is greater than that given to the target nucleus by a factor roughly equal to the nucleus-electron mass ratio. For example, by using a simple model for the spatial distribution of the electrons, one may estimate that a 1 MeV/amu +10 projectile passing .2 Å from the center of a neon atom will deposit about 1.5 keV in the electronic degrees of freedom while transmitting only about 3 eV to the nucleus. Thus the recoil is a highly-charged ion (+6) with a very-low recoil velocity (5 x $10^6$ cm/sec). It is interesting to note that, in the more common situation of foil or gas stripping a fast ion beam to achieve a high charge state, the electronic energy obtained is down from the ion's center-of-mass energy by approximately the same nucleus-electron mass ratio which acts in our favor above.

In order to investigate the possibilities of using these LEHQ recoils as a source of ions, we have invested some effort in investigating with what cross sections and what recoil velocities are these ions produced. These two characteristics are closely related, since the recoil energy may be calculated if one knows the impact parameter for the collision. We have used several experimental geometries to record "singles" charge state spectra, one of which is shown in

Fig. 1. We use a pulsed "pump" beam from an EN Tandem (pulse width < 5 nsec, repetition time 2 to 16 μsec) on a rare gas target at a pressure typically $3 - 5 \times 10^{-4}$ torr. An electric field of 100-1000 V/cm at right angles to the

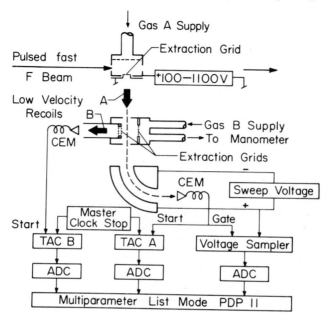

Figure 1. Schematic of Apparatus. Only Configuration II experiments use the "B" time-of-flight spectrometer and associated electronics.

beam extracts the beam and directs it through an electrostatically steered path onto a channel electron multiplier (CEM) or channelplate located 4 to 20 cm away. The ion energy is determined by the effective extraction voltage $V_1$, and thus the flight time is proportional to $\sqrt{m/q}$. A singles spectrum obtained from our current arrangement for $F^{+4}$ on Ar shown in Fig. 2. Charge states through $Ar^{+10}$ are readily identified, as well as several contaminant peaks. We note that subtracting contaminants is easy with this source, as one can simply remove the source gas without otherwise altering the source conditions. The recoils are produced primarily by direct ionization[9] of the target by the projectile although electron capture- does play a role in generating the highest charge-state recoils.[10] By way of illustration, we give in Table I a set of measured- cross sections for the production of various charge states of Ne by a $F^{+9}$ pump beam at 1 MeV/amu.

--------------------------------------------------------------

Table I   Cross sections in units of $10^{-17}$ $cm^2$ for producing $Ne^{+q}$ by $F^{+9}$ at 1 MeV/amu.

| q | 1 | 2 | 3 | 4 | 5 | 6 | 7 | 8 |
|---|---|---|---|---|---|---|---|---|
| σ | 140 | 48 | 24 | 13 | 6.0 | 1.4 | 0.80 | 0.35 |

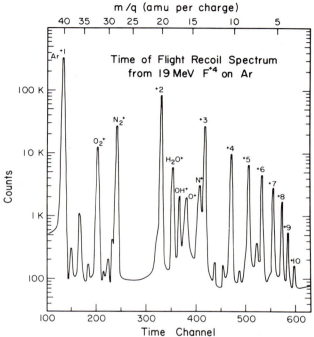

Figure 2. A time-of-flight ("A") spectrum showing the charge state spectrum generated from a slightly contaminated Ar target.
From these one can calculate that, for a target pressure of $3 \times 10^{-4}$ torr and a pump beam current of 0.5μa $F^{+9}$ (electrical) an extractor viewing a 2 mm path length should extract $2.6 \times 10^6$ $Ne^{+8}$/sec (3 pa electrical). This is more than adequate for total electron capture cross section measurements, but is weak for experiments which sample only a small fraction of the LEHQ secondary beam (e.g., electron or photon spectroscopy). In the measurements we present later, we typically run with only 100 pa of beam to keep down the count rate in the CEM.

The energies of the recoils are too low to be seen directly in spectra such as Fig. 2. However, given the cross sections for recoil production, one can obtain a characteristic impact parameter b from $\sigma = \pi b^2$ and the recoil energy $E_R$ from the small angle result

$$E_R = (Z_1^2 Z_2^2 e^4 M_1)/(b^2 E_0 M_2),$$ where $(Z_1, M_1)$ and $(Z_2, M_2)$ are target and projectile charges and masses and $E_0$ is the bombarding energy. For the $Ne^{+8}$ case discussed above, this gives $E_R = 7.6$ eV.

This result emphasizes one of the major advantages of the source, which is to go to very low ion energies. In references 9 and 11 discussions are given of detailed models for the primary collision from which the impact parameter dependence of the probability for producing each recoil charge state and thus the corresponding velocity distribution, may be calculated. In our use of the source we accelerate the ions, and have not gone below 80 eV/q. One can go lower by using $E_R$ alone as the ion energy and allowing the secondary collision to occur in the primary gas cell.[12-14]

The production cross sections for the highest q recoils increase rapidly with the charge state of the projectile, and decrease slowly with bombarding energy. We have used only the moderately charged pump beams available from our EN Tandem. By using bigger hammers, up to bare nuclei of Ne ($Xe^{+38}$, Superhilac[14]; $U^{+44}$, GSI[15]; $Br^{+27}$, Heidelberg MP-Linac[16]), as well as heluimlike Ar , $Kr^{+16}$ and $Xe^{+20}$ (ref. 15) have been produced.

## Results:  Low Energy Electron Capture and Ionization

We have to date studied only rare gas projectiles on rare gas targets.  Our data extend into an energy regime lower than that studied previously.  We have also used an unusual method of final charge state differentiation which allows us to study ionization processes as well as capture.  We have two experimental con- figurations:  Config. I allows measurements of projectile charge changing cross sections only, while config. II provides the charge differentiation results.

### A.  Configuration I

In this configuration the LEHQ ions are simply passed through a secondary gas cell and subsequently analyzed for charge change in this cell (see Fig. 1).  Two parameters are recorded for each event:  the time of flight of the LEHQ ion to the CEM (TOFA), which gives its initial charge state, q; and the analyzer voltage $V_a$, which gives its post collision charge state, q'.  A two dimensional spectrum thus obtained for $Ar^{+q}$ on Ne is shown in Fig. 3.  Events for which no charge- change occurs lie along a line at $V_a = kV_1$,  where the analyzer passes ions with

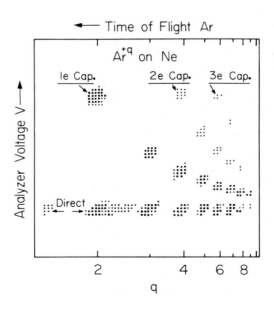

Figure 3.  A typical two-dimensional configuration I spectrum.

energy-to-charge ratio $V_a$. Capture events come at $V_a = (q/q')k\ V_1$ and give rise to the events on the single, double and triple capture lines in Fig. 3. Many cross sections are thus measured in a single run. The cross sections are extracted either from the slope of the pressure dependence curve for $0 < P_B < 5 \times 10^{-4}$ torr, or by using the nearly complete network of cross sections to correct fixed pressure data for multiple collision effects. The absolute scale is assigned from the known target length and pressure.

In Fig. 4 we show the velocity dependence of cross sections for $Ar^{+q}$ on Ne down to 80 eV/q. As has been seen previously (See, e.g., refs. 17,18) the behavior of the low-q (2 and 3) single capture cross sections is qualitatively different

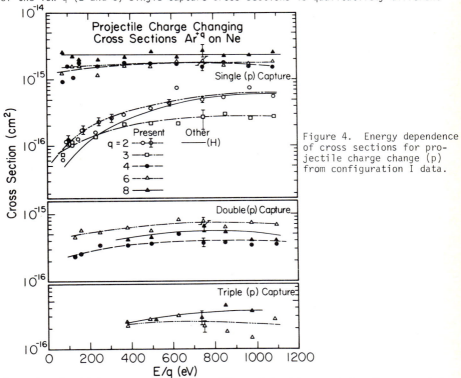

Figure 4. Energy dependence of cross sections for projectile charge change (p) from configuration I data.

from that for higher q, and can be readily explained by examining the separated-atom binding energies for the relevant $Ar^{+q}$ - Ne systems.[19] The capture is mediated by curve crossings between the incident channel (involving a neutral Ne target) and an exothermic exit channel (involving a singly charged Ne ion). The crossing radius, $R_c$, at which this occurs is given approximately (ignoring polarization) by $R_c = (q-1)e^2/\Delta E$ for a projectile of charge q feeding a channel of exothermicity $\Delta E$. In Fig. 5 we show plots of $R_c$ obtained from this expression for the feeding of various final states of the Ar ion having the initial parent configuration plus one electron in the 3p or 4s-4f shell. For q of 2 or 3, only 3p levels may be fed, at a rather well defined crossing radii. A small velocity-dependent cross section is expected, and this is obtained. For q of 4 or more,

Figure 5.  Crossing radii
for single capture (see
text).  The 3d locus,
which overlaps the n=4
band, has been omitted
for simplicity.

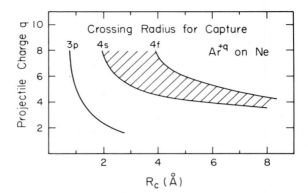

feeding of the n = 4 shell becomes possible and the energy fragmentation of the
n = 4 strength due to the core structure of the parent argon ion is sufficient
to spread the crossing radius locus into a band.  Somewhere within this band the
coupling between entrance and exit channels becomes large enough to prevent
diabatic behavior, and this roughly determines the effective radius at which
capture occurs.  Since the coupling increases rapidly with decreasing R, a
change in velocity means only a small shift in R and thus one expects large
but only weakly velocity dependent cross sections, as is seen to be the case
experimentally.  To find velocity dependence in the low velocity cross sections
at high q one will have to eliminate the projectile core structure, which means
using either bare nuclei or at least closed shell projectiles.

    In Fig. 6 we show velocity averaged cross sections versus q, and note that,
above q = 3, our results, which extend down to 80eV/q, are nearly the same as
those of Salzborn et al.,[17] at 30 keV.  Total capture cross sections are compared
with results of the absorbing sphere model of Olson and Salop,[20] and the
tunneling model of Grozdanov and Janev,[21] both of which predict cross sections
too large by about a factor of two.  The classical barrier model discussed in
references 12, 13 and 22 goes too low.  The assumption of all of these models that
the projectile behaves like a point charge is poor for Ar ions with 2 < q < 10.

    B.  Configuration II

For initial and final charge states q and q', respectively, several reaction
channels may contribute, including

$$Ar^{+q} + Ne \rightarrow Ar^{+q'} + Ne^{+(q-q')} \qquad\qquad \text{normal capture}$$

$$\rightarrow Ar^{+q'} + Ne^{+(q-q'+k)} + ke^{-1} \begin{cases} q = q', \text{ direct ionization} \\ q > q', \text{ transfer ionization} \end{cases}$$

We adopt the terminology that transfer ionization (TI) refers to any event in
which electron capture occurs and one or more electrons are liberated from the
system.[23]  To differentiate among these channels, we installed a second recoil
spectrometer in the secondary cell, and recorded TOFB as well as TOFA and $V_a$ for
each event, giving us the additional charge state of Ne as well as q and q'(Fig.1).

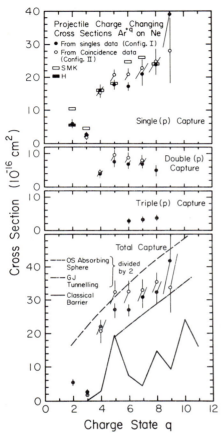

Figure 6. Velocity-averaged cross sections for projectile charge change. Other data are: SMK, Ref. 17; H, Ref. 18. Theory: OS, Ref. 20; GJ, Ref. 21; Classical barrier, Refs. 12, 13, 22. Config. II data are summed to give projectile charge changing cross sections.

In Fig. 7 we show a sample of the doubly charge differentiated data we obtained for Ar on Ne. For $Ar^{+4}$ on Ne, normal capture dominates. The small amount of direct ionization seen turns out to be due to random coincidences which are later removed in the data analysis. The weakness of this process is no surprise, since it requires feeding an endoergic channel. A surprise was encountered for projectiles with closed-shell ground state configurations, however. In Fig. 8 we show cross sections deduced from configuration II data for various projectiles on He targets. The apparent rise in the direct-ionization for $Ar^{+8}$ is, we believe, due to metastable components, probably $(2p^5 3s)\ ^3P_{0,2}$, in the LEHQ beam. The ratio of direct ionization to single capture is about 14%, independent of the target, leading us to conclude that the primary collision and subsequent electronic rearrangements leave about a 14% metastable component in the $Ar^{+8}$ beam. A similar and even stronger effect is seen in $Kr^{+8}$ and $Xe^{+8}$ beams, having low-lying $3d^9\ 4s$ and $4d^9\ 5s$ metastable states.

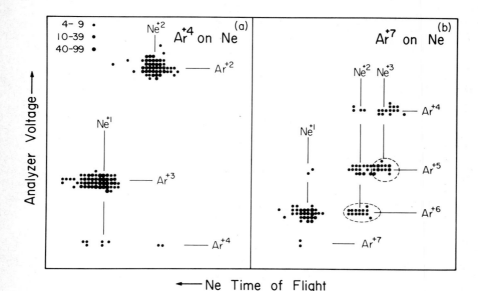

← Ne Time of Flight

Figure 7. A two-dimensional slice of config. II data, showing TOFA (dispersing Ar charge states) versus TOFB (dispersing Ne charge states). TI population for Ar$^{+7}$ is indicated by dashed ovals.

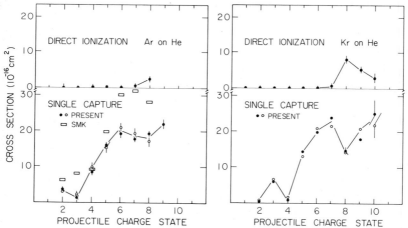

Figure 8. Apparent cross sections for Ar and Kr on He from config. II data, showing rise in direct ionization due to metastable beam components.

The configuration II data show important characteristics of the low-energy capture process itself as well as of the LEHQ source. In Fig. 7, the case of Ar$^{+7}$ on Ne shows that the TI channel is strongly fed as placed in evidence by events at the (Ar$^{+6}$, Ne$^{+2}$) and (Ar$^{+5}$, Ne$^{+3}$) intersections. We have converted these and other yields for argon on other targets to cross sections which we display in Fig. 9. The "double capture" process by which we mean here events

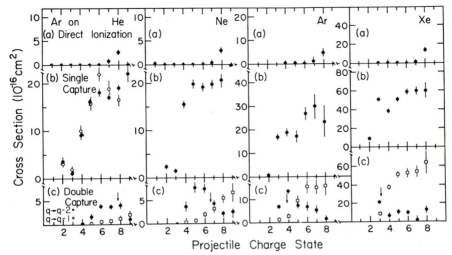

Figure 9. Config. II cross sections for $Ar^{+q}$ on rare gas targets at 500 eV/q. The double capture, $q \to q - 1$ is a TI channel.

generating +2 ions of He, Ne, Kr or Xe, represent normal ionization if $q \to q-2$, or TI if $q \to q-1$. The TI appears to have a sort of threshold q above which it becomes increasingly dominant. This threshold q is very near that for which we calculate that doubly excited states on the argon projectile, with configuration $3s^n 3p^m 4s^2$, on the argon projectile are exothermically populated (arrows in lowest part of figure). These states would autoionize before charge analysis, thus feeding, in the simplest way, TI channels. This interpretation has been previously invoked[20] and confirmed by electron spectroscopy[27] for lower charged projectiles at higher energy.

The presence of TI alters substantially our picture of multiple capture from complex targets. For a Xe target, we have found that more than half of the "single capture" identified through projectile charge change comes through the TI channel and may be more correctly called a double capture event. Multiple capture on such targets is strong, and a large fraction of the reaction ultimately results in free electron production. Results very similar to ours have recently been reported by Groh et al.,[28] who show multiple ionization of the system to be a common event accompanying capture in $Xe^{+q}$ + Xe collisions at 70 eV· q/amu. Fortunately, for the simpler targets, such as He, the importance of TI, as least in interpreting single capture, seems not to be as large as for targets of lower ionization potential. This is convenient, since such targets are the only ones which specific detailed theory is being done at present.

In a closing remark, we point out that these are alternative ways to use the LHEQ recoils generated by big hammers which already are bearing fruit. These include the selective capture experiments at GSI[12,13] and the ion trapping experiments at Berkeley[11] and now underway at Oak Ridge.[29]

Acknowledgements

We thank B. Waggoner, J. Giese, C. Schmeissner and S. L. Varghese for help in data taking. This work was supported in part by the USDOE Chemical Sciences Division.

## References

1) R. Olson, Electronic and Atomic Collisions, Invited Papers and Progress Reports, p. 391, ed. Oda and Takayanagi, (North-Holland, 1980).

2) C. F. Barnett, The Physics of Electronic and Atomic Collisions, p. 846, ed. Risley and Geballe (U. Wash. Press, Seattle, 1975).

3) G. Steigman, Astrophys. J. $\underline{199}$, 642 (1975); D. W. Rule and K. Omidvar, Astrophys. J. $\underline{229}$, 1198 (1979).

4) A. V. Vinogardov, Sobel'man, Sov. Phys. - JETP $\underline{36}$, 1115 (1973); W. H. Louisell, M. O. Scully and W. B. McKnight, Phys. Rev. A $\underline{11}$, 989 (1975).

5) M. D. Brown, J. R. Macdonald, P. Richard, J. R. Mowat and I. A. Sellin, Phys. Rev. A $\underline{9}$, 1470 (1974).

6) R. L. Kauffman, C. W. Woods, K. A. Jamison and P. Richard, ICPEAC IX, contributed papers, p. 939, ed. Risely and Geballe, U. Wash. Press (1975).

7) N. Stolterfoht, D. Schneider, R. Mann and F. Folkmann, J. Phys. B $\underline{10}$, L 281 (1977).

8) I. A. Sellin et al., Z Physik A $\underline{283}$, 329 (1977).

9) C. L. Cocke, Phys. Rev. A $\underline{20}$, 749 (1979).

10) T. J. Gray, C. L. Cocke, and E. Justiniano, Phys. Rev. A $\underline{22}$, 849 (1980).

11) R. Olson, J. Phys. B $\underline{12}$, 1843 (1979).

12) R. Mann, F. Folkmann and H. F. Beyer, J. Phys. B $\underline{14}$, 1161 (1981); R. Mann, H. F. Beyer and F. Folkmann, Phys. Rev. Lett. $\underline{46}$, 646 (1981).

13) H. F. Beyer, K.-H. Schartner and F. Folkmann, J. Phys. B $\underline{13}$, 2459 (1980).

14) R. Vane, M. H. Prior and R. Marrus, Phys. Rev. Lett. $\underline{46}$, 107 (1981).

15) R. Mann, A. Schlachter, W. Groh, A. Müller and C. Achenbach, Private comm. (1981).

16) R. Schuch, H. Schmidt-Böcking, H. Ingwersen, E. Justiniano and C. L. Cocke, private communication (1981).

17) E. Salzborn, and A. Müller, ICPEAC XI, Invited Papers and Progress Reports, p. 407, ed. Oda and Takayanagi, North-Holland (1980).

18) B. A. Huber and H. J. Kahlert, J. Phys. B $\underline{13}$ L 159 (1980).

19) C. E. Moore, N.B.S. Circular No. 467 (US GPO, Wash. D.C. 1952).

20) R. E. Olson and A. Salop, Phys. Rev. A $\underline{14}$, 579 (1976).

21) T. P. Grozdanov and R. K. Janev, Phys. Rev. A $\underline{17}$, 880 (1978).

22) H. Ryufuku, K. Sasaki and T. Watanabe, Phys. Rev. A $\underline{21}$, 745 (1980).

23) A. Niehaus, Comments on Atomic and Molecular Physics $\underline{9}$, 153 (1980).

24) P. H. Woerlee, T. M. El Sherbini, F. J. deHeer and F. W. Saris, J. Phys. B $\underline{12}$ L 235 (1979).

25) I. P. Flak, G. N. Ogurtsov and N. V. Fedorenko, J. Exptl. Theor, Phys. (USSR) $\underline{41}$, 1438 (1961); Sov. Phys. JETP $\underline{14}$, 1027 (1962).

26) W. Groh, A. Müller, C. Achenbach, A. Schlachter and E. Salzborn, Phys. Lett. to be published (1981).

27) D. Church, I. Sellin and S. B. Elston, private comm. (1981).

PHYSICS OF ELECTRONIC AND ATOMIC COLLISIONS
S. Datz (editor)
© North-Holland Publishing Company, 1982

EXCITED STATES FORMED BY ELECTRON CAPTURE IN HIGHLY
CHARGED RECOIL IONS

R. Mann, H. F. Beyer, and F. Folkmann*

GSI, D-6100 Darmstadt, West Germany

*University of Aarhus, DK-8000 Aarhus C, Denmark

Auger electron and x-ray spectra from highly stripped Ne,
Ar atoms and molecules ($NH_3$, $H_2O$, $CO_2$, $SF_6$) bombarded by
heavy projectiles (1.4 - 9 MeV/u, impact charge 12+ to 66+)
have been studied. Specific excited states carrying inner
shell vacancies result from selective electron capture from
target neutrals into outer shell orbitals of recoil- and
molecular fragment ions. Time discriminating spectroscopy
was used to separate between directly excited states from
the projectile impact and those resulting from subsequent
capture collisions.

INTRODUCTION

Spectroscopic studies of highly ionized atoms are of general interest for the
understanding of fundamental atomic physical processes being important for research in the plasma and astro physics. The K-x-ray and Auger electron emission
give information on the collisional, excitation, and deexcitation processes
allowing interpretations of the emission spectra of neutral and artificial plasma
sources. Thereby, the production of low-velocity recoil ions by bombarding rare-
gas targets with high energetic projectiles from heavy ion accelerators calls for
attention. The high degree of target ionization, e.g. the production of few
electron states, in a single collision within $t < 10^{-17}$ s combined with a small
recoil energy (Er $\lesssim$ 5 eV) and a well defined recoil direction $\theta \sim 90°$ with
respect to the projectile impact favour these recoil ions as objects for high
resolution spectroscopic studies. In particular Auger electrons which are sensitive to the movements of the emitting ion are observed with a very small
kinematical line broadening. As a consequence from the high degree of target
ionization metastable states carrying inner shell vacancies are strongly produced. They are the base for specific excited states formed by subsequent electron
capture collisions of the recoil ions with the surrounding neutral target particles. There it is a peculiarity that these recoil ions resembling the thermodynamic state of a hot plasma are present in a cold gaseous atmosphere. This
opens the possibility of studying charge exchange processes between highly
ionized but slowly moving atoms and neutrals not easily accessible so far. Besides the spectroscopic studies of such ions (1,2,3) carrying inner shell vacancy
states there are also some investigations of the production of recoil ions. More
recently C.L. Cocke (4) investigated the recoil ion production in dependence on
the energy and charge of the projectile. However, only few data exist for very
heavy and highly charged projectiles. Here we will report on measurements of
x-ray and Auger electron emission from rare gas targets and molecular systems
bombarded by heavy ions with energies $1.4 \leq E_p \leq 9$ MeV/u and charge states
$(12 \leq q \lesssim 66)$.

## HEAVY ION IMPACT ON MONATOMIC TARGETS

In the asymmetric collision of impinging heavy ions on light target atoms
($Z_t \leq 18$) most of the outer shell electrons are removed whenever the K-shell
gets ionized. Thus, a limited number of satellite lines occurs in the K-Auger
electron and K-x-ray spectra which can be studied individually (1,5). Fig. 1
displays Ne-Auger spectra induced by $Kr^{16+}$ projectiles with different energies
$E_p$ = 1.4 - 6 MeV/u. In case of $E_p$ = 1,4 MeV/u, Li-like configurations dominate
which are attributed to the decay of $(1s2s,p)n\ell$ ($n$ = 2, ... 7, $\ell$ = o ... n-1)
configurations (6). For increasing $E_p$ the degree of L-shell ionization drastically
decreases and strong contributions from Be-, B- and C-like configurations start to
complicate the spectra. However, the projectile charge q is as well an important
parameter for the multiple ionization increasing for higher q values as demonstrated for 9 MeV/u $Xe^{q+}$ projectiles (q = 29 and q $\sim$ 45 $\sim$ equilibrium charge
state after penetrating through a carbon foil) in Fig. 2.

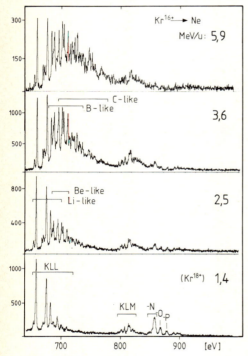
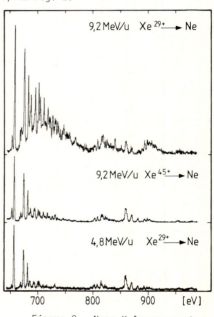

Figure 2.  Neon K-Auger spectra
for different projectile charges
and energies.

Figure 1.  K-Auger electron spectra from Ne targets induced by $Kr^{16+}$ and $Kr^{18+}$
projectiles at different energies.

Various collision systems were investigated in the past mostly for lighter projectiles and systematic studies of multiple outer shell ionization based on BEA
and PWBA scaling functions have been reported (7,8).

From the centroid energy of the KLL Auger group one may conclude on the averaged
number of L-vacancies <$\ell$> using the method of Stolterfoht et. al. (9) and an
empirical scaling function is obtained plotting <$\ell$> versus $v_p$/q ($v_p$ = projectile

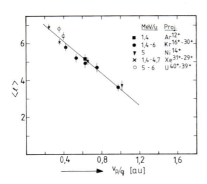

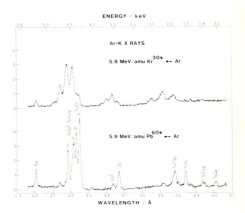

Figure 3. Averaged number of Ne-L shell vacancies $\langle \ell \rangle$ versus $v_p/q$ ($v_p$ = projectile velocity, q = charge state). The straight line fits the experimental data points.

Figure 5. Ar-K-x-ray spectra induced by 6 MeV/u $Pb^{60+}$ and $Kr^{30+}$ projectiles. The lines mainly arise from the decay of H-, He-, and Li-like configurations.

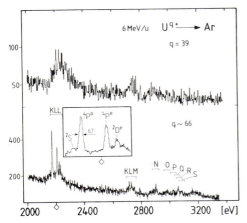

Figure 4. Ar-K-Auger electron spectra induced by 6 MeV/u projectiles for two different charge states q. The insert figure shows an expanded part of the K-LL Auger group demonstrating some prominent lines from Li-like configurations.

velocity) as shown in Fig. 3. This plot may be useful to estimate $\langle \ell \rangle$ roughly when q and $v_p$ are given.

When the target atomic number increases ($Z_t$ = 18) the dominant production of 1 to 4 electrons systems in a single collision accompanied with a K vacancy becomes very difficult not only due to the enlarged number of target electrons to be removed, but also because the projectile velocities required for a maximum ionization are quite different for each shell. This lack might be overcome by a very high projectile charge. In case of 6 MeV/u $U^{66+}$ impact on Ar it is possible to produce dominatly 1, 2, and 3 electrons systems. This can be seen from the Ar-K-Auger electron spectrum (Fig. 4) exhibiting lines from almost purely Li-like configurations and from the K-x-ray spectra (Fig. 5) demonstrating contributions from H-, He-, and Li-like states. This changes drastically going to lower values q = 39 (Fig. 4) and q = 30 (Fig. 5), respectively. The

Auger spectrum exhibits many overlapping lines reflecting the increased contribution from multi electron configurations e.g. 4 and 5 electron systems. The Li-like Auger spectrum is identified by comparison with Hartree Fock calculated transition energies (10).

The width of the prominent lines $^4p^o$ and $^4p^e$ (see insert in Fig. 4) from the decay of the metastable $(1s2s2p)^4p^o$ and $(1s2p^2)^4p^e$ states to the ground state $(1s^2)+e^-$ shows a kinematically line broadening being larger than the instrumental resolution ($\Delta E/E \sim 1.6 \times 10^{-3}$). From this a mean recoil energy $E_r = 60 \pm 6$ eV is obtained corresponding to a mean impact parameter b $\sim 0.15$ Å ($\sim 5$ times of the K-shell radius of Ar assuming Rutherford trajectories of the scattered particles with $Z_p = 66$ and $Z_t = 17$, respectively. The geometrical cross section $\pi b^2 = 7 \times 10^{-18}$ cm$^2$ agrees with the measured absolute cross section $\sigma_{KA}$ of the Auger electron emission listed in table 1 for different projectile impact on Ar.

Table 1   Cross sections $\sigma_{KA}$ of Argon K-Auger electron production

MeV/u Projectile $\sigma_{KA}$ $(10^{-18}$ cm$^2)$

| | | |
|---|---|---|
| 5.9 | $U^{66+}$ | $6.9 \pm 2$ |
| 5.9 | $U^{39+}$ | $4.8 \pm 1.5$ |
| 5.9 | $Kr^{30+}$ | $1.6 \pm 0.8$ |
| 9.0 | $Xe^{45+}$ | $2.0 \pm 1.0$ |

In case of lighter target atoms as for Ne the line width approaches to the instrumental resolution and an upper limit $E_r \lesssim 5$ eV is estimated.

STATES FORMED BY ELECTRON CAPTURE IN SLOW RECOIL IONS

When the target ionization is very high, a large fraction of metastable K-vacancy states is produced. These states undergo secondary collisions with the target neutrals and certain outer shell levels are populated by selective electron capture. Consequently, specific spectral lines become prominent as detected in Auger electron spectra from neon targets (2, 11) and confirmed by the corresponding x-ray emission (5).

The capture process may be described in a semiclassical one-electron potential curve picture (12,6) which gives the most probable main shell n populated in dependence on the recoil ion charge q and on the ionization potential $I_p$ of the neutral particle as

$$n \sim q(2|I_p|(1+(q-1)/(2q^{\frac{1}{2}}+1)))^{-\frac{1}{2}} \quad (1)$$

and giving a crossing distance

$$R_c = (q-1)/(q^2/2n^2 - |I_p|) \quad (2) .$$

From the charge exchange an exothermic energy defect $E_{exo} = |E_b - I_p|$ ($E_b$ = binding energy of the captured electron) is converted into kinetic energy shared between the collision partners according to their masses. It should be noticed that $E_{exo} \gg E_r$ is valid for the most cases studied here and a molecular orbital picture might be applied to describe these adiabatic charge exchange collisions. In the semiclassical one electron picture neglecting any dependence on the collision energy a strong oscillatory behavior of $R_c$ as a function of both, $I_p$ and q is predicted (6,12).

The selective population of n according to equation 1 has been observed in previous experiments for various $I_p$ and q values where $I_p$ was varied by mixing different gases to the target gas under investigation (Table 2). Time

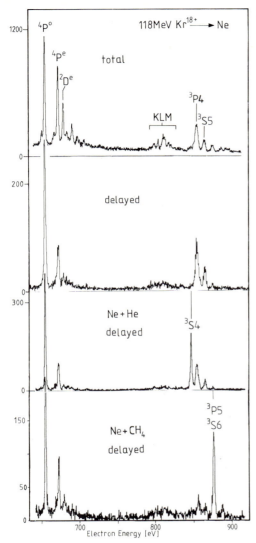

Figure 6. Total and delayed ($\lesssim$ 10 ns) Ne K-Auger spectra from different gas mixtures. The lines $^3S4$, $^3S5$, $^3S6$, $^3P4$, $^3P5$ indicate states from electron capture into main shells n = 4, 5, 6 of metastable $(1s2s)^3S$ and $(1s2p)^3P_{0,2}$ core ions.

Table 2   Calculated main shell n from equation 1 for electron capture and experimental values $n_{exp.}$ for different collision systems.

| Collision Energy [eV] | System | n | $n_{exp.}$ |
|---|---|---|---|
| $\sim$ 5 | $C^{4+}$ + $CH_4$ | 3.2 | 3 |
| $\sim$ 10 | $N^{5+}$ + $NH_3$ | 4.1 | 4 |
| $\sim$ 10 | " + He | 2.8 | 3 |
| $\sim$ 100 | " + $N_2$ | 3.4 | 3 (4) |
| $\sim$ 120 | $N^{6+}$ + $N_2$ | 4.0 | 4 |
| $\sim$ 10 | $O^{6+}$ + $H_2O$ | 4.6 | 4 (5) |
| $\sim$ 10 | " + He | 3.3 | 3 |
| $\sim$ 210 | " + $CO_2$ | 3.7 | 4 (3) |
| $\lesssim$ 5 | $Ne^{8+}$ + He | 4.2 | 4 |
| $\lesssim$ 5 | " + Ne | 4.4 | 4 (5) |
| $\lesssim$ 5 | " + $CH_4$ | 5.8 | 6 (5) |
| $\lesssim$ 10 | $Ne^{10+}$ + He | 5.0 | 5 |
| $\lesssim$ 10. | " + Ne | 5.3 | 5 (6) |
| $\lesssim$ 10 | " + $CH_4$ | 7.0 | 7 |

discriminating measurements of Auger spectra have been successfully conducted (13) separating between prompt and delayed emission in the nanosecond range using the pulsed beam structure of the UNILAC (period: 37 ns, width: 1 ns).

Fig. 6 displays total and delayed (t $\geq$ 10 ns) Ne Auger spectra from 1.4 MeV/u $Kr^{18+}$ impact on a pure Ne target and mixtures with He and $CH_4$, respectively. The pronounced lines in the delayed spectra at E $\gtrsim$ 840 eV are attributed (6) to the decay of states having metastable cores $(1s2s)^3S$ (lifetime $\tau \sim$ 9 $\mu$s) and $(1s2p)^3P_{0,2}$ ($\tau \sim$ 9 ns), and an additionally electron in outer shells n = 4 to 6 of which the multiplets from different subshell populations are not resolved. These states are produced by electron capture demonstrating a strong selectivity on the ad-

mixed target gas ($I_p$) in agreement with equation 1. The lines $^4P^o$ and $^4P^e$ at the low energy part of the spectra keep unaffected from $I_p$. They are explained by cascade feedings of the innermost quartet states $(1s2s2p)^4P^o$ and $(1s2p^2)^4P^e$ from outer shell quartets formed by electron capture. The $^4P^o$ configuration containing a long living $^4P^o_{5/2}$ component ($\tau \sim 8.4$ ns) may be partly produced by the primary collision appearing in the delayed spectrum as a pure lifetime effect (14). This was also demonstrated (13) in delayed spectra measured with low target pressure (1 mtorr) where the capture collision becomes unlikely to happen in the viewing region of the spectrometer or within the lifetime of the K vacancy core state. Then the "capture"- and "cascade"- lines disappear, only a residual intensity from the $^4P^o_{5/2}$ component is observed.

In the total spectrum some other prominent lines are seen in the KLL and KLM groups which are drastically suppressed in the delayed spectrum indicating configurations which are directly excited by the heavy ion impact.

It is possible to extract information on the dynamics of the capture collision from the time discriminated Auger electron spectra. It is based on the kinematical line broadening of certain Auger lines resulting from the exothermicity of the charge exchange. Because the outer shell capture lines from Ne are composed of several overlapping lines from different multiplets and because post collision interactions (PCI) (15) with the receding ion may cause additional line broadenings, it is difficult to measure the kinematic effect from those. However, the inner shell $^4P$-lines built up by capture and cascades decay isolated from the charge exchange partners and, therefore, they are most suitable to obtain the energy defect $E_{exo}$ using the simple relation

$$\Delta E = 4 \left( E_{exo} E_A m_e m_2 / ((m_1 + m_2) m_1) \right)^{1/2} \qquad (3)$$

which gives the kinematically line broadening ($E_A$ = Auger energy, $m_e$ = electron mass, $m_1$ = recoil ions mass, $m_2$ = mass of the neutral).

Table 3 compares line width from the prompt and delayed spectra for different charge exchange systems. From the broadening of the delayed $^4P^e$-line a value

Table 3  Experimental line widths in eV for Ne(K-LL) $^4P$-lines in the prompt and in the delayed spectra. The instrumental resolution was around 1.4 eV and the uncertainty for the widths $\pm$ 0.2 eV. $\Delta E = (\Delta E_{del.}^2 - E_{prompt}^2)^{1/2}$.

| Target | Ne | Ne + $CH_4$ | Ne + He | Ne + $H_2$ | Ne |
|---|---|---|---|---|---|
| Pressure (mtorr) | 20 | 20 + 20 | 40 + 40 | 25 + 25 | 1 |
| $^4P^o$ prompt | 1.3 | 1.4 | 1.2 | 1.6 | - |
| $^4P^o$ delayed | 2.1 | 2.0 | 1.9 | 2.2 | 1.8 |
| $^4P^e$ prompt | 1.9 | 1.9 | 1.9 | 2.3 | - |
| $^4P^e$ delayed | 2.8 | 2.4 | 2.4 | 2.5 | - |
| $^4P^e$ $\Delta E$ | 2.0 | 1.5 | 1.4 | 1.0 | - |

$E_{exo} \sim 30$ eV is obtained for $Ne^{8+} + Ne \rightarrow Ne^{7+} + Ne^+$ neglecting the smaller primary recoil energy $E_r \lesssim 5$ eV. A capture cross section $\sigma_c \sim \pi R_c^2$ can be estimated from the Coulomb repulsion $E_{exo} = (q-1)/R_c$ (6) of the charge exchanging partners. The obtained value $\sigma_c \sim 3 \times 10^{-15}$ cm$^2$ agrees with recent measurements of capture cross sections (16,21,26) of quite similar systems.

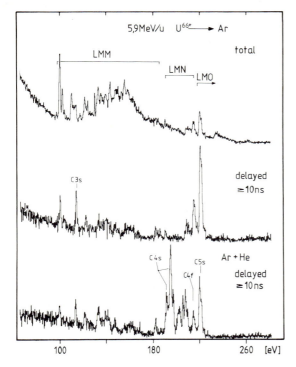

5,9 MeV/u $U^{66+}$ ⟶ Ar

total

LMM

LMN

LMO

C3s

delayed
≅10ns

C4s

C5s

C4f

Ar + He

delayed
≅10ns

100          180          260  [eV]

Figure 7. Total and delayed L-Auger spectra from Na-like argon induced by 6 MeV/u $U^{66+}$ impact on Ar and Ar+He mixture. The "delayed lines" above 180 eV result from selective electron capture into n = 4 or n = 5 of metastable cores c = $(1s^2 2s^2 2p^5 3x^1)(x=s,p,d)$.

The selective electron capture into metastable core ions resulting in auto-ionizing states appears to be fundamental not only for low atomic numbers with two electron core states but also for higher atomic numbers with a many electron core. This is demonstrated in figure 7 displaying total and delayed L-Auger spectra from Ar and Ar+He targets bombarded with $U^{66+}$ projectiles. Strong intensities of characteristic lines are seen in the delayed spectra. A selective dependence on $I_p$ (admixture of He) indicate a production by electron capture. Comparing to calculated transition energies some prominent lines can be attributed to the decay of C3s, C4s, C4f, C5s states labeled in Figure 7 where C is a Ne-like $(1s^2 2s^2 2p^5 3p^1)$ core configuration. This core is assumed to be metastable and produced either by the primary collision or by electron capture into $(1s^2 2s^2 2p^5)$ $Ar^{9+}$ ions populating the 3p level via transitions from outer "capture shells". Also metastable $(2p^{-1}3s)$ and $(2p^{-1}3d)$ core-configurations may be important and further calculations of state energies (27) are necessary for a definite line identification. Many lines at 190 - 207 eV in case of Ar+He are attached to different multiplets resulting from capture into different subshells of n = 4. Here the multiplet splitting is much larger than in case of Ne because the many electron core couples much closer to the captured electron in the neighboured shell. In this case the Auger lines can be partly resolved which allows to study the subshell population by electron capture. A similar observation was made earlier in case of $N^{5+}$ recoil ions (see Fig. 9).

Additional Auger intensities in the total spectrum at 120 - 185 eV may be attributed to L-MM transitions of directly produced $K^2 L^7 M^x$ (x = 2,3,4) configurations. A similar spectrum induced by $Cl^{12+}$ → Ar impact was discussed earlier by Schneider et al. (24).

## MOLECULAR TARGETS

Molecular targets bombarded by energetic heavy ions are of particular interest as they represent closely neighboured multi atom systems which have lost many electrons suddenly ($10^{-17}$ s) by the heavy ion impact. The highly ionized compounds will rapidly dissociate ($\sim 10^{-14}$ s, "Coulomb explosion" (17)) and remaining electrons may redistribute into the strongly disturbed system. The Auger electrons or x-rays emitted from the fragments may then carry some signatures of the molecular relaxation processes.

Light molecules as $CH_4$, $NH_3$, $H_2O$ being Ne-like with regard to their number of electrons reach similar states of ionization as for neon when bombarded by heavy ions. As an example, figure 8 depicts oxygen Auger spectra induced by $Kr^{16+}$ impact on $H_2O$ at different $E_p$. The multiple ionization in a single collision depends on $E_p$ in a similar way as for Ne, the highest degree of ionization occurs for the lowest impact energy $E_p$ = 1.4 MeV/u used in this experiment. In this case the observed Auger spectrum tends to be purely Li-like exhibiting a characteristic pattern of few strong lines in the KLL group and lines from the decay of certain outer shell states. This may indicate the fundamental nature of the production of states by electron capture into recoil ions which bear long living K-vacancy cores. Experiments with gas mixtures and using time discriminated spectroscopy may enlighten these processes.

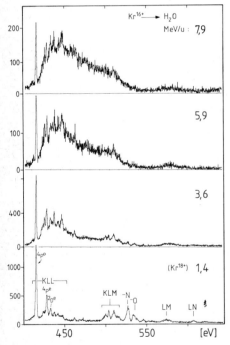

Figure 8. Oxygen K-Auger spectra from bombardment of $H_2O$ molecules with Kr projectiles of different energies.

Figure 9 displays prompt ($\pm$ 5 ns) and delayed ($\geq$ 10 ns) nitrogen Auger spectra from $NH_3$ and $NH_3$+He targets bombarded by 1.4 MeV/u $Ar^{12+}$ ions. Although $Ar^{12+}$ projectiles do not give the highest ionization significant contributions of specific lines are noticed in the delayed spectra. The line $^3$S4 is explained to result from the decay of $(1s2s(^3S)4s,p,d,f)$ states formed by electron capture from $NH_3$ molecules into the fourth shell of $N^{5+}$ $(1s2s)^3S$ core ions. The delayed intensity in the KLL group is assumed to result from states which are populated by x-ray cascades from capture states (for example $(1s2s(^3S)4d)^4D^e \rightarrow (1s2s2p)^4P^o)$ where the metastable $^4P^o$ configuration partly originates from the primary encounter. The admixture of He buffer gas having a higher ionization potential than $NH_3$ causes characteristic lines arising from capture into stronger bound states (n = 3). In this case it is possible to resolve some different multiplet states giving information on the distribution of the subshell population (s,p,d) by the capture process. The p shell population seems to be somewhat favoured but a strong selectivity of capture into a certain subshell is not found. This was already concluded from the broad structures of the capture lines in Ne-Auger spectra and supported by completing K-x-ray spectra indicating capture into p states (6).

The prompt nitrogen Auger spectrum of figure 9 is governed by many overlapping lines originating from Li- and Be-like configurations populated by the primary collision. The overlapping is not only due to a large number of multiplets lying

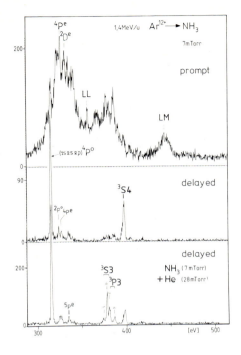

Figure 9. Prompt (+ 5 ns) and delayed (≥ 10 ns) nitrogen Auger spectra from 1.4 MeV/u $Ar^{12+}$ impact on $NH_3$ and $NH_3$+He mixtures. The delayed spectra exhibits lines $^3S4$ and $^3S3$ from electron capture into n = 4 and n = 3 of a $(1s2s)^3S$ core ion.

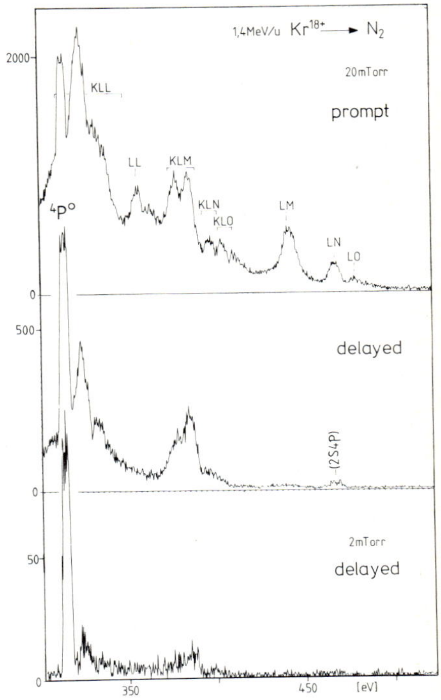

Figure 10. Prompt and delayed nitrogen Auger spectra from 1.4 MeV/u $Kr^{18+}$ impact on $N_2$ molecules at high and low pressures.

closer together for lower atomic numbers but it may result also from the PCI effect. The PCI should be important for promptly decaying states (t ≤ $10^{-14}$ s) (18) where the Auger electrons feel the electric field of the receding fragment ions. A kinematic line broadening caused by the "Coulomb explosion" of the molecule is small in case of the light molecules.

For somewhat heavier molecules like $N_2$ a strong population of few electron states especially of metastable K vacancy states by heavy ion impact is still very likely. Consequently, characteristic lines from secondary capture collisions are to be expected. However, in contrast to Ne or light molecules the capture occurs at higher collision energies (a few 100 eV) according to the kinetic energy $E_D$ obtained from the strong molecular dissociation (see table 2). In this case, $E_D$ is much larger than the energy defect $E_{exo}$ for a one-electron capture. For head-on collisions, smaller internuclear distances, and therefore, more potential curve crossings of the entrance channel and different exit channels of the charge exchange system can be reached before the Coulomb repulsion will separate the particles. Thus, contributions from double electron capture as observed elsewhere for similar colliding systems (19,20,21) cannot be excluded, however, their cross sections can be expected to be generally smaller than for a one electron capture (19,21).

Figure 10 displays prompt and delayed nitrogen K-Auger electron spectra from $N_2$ and 1.4 MeV/u $Kr^{18+}$ impact at 20 mtorr and 2 mtorr target pressure. A considerable line broadening from the Coulomb explosion kinematic covers the spectral features. The delayed spectra exhibit capture lines ($^3S3$, $^3P3$, $^3S4$) analogous to figure 9 and indicating a dominant capture into n = 3 corresponding to $I_p$ ($N_2$). These contributions and KLL states from cascades are strongly reduced at low pressure confirming to originate from secondary collisions. A structure around 470 eV to be seen in the delayed spectrum at higher pressure is attributed to a hyper-satellite (2s4$\ell$) ($\ell$ = 0,1,2,3) formed by a capture into n = 4 of a $N^{6+}$ (2s) core ion in agreement with equation 1.

It should be noticed that not only the electron background associated with the primary ionization is drastically suppressed in the delayed spectra but also the $^4P^0$-line induced by the primary collision appears well isolated at low pressure. This enables to study the line broadening by the Coulomb explosion more accurately than in earlier experiments (22), because contributions from secondary capture collisions with a subsequent cascade feeding of the $(1s2s2p)^4P^0$ state which may alter the kinematic of the molecular fragment ion can be controlled using time discriminating spectroscopy.

In the prompt spectrum of figure 10 remarkable intensities of the He-like hypersatellite sequence LL, LM, LN, and LO are observed. Two reasons may account for that. An enhanced probability for double K-vacancy production for lower target atomic numbers is expected from direct Coulomb ionization theories (23). Additionally, a relatively strong population of outer shell states (n = 2 .. 5) suggests a fast redistribution of remaining molecular electrons. Here, the molecules of low atomic number compounds ($Z_t \lesssim 10$) containing a larger number of more loosely bound electrons which have an extended spatial distribution as compared to monatomic targets introduce additional parameters for the investigation of multiple target ionization and the production of certain configurations of few electron states.

When the number of molecular electrons further increases a fast electronic rearrangement into highly ionized atoms becomes important. The metastable K-vacancy states become quenched out and the production of specific states from secondary capture collisions is reduced. Thus, a decreasing intensity of delayed Auger electron emission compared to the prompt emission is observed (Table 4).

| R | Target | Observed Atom |
|------|--------|---------------|
| 0.35 | Ne | Ne |
| 0.20 | $N_2$ | N |
| 0.08 | $CO_2$ | O |

Table 4  Relative intensity R of delayed Auger electron emission normalized to the prompt emission for different target systems bombarded by 1.4 MeV/u $Kr^{18+}$ projectiles and measured with equal time windows and target pressures.

Figure 11 displays total and delayed oxygen Auger spectra from $CO_2$ molecules and different projectile impact. Most pronounced structures characterising two and three electron states are observed in the total spectrum when the impinging projectile carries a very high charge state like 5.9 MeV/u $U^{66+}$ produced by a stripper foil. The high charge state may overcompensate the decrease of multiple ionization connected with a higher projectile velocity.

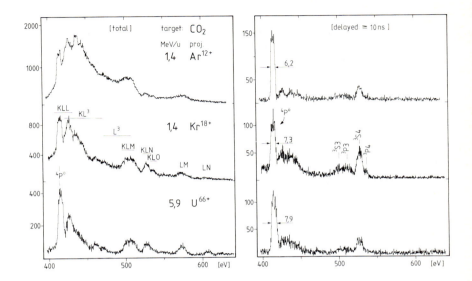

Figure 11.  Total and delayed oxygen K-Auger spectra from $CO_2$ molecules bombarded with different projectiles.  The increasing widths of the $^4p^0$-line reflect an increasing degree of molecular ionization.

The delayed spectra exhibit contributions from states emerging from selective electron capture.  From the strong broadening of the directly produced $^4p^0$-line, increasing for higher charged projectile impact, one may conclude on the mean charge $q_2$ of the molecular rest fragment CO (22).  Using in a first approach $E_M = q_1 q_2/R$ ($E_M$ = total kinetic energy of molecular fragments calculated from the line broadening (22), $q_1$ = 5 according to $(1s2s2p)^4p^0$, and R $\sim$ 1.9 Å is the internuclear distance between $0^{5+}$ and the center of gravity of the $(CO)^{q_2+}$ rest) values $q_2 \sim$ 7, 10, 12 result for $Ar^{12+}$, $Kr^{18+}$, and $U^{66+}$ impact.

Finally, for very complex molecules as $SF_6$ the rearrangement processes (25) predominates the population of states giving limits for the production of characteristic delayed lines.  The fluorine K-Auger spectrum of $SF_6$ bombarded with 1.4 MeV/u $Ar^{12+}$ ions shows practically no delayed lines (13).  However, $U^{66+}$ (6 MeV/u) projectiles are still sufficient to break up the molecule and producing highly stripped fluorine fragments (figure 12).  A remarkable intensity of the $^4p^0$-line is noticed in the delayed spectrum.  The strong kinematically line broadening corresponds to a fragment energy $E_D \sim$ 560 eV.

## CONCLUDING REMARKS

Spectroscopic studies of Auger electron and x-ray emission from highly stripped target recoils give not only information on the direct ionization and excitation due to the heavy ion impact but also information in detail on electron capture processes between neutrals and high charged ions in the near thermal region.  The observed capture states indicate that a one-electron transfer is dominant, occurring selectively into certain outershells in accordance with a classical model.  As far

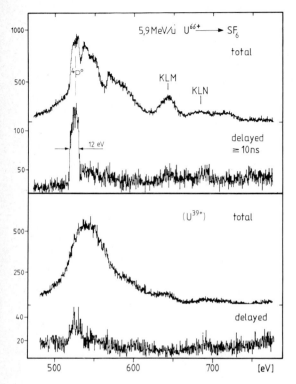

the captured electron is well separated from an inner shell electron core, the energy balance of the charge exchange process determines clearly the populated main shell. However, for capture states with a close coupling to core electrons, the subshells and the multiplets formed are more important for the energy balance, which is not in the scope of the classical barrier model in its present form.

Figure 12. Fluorine K-Auger spectra from $SF_6$ molecules bombarded with $U^{66+}$ and $U^{39+}$ projectiles.

## REFERENCES

[1] Stolterfoht, N., Schneider, D., Mann, R., and Folkmann, F., J. Phys. V: Atom. Molec. Phys. 10 (1977) L281-5.

[2] Mann, R. and Folkmann, F., J. Physique 40 (1979) C1-236.

[3] Beyer, H.F., Schartner, K.H., Folkmann, F., and Mokler, P.H., J. Phys. B: Atom. Molec. Phys. 11 (1978) L363-6.

[4] Cocke, C.L., Phys. Rev. A20 (1979) 749.

[5] Beyer, H.F., Schartner, K.H., and Folkmann, F., J. Phys. B: Atom. Molec. Phys. 13 (1980) 2459.

[6] Mann, R., Folkmann, F., and Beyer, H.F., J. Phys. B: Atom. Molec. Phys. 14 (1981) 1161.

[7] Schmiedekamp, C., Boyle, B.L., Gray, T.J., Gardner, R.K., Jamison, K.A., and Richard, P., Phys. Rev. A18 (1978) 1892.

[8] McGuire, J.H. and Richard, P., Phys. Rev. A8 (1973) 1374.

[9] Stolterfoht, N., Schneider, D., Richard, P., and Kauffman, R.L., Phys. Rev. Lett. _33_ (1974) 1420.

[10] Bar-Shalom, A. and Klapisch, M., private communication.

[11] Folkmann, F., Mann, R., and Beyer, H.F., Conference on "X-Ray Processes and Inner Shell Ionization", editors: Fabian, Watson, Kleinpoppen (Plenum Press, New York 1981).

[12] Ryufuku, H., Sasaki, K., and Watanabe, T., Phys. Rev. _A21_ (1980) 745.

[13] Mann, R., Beyer, H.F., and Folkmann, F., Phys. Rev. Lett. (1981) 646-9.

[14] Sellin, I.A., Mann, R., Frischkorn, H.J., Rosich, D., Schumann, S., and Szabó, Gy, Bull. Am. Phys. Soc. _22_ (1977) 1320.

[15] Stolterfoht, N., Brand, D., Prost, M., Phys. Rev. Lett. _43_ (1979) 1654.

[16] Vane, C.R., Prior, M.H., and Marrus, R., Phys. Rev. Lett. _46_ (1981) 107.

[17] Mann, R., Folkmann, F., and Groeneveld, K.O., Phys. Rev. Lett. (1976) 1674.

[18] Mann, R., Sellin, I.A., Folkmann, F., Frischkorn, H.J., Groeneveld, K.O., Schumann, S., and Szabó, Gy, Bulletin of the American Physical Society _22_ (1977) 1320.

[19] Huber, B.A. and Kahlert, H.J., J. Phys. B: Atom. Molec. Phys. _13_ (1980) L159-63.

[20] Crandall, D.H., Olson, R.E., and Shipsey, E.J., and Browne, J.C., Phys. Rev. Lett. _36_ (1976) 858-60.

[21] Cocke, C.L., Dubois, R., Gray, T.J., and Justiniano, E., and Can, C., Phys. Rev. Lett. _46_ (1981) 1671-4.

[22] Mann, R., Folkmann, F., Peterson, R.S., Szabó, Gy, and Groeneveld, K.O., J. Phys. B: Atom. Molec. Phys. _11_ (1978) 3045.

[23] McGuire, J.H., and Richard, P., Phys. Rev. _A8_ (1973) 1374.

[24] Schneider, D., Johnson, B.M., Hodge, B., and Moore, C.F., Physics Letters _A59_ (1978) 1302.

[25] Demarest, J.A., and Watson, R.L., Phys. Rev. _A17_ (1978) 1302.

[26] Beyer, H.F., Mann, R., and Folkmann, F., J. Phys. B: Atom. Molec. Phys. _14_ (1981) L377-81.

[27] Bhalla, C.P., private communication and Folkmann, F. et al., to be published.

PHYSICS OF ELECTRONIC AND ATOMIC COLLISIONS
S. Datz (editor)
© North-Holland Publishing Company, 1982

CROSS SECTIONS FOR ONE-ELECTRON CAPTURE FROM He
BY HIGHLY STRIPPED IONS OF C, N, O, F, Ne AND S BELOW 1 keV/amu

Y. Kaneko*, T. Iwai*, S. Ohtani, K. Okuno*, N.Kobayashi*,
S. Tsurubuchi+, M. Kimura‡, H. Tawara# and S. Takagi##

Institute of Plasma Physics, Nagoya University,
Nagoya 464, Japan

The cross sections for one-electron transfer from He atom into
the fully stripped, hydrogen-like, helium-like and lithium-like
$C^{q+}$, $N^{q+}$, $O^{q+}$, $F^{q+}$, $Ne^{q+}$ ions and also highly stripped $S^{q+}$ ions
have been measured at the energy range of 0.5q - 4.0q keV. It
is found that the measured cross sections are nearly independent
of the collision energy with a few exceptions. When plotted as
a function of the ionic charge q of ion, strong oscillations
in the cross sections are observed which are very similar in
phase but different in absolute values for ions with different
isoelectronic sequence. On the other hand, the measured cross
sections come together on a single curve when plotted as a
function of the effective core charge $Z_1^*$ of ion by taking into
account the screening by electrons. This oscillatory behavior
can be explained reasonably well through the modified classical
one electron model of Ryufuku-Sasaki-Watanabe.

1. INTRODUCTION

The electron transfer process between highly stripped heavy ion with charge q and
atomic hydrogen at low energies

$$A^{q+} + H \rightarrow A^{(q-1)+} + H^+ \tag{1}$$

is important not only in basic collision physics but also in many applications such
as astrophysics and high temperature plasma physics. In particular, the process(1)
involving impurity ions plays a key role in the energy and particle loss from the
Tokamak plasma[1]. Because only a single electron is involved in the collision
process of the fully stripped ion, theoretical treatment is considerably simple and
a number of theoretical calculations have been reported. Most of the theories are
based upon the concept of formation of the quasi-molecule $(A-H)^{q+}$ during collision.
Progress in theories is summarized by Olson[2]. On the other hand, it is difficult

* Department of Physics, Tokyo Metropolitan University, Setagaya-ku, Tokyo 158
‡ Department of Liberal Arts, Kansai Medical University, Hirakata, Osaka 573
+ Department of Applied Physics, Tokyo University of Agriculture and Technology,
  Koganei, Tokyo 184
‡ Department of Physics, Osaka University, Osaka 560
# Nuclear Engineering Department, Kyushu University, Fukuoka 812
# Department of Electric Engineering , Doshisha University, Kyoto 602

to obtain the highly ionized heavy ions at low energies and, therefore, experimental results are particularly scarce for the fully stripped ions. Data up to
early 1980 have been compiled[3]. Presently both theoretical[2] and experimental
works[4-5] are concentrated on investigations of the dependence of the cross
sections on the ionic charge of ion q and its nuclear charge $Z_1$ and on the collision
energy. Most of theories predict that the cross sections change monotonically
with the ionic charge q and its dependence is given as $q^\alpha$, $\alpha$ being roughly equal
to 2 but slightly depending on the model used, and the cross sections are nearly
independent of the collision energy below energies corresponding to the velocity of
1 a.u. with a few exceptions.

Experimental aspects including targets other than atomic hydrogen are reviewed by
Salzborn and Müller[6]. Most of the data have been obtained at energies higher
than a few keV/amu for partially ionized heavy ions. Again, almost all the experimental data show the monotonic dependence of the cross sections on q. However,
there is also experimental evidence that the cross sections do not change monotonically but some bumps or dips exist in some collision systems. For example, Müller[7]
and Crandall et al.[8] reported the cross sections for $Xe^{q+}$ ions show a significant
bump at q=5 in collisions with H and Xe targets. Very recently, Bliman et al.[9]
also reported the non-monotonic variation of the cross sections for $C^{q+}$, $N^{q+}$, $O^{q+}$
and $Ar^{q+}$ ions incident on $D_2$ and Ar gas targets at the energies of 1q - 10q keV.
They concluded that such an oscillatory variation of the cross sections is not due
to the presence of the metastable ions but due to the electronic structure of the
projectile ions. Similar variations have also been observed by Cocke et al.[10]

Meanwhile, Ryufuku, Sasaki and Watanabe (RSW)[11], based on their unitarized
distorted-wave approximation (UDWA)[12], predict that such an oscillation of the
cross sections at low energies occurs due to the crossings of the diabatic potential
curves and that the amplitude of the oscillation is large at lower energies and the
oscillation disappears at intermediate energies ($\simeq$ 25 keV/amu). They also showed
that at low energies the UDWA treatment is equivalent to the classical treatment
(see 3.3).

The present paper describes our effort in measuring the cross sections for one-
electron capture processes in highly stripped C, N, O, F, Ne and S ions including
the fully stripped ions in collisions with He gas target :

$$A^{q+} + He \rightarrow A^{(q-1)+} + He^+ \tag{2}$$

at the energy range of 0.5q - 4.0q keV. This is, to our knowledge, the first
systematic measurements of the cross sections for highly stripped heavy ions with
the isoelectronic sequence.

2. Experimental

2.1 Ion source and charge state distribution of ions

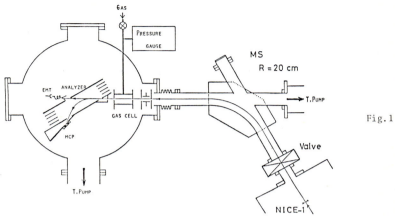

Fig.1

In the present work, ions are produced in NICE-1[13]), an electron beam type ion source (EBIS), which has a superconducting magnet to confine the high density electron beam. The surface of the superconducting magnet container at liquid He temperature works as a cryogenic pump to reduce background gas pressure in the ionization region. The background pressure measured at the outer vacuum vessel is usually $2 \times 10^{-10}$ Torr. The present experimental set-up is schematically shown in Fig.1. Ions, accelerated to a desired energy, are mass-analyzed and injected into a collision chamber. To make separation and identification of the charge and mass of ions easy and sure, the stable isotope gases, $^{13}CO$, $^{15}N_2$ and $^{18}O_2$, are used for C, N and O ions. Ne and $SF_6$ gases are used for Ne, F and S ions.

A typical charge state distribution of $^{15}N$ ions is shown in Fig.2 which is observed with a continuous electron multiplier (EMT). In contrast to the ordinary EBIS[14)], the present NICE-1 is operated in a mode where gas and electron beam are continuous-

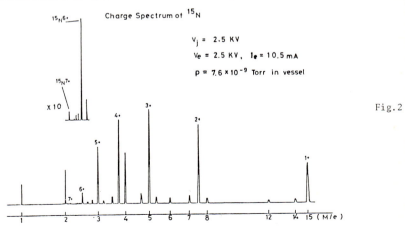

Fig.2

*Y. Kaneko et al.*

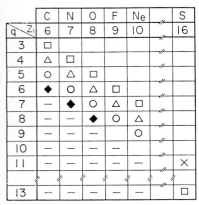

| q\Z | C 6 | N 7 | O 8 | F 9 | Ne 10 | | S 16 |
|---|---|---|---|---|---|---|---|
| 3 | □ | | | | | | |
| 4 | △ | □ | | | | | |
| 5 | ○ | △ | □ | | | | |
| 6 | ◆ | ○ | △ | □ | | | |
| 7 | — | ◆ | ○ | △ | □ | | |
| 8 | — | — | ◆ | ○ | △ | | |
| 9 | — | — | — | ○ | | | |
| 10 | — | — | — | — | | | |
| 11 | — | — | — | — | — | | × |
| 13 | — | — | — | — | — | | □ |

◆ fully stripped ion
○ hydrogen-like ion
△ helium-like ion
□ lithium-like ion
× boron-like ion

Fig.3

ly supplied[15]. Therefore, the charge of ions produced is fairly widely distributed over from q=1 to q=7 for N ions ; their distribution is strongly dependent on the gas pressure in the ion source and the electron energy. The intensity of the fully stripped $N^{7+}$ ions shown in Fig.2 is typically $2 \times 10^3$ counts per second (cps). Because of such a wide charge distribution, ions with different charge state are obtained without changing the ion source parameters.

2.2 Cross section measurements

To reduce background signals, the present collision chamber is evacuated down to $10^{-8}$ Torr with a 500 ℓ/s turbo-molecular pump. The target density of the collision cell containing He gas atoms is estimated through the pressure in a gas reservoir measured with a capacitance manometer BAROCELL and the calculated conductance of the capillary-aperture system used. Ions which pass through the collision cell are charge separated with a parallel plate electrostatic analyzer and detected with a multichannel plate detector (MCP) which works in a single particle counting mode. In this detection system, it is assumed that the sensitivity of the MCP is identical for all ions with different charge state because the ion impact energy on the MCP is always higher than a few keV where the coefficient of the secondary electron emission is usually larger than unity. It is found that the pulse height distribution from MCP used is dependent on the count rate. Therefore, in the course of measurements, care is taken to minimize the counting loss due to reduction of the pulse height by monitoring the pulse height distribution from MCP through a multichannel pulse height analyzer and an oscilloscope. The intensity of the primary ion beams is always kept less than $1.5 \times 10^4$ cps.

The cross sections for electron capture processes are determined through the initial growth of the charge-changed ions. The errors of the measured cross sections are estimated to be about ± 30 % where most uncertainties come from the determination of the growth rate, the target thickness and reproducibility.

In Fig.3 is shown a matrix of the ion and charge state which has been investigated in the present work. As seen in Fig.3, we are concentrating ourselves on measurements of the cross sections of the fully stripped, hydrogen-like, helium-like and lithium-like ions.

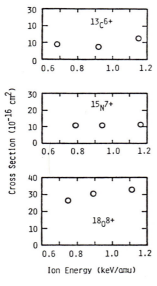

Fig. 4

## 3. Results and discussions

### 3.1 Energy dependence of the cross sections

As those in previous works, the measured cross sections for one-electron capture of multiply-charged heavy ions are nearly independent of the collision energy over $0.5q$ - $4.0q$ keV investigated in the present work, except for a few collision systems such as $C^{3+}$, $F^{8+}$, $Ne^{8+}$ and $S^{13+}$ ions where the cross sections increase slightly with collision energy. As a typical example, the cross sections for the fully stripped $C^{6+}$, $N^{7+}$ and $O^{8+}$ ions are shown in Fig.4. Full details of these results will be published elsewhere.

It is found from these data that the cross sections are varied with the charge state $q$ and also with the nuclear charge $Z_1$ of the projectile ions at the present energy range. The variation of the cross sections is particularly large for highly stripped low $Z_1$ ions and seems to be not so simple and monotonic as predicted by theories but depends on both $q$ and $Z_1$.

### 3.2 Ionic charge dependence

In Fig.5 are shown the cross sections at 0.8 keV/amu as a function of the initial charge state for all ions investigated. The lines are drawn to connect the initial charge of the isoelectronic sequence. As seen in Fig.5, the cross sections oscillate strongly with $q$ for all ions. These oscillations are particularly significant at low $q$. For example, the cross sections for $q=3$ and $5$ are almost

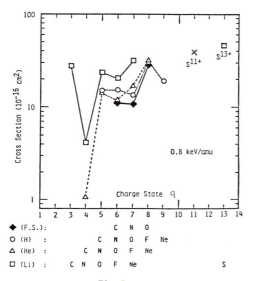

Fig.5

one order of the magnitude larger than those for q=4. Also the oscillation of the cross sections as a function of q is very similar for ions with different isoelectronic sequence. Further, for the same q, the cross sections depend on the atomic number $Z_1$ of the projectile ions. These oscillation and variation with q and $Z_1$ tend to disappear with increasing q and $Z_1$. In fact, the measured cross sections in the present work, averaged over the oscillation, are very similar to those obtained from an empirical formula of Müller and Salzborn[16].

3.3 The classical one-electron model with effective charge

As mentioned already, similar oscillations of the cross sections as a function of $Z_1$ are predicted by RSW[11] for the electron capture process between the naked ion and atomic hydrogen where only a single electron is involved. However, in the present case, the target of He has two electrons and the ion, partially ionized, also has a few electrons. Therefore, both nuclei of the target and projectile ion are screened by electrons and, then, the electron involved in the capture process feels a potential by such screened nuclei. The effective core charge, $Z_1^*$, of the ion, as seen by the electron to be transfer, is not the same as the ionic charge of the ion q.

In order to understand the oscillation phenomena observed in the present work, we follow the classical one-electron model in the electron capture process by RSW with the following modifications :

1.  It is assumed that the partially stripped projectile ion consisting of the core with the nuclear charge $Z_1$ and the screening electrons is equivalent to a naked ion with the effective core charge $Z_1^*$ and the target He atom consists of the hydrogen-like nucleus with the effective charge $Z_2^*$ and an electron which is transfered into the projectile ion.

2.  Such a core + electron system behaves hydrogenically , that is, the energy of the level of ion with the effective charge $Z^*$ is given by $-(Z^*)^2/2n^2$ where n is the principal quantum number of the level concerned.

3.  The effective charge $Z^*$ of such a partially stripped ion and helium atom is determined through the ionization potential $I_g$ in the ground state $(n_g)$ of the core + electron system :

$$Z^* = n_g(I_g/I_H)^{1/2} \qquad\qquad (3)$$

where $I_H$ represents the ionization potential of hydrogen atom in the ground state. For He target atom, $Z_2^* = 1.34$.

Then, the level energy for the excited state is calculated as follows :

$$-(Z^*)^2/2n^2 = -I_g n_g^2/(2n^2 I_H). \qquad\qquad (4)$$

As the ionization potential $I_g$ of the ground state ion with the nuclear charge Z, empirical values of Lotz[17] are used.

According to the classical one-electron model of RSW, the electron transfer to multiply charge ion at low collision energies occurs when the energy levels of the

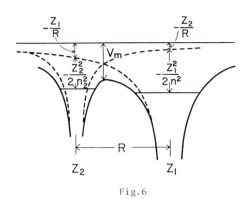

Fig.6

collision system before and after collision cross diabatically, that is,

$$-Z_1^*/R - (Z_2^*)^2/2n_2^2 = -(Z_1^*)^2/2n_1^2 - Z_2^*/R \qquad (5)$$

where R is the nuclear distance between the projectile ion and target atom. In the present case, $n_2 = 1$ as the electron is in the ground state of He atom. The left-hand side of eq.(5) corresponds to the diabatic potential energy of electron in the He target with the effective charge $Z_2^*$ perturbed by the Coulomb potential of the projectile ion with the effective charge $Z_1^*$ before collision and the right-hand side of eq.(5) does to that in the $n_1$ state of the projectile ion with the effective charge $Z_1^*$ perturbed by the Coulomb potential of the He$^+$ ion with the effective charge $Z_2^*$ after collision (see Fig.6).

The electron transfer becomes possible when the diabatic potential energy before collision (the left-hand side of eq.(5)) exceeds the maximum value of the potential barrier formed between the projectile ion and target atom $V_m$ :

$$-Z_1^*/R - (Z_2^*)^2/2n_2^2 \geq -V_m, \qquad (6)$$

$$V_m = \{(Z_1^*)^{1/2} + (Z_2^*)^{1/2}\}^2/R. \qquad (7)$$

From two equations (5) and (6), the integer n corresponding to the level where the electron is transfered can be determined as follows :

$$n \leq n_1 , \qquad (8)$$

$$n_1 = (Z_1^*/Z_2^*)\{(Z_2^* + 2(Z_1^* Z_2^*)^{1/2})/(Z_1^* + 2(Z_1^* Z_2^*)^{1/2})\}^{1/2}.$$

Then, the distance $R_n$, corresponding to the crossing point of the diabatic potential curves, is given by the following equation :

$$R_n = (Z_1^* - Z_2^*)/\{(Z_1^*)^2/2n^2 - (Z_2^*)^2/2n_2^2\}. \qquad (9)$$

Therefore, the classical one-electron transfer cross section $\sigma_{q,q-1}$ is given as follows :

$$\sigma_{q,q-1} = (1/2)\pi R_n^2 . \qquad (10)$$

3.4 Comparison between the measured cross sections and the classical model
In Fig. 7 are shown the measured cross sections plotted as a function of the effective charge $Z_1^*$ calculated from eq.(3), instead of the charge state q, together with those calculated from eq.(10) based on the classical model (dotted line). The number of n in Fig.7 represents the principal quantum number of the level of the projectile ion into which the electron is captured. It is remarkable that almost all the measured cross sections come close together on a single curve. The oscillation is large for low $Z_1^*$ and tends to vanish for higher $Z_1^*$. This oscillatory

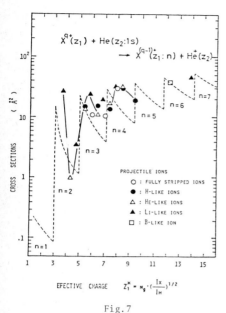

$$X^{q+}(z_1) + He(z_2:1s)$$
$$\longrightarrow X^{(q-1)+}(z_1:n) + \overset{+}{He}(z_2)$$

PROJECTILE IONS
O : FULLY STRIPPED IONS
● : H-LIKE IONS
△ : He-LIKE IONS
▲ : Li-LIKE IONS
□ : B-LIKE ION

EFECTIVE CHARGE     $Z_1^* = N_g \cdot (\frac{Ix}{IH})^{1/2}$

Fig. 7

behavior is quite similar to the calculated
one, though the agreement in the phase of
the oscillation is not so good.

Such a poor agreement in the phase of the
oscillation seems to be understandable from
the following reasons :

1. In the present classical model, the
   tunnel effect is neglected and the
   the possibility of the electron being
   captured into more than a single levels
   is neglected.  If these effects are
   taken into account, the oscillation
   should dump.

2. The charge cloud of the target atom and
   of the partially stripped ion will be
   deformed during collision because of
   their finite size of the charge distribution.  This polarization effect may
   result in the change of the electron
   transfer probability or in the change

of the effective charges $Z_1^*$ and $Z_2^*$.  In either case, this effect may be large
for higher $Z_1^*$ ions.

3. Because the energy level of the excited states is not purely hydrogenic in
   character, some corrections are necessary to obtain the accurate level energy.
   These corrections    cause the change of the effective charge $Z_1^*$ which, in
   turn, may give rise to some systematic deviations between different isoelectronic sequences.

In essence, however, the oscillatory behavior of the measured cross sections
observed in the present work is a good indication that, in highly stripped heavy
ion collision at low energies, the electron is captured dominantly into a particular
single level of the projectile ion through the crossing of the diabatic energy
levels in the collision system.  The oscillation is particularly significant for
low Z ions.  For low Z ions, the energy level responsible for the electron capture
has a small value of n and then the adjacent  levels are largely separated.  This
causes a great increase in the cross section even if the n-value is promoted by one.
On the other hand, for high Z ions, the electron is captured into a level having
a large n, where the energy levels are densely located, and then more than a single
levels have a chance to capture electron from the target atom.  This may be a
reason why the amplitude of the oscillation in the cross sections tends to diminish
toward higher Z ions.

## 4. Conclusions

We have measured the cross sections for one-electron transfer from He atom into highly stripped $C^{q+}$, $N^{q+}$, $O^{q+}$, $F^{q+}$, $Ne^{q+}$ and $S^{q+}$ ions produced in an electron beam ion source at energies less than 1 keV/amu. The measured cross sections plotted as a function of the ionic charge of ion q show significant oscillations with q which tend to disappear at large q. These oscillations are very similar for ions with different isoelectronic sequence but the observed cross sections are considerably different from each other. On the other hand, when plotted as a function of the effective charge $Z_1^*$ of ion, the cross sections measured come close together on a single curve with an oscillation which is reasonably well reproduced with the classical one-electron model, though their phase of the oscillation is not in good agreement with each other.

In order to understand the oscillatory phenomena in the cross sections for electron transfer precesses, measurements of the cross sections for lower $Z_1$ ions such as $B^{q+}$, $Be^{q+}$ and $Li^{q+}$ ions seem to be important at the low energy range and also more sophisticated calculations of the cross sections would be of great help.

## 5. References

1) M. P. C. McDowell (ed.), Atomic and Molecular Processes in Controlled Thermonuclear Fusions (Plenum Press, 1979).
2) R. E. Olson, Electronic and Atomic Collisions (ed. O. Oda and K. Takayanagi, North-Holland, 1980) p.391.
3) Y. Kaneko, T. Arikawa, Y. Itikawa, T. Iwai, T Kato, M. Matsuzawa, Y. Nakai, K. Okuno, H. Ryufuku, H. Tawara and T. Watanabe, IPPJ-AM-15 (Inst. Plasma Physics, Nagoya Univ., 1980).
4) H. J. Kim, R. A. Phaneuf, F. W. Feyer and P. H. Stelson, Phys. Rev. A17,854 (1978).
5) L. D. Gardner, J. E. Bayfield, P. M Koch, I. A. Sellin, D. J. Pegg, R. S. Peterson and D. H. Crandall, Phys. Rev. A21, 1397 (1980).
6) E. Salzborn and A. Müller, p.407 in ref.2).
7) A. Müller, PhD thesis (Univ. Giessen, 1977).
8) D. H. Crandall, R. A. Phaneuf and F. W. Meyer, Phys. Rev. A22,379 (1980).
9) S. Bliman, J. Aubert, R. Geller, B. Jacquot and D. van Houttte, Phys. Rev. A23, 1703 (1981).
10) C. L. Cocke, R. D. DuBoir, T. J. Gray and E. Justiniano, IEEE NS=28,1032 (1981)
11) H. Ryufuku, K. Sasaki and T. Watanabe, Phys. Rev. A21,745 (1980).
12) H. Ryufuku and T. Watanabe, Phys. Rev. A19,1538 (1979).
13) T. Iwai, Y. Kaneko, S. Ohtani, K. Okuno, N. Kobayashi, S. Tsurubuchi, M. Kimura, H. Tawara and S. Takagi, Proc. Sym. Ion Sources and its Application Technology (ISIAT-81,1981) p.167. Further details will be published soon.
14) E. D. Donets, IEEE NS-23,827 (1976).
15) H. Imamura, Y. Kaneko, T. Iwai, S. Ohtani, N. Kobayashi, K. Okuno, S. Tsurubuchi, M. Kimura and H. Tawara, Nucl. Instr. Meth. (in press).
16) A. Müller and E. Salzborn, Phys. Letters 62A,391 (1977).
17) W. Lotz, J. Opt. Soc. Am. 57, 873 (1967).

PHYSICS OF ELECTRONIC AND ATOMIC COLLISIONS
S. Datz (editor)
© North-Holland Publishing Company, 1982

ELECTRON CAPTURE IN SLOW $C^{+4}$/H COLLISIONS

M. Gargaud, J. Hanssen, R. McCarroll and P. Valiron

Laboratoire d'Astrophysique[*], Université de Bordeaux I
33405 Talence, France

The cross sections for charge exchange in $C^{+4}$/H collisions
are computed using a molecular treatment for energies in
the range 0.01 to 500 eV. The potential surfaces of all
molecular states which interact via avoided crossings,
either directly or indirectly with the entry channel
have been calculated using a model potential method.
Radial coupling matrix elements between all interacting
states of the same symmetry are determined with precision.
A quantum mechanical method is used to treat the collision dynamics. The results are in excellent agreement
with recent experimental measurements.

INTRODUCTION

Charge exchange reactions of multiply charged ions with neutral atomic hydrogen play an important role in many astrophysical plasmas
under coronal conditions[1,2]. Since hydrogen is effectively ionized
by electron collisions at temperatures exceeding $10^5$ K, the astro-
physically interesting cross sections are those for collision energies ranging from 0.01 to 10 eV. Experimental measurements in the
low eV energy range have been lacking in the past and cross section
estimates are to a large extent based on theoretical considerations.
Recent measurements by Phaneuf[3] of charge exchange in $C^{+q}$/H collisions now make a comparison between theory and experimental possible.

In this work we report theoretical calculations for the $C^{+4}$/H system
where charge exchange takes place via a multiple crossing network
involving $\Sigma$, $\Pi$ and $\Delta$ states of the $CH^{+4}$ molecule. Elementary considerations[4] show that only three reaction channels are likely

$$C^{+4}(1s^2)\ ^1S + H \rightarrow C^{+3}(1s^2\ n\ell) + H^+ \qquad n\ell = 3s, 3p, 3d$$

involving four $\Sigma$ states, two $\Pi$ states and one $\Delta$ state. At low energies, the influence of rotational coupling is weak compared with

[*]ER 137 du CNRS

that due to radial coupling. For that reason in most of our calculations only radial coupling has been taken into account. However, the
influence of rotational coupling has been investigated in a number
of test cases.

Although the formulation of the charge exchange process in terms of
the molecular model is qualitatively well understood, the practical
application of the model raises many problems. The dynamic coupling
matrix elements between different molecular states depend on the origin of the electron coordinates. If the basis set used is inadequate,
the correponding scattering equations may not be invariant with respect to a Galilean transformation even when trajectory effects are
small. This defect can in principle be removed by the introduction
of translation factors [5,6,7]. Unfortunately, this leads to considerable complications. Besides since the form of the translation factor in the interaction region is arbitrary [6] (only its asymptotic
form is well defined), the virtues of the translation factor approach
may be more formal than real.

In our work, explicit introduction of translation factors is avoided
by confining the collision to an interaction region of finite dimensions. The results of our calculations will show to what extent this
procedure is justified.

To investigate translation effects, it is important to dispose of
reliable coupling matrix elements, whose accuracy can be guaranteed
at all internuclear distances and for any origin of coordinates.

For these reasons, we have adopted a mono-electronic model-potential
method [8] to describe the $CH^{+4}$ ion. Since the eigen energies and vectors of the model Hamiltonian can be calculated to high precision,
it is then possible to test the validity of the molecular model in
the treatment of the collision dynamics independently of other approximations within the framework of the model-hamiltonian description
of $CH^{+4}$. The reliability of the model potential approach itself can
be ascertained by a comparison with the ab-initio calculations of
Olson et al. [9].

POTENTIAL ENERGY CURVES

Since charge exchange in $C^{+4}/H$ collisions takes place primarily via
electron capture into excited states, we require reliable potential
energies and dynamic coupling matrix elements of the first few

Rydberg states of the $CH^{+4}$ molecular ion. This problem is ideally suited for treatment by a model potential method since the ionic cores are of spherical symmetry.

The model potential method has several advantages over many ab-initio methods in that it guarantees the correct dissociation limits and long range interactions ; these considerations are of vital importance in the investigation of reactions at thermal energies. The model Hamiltonian is generated from the model potentials of the atomic cores using the prescription of Bottcher and Dalgarno[10] and Valiron et al.[8]. For the $C^{+4}$ core we have constructed a potential using the method of Valiron [11]

$$V_{C+4}(r) = - \{ 2 (1 + 4.1803 r) \exp (- 7.726 r) + 4 \} / r \qquad (1)$$

The corresponding eigen values are compared with the experimental energy levels of $C^{+3}$ on Table 1. The precision is more than sufficient for our purposes.

Table 1 - Comparison of calculated and experimental energy levels (in eV) of $C^{+3}$

| Level | Experiment | Model potential |
|-------|-----------|-----------------|
| 2s | 64.495 | 64.495 |
| 2p | 56.50 | 56.47 |
| 3s | 26.95 | 26.98 |
| 3p | 24.81 | 24.82 |
| 3d | 24.21 | 24.21 |

The model Hamiltonian of the molecular ion is then written as

$$H = T + V_{C+4} (r_b) - 1 / r_a + U_{core} \qquad (2)$$

where T is the electronic kinetic energy, $r_a$ and $r_b$ the position vectors of the Rydberg electron with respect to the H and C nuclei and $U_{core}$ is the core-core interaction. Since core polarization is relatively unimportant, it is adequate to take $U_{core}$ as equal to the Coulomb repulsion of the ionic cores.

The eigen values $\varepsilon_i$ and eigen vector $\chi_i$ for a given internuclear distance R are determined by standard variational techniques using a basis set of two centre Slater-type orbitals in prolate spheroidal coordinates.

The molecular states effective in the charge exchange process are
designated according to their dissociation limit

| | | |
|---|---|---|
| 1 | $^2\Sigma$ | $C^{+4}$ + H |
| 2 | $^2\Sigma$ | $C^{+3}(1s^2\,3d)$ + $H^+$ |
| 3 | $^2\Sigma$ | $C^{+3}(1s^2\,3p)$ + $H^+$ |
| 4 | $^2\Sigma$ | $C^{+3}(1s^2\,3s)$ + $H^+$ |
| 5 | $^2\Pi$ | $C^{+3}(1s^2\,3d)$ + $H^+$ |
| 6 | $^2\Pi$ | $C^{+3}(1s^2\,3p)$ + $H^+$ |
| 7 | $^2\Delta$ | $C^{+3}(1s^2\,3d)$ + $H^+$ |

The calculated potential energy curves are presented in figure 1.
They are in excellent agreement with the calculation of Olson et
al.[9]

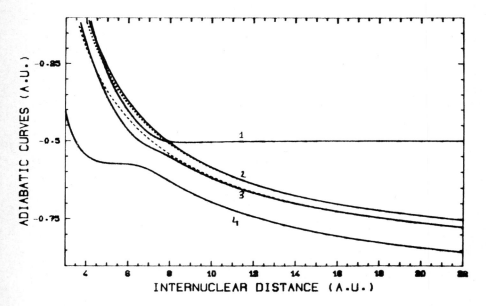

Figure 1 - Adiabatic potential energy curves of the $CH^{+4}$ molecular
          ion. The solid curves refer to the $\Sigma$ states $\sigma 1s_H$, $\sigma 3d_c$,
          $\sigma 3p_c$, $\sigma 3s_c$, labelled 1, 2, 3, 4. The dashed curves refer
          to the $\pi 3p_c$ and $\delta 3d_c$ states. The energy of the $\pi 3d_c$ is
          indistinguishable on the present scale from that of the
          $\sigma 1s_H$ state at internuclear distances less than $8a_o$.

RADIAL COUPLING MATRIX ELEMENTS

To test the various theoretical models of the collision, great care
has been taken to obtain the radial matrix elements to high precision within our model-potential approach.

Two independent methods are used. The first of these is based on a
direct numerical differentiation of the expansion coefficients of the
wave function $\chi_i$. This requires the calculation of the eigen values
and vectors at three consecutive internuclear distances. The accuracy of the numerical procedure is checked for each R value by the departure from the antisymmetry of the $\partial/\partial R$ matrix ; the accuracy is
very satisfactory, being about one part in $10^5$.

The second method is based on a variant of the Hellman-Feynman (H.F.)
theorem. (It is convenient to use the theorem in a form adapted to
prolate spheroidal coordinates, which allows all the molecular integrals to be expressed in analytic form). Since the H.F. theorem is
based on the assumption that the eigen vectors are exact, the results
provide a simple way of verifying the accuracy of our computations.
The agreement between the two methods is very satisfactory, better
than one part on $10^3$.

Typical results are presented in figures 2 and 3 showing the dependence of the matrix elements on the origin O of electronic coordinates. The origin O is assumed to lie on the internuclear axis, its
position being defined by the parameter $\gamma$

$$\underset{\sim}{r} = \underset{\sim}{r}_a - \lambda \underset{\sim}{R} = \underset{\sim}{r}_b + (1 - \gamma) \underset{\sim}{R} \tag{3}$$

where $0 < \gamma < 1$. Then

$$\left(\frac{\partial}{\partial R}\right)_{\underset{\sim}{r}} = \left(\frac{\partial}{\partial R}\right)_{\underset{\sim}{r}_a} + \gamma \left(\frac{\partial}{\partial z}\right) \tag{4}$$

where z is the component of $\underset{\sim}{r}$ along the internuclear axis and

$$< \chi_i \mid \left(\frac{\partial}{\partial R}\right)_{\underset{\sim}{r}} \mid \chi_j > = < \chi_i \mid \left(\frac{\partial}{\partial R}\right)_{\underset{\sim}{r}_a} \chi_j > + \gamma (\varepsilon_i - \varepsilon_j) < \chi_i \mid z \mid \chi_j > \tag{5}$$

We have verified  by direct calculation of the dipole matrix element
that (5) is satisfied to one part in $10^3$. This relation provides a
very sensitive test of the accuracy of our wave functions.

In general the origin dependence of the matrix elements is weak in
the vicinity of an avoided crossing. Elsewhere, the dependence on

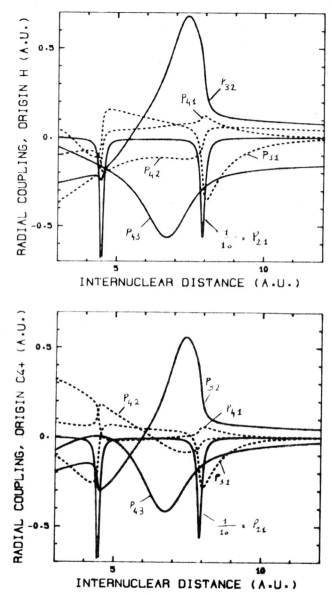

Fig. 2 and 3 - Radial coupling matrix elements of $CH^{+4}$, with the origin of
electronic coordinates on the C nucleus (fig. 2) and on the
H nucleus (fig. 3). The solid curves designate the coupling
terms 21, 32 and 43. The dashed curves refer to the terms 31,
41 and 42. Note that the plotted values of the 21 coupling
term have deen divided by a factor of 10.

origin can be considerable. Our calculated values of the radial
matrix elements differ considerably from those of Olson et al. The
reason for such a strong discrepancy is at present unclear.

SCATTERING EQUATIONS

The collision dynamics are treated in a quantum mechanical formulation using a conventional partial wave decomposition. In our work
we have assumed that at low energies the influence of rotational coupling is small compared with that due to radial coupling. The number
of coupled states is then reduced from seven to four. A quantitative
assessment of rotational coupling will be given later.

The corresponding radial wave-function $f_n^{(K)}$ for a given total angular
momentum K may be written in matrix form as

$$\frac{d^2}{dR^2} \underline{f}^{(K)} + 2 \underline{A} \frac{d\underline{f}^K}{dR} + (\underline{V} + \underline{W}) \underline{f}^K = 0 \tag{6}$$

where the matrix elements of $\underline{A}$ and $\underline{W}$ are defined as

$$A_{ij} = < \chi_i \mid \partial/\partial R \mid \chi_j > \tag{7}$$

$$V_{ij} = \left[ 2\mu (E - \varepsilon_{ij}) - \frac{K(K+1)}{R^2} \right] \delta_{ij} \tag{8}$$

$$W_{ij} = < \chi_i \mid \partial^2/\partial R^2 \mid \chi_j > \tag{9}$$

This equation is solved by first transforming to a diabatic representation[12]

$$\underline{g}^K = \underline{C} \ \underline{f}^{(K)} \tag{10}$$

where $\quad \frac{d}{dR} \underline{C} + \underline{A}\,\underline{C} = 0 \qquad\qquad \underline{C}(\infty) = \underline{I} \tag{11}$

Equation (6) then transforms to

$$\frac{d^2}{dR^2} \underline{g}^{(K)} - 2\mu \ \underline{V}^d \ \underline{g}^{(K)} + \left[ 2\mu \ E - \frac{K(K+1)}{R^2} \right] \underline{g}^{(K)} = 0 \tag{12}$$

where $\quad \underline{V}^d = \underline{C}^{-1} \ \varepsilon \ \underline{C} \tag{13}$

In the case of a two-state problem the determination of $\underline{V}^d$ is simple;
in the more general case special numerical techniques have been developed[13].

The elements of the diabatic matrix are represented in figures 4, 5.

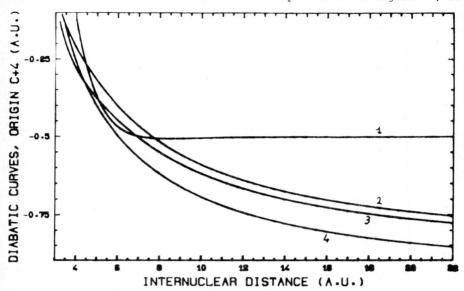

Fig. 4 - Diagonal elements of the diabatic potential matrix $\underline{V}^d$ with origin of coordinates on the C nucleus

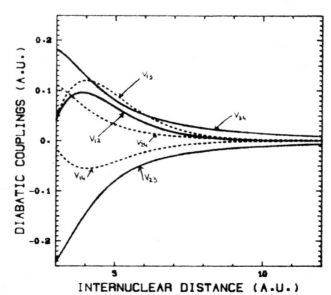

Fig. 5 - Off diagonal elements of $\underline{V}^d$ with origin of electron coordinates on the C nucleus. Solid curves refer to $V^d_{12}$, $V^d_{23}$ and $V^d_{34}$, dashed curves to $V^d_{13}$, $V^d_{14}$ and $V^d_{24}$.

Since the radial matrix elements depend on origin, so does $\underline{v}^d$. Although certain diabatic crossings correspond to well localized transition regions it is not in general possible to give a strict physical interpretation of the diabatic matrix.

The solution of (12) and the subsequent extraction of the scattering S-matrix are carried out using an extension of the log derivalue method[14] to the core of a repulsion Coulomb potential in one or more of the scattering channels. This method has proved to be particularly stable and advantageous to use at low energies for problems of this type. At higher energies (> 10 eV) rapid oscillations of the radial function require a smaller step size and the method becomes time consuming. However, this drawback is not too serious, since at these energies the partial cross sections vary smoothly with K and it becomes possible to interpolate over a large range of K.

RESULTS AND DISCUSSION

The cross sections summed over all charge exchange channels are denoted by $Q_T$ ; the capture into a specified state by $Q_{n\ell}$. The results are obtained using the radial matrix elements with the origin of coordinates at the centre of mass ; as a consequence the matrix elements $A_{23}$, $A_{34}$ do not vanish asymptotically. To circumvent the problem we determine $\underline{C}$ with the condition $\underline{C}(R_o) = \underline{I}$ where $R_o$ is some arbitrary large value of R. Sensitivity of the cross sections to particular values of $R_o$ provides a simple way of estimating the effect of these spurious couplings on $Q_{n\ell}$. (They do not, of course, have any effect on the total cross section $Q_T$).

At energies greater than a few tens of eV when the impact parameter method is valid, we may expect that $Q_T$, considered as function of collision velocity, should not depend on the C/H mass ratio. If the molecular basis set used is adequate, our results should be independent of the origin of coordinates used in computing $A_{ij}$. A number of test cases have been carried out with the H nucleus as the coordinate origin. The corresponding cross sections are designated by a superscript H.

The calculated cross sections are presented in figure 6. The contributions to $Q_T$ from the different reaction channels vary markedly with energy. At energies below 3 eV, electron capture proceeds primarily via the 3d state of $C^{+3}$ while at higher energies capture in the 3p state dominates. Capture into the 3s state does not contribute appreciably to $Q_T$ in the energy range investigated.

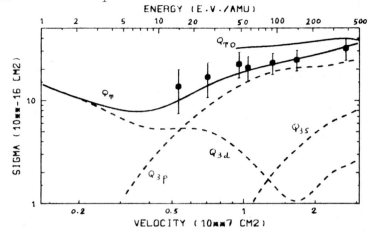

Figure 6 – Charge exchange cross sections for $C^{+4}$/H. The curves $Q_T$, $Q_{3s}$, $Q_{3p}$, $Q_{3d}$ are as defined in text. The curve $Q_{TO}$ refers to the calculations of Olson et al. The dark circles with their error bars refers to the experiments of Phaneuf

The calculated $Q_T$ are in excellent agreement with the experiments of Phaneuf. On the other hand, the theoretical cross sections of Olson et al. computed for energies greater than 50 eV/amu exceed our results by as much as 50 %. However, the substantial differences between our radial matrix elements and those of Olson et al. may explain such a discrepancy.

A comparison of $Q_T$ and $Q_{n\ell}$ with $Q_T^H$ and $Q_{n\ell}^H$ is given in Table 2. We find $Q_T$ and $Q_T^H$ do not differ significantly at energies below 100 eV ; at energies exceeding 100 eV $Q_T$ and $Q_T^H$ differ by less than 10 %.

Table 2                  Cross sections in units of $10^{-16}$ $cm^2$

| E (eV) | $Q_{3s}$ | $Q_{3b}$ | $Q_{3d}$ | $Q_T$ | $Q_{3s}^H$ | $Q_{3p}^H$ | $Q_{3d}^H$ | $Q_T^H$ |
|---|---|---|---|---|---|---|---|---|
| 24.5 | 0.12 | 8.82 | 5.06 | 13.99 | 0.18 | 8.90 | 4.99 | 14.07 |
| 81.6 | 1.83 | 18.58 | 1.64 | 22.05 | 2.50 | 18.57 | 1.99 | 23.06 |
| 136.1 | 3.39 | 20.70 | 1.06 | 25.14 | 3.70 | 21.71 | 1.07 | 26.48 |
| 27.2 | 6.29 | 22.00 | 2.00 | 30.29 | 8.44 | 20.14 | 3.41 | 31.99 |
| 54.4 | 9.68 | 23.41 | 3.43 | 36.52 | 12.57 | 19.77 | 7.58 | 39.91 |

This weak sensitivity to the origin of coordinates would indicate that the basis set of states of $\Sigma$ symmetry is an adequate represen-

tation. Support for this conclusion is provided by our calculations
of the rotational matrix elements $<1 \mid iL_y \mid 5>$ and $<2 \mid iL_y \mid 5>$ where
$L_y$ is the component of the electronic angular momentum perpendicular
to the collision plane. In estimating the influence it should be borne in mind that the matrix elements of figure 7 must be multiplied
by $K/R^2$. A transformation to the diabatic representation (see figure
7) shows that rotational coupling is effectively of little importance compared with radial coupling in the vicinity of the curve crossings.

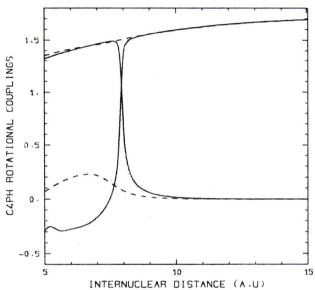

Figure 7 - Rotational coupling matrix elements $<i \mid L_y \mid 5>$ and
$<2 \mid L_y \mid 5>$ in the adiabatic representation (solid curves)
and in the diabatic representation (dashed curves)

On the other hand, the differences between $Q_{n\ell}$ and $Q_{n\ell}^H$ are more pronounced ; these may indeed be due to the neglect of rotational coupling with the $\Pi$ states 5, 6 and the $\Delta$ state 7. Calculations are in
progress to verify this point.

REFERENCES

[1] Péquignot D 1980 Astron. and Astrophys. 81, 356-8
[2] Field GB 1975 Atomic and Molecular Physics and the Interstellar
    Matter, ed. by Balian R, Encrenaz P and Lequeux J, 467-531
    (North Holland)

[3] Phaneuf R 1981 Phys. Rev. Lett. to appear

[4] McCarroll R and Valiron P 1978 J. Physique $\underline{39}$, C1, 52-4

[5] Bates DR and McCarroll R 1958 Proc. Roy. Soc. A $\underline{245}$, 175-83

[6] Schneiderman SB and Russek A 1969 Phys. Rev. A $\underline{181}$, 311-21

[7] Thorson WR and Delos JB 1978 Phys. Rev. A $\underline{18}$, 117-34

[8] Valiron P, Gayet R, McCarroll R, Masnou-Seeuws F, Philippe M
    1979 J. Phys. B : Atom. Molec. Phys. $\underline{12}$, 53-68

[9] Olson RE, Shipsey EJ and Browne JC 1978 J. Phys. B : Atom. Molec.
    Phys. $\underline{11}$, 699-708

[10] Bottcher C and Dalgarno A 1974 Proc. Roy. Soc. A 340, 187-98

[11] Valiron P 1976 Thèse de 3e cycle, Université de Bordeaux I,
     N° d'ordre 1279, unpublished

[12] Smith FT  1969 Phys. Rev. $\underline{179}$, 111-23

[13] Gargaud M, Hanssen J, McCarroll R and Valiron P 1981 J. Phys.
     B : Atom. Molec. Phys. to be published

[14] Johnson BR 1973 J. Comp. Phys. $\underline{13}$, 445-9

PHYSICS OF ELECTRONIC AND ATOMIC COLLISIONS
S. Datz (editor)
© North-Holland Publishing Company, 1982

THE CURRENT STATUS OF REACTIVE SCATTERING EXPERIMENTS
– A BRIEF INTRODUCTION TO THE SYMPOSIUM ON REACTIVE SCATTERING –

Y. T. Lee

Materials and Molecular Research Division
Lawrence Berkeley Laboratory
and Department of Chemistry
University of California
Berkeley, CA 94720

Molecular beam studies of reactive scattering between neutral atoms and molecules is the main theme of this symposium. The polarization of product angular momentum, the dynamics of endothermic reactions between atoms and polyatomic molecules and the effect of reactant internal excitation on the product state distribution will be discussed in three lectures given by Rettner, Krajnovich and Pruett. These three lectures represent some of the current activities of molecular beam studies of reactive scattering. It might be worthwhile to touch on some of the recent developments briefly before we start the main part of the symposium.

STATE–TO–STATE CHEMISTRY

Lasers are playing increasingly more important roles in the state preparation of reactants and state identification of products, especially for small molecules. Pruett's lecture entitled "State Resolved Products from High Reagent Vibrational Levels" exemplifiies the utility of these methods. The recent development of various vacuum UV laser has started to considerably expand the chemical scope of state resolved product detection by laser induced fluorescence.[1] The availability of high power visible and UV lasers has also made the multiphoton ionization process a practical alternative to laser induced fluorescence in some of the experiments. Using a plane polarized laser beam, it has been shown to be possible to determine the extent of the polarization of product angular momentum. This subject will be dealt with in Rettner's lecture on "Product Polarization in Reactive Scattering in Beam Gas Systems."

One aspect of state–to–state reactive scattering which is of special importance and which has been attracting both theoretical and experimental attention is the dynamic resonance in the reaction of atom–diatomic molecule systems containing hydrogen atoms. This quantum phenomenon involving the motion of three "atomic" particles is extremely sensitive to the details of the potential energy surface, and should provide a most stringent and meaningful comparison between theory and experiment in the coming years. Some encouraging progress has already been made on the $F + H_2 \rightarrow HF(v=n) + H$ reaction both experimentally[2] and in three dimensional quantum mechanical calculations.[3,4] At this conference, Kuppermann will review the theoretical aspects of this phenomenon in a lecture entitled "Reactive Scattering Computations; Dynamic Resonances in Chemical Reactions."

INTERACTION OF LASERS AND REACTION COMPLEXES

Extensive theoretical studies have been carried out on the interaction of high power lasers with reaction complexes. The promotion of the reaction probability, the opening up of new and desirable reaction channels and obtaining spectral information of reaction complexes are some of the goals of the interaction of the lasers and reaction complexes.[5,6] Some vigorous experimental activity is also expected in this interesting area in the near future, but it will remain a difficult experimental problem. The deposition of photons in the short lived reaction complexes is far less efficient than the resonant excitation of reagent atoms or molecules and the absorption spectra of reaction complexes in the "continuum" are expected to be broad. It remains to be seen whether the interaction of high power lasers and reaction complexes will become a useful tool for either the modification of chemical reactions or obtaining information on the structure and the potential energy surface of reaction complexes. Polanyi's progress report in this conference entitled "Probing the trnsition state in Chemically Reactive Collisions" is relevent to this subject.

## ENDOERGIC REACTIONS BETWEEN ATOMS AND POLYATOMIC MOLECULES

The relation between the location of the potential energy barrier and the effectiveness of vibrational and translation excitation of the reagents in promoting endothermic reactions of atoms with diatomic molecules has been studied quite extensively in a series of classical trajectory calculations carried out by Polanyi and coworkers.[7] Recently, several experiments on the reactions of alkali atoms and alkaline earth atoms with hydrogen halides have been carried out for the purpose of making a detailed comparison between translational and vibrational excitation.[8] Krajnovich's lecture entitled "The Effects of Reagent Translational and Vibrational Energy on the Dynamics of Endothermic Reactions" will summarize some recent studies on reaction atoms with polyatomic molecules, $Br + CH_3I \rightarrow CH_3 + IBr$ and $Br + CF_3I \rightarrow CF_3 + IBr$. The IR multiphoton excitation process was used to excite a beam of $CF_3I$ to a very high vibrationally excited state in these experiments.

## ELUCIDATION OF REACTION MECHANISM FROM DYNAMIC AND KINEMATIC RELATIONS

Crossed molecular beams experiments have been used mainly for the study of reaction dynamics of known chemical reactions. In recent crossed molecular beams studies of oxygen atoms with unsaturated hydrocarbons, it has been shown that the difficult task of determining the reaction mechanism and primary products of these reactions can be solved by using dynamic information and kinematic relations of primary products. The identification of radical products in these reactions has been frustrated in the past due to the lack of parent ions or excessive fragmentation of radical products in the mass spectrometric detection. By carrying out measurements of product angular and velocity distributions with sufficiently high resolution and by monitoring all the ions which are observable, it has been possible to convincingly derive reaction mechanisms for a series of reactions involving oxygen atoms with unsaturated hydrocarbons.[9] For example, the substitution of oxygen for a methyl radical or a hydrogen atom were found to be the major channels in the reaction of an oxygen atom with a propylene molecule.[10]

These are some of the current molecular beam studies of reactive scattering which attracted my attention. Omission is inevitable, no attempt was made to cover all important areas and I am sure the situation will change rapidly in the coming years.

## REFERENCES

[1]  Hepburn, J.W., Klimek, D., Liu, K., MacDonald, R.G., Northrup, F.J., and Polanyi, J.C., J. Chem. Phys. 74 (1981) 6226.
[2]  Sparks, R.K., Hayden, C.C., Shobatake, K., Neumark, D.M., and Lee, Y.T., in: Fukui, K. and Pullman, B. (eds.), Horizons of Quantum Chemistry (D. Reidel Publishing Co., Dordrecht, 1980).
[3]  Redmon, M.J., and Wyatt, R.E., Chem. Phys. Lett. 63 (1979) 209.
[4]  Baer, M., and Kouri, D.J., private communication.
[5]  Yuan, J.M., and George, T.F., J. Chem. Phys 68 (1978) 3040.
[6]  Lau, Albert M.F., Phys. Rev. A22 (1980) 614.

[7] Perry, D.S., Polanyi, J.C. and Wilson, Jr., C.W., Chem. Phys 3 (1974) 317.
[8] (a) Geis, M.W., Dispert, H., Budzynski, T.L. and Brooks, P.R., J. Am. Chem. Symp. 56 (1977) 103; (b) Gupta, A., Perry, D.S. and Zare, R.N., J. Chem. Phys. 72 (1980) 6250; (c) Heismann, F. and Loesch, H.J., Chem. Phys. (1981) to be published.
[9] Sibener, S.J., Buss, R.J., Casavecchia, P., Hirooka, T. and Lee, Y.T., J. Chem. Phys. 72 (1980) 4341.
[10] Buss, R.J., He, G., Baseman, R.J. and Lee, Y.T., to be published.

PHYSICS OF ELECTRONIC AND ATOMIC COLLISIONS
S. Datz (editor)
© North-Holland Publishing Company, 1982

PRODUCT POLARIZATION IN REACTIVE SCATTERING IN BEAM-GAS SYSTEMS

Michael G. Prisant, Charles T. Rettner and Richard N. Zare

Department of Chemistry
Stanford University
Stanford, CA   94305, U.S.A.

A procedure is developed for determining the product
rotational alignment in the center of mass frame from
polarization measurements of the chemiluminescent atom-
diatom exchange reaction, A+BC→AB*+C, under beam-gas
conditions.  The beam-gas relative velocity vector
distributions required for this inversion are generated
numerically using Monte Carlo sampling.  Beam-gas polarization measurements are presented for the reaction
$Ca(^1D)+HCl \rightarrow CaCl(B^2\Sigma^+)+H$ and analysis reveals a close-to-
limiting center of mass rotational polarization.

The dynamics of reactive collisions can be studied through a wide variety of
measurements.[1-3]  Recent advances in molecular beams, lasers and support technology have substantially enlarged the list of observables.  However, not all
measurements are equally facile or have equal value.  Here we describe a relatively simple method for examining an important dynamical quantity, the rotational
polarization of reaction products.  Before considering this in detail, let us survey the scope of current measurements.

Most of what is known about the dynamics of chemical reactions stems from determinations of scalar quantities, such as rate constants, cross sections, product
state distributions, etc.  In particular, studies at the state-to-state level
have helped to unfold the detailed balance sheet of conservation of energy,
revealing the effectiveness of reagent energy (in its various forms) in promoting
reaction and the partitioning of the excess energy of reaction amongst products.[1-3]
The balance sheet for conservation of angular momentum can be equally informative.
Here determinations of vector or directional properties of chemical reactions are
required.  Thus far such studies have emphasized the determination of product
angular distributions.[1-3]  These measurements relate the direction of the orbital
(external) angular momentum of the products $\underline{L}'$ to the initial relative velocity
of the reagents $\underline{k}$.  The corresponding relation of the direction of the internal
angular momentum $\underline{J}'$ of the products to $\underline{k}$ has received much less attention,
although its importance as a means for characterizing reaction dynamics was
recognized by Herschbach[4] early in the development of molecular beam scattering
studies.  What experimental information that is available has been obtained from
the measurement of polarized chemiluminescence,[5-9] the electric field deflection
of polar reaction products,[10-13] and laser induced fluorescence of product molecules.[14]  In anticipation that improved techniques would make these and related
vector correlations more experimentally accessible, a number of theoretical
treatments have appeared.[15-21]  This paper examines in detail the polarized
chemiluminescence method as applied to beam-gas scattering experiments.

For the prototype atom transfer reaction A + BC → AB + C, where BC and AB are
diatomic (or quasi-diatomic) molecules, the rotational polarization of the AB
product may be assessed in a number of ways.  Herschbach and coworkers[10-13] used
inhomogeneous two-pole electric fields to selectively deflect product molecules
in different spatial orientations.  Although valuable, this approach determines

the alignment averaged over all product states. Moreover, this approach has so far been restricted to the study of highly polar alkali halide products. More recently, it has been suggested[16,22,23] that the technique of laser-induced fluorescence (LIF) may provide a sensitive probe of alignment or orientation; in theory, as many as twelve independent vector moments can be obtained using a crossed beam reaction geometry,[16] and individual vibrational-rotational product states can be interrogated. Preliminary results in this laboratory[14] indicate that the successful application of this technique requires a judicious choice of reaction system and laser source. In particular, open-shell diatomic products may become disoriented by the presence of ambient magnetic fields that cause Larmor precession between the time of formation and the time of detection.[14] Moreover, cw laser sources may prove preferable over pulsed laser sources because the former present less severe optical pumping problems.[24,25] A third approach utilized the intrinsic fluorescence produced by reactions yielding electronically excited products.[5-9] Jonah et al.[5] demonstrated that the degree of (linear) polarization of the molecular chemiluminescence (CL) can be related to the alignment of the product molecules in the laboratory frame. Resolution of the CL spectrum can readily provide polarization measurements for individual vibrational bands. Although this method is blind to the vast majority of (dark) reactions, the comparative experimental simplicity of analyzing the degree of polarization of chemiluminescent emission makes it appealing, when applicable.

This approach has been utilized recently by Simons and co-workers in crossed supersonic molecular beam experiments.[6-9] Their studies have provided the first information about the energy dependence of angular momentum alignment in reactive collisions. Moreover, they have revealed a wide variety of behavior for different chemical systems, complementary in many ways to the earlier electric deflection results.[11,12] Although these studies have employed a crossed beam geometry in which the direction of $\underset{\sim}{k}$ is well defined, single-collision CL reactions are in fact most commonly studied by firing a reagent (metal) beam into a low pressure target (oxidant) gas.[1,26] This so-called beam-gas geometry provides essentially maximum single-collision signal intensities, and hence maximum product state selection, at the sacrifice of definition of $\underset{\sim}{k}$ for each reactive collision. With this in mind and in light of the renewed interest in CL polarization studies,[6-9] we have reexamined the problems associated with extracting meaningful information about angular momentum disposal from CL polarization measurements using a beam-gas scattering geometry.

At first glance the intrinsically poor definition of initial relative velocity per reactive collision might be thought to prohibit the accurate measurement of vector properties of the reaction. However, further inspection shows that the averaging process may be turned into an advantage. The distribution of relative velocity vectors is so well defined that the averaging process can be quantitatively calculated to high accuracy. Although the averaging over all possible target molecule velocities diminishes the apparent degree of CL polarization, we shall show that in most cases of practical interest the degree of polarization is reduced by less than 50 percent from that of a crossed beam geometry.

In what follows, we fully develop the analysis begun by Jonah et al.[5] We show how to invert beam-gas CL polarization measurements to provide center-of-mass information. We illustrate this approach with new CL polarization data for the reaction

$$Ca(^1D_2) + HCl \rightarrow CaCl(B^2\Sigma^+) + H .$$

We also apply this analysis to the results of the reaction $Ba + NO_2 \rightarrow BaO^* + NO$ previously studied.[5]

II.    THEORY

A.    From Chemiluminescent Polarization to Rotational Alignment

For a beam-gas configuration the chemiluminescence is traditionally viewed at right angles to the beam axis. Let $I_{\parallel}$ and $I_{\perp}$ denote the CL intensities polarized parallel and perpendicular to the beam axis. Then the degree of polarization, P, of the chemiluminescence is defined by

$$P = (I_{\parallel} - I_{\perp})/(I_{\parallel} + I_{\perp}) \ . \tag{1}$$

Since there is cylindrical symmetry about the beam axis, the product $\hat{J}'$ distribution can be expressed as a Legendre expansion

$$f(\hat{J}'\cdot\hat{Z}) = \sum_{\ell} a_{\ell} \ P_{\ell}(\hat{J}'\cdot\hat{Z}) \ . \tag{2}$$

Here $\hat{J}'$ is the unit vector along $J'$, $\hat{Z}$ is the unit vector along the beam axis,

$$\hat{J}'\cdot\hat{Z} = \cos\theta \tag{3}$$

is the cosine of the angle included by $\hat{J}'$ and $\hat{Z}$, and the coefficients

$$a_{\ell} = \frac{2\ell+1}{2} \int_{-1}^{1} P_{\ell}(\hat{J}'\cdot\hat{Z}) \ f(\hat{J}'\cdot\hat{Z}) \ d(\hat{J}'\cdot\hat{Z})$$

$$= \frac{2\ell+1}{2} <P_{\ell}(\hat{J}'\cdot\hat{Z})> \tag{4}$$

are the $\ell^{th}$ Legendre moments of the distribution $f(\hat{J}'\cdot\hat{Z})$. The degree of polarization is then given by[5]

$$P(P,R \text{ line}) = -3\alpha/(20 - \alpha) \tag{5}$$

for both a parallel-type transition in which the transition moment, $\mu$, lies along the internuclear axis and for a perpendicular-type transition in which $\mu$ lies in the plane of rotation perpendicular to the internuclear axis. Similarly,

$$P(Q \text{ line}) = 3\alpha/(10 + \alpha) \tag{6}$$

for a perpendicular-type transition in which $\mu$ lies along $J'$. Here $\alpha$ is the alignment parameter

$$\alpha = a_2/a_0$$

$$= \frac{\frac{5}{2} <P_2(\hat{J}'\cdot\hat{Z})>}{\frac{1}{2} <P_0(\hat{J}'\cdot\hat{Z})>}$$

$$= 5 <P_2(\hat{J}'\cdot\hat{Z})> \tag{7}$$

which is proportional to the second Legendre moment of $f(\hat{J}'\cdot\hat{Z})$. In Eq. (7) we have set $<P_0(\hat{J}'\cdot\hat{Z})> = 1$.

Thus determination of the degree of polarization of the CL leads to a knowledge of $\alpha$ which in turn yields $<P_2(\hat{J}'\cdot\hat{Z})>$. Note that this is the sole Legendre moment that can be extracted from dipole emission polarization.

B.  From Rotational Alignment in the Laboratory Frame to Rotational Alignment in the Center of Mass

The center of mass quantity we seek is $<P_2(\hat{J}'\cdot\hat{k})>$, the second Legendre moment of the (cylindrically symmetric) distribution of $\hat{J}'$ about $\hat{k}$, $f(\hat{J}'\cdot\hat{k})$. The laboratory distribution of $\hat{J}'$ about $\hat{Z}$ results from a convolution of the distribution of $f(\hat{J}'\cdot\hat{k})$ with the distribution of initial laboratory relative velocity vectors, $f(\hat{k}\cdot\hat{Z})$. It can be shown[27,28] that provided the two distributions are uncorrelated,

$$\langle P_2(\hat{\underset{\sim}{J}}' \cdot \hat{\underset{\sim}{Z}}) \rangle = \langle P_2(\hat{\underset{\sim}{J}}' \cdot \hat{\underset{\sim}{k}}) \rangle \ \langle P_2(\hat{\underset{\sim}{k}} \cdot \hat{\underset{\sim}{Z}}) \rangle \tag{8}$$

Therefore extraction of $\langle P_2(\hat{\underset{\sim}{J}}' \cdot \hat{\underset{\sim}{k}}) \rangle$ from an experimental determination of $\langle P_2(\hat{\underset{\sim}{J}}' \cdot \hat{\underset{\sim}{Z}}) \rangle$ requires a knowledge only of $\langle P_2(\hat{\underset{\sim}{k}} \cdot \hat{\underset{\sim}{Z}}) \rangle$, the second Legendre moment of the distribution of $\hat{\underset{\sim}{k}}$ about $\hat{\underset{\sim}{Z}}$.

C.  Calculation of $\langle P_2(\hat{\underset{\sim}{k}} \cdot \hat{\underset{\sim}{Z}}) \rangle$

The analytic calculation of $\langle P_2(\hat{\underset{\sim}{k}} \cdot \hat{\underset{\sim}{Z}}) \rangle$ requires evaluation of the distribution $f(\hat{\underset{\sim}{k}} \cdot \hat{\underset{\sim}{Z}})$. This calculation is easily formulated: we must integrate over all quantities which affect $\hat{\underset{\sim}{k}} \cdot \hat{\underset{\sim}{Z}}$ and which are not uniquely specified by the experiment. For a beam-gas configuration these quantities include: the distribution of magnitudes and directions of the beam velocities, the distribution of magnitudes and directions of the gas velocities, and the distribution of relative velocities that react. It has not proven possible to obtain a solution in a closed form.[5] We have instead sought a numerical solution based on the Monte-Carlo method,[29,30] details of which will be published elsewhere.[31,32] Briefly, $f(\hat{\underset{\sim}{k}} \cdot \hat{\underset{\sim}{Z}})$ is obtained by accumulation of events determined by the quantities averaged in the experiment - effectively mimicking the experiment itself. Beam and gas velocities are selected from appropriate Maxwellian distributions and the angle between them is selected from a cosine weighted distribution. The final event count is weighted by $|\underset{\sim}{k}|$, reflecting the greater flux of events with higher relative velocities (assuming the cross section to be independent of $|\underset{\sim}{k}|$). After a pre-set number of events the program calculates $\langle P_2(\hat{\underset{\sim}{k}} \cdot \hat{\underset{\sim}{Z}}) \rangle$ from the accumulated $f(\hat{\underset{\sim}{k}} \cdot \hat{\underset{\sim}{Z}})$ distribution.

We note the following reduction of parameters in the calculation. The value of $(\hat{\underset{\sim}{k}} \cdot \hat{\underset{\sim}{Z}})$ depends on the speed ratio $v_b/v_g$ and not on the individual magnitudes of these velocities. The distribution of this ratio is uniquely determined by the ratio

$$\rho = T_g M_b / T_b M_g \tag{9}$$

where T and M refer to temperatures and masses and the subscripts g and b indicate the gas and beam species, respectively. Hence $\langle P_2(\hat{\underset{\sim}{k}} \cdot \hat{\underset{\sim}{Z}}) \rangle$, can be parameterized by $\rho$ for all effusive beam-gas experiments. We call $\rho$ the universal beam-gas parameter.

D.  Results of the Monte Carlo Calculation

Table I lists the computed values of $\langle P_2(\hat{\underset{\sim}{k}} \cdot \hat{\underset{\sim}{Z}}) \rangle$ obtained after accumulation of 5000 events for values of the universal parameter $\rho$ over the range 0.0 to 2.0 at intervals of 0.1. As before, we have chosen the normalization $\langle P_0(\hat{\underset{\sim}{k}} \cdot \hat{\underset{\sim}{Z}}) \rangle = 1$. This data is presented in Fig. 1. As $\rho$ increases, $\langle P_2(\hat{\underset{\sim}{k}} \cdot \hat{\underset{\sim}{Z}}) \rangle$ asymptotically approaches zero. However, values of $\rho$ greater than unity are rarely encountered in practice. Of course, $\rho$ refers strictly to an effusive beam; for a nozzle beam Fig. 1 must be appropriately modified.

Fig. 2 shows actual $f_{LAB}(\hat{\underset{\sim}{k}} \cdot \hat{\underset{\sim}{Z}})$ distributions computed for $\rho$ = 0.0, 0.1, 0.2, 0.4, 0.8 and 1.6. The plots have been arbitrarily scaled to the peak heights of each plot, rather than to the total number of counts. The noise on these distributions results from the statistical nature of the Monte Carlo averaging procedure.

III.  MEASUREMENT OF CL POLARIZATION FROM THE REACTION Ca($^1$D) + HCl →
      CaCl*(B$^2\Sigma^+$) + H

Brinkman et al.[33] have shown that the reaction of hydrogen chloride gas with calcium atoms in the metastable 4s3d $^1$D state produces visible chemiluminescence originating from the CaCl A$^2\Pi$ and B$^2\Sigma^+$ states and terminating on the X$^2\Sigma^+$ ground state. We choose to study the CL polarization of the reaction Ca($^1$D) + HCl → CaCl(B$^2\Sigma^+$) + H because the mass combination exemplifies a kinematically

Table I.  Monte Carlo values[a] of $<P_2(\hat{\underset{\sim}{k}} \cdot \hat{\underset{\sim}{Z}})>$ as a function
of the universal beam-gas parameter $\rho$.[b]

| $\rho$ | $<P_2(\hat{\underset{\sim}{k}} \cdot \hat{\underset{\sim}{Z}})>$ | | $\rho$ | $<P_2(\hat{\underset{\sim}{k}} \cdot \hat{\underset{\sim}{Z}})>$ |
|---|---|---|---|---|
| 0.0 | 1.000 | : | 1.1 | 0.407 |
| 0.1 | 0.861 | : | 1.2 | 0.388 |
| 0.2 | 0.772 | : | 1.3 | 0.368 |
| 0.3 | 0.698 | : | 1.4 | 0.352 |
| 0.4 | 0.636 | : | 1.5 | 0.312 |
| 0.5 | 0.593 | : | 1.6 | 0.308 |
| 0.6 | 0.541 | : | 1.7 | 0.299 |
| 0.7 | 0.513 | : | 1.8 | 0.294 |
| 0.8 | 0.476 | : | 1.9 | 0.271 |
| 0.9 | 0.451 | : | 2.0 | 0.270 |
| 1.0 | 0.423 | : | | |

(a)  Computed by averaging over 5000 events.  One standard deviation corresponds to 1.4%.

(b)  Defined as $T_g M_b / T_b M_g$, where T and M are temperatures and masses and g and b refer to gas and beam species.

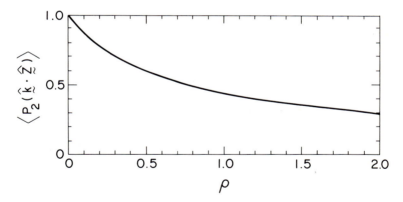

Figure 1.  Variation of $<P_2(\hat{\underset{\sim}{k}} \cdot \hat{\underset{\sim}{Z}})>$ with $\rho$

constrained system,[4,12,18] for which considerable rotational polarization is inevitable.  Furthermore, the $B^2\Sigma - X^2\Sigma$ system is a parallel transition, having only P and R lines, permitting analysis through Eq. (5) alone, without complications due to overlapping Q lines.

A.  Experimental Procedure

The beam-gas apparatus is similar to that described previously.[5,34]  Depending on configuration it can take chemiluminescence spectra, measure the polarization of

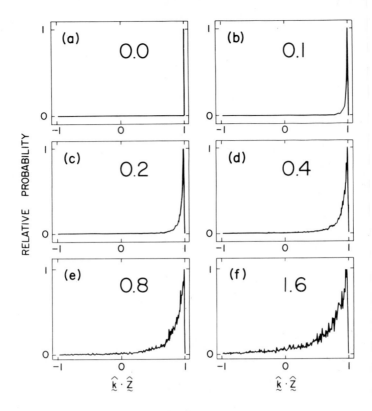

RELATIVE PROBABILITY

$\hat{k} \cdot \hat{\underset{\sim}{Z}}$

Figure 2.   The $f_{LAB}(\hat{\underset{\sim}{k}} \cdot \hat{\underset{\sim}{Z}})$ distributions computed by the
Monte Carlo method for a range of $\rho$ values.
The value of $\rho$ is indicated in each case.

chemiluminescent features, and obtain the dependence of a chemiluminescent feature
on beam velocity.  Only the main points are summarized here.

The calcium sample is loaded into a graphite oven suspended inside a resistively-
heated graphite tube.  Metastable calcium atoms are produced by an intense discharge which is struck between the oven orifice and the heating tube.  Normally a
DC discharge of 1.0 A is employed.  In TOF experiments the discharge is pulsed on
for 10 µs at a repetition rate of 2000 Hz.

The static HCl gas is at a pressure of $10^{-1}$ Pa($10^{-4}$ Torr). Its temperature can be varied by cooling a 10 cm diameter copper box that almost completely surrounds (~11 sr) the reaction zone.

The chemiluminescence spectrum is taken with a 3/4 m monochromator using a 1200 groove/mm grating blazed at 600 nm. An 8 cm focal length lens collects ~0.1 sr of the chemiluminescence and focuses the reaction zone on the entrance slit of the monochromator. For polarization measurements, a polarization scrambler is placed before the entrance slit. It consists of two commercially available (Karl Lambrecht) birefringent wedges held with their faces parallel and their optic axes making an angle of 45°. With this device the response of the detection system is found to be independent of polarization of the incident light. Between the polarization scrambler and the lens is inserted a polarization analyzer, consisting of a sheet of polaroid.

TOF spectra were recorded using a boxcar signal averager (PAR model 162 with a model 165 plugin). A gate width of 10 μs is used in the "RUN" mode.

B. Experimental Results

The $\Delta v = 0$ sequence portion of the CaCl B-X chemiluminescence spectrum was recorded with a resolution of 0.02 nm (FWHM) using a Ca oven temperature of 1200 K and an HCl gas temperature of 293 K. As is characteristic of the visible band systems of the alkaline earth monohalides, the $\Delta v = 0$ sequence has almost all the intensity. Band heads are formed in the $P_1$ and $P_2$ branches. The R branch lines contribute to an underlying background. The spectrum we have obtained is essentially identical to that observed by Brinkmann et al.[33]

Having verified the identity of the emitter, subsequent experiments were carried out under the same conditions but with the monochromator set to a resolution of 0.1 nm (FWHM) and fixed on the peak found at 593.52 nm. This feature has been assigned[35] to the overlap of the $P_1$ head of the (0,0) band and the $P_2$ head of the (1,1) band, corresponding to emission from v' = 0 and 1. The CL polarization of this CaCl B-X feature was measured for two temperatures of the HCl gas. We find for $T_g = 293 \pm 2$ K

$$P_{CL} = 0.22 \pm 0.01 \tag{10}$$

and for $T_g = 150 \pm 10$ K

$$P_{CL} = 0.25 \pm 0.01 . \tag{11}$$

Similar values of $P_{CL}$ are found throughout the $\Delta v = 0$ sequence. The degree of polarization is positive and sizeable, as anticipated for this kinematically constrained reaction system. Moreover, $P_{CL}$ increases with decreasing gas temperature. We note also that a time of flight spectrum, taken to check for the existence of a translational barrier in the reaction entrance channel, showed no indication of such.[32]

IV. DISCUSSION

The interpretation of vector correlations in beam-gas scattering experiments requires a knowledge of the distribution $f_{LAB}(\hat{k} \cdot \hat{Z})$ of relative velocity directions $\hat{k}$ with respect to the beam direction $\hat{Z}$. We have shown that $f_{LAB}(\hat{k} \cdot \hat{Z})$ can be calculated to any desired degree of accuracy by a Monte Carlo sampling procedure. Moreover, we have found that the second Legendre moment, $\langle P_2(\hat{k} \cdot \hat{Z}) \rangle$, of this distribution can be expressed as a function of the single universal kinematic parameter $\rho = T_g M_b / T_b M_g$.

The values of $\langle P_2(\hat{k} \cdot \hat{Z}) \rangle$ given in Table I and shown in Fig. I are a measure of the width of the angular spread in $f_{LAB}(\hat{k} \cdot \hat{Z})$. It is apparent from these values and from the simulated distributions presented in Fig. 2 that $f_{LAB}(\hat{k}\cdot\hat{Z})$ remains highly anisotropic over the range of $\rho$ considered ($0 \leq \rho \leq 2$).

From the measured degree of chemiluminescence polarization, $P_{CL}$, one determines, using Eq. (5) or (6), the alignment parameter $\alpha = 5 \langle P_2(\hat{J}'\cdot\hat{Z})\rangle$ which is proportional to the second Legendre moment of the distribution $f_{OBS}(\hat{J}'\cdot\hat{Z})$ of product angular momentum directions about the beam axis. Once $\langle P_2(\hat{k}\cdot\hat{Z})\rangle$ is known (from the above Monte Carlo calculations), then the dynamical quantity $\langle P_2(\hat{J}'\cdot\hat{k})\rangle$, i.e., the second Legendre moment of the center-of-mass distribution $f_{CM}(\hat{J}'\cdot\hat{k})$ of product angular momentum directions about the relative velocity vector, can be extracted by dividing $\langle P_2(\hat{J}'\cdot\hat{Z})\rangle$ by $\langle P_2(\hat{k}\cdot\hat{Z})\rangle$. Therefore one can picture $\langle P_2(\hat{k}\cdot\hat{Z})\rangle$ as an effective "resolution" with which $f_{CM}(\hat{J}'\cdot\hat{k})$ can be studied in a beam-gas chemiluminescent reaction.

The polarization values reported in the previous section for the reaction $Ca(^1D) + HCl \rightarrow CaCl(B^2\Sigma^+) + H$ have been inverted using this recipe. Values for $\rho$ are calculated as 0.267 ($T_g$ = 293 K, $T_b$ = 1200 K) and 0.137 ($T_g$ = 150 K, $T_b$ = 1200 K). Table II lists the values of $\langle P_2(\hat{J}'\cdot\hat{k})\rangle$ determined from $P_{CL}$ and $\rho$ in each case. In addition, data are given for the reaction $Ba + NO_2 \rightarrow BaO* + NO$, corresponding to the reported[5] value of $P_{CL}$ = 0.020 ± 0.005 and our estimation of $\rho$ = 0.78.

Table II. Values of $\langle P_2(\hat{J}'\cdot\hat{k})\rangle$ for beam-gas chemiluminescent reactions.

| Emitter | $P_{CL}$ | $\rho$ | $\langle P_2(\hat{J}'\cdot\hat{Z})\rangle$ | $\langle P_2(\hat{k}\cdot\hat{Z})\rangle$ | $\langle P_2(\hat{J}'\cdot\hat{k})\rangle$ |
|---|---|---|---|---|---|
| CaCl(B) | 0.22 ± 0.01 | 0.267 ± 0.005 | −0.316 ± 0.014 | 0.72 ± 0.02 | −0.44 ± 0.02 |
| CaCl(B) | 0.25 ± 0.01 | 0.137 ± 0.009 | −0.364 ± 0.016 | 0.83 ± 0.03 | −0.44 ± 0.03 |
| BaO(A) | 0.02 ± 0.005 | 0.78 ± 0.005 | −0.027 ± 0.008 | 0.48 ± 0.01 | −0.06 ± 0.02 |

It is seen that the two different values of $P_{CL}$ for the $Ca(^1D) + HCl$ reaction obtained at different gas temperatures yield the same value of $\langle P_2(\hat{J}'\cdot\hat{k})\rangle$ within the experimental uncertainty. As expected, the value of $\langle P_2(\hat{J}'\cdot\hat{k})\rangle$ for the kinematically unconstrained reaction $Ba + NO_2$ is much smaller, indicating a more nearly isotropic distribution of $\hat{J}'$ about $\hat{k}$.

The value of $\langle P_2(\hat{J}'\cdot\hat{k})\rangle$ obtained for the $Ca(^1D) + HCl$ chemiluminescent reaction indicates a highly polarized distribution of product angular momentum directions in the center of mass frame. A value of $-0.45$ for $\langle P_2(\hat{J}'\cdot\hat{k})\rangle$ is equivalent to a value of 0.033 for $\langle|\hat{J}'\cdot\hat{k}|^2\rangle$, i.e., the average angle between $\underline{J}'$ and $\underline{k}$ is of the order of 80°. Thus $\underline{J}'$ is confined to a fairly narrow disk perpendicular to $\underline{k}$. Similar results have been obtained by Hsu et al.[12] for the reactions Cs + HI, Cs + HBr, and K + HBr; they obtain values of $\langle P_2(\hat{J}'\cdot\hat{k})\rangle$ = −0.44 ± 0.04, −0.38 ± 0.04, and −0.42 ± 0.06, respectively. Thus the values for Cs + HI and K + HBr agree with one another and with our value for $Ca(^1D) + HCl$ to within experimental errors. The result for Cs + HBr is slightly smaller but in all cases $\langle P_2(\hat{J}'\cdot\hat{k})\rangle$ is close to its limiting value of −0.5. This supports the contention[4,32] that kinematics dominate all four systems.

ACKNOWLEDGMENT

This work is supported by the National Science Foundation under NSF CHE 78-10019.

REFERENCES:

[1]  Levy, M. R., Dynamics of Reactive Collisions, in: Prog. Reaction Kinetics, 10 (1979) 1.

[2]  Smith, I. W. M., Kinetics and Dynamics of Elementary Gas Reactions (Butterworths, London, 1980).

[3]  Bernstein, R. B., Adv. Atomic Mol. Phys. 15 (1979) 167.

[4]  Herschbach, D. R., Disc. Faraday Soc. 33 (1962) 283; Adv. Chem. Phys. 10 (1966) 319.

[5]  Jonah, C. D., Zare, R. N., and Ottinger, Ch., J. Chem. Phys. 56 (1972) 263.

[6]  Rettner, C. T. and Simons, J. P., Chem. Phys. Lett. 59 (1978) 178.

[7]  Rettner, C. T. and Simons, J. P., Faraday Disc. Chem. Soc. 67 (1979) 329.

[8]  Hennessy, R. J. and Simons, J. P., Chem. Phys. Lett. 75 (1980) 43.

[9]  Hennessy, R. J., Ono, Y, and Simons, J. P., Mol. Phys. (in press); Chem. Phys. Lett. 75 (1980) 47.

[10] Maltz, C, Weinstein, N. D., and Herschbach, D. R., Mol. Phys. 24 (1972) 133.

[11] Hsu, D. S. Y. and Herschbach, D. R., Faraday Disc. Chem. Soc. 55 (1973) 116.

[12] Hsu, D. S. Y., Weinstein, N. D., and Herschbach, D. R., Mol. Phys. 29 (1975) 257.

[13] Hsu, D. S. Y., McClelland, G. M., and Herschbach, D. R., J. Chem. Phys. 61 (1974) 4927.

[14] Perry, D. S., Gupta, A., and Zare, R. N., Electro-Optic Laser '80, Proceedings, Industrial and Scientific Conference Management Inc., Chicago, 1981 (to be published).

[15] Case, D. A. and Herschbach, D. R., Mol. Phys. 30 (1975) 1537; J. Chem. Phys. 64 (1976) 4212; J. Chem. Phys. 69 (1978) 150.

[16] Case, D. A., McClelland, G. M., and Herschbach, D. R., Mol. Phys. 35 (1978) 541.

[17] McClelland, G. M. and Herschbach, D. R., J. Phys. Chem. 83 (1979) 1445; Faraday Disc. Chem. Soc. 67 (1979) 360.

[18] Hijazi, N. H. and Polanyi, J. C., J. Chem. Phys. 63 (1975) 2249; Chem. Phys. 11 (1975) 1.

[19] Kafri, A., Shimoni, Y., Levine, R. D., and Alexander, S., Chem. Phys. 13 (1976) 323.

[20] Schulten, K. and Gordon, R. G., J. Chem. Phys. 64 (1976) 2918.

[21] Alexander, M. H., Dagdigian, P. J., and DePristo, A. E., J. Chem. Phys. 66 (1977) 59.

[22] Sinha, M. P., Caldwell, C. D., and Zare, R. N., J. Chem. Phys. 61 (1974) 491.

[23] McCaffery, A. J., Jeyes, S. R., and Rowe, M. D., Ber. Bunsenges 81 (1977) 225.

[24] Drullinger, R. E. and Zare, R. N., J. Chem. Phys. 51 (1969) 5532; J. Chem. Phys. 59 (1973) 4225.

[25] Rettner, C. T., Wöste, L., and Zare, R. N., Chem. Phys. (in press).

[26] Menzinger, M., Adv. Chem. Phys. 42 (1980) 1.

[27] Soleillet, P., Ann. Phys. (Paris) 12 (1929) 23; Perrin, F., J. Phys. Radium 7 (1926) 390; Ann. Phys. 12 (1929) 169.

[28] Weber, G., Adv. Protein Chem. 8 (1953) 415.

[29] Porter, R. N. and Raff, L. M., Classical Trajectory Methods in Molecular Collisions, in: Modern Theoretical Chemistry (Plenum Press, New York, 1976) Vol. 2, pp. 1-52.

[30] Prisant, M. G., Ollison, W. M., and Cross, Jr., R. J., J. Chem. Phys. 69 (1978) 4797.

[31] Prisant, M. G., Ph.D. thesis, in preparation.

[32] Prisant, M. G., Rettner, C. T., and Zare, R. N., J. Chem. Phys. (in press).

[33] Brinkmann, U. and Telle, H., Mol. Phys. 39 (1980) 361; Chem. Phys. Lett. 33 (1980) 530.

[34] Kiang, T., Estler, R. C., and Zare, R. N., J. Chem. Phys. 70 (1979) 5925.

[35] Berg, L. E., Klynning, L., and Martin, H., Physica Scripta 22 (1980) 216.

PHYSICS OF ELECTRONIC AND ATOMIC COLLISIONS
S. Datz (editor)
© North-Holland Publishing Company, 1982

THE EFFECTS OF REAGENT TRANSLATIONAL AND VIBRATIONAL ENERGY
ON THE DYNAMICS OF ENDOTHERMIC REACTIONS

D. Krajnovich, Z. Zhang, F. Huisken, Y. R. Shen
and Y. T. Lee

Materials and Molecular Research Division
Lawrence Berkeley Laboratory
Berkeley, CA 94720

The endothermic reactions $Br + CH_3I \rightarrow CH_3 + IBr$ ($\Delta H_0^O = 13$ kcal/mole) and $Br + CF_3I \rightarrow CF_3 + IBr$ ($\Delta H_0^O = 11$ kcal/mole) have been studied by the crossed molecular beams method. Detailed center-of-mass contour maps of the IBr product flux as a function of recoil velocity and scattering angle are derived. For both systems it is found that the IBr product is sharply backward scattered with respect to the incident Br direction, and that most of the available energy goes into product translation. Vibrational enhancement of the $Br + CF_3I$ reaction was investigated by using the infrared multiphoton absorption process to prepare highly vibrationally excited $CF_3I$. At a collision energy of 31 kcal/mole (several times the barrier height), reagent vibrational energy appears to be less effective than an equivalent amount of (additional) translational energy in promoting reaction. More forward scattered IBr is produced in reactions of Br with vibrationally hot $CF_3I$.

I.   INTRODUCTION

There have not been many direct experimental studies of subtantially endothermic chemical reactions. In the early days, using effusive beam sources, reactive scattering experiments were generally limited to collision energies in the thermal energy range. The development of seeded supersonic beams,[1] however, has largely removed this restriction. For reasonably heavy systems, it is easy to achieve collision energies comparable to, or greater than, typical bond dissociation energies.

In an early application of the seeded beam method to the study of endothermic reactions, Jaffe and Anderson[2] attempted to determine the cross section for the reaction $HI + DI \rightarrow HD + I + I$. For collision energies between 20 and 109 kcal/mole, no HD attributable to reaction was detected ($\sigma_r < 0.04$ Å$^2$) suggesting that internal excitation of one or both reagents is required for this reaction to occur. The application of microscopic reversibility[3] to the detailed rate constants obtained in infrared chemiluminescence studies of exothermic reactions has also been used to suggest that reagent vibrational energy (V') is vastly more effective than reagent translational energy (T') in crossing the endothermic energy barrier. Of course, since the exothermic reactions were typically studied at thermal energies, the results of these microscopic reversibility arguments are only applicable to reaction in the endothermic direction at energies just above threshold. Under these conditions, the preference for V' over T' will at least partially result from the requirement to conserve angular momentum in the collision. That is, for reaction just above threshold, if the reagent energy is present mainly as translation an appreciable fraction of the translational energy will transform into rotational energy of the "collision complex" (except in very small impact parameter collisions), such that the remaining energy is insufficient to overcome the potential energy barrier. For collision energies well in excess of the barrier height, this angular momentum constraint will play a less important role, and it is unclear whether or not V' will retain its overwhelming superiority in most endothermic reactions.

In fact, some endothermic reactions appear to proceed quite readily when the reagent energy is vested almost exclusively in translation. For example, it has been shown[4] that the collision-induced dissociation of alkali halides by rare gas atoms becomes very efficient when the collision energy is substantially higher than the endothermicity. Preliminary results on the endothermic reactions of I atoms with $CH_3Br$,[5] reported at VII ICPEAC, provide additional evidence of the adequacy of translational energy in prómoting endothermic reaction.

It has been recognized from early on[6] that it woud be advantageous to have a direct comparison of the relative effectiveness of V' and T' in promoting substantially endothermic chemical reactions. This problem has been attacked both theoretically and experimentally. Polanyi and coworkers have performed an extensive series[7],[8] of classical trajectory calculations on what are thought to be "typical" endothermic potential energy surfaces. In one of their studies of AB + C reactions,[7] they employed a potential energy surface which was endothermic by 35.7 kcal/mole and varied T' and V' in the range up to T' + V' = 90 kcal/mole. They found an orders-of-magnitude increase in the reaction cross section when energy was transferred from T' to V', until an optimal distribution over T' and V' was achieved with V' >> T'. When V' was close to zero, practically no reactive trajectories were observed. The greater efficiency of V' over T' in promoting endothermic reaction was marked even at the highest reagent energy studied. One other interesting effect they observed is that while the molecular (BC) product was mainly backwards or sideways scattered, enhanced V' at constant T' gave rise to more forward scattering. More recently, Polanyi and Sathyamurthy[8] have shown that on surfaces which exhibit a more gradual rise to the barrier crest (i.e., surfaces with more of the barrier climb in the coordinate of approach), T' may compete much more effectively with V'. On extremely gradual endothermic surfaces they found that the best mix of V' and T' may comprise more than 50% translation.

Experimentally, there have only been three attempts to directly compare the effect of V' and T' at the same total reagent energy. All of these involved reactions of alkali or alkaline earth atoms with hydrogen halide molecules and employed either pulsed chemical lasers or thermal heating to vibrationally excite the HX. Brooks and coworkers[9] studied the approximately thermoneutral K + HCl reaction. They found that when HCl was excited to v = 1, the reaction cross section was around 10 times greater than when the same total reagent energy was supplied as translation. Gupta, Perry and Zare[10] studied the Sr + HF reaction, which is endothermic by 6 kcal/mole. They found that reagent translation was quite effective in promoting this reaction. The reaction cross section for HF(v = 1) was only 1-10 times greater than when the same total energy was supplied as translation. Using thermal heating to excite vibrational motion, Heismann and Loesch[11] found that although vibrational excitation is more effective than translational energy in promoting the K + HCl reaction at lower collision energies, at collision energies higher than 0.5 eV translational energy becomes as effective as vibrational excitation. On the other hand, in the more endothermic K + HF reaction, vibrational excitation was found to be much more effective than translation even at collision energies as high as 1.7 eV.

In this paper we will present some results of crossed molecular beams experiments on the following endothermic reactions:

$$Br + CH_3I \rightarrow CH_3 + IBr \qquad \Delta H_0^o = 13 \text{ kcal/mole}$$

$$Br + CF_3I \rightarrow CF_3 + IBr \qquad \Delta H_0^o = 11 \text{ kcal/mole} \quad .$$

For each reaction, laboratory angular and velocity distributions of the IBr product (measured at a collision energy of approximately 60 kcal/mole) were used to deduce the center-of-mass (CM) product translational energy and angular distributions. The results, besides proving that both reactions proceed readily at enhanced collision energies without the aid of reagent vibration, provide a fairly detailed picture of the reaction dynamics. In addition, studies of vibrational enhancement of the Br + $CF_3I$ reaction were performed by using a $CO_2$ TEA laser to prepare highly vibrationally excited $CF_3I$ molecules via the infrared multiphoton absorption process. These experiments revealed the qualitative effect of reagent vibration on the reaction dynamics, and allowed a comparison to be made of the relative efficiency of reagent translational and vibrational energy in promoting this endothermic reaction.

## II. EXPERIMENTAL

The molecular beam apparatus used in these experiments has been described in detail previously.[12] Briefly, seeded supersonic beams of Br atoms and $CH_3I$ ($CF_3I$) molecules were crossed at right angles in a liquid nitrogen cooled interaction chamber maintained at ~5 x $10^{-7}$ torr. The products were detected in the plane of the atomic and molecular beams by a rotatable, ultra-high vacuum mass spectrometer.

The atomic bromine beam was produced by thermal dissociation of a $Br_2$-rare gas mixture in a resistively heated graphite oven[13] which was operated at about $1500°K$. The hot gases expanded through a 0.18 mm diameter hole in the end of the oven. The $CH_3I$ and $CF_3I$ beams were produced by expanding He- or Ar-seeded mixtures through a 0.13 mm diameter stainless steel nozzle. Both beam sources utilized two stages of differential pumping.

Beam velocity distributions were determined by standard time-of-flight (TOF) measurements using an extended (80 cm) flight path. The $CH_3I$ and $CF_3I$ beams had FWHM velocity spreads of about 15%, while the velocity spread of the Br beam varied between 15 and 25%, depending on the particular conditions. In general, the $CH_3I$ ($CF_3I$) beam velocity was held fixed, and the collision energy, $E_c$, was varied by seeding the bromine in different carried gases or by changing the bromine and carrier gas partial pressures. The maximum Br velocity was obtained by passing 600 torr of He through a $Br_2$ reservoir held at $-30°C$ (bromine partial pressure ≈ 5 torr). This resulted in a peak Br velocity of 2.90 km/s.

The Br + $CH_3I$ → $CH_3$ + IBr and Br + $CF_3I$ → $CF_3$ + IBr reactions were studied in the collision energy range 15–65 kcal/mole. The cross section for both reactions increased by more than two orders of magnitude over this energy range. Laboratory angular and velocity distributions of the IBr product were measured at three collision energies for the Br + $CH_3I$ reaction and at two collision energies for the Br + $CF_3I$ reaction. Here, only the best data for each reaction (obtained at the highest collision energy) will be presented, since this data allowed a direct deconvolution to obtain the CM angular and velocity distributions (see Sec. III).

Angular distributions were obtained by modulating the Br beam with a 150 Hz tuning fork chopper. For each laboratory angle, $\Theta$, data was collected for equal times during the open and closed portions of the tuning fork cycle by a dual channel scaler. The difference of the counts in the "open" and "closed" channels constituted the angular scan signal, $N(\Theta)$. For the experiments reported here, counting times of 1–2 minutes at each angle were sufficient to reduce the error bars (resulting from statistical counting error) to less than 1%. The $IBr^+$ signal rate at the peak of the angular distribution was 4800 counts/s for Br + $CH_3I$ at $E_c$ = 61 kcal/mole and 1300 counts/s for Br + $CF_3I$ at $E_c$ = 65 kcal/mole, with a background count rate of 200 counts/s.

Velocity analysis for the Br + $CH_3I$ reaction was performed by the standard TOF method. The 17.8 cm diameter wheel had four 0.75 mm slots on its perimeter. A 0.58 mm high reducing slit was mounted in front of the first 3 mm x 3 mm detector aperture. The wheel was rotated at 450 Hz, resulting in a shutter opening time of 5.3 μs. Following each wheel opening the TOF distribution, N(t), of reactively scattered IBr was collected by a 255-channel multichannel scaler using a 2 μs dwell time per channel. The distance from the wheel to the detector was 18.0 cm. Counting times were 1-3 hours at each angle.

For the Br + $CF_3I$ reaction, IBr velocity distributions were obtained by the cross-correlation TOF method.[14] Our cross correlation wheel utilizes a 255-channel sequence. The wheel was rotated at 326.8 Hz, corresponding to a dwell time per channel of 12 μs. The distance from the wheel to the detector was 18.3 cm. Counting times were 30-60 minutes at each angle.

The studies of vibrational enhancement of the Br + $CF_3I$ reaction were performed using a high repetition rate $CO_2$ TEA laser (Gentec DD-250). The laser beam crossed the $CF_3I$ beam at 90° inside the second differential pumping chamber of the $CF_3I$ beam source. (I.e., the molecular beam was irradiated after it passed the source skimmer but before it entered the main interaction chamber.) To improve the duty cycle, a cylindrical lens was used to focus the laser beam to a 2 mm high x 21 mm long rectangular spot at the point where it intersected the molecular beam. All of the experiments were performed at a laser frequency of either 1074.6 or 1076.0 $cm^{-1}$ and an energy fluence of 0.5-0.7 $J/cm^2$. The laser was typically operated at a repetition rate of 70 Hz. By pointing the mass spectrometer directly into the $CF_3I$ beam and measuring the amount of depletion of the $CF_3I^+$ signal following each laser pulse, we inferred that approximately 12-15% of the $CF_3I$ molecules in the irradiated zone were being dissociated. Most of the remaining $CF_3I$ molecules are expected to be vibrationally hot.

The reaction of Br atoms with these hot $CF_3I$ molecules was studied at two collision energies, 31 kcal/mole and 11.7 kcal/mole. At each collision energy, TOF distributions of $IBr^+$ were recorded at various lab angles by using a reference pulse from the $CO_2$ laser to trigger the multichannel scaler. The dwell time per channel was 10 μs. The distance from the interaction center to the detector was 21.2 cm. These TOF distributions were then integrated and the background subtracted to obtain the laboratory angular distribution of the laser-correlated IBr product. At $E_c$ = 31 kcal/mole, signal was collected for 6 x $10^5$ laser pulses at each of 10 angles. The maximum $IBr^+$ signal level was 0.018 counts/pulse. At $E_c$ = 11.7 kcal/mole, signal was collected for 9 x $10^5$ laser pulses at each of 13 angles, and the maximum $IBr^+$ signal level was 0.012 counts/pulse. No laser-correlated $CF_3Br$ or BrF product was detected at either collision energy.

III. RESULTS AND ANALYSIS

A.    Br + $CH_3I$ → $CH_3$ + IBr

The laboratory angular distribution of $IBr^+$, obtained at a collision energy of 61 kcal/mole, is shown in Fig. 1 together with the nominal Newton diagram. The circle on the Newton diagram corresponds to the maximum CM velocity of the IBr product allowed by energy conservation. The TOF distributions of $IBr^+$ at 6 lab angles are shown in Fig. 2. The time scale in Fig. 2 has not been corrected for the flight time of the $IBr^+$ ions through the mass spectrometer (about 45 μs).

A deconvolution program[15] was used to directly obtain the CM differential cross section from the experimental results. Briefly, this program proceeds as follows: (i) A laboratory velocity-angle flux map is constructed from the measured angular and TOF distributions. (ii) An iterative ratio method developed by Siska[16] is used to deduce a "monochromatic" laboratory flux map, which is supposed to represent the result of an ideal experiment performed with monoenergetic molecular beams. (iii) This monochromatic LAB flux map is directly inverted to CM coordinates using a single (nominal) Newton diagram. The TOF distributions were smoothed by a 15-point filter before the lab flux map was constructed. Also, the collision energy dependence of the reaction cross section was taken into account when weighting the various Newton diagrams for the deconvolution in step (ii).

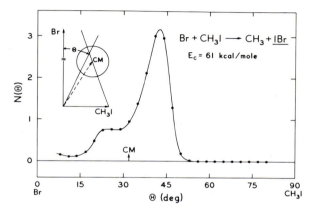

Fig. 1. Laboratory angular distribution of the IBr product from the Br + $CH_3I$ reaction at 61 kcal/mole collision energy. The curve is drawn free-hand through the points.

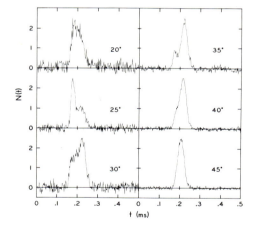

Fig. 2. Time-of-flight analysis of the IBr product from the Br + $CH_3I$ reaction at 61 kcal/mole.

A contour map of the resulting CM differential cross section in velocity space, $d^2\sigma/dud\omega$, as a function of IBr scattering angle, $\theta$, and recoil velocity, u, is shown in Fig. 3, superimposed on the nominal Newton diagram. (Note that $d\omega$ represents the unit of solid angle in CM coordinates, and that $\theta = 0°$ is defined to be the direction of the incident Br atom.) It is usually more meaningful to talk about the differential cross section in energy space, $d^2\sigma/dE_T^\dagger d\omega$, where $E_T^\dagger$ is the total CM translational energy of both products. The computer program also calculates this quantity at various IBr scattering angles. The results suggest that there is strong coupling of the product energy and angular distributions. The translational energy distribution broadens and shifts to lower energies as the IBr scattering angle decreases. In Fig. 4 are shown the average CM translational energy distribution (obtained by averaging $d^2\sigma/dE_T^\dagger d\omega$ over all solid angles) and the CM angular distribution (obtained by integrating $d^2\sigma/dE_T^\dagger d\omega$ over $E_T^\dagger$ at each scattering angle $\theta$). These distributions are regarded as tentative, especially in view of the fact that the translational energy distribution is not entirely successful at conserving energy! The high energy tail which extends beyond the maximum available energy must be an artifact of the deconvolution procedure. This tail contains about 10% of the total product flux. Nevertheless, for the energy distribution shown, the mean product translational energy is 31 kcal/mole, or 64% of the available energy. The true values probably lie slightly lower. By applying the sine weighting factor to $d\sigma/d\omega$ (to account for out-of-plane scattering), we calculate that 80% of the IBr product is scattered into the backward hemisphere.

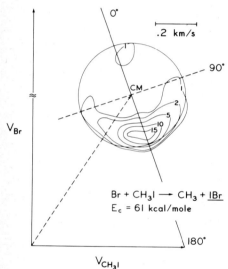

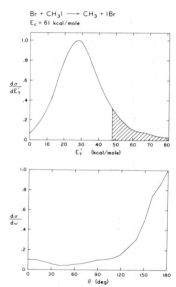

Fig. 3. Center-of-mass IBr product flux contour map for the Br + $CH_3I$ reaction at 61 kcal/mole, obtained by direct deconvolution of the data in Figs. 1 and 2.

Fig. 4. Center-of-mass product translational energy and angular distributions for the Br + $CH_3I$ reaction at 61 kcal/mole, obtained by direct deconvolution. The shaded portion of the translational energy distribution is energetically forbidden.

B.    Br + $CF_3I$ → $CF_3$ + IBr

Laboratory angular and TOF distributions of $IBr^+$, obtained at a collision energy of 65 kcal/mole, are shown in Figs. 5 and 6. TOF distributions were also measured at $\Theta = 40°$ and $50°$, but these are not shown. The time scale of Fig. 6 has not been corrected for ion flight time. These data were input to the deconvolution program, after smoothing all of the TOF distributions with a 5-point filter. The calculated contour map of the CM differential cross section in velocity space is shown in Fig. 7. The angle-averaged CM translational energy distribution and the CM angular distribution are shown in Fig. 8. (Since the total CM flux could not be sampled in the laboratory, only the data on the "far" side of the relative velocity vector was used to calculate $d\sigma/dE_T^{\dagger}$ and $d\sigma/d\omega$.) The mean product translational energy is 35 kcal/mole, or 65% of the available energy. The CM angular distributions falls off more gradually here than for the $Br + CH_3I$ reaction. Also, there is no relative maximum in $d\sigma/d\omega$ at $\theta = 0°$. About 74% of the IBr product is scattered into the backward hemisphere.

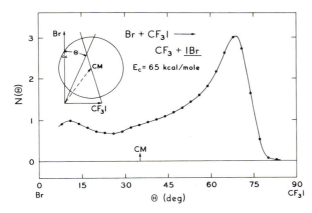

Fig. 5. Laboratory angular distribution of the IBr product from the $Br + CF_3I$ reaction at 65 kcal/mole. The curve is drawn free-hand through the points.

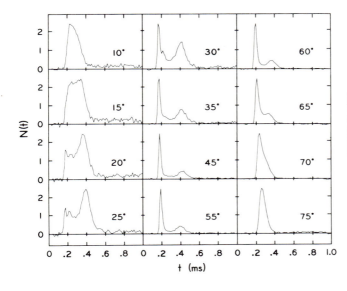

Fig. 6. Time-of-flight analysis of the IBr product fom the $Br + CF_3I$ reaction at 65 kcal/mole.

$$Br + CF_3I \longrightarrow CF_3 + \underline{IBr}$$
$$E_c = 65 \text{ kcal/mole}$$

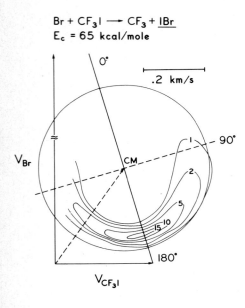

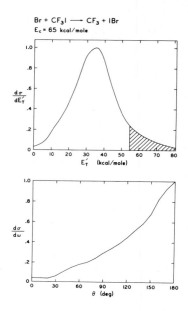

Fig. 7. Center-of-mass IBr product flux contour map for the Br + CF3I reaction at 65 kcal/mole, obtained by direct deconvolution of the data in Figs. 5 and 6.

Fig. 8. Center-of-mass product translational energy and angular distributions for the Br + CF3I reaction at 65 kcal/mole, obtained by direct deconvolution. The shaded portion of the translational energy distribution is energetically forbidden.

C.    Br + $CF_3I^{\dagger}$ → $CF_3$ + IBr

The angular distribution of laser–correlated IBr product obtained at a collision energy of 31 kcal/mole is shown in Fig. 9. This signal is a result of reactive scattering events between Br atoms and vibrationally hot CF3I molecules produced by the infrared multiphoton absorption process. For comparison, the angular distribution of IBr obtained at the same collision energy but without the laser is also shown. As was the case at the higher collision energy, most of the IBr produced in collisions of Br atoms with cold CF3I molecules is scattered backwards. However, Fig. 9 shows that there is much more forward scattered IBr produced in the reactions with vibrationally hot CF3I. Fig. 10(a) shows the TOF distribution of laser–correlated IBr measured at Θ = 60°, which is close to the CM angle. Apart from a minor offset, t = 0 corresponds to the firing of the laser. The solid line in Fig. 10(a) is just a 5–point smooth of the raw data. Fig. 10(b) shows the corresponding TOF distribution of IBr from the "cold" reaction as obtained by cross–correlation TOF. (Actually, this data was taken at a slightly lower collision energy, $E_c$ = 27 kcal/mole, but at a lab angle of Θ = 50°, which was near the CM angle for this experiment. Therefore, to the qualitative extent intended, Figs. 10(a) and (b) may be directly compared.) The similarity between the two TOF distributions suggests that the product translational energy distribution is not drastically altered when the CF3I reagent is vibrationally excited.

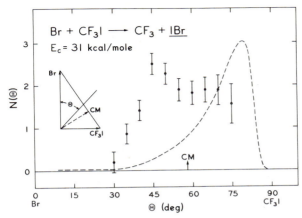

Fig. 9. ● Laboratory angular distribution of IBr produced in collisions of Br with vibrationally hot $CF_3I$. Error bars represent one standard deviation. --- Laboratory angular distribution of IBr produced in collisions of Br with vibrationally cold $CF_3I$. Both distributions were obtained at a collision energy of 31 kcal/mole.

At the peak of the laser-on angular distribution, the $IBr^+$ signal rate was 0.018 counts/laser pulse. In order to compare the signal levels from the "hot" and "cold" $CF_3I$, we must correct for the duty cycle of the laser excitation. Each laser pulse excites a 2.1 cm long segment of the $CF_3I$ molecular beam, and the $CF_3I$ beam velocity in this experiment was 11.7 x $10^4$ cm/s. Therefore, the signal rate from the "hot" $CF_3I$ may be calculated to be be

$$\text{IBr signal rate (laser on)} = \frac{0.018 \text{ counts/pulse}}{2.1 \text{ cm/pulse}} \times 11.7 \times 10^4 \text{ cm/s}$$

$$= 1000 \text{ counts/s.}$$

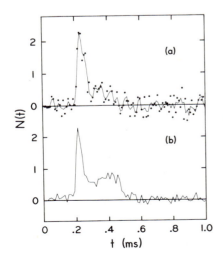

Fig. 10. (a) Time-of-flight distribution of laser-correlated IBr product from the Br + $CF_3I$ reaction at a collision energy of 31 kcal/mole. This distribution was measured at $\Theta$ = 60°. The solid curve is a 5-point smooth of the measured points. (b) Time-of-flight distribution of IBr product from the Br + $CF_3I$ reaction at 27 kcal/mole. This distribution was measured at $\Theta$ = 50° by the cross-correlation method.

At the peak of the laser-off angular distribution, the IBr signal rate (after correcting for the duty cycle of the tuning fork chopper) was 2400 counts/s. The 1000 counts/s enhancement due to vibrational excitation is quite small; an equivalent enhancement could be achieved by merely increasing the collision energy by 2 or 3 kcal/mole. Due to our imprecise knowledge of the $CF_3I$ vibrational population distribution created by the $CO_2$ laser pulse, we cannot make an accurate quantitative statement concerning the relative effectiveness of reagent vibrational vs. translational energy in promoting this endothermic reaction. However, we can still make some reasonable estimates. We know that 12–15% of the $CF_3I$ molecules are being dissociated by the laser pulse. It seems likely that the average vibrational energy content of the remaining $CF_3I$ molecules is around 30–40 kcal/mole. Even if we consider only that fraction of the total vibrational energy which lies in the reaction coordinate, i.e., the C–I bond, it still appears that, at this collision energy (which is ~3 times the endothermic barrier height), reagent vibrational energy is less effective than an equivalent amount of additional reagent translational energy in promoting reaction.

The angular distribution of laser-correlated IBr product obtained at a collision energy of 11.7 kcal/mole is shown in Fig. 11. (The laser conditions were unchanged from the previous experiment, so the $CF_3I$ vibrational energy distribution should be identical.) For this experiment, the peak IBr signal rate was 0.012 counts/pulse and the $CF_3I$ beam velocity was $6.5 \times 10^4$ cm/s. The IBr signal rate was calculated as before:

$$\text{IBr signal rate (laser on)} = \frac{0.012 \text{ counts/pulse}}{2.1 \text{ cm/pulse}} \times 6.5 \times 10^4 \text{ cm/s}$$

$$= 370 \text{ counts/s.}$$

Since the collision energy was essentially equal to the barrier height in this experiment, there was almost no IBr signal with the laser off. We measured $45 \pm 11$ counts/s at the CM angle. To achieve a signal rate of 370 counts/s with translational energy alone, the collision energy would have to be increased from 11.7 kcal/mole to around 20 kcal/mole. Therefore, vibration competes much more effectively with translation when the collision energy is comparable to the barrier height.

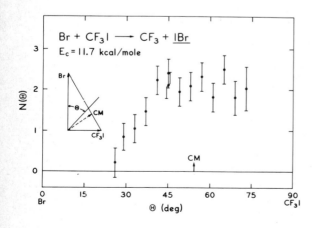

Fig. 11. Laboratory angular distribution of laser-correlated IBr product from the Br + $CF_3I$ reaction at a collision energy of 11.7 kcal/mole.

## IV.  SUMMARY AND DISCUSSION

The dynamics of the Br + CH$_3$I and Br + CF$_3$I reactions appear to be quite similar.  In both cases:

(1)  Reagent translational energy alone can adequately promote reaction.

(2)  The product energy and angular distributions are coupled.  This is expected to be the rule rather than the exception in direct reactions.  On the average, the products carry away about 2/3 of the available energy as translation.

(3)  The IBr product is mainly backward scattered with respect to the incident Br direction; about 3/4 of the IBr scatters into the backward hemisphere, and the IBr angular distribution peaks at $\theta$ = 180°.

Although the percentage of IBr scattered into the forward and backward hemispheres is about the same for both reactions, the shapes of the CM angular distributions are rather different (see Figs. 4 and 8).  The angular distribution of IBr from Br + CF$_3$I falls off rather gradually and monotonically on moving frm $\theta$ = 180° to $\theta$ = 0°.  In contrast, the IBr product from Br + CH$_3$I is more strongly peaked on the relative velocity vector, and there is even a relative maximum in d$\sigma$/d$\omega$ at $\theta$ = 0°.  This small concentration of flux in the forward direction also appears in the contour map (Fig. 3).  The fast peak in the TOF distribution at $\theta$ = 25° (see Fig. 2) is mainly responsible for this feature.  Sharp product peaking on the relative velocity vector suggests[17] that there may be significant polarization (i.e., parallel or antiparallel alignment) of the initial and final orbital angular momenta, $\vec{L}$ and $\vec{L}'$, even though with CH$_3$ departing L' must be rather small.  If this interpretation is correct, then it implies that there is a stronger preference for coplanar reaction geometries in the Br + CH$_3$I reaction (treating CH$_3$ as a point) than in the Br + CF$_3$I reaction.  Similar orbital angular momentum polarization effects have been noted in the Li + HF[18] and F + D$_2$[19] reaction systems, which also have very light leaving groups.

The experiments on the vibrational enhancement of the Br + CF$_3$I reaction reveal that:

(1)  Much more (around 50%) of the IBr product is forward scattered when Br reacts with vibrationally hot CF$_3$I.  This trend has also been observed in trajectory calculations.[7]  A large amount of vibrational energy in the bond under attack may allow larger impact parameter collisions to react, giving rise to more forward scattering.

(2)  The product translational energy distribution is not greatly perturbed by the addition of large amounts of reagent vibrational energy; most of the reagent vibrational energy is retained as vibrational energy in the products.

(3)  When the collision energy is several times the barrier height, reagent vibrational energy appears to be less effective than an equivalent amount of additional translational energy in promoting reaction.  This suggests that the reaction proceeds on an extremely gradual endothermic potential energy surface,[8] with most of the barrier lying along the coordinate of approach.

(4)  When the collision energy is comparable to the barrier height, vibrational energy in the bond under attack is more effective than additional translational energy in crossing the endothermic barrier.  This preference for V' over T' at low collision energies results, at least in part, from simple angular momentum constraints.

ACKNOWLEDGMENTS

This work was supported by the Assistant Secretary for Nuclear Energy, Office of Advanced Systems and Nuclear Projects, Advanced Isotope Separation Division, U.S. Department of Energy under Contract No. W-7405-Eng-48.

REFERENCES

[1]  Abuaf, N., Anderson, J.B., Andres, R.P., Fenn, J.B. and Marsden, D.G.H., Science 155 (1967) 997-999.
[2]  Jaffe, S.B. and Anderson, J.B., J. Chem. Phys. 49 (1968) 2859-2860; J. Chem. Phys. 51 (1969) 1057-1064.
[3]  Anlauf, K.G., Maylotte, D.H., Polanyi, J.C. and Bernstein, R.B., J. Chem. Phys. 51 (1969) 5716-5717; Polanyi, J.C. and Tardy, D.C., J. Chem. Phys. 51 (1969) 5717-5719; Perry, D.S., Polanyi, J.C. and Wilson, Jr., C.W., Chem. Phys. Lett. 24 (1974) 484-487.
[4]  Tully, F.P., Lee, Y.T. and Berry, R.S., Chem. Phys. Lett. 9 (1971) 80-84; Tully, F.P., Cheung, N.H., Haberland, H. and Lee, Y.T., J. Chem. Phys. 73 (1980) 4460-4475.
[5]  Lee, Y.T., in: Branscomb, L.M., and others (eds.), Physics of Electronic and Atomic Collisions, VII ICPEAC, 1971 (North-Holland, Amsterdam, 1972).
[6]  Douglas, D.J., Polanyi, J.C. and Sloan, J.J., Far. Disc. Chem. Soc. 55 (1973) 310-311.
[7]  Perry, D.S., Polanyi, J.C. and Wilson, Jr., C.W., Chem. Phys. 3 (1974) 317-331.
[8]  Polanyi, J.C. and Sathyamurthy, N., Chem. Phys. 33 (1978) 287-303; Chem. Phys. 37 (1979) 259-264.
[9]  Odiorne, T.J., Brooks, P.R. and Kasper, J.V.V., J. Chem. Phys. 55 (1971) 1980-1982; Pruett, J.G., Grabiner, F.R. and Brooks, P.R., J. Chem. Phys. 63 (1975) 1173-1183.
[10] Gupta, A., Perry, D.S. and Zare, R.N., J. Chem. Phys. 72 (1980) 6250-6257.
[11] Heismann, F. and Loesch, H.J., submitted to Chem. Phys.
[12] Lee, Y.T., McDonald, J.D., LeBreton, P.R. and Herschbach, D.R., Rev. Sci. Instrum. 40 (1969) 1402-1408.
[13] Valentini, J.J., Coggiola, M.J. and Lee, Y.T., Rev. Sci. Instrum. 48 (1977) 58-63.
[14] Hirschy, V.L. and Aldridge, J.P., Rev. Sci. Instrum. 42 (1971) 381-383.
[15] Valentini, J.J., Ph.D. Thesis, University of California, Berkeley, 1976.
[16] Siska, P.E., J. Chem. Phys. 59 (1973) 6052-6060.
[17] Miller, W.B., Safron, S.A. and Herschbach, D.R., Far. Disc. Chem. Soc. 44, (1967) 108-122.
[18] Becker, C.H., Casavecchia, P., Tiedemann, P.W., Valentini, J.J. and Lee, Y.T., J. Chem. Phys. 73 (1980) 2833-2850.
[19] Sparks, R.K., Hayden, C.C., Shobatake, K., Neumark, D.M. and Lee, Y.T., in: Fukui, K. and Pullman, B. (eds.), Horizons of Quantum Chemistry (D. Reidel Publishing Co., Dordrecht, 1980).

PHYSICS OF ELECTRONIC AND ATOMIC COLLISIONS
S. Datz (editor)
© North-Holland Publishing Company, 1982

STATE RESOLVED PRODUCTS FROM HIGH
REAGENT VIBRATIONAL LEVELS

Afranio Torres-Filho and J. Gary Pruett[*]

Department of Chemistry
University of Pennsylvania
Philadelphia, Pennsylvania
U.S.A.

INTRODUCTION

Reactions of vibrationally excited molecules have now been studied
for over ten years. Beginning with the work of Odiorne, Brooks, and
Kasper in 1970, infrared chemical lasers have been the primary source
of vibrational excitation radiation.[1] The use of these fixed line
resonant lasers follows logically from their initial use as
excitation sources in infrared fluorescence studies of hydrogen
halide vibrational relaxation.[2] Their obvious advantages are a
resonant match with the hydrogen halide absorption frequency and
easily generated high spectral brightness. The disadvantage of
these sources is their propensity for poor lasing on low, thermally
populated, rotational lines, and relatively poor lasing on fundamental rather than hot band vibrational transitions. The net result
has been that all use of such lasers has been exclusively to excite
v=1 of hydrogen halide reagents, primarily HF and HCl. Even with
this resonant excitation, only a small fraction of the thermally
populated v=0 molecules are excited. Other methods of generating
vibrationally excited reagents have been tried with some success.
Polanyi et al. have used chemical generation of vibrationally
excited hydrogen halides in their chemical depletion experiments,[3]
and $CO_2$ lasers have been used to excite a few non-halide reagents.
All of these studies have motivated the scientific community with
the desire of studying reaction rates as a function of reagent
vibration.

The combination of I.R. laser excitation of reagents with visible
laser detection of vibrationally state resolved products began in
the laboratory of Dr. R. N. Zare with the study of the exothermic
reaction Ba + HF (v=1) → BaF (v=0-12) + H.[4] Karney, Estler, and
Zare[5] and Gupta, Perry, and Zare,[6] have expanded these studies to
include other reagents and to begin studying orientation effects
using polarized laser sources. We have developed an apparatus
particularly suited for these kinds of experiments which allows
operation of the molecular beam reaction in the chemical laser
cavity, and sets up an accurate dual beam sample to allow subtraction of v=0 reagent contributions to the observed products. Use of
this apparatus initially allowed state-to-state determination of
rates of product formation from v=1 reagents into levels dominantly
populated by v=0 reagents.[7]

In all of these studies, excited reagents have been limited to the
v=1 level because of the fundamental line excitation schemes used.

For some reactions which are endothermic at v=0 but exothermic from
v=1, this results in a large change in reaction probability, but
only gives state-to-state information from one reagent state (v=1).
For reactions which are exothermic at v=0, excitation to v=1 allows
study of two reagent vibrational levels, but requires the accurate
handling of subtraction methods.  For some reactions, even v=1 is
non-reactive, and requires reagents at v=2 or higher to proceed at
measurable rates.  In principle, it seems desirable to study as many
reactive vibrational levels as possible.

Excitation to higher vibrational levels of the hydrogen halides is
not new to those studying energy transfer collisions in bulk phases.
Osgood and Javan excited HF up to v=3 using multiple resonance
techniques,[8] and even higher vibrational levels are accessible using
over tone transitions pumped by high power dye lasers.[9]  We have
recently incorporated a variation on the multiple resonance technique
to excite sufficient HF molecules to v=2 to allow state resolved
products to be observed under molecular beam conditions.

EXPERIMENTAL

Figure 1, reproduced from reference 7 shows the basic apparatus used
in these experiments.

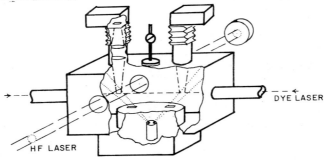

Figure 1.  Dual beam intracavity apparatus.

We utilize a dual beam + gas arrangement in which two identical
reaction zones are created by forming two metal atom beams from a
single, resistively heated and differentially pumped oven.  Each
beam enters a common reagent gas scattering chamber which is typically operated at $5-10 \times 10^{-5}$ torr reagent gas pressure.  The reaction
zones are separated by 25cm as the beams enter the scattering
chamber.  One reacting zone is irradiated intracavity by the HX
chemical laser, pulsed at 10hz with a pulse duration of .5 - 1.0
μsec.  After a 10-20 μsec delay, a $N_2$ laser pumped tunable dye laser
is fired to excite electronic fluorescence in the metal monohalide
reaction products.  The dye laser pulse (10 nsec duration) passes
through both reaction zones, using the "same" photons to excite
both product samples.  Because of the angles between the metal atom
beams, reagents and products interpret the dye laser frequency
differently due to the Doppler shift, so the probe beam must be
equally split and counterpropagated through the samples to avoid a
Doppler-induced difference between the excitation spectra.  Earlier
published results obtained with this system show its ability to
extract small effects in the product distribution due to the small
fraction of excited reagents.[7]

Excitation of reagent molecules to v=2 is accomplished by arranging the HF laser cavity to be resonant on a selected fundamental line as well as a hot band (v=2 → v=1) line whose lower rotational level is the upper level of the fundamental line. This is accomplished as shown in Figure 2.

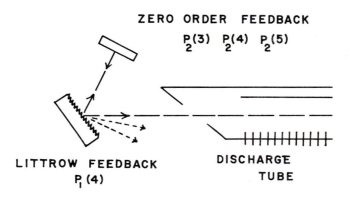

Figure 2. Multiple resonance feedback laser.

The grating at one end of the laser cavity is fixed, as before, to Littrow reflect a selected fundamental ($P_1(4)$) line whose lower level is significantly populated in the room temperature reaction chamber. Infrared fluorescence on all insipient fundamental and hot-band lines is reflected at zero order of the grating to a plane mirror, and returned through zero order back into the cavity. The high gain of the hot-band transitions allows the strongest $P_2(3)$, $P_2(4)$ and $P_2(5)$ lines to lase as seen in Figure 3.

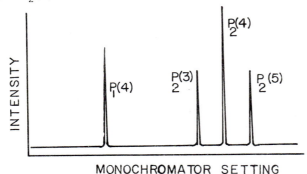

Figure 3. Dispersed laser emission from IR pump laser.

Hot-band lasing actually reaches threshold earlier than the
fundamental lasing, but the two occur simultaneously for a significant portion of the lasing duration, allowing double resonance
excitation to v=2, J=2 from v=0, J=4.

In a typical experiment, a dual beam v=0 and v=1 reaction pair is
run to allow subtraction of the v=0 reagent contribution, if any.
Then the hot-band feedback is introduced and another excitation
spectrum is obtained.  Because the products now result from v=0,
v=1, and v=2 reagents, the subtraction problem is highly compounded.
The essential piece of information, which can only be inferred from
saturation limits and previous results, is the fraction of reagent
molecules in v=0, 1 and 2.  Reasonable estimates for the fractions
in HF are 94% v=0, 4% v=1 and 2% v=2.  These are close to the
saturation limits, but could be in error by ± 50% of their respective
values.

Figure 4 shows on energy level diagram for the HF reactions indicating
accessible product vibrational levels from HF v=0, 1, and 2.

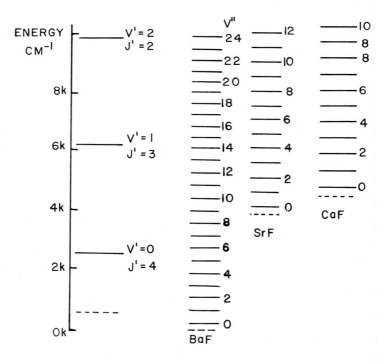

Figure 4.   Energy level diagram of reagents and products.


Ba + HF (v=0, 1, 2)

Figure 5 shows three excitation spectra, offset vertically to avoid
congestion, of the $^2\pi_{\frac{1}{2}}$ - $^2\Sigma^+$ system in BaF.  The lowest trace occurs

with the v=0 "blank" reaction zone and shows rotationally unresolved
Δv=0 bands with pronounced Q branch peaks above P and R backgrounds.
The height of the peaks above background is used to infer relative
vibrational level populations and thus relative state-to-state
formation cross sections.

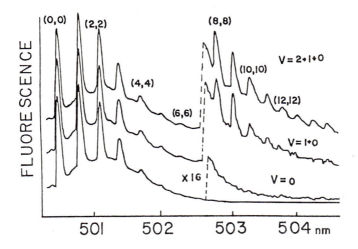

Figure 5.   Excitation spectra of BaF products.

For these bands, all Franck-Condon factors are nearly equal so that
relative cross sections are closely displayed in the raw data.   The
second trace shows the combined v=0 + 1 reaction zone excitation
spectrum, which clearly exhibits increased product vibration, while
still dominated by the predominant v=0 reagents.   Finally with the
infrared double resonance excitation, the upper figure shows the
products from a combined v=0, 1, and 2 reagent population.   Additional vibrational peaks occur beyond the figure out to BaF (v=20).

Adjusting the signals to account for the reagent excitation fraction
allows extraction of state-to-state relative cross sections for all
reagent and product vibrational levels.   These relative cross
sections are subject to interdependent error relations which tend
to cause groups of cross sections to rise or fall while keeping
relative values within those groups fairly stable.   Detailed accounts
of these effects can be seen elsewhere.[10,11]   It is clear that all
products are predominantly formed in low reagent vibrational levels
with population extending to nearly the exothermic limit.   Figure 6
displays the resulting relative state-to-state reaction cross
sections.

## Sr + HF (v=0, 1, 2)

As seen in Figure 4b, the reaction of Sr with HF (v=0) is endothermic
by over $1000cm^{-1}$.   This probably allows a small amount of reaction
to occur via the high kinetic energy tail of the Boltzman velocity
distribution, but this reaction is quite minor.   From v=1 however,
SrF may be formed out to v=5 exothermically, and from HF (v=2)
products out to v=12 may be produced.   Figure 7 shows three

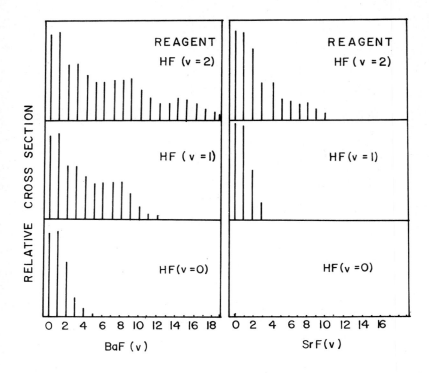

Figure 6.   State-to-state cross sections.

excitation spectra of the SrF $A^2\pi_{3/2}$ - $X^2\Sigma^+$ band system.   Again,
only $\Delta v=0$ bands are observed and $Q$-type heads are prominant above
P + R type backgrounds.   Reaction from v=0 is observable but very
weak.   Excitation to v=1 results in a product distribution first
reported by Karney et al.[5]   Double resonance excitation to v=2
shows clear evidence of increased product energy out to v=10.   In
this experiment, since v=0 does not react, all observed products
are from either v=1 or v=2 reagents.   If the probe laser is fixed
on any product peak and the double resonance switched off, leaving
only v=1 states reactive, the signal always decreases.   This could
be due to a reduced net population of excited states of equal
reactivity and/or a loss of more reactive v=2 reagents.   Assuming
the hot band excitation nears saturation allows separation of the
effects.   Figure 6b displays the resulting relative cross sections.
Again we see that production of low product vibrational levels
dominates the process, giving a small product vibrational energy
release, but with products populated to nearly the exothermic limit.

Ca + HF (v=0, 1, 2)

Figure 5c reveals that Ca + HF (v=0) is also endothermic but that

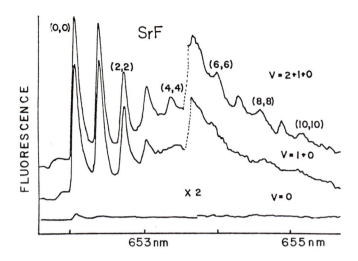

Figure 7.   Excitation spectra of SrF.

v=1 and v=2 are exothermic.  Figure 8 displays the excitation spectra of the CaF $A^2\pi_{\frac{1}{2}}$ - $X^2\Sigma^+$ system.

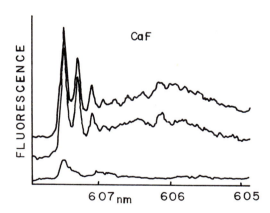

Figure 8.   Excitation spectra of CaF.

Again very little v=0 reaction is observed and v=1 is reactive.  The excitation spectrum of the double resonance pumped experiment shows some evidence of additional product state populations but signal-to-noise limitations preclude meaningful interpretation.

CONCLUSIONS

The state-to-state resolution of these experiments can be averaged
to obtain overall fractional energy release figures for the reactions
studied.  For Ba + HF, of the total energy available to products,
v=0 reagents give 21% product vibrational energy release, v=1
reagents give 30%, and v=2 reagents drop back down to 28% vibrational
energy release.  If one considers vibrational energy adiabaticity,
then 35% of the additional vibration, while only 25% of the incremental vibrational increase in HF (v=2) is carried to additional
product vibration.  In addition the total reaction cross section
increases by factors of 2 and 0.25 going from v=0 to v=1 then v=2.
Since the value of the rotational quantum number J is also changing
in these experiments, the change in total cross section cannot be
described intirely to the change in v.[10]

Similar effects are seen in the Sr + HF results.  For v=1 reagents
only 22% of the exothermicity appears as product vibration and for
v=2 reagents again only 23% appears as product vibration while the
total cross section increases only modestly by about 50%.  It must
be emphasized that these results are quite crude, since we have no
direct method of determining precise excitation fractions.  The
situation will be simplified somewhat when high powered infrared
dye lasers allow direct pumping of overtone levels to directly access
higher vibrational states.  With present experimental techniques we
can distinguish results from excitations of as little as 0.5% of the
reagent gas.  The future of these experiments is surely to include
jet cooled polyatonic reagents which can be directly excited by
tunable laser sources to a variety of isoenergetic but dynamically
different reagent vibrations.

ACKNOWLEDGEMENTS

This research was supported by the National Science Foundation under
NSF CHE 76-11468.  Part of the optical support of this work was
provided through the University of Pennsylvania, Regional Laser
Laboratory, NSF CHE 78-18719.

REFERENCES

1.  T. J. Odiorne, P. R. Brooks, and J. V. V. Kasper, J. Chem. Phys.,
    55, 1980 (1971).

2.  H. L. Chen, J. C. Stephenson, and C. B. Moore, Chem. Phys., Lett.
    2, 593 (1963).

3.  D. J. Douglas, J. C. Polanvi, and J. J. Sloan, J. Chem. Phys.,
    59, 6679 (1973).

4.  J. G. Pruett and R. N. Zare, J. Chem. Phys., 64, 1774 (1976).

5.  Z. Karny, R. C. Estler, and R. N. Zare, J. Chem. Phys., 69, 5199
    (1978).

6.  A. Gupta, D. S. Perry, and R. N. Zare, J. Chem. Phys., 72, 6250
    (1980).

7.  A. Torres-Filho and J. Gary Pruett, J. Chem. Phys., 72, 6736
    (1980).

8. R. M. Osgood, Jr., P. B. Sackett, and A. Javan, J. Chem. Phys., 60, 1464 (1974).

9. G. M. Jursich and F. F. Crim, J. Chem. Phys., 74, 4455 (1981).

10. A. Torres-Filho, Ph.D. Thesis, University of Pennsylvania 1981.

11. A. Torres-Filho and J. Gary Pruett, J. Chem. Phys. (accepted for publication).

PHYSICS OF ELECTRONIC AND ATOMIC COLLISIONS
S. Datz (editor)
© North-Holland Publishing Company, 1982

ATOMIC PHYSICS ISSUES IN FUSION

D. E. Post


Princeton University, Plasma Physics Laboratory
Princeton, New Jersey 08544
U.S.A.

Atomic physics issues have played a large role in controlled
fusion research.  A general introduction to the present role
of atomic processes in both inertial and magnetic controlled
fusion work is presented.

## I.  INTRODUCTION

Atomic collision processes have played a large role in both magnetic and inertial
confinement fusion research.   There are four primary areas of fusion plasma
research where atomic processes are important: (1) the hot central plasma where
the fusion reactions occur, (2) the plasma edge where the plasma interacts with
the external environment, (3) plasma heating methods used to initially produce
the hot plasma, and (4) diagnostic techniques used to measure the physical
properties of the plasma.  In magnetic fusion research, atomic processes such as
line radiation and charge-exchange have contributed significantly to the energy
and particle balance in the hot center of the plasma.  Atomic processes involving
low $Z$ ions, hydrogen and non-hydrogen molecules, and ion-solid collisions are
important at the plasma edge.  Neutral beam heating, which involves much atomic
physics, has been crucial to the success of the tokamak and mirror programs.
More than half of the techniques used to diagnose magnetic fusion experiments
rely on electron and ion collision processes.

The same is true of inertial confinement research.  Equation of state and opacity
issues are crucial to designing fusion target pellets and to understanding their
behavior.  The absorption and stopping of laser light and energetic ions in the
outer layers of an imploding pellet, the design and construction of high energy
lasers and ion beams, and the use of spectroscopy and laser scattering to
diagnose pellet implosions all depend on atomic collisions of one type or
another.

This symposium will explore each of these areas.  It is useful, however, to first
describe the conditions (electron and ion temperature, density, etc.) that
characterize inertially and magnetically confined plasmas.  These conditions can
be derived from a few simple assumptions.

Most currently envisioned fusion schemes rely on the nuclear reaction $D + T \rightarrow n +$
$He + 17.5$ MeV.  The neutron has $\sim 14$ MeV of kinetic energy.  The mean free path
for neutrons is much larger than the dimensions of commonly discussed fusion
plasmas, and thus, the neutron energy is absorbed in a "blanket" structure
surrounding the fusion chamber.  The alpha particle is charged and is confined
and used to heat the plasma directly, that is to resupply the heat losses of the
fusing plasma.  The reaction rates of D-T fusion [1] are large only for $T \gtrsim 10$
keV (Fig. 1).  This sets a lower limit on the temperature to be expected at the
center of a fusion plasma.  The requirement that the alpha particles deposit
their energy in the plasma, and thus keep the reaction going against losses

introduces a second condition, first popularized by Lawson [2]. A simple energy balance equation for a DT plasma would be

$$\frac{3}{2} k \frac{\partial(\sum_i n_i T_i)}{\partial t} \approx \frac{3}{2} k \frac{(n_D T_D + n_T T_T + n_e T_e)}{\tau_E} \lesssim n_D n_T \langle \sigma v \rangle_{DT} E_\alpha$$

where the sum is over the different ion species and the electrons, $\tau_E$ is the energy confinement time, $n_i$ and $T_i$ are the density and temperature of plasma species i, $\langle \sigma v \rangle_{DT}$ is the fusion reaction rate of DT (Fig. 1), and $E_\alpha$ is the kinetic energy of the alpha particle (3.5 MeV). Assuming that $n_D = n_T = (n_e/2)$ and that $T_D = T_T = T_e$, one obtains the condition $n_e \tau_e \gtrsim (12kT/\langle \sigma v \rangle E_\alpha)$ for ignition. $n_e \tau_e$ is a strong function of temperature (Fig. 2), and has a minimum at $T \sim 25$ keV, with $n_e \tau_e \gtrsim 1.5 \times 10^{14}$ sec/cm$^3$. Thus, the goal of controlled fusion research is the attainment of $T \gtrsim 10$ keV and $n_e \tau_e \gtrsim 10^{14}$ sec/cm$^3$.

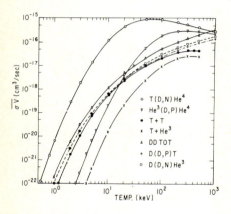

Figure 1

Reaction rates for a variety of fusion reactors averaged over a Maxwellian ion temperature distribution [1].(762346)

Figure 2

$n_e \tau_E$, the product of the plasma density and confinement time, as a function of temperature.(81P0127)

## II. MAGNETIC CONFINEMENT ATOMC PHYSICS ISSUES

Since a temperature of 10 keV is far hotter than can be tolerated by any material, the hot plasma must be kept out of contact with any containment vessel. One approach is use the pressure of high strength magnetic fields (1-10 Tesla) to confine the plasma. The plasma pressure is limited by the requirement that it cannot be very much greater than the magnetic field pressure. Since most engineering structures can support at best pressures of $\sim 1,000$ atmospheres, it is reasonable to expect the plasma pressure to be on the order of 1-1,000 atmospheres. For a temperature of 10 keV, p = nT yields a density of $6 \times 10^{13}$cm$^{-3}$ up to $6 \times 10^{16}$cm$^{-3}$. For realistic systems, this works out to densities in the $10^{14}$ to $10^{15}$cm$^{-3}$. For the confinement condition $n_e \tau_e \sim 10^{14}$sec/cm$^3$, these densities imply $\tau_e \sim 0.1$ to 1 sec. Since the plasma consists of charged particles (ions and electrons), the Lorentz force $[(e/c) \vec{v} \times \vec{B}]$ will be perpendicular to $\vec{v}$ and no work will be done on the plasma by the magnetic field. The particles gyrate about the field lines, and thus are confined in two-dimensions. However, they are free to move along the field lines with velocities $v \sim (2E/m)^{1/2} \sim 10^8 cm/sec$. Since the confinement time is of the order of one second, a system with good confinement across the magnetic field would have to be $\sim 10^8$ cm long. Thus confinement along the field lines is important. There are two basic approaches to this problem (Fig. 3). One is to use some method to limit the transport of particles along the field lines at the ends of the confinement device. The most successful of these "open" confinement schemes is the "mirror" machine. The magnetic field at each end of the experiment is made larger than the central field. This causes many of the plasma ions to be reflected from the high field region at ends back to the central plasma. This work is primarily being done at the Lawrence Livermore Laboratory in the United States [3] and other labs abroad.

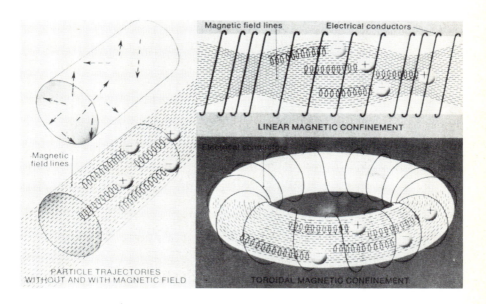

Figure 3
Schematic drawing of charged particle confinement by a magnetic field. (786452).

The other approach is to bend the field lines around into a torus so that the particles can travel freely along the field lines, and not escape the plasma. Most magnetic fusion research follows this general approach with tokamaks, stellarators, and toroidal pinches. Tokamaks [4,5] are formed by using large solenoidal coils to produce a toroidal field, and obtaining a stable plasma, and some heating by inducing an electrical current in the plasma by making the plasma the secondary of a transformer circuit (Fig. 4). The plasma current in the tokamak makes a magnetic field around the plasma. The combination of the magnetic field in the toroidal direction (long direction) and poloidal field (short direction around the plasma) makes the field lines helical. The field lines map out closed surfaces around the magnetic axis (roughly the center of the plasma current).  The center of the plasma is at the magnetic axis, and the edge is near the walls.  The highest temperatures coupled with the largest confinement times have been achieved in tokamaks, so I will concentrate on them.  As other fusion approaches reach tokamak parameters, they should have many of the same atomic physics issues.

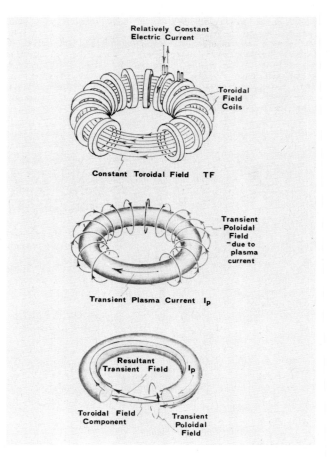

Figure 4
Schematic illustration of a tokamak showing the toroidal and poloidal magnetic fields. (754023)

Atomic processes can play a key role in the behavior of the central plasma of a tokamak.  Their major effect is usually to degrade the energy confinement by radiating energy from the center.  At temperatures of 1 keV, iron is only partially ionized.  At higher temperatures, such as 10 keV, higher Z elements such as molybdenum and tungsten are not fully ionized.  These partially ionized elements lose much of the energy in the central plasma by line radiation since the plasma is thin to visible and higher energy photons at $n_e \sim 10^{14}/cm^3$. The typical energy balance in tokamaks (Fig. 5) is that heat is supplied to the center plasma by ohmic heating due to the plasma current and auxiliary heating by neutral beams or radio-frequency waves. That heat is conducted and convected from the center plasma to the plasma edge. The plasma heat transport is due to classical collisions among the plasma ions and electrons, and also due to small scale turbulence produced by instabilities in the plasma. These plasma heat loss processes produce a minimum heat loss rate. Since the losses are probably diffusive, they can be reduced by making the plasma larger. Energy losses from the central plasma by impurity line radiation essentially short circuit the plasma physics heat losses, and thus can severely degrade the energy confinement. Higher Z impurities such as Mo and W are not fully ionized at the plasma center, and thus can radiate strongly [6] (Fig. 6). If we balance the alpha heating rate against impurity radiation and plasma transport losses, we find that above a certain impurity level, the impurities can radiate all of the alpha heating energy, and even having no plasma transport losses does not help (Fig. 7).

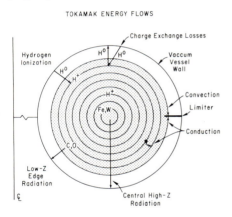

TOKAMAK ENERGY FLOWS

Figure 5
Diagram of the energy flows in a tokamak. (782058)

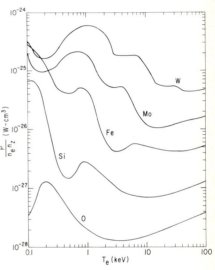

Figure 6
Emissivities of various elements in coronal equilibrium [6]. (772182)

From the requirement that $n_e \sim 10^{14}/cm^3$ and $T \sim 10$ keV, we can specify the conditions appropriately for the atomic processes in the central plasma. At those densities, the plasma is thin to photons. The radiative lifetime for most excited states ($A \sim 10^{10}sec^{-1}$ => $\tau_{rad} \sim 10^{-10}sec$) is short compared to the electron excitation collision time ($v \sim 10^5 sec^{-1}$ => $\tau_{ei} \sim 10^{-5}sec$). The ions are thus in the ground state before excitation. Dielectronic recombination is usually the dominate recombination mechanism [6]. For high Z atoms such as tungsten (Z = 74), adding dielectronic recombination to radiative recombination shifts the dominant charge state from +50 to +28 for a temperature of 1 keV. Recombination due to charge transfer between neutral hydrogen and multiple-charged ions can also be important in neutral beam heated plasmas [7,8].

Atomic processes play an important role in the behavior of the tokamak edge plasma. The particle balance (Fig. 8) in a tokamak consists of neutral hydrogen atoms and molecules going into the plasma and undergoing ionization and charge exchange. Some of the hydrogen atoms gain enough energy by charge-exchanging with hot plasma ions to increase their mean free path for ionization [$\lambda = (v_o/n_e\langle\sigma v\rangle)$] to penetrate to the center before they are ionized [9]. The plasma

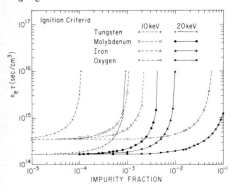

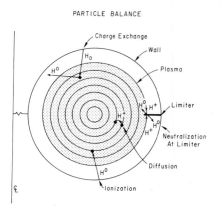

PARTICLE BALANCE

Figure 7

The $n_e\tau_E$ due to plasma transport losses required for ignition as a function of the impurity fraction $n_z/n_i$ for four different impurities and two temperatures $n_i$ is the total ion density [6].(772101)

Figure 8
Diagram of the particle flows in a tokamak. (782057)

ions then diffuse across the flux surfaces until they reach the edge where they recombine at the wall or at a limiter. The ions and neutral atoms strike the wall and limiter and can sputter off wall material so that some of it enters the plasma and becomes an impurity. Light impurities such as oxygen are also desorbed from the wall due to electron and ion impacts with the wall.

The density in the plasma edge varies from $10^{10}$ to $10^{14}cm^{-3}$, and the temperature ranges from 1 eV to 1,000 eV [10]. Thus, atomic processes that would play a role involve the plasma-wall interaction with the reflection of ions and neutrals from the wall and wall sputtering. The formation and breakup of molecules of hydrogen and even impurities such as $CH_4$ are important. The transport of helium in the edge is important as well since helium will be formed by the fusion of D and T and must be exhausted and pumped [11]. A good understanding of the plasma edge will grow in importance as the fusion community begins to design realistic reactor level experiments which require the handling of large heat and particle fluxes.

A third area of fusion work involving atomic physics is plasma heating. Ohmic heating is difficult to use for high temperatures because $\eta$, the plasma resistivity, is proportional to $T^{-3/2}$ and is small for high temperatures. The most successful heating method for tokamaks and mirrors used so far has been the injection of high current beams ($\sim 160$ amps) of high energy neutral atoms ($\sim 50$ keV) into the plasma, where they are ionized [12]. The fast ions thus formed are confined by the magnetic field and slow down by colliding with the background plasma, thus heating the background plasma.

The neutral beams now used are made by forming $H^+$ or $D^+$ in a plasma discharge, extracting and accelerating the $H^+$ or $D^+$ to 20-120 keV electrostatically, and neutralizing the $H^+$ or $D^+$ beam by charge-exchange in a $H_2$ or $D_2$ gas. The powers used range from 3 MW now to 60 MW for planned experiments (Table I).

To allow the neutrals to penetrate to the center of the larger plasmas, one needs to raise the energy of the neutral beams since $\sigma \propto 1/E$, and 120 keV $D^0$ is useful only for tokamaks with minor radii of $\sim$ 100 cm or less. However, the charge-exchange cross-section for $H^+ + H_2 \rightarrow H^0 + \ldots$ is dropping rapidly for $E > 40$ keV

TABLE I.
Neutral Beam Systems

| Energy | Ion | Power | Experiment | Date |
|--------|-----|-------|------------|------|
| 40 keV | $D^0, H^0$ | 3 MW | PLT | (1978) |
| 50 keV | $D^0, H^0$ | 8 MW | PDX | (1981) |
| 80 keV | $D^0, H^0$ | 7 MW | D-III | (1983) |
| 120 keV | $D^0$ | 20 MW | TFTR | (1984) |
| 150 keV | $D^0$ | 60 MW | FED | (1990) |

[13]. 80-90 keV/amu is the reasonable upper limit on the energy set by the dropping neutralization efficiency. One method for raising the beam energy while obtaining a reasonable neutralization efficiency is to form $D^-$ accelerate it, and neutralize it either with a thin gas cell ($\sim$ 65% production of $D^0$) or by photodetachment using a laser [14]. The chief difficulty has been making 10-100 amp sources of $D^-$. An alternative suggestion [15] has been to form beams of $Li^-$, $C^-$, $O^-$, or $Si^-$, accelerate them to $\sim$ 1 MeV/amu, neutralize them by collisions in a gas cell or by photodetachment and inject them into the plasma. Recent measurements by groups at Belfast and Brookhaven indicate that $\sim$ 50% of a $Li^-$ beam (Fig. 9) could be neutralized, but that 20-30% was the best that could be done with $C^-$, $O^-$, or $Si^-$. Thus, high energy $C^-$, $O^-$, or $Si^-$ neutral beams are feasible only if photodetachment is used.

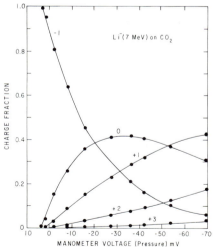

Figure 9
The charge state fraction in a 7 MeV $Li^-$ beam after passing through a $CO_2$ gas cell as a function of pressure in the gas cell for an experiment at the Brookhaven National laboratory [17]. (81P0121)

Some early work (∼ 1974) indicated that neutral beams for large tokamaks would
not penetrate if there were a significant impurity level in the plasma [18].
This work indicated that the cross-section for $H^o + A^{+q} \rightarrow H^+ + \ldots$ should scale
as $Z^2$ times the hydrogen cross-section. This has led to a lot of work on the
interaction of $H^o$ and multi-charged ions in the last five years, with the result
that $\sigma \propto \sigma_H Z^{\sim 1.2}$, due to the dominance of charge-exchange instead of ionization
for $Z \gtrsim 8$, and hydrogen energies ∼ 50 keV/amu [19].

Probably a majority of plasma diagnostics rely on atomic processes [20].
Diagnostic techniques can be divided into two categories, passive and active.
Passive diagnostics measure the energy spectra of photons and particles that
naturally leave the plasma. Active diagnostics introduce probing beams of
photons or particles to produce an outflux of particles or photons which can be
measured. Spectroscopy in the visible, ultra-violet and x-ray is the most common
passive diagnostic. The frequencies and intensities of emission lines is used to
measure the location and density of the different charge states of plasma
impurities. This data not only tells us the impurity level in the experiment and
how much energy is being lost in radiation from a particular region in the
plasma, but can also be used in conjunction with information about the ionization
and recombination rates to deduce the particle transport rates. Measuring the
doppler broadening of the emission line gives the local ion temperature [21].
The electron temperature can be determined from the energy spectrum of the
bremsstrahlung emission in the soft x-ray region [4]. The ion temperature can
also be measured from the energy distribution of the neutral atoms escaping from
the plasma. The high energy neutral atoms are formed by the charge-exchange of
low energy neutral atoms with the hot ions.

Neutral beams are often injected to increase the local neutral density so that
the charge-exchange rate is increased. Diagnostic neutral hydrogen beams are
used to "neutralize" hot ions and fast ions from beam injection [22]. Not only
can the energy spectrum of the fast ions be measured this way, but plasma
parameters such as Zeff and q, related to the plasma current distribution, can be
determined as well. A recent proposal has been made to measure the fast alpha
particle distribution by using a high energy (3-6 MeV) neutral lithium beam to
neutralize the fast alphas by double charge-exchange ($Li^o + He^{++} \rightarrow He^o + \ldots$)
[23]. This cross-section has recently been measured by the Belfast group (Fig.
10) [16].

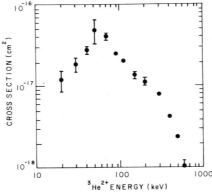

Figure 10
Cross-sections for charge-exchange of $He^{++}$ and $Li^o$ as a function of energy [16].
(81P0190)

To date, tokamaks have achieved central temperatures of $\sim 7$ keV [12], densities of $10^{14} - 10^{15}$cm$^{-3}$ [4], and $n\tau \sim 3 \times 10^{13}$sec/cm$^3$ [4]. Mirror experiments (TMX) [24], have achieved temperatures of $\sim 200$ eV, densities of $\sim 5 \times 10^{13}$ and $n\tau \sim 7 \times 10^{10}$sec/cm$^3$. The growth in the United States tokamak fusion program is illustrated in Table II. Similar sized tokamaks to TFTR are being constructed throughout the world (JT-60, Japan; JET, Europe; and T-15, USSR). The mirror program is building a large experiment, $\sim \$200$ million, to be ready in about 1984. The United States magnetic fusion budget is about $\$450,000,000$ a year, and the worldwide budget is about $\$10^9$/year.

TABLE II.
Examples of the United States Tokamak System

|  | PLT | $\rightarrow$ | TFTR | $\rightarrow$ | FED |
|---|---|---|---|---|---|
| R (major radius) | 140 cm | | 250 cm | | 520 cm |
| a (minor radius) | 40 cm | | 85 cm | | 130 cm |
| Date | 1978 | | 1984 | | 1991 |
| Cost (millions of dollars) | 30 | | 350 | | 1,500 |

The invention of lasers and high energy particle beams in the 1960's opened the possibility of compressing small pellets of DT and other fusionable materials to high enough densities and temperatures so that the pellet would confine the fusion alphas, and thus ignite in a very much scaled down version of a hydrogen bomb. The basic idea is that the driven energy (laser, beams) is focused on the pellet surface (Fig. 11). The surface is rapidly heated and ablated. The continued ablation drives the pellet material inward with a spherically conver-gent shock wave. Typical pellet dimensions (consistent with yields less than 1 ton of TNT) are $0.01 - 1$ cm for the uncompressed pellet and $10^{-1}$ to $10^{-3}$cm for the compressed pellet. Since the temperature must be in the 1–10 keV range for fusion to occur, the thermal velocity of the particles is $\sim 10^8$cm/sec. The pellet will thus fly apart in $10^{-9} - 10^{-10}$ seconds. This time can roughly be interpreted as a confinement time so $n_e \tau_c \sim 10^{14}$sec/cm$^3$ implies that $n_e \sim 10^{23} - 10^{25}$/cm$^3$. Thus, the conditions for inertial confinement fusion are $T \sim 10$ keV, $\tau_c \sim 10^{-9}$sec, and $n_e \sim 10^{24}$cm$^{-3}$.

The division of issues into those related to the central plasma, edge plasma, heating and diagnostics is again appropriate. In the central plasma, many materials beside hydrogen are present since the pellets contain layers of many different materials. Adequate understanding of the pellet dynamics requires a knowledge of hydrodynamics and radiation transport in the pellet. Thus knowledge of the equation of state and opacity for dense, hot plasmas is essential [26].

The inter atom spacing $\ell = n^{-1/3} \sim 10^{-8}$cm is of the order of the size of the electron orbits so the usual assumptions about binary collisions do not apply. One has conditions more appropriate to "solid-state" type physics except that the system is not periodic. The collision rates are affected. The energy levels are shifted, and some of the higher levels are eliminated.

The edge plasma is extremely important to inertial confinement since the energy
needed to compress the pellet is deposited there.  Typical densities run from
$10^{14} - 10^{22} cm^{-3}$ and typical temperatures run from 10 eV to 1 keV.  The key issues
are the absorption and stopping of laser light, and the absorption and stopping
of fast ions and electrons.  Magnetic fields up to the megagauss level are

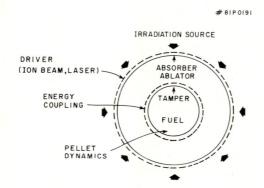

Figure 11
Schematic illustration of an inertial confinement pellet. (81P0191)

generated which will affect collision rates.  As in the magnetic fusion case, the
interaction of the pellet debris with the reactor vessel is important.  The
emphasis is on the survival of wall to pulsed heat and particle loads ($\sim 1,000$
joules/cm$^2$ in $10^{-9}$ seconds).

The main approaches to inertial confinement are defined by the type of beam used
to compress the pellets (lasers, light ions, heavy ions, and electrons).  The
original approach was to use high energy lasers [27].  The primary difficulties
with lasers is their low electrical efficiency and poor coupling of the laser
power to the pellet.  An additional problem is the generation of high energy
electrons which penetrate to the pellet core and heat it before the final
compression stage, thus increasing the power required to reach the required high
density.  Current experiments indicate that the laser-plasma coupling is improved
and the hot electron preheat is reduced as the laser wavelength is reduced.
Large $CO_2$ laser (10 μ) are being used at the Los Alamos National Laboratory.
Neodymium glass lasers (1 μ) are the most common with programs at Lawrence
Livermore National Laboratory, KMS, Inc., and the University of Rochester, in the
U.S., at the Lebedeev Institute in USSR, at Osaka University in Japan, at LeMeil
in France, and at the Max Planck Institute in Garching, West Germany.  There is
intense interest in going to shorter wavelengths using exotic lasers such as KrF
lasers or by frequency doubling.  Livermore is constructing a frequency doubled
laser ($\sim 0.5$ μ) beginning with a Neodymium glass laser ("Novette").

Given the apparent simplicity of the stopping of high energy ions in a plasma,
the short range of the ions, and the possibility of high electrical efficiency
for ion beams, there is an active program to use both light and heavy ion beams
for compressing pellets.  The light ion beam program is being done in the United

States at Sandia Laboratories in Alburquerque, New Mexico, and at the Kurchatov Institute in Moscow. The heavy ion beam fusion work is being done at the Argonne National Laboratory, and the Lawrence Berkeley Laboratory in the United States, and the Rutherford Laboratory in the United Kingdom.

One candidate beam is $\sim 1$ GeV $Cs^+$. Recent work on the reaction $Cs^+ + Cs^+ \rightarrow Cs^0 + Cs^{+2}$ has had a bearing on the feasibility of such beams, since this reaction could destroy the beam.

Since lasers involve electronic transitions in molecules and atoms, atomic processes will continue to play a large role in the search for high power, efficient lasers. The thrust toward shorter wavelengths ($< 1\mu$) will be an important research area. Significant improvements in both these areas will have to be made for lasers to be feasible as drivers for a inertial confinement fusion reactor. At this moment ion beams seem more promising for reactor applications than lasers. Thus, there will be an increasing emphasis on the production and acceleration of efficient, high energy, and high power ion beams.

Diagnostic techniques for inertial confinement experiments rely heavily on atomic processes. Spectroscopy in the UV, x-ray, and visible is used [28]. One common technique is to seed the DT gas with neon or argon and look at the emission lines for line profile changes, line shifts, and other high density effects [29]. Many pellets have layers of high Z materials, and time and space resolved line radiation, and x-ray continua from these high Z layers allows one to determine the time history of the temperature and density of the layers in the compressing and exploding pellet. Spectral information coupled with rate equation modeling of ionization and recombination in the strong temperature and density gradients at the ablation layer where the laser is deposited provides a measure of the laser plasma interaction. Optical laser probes are used with interferometry, faraday rotation, and polarization to measure the density profiles in the ablating surface of the pellet.

The best conditions achieved so far by the inertial confinement program are $n \sim 3 \times 10^{24} cm^{-3}$, $T \sim 1$ keV, and $\tau_c \sim 1.2 - 1.5$ nsec. Pellets of initial diameters of 200 to 400 microns have been compressed to 10-20 microns. The major funding commitment is for large laser systems at Livermore and Los Alamos in the United States (Table III). The United States budget is $215 million/year.

The programmatic needs of the fusion community have provided motivation and resources for many of the current research topics in atomic physics. The interest in charge-exchange of multi-charged ions with hydrogen was spurred on by the concerns that the fusion community that had neutral beams would not penetrate plasmas with even relatively small impurity levels. In addition, fusion research has, as a by-product, produced new experimental and theoretical tools to explore atomic physics in brand new regimes. Plasmas with temperatures as high as 7 keV

TABLE III.
Large United States Inertial Confinement Experiments

| Name | Type | Wavelength | Site | Date | Power | Engergy |
|------|------|------------|------|------|-------|---------|
| Helios | $CO_2$ | $10\mu$ | LASL | 1978 | 30 TW | 6 kJ |
| Shiva | Ni glass | $1\mu$ | LLL | 1978 | 30 TW | 6 kJ |
| Antares | $CO_2$ | $10\mu$ | LASL | 1982 | 50 TW | 40 kJ |
| Nova | Ni glass | $1\mu$ | LLL | 1985 | 60 TW | 30 kJ |

and densities up to $3 \times 10^{24} \mathrm{cm}^{-3}$ are produced in laboratories. These have led to the identification of the energy levels of highly stripped ions (such as $W^{+50}$) [30]. Large spectroscopy groups have been set up at the major fusion laboratories with the attendant improvements in spectrometers and detector technology. These groups have observed effects seen only in astrophysical plasmas before, such as dielectronic satellite lines, and recombination edges. The work on excitation-autoionization is of great interest to the fusion community. The work on dense matter atomic physics is a qualitatively new area that will almost certainly grow in interest and importance.

A lot of extremely useful work is being done by the various data center groups throughout the world to assemble the data relevant to fusion and put it in a digestible form so that plasma physicists can easily use the data. In the United States, there are data centers at the Oak Ridge National Laboratory in Oak Ridge, Tennessee; at the Joint Institute Laboratory for Astrophysics in Boulder, Colorado, and at the National Bureau of Standards in Washington, D.C. In Japan, there are data centers at the Plasma Physics Institute at Nagoya University in Nagoya, and at the Japanese Atomic Energy Research Institute at Tokai. In Europe, there are groups at the International Atomic Energy Agency in Vienna, Austria; at Queen's University, Belfast, Northern Ireland, and at the University of Paris in France. In the USSR, there is a group at the Kurchatov Institute in Moscow.

The rest of the symposium will consist of four speakers who will cover the theoretical and experimental needs and applications of atomic data in magentic and inertial confinement fusion research. The final paper will discuss the observation of excited states of $O^{+7}$ produced in a tokamak by charge-exchange of $O^{+8}$ and $H^0$. References 3, 4, 5, 10, 25, and 26 are good reviews of their topics. Volume 23, pp. 70-214 of Physica Scripta provides a good, overall review of the atomic physics needs of magnetic fusion.

ACKNOWLEDGMENTS

The author is grateful for discussions with Dr. R. Hulse, Dr. Jon Weisheit, and the symposium speakers. The work done at Princeton is supported by the U. S. Department of Energy Contract No. DE-AC02-76-CHO-3073.

REFERENCES

[1]   Greene, S., UCRL-70522 (1967), Lawrence Livermore Laboratory (1967).
[2]   Lawson, J. D., Proceedings of the Physical Society, London 70B (1957) 6.
[3]   Cohen, B. I., Editor, Lawrence Livermore Laboratory Report UCAR-10049-80 (1980).
[4]   Rawls, J., Status of Tokamak Research, DoE/ER-0034 United States Department of Energy, Washington, D.C. (1979).
[5]   Furth, H. P., Nuclear Fusion 15 (1975), 487.
[6]   Jensen, R. V., et al., Nuclear Fusion 17 (1977), 1187.
[7]   Hulse, R. A., et al., J. Phys. B. 13 (1980), 3895.
[8]   Krupin, V. A., et al., JETP Lett. 24 (1979), 318.
[9]   Hughes, M. H., et al., J. Comp. Phys. 28 (1978), 43.
[10]  McCracken, G. and Stott, P., Nuclear Fusion 19 (1979), 889.
[11]  INTOR, International Tokamak Reactor, Zero Phase, Intenational Atomic Energy Agency, Vienna (1980).
[12]  Eubank, H., et al., Phys. Rev. Lett. 43 (1979), 270.
[13]  Freeman, R., et al., CLMR-137, Culham Laboratory (1974).
[14]  Sluyters, Th., BNL 51304, Brookhaven National Laboratory (1980).
[15]  Grisham, L., et al., PPPL-1759, Princeton University Plasma Physics Laboratory (1981).

[16]   McCullough, R., Goffe, T., Shah, M., Lennon, M., Gilbody, H., private communication.

[17]   Johnson, B., private communication.

[18]   Moriette, P., Proceedings Seventh Yugoslav Symposium on the Physics of Ionized Gases (1974), p. 43.

[19]   Olson, R. E., et al., Phys. Rev. Lett. 41 (1978), 163.

[20]   Equipe TFR, Nuclear Fusion 18 (1978), 647.

[21]   Suckewer, S., Physica Scripta 23 (1981), 73.

[22]   Goldston, R., Phys. Fluids 21 (1978), 2346.

[23]   Post, D., et al., J. Fusion Energy 1 (1981), 129.

[24]   Coenogen, F., et al., Phys. Rev. Lett. 44 (1980), 1132.

[25]   Proceedings of the IEEE Minicourse on Inertial Confinement Fusion, Montreal, Canada, June, 1979.

[26]   Weisheit, J., Princeton University Plasma Physics Laboratory Report PPPL-1765 (1981).

[27]   Nuckolls, J., et al., Nature 239 (1972), 139.

[28]   Yaakobi, B., et al., Phys. Rev. Lett. 39 (1977), 1526.

[29]   Bristow, T., et al., IAEA-CN-37/B-4 (1979), International Atomic Energy Agency, Vienna, Austria.

[30]   Reader, J. and Luther, G., Phys. Rev. Lett. 45 (1980), 609.

PHYSICS OF ELECTRONIC AND ATOMIC COLLISIONS
S. Datz (editor)
© North-Holland Publishing Company, 1982

THE EFFECTS OF LOW ENERGY ELECTRON CAPTURE COLLISIONS

$(H_o + C^{n+})$ ON THE PARTICLE AND ENERGY BALANCE
OF TOKAMAK PLASMAS

John T. Hogan

Oak Ridge National Laboratory
Fusion Energy Division
Oak Ridge, Tennessee
U.S.A.

## 1. INTRODUCTION

The particle and energy balance in magnetic fusion plasmas is a complex subject in which atomic and molecular collisions data often plays a determining role. Appreciation for the many processes involved may be gained from the Proceedings of the NATO Advanced Study Institute on Atomic and Molecular Processes in Controlled Fusion Research [1], from the published proceedings of the IAEA Advisory Committee Meetings on Atomic and Molecular Data for Fusion [2,3], and from a recent topical article on atomic physics processes in fusion [4].

To illustrate the way in which atomic data provides enlightenment in the search for understandable (and thus extrapolable) confinement models, we restrict our scope to electron capture collisions involving $H_o$ and multiply-charged ions. Many such foreign (impurity) multiply-charged ion species are found in plasma discharges, as a result of gas recycling and damage to the surrounding surfaces by energetic plasma particles. Typical 'low-Z' ions are carbon and oxygen; the major constituents of the stainless steel wall (Fe, Ni, Cr) are intermediate impurities, while 'high-Z' impurities (Mo, W) enter from 'limiter' plates which constrict the hot plasma zone to reduce direct plasma-wall contact. In this discussion, however, attention will be given only to applications of data involving $H_o + C^{n+} \to H^+ + C^{(n-1)+}$ reactions with energy 10 eV to 2 keV. This energy range is typical of the plasma edge in present devices. The reasons for choosing only carbon for discussion are:

   a) As a result of recent theoretical and experimental progress, some reported at this conference, the electron capture rates for the full suite of charge states $(C^{2+} \to C^{6+})$ of interest in the plasma edge may be estimated for the first time.

   b) Electron impact ionization cross-sections have been measured for these ions, and the use of semi-empirical ionization rates is supported by direct measurement [5,6].

   c) The cross-sections for electron capture have been found to be sufficiently large that, taken together with the typical measured $H^o$ concentrations in the edge region, this process is the dominant recombination mechanism for $C^{n+}$ by a wide margin: hence both the dominant ionization and recombination processes may be estimated for the full suite of $C^{n+}$ ions of interest.

   d) This situation does not exist for any other impurity ion of fusion interest: a need exists for data in this energy range for impurity ions found in fusion plasmas.

While it is encouraging to be able to make quantitative estimates for a particular process for the first time, it is not necessarily true a priori that the results

will be relevant to fusion.  In this case, however, the results have a major im-
act in several topical areas:

a) An attempt is being made to fashion a 'photosphere,' [7] or 'cool plasma
mantle' [8] around future large fusion plasmas.  It would be desirable to create,
and maintain, a radiating zone on the plasma periphery in which power deposited in
the plasma core by thermalizing α-particle reaction products is lost to low-Z impurity ions and then radiated from the plasma. Wall damage (and subsequent introduction of a large number of impurities) would thus be minimized.  Since $H_o$ + $H^+$
charge exchange in the edge region creates an efflux of energetic neutral atoms
which can sputter wall atoms into the discharge, it is desired to keep the temperature of the radiative zone well below the maximum in the sputtering yield.  (As a
typical value, $S_{max}$ occurs at ∿2 keV for $F_e$.)  Thus, we will adopt 350 eV as the
upper limit of our temperature range.

b) Recent high-power heating results have been carried out with graphite lim-
iters.  The dominant impurity ion in these cases is carbon, and the contribution
of carbon radiation to the overall energy balance is significant [9].

c) Study of the dynamics of impurity motion has disclosed asymmetries in the
emitted light, which could confound collisional (so-called 'neoclassical') theory.
While a theoretical analysis predicts small asymmetries to result from the theory,
quite large asymmetries are measured.  The possible role of charge-exchange of
$H_o$ with multi-charged impurities in producing this asymmetry has been suggested
[10], but the lack of data has prevented calculation till now.


2.   ATOMIC DATA USED IN THE CALCULATIONS

Calculation of the impurity charge state and radiative cooling power requires the
solution of the ionization-recombination balance equations for multiply-charged
ions in a plasma with specified temperature and density.  These calculations are
routine, and the usual estimates for <Z> (charge state) and $P_{RAD}$ (radiative
cooling power) are available in tabular form [11].  However, these conventional
estimates cannot be used for the plasma edge, because charge exchange recombination has not been included in them.  Hence we recompute the so-called coronal
equilibrium relative abundances and radiative cooling rates for typical plasma
conditions.

2.1  Balance Equations

The steady-state equations for the ionization-recombination balance are:

$$0 = I_{k-1}^i n_{k-1}^i + R_{k+1}^i n_{k+1}^i - (I_k^i + R_k^i)n_k^i \tag{1}$$

$I_k$, $R_k$ are the total ionization rates for the ion with charge $Z_k$ = k - 1
(k = 1, ..., Z); $n_k^i$ is the spatial density of that ion; i denotes the spatial
zone.

Plasma transport has a large effect on radiation and in the diagnostic applications
of spectroscopic data [12].  Because the calculation of cooling rates without including transport processes, but comparing cases with and without charge exchange
recombination, will give a fair picture of the changes made when the new data are
taken into account we will neglect plasma transport processes.

## 2.2 Ionization and radiative and dielectronic recombination rates

Restricting attention to the case of carbon, we use the semi-empirical Lotz formulae to calculate the electron impact ionization rates [5]. The justification for using this rate has been presented by Crandall [6]. Uncertainties of up to 40% with respect to the data we use are suggested, and so the sensitivity of our results to variation in this rate is examined in Section 6. The total recombination rate for plasmas in the density and temperature range we consider has contributions from radiative, dielectronic, and charge exchange-induced recombination. The expressions which we use for radiative and dielectronic recombination have been described by Mattioli [13]. They embody estimates by Von Goeler et al. [14] for radiative recombination and the results of Burgess [15] are used for the dielectronic recombination rate. Again, the sensitivity to variations in the assumed data is treated in Section 6.

## 2.3 Charge-exchange Induced Recombination

The rate of recombination due to electron capture reactions is

$$\alpha_{cx}^{Z} = n_o n_Z \sigma_{cx}^{Z} V_{th}(H_o)$$

where $\sigma_{cx}^{Z}$ is the cross-section for electron capture from $H_o$, $n_o$ is the neutral hydrogen density, and $V_{th}(H_o)$ is the hydrogen thermal speed. In Monte Carlo calculations of $n_o$, the exact particle thermal speed is used. However, for steady state radiative cooling calculations we assume $V_{th}(H_o)$ is equal to the hydrogen ion thermal speed, and that $T_e = T_i$. Low $T_e$ in the plasma edge, and the good electron-proton energy coupling which this implies, serves to justify this simplification.

The atomic data used for $\sigma_{cx}^{Z}$ are as follows: Theoretical values for $C^{2+}$, $C^{3+}$ (electron capture to $C^{+}$, $C^{2+}$, respectively) have been obtained by Heil and Bottcher [16] extending previous results of Heil, Dalgarno and Butler [17], for energies up to ~250 eV/amu. Experimental results for $C^{2+}$ electron capture [18] are used for energies above 500 eV/amu, and (approximately coincident) theoretical and experimental [19] results are used for $C^{3+}$ for all energies considered.

Theoretical results for $C^{4+,5+}$ presented by Olson, Shipsey, and Browne [20,21] are assumed. Other theoretical work [22] and recent experiments [19] suggest that factor 2 variations about these values should be studied. The sensitivity of the results to these uncertainties will be discussed in Section 6.

## 2.4 Discussion

The most significant aspect of the cross-section data for charge exchange is the large difference between the $C^{3+}$ and $C^{4+}$ cross sections. $C^{4+}$ will readily recombine to $C^{3+}$, but there will be a sharply reduced flow from $C^{3+}$ downwards. The relative population of this strongly radiating Li-like ion will thereby be enhanced by charge exchange effects.

The existence of a large $C^{6+}$ cross-section implies that fully stripped carbon will only be found in plasmas at much higher electron temperatures than heretofore calculated.

## 3.  PARTICLE BALANCE: NEUTRAL DENSITY

### 3.1  Particle Balance

The particle balance in the plasma may be viewed as a form of book-keeping, to ensure that sources and losses of particles add up to the observed rate of change of density.  Balance equations for particles are given in Ref. [1].  The source of particles due to impurity stripping, from beam injection, and from ionization of neutral hydrogen atoms must be computed; hence neutral gas transport should be calculated.

### 3.2  Neutral Transport

Charge exchange processes can influence the spatial distribution of $H_o$ in the plasma (and thus the particle balance through the source term).  Neutral particles travel freely, of course, with respect to the magnetic field.  Their transport is governed by the Boltzmann equation, with impact ionization and (resonant) $H_o + H^+$ charge exchange the dominant atomic collisions effects which must be considered. The transport equation is described, e.g., in [23].

The transport of neutrals is affected by electron impact ionization (annihilation) by the exchange of velocities with the background hydrogenic ion population, and by annihilation of the neutral by $H_o + C^{n+}$ electron capture.  Although the electron capture cross sections for $H_o$ with multiply charged ions are larger than that for $e^- + H_o$ impact ionization, the much larger electron thermal speed leads to the domination of impact ionization over the impurity charge exchange process as an annihilation mechanism for many cases of interest.

Neutral densities in the plasma edge in the range $10^{10}$–$10^{11}$ cm$^{-3}$ are typically estimated.  However, the asymmetric nature of the spatial distribution of external sources of neutrals can lead to much higher local densities.  The sources are cold neutrals injected for fueling, (1–10 eV); Franck-Condon dissociative ionization of $H_2^o$ on the walls of the device, diffusing there from the bulk metal (1–10 eV) [24]; ions, neutralized at the limiter, and re-entering with energies typical of plasma on the outermost flux surface (10 eV $\sim$ 100 eV); and emergent charge exchange neutrals reflected from the wall with intermediate energies close to ion energies in the plasma edge (100 eV $\sim$ 2 keV).

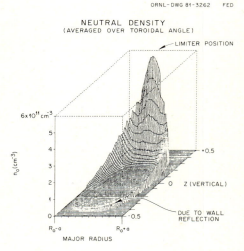

ORNL-DWG 81-3262   FED

NEUTRAL DENSITY
(AVERAGED OVER TOROIDAL ANGLE)

Figure 1
Spatial density of $H_o$: $n_o(r,\phi,Z)$ ($\phi$ the toroidal angle) is averaged with respect to $\phi$ and shown as a function of major toroidal radius, r, and the vertical dimension, Z.  The source of particles in this Monte Carlo calculation is assumed to be the top center of the box: this is typical of emission from a horizontal rail limiter. Parameters are chosen to be typical of an ohmic discharge in the ISX-B tokamak.

CARBON IV RELATIVE ABUNDANCE

ORNL-DWG 81-3258    FED

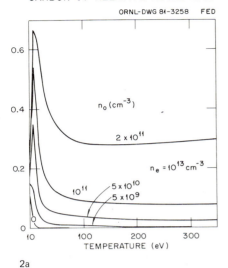

2a

CARBON VI RELATIVE ABUNDANCE

ORNL-DWG 81-3260    FED

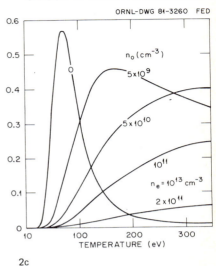

2c

CARBON V RELATIVE ABUNDANCE

ORNL-DWG 81-3243    FED

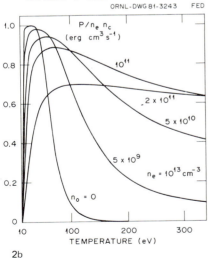

2b

Figure 2
The relative abundance of CIV, CV and CVI ions with respect to the total carbon density. In these, as in all subsequent figures showing neutral density effects (i.e., Figures 3, 4, 5 and 7) the values displayed are valid for other cases with the same ratio of $n_o/n_e$. Here $n_e = 10^{13}$ cm$^{-3}$ has been chosen for specificity, hence the results displayed for $n_o = 2 \cdot 10^{11}$ cm$^{-3}$, say, are also true for $n_o = 10^{11}$ cm$^{-3}$, $n_e = 5 \cdot 10^{12}$ cm$^{-3}$. The temperature range is 10 eV to 350 eV.

a. relative abundance of CIV
b. relative abundance of CV
c. relative abundance of CVI

Two of the sources mentioned are strongly localized in space: gas injection (commonly called 'puffing') and limiter recycling. Figure 1 shows a typical calculated distribution of neutral hydrogen density around the cross section of the torus. Gas is assumed to enter from a rail limiter, near the top of the cross-section. Particles are then followed with a 3-dimensional Monte Carlo code, which includes the processes needed for the Boltzmann equation description [25].

774 *J.T. Hogan*

The code assumes that plasma parameters (density, temperature, e.g.) are constant on magnetic flux surfaces, because of the rapid communication around flux surfaces due to motion along field lines.

## 3.3 DISCUSSION

As seen in Fig. 1 there may be a factor of 3-5 difference between local values of the neutral density and the flux surface average (which enters the particle balance). The local density, of course, must be used for calculations of radiative cooling enhancement and charge state depression due to charge exchange with impurities. However, this lack of symmetry is an expected effect arising from the spatial localization of the neutral particle source and the straight-line flight of the neutral atoms.

## 4. RELATIVE ABUNDANCES AND RADIATIVE COOLING RATES

### 4.1 Relative Abundance

The ratio of the number of ions with charge Z to the total number of carbon ions at a spatial location is computed from the steady-state balance equation. The results for $C^{3+} \rightarrow C^{5+}$ are shown in Figure 2a-c. Several features of fusion interest emerge.

The coronal equilibrium dependence of relative abundance solely on $T_e$ is destroyed with the correct treatment of $C^{n+}$ charge exchange. For $\sigma_{cx}^Z = 0$, there is a sharp localization of relative abundance with respect to temperature (as seen in the $n_o = 0$ cases in Figures 2a-c).

The existence of a peak in, e.g., CV emission in the plasma, has in the past been used to give the temperature value at that position. This diagnostic application is only possible, however, when the ratio $n_o/n_e$ is less than ~$2 \cdot 10^{-4}$. This can be a stringent requirement during high-power heating experiments when intense 'gas puffing' is used to fuel the discharge.

The most apparent effect of the charge exchange reactions is to depress the average charge of the carbon at a given temperature. As seen in Figure 3, with $n_o = 0$ (or, equivalently $\sigma_{cx}^Z = 0$) the $<Z> \equiv \Sigma Z_k n_k / \Sigma n_k$ quickly rises to the fully ionized value above a temperature of ~100 eV. With $\sigma_{cx}^Z \neq 0$, however, the fully stripped stage is not attained in the 10-350 eV temperature range of interest for plasma edge calculations.

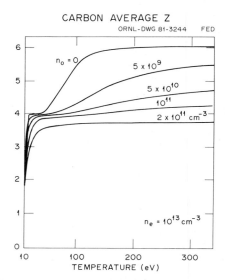

CARBON AVERAGE Z

ORNL-DWG 81-3244    FED

$n_o = 0$

$5 \times 10^9$

$5 \times 10^{10}$

$10^{11}$

$2 \times 10^{11}$ cm$^{-3}$

$n_e = 10^{13}$ cm$^{-3}$

TEMPERATURE (eV)

Figure 3
The average ionic charge of carbon ions for $T_e$ = 10-350 eV, $n_o$ = 0-2.10$^{11}$ cm$^{-3}$

## 4.2 Radiative Cooling Rates

The calculation of radiative emission can proceed once all the relative abundances have been found. We include contributions to radiation arising from radiative and dielectronic recombination, and from line radiation, using estimates calculated by Breton et al [26,27]. The major effect of charge exchange collisions is to enhance the recombination rate, and hence to increase the abundance of radiating ions. The electron energy added to the plasma with $H_o$, transferred to excited states of $C^{n+}$ and subsequently radiated is not counted as a direct cooling loss from the plasma electrons.

Typical results, indicating both the relative contributions of radiative and di-electronic recombination, and line radiation losses are shown in Figures 4a-c. We have chosen CV as a typical ion spectroscopically observed in the plasma edge. The cooling rates are expressed as power per electron per CV ion. The total radi-ative and dielectronic recombination rate, shown in Figure 4a, has a peak value of $\sim 2.10^{-21}$ erg cm$^3$ s$^{-1}$, while the line radiation has a peak some 40 times higher. For both, though, there is a very large enhancement due to charge exchange recom-bination. The total CV radiation, shown in Figure 4c, peaks at $\sim 70$ eV with a charge exchange, but with $\sigma_{cx}^{Z} \neq 0$ the radiative cooling rate rise throughout the plasma edge energy range.

For application to 'cool plasma mantle', or 'photosphere' techniques the total carbon radiation is of interest. This is shown for the relevant energy range in Figure 5. If 'mantle' temperatures can be kept in the 10-20 eV range, then radi-ative rates of 2-3 W/cm$^3$ can be achieved for an electron density of $10^{13}$ cm$^{-3}$ and a carbon density $2.10^{12}$ cm$^{-3}$. This cooling rate could radiate most of the power emission in the presently considered design for the Fusion Engineering Device [28], with a radiating layer only $\sim 10$ cm thick, as compared with the plasma minor radius of 125 cm. (The goals of the FED are to produce sufficient thermonuclear power to demonstrate the engineering features of a fusion power plant: neutron shielding, tritium breeding, blanket, etc.). As the temperature is increased, the width of the radiating layer would have to be increased as well.

The discrepancy between the rate calculated including charge exchange, and that without, is approximately two orders of magnitude at $T_e$ = 100 eV!

## 5. IMPURITY DYNAMICS

An important experimental finding, related to the validity of neoclassical theory, has been the observation of spatial asymmetry in impurity light emission [10,29, 30]. The assumption that plasma parameters should be approximately constant on magnetic flux surfaces, and hence symmetric in space, is a typical assumption in theoretical models. Corrections to the collisional diffusion theory have been proposed [31], but the maximum asymmetry allowed is approximately 25%. Much larger discrepancies (typically a factor 2 variation in light emission) are observed.

The asymmetry in the neutral hydrogen distribution, discussed in Section 3, yields a natural explanation for the observations. The predicted spatial distribution of CV is shown in Figure 6a, with $\sigma_{cx}^{Z}$ = 0 assumed. Figure 6b shows the case for $\sigma_{cx}^{Z} \neq 0$, with only the left half of the CV spatial distribution ($\phi < \Pi$) presented for clarity. The distribution is symmetric about the $\phi = \Pi$ plane, and shows a pronounced dip where the neutral hydrogen density has a peak near the limiter. This model thus traces the asymmetry in impurity light emission to the naturally asymmetric $H_o$ spatial distribution. The spatial localization of the $H_o$ distribu-tion (and hence of carbon light emission) will depend on the details of placement of the limiter and gas injection parts for specific experiments.

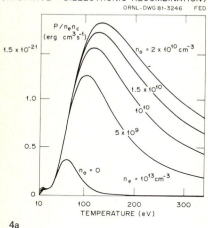

4a

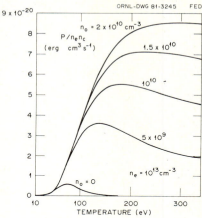

4c

**Figure 4**
Radiative rates for CV ions, for $T_e$ =
10–350 eV, $n_o$ = 0–2.$10^{11}$ cm$^{-3}$.

a. The contributions from radiative
   and dielectronic recombination
   events.
b. Total line radiation emission.
c. The overall radiative rate,
   summing contributions in 4b and 4c.

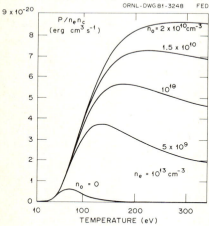

4b

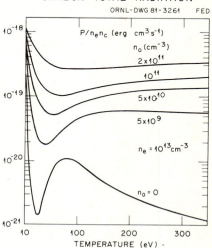

**Figure 5**
Total radiative cooling for all carbon ions,
for $T_e$ = 10–350 eV, $n_o$ = 0–2.$10^{11}$ cm$^{-3}$. Results are expressed as the power radiated
per electron per carbon ion.

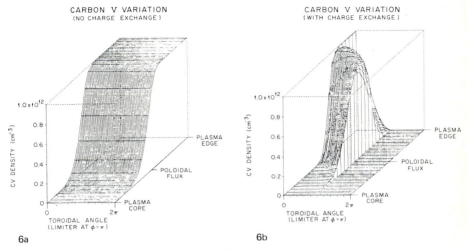

ORNL-DWG 81-3264   FED                    ORNL-DWG 81-3263   FED

Figure 6

Effects of charge transfer between H$_0$ and multiply charged carbon impurities on the spatial distribution of a typical carbon ion, CV. The results are shown as functions of the angle around the torus (with particles entering the plasma at $\phi = \Pi$), and of the poloidal magnetic flux, which serves to label surfaces with constant plasma density, temperature, etc.

a.  $\sigma_{cx}^Z = 0$ assumed, the CV relative abundance depends only on $T_e$, hence it is depressed in the plasma core ($T_e$ = 700 eV) and rises toward the plasma edge ($T_e$ = 50 eV). No dependence on $\phi$ is present since $T_e$ depends only on the value of poloidal magnetic flux, which does not change as $\phi$ changes.

b.  Same conditions as in 6a, but now the correct cross-sections for $\sigma_{cx}^Z$ are included. The CV distribution is strongly asymmetric. For clarity in viewing, the right half of the CV distribution has been excised in this Figure ($\phi > \Pi$). The computed distribution is mirror-symmetric about $\phi = \Pi$.

## 6.  SENSITIVITY ANALYSIS

The atomic data which have been used for these calculations are susceptible to improvements in accuracy. The specific needs discussed here, which bear on obtaining an accurate particle and energy balance for some magnetically confined plasmas, are met by the available data. Improvements would be welcome for diagnostic purposes, and the details of 'cool plasma mantle', or 'photosphere', power emission scheme will require a narrower range for some cross-sections. To discuss the accuracy of the data, we compute the sensitivity of the results to variations in the assumed atomic data. The reference values used are those described in Section 2.

### 6.1  Electron Impact Ionization

The uncertainty in using the Lotz semi-empirical ionization rates is often quoted as 40%. We have found that the variation in total radiated power produced by enhancing the Lotz rate by a factor 1-1.4 is a decrease by 30% in the 30 eV region, and there is a smaller decrease elsewhere.

## 6.2  Radiative and Dielectronic Recombination

The carbon charge exchange rate is the dominant recombination process, hence there is only a small sensitivity to variation of the dielectronic and radiative recombination rates. Figure 7a shows the results of a 1-400X enhancement of the dielectronic recombination rate on the radiated power. A significant difference is seen only where the Li-like ion is active ($T_e$ < 50 eV) and these by a factor 2. The same variation in the radiative recombination rate produces an order of magnitude smaller change.

## 6.3  Charge Exchange Cross-sections

The cross-sections for $C^{3+}$, $C^{4+}$, and $C^{5+}$ collisions have been measured over much of our relevant energy range and are typically within a factor 2 of theoretical calculations. Using the base data described in Section 2, then, we study the sensitivity in the radiative cooling rate to changes in the $C^{3+,4+,5+}$ cross-sections, and the $C^{6+}$ cross-sections separately. Since the $C^{6+}$ cross-sections has not been measured over the whole energy range, the results of the parameters variation are somewhat more critical.

For the $C^{3+,4+,5+}$ variation (Figure 7b) the uncertainty in cross-section strongly affects the variation in the cooling rate. Hence, while the general features of the cooling rate curves are similar, if the cross-sections were to be a factor 2 lower than the center of the present experimental range, then a larger portion of the edge plasma would be needed to provide the required radiative emissions.

For $C^{6+}$ the sensitivity has been found to be smaller since the lower ionization stages radiate more effectively. The presence of significant radiation levels at high temperature is maintained (Fig. 7c).

## 7.  DISCUSSION

The particle and energy balance in magnetically confined fusion plasmas must inevitably be affected by atomic collisions processes, if only because perfect isolation from material surfaces is impossible. The particular process considered here, $H_o$ + $C^{n+}$ electron capture reactions in the 10 eV - 2 keV plasma edge thermal range, illustrates the typical role of collisions data in magnetic fusion particle and energy balance.

The inclusion of charge exchange cross-section data for the suite of carbon ions expected at these temperatures produces the following results:

· The dependence of relative abundance on $T_e$, characteristic of steady-state conditions without significant neutral hydrogen density ($n_o/n_e$ < 2.10$^{-4}$) is replaced by a dependence on $n_o$ and $T_e$. Further, the localization of relative abundance in a narrow temperature range is lost. Incompletely stripped carbon ions will exist, for typical neutral densities, at temperatures above 300 eV.

· Radiative cooling rates are enhanced by several orders of magnitude over those calculated with $\sigma_{cx}^z$ = 0. The levels are sufficient to encourage further attempts to fashion a radiative layer for transmission of large fusion power fluxes to external surfaces in future reactors. The large difference between charge exchange cross-sections for $C^{4+}$ and $C^{3+}$ leads to efficiently radiating charge states being well populated.

• Spatial asymmetries in light emission, viewed as a possible paradox with regard to drift-collision theory, are seen as a natural consequence of the asymmetry of the neutral hydrogen spatial distribution and the large values of the charge exchange cross-sections for $C^{4+}$, $C^{5+}$.

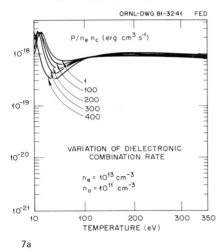

CARBON TOTAL RADIATION

7a

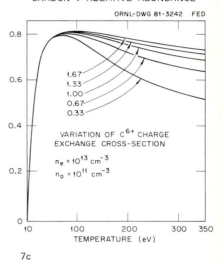

CARBON V RELATIVE ABUNDANCE

7c

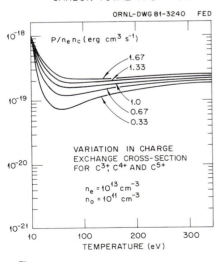

CARBON TOTAL RADIATION

7b

Figure 7
Sensitivity to variations in atomic data. For these cases $n_o/n_e = 10^{-2}$ has been assumed ($n_o = 10^{11}$ cm$^{-3}$, $n_e = 10^{13}$ cm$^{-3}$). $T_e = 10$ eV – 350 eV.

a. Variation of the dielectronic recombination rate. Cases for 1–400X the reference rate are shown.

b. Variation in the base cross-section for $C^{3+}$, $C^{4+}$, $C^{5+}$ + $H_o$ reactions (0.33–1.67X).

c. Variation in the $C^{6+}$ + $H_o$ cross-section (0.33 → 1.67 X).

REFERENCES

[1]  Atomic and Molecular Processes in Controlled Thermonuclear Fusion, NATO Advanced Study Institutes, Series B: Physics, Vol. 53 (Eds., M.R.C. McDowell, A. M. Ferendici) Plenum Press, New York, 1980.

[2]  Proceedings of the IAEA Advisory Group Meeting on Atomic and Molecular Data for Fusion, held at Culham Laboratory (UKAEA) 1976. Proceedings published by IAEA, Vienna, as IAEA-199 (1977). Invited papers published in Physics Reports, J. Phys. $\underline{37C}$ (1978).

[3]  Proceedings of the Second IAEA Advisory Group Meeting on Atomic and Molecular Data for fusion, held at CEN/Fontenay-aux-Roses Laboratory (CEA) 1979. Invited papers published in Physica Scripta $\underline{23}$ (1981).

[4]  K. Dolder, J. Hogan, Comments on Plasma Physics and Controlled Fusion, $\underline{6}$ 1 (1980).

[5]  W. Lotz, Z. Phys. $\underline{216}$ 241 (1968), Z. Phys. $\underline{220}$ 466 (1969).

[6]  D. H. Crandall, Physica Scripta $\underline{23}$ 154 (1981).

[7]  K. Lackner, private communication.

[8]  A. Gibson, M. Watkins, Proc. 8th European Conference on Controlled Nuclear Fusion and Plasma Physics, Prague, 1977, Vol. I, p. 31.

[9]  PLT Group, Proc. 7th International Conference on Plasma Physics and Controlled Nuclear Fusion, Innsbruck, 1978, IAEA Vienna $\underline{1}$ (167) 1979.

[10] S. L. Allen, H. W. Moos, R. K. Richards, J. L. Terry, E. S. Marmar, Nuclear Fusion $\underline{21}$ (251) 1981.

[11] R. V. Jensen, D. E. Post, C. B. Tarter, W. H. Grasberger, W. A. Lokke, Atomic and Nuclear Data Tables $\underline{20}$ 397 (1977).

[12] S. Suckewer, E. Hinnov, M. Bitter, R. Hulse, D. Post, Phys. Rev. A $\underline{22}$ 725 (1980).

[13] M. Mattioli CEN/Fontenay-aux-Roses Report EUR-CEA-FC 761 (1975).

[14] S. Von Goeler, W. Stodiek, H. Eubank, H. Fishman, S. Grebenshikov, E. Hinnov, Nucl. Fusion $\underline{15}$ 301 (1975).

[15] A. Burgess, Astrophys. Jour. $\underline{141}$ 1588 (1965).

[16] T. Heil, C. Bottcher, Proc. XII ICPEAC and private communication.

[17] T. Heil, A. Dalgarno, J. Phys. B. Lett. $\underline{12}$ L557 (1979); S. E. Butler, Phys. Rev. A $\underline{20}$ 2317 (1979).

[18] W. L. Nutt, R. W. McCullough, H. B. Gilbody, J. Phys. B Letters $\underline{11}$ L181 (1978).

[19] R. Phaneuf., Phys. Rev. A. $\underline{24}$ 1138 (1981).

[20] R. E. Olson, E. J. Shipsey, J. C. Browne, J. Phys. B $\underline{11}$, 699 (1978).

[21] E. J. Shipsey, J. C. Browne, R. E. Olson, J. Phys. B. $\underline{14}$ 869 (1981).

[22] M. Gargand, H. Hassen, R. McCarroll, P. Valiron, Proc. XII ICPEAC.

[23] J. T. Hogan, Methods in Computational Phys., V. 16, Applications to Controlled Fusion Research, ch. 3, (B. Alder, S. Fernbach, J. Killeen, eds.) Academic Press, New York, 1976.

[24] H. C. Howe, J. Nucl. Materials 93 and 94 17 (1980).

[25] G. G. Kelley (private communication).

[26] C. Breton, C. DeMichelis, M. Mattioli, Nucl. Fusion 16 891 (1976).

[27] C. Breton, C. DeMichelis, M. Mattioli, Nucl. Fusion CEN/Fontenay-aux-Roses Report EUR-CEA-FC 853 (1976).

[28] C. A. Flanagan, D. Steiner, G. E. Smith and the FED Design Center Staff, Oak Ridge National Laboratory Report ORNL/TM-7777, June 1981.

[29] S. Suckewer, E. Hinnov, J. Schivell, Bull. Amer. Phys. Soc. 23 697 (1978), also Princeton Plasma Physics Laboratory Report PPPL1430 (March 1978).

[30] J. L. Terry, E. S. Marmar, K. I. Chen, H. W. Moos, Phys. Rev. Lett. 39 1615 (1977).

[31] K. H. Burrell, S. K. Wong, Nucl. Fusion 19 1571 (1979).

PHYSICS OF ELECTRONIC AND ATOMIC COLLISIONS
S. Datz (editor)
© North-Holland Publishing Company, 1982

ATOMIC PROCESSES FOR DIAGNOSTICS OF MAGNETICALLY
CONFINED PLASMAS

S. Suckewer and E. Hinnov

Princeton Plasma Physics Laboratory
Princeton University
Princeton, New Jersey   08544
U.S.A.

In this paper the following subjects and related atomic
processes will be addressed: (a) the role of radiation in
magnetically confined plasmas; (b) deviation of the ion
species abundance from the coronal equilibrium caused by the
transport and charge exchange processes; and (c) applications
of spectral lines from the magnetic dipole transitions in
ground configurations of highly ionized atoms for plasma
diagnostics.

I.  INTRODUCTION

Magnetically confined plasmas comprise a large range of plasma parameters, the
most important of which, for atomic processes, is the electron temperature.
Since the plasmas produced in large tokamaks cover the largest range of
temperatures from a few tens of eV near the periphery to several keV in the
center in presently operating devices [1,2], with prospects of 5-10 keV in a few
years in the devices now under construction, it is sufficient to consider atomic
processes in tokamak plasmas only.

Even with this restriction, the scope of atomic processes and the diagnostics
based on such processes is much too large to aim at anything resembling
comprehensive treatment in the space limits of the present paper.  We have
therefore chosen for review three aspects of the problem, common to magnetically
confined plasmas:  1) the effect of plasma composition, i.e., impurity
concentration, on the radiative energy losses and energy balance in the plasma,
2) the prospect of using localized impurity ion density measurements to deduce
cross-field ion transport rates, and 3) the application of the relatively long
wavelength magnetic dipole lines from the ground configurations of highly-
ionized impurity atoms for localized diagnostics in the high-temperature plasmas.

These topics, and particularly the last two, are of course, intimately
connected.  Measurement of ion transport at a particular location obviously
requires the presence of an ion species at that location and the means of its
quantitative detection.  Along the (minor) radius, i.e., along a temperature
gradient, the various ionization states of an element would be distributed
according to the coronal ionization equilibrium, collisional ionization balanced
by electron-ion recombination, if the cross-field transport rate were
sufficiently slow.  The maximum densities of given ion species in this state tend
to occur at temperatures such that $T_e \approx (1/2 - 1/3) E_i$, the ionization potential
of the ion [3,4].  In tokamak plasmas where the neutral hydrogen density is
generally far above the equilibrium value, particularly during intensive netural-
beam heating, the recombination rate of highly ionized atoms may be significantly
augmented by charge-exchange with these H° atoms [5].  The result of this is to
move the ion density peaks to higher temperature, i.e., to smaller radii.  In
addition, if the radial particle transport rate is sufficiently rapid to move the
ion to a significantly different temperature location during an ionization or
recombination time, the radial ion density profiles are further modified, and it

is the latter effect that allows in principle the deduction of the ion transport
rates from measured density profiles [4,6]. The success of such a program
evidently depends on accurate measurements of the ion density and electron
temperature profiles, the rate coefficients of ionization and recombination, the
distribution of central hydrogen and the requisite charge-exchange rates, but
first of all the presence of the appropriate ion at a time and place where it is
to be measured.

Experiments have shown that ions in tokamak discharges generally occur at
temperatures higher than the coronal maximum, in the neighborhood of $E_i \approx T_e$
[5,7]. This criterion thus establishes the appropriate diagnostic elements.
Although the n = 1 shell or K-shell radiation offers various important diagnostic
possibilities, it is the n = 2 or L-shell that is of primary diagnostic interest,
being sufficiently simple for theoretical treatment and yet sufficiently complex
to allow a variety of measurement techniques. Thus, the preferred diagnostic
ions are those that have the ionization potentials of the lithiumlike states
roughly comparable to the peak electron temperature.

Some of these elements occur spontaneously in tokamak plasmas, e.g., Ti, Cr, Fe,
Ni. Others may be added deliberately for diagnostic purposes, e.g., Sc, Mn,
Cu. In the latter event it is important that the amount is sufficiently small so
as not to affect the plasma conditions appreciably and still sufficiently large
for quantitative detection. Fortunately such a range exists, at least in the
medium-Z elements ($Z \approx 20-30$).

For the next generation tokamaks, with peak electron temperatures hopefully
approaching 10 keV, this range of diagnostic elements must be expanded up to
about Zr or Mo. It is particularly in this range, $Z \sim 30-40$, where large amounts
of work in atomic structure and line identification is still ahead of us, in
addition to the perennial problems of accurate ionization and recombination
rates.

II.  RADIATED ENERGY AND IMPURITY ION CONCENTRATIONS

In a successful fusion reactor radiative energy losses must be small compared to
fusion energy production, at least in the high-temperature core of the plasma.
The conceptual designs thus require that radiation emmissivity be not much larger

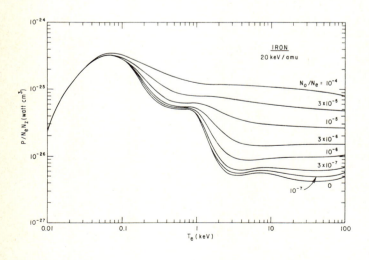

Fig. 1.
Calculated
changes in iron
radiation
losses per
particle due to
charge exchange
of iron with
fast neutral
hydrogen (by R.
Hulse, et al.,
[9])

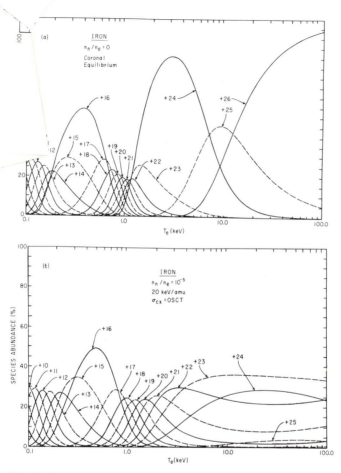

Fig. 2. Calculations of ionic species abundance for iron as a function of temperature for coronal equilibrium (a) and with additional charge-exchange recombination with 20 keV/amu H° neutrals with concentration $n_0 = 10^{-5} n_e$ (by R. Hulse et al., [9]).

than the hydrogen bremsstrahlung, or $\sim$ 20mW/cm$^3$. This condition sets limits to the allowable impurity concentration in the plasma, depending on the atomic weight of the impurity, the plasma temperature, and other conditions that may affect the ionization balance, i.e, deviation from coronal equilibrium.

In general the radiative efficiency of an element increases with Z and decreases with temperature as the atom becomes more completely stripped [3,8]. There is a broad, flat minimum of temperatures where a large fraction of the ions have reached heliumlike or higher states, and a slow increase at still higher temperatures where bremsstrahlung continuum emission becomes dominant. The lowest

The ultimate limits [13,14] for the concentrations of several elements under reactor-like conditions, a 50-50% mixture of deuterium and tritium with total density $10^{14}$ $cm^{-3}$ and electron-ion temperatures 10 keV or 20 keV, are shown in Fig. 3. For each element there evidently exists a fairly narrow range of concentrations in which its effect ranges from negligible to prohibitive for this purpose of achieving energy break-even conditions. This limit is lower with increasing atomic number Z, and with decreasing temperature. The data in Fig. 3 are calculated [14] for ordinary coronal equilibrium distributions. Charge-exchange recombinations or other deviations from the equilibrium will lower these limits further, because of the increased radiation efficiency, as described above.

III. PARTICLE TRANSPORT

The principal problem in magnetic confinement, the cross-field particle transport rate, is still generally called "anomalous," which is a euphemism for "not understood." The reason for this dismaying state of affairs is undoubtedly the dearth of adequate measurements, and it is in this area where plasma spectroscopy, employing diagnostic "impurities," must face its greatest challenge.

Available circumstantial evidence [15-18] appears to indicate that in tokamak discharges various ions can move both in and out with relative ease, and as a result, the plasma composition is fairly homogeneous. However, there is no clear consensus even on such a question, whether significant deviations from coronal ionization equilibrium exist. We shall present here samples of various representative observations.

Figure 4 shows measured radial profiles of several oxygen ion emissivities in the outer part of the TFR tokamak [15] (limiter radius 18 cm), and the electron temperature deduced from Thomson scattering measurements. Also on the upper graph are marked the locations of the emissivity peaks of these oxygen ions as well as two nitrogen and two carbon ions. (The emissivity peaks are quite close to the corresponding ion density peaks, except for the OVII ion, where the ion density profile is probably noticeably further out.) The peak emissivities occur at locations where $T_e(r) = 1/3$ $E_i$, which positions are indicated by crosses (+). Similar measurements in PLT and ST tokamaks have consistently shown higher-temperature locations, in the range $T_e(r) = (0.5-1.0)$ $E_i$, for the oxygen and carbon ions. Thus, although both sets of measurements show some deviation from coronal equilibrium, the deviation is much larger in the Princeton tokamaks. The reason for this different behavior is not know, but it may be related to another observation. The PLT discharges often exhibit pronounced poloidal asymmetries [19] in the peripheral light intensities (e.g., hydrogen, carbon, and oxygen light). Although their origin is also undetermined, they may be a symptom of a rapid nondiffusive radial transport of plasma.

However, closer to the center the poloidal asymmetry disappears, but the difference in observed deviations from equilibrium appear to persist. Figure 5 shows the measured [16] radial profiles of several Cr, Fe, and Ni ions in the TFR tokamak, compared to expected coronal equilibrium distributions (curves labelled 2). Curves 1 and 3 show the possible variation of the calculated coronal equilibrium curves due to the authors' estimate of the uncertainties in the ionization and recombination rates. The conclusion is that within the uncertainties, these ions may be close to coronal equilibrium.

A representative result from PLT tokamak [5,7] is shown in Fig. 6. There is a clear systematic shift of the observed ion density profiles toward higher temperatures, i.e., smaller radii. Although it is possible, by invoking systematic uncertainties in ionization and recombination rates and the temperature profile, to make the calculated profiles more or less coincide with observations, such an explanation seems rather forced (it is in fact easier to make the deviations larger than shown in the figure). The deviation is ascribed

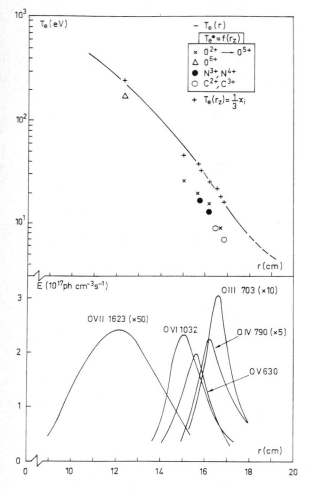

Fig. 4. Radial intensity
distribution of lines of
five stages of ionization
of oxygen in TFR tokamak
(lower part). In the
upper part the electron
temperature profile
(continuous curve) is
shown, and different
symbols indicate the
position of peak radiation
of the ions and the
correspondent temperature
of the maximum abundance
from coronal
equilibrium. Crosses mark
temperatures for these
peaks which are equal to
1/3 of ionization energy
$E_i = \chi_i$ (by TFR Group
[15]).

primarily to a net inward drift of these ions, with speed in the neighborhood of
3-4 cm/msec.

In order to reconcile such inward drift with observed lack of systematic
accumulation of any ion, there must be an outward drift, and this must be carried
to a large extent by ions in heliumlike and higher states of ionization.

Figure 7 shows measured [17] singlet (resonance) and triplet (intercombination)
line intensities of the heliumlike OVII ion at various chord distances in the
DITE tokamak. The relative increase of the triplet line with increasing
distance, r, [decreasing $T_e(r)$] is interpreted as being caused by recombination
of the hydrogenlike OVIII ion, which in turn must be produced at high
temperatures near the center. This measurement thus allows the deduction of the
distance an OVIII ion moves radially during a recombination time.

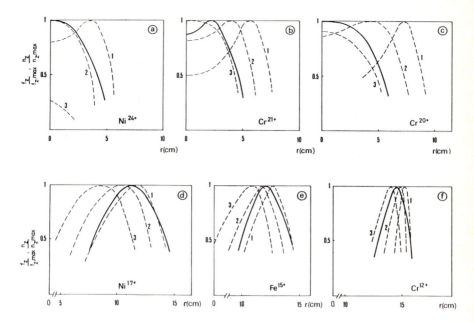

Fig. 5. Comparison of ion density profiles (solid lines) with its coronal equilibrium distribution (dashed lines) for six metal ions in TFR; curve 2 corresponds to the calculation of coronal equilibrium by Breton et al., whereas curves 1 and 3 are obtained by multiplying and dividing, respectively, this ratio by a factor of two (by TFR Group [16]).

An analogous measurement in argon [18] (added to the discharge for diagnostic) in the PLT tokamak is shown in Fig. 8. The soft-x-ray spectrum measured at different chords, shows the appearance of the recombination continuum of the hydrogenlike ArXVIII ion, at a distance (22.5 cm) where the electron temperature was well below the equilibrium temperature of ArXVII. These measurements imply outward drifts of ions at the rate of several cm/msec, commensurate with the inward drifts deduced from the lower ionization state measurements.

Similar conclusions about ion drifts, both inward and outward, have been reached in many other experiments, e.g., silicon injection in Alcator [20], scandium injection in PDX [21], disappearance of tungsten in PLT [22] upon edge cooling by neon and of Fe and Ti by hydrogen puffing [23], iron K-line fine-structure [24] measurements, etc.

There is also an experiment on ISX [25] tokamak that is interpreted to exhibit only inward drift and resulting accumulation of injected silicon.

Thus, there is a lot of fragmentary evidence showing various tantalizing aspects of particle transport. What is needed is systematic experiments with improved measurement techniques, and refined evaluation of the ionization and recombination processes.

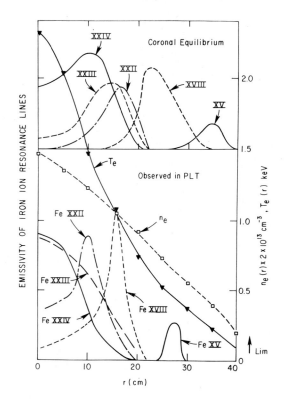

Fig. 6. Observed radial distribution of highly ionized iron lines and the distribution expected in coronal equilibrium in a PLT discharge [5,7].

DITE GRAZING INCIDENCE GRATING SPECTRA

Demonstrating the variation of O VII $1s^2$-$1s2p$ $^1P_1$ $^3P_1$
ratio, R, with radial position ,r.    100kA discharge

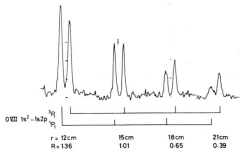

Fig. 7. Ratio of intensities of intercombination and allowed lines $1s^2$ $^1S_0$-$1s2p$ $^3P_1$/$^1P_1$ of OVII versus plasma radius r in DITE tokamak [17].

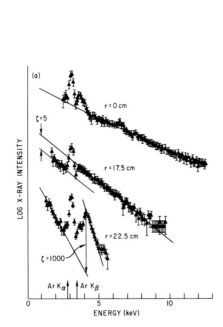

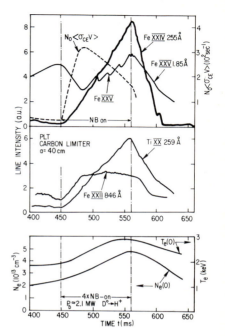

Fig. 8. Soft x-ray continua at different chords, showing argon XVIII recombination edge at ~22.5 cm in PLT tokamak [18].

Fig. 9. Time evolution of FeXXIV, FEXXV, FeXII and TiXX line intensities during neutral beam injection into PLT [12]. Dashed curve represents rate of charge exchange of FeXXV with D°. Central electron temperature $T_e(o)$ and density $n_e(o)$ are shown in the lower part of the figure.

## IV. CHARGE-EXCHANGE RECOMBINATION

One of the atomic processes that evidently needs to be taken into account quantitatively is charge-exchange recombination between highly-ionized atoms and neutral hydrogen. Clearly, if the H° charge-exchange rate coefficient is about $10^{-6}$ cm$^{-3}$ sec$^{-1}$ (which may be a conservative figure for 20-30 times ionized atoms) and the collisonal ionization or recombination rate is ~$10^{-11}$ cm$^{-3}$ sec$^{-1}$ (as it is, e.g., for Fe XXII - XXIV), a H°/electron ratio ~$10^{-5}$ is sufficient to influence strongly the ionization state balance, as shown [9] in Fig. 2. This not only changes the radiative efficiency of an element at high temperatures, as was shown in Fig. 1, but also influences the interpretation of radial transport rates as deviation from coronal equilibrium. During high power neutral beam heating experiments in PLT it was indeed observed [12] that the iron (and titanium) ionization states were substantially lowered (Fig. 9) in spite of increasing electron temperature, and the change was quantitatively consistent with the expected charge-exchange recombination, although many important details concerning the observed radial distributions have remained unexplained.

The effect on recombination becomes relatively less important in lighter elements, where charge-exchange rate coefficients are smaller and ionization rates larger. However, in light elements, e.g., carbon and oxygen, the charge-exchange effect may be useful in diagnostics, either for measurement of the very important distribution of otherwise unobservable completely stripped atoms [26,27,28], or possibly, with the help of quantitative diagnostic H° beams, measurement of the intrinsic H° density profiles, especially during neutral beam heating. With increasingly improved charge-exchange data becoming available [29,30,31], the most critical part of the problem may indeed be the quantitative determination of the H° distribution.

## V. MAGNETIC DIPOLE LINES OF HIGHLY IONIZED ATOMS

Experiments such as radial ion distribution measurements require many observations in a given discharge for a sufficiently complete set of experimental data. Furthermore, various uncertainties in the atomic rate coefficients make it desirable, or even necessary, that the measurements be redundant, i.e., by observing emission of different elements at the same locations in plasma. Such detailed measurements in the extreme u.v. with grazing incidence or equivalent spectrometers tend to be very cumbersome, or very expensive.

Fortunately, various magnetic dipole transitions in the ground configurations of highly ionized atoms have often sufficient intensity for measurement, in a wavelength region where normal incidence optics and multiple reflections are still practicable [32,33]. The use of such lines, in conjunction with laser blow-off injection [34] of the suitable diagnostic elements into the plasma greatly extends the facility and precision of localized spectroscopic measurements [35].

Figure 10 shows a set of scandium ion lines that were identified [35] in the PDX tokamak. Note the ionization potentials which give the approximate temperature

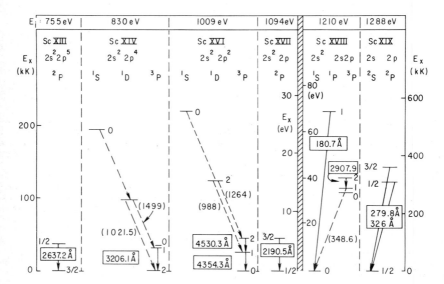

Fig. 10. Energy levels and wavelengths of scandium forbidden transitions in $2s^2 2p^n$ configurations and allowed 2p-2s transitions of ScXVIII and ScXIX. Wavelengths in boxes are those observed in PDX tokamak [35].

at which the lines are applicable. Recently, we have undertaken a systematic measurement of analogous lines in Cr, Fe, Ni and Cu in the PLT tokamak, using the laser blow-off for eased identification. During the course of this measurement it was found that two of the lines, corresponding to the $^3P_1 \rightarrow {}^3P_0$ transition in CrXIX and $^3P_2 \rightarrow {}^3P_1$ transition in FeXXI, had been misidentified [36] (2100.$\overset{\circ}{A}$ and 2304.5$\overset{\circ}{A}$ respectively). The correct wavelengths, together with other transitions of these ions observed in PLT or PDX tokamaks, are given in Table 1 (line TiXIX 2344.5$\overset{\circ}{A}$ was first observed in DITE [37]). The corresponding copper wavelengths and lines arising from intercombination (change of multiplicity) transitions are still being evaluated. Lines in the nitrogen sequence, except the 2665 $\overset{\circ}{A}$ line of FeXX, have not yet been positively identified.

TABLE 1. WAVELENGTHS (Å)* OF MAGNETIC DIPOLE TRANSITIONS OBSERVED IN PLT AND PDX TOKAMAK DISCHARGES

| TERM | $2s^2 2p^5$ $^2P$ | $2s^2 2p^4$ $^3P$ | $2s^2 2p^2$ $^3P$ | | $2s^2 2p$ $^2P$ | $2s^2 2p$ $^3P$ |
|---|---|---|---|---|---|---|
| ΔJ | $1/2 \rightarrow 3/2$ | $1 \rightarrow 2$ | $2 \rightarrow 1$ | $1 \rightarrow 0$ | $3/2 \rightarrow 1/2$ | $2 \rightarrow 1$ |
| SEQUENCE | FI | OI | CI | | BI | BeI |
| Sc | 2637.2 ±0.2 | 3206.1±0.3 | 4530.3±0.4 | 4354.3±0.4 | 2190.5 ±0.2 | 2907.9±0.3 |
| Ti | 2117.1 ±0.2 | 2544.8±0.1 | 3834.4±0.2 | 3370.8±0.2 | 1778.1 ±0.1 | 2344.6±0.2** |
| Cr | 1410.6 ±0.3 | 1656.3±0.2 | 2885.4±0.2 | 2090.9±0.2 | 1205.9 ±0.3 | |
| Fe | 974.85±0.15 | 1118.2±0.15 | 2298.0±0.2 | | 845.55±0.1 | 1079.3±0.3 |
| Ni | 694.55±0.15 | 2818.2±0.3 | 1914.7±0.3 | 911.0±0.3 | | |

*WAVELENGTHS ABOVE 2000Å ARE IN THE AIR.

**THIS LINE WAS FIRST OBSERVED IN DITE TOKAMAK AND PRACTICALLY COINCIDES WITH OUR MEASUREMENTS.

The magnetic dipole lines have been applied to a variety of local diagnostics, such as ion Doppler temperature, plasma rotation, and local (sawtooth) fluctuation measurements [38,39,40]. An interesting case study of partially determined $T_i(r)$ profile is given in Fig. 11, which shows the ion temperature change caused by application of 340 kW ICRF power. Both the Doppler widths and the radial locations were directly measured for the 2545 Å line of TiXV ($E_i$ = 0.94 keV) and the 2885 Å line of CrXIX ($E_i$ = 1.4 keV). Both the spontaneously present TiXV and the (laser blow-off) injected Cr XIX lines were centrally peaked, the latter more narrowly. With the central $T_e(0) \sim$ 800 eV, the TiXV line fulfills the approximate criterion $E_i \sim T_e$, but the CrXIX $E_i$ nearly twice this figure. The implication, although not directly confirmed in this case because of insufficient operation time, is that CrXVI-XVII were also centrally peaked and presumably more abundant than CrXIX. This indicates that the injected amount of chromium, although it did not noticeably affect the temperature (or other plasma condition) behavior, was as well above the minimum necessary for the measurement. On the other hand, for an ion temperature profile measurement at the given $T_e(r)$, one should have chosen a lighter element such as Sc or Ca. Of course, it would have been of great interest to have direct experimental demonstration of this conjecture.

The approximate $T_e(r)$ range of applicability of the various ions in their n = 2 shells are shown in Fig. 12, for two hypothetical electron temperature profiles (together with some often used carbon and oxygen lines, and also an n = 3 shell line of Kr for comparison). Clearly for the $T_e(0)$ = 2 keV profile, typical of present day large tokamaks, elements with Z ≈ 20-30 are adequate, but at $T_e(0) \sim$ 5 keV even krypton .is becoming marginal. Thus, for the prospective 5-10 keV plasma diagnostics, it is high time to establish the atomics structure and wavelengths, as well as the appropriate ionization and recombination rates for the elements in the range Z ~ 30-40.

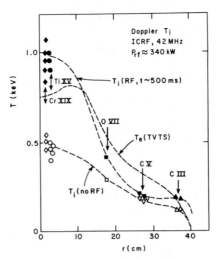

Fig. 11. Application of TiXV and CrXIX forbidden lines for ion temperature profile measurement during ICRF(42MHz) heating experiment in PLT.

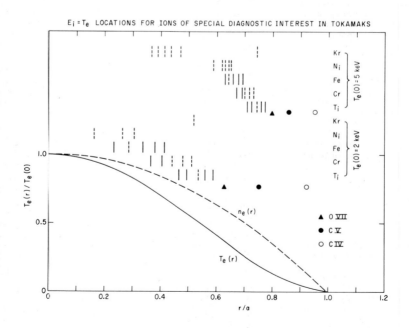

Fig. 12. Approximate positions of radiation maxima, assuming $E_i = T_e$, of the various magnetic dipole lines and some popular carbon and oxygen lines in tokamak discharges with central temperatures 2 keV and 5 keV.

CONCLUSIONS

There exists a range of densities for any particular element that may be introduced into a tokamak discharge, that is sufficiently large for adequate local diagnostics, and yet does not change the plasma behavior significantly.

Detailed systematic measurements of various ion density profiles of successive state of different elements are necessary for adequate description and ultimate understanding of particle transport in tokamaks and other confinement devices.

The most appropriate (although not the only valuable) diagnostic elements are those for which the lithiumlike state ionization potential is comparable to the peak electron temperature. The laser blow-off method has proved valuable for controlling the amount and the time of injection of the diagnostic tracer elements.

Gathering the necessary large quantity of data for adequate ion density profiles is greatly facilitated by the use of the relatively long wavelength magnetic dipole lines of highly ionized atoms.

Continued refinement of atomic rate coefficients, of ionization and recombination, including charge exchange recombination with neutral hydrogen (isotopes), is essential for adequate interpretation of measured ion density profiles in terms of particle transport. Equally important is further development of electron temperature measurement. (We note that 90° Thomson scattering becomes quite problematic in the 5-10 keV range, expected in the near future.)

Determination of particle transport rates by interpreting measured ion density profile is clearly a formidable task, but it must be kept in perspective by considering the time and effort already spent on this problem without very noticeable success.

ACKNOWLEDGMENTS

Most of the results presented here are from PLT and PDX and reflect effort of entire groups working on these devices. Figures 1,2,3 are taken from the works of D. Post, R. Hulse et al., and R. Jensen et al., Fig. 4,5 from the TFR Group, and Fig. 7 from DITE tokamak.

We are indebted to Professor B. Edlen for casting doubt on the identity of the 2100.6 Å line as originating from CrXIX.

This work is supported by the United States Department of Energy Contract No. DE-AC02-76-CHO-3073.

REFERENCES

[1]   Artsimovich, L.A., Nuclear Fusion 12 (1972) 215.
[2]   Furth, H.P., Nuclear Fusion 15 (1975) 487.
[3]   Breton, C., DeMichelis, C., and Mattioli, M., Nuclear Fusion 16 (1976) 891.
[4]   Drawin, H.W., Review Paper at IAEA Advisory Group Meeting on Atomic and Molecular Data for Fusion, Culham Laboratory, England (1976); Physics Reports 37 (1978) 125.
[5]   Hinnov, E., Diagnostics for Fusion Experiments, Proc. of the Course, Varenna, Italy, 1978, (Edited by Sindoni, E. and Wharton, C.) Pergamon, Oxford (1979).
[6]   Suckewer, S., Review Paper at IAEA Advisory Group Meeting on Atomic and Molecular Data for Fusion, Fontenay-aux-Roses, France (1980); Physica Scripta 23 (1981) 72.

[7]     Suckewer, S., and Hinnov, E., Phys. Rev. A20 (1979) 578.
[8]     Post, D.E., et. al, Atomic Data and Nuclear Data Tables 20 (1977) 397.
[9]     Hulse, R.A., Post, D.E., and Mikkelsen, D.R., J. Phys. B. 13 (1980) 3895.
[10]    Olson, R.E., and Salop, A., Phys. Rev. A14 (1976) 579.
[11]    Olson, R.E., and Salop, A., Phys. Rev. A16 (1977) 531.
[12]    Suckewer, S., Hinnov, E., Bitter, M., Hulse, R., and Post, D., Phys. Rev. A22 (1980) 725.
[13]    Meade, D.M., Nuclear Fusion 14 (1974) 289.
[14]    Jensen, R.V., et al., Nuclear Fusion 17 (1977) 1187.
[15]    TFR Group, Plasma Physics 20 (1978) 207.
[16]    TFR Group, Plasma Physics 22 (1980) 851.
[17]    Peacock, N.J., et al., Plasma Physics and Controlled Fusion Research (Innsbruck, 1978), Nuclear Fusion Supplement, Publ. IAEA Vienna, Vol. 1 (1979) 303.
[18]    Brau, K., von Goeler, S., et al., Phys. Rev. A22 (1980) 2769.
[19]    Suckewer, S., Hinnov, E., and Schivell, J., Princeton University, Plasma Physics Laboratory Report PPPL-1430 (1978).
[20]    Marmar, E.S., Rice, J.E. and Allen, S.L., Phys. Rev. Lett. 45 (1980) 2025.
[21]    Cecchi, J., et al., Bull. Am. Phys. Soc. 25 (1980) 927.
[22]    Hinnov, E., et al., Nuclear Fusion 18 (1978) 1305.
[23]    Hinnov, E., et al., Bull. Am. Phys. Soc. 24 (1979) 1054.
[24]    Bitter, M., von Goeler, S., et al., Phys. Rev. Lett. 42 (1979) 304.
[25]    Burrell, K., et al., Nuclear Fusion (1981) (to be published).
[26]    Isler, R.C., Phys. Rev. Lett. 38 (1977) 1359.
[27]    Isler, R.C. and Crume, C., Phys. Rev. Lett. 41 (1978) 1296.
[28]    Afrosimov, V.V., et al., JETP Lett. 28 (1978) 500.
[29]    Gilbody, H.B., Physica Scripta 23 (1981) 143.
[30]    Bliman, S., et al., Phys. Rev. A23 (1981) 1703.
[31]    Salzborn, E. and Muller, A., in Invited Papers and Progress Reports of the XI ICPEAC, Kyoto, 1979 (Edited by Oda, N. and Takayanazi, K.), North-Holland, Amsterdam (1980), p. 407.
[32]    Suckewer, S. and Hinnov, E., Phys. Rev. Lett. 41 (1978) 756.
[33]    Hinnov, E., Fonck, R., and Suckewer, S., Princeton University, Plasma Physics Laboratory Report PPPL-1669 (1980).
[34]    Cohen, S.A., Cecchi, J.L., and Marmar, E.S., Phys. Rev. Lett. 35 (1975) 1507.
[35]    Suckewer, S., Cecchi, J., Cohen, S., Fonck, R., and Hinnov, E., Phys. Lett. A80 (1980) 259.
[36]    Hinnov, E. and Suckewer, S., Phys. Lett. A79 (1980) 298.
[37]    Lawson, D.K., Peacock, N.J., and Stamp, M.F., J. Phys. B. (to be published).
[38]    Eubank, H., et al., Phys. Rev. Lett. 43 (1979) 270.
[39]    Suckewer, S., et al., Nuclear Fusion (to be published).
[40]    Hinnov, E., et al., Bull. Am. Phys. Soc. 25 902 (1980).

PHYSICS OF ELECTRONIC AND ATOMIC COLLISIONS
S. Datz (editor)
© North-Holland Publishing Company, 1982

# SURVEY OF ATOMIC PHYSICS ISSUES IN EXPERIMENTAL
# INERTIAL CONFINEMENT FUSION RESEARCH*

Allan Hauer

University of California
Los Alamos National Laboratory


E. J. T. Burns
Sandia National Laboratory

Atomic processes impact many of the interactions important to
inertial confinement fusion. For example, ionization of bound
electrons represent one of the primary stopping mechanisms for
both ions and electrons. Atomic processes are also important in
predicting the behavior of the very dense plasmas which are central
to the success of inertial fusion. Atomic spectroscopy is an
important diagnostic of ICF plasmas.


## INTRODUCTION

Inertial confinement fusion (ICF) requires the production of high temperature,
high density plasma. Laser driven implosions have, for example, been able to
produce plasma densities of the same magnitude as ordinary solids (a few
$g/cm^3$) with temperatures of the order of a kilovolt. Atomic processes play
an important role in the generation of such plasma conditions. In addition,
atomic spectroscopy has become an important diagnostic of ICF plasma conditions.

Atomic processes are also very important in the physics of ICF drivers. Design
of ion sources and accelerators, beam propagation, more efficient and variable
wavelength lasers, are crucial problems in the ICF field. Space does not,
however, permit dealing with this very important component of the ICF field.

In Fig. 1, we illustrate the general nature of the interactions involved in
particle and laser driven compression.

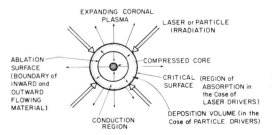

Figure 1.
Schematic of Laser/Particle
Target Interaction

In the case of laser heating, energy is absorbed at an electron density near
the critical value for the laser involved ($10^{19}$ cm$^{-3}$ for $CO_2$, $10^{21}$
cm$^{-3}$ Nd:glass). Atomic processes can be quite important in the ionization
and recombination dynamics in the absorption region.

In the case of ion beam drivers, the absorption process is, of course, quite
different. The ionization and excitation of the target material by ions is a
major contribution to the stopping power. Atomic processes are important, not
only in the calculation of ionization cross sections, but evaluating the
ionization dynamics of the ion heated target.

The ablation of material from the outer portion of the target drives an
implosion. When material converges and stagnates in the center, a heated and
compressed core is formed. Spectral lines emitted during this compressed phase
are the chief source of information on implosion dynamics.

In the absorption of laser light, a significant portion of the laser energy may
be deposited in a hot electron distribution. The transport and deposition of
this hot electron population involves several important atomic physics
questions, such as bound ionization stopping power.

In Table I, we briefly review some of the major ICF processes that will be
discussed and their correlative atomic processes.

TABLE I

| ICF RELATED PROCESS | ATOMIC PROCESSES INVOLVED | THEORY AND MODELING | DIAGNOSTIC TECHNIQUES |
|---|---|---|---|
| Absorption of laser light | Ionization and re-combination dynamics in presence of large density gradients and electric fields | Rate equation models coupled to hydro codes or other plasma modeling | (1) Spatially re-solved spectral line observation (2) Charged particle ionic species and energy spectrum |
| Transport and deposition of fast ions and electrons | (1) Collisional ionization and excitation (2) Bremsstrahlung (3) electron scattering | Stopping power formulations needed for input to comprehensive hydro modeling | Inner shell and bremsstrahlung radiation used as indicator of particle deposition |
| Thermal, particle, and radiation transport | (1) Bound contri-butions to absorption (2) Thermal and electrical con-duction in partially ionized plasma | (1) Opacity cal-culations needed for radiation treatment in com-prehensive models (2) Calculation of transport coefficients | (1) Line radia-tion from lay-ered targets as indicator of thermal transport (2) Absorption spectra |
| Production of high density imploded plasma | (1) Quantum effects in the equation of state (EOS) of dense plasmas (2) Line radiation from higher Z plasmas | (1) Thomas Fermi Dirac EOS as starting point (2) Spectral model-ing with rate equations coupled to hydro codes | (1) Observation in dense plasma of (a) line pro-files (b) line shifts (c) continuum edge structure (2) High intensity shock studies |

ENERGY ABSORPTION AND TRANSPORT IN ICF PLASMA

The transport and deposition of hot electrons and fast ions involve similar atomic problems. Questions such as bound ionization stopping power are common to both problems. Hot electron transport is a crucial question in laser fusion while fast ion stopping is, of course, central to ion beam fusion.

A. Ion Beam Deposition

Most experimental data on ion beam energy loss has been obtained only for cold materials.[1] A central question is thus the effect of dense high temperature plasma conditions on ion beam stopping. The experimental stopping power of light ion beams, as a function of ion beam energy and target atomic number for cold matter, are summarized in a number of references.[2,3] The energy loss of ions in cold material is primarily due to the processes of ionization and excitation of bound electrons. At low ion energies, inelastic nuclear scattering can also result in appreciable energy loss.

A number of authors[4,5,6,7] have calculated the stopping power of light and heavy ions due to bound electrons for various elements. Mehlhorn[7] has calculated stopping power for bound electrons using the LSS[8] model at low incident ion beam velocities and the Bethe[9] model at high incident ion beam velocities. The dividing line between the two models is for the incident particle velocity parameter $\beta$ (= $v/c$),

$$\beta \sim \frac{Z_1^{1/3}}{137} , \qquad (1)$$

where $Z_1$ is the atomic number of the projectile. In the Bethe model, the stopping power is proportional to $Z_{eff}^2$, where $Z_{eff}$ is the effective charge of the projectile given by the expression[9]:

$$Z_{eff}/Z_1 = 1 - 1.034 \exp(- 137.04 \ \beta/Z_1^{0.69}). \qquad (2)$$

Rogerson[5] uses the "local oscillator model" (LOM) to calculate the stopping power of bound electrons. The local bound electron density in the target atom is taken from the Thomas-Fermi (TF) model proposed by Zink.[10] For ionized targets, the Saha equation is used to find the average ionization state of the target material as a function of the target temperature and density. For lower atomic number materials it may be necessary to perform non-LTE calculations of the average ionization state.

McGuire[11] uses a generalized oscillator strength (GOS) formulation of the Born approximation to calculate the excitation and ionization contribution to the proton stopping power by individual subshells. McGuire[11] presents stopping powers for individual subshells for Au I – AU XII for protons with energy between 0.1 and 10 MeV.

Figure 2 shows a comparison between proton stopping powers (incident on Au XII) as a function of incident proton energy for the detailed subshell stopping powers of McGuire[11] and the Bethe model as described by Mehlhorn.[4] For plasma temperatures on the order of 100 eV in gold, bound electrons make significant contributions to stopping power.

The free electron stopping power in plasmas are described by either short range binary collisions plus long range collective effects[12] or by dielectric function models.[4,7]

Generally, free electrons are more effective in stopping ions than bound electrons. However, if the thermal velocity of the free electrons greatly exceeds the velocity of the incoming ion, the stopping power actually decreases. Bound electrons are very important in ion stopping, particularly where the target plasma is not fully stripped. Obviously, bound electron stopping powers are very important at low temperatures where the ratio of bound

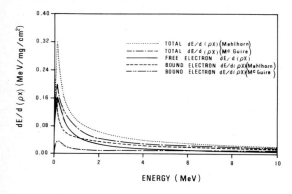

Fig. 2

Comparison of
Ion Stopping Power
Formulations

electrons to free electrons is great. Also, the stopping power is higher for low energy ions, so long as the ion has sufficient energy to ionize the electrons in the target. Detailed calculations of subshell stopping powers of various elements are very important for ions with velocities equal to the orbital velocities of the target subshells.

B. Hot Electron Transport in Laser Plasmas

When laser light is absorbed in a plasma (near the critical surface), a portion of the absorbed energy is usually contained in a suprathermal electron distribution. Large electric fields arise because of the charge separation that occurs when high energy electrons initially escape the target. The presence of these fields may cause the electron transport to differ from the classical picture of electron penetration in cold solids. In addition, it has been suggested that other mechanisms such as magnetic fields and return current inhibition[13] may alter the transport of these hot electrons. The energy deposition and transport of these hot electrons is of particular importance in laser driven fusion.

As with heavy ions, the stopping of fast electrons in dense matter is due to ionization and excitation of bound electrons and collective loss to free electrons.[12] Evaluation of the bound contribution begins with the well known Bethe form for the cross sections for ionization of electrons in various shells.[14]

$$\sigma_{n\ell} = \frac{\pi e^4 \, Z_{n\ell}}{E \, E_{n\ell}} \, A_{n\ell} \; \ell n \, [\frac{4E}{B_{n\ell}}] \tag{4}$$

where $E$ = electron energy, $Z_{n\ell}$ = number of electrons in $n\ell$ shell, $E_{n\ell}$ = ionization potential of $n\ell$ shell. $A_{n\ell}$ and $B_{n\ell}$ are constants dependent on atomic structure. The simplest form of the Bethe–Bloch energy loss formula is obtained by summing Eq. (4) over all shells. As in the ion stopping power formulation, an average excitation potential is usually used in place of a rigorous summation over atomic shells. It can be shown that the total stopping power due to both bound and free contributions can be written approximately[15] as:

$$\frac{dE}{dx} = - \frac{2\pi N_{ion} e^4}{E} \left[ Z_B \ln \left(\frac{1.1E}{\bar{I}}\right) + Z_f \ln \left(\frac{1.1E}{I_0}\right) \right] , \qquad (5)$$

where $\bar{I}$ = average bound excitation energy, $I_0 = (0.53)/\hbar\omega_p$ ($\omega_p$ = plasma frequency), $N_{ion}$ = ion density. $Z_f, Z_B$ are the number of free and bound electrons per ion, respectively. The second term in brackets is the stopping contribution due to free electrons. This contribution thus depends on the plasma density and average charge state. Spectroscopic techniques exist for measuring the bound contribution to stopping, as is discussed in the next section.

In addition to the energy loss mechanisms, atomic processes also impact the overall models of energy transport. Hot electron transport has been modeled with a variety of techniques[16] including Boltzmann equations[17] (in the presence of electric fields) and Monte Carlo[18] codes. In formulating these models more accurate experimental measurements of electron angular scattering distributions would be desirable.

C. Inner Shell and Bremsstrahlung Radiation Diagnosis of Particle Deposition

Inner shell ($K_\alpha$) radiation is an important diagnostic of both suprathermal electron and ion beam deposition in dense materials.

In Fig. 3, we show the basic configuration of experiments in inner shell electron transport diagnosis. Hot electrons generated near the laser critical surface propagate through a low Z material and are incident on the $K_\alpha$ radiator. The basic question is "what is the relation between observed $K_\alpha$ intensity and electron energy deposited in any material " As has been pointed out by Hares, et al.,[19] when the average potential formulation is used, the ratio of K shell to total stopping power is relatively insensitive to incident electron energy. If the average excitation picture is valid, a simple proportionality between $K_\alpha$ intensity and deposited fast electron energy can be established.

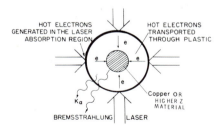

HOT ELECTRONS GENERATED IN THE LASER ABSORPTION REGION

HOT ELECTRONS TRANSPORTED THROUGH PLASTIC

Copper OR HIGHER Z MATERIAL

$K_\alpha$

BREMSSTRAHLUNG     LASER

Fig. 3
Inner shell and Bremsstrahlung measurements of electron transport

Taking the ratio of K shell to total stopping power one obtains:

$$\frac{dE/dx|_{tot}}{dE/dx|_k} = c \frac{\ln(4E/\bar{I})}{\ln(E/Ek)} . \qquad (6)$$

Where c = constant. In many cases Eq. (6) is probably a good general approximation. Several complications should, however, be considered.

In the case of $CO_2$ laser irradiation, the hot electrons carry a major fraction of the absorbed energy. We must, therefore, consider the heating of the $K_\alpha$ emitting material. As the electrons are ionized from the emitter the

$K_\alpha$ transition is shifted to higher energies. Substantial shifts begin to occur when electrons are removed from the L shell.[20]

In Fig. 4, we show such a shifted satellite emission spectrum. A slit perpendicular to the dispersion direction provided coarse spatial resolution (~ 100 μm). The line marked $K_\alpha$ is the usual cold inner shell transition. Satellite lines on the high energy side are the $K_\alpha$ lines shifted by ionization. The $1s^2-1s2p$ helium-like line is also present in the spectrum. A densitometer trace of the spectrum is also shown in Fig. 4. Satellite line identifications agree with the calculations of House[20] to within a few percent.

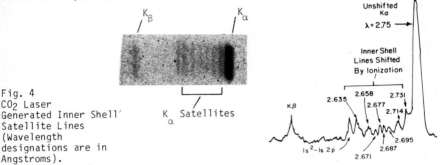

Fig. 4
$CO_2$ Laser
Generated Inner Shell
Satellite Lines
(Wavelength
designations are in
Angstroms).

The lateral shift in the position of the satellite lines (with increasing energy) is due to the fact that the lines are generated in different regions of the plasma. In these tests, the slit was too large to obtain detailed spatial information. The lateral shift does, however, indicate that the more highly ionized lines are emitted from regions farther from the original target surface. This agrees with the general picture of a hot plasma expanding from a laser irradiated solid surface.[21] In order to use the inner shell emission as an energy deposition indicator, the intensity from all shifted components of $K_\alpha$ must be considered. This, in turn, requires that careful consideration be given to the transport of $K_\alpha$ radiation through the heated material.

When a vacancy in the L shell occurs, the possibility of resonant line absorption is present.[22] This must be taken into account in calculating the observed $K_\alpha$ intensity. Calculations of the complete $K_\alpha$ spectrum from heated material have been undertaken in several laboratories. The study of inner shell satellite generation should be a valueable diagnostic of plasma conditions in addition to being an overall indicator of electron energy deposition.

An additional complication with inner shell diagnosis of energy deposition is the contribution to stopping power from collective loss to free electrons (plasma oscillations). This loss mechanism is important in the stopping of both electrons and heavy ions. As we have seen in Eq. (5), the loss due to free and bound electrons can be written in a similar way. If a rough indication of the temperature of the $K_\alpha$ emitting material is available, the bound contribution to deposition (as indicated by $K_\alpha$ intensity) can be used to estimate the free electron loss.

Another consideration in the use of $K_\alpha$ diagnostics of hot electron deposition is the variation in fluorescent yield with ionization state of the $K_\alpha$ emitting material. Ionization causes the ratio of $K_\alpha$ line emission to Auger emission to vary. Calculations of this effect have been made by McGuire.[23]

For example, $Al^{+5}$ would have a fluorescent yield that differs from the neutral value by a factor of 1.5.

In $CO_2$ laser work, the hot electron energies can be very high in some cases having a significant portion of the population with energies of several hundred kilovolts. At these energies, one should consider whether relativistic corrections (to the "Bethe" form as used in Eqs. (4) and (6)) to the inner shell cross sections are in order. Good general guides to this question are the works of Kolbenstvedt[24] and Arthurs and Mosiewitsch[25] (see also 26). For high Z $K_\alpha$ emitters (Z > 40) and high energies (> 200 keV) errors of the order of 2 might be made in the deposited electron energy by using the "Bethe-like" formulation used in Eq. (6).

Inner shell diagnostics are also very important in ion beam work. For example, in light ion fusion work the ion current incident on the target is often determined from inner shell ionizations.

X-ray yields must be considered thick-target yields, since most if not all of the incident ion energy is deposited in the target. As in the electron case, heating of the $K_\alpha$ emitter must be considered. In present light, ion experiments energy is deposited over 30-40 ns. The time dependent ionization of the emitter thus should be taken into account.

Inner-shell emission, via ions incident on cold target material (atoms) has been reviewed by Richard.[27] We will confine our discussion to direct coulomb interactions between the field of the incoming ion and the bound electron in the target atom and neglect electron promotion via the molecular orbital (MO) model. This picture is appropriate for present work which has been done with protons as the incident particles. The cross sections for K-shell and L-shell ionization are well characterized in the binary encounter approximation (BEA) by[28]

$$\sigma(E_i) = (nZ_1^2 \sigma_0/V^2) G(v_i/v_e) ,\qquad (7)$$

where $E_i$ is the incident ion energy, n = 2 for K-shell, n = 3 for L shell, etc., $Z_1$ is the projectile atomic number and V is the binding energy of the ionized electron shell in the target, $G(v_i/v_e)$ is a function of the ratio of the incident-projectile velocity $v_i$ and the average ionized orbital-electron velocity $v_e$. For ionized targets, then, the cross section would decrease as the binding energy of the K- or L-shell electrons increases and is thus a function of the ionization state of the plasma.

As in the electron case, a more complete model of $K_\alpha$ satellite production would be desirable for energy deposition studies. Nardi and Zinamon (See ref. 22) calculate the line absorption for a given ionization species by merging all of the possible L-S couplings of the excited inner-shell states into a single satellite line and assign it an average oscillator strength. These transitions are then transported through the plasma. They find significant line emission differences with and without line opacity.

Inner shell spectroscopy discussed here has the most commonality between ion beam and laser work. In laser ICF experiments, "optical" spectroscopy using transitions in highly stripped (usually hydrogen- and helium-like species) ions has become well developed as a diagnostic of thermal plasma conditions. This is discussed at length in other references.[29]

The other major diagnostic of hot electron temperature and transport is Bremsstrahlung radiation.

It is not itself an important stopping, loss mechanism in the electron energy range of interest to laser fusion. Birkhoff[30] gives the following criterion for the relative importance of collisional ionization and radiative losses.

$$\frac{(dE/dx)RAD}{(dE/dx)COLL} = \frac{EZ}{800} \quad (E \text{ in MeV}). \tag{8}$$

One MeV electrons in gold would lose about 10% of their energy by Bremsstrahlung radiation.

In ICF experiments, the slope of the Bremsstrahlung spectrum has typically been taken to be indicative of the hot electron temperature (or temperatures). In some cases, however, interpretation may be more complex. In some experiments of this sort, hot electron temperature can be measured by ion charge collectors. Often, this value does not agree with the Bremsstrahlung slope. This discrepancy could be due to atomic physics issues such as the Bremsstrahlung cross sections used or plasma-hydro considerations such as anomalous transport mechanisms – at present this question remains unsettled.

Over the range of applicability of Eq. (8), the integrated hard x-ray yield can be taken as a measure of the collisionally deposited hot electron energy. In recent experiments, this method has been employed to study the energy deposition in laser irradiated shells. In these tests, hot electrons can make multiple passes through the shell (return paths being caused by electric fields mentioned above). Estimates of total deposited energy agree with other measurements of deposition.

Detailed evaluations of Bremsstrahlung and $K_\alpha$ radiation are being included in the comprehensive modeling of ICF target experiments. The primary modeling of laser and ion driven fusion experiments is done with large hydrodynamic codes such as LASNEX.[31] Hot electron transport in such codes is basically handled as flux limited diffusion in the presence of electric fields.

Generation of inner shell radiation and Bremsstrahlung are calculated using the known electron spectrum as a function of space for each time step of the hydro simulation. The $K_\alpha$ cross section presently employed is the "Bethe like" cross section mentioned earlier. Absorption of $K_\alpha$ radiation is realistically handled using ray trace transport. Bremsstrahlung is calculated using what are believed to be the best available high energy Bremsstrahlung cross sections.[32]

In ion beam fusion work, electron currents are a loss mechanism. In other words, the optimum irradiation would have 100 of the diode current in the form of ions. Bremsstrahlung is used to evaluate these electron loss currents.

## COMPRESSION DIAGNOSTICS AND THE PHYSICS OF DENSE PLASMAS

As was mentioned above, laser driven implosions have produced the highest density plasmas available in the laboratory. Present understanding of the physics of ICF indicate that practical fusion conditions can only occur at very high compressed densities. Atomic physics plays an extremely important role in the understanding of these very dense plasmas. For example, quantum-mechanical effects are prominent in the formulation of equations of state (EOS) of dense plasmas.[6] Atomic physics questions are important in both the diagnosis of dense plasmas and in understanding their basic physical behavior.

### A. Spectroscopic Diagnosis of High Density Implosions

Spectroscopy has played a prominent role in the diagnosis of high density laser driven implosions.[33] A typical target used in high density implosion experiments is shown in Fig. 5.

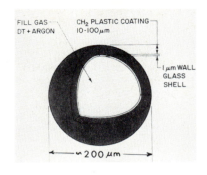

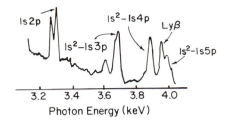

| Line | Lyα | Lyβ | $1s^2 - 1s3p$ |
|------|-----|-----|---------------|
| $\tau_0$ | 13 | 0.25 | 0.5 |

Fig. 5.
Target Used in High Density Laser Implosions and Typical Spectrum
Obtained from such an Implosion.

A glass shell is filled with a variety of gases; DT is used as the fuel for
thermonuclear burn. Spectroscopic diagnosis of plasma temperature and density
is accomplished by seeding the DT fuel with a small amount of higher Z
material, such as neon or argon. Analysis of the spectral profiles of line
emission (from hydrogen and helium—like states) from these elements is the
most direct diagnostic of compressed density and $\rho R$ (quantities of crucial
importance to ICF).[34,35]

The thick plastic layer is used to shield the fill gas from high energy
electrons that would tend to prematurely penetrate and heat the fill gas thus
degrading the implosion.

Transport of the line emission from its region of generation (in the compressed
core) is of major concern. Large optical depths usually only occur for the
first one or two members of a series. One must also, however, consider the
interaction of the compressed core radiation with the glass shell (referred to
as the pusher) containing and compressing it. In the higher density cases, the
pusher is much colder and often denser than the compressed fuel gas.
Non—resonant absorption can usually be quite accurately estimated from well
known opacity calculations (when LTE holds). When resonances and near
resonances occur, much more detailed transport calculations must be performed.
Prominent line absorption spectra are often observed from such targets.[33]

A typical spectrum from a high density imploded plasma is also shown in Fig.
5. The compressed electron density is $7 \times 10^{23}$ cm$^{-3}$. Typical optical
depths for various lines are indicated in the figure.

Helium—like profiles of moderate Z ion emission were first produced as part of
the program of analysis of these experiments.[36,37] Calculations of the
helium—like profiles are somewhat more difficult than the more widely utilized
(in laser plasmas) hydrogen—like lines. Without the orbital angular momentum
degeneracy (of H—like species) the Hamiltonian for a two electron configuration
is no longer diagonal in a single representation that is good for all field
strengths. An individual representation must thus be constructed for each
value of field.

The energy levels used in these calculations have been calculated with the Hartree-Fock-Slater atomic structure code developed by R. D. Cowan.[38]

B.  Non-LTE Radiation Modeling

In many of these high compression experiments, deviations from LTE conditions occurred.  Some simple considerations make this clear.  For a level to be in collision dominated (CD) equilibrium with higher levels,[39] we use the following criterion (for Argon):

$$N_e > 7 \times 10^{18} \frac{Z^6}{(n)^{17/2}} \left[ \frac{kTe}{Z^2 E_H} \right]^{1/2} \quad cm^{-3}, \tag{9}$$

where $Z - 1$ = charge of hydrogenic ion = 17, n = principal quantum number, for example, for n = 2 $N_e > 4 \times 10^{24}$, while for n = 3, $N_e > 1.1 \times 10^{23}$.  The state n = 2 is not in LTE in the present experiments, while n = 3 almost certainly is.  It is, thus, clear that in order to deal with questions such as the detailed transport of optically thick lines (e.g., Lyman $\alpha$ and $1s^2$-1s2p) that a more accurate radiation model must be developed.

Given this motivation, a program for accurately modeling non-LTE radiation in laser imploded pellets has been undertaken in a joint NRL LANL effort.[40]  The overall plasma conditions are determined with a one dimensional version of the lagrangian hydrodynamic code LASNEX.[31]  Detailed radiation characteristics are calculated with a collisional radiative (CR) model (developed by the plasma radiation group at the Naval Research Lab).  The CR model is, at present, used in a post-process mode where temperture and density output information (as a function of space and time) is processed separately and results are not fed back into the hydro code.  Non-LTE radiation dynamics are handled in real time (and the results fed back into the hydro calculation) using an average ion model.[41]  The average ion approximation seems adequate for handling the effects of radiation on the overall hydro behavior.

At each step in the hydrodynamic simulation information, such as temperature, density, average opacity, etc., are output and stored.  This information is then processed by the collisional radiative code.  The evaluation of the spectrum, as a function of time is calculated as well as the integrated final result.  A time integrated spectral plot is produced.  The complete time evolution of the spectrum is also displayed as a motion picture.

This model follows in detail all of the complex excitation and de-excitation processes that are occuring in the radiating ion of interest; and thus is crucially dependent on detailed atomic physics data, such as collisional cross sections and rate coefficients.  For example, some of the processes that are included are:

1.    collisional excitation and de-excitation,

2.    radiative and dielectric recombination,

3.    resonant absorption (and radiative pumping).

In Fig. 6, we show an example of such a calculation.

A careful evaluation has been made of the electron collisional rate coefficients.  Coulomb-Born calculations have been made for hydrogen-like rates and distorted wave for helium- and lithium-like.[42]

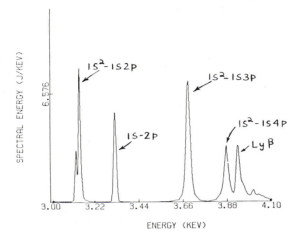

Fig. 6   Simulation of Argon Spectrum Emitted By a Laser Compressed Plasma

In addition to diagnostic modeling, these detailed collisional radiative models are also useful in analyzing the validity of average ion calculations of non-LTE conditions.  Comparisons can be made between average ion and the more detailed calculations of quantities, such as average shell populations and Z.

C. Physics of Dense Plasmas

The densities and temperatures reached in ICF implosion plasmas represent a unique state of matter that is not adequately modeled by simple gas dynamics. In order to calculate properties, such as equations of state (EOS), important quantum mechanical effects must be taken into account.[6,10]

Two potentially measurable quantities that would be of value in determining plasma potentials are the shift of spectral lines, and the shift and structure of continuum edges.

At very high plasma densities, small line shifts may occur.[43]  This is due to the free electron wavefunctions penetrating the orbits of the radiating electrons.  This effect is usually referred to as plasma polarization.  At electron densities well into the $10^{24}$ cm$^{-3}$ range shifts of a few eV are possible.

It is possible that at presently attainable laser fusion conditions, line shifts might be observable in a carefully designed experiment.  One possible candidate is the target shown in Fig. 5.

When such a target is imploded with a laser, the conditions that occur in the glass (pusher) are often quite different from those in the central fuel region.  Electron densities average about 2 x $10^{24}$ cm$^3$ with temperatures of about 300 eV.  The glass might be used as a host material for a radiating species, such as neon or sodium.  A controlled percentage of such material would be introduced as a tracer into the glass.

The first problem that must be addressed is the escape of about 1 keV (neon or sodium) radiation from the cool glass plasma containing it.  Using the opacity tables for glass calculated by Argo and Huebner,[44] we find that 1 keV will escape from depths of about 5-6 μm in the glass (the 1 μm initial wall thickens to 15-20 μm near peak compression).

At a depth of 5-6 μm, the temperature of the glass is approximately 300 eV.  An acceptable level of hydrogen-like sodium radiation would be observed under these conditions and very strong neon emission would occur.  More detailed calculations of line emission have been performed using the temperature and density from hydrodynamic simulations.  These indicate that measurable radiation (a favorable line to continuum ratio) can be obtained with a moderate dopant ratio, thus keeping the optical depth relatively low (under 10).

Observability of small line shifts, given the prevailing plasma conditions, is probably one of the most important questions to be answered.  Finite optical depth, stark broadened width, and temporally changing plasma conditions all affect the accuracy with which the center of gravity of a spectral line can be located.

The theoretical aspects of dense plasma radiation are dealt with in another paper in this conference.[45]

CONCLUSION

Atomic processes are important in all of the major phases of ICF research. Further work on inner shell cross sections will be of value in the evaluation of both electron and ion transport in dense matter.  Calculation of non-LTE radiation conditions will continue to be important in many phases of particle and laser target interaction.

The physics of very dense plasmas will remain an important and very interesting area of research.  Experimental data is needed to correlate with the already extensive theoretical work in this field.

*Work performed under the auspices of the U. S. Department of Energy.

REFERENCES

[1] Some ion stopping measurements have been done in plasma; see, for example, H. H. Brown, New York University Workshop on Charged Particle Penetration in Matter (Jan. 3-6 1980).  Edited by S. Stern.

[2] Anderson, H. N. and Ziegler, J. F., Hydrogen-Stopping Powers and Ranges in All Elements; Ziegler, J. F., Helium-Stopping Powers and Ranges in All Elements; Ziegler, J. F., Heavy Ions-Stopping Powers and Ranges, (Perganon Press, 1977).

[3] Northcliffe, L. and Schilling, R., Nucl. Data Tables A7, 233 (1970).

[4] Mehlhorn, T. A., A Finite Material Temperature Model for Ion Energy Deposition in Ion-Driven ICF Targets, Sandia Laboratories, Albuquerque, NM, Report No. SAND80-0038, May 1980, accepted for publication J. Appl. Phys.

[5] Rogerson, J. E., Fast  Ion Stopping Power in Dense Ionized Plasma, Naval Research Laboratory Memorandum Report No. 4485, April 1981.

[6] More, R. M., Materials at Extreme Conditions: ICF Targets, Lawrence Livermore National Laboratory-Report UCRL-84115, April 7, 1980.

[7] Nardi, E., Peleg, E., and Zinamon, Z., Phys. Fluids 21, 574 (1978).

[8]     Linhard, J. Scharff, M., and Schiott, H. E., Kgl. Danske Videnskab. Selskab, Mat. Fys. Medd. 33, No. 14 (1963).

[9]     Janni, J. F., Air Force Weapons Laboratory Technical Report No. AFWL-TR-65-150, 1966 (unpublished).

[10]    Zink, J. W., Phys. Rev. 176, 279 (1968).

[11]    McGuire, E. J., Phys. Rev. A 22, (3) 868 (1980).

[12]    Jackson, J. D., Classical Electrodynamics, John Wiley and Sons, Inc., 2nd edition, (1975) p. 643.

[13]    Bond, D. J., Hares, J. D., Kilkenny, J. D., Phys. Rev. Lett., 45, (4) 252 (1979).

[14]    Powell, C. J., Rev. Mod. Phys. 48, (1) 33 (1976).

[15]    This form of the stopping power is based on Reference 12 and was suggested by C. Cranfill, Los Alamos National Laboratory.

[16]    Harrach, R. J., Kidder, R. C., Phys. Rev. A, 23, (2) Feb. 1981.

[17]    Kershaw D., Lawrence Livermore National Laboratory Report: No. UCRL 8394 (1979).

[18]    Halblieb, J. A., Sandia Labs Report No. SAND 79-0415, May, 1979.

[19]    Hares, J. P., Kilkenny, J. D., Key, M. H., and Lunney, J. G., Phys. Rev. Lett. 42, (18) p. 1216 (1979).

[20]    House, L., Astro, J., Supp. 155, (18) p. 21 (1969).

[21]    Hughes, D., Plasma and Laser Light, p. 273, Wiley (1973).

[22]    Analysis of this resonant absorption is being pursued by the present authors and Dr. Z. Zinamon. See Weizmann Institute (Isreal) Report WIS-80/201. See, also, A. Hauer, W. Priedhorsky, and D. van Hulsteyn, Los Alamos National Laboratory Report LA-UR 81-1788, submitted to Applied Optics.

[23]    McGuire, E. J., Sandia Laboratories Report No. SAND74-03245, Dec. 1974.

[24]    Kolbenstvedt, H. J., Appl. Phys. 38 (12) p. 4785 (1967).

[25]    Arthurs, A. M. and Moiseiwitsch, B. L., Proc. Roy. Soc. Con. A247, p. 550 (1958).

[26]    Madison, D. A. and Mertzbacher, B. A., in Atomic Inner Shell Processes, Academic (1975).

[27]    Richard, P., "Ion-Atom Collisions" in Atomic Inner-Shell Processes Vol. I, edited by B. Crasemann, Academic Press, New York, pp. 73-158 (1975).

[28]    Garcia, J. D., Fortner, R. J., and Kavanagh, T. M., Rev. Mod. Phys. 45, 111 (1973).

[29]    Key, M. H., Hutcheon, R. J., "Spectroscopy of Laser Produced Plasmas" in Advances in Atomic and Molecular Physics, Vol. 16 (Academic 1981).

[30]    Birkhoff, G. D., Handbuch der Physik Vol. XXXIV, p. 63, Springer Verlag (1958).

[31]    Zimmerman, G. and Kruer, W., Comm. Plasma Phys. 2, 85 (1975).

[32]    Pratt, R. H., Tseng, H. K., Lee, C. M., Kissel, L., MacCallum, C., and Riley, M., Atomic and Nuclear Data Tables, Vol. 20 (2) p. 175 (1977).

[33]    Hauer, A., "Diagnosis of High Density Laser Compressed Plasmas Using Spectral Line Profiles," p. 295 in Spectral Line Profiles, Walter de Gruyter (1981).

[34]    Yaakobi, B., et al., Phys. Rev. Lett. 44, 1072 (1980).

[35]    Hauer, A., et al., Phys. Rev. Lett. 45, (18) p. 1495 (1980).

[36]    Griem, H. R. and Kepple, P., in Spectral Line Profiles, Walter de Gruyter (1981).

[37]    Hooper, C. F., Univ. of Florida Dept.of Physics Rpt. Contract No. 8820309.

[38]    Cowan, R. D., J.O.S.A. 58 (6) 808 (1968).

[39]    Griem, H. R., Plasma Spectroscopy (McGraw-Hill 1964), available from University microfilms.

[40]    NRL Plasma Radiation Group:  J. Davis, P. Kepple, K. Whitney, D. Duston, J. Apruseze; LANL, A. Hauer, E. Linnebur.

[41]    Locke, W. A. and Grasberger, W. H., Lawrence Livermore National Laboratory Report No. UCRL 52276.

[42]    Jacobs, V. L., Davis, J., Kepple, P. C., and Blaha, M., Astrop. J. 211, 605 (1977).

[43]    Skupsky, S., Phys. Rev. A 21, (4) 1316 (1980).

[44]    Argo, H. and Huebner, W., J. Quan. Spec. Rad. Transf. 16, 1091 (1976).

[45]    Davis, J., Paper S4B5 this Conference.

PHYSICS OF ELECTRONIC AND ATOMIC COLLISIONS
S. Datz (editor)
© North-Holland Publishing Company, 1982

ATOMIC PROPERTIES AND PROCESSES IN DENSE PLASMAS

J. DAVIS

Plasma Radiation Group
Naval Research Laboratory
Washington, D.C.   20375

and

M. BLAHA

Laboratory for Plasma and Fusion Energy Studies
University of Maryland
College Park, Maryland   20742

A completely quantum mechanical formalism has been developed to
describe the high density plasma effects on fundamental atomic
structure and processes.  The formalism is then applied to
characterize the magnitude of the plasma polarization shift for
a hydrogenic impurity ion immersed in a fully ionized hydrogen
plasma.  Results are presented for level shifts, spontaneous
decay rates, and electron collision cross sections for $Ne^{+9}$ and
$Ar^{+17}$.

INTRODUCTION

The influence of plasma microfields on atomic properties and processes is
currently experiencing a period of intensive investigation due, in part, to its
relevance in the inertial confinement fusion program and in astrophysical
applications, particularly in the study of stellar interiors.  In addition to its
diagnostic applications, the modification of atomic properties and processes is
of fundamental interest and importance.  The knowledge that environmental effects
can affect the plasma constituents and how they interact amongst themselves has
been known for a long time.  In fact, screening effects in fully ionized plasmas
is a well known and understood phenomenon and can be described elegantly by using
"dressed" particle theories.  The simplest and best known example is Debye
screening.  However, in a plasma containing partially stripped ions having bound
electrons the situation is not so clear; current dressed particle theories deal
only with free electrons.  In principle it should be possible to treat both free
and bound electrons on an equal basis; some of the many-body approaches currently
under investigation seem to be heading in the right direction.  Our goal is less
ambitious in that we will treat bound and free electrons on an equal basis but
will limit our investigation to describe the plasma polarization shift (PPS).  By
the PPS we mean the effect that describes the interaction of the net charge of an
ion with the plasma environment which, averaging over time, produces an excess of
electrons in the vicinity of the ion.  The induced polarization charge is assumed
to derive partially as a result of the intrusion of plasma electrons within the
bound state orbital, partially shielding the nuclear charge, causing a shift of
the energy level structure of the ion.  In reality a number of processes
contribute to the shift and broadening of energy levels.  However, we will adopt
a more restrictive definition, more in line with the above comments, and consider
only those static effects associated with the plasma constituents on the

effective charge of the ion under investigation. Effects due to ion quadrupole, nearest neighbor ion sphere, and elastic scattering will be omitted here and will form the basis of a forthcoming work. Nonetheless, even in the simplest interpretation of the PPS of spectral lines it remains a basic and controversial effect.

Griem[1] tried to explain the blue shifts for the resonance lines of ionized helium by considering the perturbation of the upper levels by free plasma electrons. An alternate approach involves the solution of the Schrödinger equation for the bound electrons in the Debye-Hückel potential. This leads to red shifts of the lines because it causes large shifts of the lower levels. Experimentally, evidence has been reported for blue shifts[2] of Lyman lines and red shifts[3] of Balmer lines for He II, but shifts measured for other similar ionized emitters are much smaller or zero.

Theimer and Kepple[4] have shown that the theory using the Debye-Hückel potential is inadequate because it ignores the reaction of plasma electrons to the presence of the bound electron. Their self-consistent calculation for a hydrogen plasma indicates much smaller level shifts than those predicted by previous theories. More recently, an analogous treatment was adopted by Rozsnyai[5] and Skupsky[6] with similar results. In these theories the free electron gas was treated classically and the energy levels were obtained from the solution of the Schrödinger equation.

In the present investigation, a self-consistent quantum mechanical theory is presented in which level shifts, transition probabilities, and collision cross sections for $Ne^{+9}$ and $Ar^{+17}$ are calculated in a plasma with electron density ranging from $10^{24}$ to $6 \times 10^{24}$ cm$^{-3}$ (for $Ne^{+9}$) and from $2 \times 10^{25}$ to $8 \times 10^{25}$ cm$^{-3}$ (for $Ar^{+17}$).

DESCRIPTION OF THE METHOD

We consider an isolated impurity ion with nuclear charge Z immersed in a fully ionized hydrogen plasma. In an equilibrium situation, the whole system can be described by specifying the occupation numbers of all possible quantum states, and the corresponding wave functions are solutions of time-independent equations. This description leads to the "average atom" model for the impurity ion, in which no distinction is made between individual ionization stages and excited states of the ion. Observations of line radiation from hot dense plasmas indicate, however, that individual ions can be in specific states of excitation and ionization for periods of time longer or comparable to the lifetime of these states. This is possible because the local plasma fluctuations allow the free electrons to become temporarily trapped into bound orbits for the duration of the corresponding fluctuation. Fluctuations of local density and potential cannot be adequately described by equilibrium models and therefore it appears more appropriate to use a time-dependent theory for the study of plasma interaction with radiating atomic systems. In a time-dependent description, plasma electrons should be represented by wave packets and there will be a definite probability of finding an ion, at a given time, in a specific state depending on the initial conditions.

In the present study we will investigate plasma effects on one-electron hydrogenic ions. Due to the complexity of the problem, several simplifying assumptions were made in our model. Fluctuations of the local potential due to the ionization and recombination processes have generally a much longer time scale than the lifetime of deeply bound excited states and we will assume that the impurity ion is temporarily in equilibrium with the surrounding plasma while being in a given ionization stage. The spatial extension of wave packets corresponding to plasma electrons is usually larger than the dimension of low lying bound orbits and in our procedure they will be replaced by infinite waves. Consequently, the applicability of the present method is limited to bound orbits whose dimensions are smaller than the de Broglie wavelength of plasma electrons, evaluated at the mean thermal velocity. For simplicity we assume that all electric charges are spherically symmetric. The average local charge density due to the bound electron can be written as (atomic units are used throughout the paper, unless indicated)

$$\rho_b = - (4\pi r^2)^{-1} \sum_{n\,\ell} b_{n\ell} P_{n\ell}^2 (r), \tag{1}$$

where $b_{n\ell}$ are the occupation numbers, $P_{n\ell}(r)$ are normalized radial wave functions, and $\sum_{n\ell} b_{n\ell} = 1$. Generally, the occupation numbers represent an average state of excitation of the ion and are determined by all possible collisional and radiative processes involving the bound electron. At temperatures and densities considered in the present paper, the coupling constant $\Gamma$ for a fully ionized hydrogen plasma is less than one and therefore it is assumed that one can ignore the discrete character of positive charges carried by protons and use instead an average positive charge density $\rho_p$.

If $\rho_e$ and $\rho_p$ represent local charge densities due to the free electrons and protons, the electrostatic potential $V(r)$ at the distance $r$ from the nucleus with charge Z is then given by

$$V(r) = Z r^{-1} + 4\pi \left[ r^{-1} \int_0^r r^2 (\rho_e + \rho_p + \rho_b)\, dr \right. $$
$$\left. + \int_r^\infty r (\rho_e + \rho_p + \rho_b)\, dr \right]. \tag{2}$$

At large distances from the ion, $\rho_b = 0$ and the plasma is neutral, so that

$$- \rho_e = \rho_p \equiv \rho_\infty \equiv N_e. \tag{3}$$

The positive charge distribution is described by classical Boltzmann statistics, viz.

$$\rho_p = \rho_\infty \exp\, (-V/k_B T), \tag{4}$$

where $V(r)$ is given by (2), T is the temperature and $k_B$ the Boltzmann constant.

The free electrons are represented by a Fermi-Dirac energy distribution outside a spherical boundary corresponding to the distance $r_o$ from the nucleus; however, inside this boundary they can be treated quantum-mechanically and are described by wave functions which are solutions of the time independent Schrödinger equation. In practical calculations the distance $r_o$ is taken to be large enough so that plasma at the boundary may be considered neutral and condition (3) valid within numerical accuracy. The spherically averaged charge density due to the free electrons at any r is given by

$$\rho_e = - \int_0^\infty W(k) (kr^2)^{-1} \sum_{\ell=0}^\infty (2\ell+1) F_{k\ell}^2 (r)\, dk, \tag{5}$$

where

$$W(k) = \pi^{-2} k^2 \left[ 1 + \exp\, \left\{ (\tfrac{1}{2}k^2 - \mu)/k_B T \right\} \right]^{-1}, \tag{6}$$

and $1/2k^2$ is the electron kinetic energy at $r > r_o$. $\mu$ is the chemical potential of the free electron gas determined by the condition

$$\int_o^\infty W(k) \, dk = \rho_\infty \tag{7}$$

and the functions $F_{k\ell}$ are solutions of the equation

$$\left[\frac{d^2}{dr^2} - \frac{\ell(\ell+1)}{r^2} + 2V(r) - 2V_{ex}(r) - 2V_{corr}(r)\right. \tag{8}$$

$$\left. + 2V_{ex}(\infty) + 2V_{corr}(\infty) + k^2\right] F_{k\ell}(r) = 0$$

with the asymptotic form

$$F_{k\ell}(r) \sim k^{-1/2} \sin (kr + \delta_{k\ell}). \tag{9}$$

For small $r$, $V(r)$ behaves like $r^{-1}$ and eq. (8) has solutions $F_{k\ell}$ which are bound at the origin and which behave like $r^{\ell+1}$. It follows from (5) that $\rho_e$ is finite at $r = 0$.

The electrostatic potential V is given by (2) and the exchange energy $V_{ex}$ and the correlation energy $V_{corr}$ of a free electron in a plasma were taken from Dharmawardana[7,8] and Taylor[8] and are given in terms of temperature T and local density of free electrons $n_e \equiv - \rho_e$:

$$V_{ex} = - 0.4073 \, r_s^{-1} \tanh (\tau^{-1}), \tag{10}$$

$$V_{corr} = - 0.6109 \, r_s^{-1/2} (-0.0081+1.127\tau^2+3.756\tau^4) \tag{11}$$

$$\times (1.0+1.291\tau^2+3.593\tau^4)^{-1} \tanh (\tau^{-1/2}),$$

$$\tau = 2(\frac{4}{9\pi})^{2/3} r_s^2 k_B T,$$

$$r_s^{-1} = (\frac{4}{3} \pi n_e)^{1/3} .$$

In practice, the integral in (5) is replaced by a summation. At large distances, the convergence of the sum over partial waves in (5) is rather slow and we found it necessary to include up to 50 or 70 partial waves, depending on the value of k, to achieve sufficient accuracy for $r < r_o$.

The effect of Fermi degeneracy is taken into account by the distribution function W(k) in (5) for densities equal to $\rho_\infty$. However, we have ignored the possible additional effects of Fermi degeneracy inside $r_o$. Also all effects due to relativistic corrections have been ignored since they are probably negligible for light nuclei.

Radial functions $P_{n\ell}$ of the bound electron satisfy the equation

$$\left[\frac{d^2}{dr^2} - \frac{\ell(\ell+1)}{r^2} + 2V_b(r) + 2E_{n\ell}\right] P_{n\ell}(r) = 0. \tag{12}$$

The potential $V_b$ acting on the bound electron is given by (2) with $\rho_b = 0$. The energy shift of a hydrogenic level $n\ell$ due to the plasma polarization is

$$\Delta E_{n\ell} = E_{n\ell} + Z^2(2n^2)^{-1}.$$  (13)

Equations (1), (2), (4), (5), (8) and (12) are solved self-consistently with the boundary conditions

$$rV(r) \to 0, \; P_{n\ell}(r) \to 0$$  (14)

for $r \to \infty$. In practice these conditions should be satisfied at $r = r_o$. The self-consistent solution yields the eigenvalues $E_{n\ell}$, and the level shifts are then obtained from (13).

According to the definition of $V_b$ and to the boundary conditions (14), $rV_b(r) \to 1$ as $r \to \infty$. However, it would be incorrect to conclude that the present method leads to an infinite number of bound states. The validity of our procedure is restricted to deeply bound states, as mentioned earlier, and it would be unphysical to calculate energies of high levels using the present method, even if the self-consistent solutions existed.

The potential $V$ is different from $V_b$ and therefore the functions $F_{k\ell}$ and $P_{n\ell}$ corresponding to the same angular momentum are not generally orthogonal. The lack of orthogonality may lead to serious errors in the calculation of exchange energy, but in our procedure the effects of exchange between the bound and free electrons are not considered to be important and are ignored altogether.

## RESULTS FOR $Ne^{+9}$ and $Ar^{+17}$

It is of interest to compare the self-consistent potential $V$ with the potential derived from the Debye-Hückel theory of plasma screening. Fig. 1 shows two typical examples of the quantity $rV$ for the argon nucleus immersed in a fully ionized hydrogen plasma according to our self-consistent calculations. In this

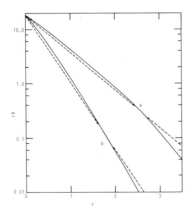

Figure 1.  The quantity $rV$ for $Ar^{+17}$ in a fully ionized hydrogen plasma. Full curve - present results; dashed curve - Debye-Hückel theory.

$$a - N_e = 5 \times 10^{25} \; cm^{-3}, \quad T = 2000 \; eV;$$
$$b - N_e = 8 \times 10^{25} \; cm^{-3}, \quad T = 1000 \; eV.$$

case, $\rho_b = 0$ and the procedure is the same as described above.  The Debye-Hückel
potential is lower than the self-consistent potential up to the distance
$4\Lambda$, where $\Lambda$ is the Debye screening length for a fully ionized hydrogen.

The calculations for hydrogen-like $Ne^{+9}$ and $Ar^{+17}$ were performed with the
simplifying assumption that the levels 1s, 2s and 2p are populated according to
Boltzmann statistics and that all occupation numbers $b_{n\ell}$ with $n > 2$ are zero.

Table 1 presents the atomic level shifts (13) for two temperatures and three
electron densities $N_e$.  Resulting line shifts of the Lyman-$\alpha$ radiation are shown
in Figs. 2 and 3.  The line shifts are assumed to be equal to the difference of
level shifts.  The present values of line shifts are slightly larger than the
results of Skupsky[6] who treated the free electrons classically with quantum
mechanical corrections to the potential.

Table 1

LEVEL SHIFT (eV)

| | Ne X | | | | Ar XVIII | | | |
|---|---|---|---|---|---|---|---|---|
| T(eV) | $N_e(cm^{-3})$ | 1s | 2s | 2p | T(eV) | $N_e(cm^{-3})$ | 1s | 2s | 2p |
| 200 | $10^{24}$ | 122.8 | 115.6 | 117.5 | 1000 | $2\times10^{25}$ | 475.8 | 439.9 | 449.1 |
| | $3\times10^{24}$ | 194.0 | 173.3 | 178.5 | | $5\times10^{25}$ | 702.0 | 616.4 | 637.5 |
| | $6\times10^{24}$ | 225.5 | 189.8 | 198.1 | | $8\times10^{25}$ | 865.0 | 736.0 | 766.4 |
| 500 | $10^{24}$ | 90.0 | 85.5 | 86.7 | 2000 | $2\times10^{25}$ | 375.8 | 351.0 | 357.3 |
| | $3\times10^{24}$ | 147.3 | 134.2 | 137.5 | | $5\times10^{25}$ | 563.8 | 504.5 | 518.8 |
| | $6\times10^{24}$ | 199.6 | 174.3 | 180.3 | | $8\times10^{25}$ | 695.0 | 602.6 | 624.3 |

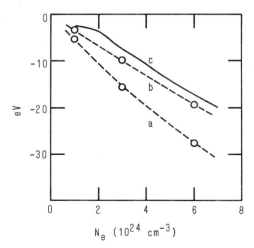

Figure 2.   Density dependence of the Ne X Lyman-$\alpha$ frequency shift.[6]
a - T = 200 eV;  b - T = 500 eV;  c - results of Skupsky[6] (T=500 eV).

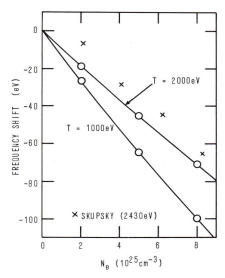

Figure 3.   Density dependence of the Ar XVIII Lyman-$\alpha$ frequency shift.
Crosses - results of Skupsky[6].

Table 2 shows how the Einstein coefficients of transition probability A for
Lyman-$\alpha$ line are affected by plasma conditions.   $A_{Coul}$ is the coefficient
corresponding to the pure Coulomb field, i.e. to $N_e = 0$.   The change of
transition probabilities is a consequence of the line shift combined with the
change of radial functions $P_{n\ell}$.   In order to ascertain the sensitivity of line
shifts to the choice of $b_{n\ell}$, we carried out additional calculations for $Ne^{+9}$ with
the assumption that only the 1s ground level is occupied.   The resulting line
shifts were found to be only about 1% smaller than those shown in Fig. 2.

Table 2

$$A(2p \rightarrow 1s)/A_{Coul} \ (2p \rightarrow 1s)$$

|          | T(eV)        | $N_e \ (cm^{-3})$ | | |
|----------|--------------|-----------------|--------------------|------------------|
|          |              | $10^{24}$       | $3 \times 10^{24}$ | $6 \times 10^{24}$ |
| Ne X     | 200          | 0.967           | 0.904              | 0.840            |
|          | 500          | 0.981           | 0.942              | 0.887            |
|          |              | $N_e \ (cm^{-3})$ | | |
|          |              | $2 \times 10^{25}$ | $5 \times 10^{25}$ | $8 \times 10^{25}$ |
| Ar XVIII | 1000         | 0.950           | 0.879              | 0.816            |
|          | 2000         | 0.967           | 0.919              | 0.873            |

In the present calculations we have ignored all effects related to the time dependent perturbations by plasma electrons and protons. In particular, a possible redistribution of electron velocities inside the boundary $r_o$ due to electron-electron and electron-proton collisions was not taken into account and effects leading to broadening of spectral lines were also omitted. Thus the results indicate the magnitude of effects caused by plasma polarization alone. In situations where there are only few electrons inside the Debye sphere and collisions do not happen very often, the present treatment should be preferable to the procedures based on statistical considerations for plasma electrons.

The calculation of electron collision excitation cross sections and rates in a dense plasma should proceed, in principle, along the lines of many-body quantum theory. However, the two-body approximation may still be useful not only for approximate evaluation of excitation rates, but especially because its simplicity allows us to separate various plasma effects.

The change in the value of excitation cross sections and rates in dense plasmas is the result of several causes: (a) changed excitation energy, (b) changed wave functions of atomic electrons, (c) distortion of the incident and scattered waves by the charged cloud around the target ion, (d) modification of the Coulomb interaction potential $|\vec{r}_1 - \vec{r}_2|^{-1} \equiv r_{12}^{-1}$ for the colliding and bound electron by plasma effects, and (e) other effects of many-body interactions.

We have calculated electron collision strengths for the 1s – 2s, 1s – 2p, and 2s – 2p transitions of $Ne^{+9}$ and $Ar^{+17}$ using the formalism of the distorted-wave method without exchange. Excitation energies and atomic wave functions were taken from our self-consistent calculations and the equation for the radial functions of the colliding electron were solved using the self-consistent potential (2). However, the modification of the potential of mutual interaction $r_{12}^{-1}$ cannot be properly described within the formalism of the distorted-wave approximation, unless the wave functions of plasma electrons take into account correlation effects between the bound and the free electrons. The variational expression for the element of the reactance matrix then contains an additional term that partially cancels the term of direct Coulomb interaction $r_{12}^{-1}$ as a result of decreased local density of plasma electrons in the vicinity of the bound electron. One can approximate this effect by using the free electron wave functions without correlation and replacing $r_{12}^{-1}$ in the matrix element by $r_{12}^{-1} f(r_{12})$, where $f(r_{12})$ is a properly defined screening function. $r_{12}^{-1} f(r_{12})$ can be expanded in the form

$$r_{12}^{-1} f(r_{12}) = r_{>}^{-1} \sum_{\lambda=0}^{\infty} P_{\lambda}(\cos \omega) \sum_{\mu=0}^{\infty} A_{\lambda\mu}(r_{>}) (r_{<}/r_{>})^{\lambda + 2\mu}, \quad (15)$$

where $\omega$ is the angle between $\vec{r}_1$ and $\vec{r}_2$ and the coefficients $A_{\lambda\mu}$ depend on the form of $f(r_{12})$.

To demonstrate the effect of screening of the mutual interaction, we have adopted a Debye-Hückel form of the screening function. The expansion (15) has the form

$$r_{12}^{-1} \exp(-r_{12}/D) = \left[ r_{>}^{-1} + \frac{1}{6} (r_{>}/D)^2 (r_{<}^2/r_{>}^3) + \cdots \right] \exp(-r_{>}/D) P_o(\cos \omega)$$
$$+ \left[ (1 + r_{>}/D) (r_{<}/r_{>}^2) + \frac{1}{10} (1 + r_{>}/D) (r_{>}/D)^2 (r_{<}^3/r_{>}^4) + \cdots \right] \quad (16)$$
$$x \exp(-r_{>}/D) P_1 (\cos \omega) + \cdots$$

The results for typical values of T and $N_e$ are shown on Figs. 4 to 9. Also shown

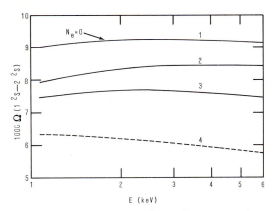

Figure 4.   Collision strength $\Omega$ ($1s\,^2S$, $2s\,^2S$) for $Ne^{+9}$.   T = 200 eV.
1 – $N_e$ = 0;  2,3,4 – $N_e$ = $3\times10^{24}$ $cm^{-3}$;  2 – D = $\infty$;   3 – D = $\Lambda$;
4 – D = $\frac{1}{2}$ $\Lambda$ ($\Lambda$ = Debye length).

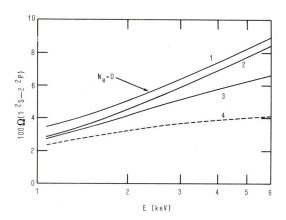

Figure 5.   Collision strength $\Omega$ ($1s\,^2S$, $2p\,^2P$) for $Ne^{+9}$.   T = 200 eV.
1 – $N_e$ = 0;  2,3,4 – $N_e$ = $3\times10^{24}$ $cm^{-3}$;  2 – D = $\infty$;   3 – D = $\Lambda$;
4 – D = $\frac{1}{2}$ $\Lambda$ ($\Lambda$ = Debye length).

*J. Davis and M. Blaha*

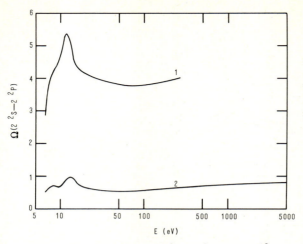

Figure 6.   Collision strength $\Omega$ $(2s^2S, 2p^2P)$ for $Ne^{+9}$.   T = 200 eV; $N_e$ = 3 x $10^{24}$ $cm^{-3}$.   1 – D = $\infty$;   2 – D = Debye length.

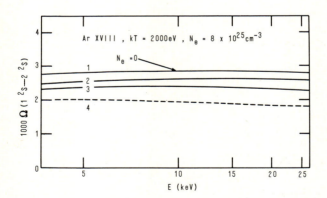

Figure 7.   Collision strength $\Omega$ $(1s^2S, 2s^2S)$ for $Ar^{+17}$.   T = 2000 eV. 1 – $N_e$ = 0; 2,3,4 – $N_e$ = $8 \times 10^{25}$ $cm^{-3}$; 2 – D = $\infty$; 3 – D = $\Lambda$; 4 – D = $\frac{1}{2}$ $\Lambda$   ($\Lambda$ = Debye length).

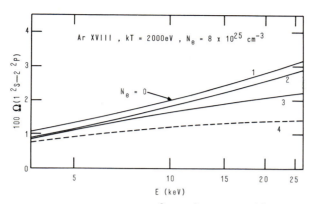

Figure 8.  Collision strength $\Omega$ ($1s^2S$, $2p^2P$) for $Ar^{+17}$. T = 2000 eV.
1 – $N_e$ = 0; 2,3,4 – $N_e$ = $8\times10^{25}$ $cm^{-3}$; 2 – D = $\infty$; 3 – D = $\Lambda$;
4 – D = $\frac{1}{2}$ $\Lambda$  ($\Lambda$ = Debye length).

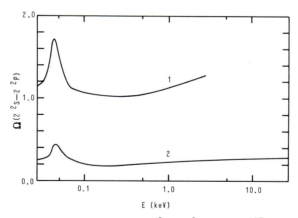

Figure 9.  Collision strength $\Omega$ ($2s^2S$, $2p^2P$) for $Ar^{+17}$.  T = 2000 eV;
$N_e$ = $8\times10^{25}$ $cm^{-3}$.  1 – D = $\infty$;  2 – D = Debye length.

are the collision strengths without any plasma effects ($N_e$ = 0) for the transitions 1s - 2s, 1s - 2p. The change of atomic wave functions by plasma polarization has a very small effect on the cross sections for conditions considered in this investigation. The change of excitation energies produces a negligible effect on the 1s - 2s and 1s - 2p transitions, but it removes the degeneracy of the 2s and 2p levels and reduces the 2s - 2p cross sections to finite values. For the 1s - 2s and 1s - 2p transitions, calculations were made for 3 values of the screening parameter D in (16). D = ∞ corresponds to the situation where the modification of mutual interaction is ignored and the lowering of the collision strength is caused only by the defocusing of colliding electrons (curve 2 on Figs. 4,5,7,8). Curves 3 and 4 were obtained with D = Λ and D = 1/2 Λ, where Λ is the Debye length for the corresponding temperature and density.

The effect of screening on partial collision strengths increases with increasing angular momentum and therefore the dipole transition 1s - 2p is more affected than the monopole transition 1s - 2s. The term corresponding to μ = 1 in the expression (15) and (16) has only a very small effect on the collision strengths.

Collision strengths for the 2s - 2p transition were calculated for D = ∞ and D = Λ. The sensitivity of Ω to the value of D is much greater than for the 1s - 2s and 1s - 2p transitions, because the excitation energy is much smaller, many more partial waves contribute to Ω at a given energy and higher angular momenta are more affected by screening. Collision strengths exhibit a resonance-like behavior in the low-energy region above the excitation threshold. The enhancement of Ω is caused almost exclusively by the contribution from the ℓ = 2,3 incident partial waves in Ne$^{+9}$ and by ℓ = 3 in Ar$^{+17}$.

## ACKNOWLEDGEMENT

This work was supported, in part, by the Office of Naval Research and NASA. We would also like to acknowledge useful discussions with Drs. H.R. Griem and M.W.C. Dharma-wardana.

## REFERENCES

[1]  Griem, H.R., Proc. 7th Int. Conf. on Phenomena in Ionized Gases, Belgrade, Yugoslavia (1965).
[2]  Neiger,M. and Griem, H.R., Phys. Rev. A14 (1976) 291-299.
[3]  Van Zandt, J.R., Adcock, J.C., and Griem, H.R., Phys. Rev. A14 (1976) 2126-2132.
[4]  Theimer, O. and Kepple, P., Phys. Rev. A1 (1970) 957-965.
[5]  Rozsnyai, B.F., J. Quant. Spectrosc. Radiat. Transfer 15 (1975) 695-699; Phys. Rev. A5 (1972) 1137-1149.
[6]  Skupsky, S., Phys. Rev. A21 (1980) 1316-1326.
[7]  Dharma-wardana, M.W.C., private communication (1980).
[8]  Dharma-wardana, M.W.C. and Taylor, R., J. Phys. C: Solid St. Phys. 14 (1981) 629-646.

PHYSICS OF ELECTRONIC AND ATOMIC COLLISIONS
S. Datz (editor)
© North-Holland Publishing Company, 1982

OPTICAL EXCITATION BY CHARGE EXCHANGE*

OF $H^0$ WITH $O^{8+}$

R. C. Isler, L. E. Murray, and S. Kasai[†]

Fusion Energy Division
Oak Ridge National Laboratory
Oak Ridge, Tennessee 37830
U. S. A.

Several spectral lines produced by charge transfer
of $O^{8+}$ with atomic hydrogen from neutral heating
beams have been observed in the ISX-B tokamak.
These signals have been utilized to measure the
concentrations of fully stripped oxygen.

## I.  INTRODUCTION

Completely stripped light ions may be the dominant impurities in
the center of tokamak plasmas.  It is difficult to study their
concentrations and transport since they do not radiate spectral
lines.  But charge-transfer reactions with neutral hydrogen provide a
means for circumventing this problem.  These processes, which are
typified by

$$H^0 + O^{8+} \rightarrow H^+ + (O^{7+})^*, \qquad\qquad (1)$$

leave the product impurity ion in excited states.  The subsequent
radiation then provides a signal directly related to the $O^{8+}$
concentration.  It is crucial, of course, that this process be
distinguishable from the electron excitation of $O^{7+}$,

$$e + O^{7+} \rightarrow e + (O^{7+})^*. \qquad\qquad (2)$$

Emissions produced by the charge-transfer process were observed from
oxygen in the ORMAK device where the neutral heating beams provided
the source of hydrogen[1], and from carbon in the T-4 tokamak where an
active diagnostic beam was employed[2].  No attempt was made to
calculate the oxygen concentrations from the ORMAK observations,
mainly because the partial cross sections for exciting the individual
n- and ℓ-states were not known.  Since that time, these individual
cross sections for hydrogenic ions have become available[3,4,5], and
this paper describes how they have been utilized to measure the fully
ionized oxygen concentration in the ISX-B tokamak.

## II.  EXPERIMENTAL ARRANGEMENT AND OBSERVATIONS

Two neutral beam injectors are installed on ISX-B, each of which
can provide over 1 MW of power into the plasma with a full energy
component between 30 and 40 keV (100 A sources).  The best present
analysis indicates that the fractional currents associated with the
full, the one-half, and the one-third energy components of the beams
are 0.60, 0.20, and 0.20 respectively.  Both beams are injected in

the same direction as the plasma current (co-injection).  The
experiments discussed here have been carried out using hydrogen beams
and deuterium plasmas.

A top view of the tokamak, the beamlines, and some of the
diagnostic devices is shown in Fig. 1.  The grazing incidence
monochrometer (McPherson Model 247) used for the line integrated
spectral observations is situated so that its field of view includes
high energy hydrogen atoms from the west beam, but not atoms from the
east beam.  When the west beam is operated alone, the spectral
signals observed result from the sum of charge-exchange and electron
excitation.  But because the lifetimes of the pertinent excited
states are so short (less than 2 x $10^{-10}$ s), ions excited by
charge-exchange with the particles of the east beam decay before
transport along the toroidal magnetic field brings them into the
field of view of the spectrometer.  As a result, when only the east
beam is operating, the spectral radiation observed is due solely to
electron excitation.  This excitation should be toroidally uniform
for the highly ionized stages because the transit time around the
machine (5-8 x $10^{-5}$ s) is much shorter than the characteristic time
for electron ionization (1-2 x $10^{-3}$ s).  The experimental arrangement
shown in Fig. 1 allows the direct charge-exchange signal to be
separated from the total signal by operating the beams at the same
current and subtracting an "east beam shot" from a "west beam shot".

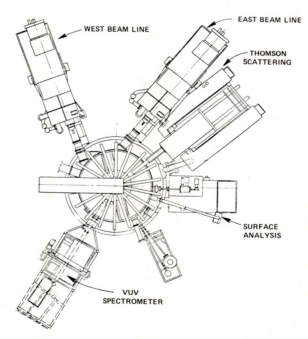

Fig. 1. Schematic view of ISX-B showing the
locations of the grazing incidence spectrometer and
the two neutral beam injectors.

The characteristics of one sequence of the discharges that were studied are shown in Figs. 2-4.   Fig. 2 illustrates the temporal dependences of the plasma current, the neutral beam current, the electron concentration, the soft x-ray radiation, and the magnetohydrodynamic activity.   The steady state plasma current is 160 kA, and the toroidal field is 12.3 kG.   Deuterium is added to the machine throughout the shot causing the electron concentration to rise steadily.   The neutral beam injectors, each of which provides 600 kW at a primary energy of 33 keV, are turned on from 80 to 180 ms.   After this period the current begins to run down, and the discharge terminates.   The soft x-ray signal arises primarily from bremsstrahlung and from radiative recombination with light impurities.

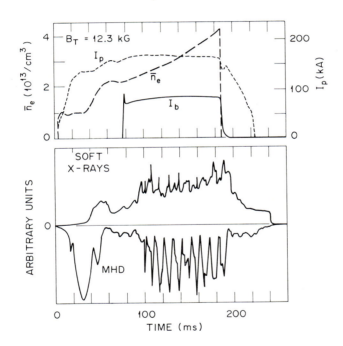

Fig. 2. Temporal behaviour of several plasma parameters.   $I_p$ and $n_e$ are the plasma current and the line averaged electron density.   The total beam current, $I_b$, is 36 amperes.

Electron temperature and density profiles were taken at 130 ms into the discharge.   These are shown in Fig. 3.   The temperature profile is relatively flat in the center with a maximum near 1150 eV. Profiles were not obtained for the pre-injected plasma, but the sequence immediately preceeding the one we describe here employed only ohmic heating, and the central temperature was 730 eV.   We have assumed that this is a close approximation for the injected plasmas just before the beam is turned on.

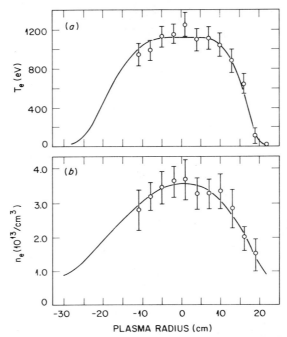

Fig. 3. Plasma temperature and density profiles.

      The emission rates[6] for lines of $O^{7+}$ are illustrated in Fig. 4.
Traces which are specified as electron excitation are "east beam
shots", and those noted as charge-exchange excitation are obtained by
a subtraction of two signals as we have described.  These data have
been numerically smoothed over 10 ms intervals except for the
charge-exchange signals around the time of injection, 78 ms, where
the normal 1 ms sampling rate was preserved in order to demonstrate
the characteristic rapid rise.  The initial spike and the following
relaxation correspond to the same features that are present in the
beam current pulse (Fig. 2).  The rising plasma density and the
concomitant attenuation of the beam before it crosses the field of
view of the spectrometer cause the decrease of the charge transfer
signals after 120 ms.  These data represent the first confirmation
that several lines of oxygen produced by charge-exchange excitation
can be observed during neutral beam injection.  As a result, our
identification of this process is greatly strengthened.  In contrast
to the 1-2 ms rise in the charge exchange signal, the electron
excited radiation requires about 30 ms to attain its peak value
following the start of the beam pulse, and if an average is taken
through the large amplitude oscillations, the signal level decreases
only slightly until the discharge is terminated.  At first sight, it
might be supposed that the increase of the electron excited signal
during injection results from charge exchange recombination changing
the ionization balance.  The relaxation time for this process is only
6 ms, however, much too short to explain the experimental data.
Detailed analysis shows that a change in the transport rate of the
oxygen is the most probable factor influencing the evolution of the
electron excited $O^{7+}$ radiation during injection.

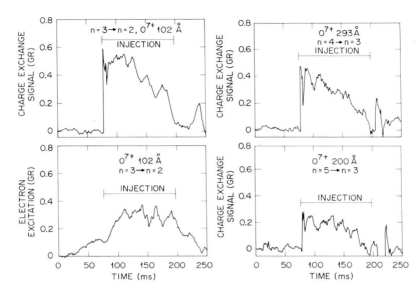

Fig. 4. Emission rates of $0^{7+}$ ions.

Signals from a second sequence for charge exchange excitation of the n=3→2 and the n=5→4 transitions of hydrogen-like oxygen are shown in Figs. 5 and 6. Again, it is clear that the onset of the charge-exchange excitation is quite distinctive and is coincident with the neutral beam injection.

## III. ANALYSIS OF CHARGE-EXCHANGE EXCITATION

The concentration of $0^{8+}$ in the center of the plasma, $n_8(0)$, is determined from the charge-exchange signals through the relationship,

$$S_c = \sum_{k=1}^{3} \int_{-a}^{a} j_k(r)\ n_8(0)\ f(r)\ \sigma_E^k\ dr. \qquad (3)$$

Here $S_c$ is the measured emission rate of a spectral line, $j_k$ is the current density of a given energy component of the neutral beam, and $\sigma_E^k$ is the cross section for excitation of the line at a collision energy $E_k$. A 20% correction is added to the nominal value of the beam current to account for the low energy neutral deuterium atoms (halo neutrals) which are formed by charge-exchange of plasma ions with the beam and which can subsequently charge-exchange with the impurities. The integral of Eq. (3) is evaluated along the optic axis of the spectrometer by using assumed radial profiles, $f(r)$, for the $0^{8+}$ profiles, and by calculating the attenuation of the beam between the source and the observation points.

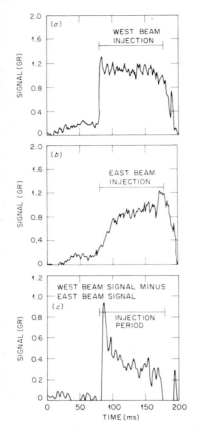

Fig. 5. Signals from the 3→2
transition of O⁷⁺.

Fig. 6. Signals from the 5→4
transition of O⁷⁺.

The total effective cross sections for charge-exchange
excitation of a given spectral line are derived from the individual
charge-transfer cross sections for populating specific n- and
ℓ-states of the hydrogen-like oxygen by taking account of all the
branching and cascading in the decay process. As an example,
effective excitation cross sections computed from Salop's charge
transfer results[3] are shown in Table I.

When necessary, Salop's calculations at 26 and at 32 keV have
been extrapolated linearly to obtain the values for the full- and the
one-half energy components of the beam. The cross sections for the
one-third energy components of the beam, are assumed the same as for
the one-half energy component. Uncertainties in the extrapolations
do not strongly influence the analysis, since about 70% of observed
excitation occurs from the full energy component.

Table I. Effective excitation cross sections in units
of $10^{-15}$ cm$^2$ for spectral lines produced by charge
transfer of hydrogen on completely ionized oxygen at 32
keV. The calculations of Salop (Ref. 3) have been used.

| n (lower level) | n (upper level) | | | | |
|---|---|---|---|---|---|
| | 2 | 3 | 4 | 5 | 6 |
| 1 | 2.52 | 0.18 | 0.12 | 0.07 | 0.06 |
| 2 | | 1.98 | 0.33 | 0.21 | 0.07 |
| 3 | | | 1.64 | 0.44 | 0.08 |
| 4 | | | | 1.28 | 0.19 |
| 5 | | | | | 0.54 |

Excitation takes place principally into the high angular
momentum states of the n=4 and n=5 levels. As a result, a large
fraction of the decay takes place through an yrast series, and the
Δn=1 transitions between low levels, such as n=3→n=2, come from
cascades. In Table II we list the comparisons between the measured
and the calculated relative intensities at 115 ms after the discharge
begins. The ratios of the 4→3 and the 5→4 transitions differ by
30-40% from the calculated values, and the relatively weak 5→3
transition differs by a factor of 2. Because of fluctuations in the
charge-exchange signal and uncertainties in the spectral
calibrations, a factor of 1.5, it is not feasible to use these data
to ascertain the accuracy of different theoretical calculations.

Table II. Comparison of measured and calculated emissions
normalized to the 3→2 transition of $O^{7+}$.

| Transition | λ(Å) | Meas. | Calc. (32 keV) | Calc. (26 keV) |
|---|---|---|---|---|
| 3→2 | 102 | 1.00 | 1.00 | 1.00 |
| 4→3 | 293 | 0.63 | 0.83 | 0.87 |
| 5→4 | 633 | 0.83 | 0.62 | 0.67 |
| 5→3 | 200 | 0.38 | 0.22 | 0.20 |

The central concentration of the fully ionized oxygen is first estimated from Eq. (3) by assuming a uniform distribution of $O^{8+}$. The result is then employed in the RECYCL code[7] which models the plasma by calculating profiles of various ionization stages of oxygen from measured electron temperature and density profiles by using ad hoc transport rates for the impurities. The calculated profiles are then employed in Eq. (3) to recompute $n_8(0)$. The two computations are iterated until they give consistent results for both the charge-exchange emission and the electron excited emission of the $O^{7+}$ lines.

Results of analyzing the data shown in Fig. 5 at 140 ms are given in Table III. The best fit using the modeling code indicates that the central concentrations of $O^{8+}$ and $O^{7+}$ are $2.4 \times 10^{11}$ and $0.9 \times 10^{11}$ respectively. Some insight into the absolute accuracy of the oxygen concentrations deduced from spectral data can be gained by comparing to other, non-specific measurements of the impurity content. The plasma resistivity provides a value of the average ion charge, $\langle Z_{eff} \rangle$, and the soft x-ray signal comes mainly from bremsstrahlung and radiative recombination into light impurities. The oxygen contribution to $\langle Z_{eff} \rangle$ is calculated from the spectroscopic data to be 0.5 in the inner half of the plasma both before and during injection. Resistivity measurements indicate that the contribution from all impurities is 1.0. In most of our discharges we estimate that other impurities contribute roughly one-half the oxygen value. The poor quality of the loop voltage signal and the uncertainity of spectrometer calibration can easily account for the 30% discrepancy between the spectroscopic and the resistivity measurements of $\langle Z_{eff} \rangle$. Similarly, the calculated soft x-ray signal, including only bremsstrahlung and recombination from hydrogen and from oxygen (as measured by spectroscopy), is one-half the observed value; the hydrogen contribution amounts to 31% of the total. But we expect bremsstrahlung and recombination from other impurities to contribute to the observed signal, and some fraction also results from x-ray line emission. The spectroscopic measurements appear to underestimate the oxygen concentration by 50% at most when compared with the resistivity and the soft x-ray signals. This accuracy is as good as can be expected from the present data; the combined uncertainties of the spectroscopic calibration, the beam parameters, and the theoretical cross sections is judged to be 60%.

In summary, the observation of several lines of $O^{7+}$ produced by charge-exchange excitation in ISX-B has strengthened the identification of this process in tokamaks. If experiments are designed so that the electron excited component of the line radiation can be distinguished from the charge exchange component, it appears feasible to monitor the fully ionized species in the discharge. Experimental uncertainties limit the measurements of the concentrations to 50% in the present experiments, and verification of the theoretical cross sections used for the analysis must await laboratory studies.

We would like to acknowledge several helpful discussions with R. E. Olson and with T. A. Green concerning charge transfer calculations. Thanks are also due to all members of the ISX-B group who operated the machine and recorded data during these experiments.

Table III. Calculated values of $O^{8+}$ and $O^{7+}$ at r=0. The inferred contributions of oxygen and hydrogen to $<Z_{eff}>$ and to the x-ray (PIN diode) signal from spectroscopic analysis are compared to other measurements of these quantities.

| | $n_8(0)$ ($10^{11}/cm^3$) | $n_7(0)$ ($10^{11}/cm^3$) | $<Z_{eff}>$ | X-Ray Signal (Volts) |
|---|---|---|---|---|
| Spectroscopy | 2.4 | 0.9 | 1.5 | 0.52 |
| Plasma resistivity | | | 2.0 | |
| PIN diode Measurement | | | | 1.00 |

## REFERENCES

*Research sponsered by the Office of Fusion Energy, U.S. Department of Energy, under contract W-7405-eng-26 with the Union Carbide Corporation.

†Visitor from the Japanese Atomic Energy Research Institute

1. R. C. Isler, Phys. Rev. Lett. 38, 1359 (1977).

2. V. V. Afrosimov, Y. S. Gordeev, A. N. Zinoviev, and A. A. Korotkov, Pis'ma Zh. Eksp. Teor. Fiz. 28, 505 (1978) [JETP Lett. 28, 500 (1978)].

3. A. J. Salop, J. Phys. B 12, 919 (1979).

4. Private communication, R. E. Olson.

5. V. A. Abramov, F. Baryshnikof, and V. S. Lisitsa, Pis'ma Zh. Eksp. Teor. Fiz. 27, 494 (1978).

6. All spectral emission rates are given in gigarayleighs (GR). 1 gigarayleigh = $10^{15}$ photons/cm2-s.

7. R. C. Isler, E. C. Crume, and H. C. Howe, Nucl. Fusion 19, 727(1979).

PHYSICS OF ELECTRONIC AND ATOMIC COLLISIONS
S. Datz (editor)
© North-Holland Publishing Company, 1982

# COHERENCE, ORIENTATION, AND ALIGNMENT
# IN EXCITED HYDROGENIC STATES FORMED AT SURFACES

*N. H. Tolk, J. C. Tully and Y. Niv*

Bell Laboratories
Murray Hill, NJ 07974
USA

An increasing number of atomic collisions in solids studies utilize experimental and theoretical techniques similar to those used in studies of two-body interactions involving ions, atoms, molecules, leptons and photons. Many inelastic processes involving solids may be simply understood in terms of microscopic quantum mechanical descriptions. To illustrate this, we describe measurements of elliptic polarization produced by both grazing incidence and tilted foil collision interactions arising from the decay of excited hydrogenic states.

## INTRODUCTION

Unquestionably, the collisional interaction of ions, photons, and electrons with solids introduces new features and complexities that are not present in simple two-body collisions. Nevertheless, electronic excitations occurring in collisions with solids and in particular solid surfaces have received an extraordinary amount of recent scientific attention. This is due partly to a practical need to solve problems associated with (a) first wall plasma interactions in controlled fusion devices (b) electronic device fabrication and (c) surface and bulk characterization involving a large variety analysis techniques. In addition, this topic has become an increasingly fruitful area of fundamental research with substantial overlapping interests in solid state physics, surface physics and chemistry, atomic physics, nuclear physics and space physics.

As amply shown in the contributions to this symposium, atomic collisions on and in solids increasingly are being understood in microscopic and even elegant terms involving molecular structure, simple electronic processes and quantum mechanical amplitude and phase considerations. As an example, in this chapter, we will discuss measurements of elliptic polarization arising from the decay of coherent oriented and aligned hydrogenic states produced by grazing incidence and tilted foil atomic collisions.

A central and fundamental challenge in the area of ion-surface interactions is to understand in detail the static and dynamic electronic behavior of hydrogen near solid surfaces. Grazing incidence and tilted foil collision configurations provide excellent tools for this study in that (a) the perpendicular velocity of the particles may be varied over a wide range and (b) the region of momentum changing collisional excitation of the surface may be separated in time and space from the final electronic interactions.

Many studies of elliptic polarization of radiation produced as a result of tilted foil (transmission) and grazing incidence collision configurations have been carried out over the past few years.[1-5] A multitude of physical models have been invoked to explain this phenomenon including asymmetric electron pickup, the transformation of alignment to orientation by surface electric fields, asymmetric electron-atom collisions and other mechanisms.[6-15] Not only is there considerable uncertainty with regard to the nature of the surface interaction, but also the role of the bulk material (particularly in the transmission geometry) has been a matter of controversy.[16,17] It is argued that the elliptic polarization is either developed completely or partially from a bulk induced alignment[6-9] or is determined exclusively by the final electronic interaction at the outgoing surface.[10-15] In order to attempt to provide answers to these questions experimental and theoretical studies have been carried out with the intent of determining the role of the bulk and also the nature of excited states created in these interactions. Measurements have been carried out at two different collision energies.

The first series of experiments was performed at 530 keV and involved comparison of polarized radiation emitted by $He^+$ ions (468.6 nm) in transmission (tilted foil) and reflection (grazing incidence) geometries, using both carbon foil and silicon single crystal targets.[1] These are the first tilted foil experiments involving a single crystal target, and the first experiments comparing tilted foil and grazing-incidence results with the same target. Our measurements suggest that the same physical mechanism is responsible for the excited state formation in both collision geometries, namely a surface interaction with no bulk contribution.

The second series of measurements was carried out at 9 keV in a ultra-high vacuum environment.[2] In these experiments, strong electric-field-dependent oscillations were observed in the polarization of $H_\alpha$ radiation following low-energy grazing incidence collisions of protons on a clean nickel crystal. Using a density matrix analysis, we have been able to account for the major features of the oscillations. This analysis demonstrates that at the instant of state formation at the surface the n=3 state is a *coherent* superposition of p,s and/or d states.

## $He^+$ MEASUREMENTS AT 530 keV

In this set of experiments, a beam of $He^+$ particles traverses a tilted foil (either a .5$\mu$m silicon single crystal (11) or a 5$\mu$g/cm carbon foil) or is reflected at a grazing incidence angle (approximately 4°) from the same silicon single crystal used in the tilted foil case. This is shown schematically in Figure 1. The elliptically polarized photons at 468.6 nm (n=4 to n=3) arising from the decay of the $He^+$ (n=4) hydrogenic state formed at the surface are detected downstream with a 0.3m monochromator, a Polaroid HN-38 polarizer, a depolarizer which removes spectrometer polarization bias, and a retardation plate with one quarter-wave retardation at 468.6 nm. Single-photon counting techniques were used to detect and process photons emitted normal to the plane defined by the incoming beam and surface normal for both tilted foil and grazing-incidence configurations as shown in Fig. 1. Crystal orientation was monitored by Rutherford backscattering. The $He^+$ ion beam was introduced into target chamber at a vacuum of $\sim 1 \times 10^{-7}$ Torr. The target area was surrounded by a metal enclosure maintained at liquid nitrogen temperature. In this set of experiments, the beam energy was adjusted to maintain a constant emergence energy of 530 keV. The results are illustrated in Fig. 2 showing the smooth behavior of $S/I = \dfrac{I_{RHC} - I_{LHC}}{I_{RHC} + I_{LCH}}$ where $I_{RHC}$ and $I_{LHC}$ are the intensities of right handed and left handed circular polarization.[18] The inset in figure 2 shows a detailed behavior as the tilt angle traverse an unspecified channel direction (which was revealed by a reduction of a factor of 5 in backscattered $He^{++}$ yield in an annular particle detector). Traversing the channeling direction, where the bulk properties change very drastically (critical angle of $\sim 0.7$), didn't produce any detectable change

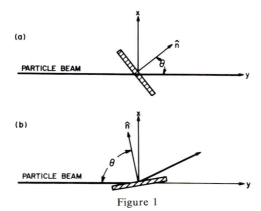

Figure 1

Schematic illustrations of the experimental configuration: (a) Tilted foil case. (b) Grazing incidence case. Photons emitted in the -z direction (out of the page) are detected.

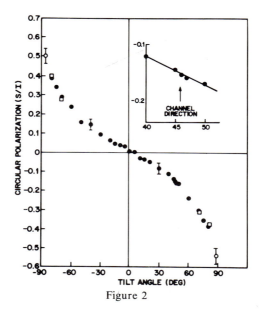

Figure 2

Circular polarization, S/I, of *HeII* (4-3) at 468.5 nm as a function of tilt angle $\theta$ at a constant emerging energy of 530 keV for both grazing incidence with a silicon crystal (open circles ○), and for beam-foil transmission with a silicon crystal (black circles ●) and a carbon foil (open squares □). The inset shows the tilt angle dependence of S/I in the vicinity of a low index planar channeling direction with channeling critical angle of approximately 0.5 degrees.

in the smooth behavior of $S/I(\theta)$. As shown in Fig. 2, different target materials (carbon and silicon) gave similar results, while using the Si crystal in a grazing angle geometry yielded a result for $S/I$ which tends to join smoothly with the transmission geometry data at corresponding emergence angle.

Even though (a) symmetry considerations allow the possible creation of oriention in a bulk three-dimensional crystal, and (b) some theoretical models predict the transformation of bulk created alignment into orientation at the surface, our observations lead us to conclude that there is no bulk contribution or memory of a bulk interaction in the production of the observed $He^+$ $n=4$ excited hydrogenic state. The excited state formation appears to occur exclusively at the exit surface.

## BALMER ALPHA MEASUREMENTS AT 9 keV

In these experiments 9 keV protons are incident at grazing incidence (4°) on a clean nickel (110) crystal. Figure 3 is a schematic representation of the experimental configuration. The experiments are carried out in an ultrahigh-vacuum chamber ($10^{-10}$ Torr). The target was cleaned by ion sputtering and heating. Following the grazing incidence interaction, beam particles enter an applied electric field region. Radiation from the decaying particles is then observed in a region free of electric field. Polarization data is acquired in the standard manner as previously described. Balmer alpha, $H_\alpha$, radiation arising from the $n=3$ to $n=2$ transition of the hydrogen atom emitted in the z and x directions (see figure 1) was detected. The three Stokes parameters $S/I$, $M/I$, and $C/I$, as defined below[18] were measured,

$$S/I = (I_{RHC} - I_{LHC})/(I_{RHC} + I_{LHC})$$

$$M/I = (I_0 - I_{90})/(I_0 + I_{90})$$

$$C/I = (I_{45} - I_{135})/(I_{45} + I_{135})$$

where $I_0$, $I_{45}$, $I_{90}$ and $I_{135}$ are the intensity of radiation with direction of polarization 0°, 45°, 90°, and 135°; and as before $I_{RHC}$ and $I_{LHC}$ are the intensities of right-handed and left-handed circular polarization. 90° is defined as being along the beam direction

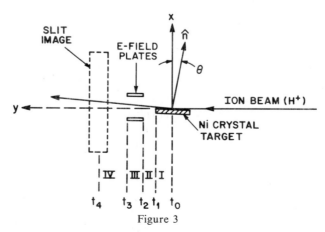

Figure 3

Schematic illustration of $H^+$ grazing incidence experimental configuration.

(y direction) as shown in figure 3.

These studies were motivated by the desire to (a) identify in detail the shape of the anisotropic oriented and aligned n=3 state produced when protons interact with a well defined surface at grazing incidence, and (b) given the actual states, to identify and describe the physical mechanism responsible for the creation of the state. As in previous measurements,[4] strong elliptic polarization was observed at zero applied field. Unique to these experiments however is the observation of pronounced oscillatory behavior in the polarization of the radiation measured as a function of applied electric field. Figure 4 illustrates measurements of S/I, C/I and M/I for photons emitted along the z direction as a function of electric field strength from -800 to 800 V/cm. Note that each of the Stokes parameters oscillates rapidly and independently as a function of field strength. In fact S/I, a measure of circular polarization, is seen to vary through zero and change sign indicating a reversal of direction.

The observed oscillations are due to quantum mechanical phase interferences arising from electric field induced Stark splittings of the nearly degenerate hydrogen n=3 levels. this oscillatory structure constitutes an important fingerprint of the initial state. Analysis of these experiments conclusively demonstrates coherence among the 3s, 3p and 3d states. these are the first observations of electric field dependent polarization oscillations and the first demonstration of coherence in excited states formed by grazing incidence.

Quantitative specification of the coherent state is important in determining the mechanism of formation. We may extract this information by means of a density matrix formulation. The details of this treatment are published elsewhere,[19] however we may outline the basic procedure.

In this case, it is necessary to make informed choices of all the elements of the $18 \times 18$ n=3 density matrix $\rho(t)$ at t=0, the instant of state creation. Symmetry considerations and spin independence guide us in these decisions. In addition, we are aided in this effort as hydrogenic levels and their behavior in electric fields are well understood. Once selected the initial density matrix is propagated forward in time by direct numerically integration of its equation of motion using external field, spin orbit,

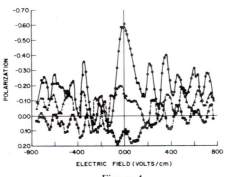

Figure 4

Experimental measurement of the variation of three relative Stokes parameters, S/I (circles, ●), M/I (squares, □) and C/I (triangles, Δ) as a function of applied electric field. Incident proton energy is 9 keV, incident grazing angle is 4°, and the target the (100) face of a nickel single crystal.

and Lamb shift Hamiltonian appropriate to the particular region traversed by excited hydrogen as shown in Figure 3. The final result is then compared with experiment to determine the best initial $\rho(0)$.

Our best calculation defined as coming closest to our experimental data is shown in Figure 5. We observe that these results are in qualitative, not quantitative agreement; the results do show significant similarities. One important similarity is that C/I is nonzero at zero applied electric field. Thus, the polarization is elliptical, with major axis at some oblique angle with respect to the surface normal. This arises in our calculation as a consequence of the evolution of the n=3 excited state in the vicinity of the surface (region I). In view of the uncertainties associated with electric field lines, cascades from higher levels, and the times at which particles enter and leave the various regions shown in fig. 3 we consider this analysis to be surprisingly satisfactory.

Although many physical mechanisms may play a role in creating oriented and aligned state at surfaces including anisotropic de-excitation or ionization,[20] scattering from surface plasmons,[15] spin-aligned pickup,[21] impact parameter selection,[14] and evolution in surface fields,[6,7,9] our results appear most consistent with anisotropic electron pickup.[4,10,12]

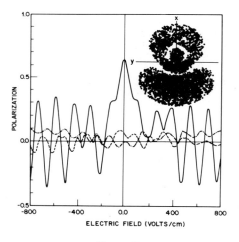

Figure 5

Variation of Stokes parameters, S/I (solid curve), M/I (dashed curve), and C/I (dashed-dot curve) with strength of applied electric field computed by time evolution of the density matrix for the same conditions as experimentally obtained in Fig. 4. The calculation assumes the initial state $\sqrt{1/3}s - \sqrt{1/2}p_x + \sqrt{1/24}d_{z^2} + \sqrt{3/24}d_{x^2-y^2} - i\sqrt{1/2}p_y + i\sqrt{1/2}d_{xy} + \sqrt{1/2}p_z + \sqrt{1/2}d_{xz} + i\sqrt{1/2}d_{yz}$. The inset illustrates this wave function. Density of dots represents absolute value of amplitude in the xy plane (z=0).

In order to obtain a more definite specification of the density matrix, in the future we intend to measure the elliptic polarization of $Ly_\alpha$ ($n=2$ to $n=1$) radiation. We also intend to measure the effect on the polarization of radiation from magnetically aligned ferromagnetic targets, to explore the possibility of picking up electrons which are partially spin aligned. Theoretical calculations show that spin-alignment of only a few percent should be readily discernible in electric field dependent measurements of S/I.

## CONCLUSIONS

The experiments described here provide information on detailed electronic interactions of particles near surfaces. Measurements of elliptically polarized light provide data which lead to specification of the excited states at the instant of creation and consequently contribute important data which may ultimately lead to the precise identification and physical description of the responsible mechanisms.

## REFERENCES

1. N. H. Tolk, L. C. Feldman, J. S. Kraus, J. C. Tully, J. Hass, Y. Niv and G. M. Temmer, Phys. Rev. Lett. *47*, 487 (1981).

2. N. H. Tolk, J. C. Tully, J. S. Kraus, C. Rau, and R. J. Morris, Submitted to Phys. Rev. Lett.

3. H. J. Andrä, R. Fröhling, and H. J. Plöhn, in *Inelastic Ion-Surface Collisions,* edited by N. H. Tolk, J. C. Tully, W. Heiland, and C. W. White (Academic Press, New York, 1977), p. 329, and references therein.

4. N. H. Tolk, J. C. Tully, J. S. Kraus, W. Heiland, and S. H. Neff, Phys. Rev. Lett. *41*, 643 (1978); *42*, 1475 (1979).

5. H. G. Berry, G. Gabrielse, A. E. Livingston, Phys. Rev. *A16*, 1915 (1977).

6. T. G. Eck, Phys. Rev. Lett. *33*, 1055 (1974).

7. M. Lombardi, Phys. Rev. Lett. *35*, 1172 (1975).

8. E. Lewis and J. D. Silver, J. Phys. B *8*, 2697 (1975).

9. Y. B. Band, Phys. Rev. *A13*, 2016 (1976).

10. H. Schröder and E. Kupfer, Z. Physik *A279*, 13 (1976).

11. Ref. 4, first entry.

12. J. Burgdörfer, H. Gabriel and H. Schröder, Z. Physiz *A295*, 7 (1980).

13. J. C. Tully, N. H. Tolk, J. S. Kraus, C. Rau, and R. Morris, in *Inelastic Ion-Surface Collisions,* ed. by W. Heiland and E. Taglauer, Springer-Verlag, Berlin, in press.

14. R. Herman, Phys. Rev. Lett. *35*, 1626 (1975).

15. A. A. Lucas, Phys. Rev. *B20*, 4990 (1979).

16. T. J. Gay and H. G. Berry, Phys. Rev. *A19* 952 (1979).

17.  H. J. Andrä, R. Fröhling, H. J. Plöhn, H. Winter and W. Wittmann, J. de Phys. *41*, C1-275 (1979).

18.  N. Born and E. Wolf, *Principles of Optics,* 4th ed. (Pergamon, Oxford, 1979), p. 30.

19.  Explicit prescriptions for the density matrices and Hamiltonian matrices are given in N. H. Tolk, J. C. Tully, J. S. Kraus, C. Rau and R. Morris, to be submitted to Phys. Rev.

20.  N. H. Tolk, J. C. Tully, J. S. Kraus, W. Heiland and S. H. Neff, Surf. Sci. *90*, 447 (1979).

21.  C. Rau, Comm. Sol. St. Phys. *9*, 177 (1980).

PHYSICS OF ELECTRONIC AND ATOMIC COLLISIONS
S. Datz (editor)
© North-Holland Publishing Company, 1982

MOLECULAR-ION STRUCTURES DETERMINED FROM COULOMB EXPLOSIONS
OF PENETRATING IONS

Donald S. Gemmell

Physics Division, D203, Argonne National Laboratory, Argonne, IL 60439

Traditional experimental techniques (e.g. studies on photon
absorption or emission) for determining the stereochemical
structures of neutral molecules are extremely difficult to
apply to molecular ions because of problems in obtaining a
sufficient spatial density of the ions to be studied. Recent
high-resolution measurements on the energy and angle distributions of the fragments produced when fast (MeV) molecular-ion
beams from an electrostatic accelerator dissociate in thin foils
and in gases, offer promising possibilities for deducing the
stereochemical structures of the molecular ions constituting
the incident beams. In this paper we describe this "Coulomb
explosion" technique and give some examples of its use in
structure determinations.

INTRODUCTION

In most experimental techniques used to derive the stereochemical structures of
molecular ions, the difficulty in obtaining a sufficient column density of the ions
to be studied presents a severe limitation. Thus, for example, photon emission or
absorption techniques usually involve searching for extremely weak and narrow lines
in the presence of background radiation. Where such methods have been successfully
applied, they result in very precise structure determinations. However, it has so
far only proved possible to measure a handful of molecular ion structures with
these "standard" techniques. This situation contrasts strongly with that for
neutral molecules for which the experimental problems are much less severe and for
which thousands of structures have been determined.

Recent studies on the dissociation of fast (MeV) molecular ion beams in thin foils
suggest a novel alternative approach to the determination of molecular ion structures. At these high beam velocities (typically a few percent of the speed of
light) each dissociating molecular-ion projectile rapidly loses many (sometimes
all) of its electrons in sudden violent collisions with electrons in the foil
target. The projectile then undergoes a so-called "Coulomb explosion" as the now
highly charged and monatomic fragments repel one another apart via their mutual
Coulomb forces. These fragments emerge downstream from the foil with their velocities shifted in both magnitude and direction from the beam velocity. Typically
the Coulomb explosion results in energy shifts of a few keV and angular shifts of
a few mrad for the dissociation fragments. High-resolution measurements of the
joint energy and angle distributions for these fragments (especially when two or
more fragments are detected in coincidence) offer promising possibilities for
deducing the stereochemical structures of the molecular ions constituting the
incident beam.

If, instead of a foil target, a dilute gas target is used, the projectile dissociations then arise mainly from relatively gentle collisions. Nevertheless additional useful structure information can be obtained, especially when taken in
conjunction with results from foil-induced dissociation.

In this paper, we briefly describe this Coulomb explosion technique and give some examples of its use in structure determinations. (For a more complete discussion see, for example, Refs. 1-4.)

## 2. COLLISION-INDUCED DISSOCIATION AT HIGH VELOCITIES

The collision-induced dissociation of molecular ions has previously been studied mostly with gaseous targets and with ion-beam energies in the keV range. There are, however, definite advantages in extending such studies to the MeV range. We summarize here some of the most significant of these advantages (for a more detailed discussion, see Ref. 5).

(1) The collision times become short compared with the times for molecular vibration and rotation and with the time for the subsequent dissociation [the cross sections for electron removal from the projectile are on the order of $10^{-16}$ $cm^2$ (see, for example, Ref. 6)].

(2) The use of thin foil targets becomes feasible. In foils, close collisions with target electrons cannot be avoided, and this leads to more energetic Coulomb explosions than those occurring with gas targets.

(3) For foil targets, higher beam energies lead to higher fragment charge states, giving more vigorous Coulomb explosions.

(4) As the projectile velocity v is raised, there is a decline not only in the absolute values of multiple scattering angles (which vary as $v^{-2}$) but also in the multiple scattering relative to the widths of the fragment angular distributions (which vary roughly as $v^{-1}$).

(5) Screening of the fragment charges by the target electrons in a foil is reduced at higher beam velocities.

For high projectile velocities ($v \gg v_o = e^2/\hbar$), collision-induced dissociation may be well approximated as a two-step process. First there occurs a rapid collision with one or more target atoms during which the projectile's nuclei do not move in their center-of-mass (cm) frame. Because of their low mass, the electrons associated with the projectile reconfigure themselves in a time comparable with the collision time (this process includes the possibility of removal by ionization of some or all of the electrons). There then follows, on a much longer time scale, a dissociation of the resultant excited molecular state into two or more fragments.

## 3. COULOMB EXPLOSIONS

To illustrate the basic considerations in Coulomb explosions, we sketch here a simplified model for the dissociation of 3-MeV $HeH^+$ ions incident on a 100-Å thick carbon foil (Fig. 1). This model can then be refined[5] and extended to more complex projectiles.[3,4]

Since the mean time to strip the two electrons from each projectile is short (a few times $10^{-17}$ s) compared with the dwell time in the foil (0.93 fs at the beam velocity $v = 1.07 \times 10^9$ cm/s) and with the times for molecular vibration ($\sim 10^{-14}$ s) and rotation ($\sim 10^{-12}$ s) of the projectile, we may make the approximation that the electron stripping and the start of the consequent Coulomb explosion both occur instantaneously at the front surface of the foil. Let the internuclear vector from the $\alpha$ particle to the proton in a given $HeH^+$ projectile have a length $r_o$ at this instant and be inclined at an angle $\phi$ relative to the beam direction. We suppose the values of $r_o$ to have a probability distribution $D(r_o)$ which depends upon the population of the various vibronic states of the projectile.

If all the $HeH^+$ projectiles were in the ground electronic, vibrational, and

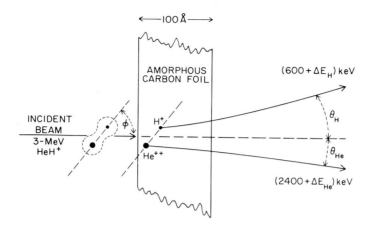

Fig. 1. A (very) schematic picture of a 3-MeV HeH[+] projectile dissociating in a 100-Å-thick amorphous carbon foil.

rotational state, the most probable value of $r_0$ would be expected to be 0.79 Å (based, for example, on the potentials calculated by Kolos and Peek[7]). Let us for the moment consider a projectile having this particular value of $r_0$ at the instant of entry into the foil. Electron stripping then produces an α-particle-proton pair with a mutual Coulomb energy $\varepsilon = 2e^2/r_0 = 36.5$ eV. This potential energy is converted into kinetic energy in the cm as the Coulomb explosion develops. Since this Coulomb energy is much larger than the energies that the fragments have due to vibration and rotation in the incoming projectile, we may make the further approximation that at the time of stripping (t = 0) the α particle and the proton are at rest in the cm frame.

If, for now, we ignore further interactions of the projectile nuclei with the target foil, then the internuclear separation as a function of time, r(t), is found by solving

$$\mu \ddot{r} = Z_1 Z_2 e^2 / r^2 \qquad (1)$$

with the initial conditions

$$r(0) = r_0, \dot{r}(0) = 0 \qquad (2)$$

where $\mu = M_1 M_2 / (M_1 + M_2)$ is the reduced mass of the two fragments and $Z_1 e$ and $Z_2 e$ are their charges. From this one finds that the time for the internuclear separation to grow to r is given by

$$t(r/r_0) = t_0 f(r/r_0) \qquad (3)$$

where

$$t_0 = [ r_0^3 / (2 Z_1 Z_2 e^2) ]^{1/2} \qquad (4)$$

and

$$f(x) = \sqrt{x}\,\sqrt{x-1} + \ln\,(\sqrt{x} + \sqrt{x-1}).$$ (5)

The final asymptotic $(t \gg t_o)$ relative velocity of recession is

$$\dot{r}(\infty) = [2Z_1 Z_2 e^2/\mu r_o]^{1/2}.$$ (6)

For our example of 3-MeV HeH$^+$ ($Z_1 = 1$, $Z_2 = 2$), $t_o = 0.84$ fs and $\dot{r}(\infty) = 9.4 \times 10^6$ cm/s (about 1% of the beam velocity).

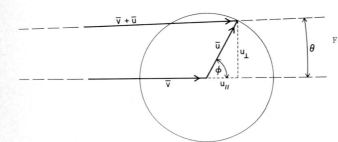

Fig. 2.  Vector diagram illustrating the relationship between cm and LAB coordinates for a fragment emerging from a Coulomb-exploded molecular ion projectile.

The final energies and angles of the fragments in the laboratory (LAB) frame are easily obtained by referring to the vector diagram in Fig. 2.  Let u be the asymptotic cm velocity acquired by the fragment of mass $M_1$ as a result of the Coulomb explosion.  That is

$$u = (\mu/M_1)\dot{r}(\infty).$$ (7)

The asymptotic Lab velocity of this fragment is then $\bar{v} + \bar{u}$.  The shift in Lab angle is

$$\theta \approx (u \sin \phi)/v = u_\perp/v$$ (8)

and the shift in LAB energy is

$$\Delta E = 1/2M_1 (\bar{v} + \bar{u})^2 - 1/2M_1 v^2$$

$$\approx M_1 vu \cos\phi = M_1 vu_{||} = 2E_1 (u_{||}/v)$$ (9)

where

$$E_1 = 1/2M_1 v^2$$ (10)

(we neglect terms with relative magnitudes on the order of u/v).

From the foregoing, one readily sees that there exist extreme values $\theta_{max}$ and $\Delta E_{max}$ corresponding to the extreme orientations ("parallel": $\phi = 0$ or $\pi$; and "transverse": $\phi = \pi/2$).  In our example of 3-MeV HeH$^+$ with $r_o = 0.79$ Å, we find for the protons $\theta_{max} = 7.0$ mrad ($0.40^\circ$) and $\Delta E_{max} = 8.4$ keV.  For the $\alpha$ particles,

$\theta_{max}$ is a factor of four smaller and $\Delta E_{max}$ is the same.

If the spatial orientations of the incident projectiles are isotropically distributed, we may think of the tip of the fector $\bar{u}$ in Fig. 2 as populating with uniform probability the surface of a spherical shell of radius u. Thus, for a given projectile orientation, there is a $(4\pi u^2)^{-1}$ weighting factor for the values of u that arise from an initial distribution $D(r_o)$ of internuclear spacings. The distribution function for u is

$$G(u) = (4\pi u^2)^{-1} D(r_o) \mid dr_o/du \mid \tag{11}$$

where $G(u)$ and $D(r_o)$ are normalized by

$$\int_o^\infty D(r_o) dr_o = 1, \qquad 4\pi \int_o^\infty G(u) u^2 du = 1. \tag{12}$$

For a simple Coulomb explosion we obtain

$$G(u) \propto u^{-5} D(r_o(u)). \tag{13}$$

Thus a measurement of the LAB variables $\theta$ and $\Delta E$ for one of the fragment species ($M_1$, say) is equivalent to a measurement of $u_\perp$ and $u_{\parallel}$ in the cm system. Knowledge of the Coulomb dissociation potential then permits determination of $D(r_o)$. Actually just one of the two measurements suffices: either the distribution of $\theta$ for $\Delta E = 0$, or the distribution of $\Delta E$ for $\theta = 0$.

A two-parameter measurement ($\theta, \Delta E$) of the yield of fragments of mass $M_1$ is expected to give a uniform "ring pattern" since it corresponds to a cut across the center of the sphere indicated in Fig. 2.

We have thus far ignored the influence of the medium apart from its role in initiating the Coulomb explosion. In practice various phenomena occurring as fast ions traverse foil targets can influence the dissociating fragments. The most important effects are multiple scattering, energy loss and energy straggling, wake effects, electronic screening, and charge exchange.

Multiple scattering can be a serious problem in that it effectively blurs the experimental angular resolution. It is minimized by using the thinnest possible target foils with low atomic number and by using high projectile velocities. Thus, most of the work discussed here has employed carbon targets about 100 Å thick. For such targets, 600-keV protons (e.g., from 3-MeV $HeH^+$) suffer multiple scattering with an average angular deflection of about 0.6 mrad.

Energy loss and straggling due to the normal stopping power of the medium are much less severe problems. For 600-keV protons traversing 100-Å carbon, the energy loss is 0.7 keV and the straggling is only a few tens of eV.

Polarization wake effects and electronic screening have been discovered[8] to have a significant influence on the motion of a cluster of fragments. (For a discussion of wake effects, see the papers by Vager et al.[9] and Ritchie et al.[10] at this conference). The wakes of a Coulomb-exploding cluster of ions traversing a foil superpose upon one another. Thus, for a diatomic projectile, the motion of one of the fragments while inside the foil is affected by the wake of its partner. This results in a tendency for trailing fragments to align more closely behind their partners.[8]

For more slowly moving fragments, screening by target electrons begins to play a significant role. The Coulomb explosion develops between charges for which the potential is exponentially screened, the screening distance being given by

$$a = v/\omega_p \tag{14}$$

where $\omega_p$ is the plasma frequency appropriate to the target electrons. For 1-MeV carbon ions, for example, traversing a carbon foil ($\hbar\omega_p$ = 25.9 eV), we find $a = 1.0$ Å.

Charge exchange (electron capture and loss) in the target can have a major effect upon the ionic fragments traversing a foil. This is particularly true of heavy fragments. Inside the foil the extremely rapid ($\sim10^{-17}$ s) capture and loss of electrons by the fragments leads to a well-defined effective charge that determines the stopping power and also the magnitude of the Coulomb explosion. At high velocities the effective charge for a given ion can be written as

$$z^{eff} = (S/S_p)^{1/2} \tag{15}$$

where S is the stopping power of the foil for the ion and $S_p$ is the corresponding value for a proton of the same velocity.[11] Outside the foil the ions adopt integral charge states.

## 4.  APPARATUS FOR HIGH-RESOLUTION MEASUREMENTS

To explore the details of ring patterns having "radii" of a few mrad and a few keV, experimental resolution widths at least an order of magnitude smaller than these values are desirable. This poses some technical problems, particularly in the angular coordinate. The solutions to these problems as implemented at Argonne[12] have been described in the literature (e.g. Refs. 1,5,13-15). Figures 3 and 4 show the experimental arrangement now in use.

Magnetically analyzed molecular-ion beams are collimated to have a maximum angular divergence of ±0.09 mrad and a spot size of 1 mm at the target position. A set of "pre-deflector" plates permits electrostatic deflection of the beam incident on the target. Similarly a set of "post-deflectors" is used to deflect charged particles emerging from the target. The pre- and post-deflectors are used in combination so as to avoid detecting particles that arise from spurious incident beams (e.g., fragments arising from dissociation of the primary beam along the long flight path between the beam collimators). A 25° electrostatic analyzer having a relative energy resolution of ±3 × $10^{-4}$ is located several meters downstream from the target. An aperture placed ahead of the analyzer accepts a 1-mm-diameter group of trajectories originating at the target position.

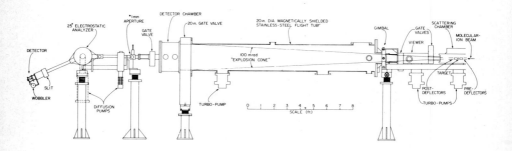

Fig. 3.  Schematic diagram of the experimental arrangement at Argonne's 4-MV Dynamitron accelerator.

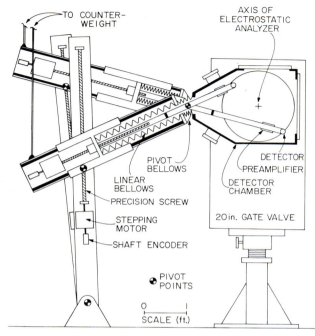

Fig. 4. Schematic diagram showing a cross-sectional view of the detector chamber and movable detector systems at Argonne's 4-MV Dynamitron accelerator.

Distributions in energy and angle are made for particles emerging from the target by varying the voltages on the horizontal pre-deflectors and/or the post-deflectors in conjunction with that on the electrostatic analyzer. The overall angular resolution is ±0.15 mrad.

The detector chamber shown in Fig. 4 is a recent addition to bhe beam line. It permits the coincident detection of dissociation fragments. The chamber houses two detectors that can be positioned to an accuracy of about 0.001 in. anywhere on a 20-in. dia. circular area subtending an angle of 100 mrad at the target.

The resolution obtainable with our apparatus is

$$\delta\theta = \pm 1.5 \times 10^{-4} \text{ rad, } \delta E/E_1 = \pm 3 \times 10^{-4}. \tag{16}$$

## 5. MEASUREMENTS ON "RING PATTERNS" FOR DIATOMIC PROJECTILES

Figure 5 shows a ring pattern measured for protons from 3-MeV HeH[+] incident upon a 195-Å thick carbon foil.[5] The ring has approximately the diameter predicted from our simple model. The most obvious feature not predicted from the model is the nonuniform distribution of proton intensity around the ring. There is an enhanced intensity for trailing protons and (to a much lesser degree) for leading protons. This redistribution of particle flux is a consequence of the interaction of the fragments with the wakes that their partners induce in the foil. It is to be observed for both heteronuclear and homonuclear projectiles.

The widths of the peaks in the measured cross are larger than expected for the HeH[+] ground state alone. This is because $D(r_o)$ is a broader distribution (excited

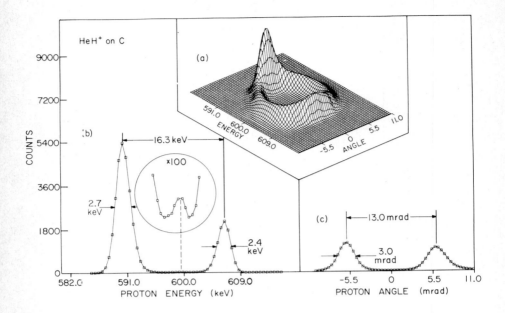

Fig. 5.   (a) "Ring pattern" and (b,c) "cross" for protons from 3.0 HeH$^+$
dissociating in a 195-Å-thick carbon foil.[5]

states of the projectile contribute) and because of multiple scattering.  The
asymmetry in the peak widths in the $\theta = 0$ energy spectrum is caused by wake effects.

Figures 6-8 show similar results for the foil-induced dissociation of beams of
H$_2^+$, $^3$He$_2^+$, and N$_2^+$.

Figure 9 shows a comparison between the measured and calculated[17] ring patterns for
a typical case of foil-induced dissociation.  When the data were fitted, the only
adjustable parameters were the most probable value of $r_o$ and the standard deviation
$\sigma$ of the distribution $D(r_o)$, assumed here to be Gaussian.  The best fit is obtained
with $\bar{r}_o = 0.79$ Å and $\sigma = 0.15$ Å.  For the vibrationless ground state of HeH$^+$, a
Gaussian approximation to $D(r_o)$ would require $\bar{r}_o = 0.79$ Å and $\sigma = 0.081$ Å.  The
increased width of the experimentally derived $\sigma$ implies the population of excited
states in the incident projectiles.  A detailed analysis[5] of data such as those in
Figs. 5 and 9 shows that the four lowest lying vibrational states in the HeH$^+$ ions
represent 53%, 22%, 11%, and 6%, respectively, of the incident beam.  The most
probable value of $r_o$ remains 0.79 Å.  (The data shown for HeH$^+$ in Figs. 5 and 9
were obtained by using a duoplasmatron ion source fed with a gas mixture of 90% He
and 10% H$_2$.  Recently, it was discovered[18] that an RF source fed with the same gas
mixture produces HeH$^+$ ions in much higher states of vibrational excitation.)  A
similar analysis has been performed for H$_2^+$ beams where it is found that the vibrational population is fairly close to that expected on the basis of Franck-Condon
factors.[5]

For slowly moving heavy-ion fragments, an improved fit to the data can be obtained
by using a wake model that takes better into account the close collisions with

target electrons.   Thus, for example, Breskin et al.[19,20] have accounted in a very precise way for the foil-induced dissociation of 11.2-MeV OH$^+$.

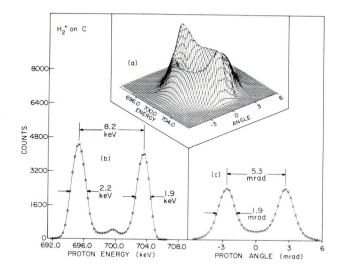

Fig. 6.   (a) "Ring pattern" and (b,c) "cross" for protons from 1.4-MeV H$_2^+$ dissociating in a 88-Å-thick carbon foil.[5]

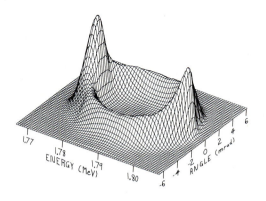

Fig. 7.   "Ring pattern" for $^3$He$^{2+}$ from 3.6-MeV $^3$He$_2^+$ dissociating in a 118-Å-thick carbon foil.[15]

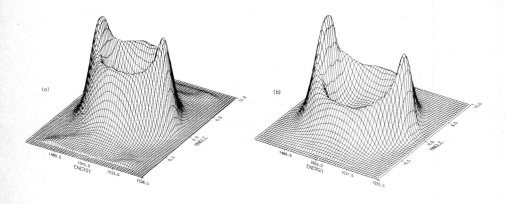

Fig. 8.  "Ring patterns" for (a) N$^+$ and (b) N$^{4+}$ fragments arising from the dissociation of 3-MeV N$_2^+$ in a 75-Å-thick carbon foil.[16]

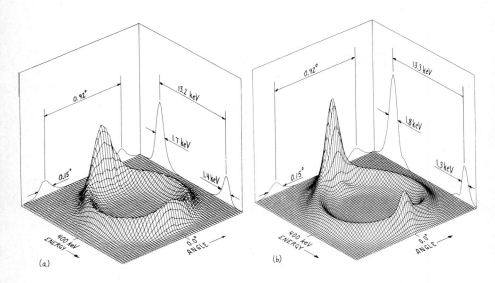

Fig. 9.  (a) Experimental and (b) calculated (based on a Coulomb explosion modified by wake effects and multiple scattering) ring patterns for protons from 2.C-MeV HeH$^+$ dissociating in an 85-Å-thick carbon foil.[17]

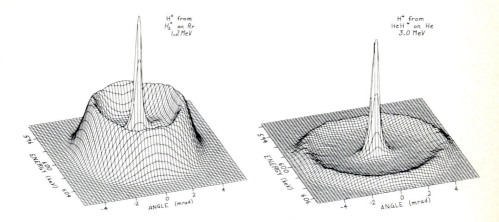

Fig. 10.  Joint energy-angle distributions for protons from 1.2-MeV $H_2^+$ and
3.0-MeV $HeH^+$ dissociating in gaseous Ar and He, respectively.[5]

When fast molecular ions dissociate in a dilute gas target, wake effects,
screening, and multiple scattering no longer play a role.  Figure 10 shows two
examples.  One sees immediately that the rings are uniformly populated.  The
diameters of the rings correspond to the dissociations $H_2^+ \rightarrow H^+ + H^+$ and $HeH^+ \rightarrow$
$He^+ + H^+$.  The patterns display dominant central peaks.  The central regions can be
quantitatively accounted for[5] in terms of collision-induced transitions of the
projectiles to excited electronic states which then decay dissociatively, yielding
one neutral and one charged fragment.  Gentle collisions of this type are possible
in gases (unlike solids) where large-impact-parameter collisions are favored.  For
such excitations the resultant cm energy acquired by the fragments can be small
(<0.1 eV, say) if the initial internuclear separation is large [i.e., on the tail
of the distribution $D(r_o)$].  These fragments, although arising in only a small
fraction of the dissociations, are detected for all initial projectile orientations
and thus give rise to the large central peak.

## 6.  IMPLICATIONS FOR STRUCTURE DETERMINATIONS

Analysis of many ring patterns and crosses has shown (see, for example, Ref. 14)
that for a great variety of diatomic molecular ion projectiles the bond length can
be determined with an accuracy of about 0.01 Å.  In the analysis the influences
(usually relatively small) of wake effects, multiple scattering, post-foil charge-
state distributions, etc., are taken into account.

At Argonne we have become interested in exploring the extent to which these high-
resolution studies with fragmentation techniques may be extended to the difficult[1,2,15]
problem of determining the geometric structures of polyatomic molecular ions.
Related studies are also under way at Brookhaven[21] and at the Weizmann
Institute.[20-22]  Although the accuracy expected from the fragmentation techniques
may be poor (e.g., ∿0.01 Å in bond lengths and ∿1° in bond angles for not-too-
complicated species) compared with that attainable with "standard" photon
techniques (when they can be applied), it should be good enough to resolve many
conflicts between predictions from various structure calculations.  Further, this

level of accuracy should suffice to assist practitioners of the photon techniques to zero in on the frequencies of their (usually very narrow) resonances. One further point: the molecular ions that are of the greatest interest in astro-chemical and fusion studies are very much the same ones produced copiously from the plasma ion sources normally used in electrostatic accelerators.

Except for the very simplest polyatomic molecular ions (e.g., $H_3^+$), high-resolution studies on single fragments yield only gross features of the stereo-chemical structures. For example, our measurements on $C^{2+}$ fragments from 3.6-MeV $C_3H_3^+$ ions dissociating in thin foils demonstrate only that the carbons sit on the corners of an approximately equilateral triangle [that is, we have a beam of cyclopropenyl ions and not propargyl ions (which are linear in carbons)]. Similarly, our measurements on single fragments from $OH_2^+$ show only that the protons are equivalent and that the oxygen is "in the middle". The accuracy in determining bond lengths and bond angles is poor because there are wide ranges of values for these parameters that combine to give about the same Coulomb explosion velocity u for any given fragment.

A further difficulty associated with this type of measurement for polyatomic molecular ions lies in analyzing the effects of vibrational excitations of the projectiles. Excitations of some modes (e.g., symmetric breathing modes) can frequently be expected to result only in apparent changes in the bond lengths determined by the Coulomb-explosion method. However, many nonsymmetric modes can result in apparent structures that differ markedly from the structure of the vibrationless ground state. It is therefore important to analyze the Coulomb-explosion data in terms of the specific modes that can be excited for each projectile species considered. The analysis can be greatly simplified if the projectiles can be prepared in their ground (or at most a small range of low-lying) vibrational states - often a difficult technical task. In a recent paper, Pratt and Chupka[23] have drawn attention to the influence of vibrational excitations upon the conclusions to be deduced from Coulomb explosion experiments. These authors also discuss methods of circumventing some of these difficulties.

Polyatomic structures can be much more precisely determined if spatial and temporal coincidences are recorded for two or more dissociation fragments from a given projectile. With this in mind, we recently revised the apparatus at Argonne so as to permit a wide variety of coincidence measurements (Fig. 4). We are now able to measure double or triple coincidences and record simultaneously information on fragment charge states, energies, and flight times from the target. The system has been tested with various simple diatomic and triatomic projectiles ($H_2^+$, $HeH^+$, $CH^+$, $NH^+$, $OH^+$, $H_3^+$, $CH_2^+$, $NH_2^+$, $OH_2^+$, etc.).

In the remaining sections we present some measurements on single fragments and then some preliminary results on coincidence measurements.

## 7. MEASUREMENTS ON SINGLE FRAGMENTS FROM POLYATOMIC PROJECTILES

$\underline{H_3^+}$. A joint study[24] of the simplest polyatomic molecular ion, $H_3^+$, was under-taken at the University of Lyon, the Weizmann Institute, and Argonne National Laboratory. Each laboratory performed a measurement based on the Coulomb explosion of fast $H_3^+$ ions. Although somewhat different techniques were used, the three measurements gave results in agreement with one another. It was experimentally demonstrated (for the first time) that $H_3^+$ is equilateral triangular in shape. The three measurements of the proton-proton bond distance yielded 0.97 ± 0.03 Å (Argonne), 0.95 ± 0.06 Å (Lyon), and 1.1 ± 0.2 Å (Weizmann Institute). Figure 11 shows a comparison of these results with a recent calculation by Carney[25] based on the vibrational-state population parameters of Smith and Futrell.[26] (The vibrational ground state of $H_3^+$ has a calculated[27] bond length of 0.91 Å.)

$\underline{CO_2^+}$ and $\underline{N_2O^+}$. A typical example of the manner in which gross structures may be

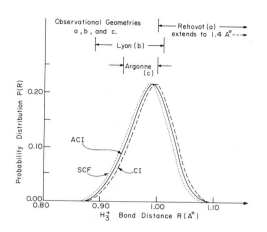

Fig. 11. Comparison between measured[24] and calculated[25] distributions in the proton-proton
bond distance in $H_3^+$ (from Ref. 25).

rapidly determined by Coulomb-
explosion techniques is to be
found in recent studies at
Argonne[28] with 3.5-MeV beams of
$CO_2^+$ and $N_2O^+$. These molecular
ions in their ground and low-
lying states are known[29] to be
linear; but while $CO_2^+$ has the
symmetric form (O-C-O), $N_2O^+$ is
asymmetric (N-N-O). Figure 12
shows $\theta = 0$ energy spectra for
$O^{4+}$ and $C^{2+}$ and $CO_2^+$ and $O^{4+}$
and $N^{3+}$ from $N_2O^+$. Similar
spectra were obtained for fragments emerging in other charge
states. Each spectrum takes
about 5 min to accumulate. The
principal structural characteristics are evident from just a
casual inspection of Fig. 12.
The existence of only two peaks
in Fig. 12a indicates that the
two oxygen atoms in $CO_2^+$ are
equivalent. The existence of
just one peak in Fig. 12b shows
that the carbon atom is central
in a linear molecule (no net
Coulomb-explosion velocity). The two peaks in Fig. 12c show that the oxygen atom
in $N_2O^+$ lies "on the outside" and the three peaks in Fig. 12d show that one nitrogen atom is "on the outside" and that one is in the center of a linear molecule.
The central peak in Fig. 12d is much more strongly populated than the two side
peaks because many more incident orientations contribute to it.

From these considerations, it can be seen that by simply counting the number of
peaks in each spectrum, one can infer that both $CO_2^+$ and $N_2O^+$ are linear with
structures O-C-O and N-N-O. A more detailed analysis is obviously required to
obtain precise values for the bond lengths and angles.

$CH_n^+$ (n = 0,4). The proton and the carbon fragments arising from the Coulomb
explosion of $CH^+$, $CH_2^+$, $CH_3^+$, and $CH_4^+$ have been studied at beam velocities corresponding[30] to 0.194 MeV/amu. The singlet proton spectra, although reflecting
vigorous Coulomb explosions, are not particularly informative concerning the projectile structures. For the carbon fragments, energy straggling and multiple
scattering in the target blur out the structure information in the energy and angle
spectra. However, the width of the peak observed in these carbon-ion spectra is
sensitive to any asymmetry in the distribution of the protons that surround the
carbon atom in the projectile.

Figure 13 shows the measured energy widths (fwhm) of outgoing ($\theta = 0$) $C^{4+}$ ions that
emerge after the incident beam strikes a foil target. The value of 6.1 keV for
incident $C^+$ represents the contribution of energy straggling convoluted with both
the beam energy spread and the resolving power of the electrostatic analyzer system.
The Coulomb explosion of the highly asymmetric $CH^+$ ions adds a large contribution
which increases the measured width to 14.6 keV. For the more symmetric $CH_2^+$ ions,
because of near cancellation of the impulses produced by each proton on the carbon
ion, the Coulomb explosion is reduced and thus we measure a width of only 10.1 keV.
If $CH_2^+$ were rigidly linear, the width would be expected to be close to the $C^+$
straggling value of 6.1 keV (the width would actually be somewhat greater than
6.1 keV because of charge-state fluctuation effects that modify the Coulomb explosion while the ion fragments are within the target foil). A similar effect is seen
in the measurement of the carbon width for the dissociation of $CH_3^+$. Again, the

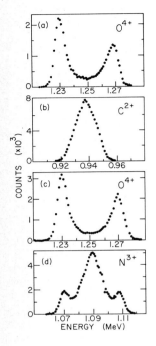

Fig. 12. Energy spectra at
$\theta = 0$ for (a) $O^{4+}$ and (b)
$C^{2+}$ resulting from 3.5-MeV
$CO_2^+$ bombarding a 113-Å-thick
C foil, and for (c) $O^{4+}$ and
(d) $N^{3+}$ resulting from 3.5-MeV
$N_2O^+$ bombarding a 160-Å-thick
carbon foil.[28] The spectra
are not normalized to one
another.

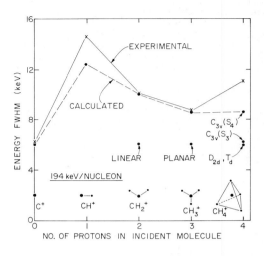

Fig. 13. Comparison of energy widths of $C^{4+}$
spectra for $CH_n^+ \to C^{4+}$. The calculations are
based on a carbon-ion effective charge of 3.5
and neglect wake forces.[30]

width is increased over the minimum that one
would expect for a rigid planar structure; however, it is smaller than either of the $CH^+$ or
$CH_2^+$ results. The data for $CH_4^+$ show a dramatic departure from this general trend. The
measured width of 11.1 keV is larger than all
but the $CH^+$ measurement. This indicates a
highly asymmetric proton distribution around the
carbon nucleus and is most likely a consequence
of the Jahn-Teller distortion of $CH_4^+$.

The dashed line in Fig. 13 shows the width calculated on the basis of a very crude
model in which it is assumed that rigid structures having carbon charges of 3.5 and
proton charges of 1.0 Coulomb-explode. No attempt was made to include the effects
of molecular vibrations, wakes, multiple scattering, charge-state distributions,
etc. These calculations thus serve only as a rough guide to the possible structures of the projectiles and are not to be interpreted as determining the actual
structures. The calculation for $CH^+$ assumes a bond length of 1.13 Å. For $CH_2^+$
a carbon-proton distance of 1.03 Å is assumed and the H-C-H angle is taken to be
$140°$ (a guess based on the expectation that the bond angle will be close to the
value of $131°$ known[29] for the isoelectronic molecule $BH_2$). For $CH_3^+$ the calculation assumes a rigid pyramidal structure with an interproton distance of 1.08 Å and
with the carbon ion 0.2 Å off the proton plane. The values calculated for $CH_4^+$
were based on the four Jahn-Teller distorted structures derived by Dixon.[31] As
noted above, taking account of vibrational excitations of the projectiles can
affect the implications of these calculations. For example, $CH_3^+$ is commonly

thought to be planar[29] and the results shown in Fig. 13 are consistent with a planar structure in which a low-frequency out-of-plane oscillation of the carbon ions exists with an amplitude of ∿0.2 Å.

$HCO^+$. In Table I we compare the diameters of the ring patterns obtained[32] for $C^{4+}$, $O^{3+}$, and $O^{5+}$ fragments from the foil-induced dissociation of $HCO^+$ and $CO^+$ of the same velocity (124 keV/nucleon). A cursory glance suffices to deduce that the tri-atomic beam ions must have the structure H-C-O and not C-O-H or C-H-O.

TABLE I

Diameters (in keV and mrad) of the ring patterns for $C^{4+}$, $O^{3+}$ and $O^{5+}$ fragments from the foil-induced dissociation of beams of 4.476-MeV $CO^+$ and 3.6-MeV $HCO^+$.

| Fragment | Diameter in keV | | Diameter in mrad | |
| --- | --- | --- | --- | --- |
| | $CO^+$ | $HCO^+$ | $CO^+$ | $HCO^+$ |
| $C^{4+}$ | 42.1 | 37.9 | 13.9 | 13.2 |
| $O^{3+}$ | 37.6 | 39.1 | 8.8 | 9.2 |
| $O^{5+}$ | 42.0 | 44.4 | 9.7 | 10.4 |

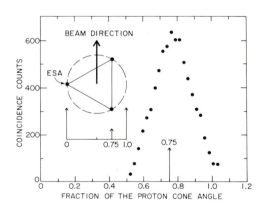

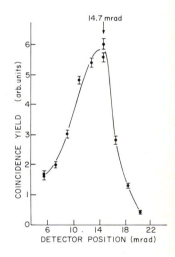

Fig. 14. Proton-proton counting rate as a function of the relative angular separation of the two detectors (the ESA and one of the movable detectors shown in Fig. 4) (Ref. 33). The protons arose from 1.8-MeV $H_3^+$ incident on a carbon foil 93-Å thick.

Fig. 15. Coincidence counting rate for protons from 3.6-MeV $OH_2^+$ dissociating in an 80-Å-thick carbon foil.[2] The rate is plotted as a function of the angle between the electrostatic analyzer (set on the low-energy, θ = 0 proton group) and one of the movable surface-barrier detectors (Fig. 4).

## 8.  COINCIDENCE MEASUREMENTS

$H_3^+$, $H_2D^+$, $HD_2^+$, and $D_3^+$.  Using the movable detector systems shown in Fig. 4 in conjunction with the electrostatic analyzer (ESA) indicated in Fig. 3, a group at Argonne[33] has recently begun a series of coincidence measurements on fragments from $H_3^+$ and its deuterated forms.  Figure 14 shows the proton-proton coincidence counting rate obtained with a beam of 1.8-MeV $H_3^+$ striking a carbon foil.  In these measurements one of the detected protons was counted in the ESA.  The deflections and the ESA voltage were set so that the ESA detected "sideways-going" (in the C.M. frame) protons.  The other detected proton was sensed in one of the movable semi-conductor counters shown in Fig. 4.  The abscissa in Fig. 14 is measured in terms of the angular separation of the two detected protons relative to the angular diam-eter of the proton ring pattern.  For an equilateral triangular $H_3^+$, one would expect the coincidence rate to peak at an abscissa value of 0.75.  The slight upward shift in the experimental value (0.77) is explainable in terms of the de-creased phase space available for larger angular separations.

$CH_2^+$, $NH_2^+$ and $OH_2^+$.  Preliminary coincidence measurements have been performed for these dihydride ions.[2]  Of the three, only $OH_2^+$ has previously had its structure determined experimentally.  From optical measurements Lew and Heiber[34] found the O-H bond length to be 0.999 Å and the H-O-H bond angle to be 110.5°.

Figure 15 shows a spatial scan of the proton-proton double coincidence rate for the foil-induced dissociation of 3.6-MeV $OH_2^+$ ions.  Note that a given combination of post-deflector and ESA voltage settings amounts to choosing for study a limited subset of the incoming projectile orientations.  For the data shown in Fig. 15, only those $OH_2^+$ ions in which one proton is trailing are selected.  The angular radius of the proton ring pattern is 16.2 mrad.  Thus the bond angle is close to $\beta = 180° - \sin^{-1} (14.7/16.2) = 115°$.  Approximate corrections for the displaced cm and for the oxygen recoil result in values of $110 \pm 2°$ and $1.0 \pm 0.04$ Å for the bond angle and bond length, respectively.  These values agree with those from the optical measurements.[34]  Measurements with other orientations chosen in the ESA give similar results.  A more detailed analysis, properly taking into account wake effects and multiple scattering, should result in a significant improvement in the level of accuracy.

Figure 16 shows a comparison of the results for foil-induced dissociation of 3.6-MeV $CH_2^+$, $NH_2^+$, and $OH_2^+$.  For these data, the deflections are chosen so that the ESA detects only those protons with the maximum transverse momentum.  The double coincidence rate for $OH_2^+$ peaks a little past the center of the proton cone - again consistent with a bond angle of 110°.  However, for the other two projec-tiles, the peak occurs at the extreme angle of the proton cone - opposite the ESA. This would be consistent with a linear structure, but again vibrational effects may be playing a large role in these projectiles.

We have recently begun triple coincidence measurements, e.g., on the pair of protons and the $N^{3+}$ fragments arising from the dissociation of 3.6-MeV $NH_2^+$ ions in a dilute Ar gas jet (Fig. 17).  Although the triple coincidence counting rates are low, the data are very clean and the analysis is simplified as compared with the results obtained with foil targets.  In Fig. 17 the detailed time-of-flight infor-mation for each point in the angular scan has not yet been taken into account in the data analysis.

To summarize, we have commenced high-resolution coincidence measurements on the fragments from foil- and gas-induced dissociation of fast polyatomic molecular ions.  Our measurements on $H_3^+$ indicate an equilateral triangular structure.  For $OH_2^+$, we are able to reproduce the bond angle and bond length found in optical experiments.  The precision of the method can now be expected to improve consider-ably as more refined data-analysis procedures are developed.

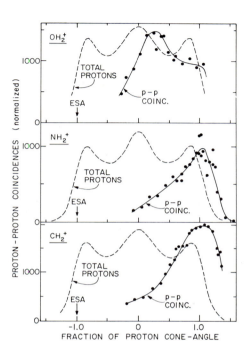

Fig. 16.  The proton-proton coincidence counting rates for 3.6-MeV beams of $CH_2^+$, $NH_2^+$, and $OH_2^+$ dissociating in carbon foils of thicknesses 98, 66, and 89 Å, respectively.[2]  The rates are plotted as functions of the fraction of the proton cone angle (16.25, 14.28, and 13.2 mrad for $CH_2^+$, $NH_2^+$, and $OH_2^+$, respectively) lying between the electrostatic analyzer (set on the protons having the maximum transverse momentum) and one of the movable surface barrier detectors (Fig. 4).  Also shown are the total (energy-summed) proton-singles rates in the movable detector.

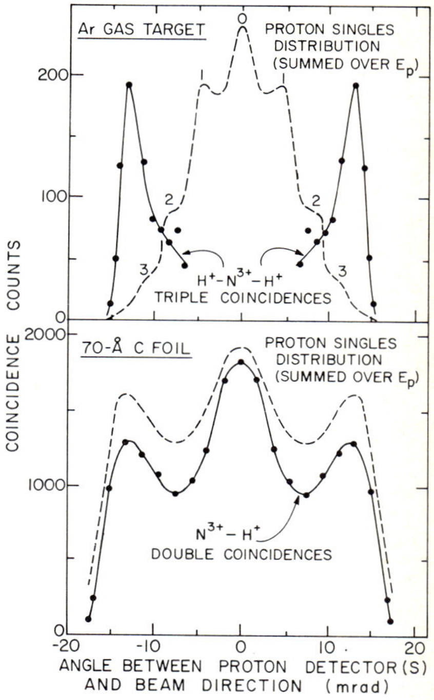

Fig. 17.  (Top) $H^+-N^{3+}-H^+$ triple coincidence counting rate for 3.6-MeV $NH_2^+$ ions dissociating in an Ar gas-jet target.  $N^{3+}$ ions are detected at $0°$ C in the electrostatic analyzer.  Protons are detected in the two movable detectors.  The coincidence rate is plotted as a function of angle between the (symmetrically placed) proton detectors and the beam direction.  (Bottom) Same, but double coincidences ($N^{3+}-H^+$) obtained with a 70-Å carbon foil target.  In both figures the total (energy-summed) proton-singles rates in the movable detectors are shown as dashed curves.  In the top figure, the numbers (0,1, 2,3) on the dashed curve refer to the corresponding nitrogen-ion charge state.[2]

## 9.  MEASUREMENTS USING IMAGING TECHNIQUES

At the Weizmann Institute, a photographic technique was developed[24] to study $H_3^+$. With this technique, all of the fragments from individual projectiles were captured and rendered visible in a photographic emulsion.  A similar method has recently been applied by a group at Brookhaven[21] to study the foil-induced dissociation of $C_n^-$ beams.  This group uses a polycarbonate plastic sheet as detector.  Figure 18 shows such a sheet after exposure to fragments from foil-dissociated $C_4^-$ projectiles.  Inferring structure information from data such as these is difficult (except perhaps to observe that $C_4^-$ is not linear).  A higher data rate would be very desirable because a statistical analysis of the fragment patterns would then become feasible.  Promising steps in this direction are now being taken at Brookhaven and at Weizmann Institute where electronic imaging techniques are being developed.  At Brookhaven[21] a fluorescent screen used in conjunction with an image intensifier is being tested.  Workers at the Weizmann Institute[22] have demonstrated that a charge-coupled semiconductor device is capable of acting as a high-resolution two-dimensional detector of charged particles.  The energy resolution in each element is expected to be adequate for purposes of distinguishing masses.  It is expected that this device will permit the rapid acquisition ($\sim$50 images/s) and storage of two-dimensional projections of individual Coulomb-exploded molecular ions.  The digital coordinates of the fragments in each image together with rough measures of their energies will be stored and processed by a computer.

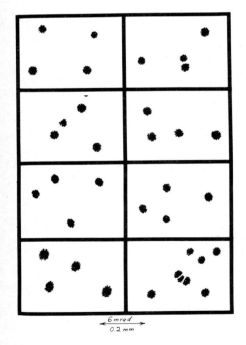

6 mrad<br>0.2 mm

Fig. 18.  Etched polycarbonate sheet after exposure to fragments from 3.5-MeV $C_4^-$ ions dissociated in a thin carbon foil.[21]

These imaging techniques and the Argonne technique are complementary approaches to the structure problem. The one approach will produce a high rate of coincidences between all fragments for all incident projectile orientations, but little detailed information on energies, charge states, flight times, etc.  The other approach selects individual charge states and projectile orientations with very precise information on parameters such as energies, flight times, direction, etc. but with low coincidence counting rates.

## CONCLUSION

We have described the main features of the Coulomb explosion of fast-moving molecular-ion projectiles and the manner in which Coulomb-explosion techniques may be applied to the problem (difficult to attack by more conventional means) of determining the stereochemical structures of molecular ions.  Examples have been given of early experiments designed to elicit structure information.  The techniques are still in their infancy, and it is to be expected that as both the technology and the analysis are refined, the method will make valuable contributions to the determination of molecular-ion structures.

ACKNOWLEDGMENTS

The research at Argonne has been made possible through the dedicated efforts of an energetic set of collaborators. In particular the author wishes to acknowledge the contributions of P. J. Cooney, N. Cue, A. K. Edwards, K.-O. Groeneveld, E. P. Kanter, W. J. Pietsch, I. Plesser, J.-C. Poizat, A. J. Ratkowski, Z. Vager, and B. J. Zabransky. The work at Argonne was conducted under the auspices of the Division of Basic Energy Sciences of the U. S. Department of Energy.

REFERENCES

[1]  Proceedings of the Workshop on Physics with Fast Molecular-Ion Beams, Argonne National Laboratory, Argonne, IL, Aug. 20-21, D. S. Gemmell, Ed., Physics Division Informal Report ANL/PHY-79-3 (Aug. 1979).

[2]  D. S. Gemmell, P. J. Cooney, and E. P. Kanter, Nucl. Instrum. Methods 170, 81 (1980).

[3]  D. S. Gemmell, Chem. Rev. 80, 301 (1980).

[4]  E. P. Kanter, Proc. of NATO International Advanced Study Institute, Kos, Greece, Sept. 30-Oct. 10, 1980 (to be published).

[5]  E. P. Kanter, P. J. Cooney, D. S. Gemmell, K.-O. Groeneveld, W. J. Pietsch, A. J. Ratkowski, Z. Vager, and B. J. Zabransky, Phys. Rev. A 20, 834 (1979).

[6]  N. Bohr, K. Dan. Vidensk. Selsk. Mat.-Fys. Medd., 18, No. 8 (1948).

[7]  W. Kolos and J. M. Peek, Chem. Phys., 12, 381 (1976).

[8]  D. S. Gemmell, J. Remillieux, J.-C. Poizat, M. J. Gaillard, R. E. Holland, and Z. Vager, Phys. Rev. Lett., 34, 1420 (1975); Nucl. Instrum. Methods, 132, 61 (1976).

[9]  Z. Vager et al. (this conference).

[10] R. H. Ritchie et al. (this conference).

[11] See, for example, L. C. Northcliffe and R. F. Schilling, Nucl. Data Tables, A7, 233 (1970).

[12] The Argonne apparatus has evolved over several years. Major contributions to its development have been made by P. J. Cooney, D. S. Gemmell, E. P. Kanter, W. J. Pietsch, I. Plesser, A. J. Ratkowski, Z. Vager, and B. J. Zabransky.

[13] Z. Vager, D. S. Gemmell, and B. J. Zabransky, Phys. Rev. A 14, 638 (1976).

[14] D. S. Gemmell, P. J. Cooney, W. J. Pietsch, A. J. Ratkowski, Z. Vager, B. J. Zabransky, A. Faibis, G. Goldring, and I. Levine, Proceedings of the 7th International Conference on Atomic Collisions in Solids, Moscow, Sept. 19-23, 1977, to be published.

[15] D. S. Gemmell, "Radiation Research" (Proceedings of the Sixth International Congress of Radiation Research, Tokyo, Japan, May 13-19, 1979), S. Okada, M. Imamura, T. Terashima, and H. Yamaguchi, Eds., University of Tokyo, 1979, pp. 132-144.

[16] D. S. Gemmell, I. Plesser, and B. J. Zabransky, private communication.

[17] Z. Vager and D. S. Gemmell, Phys. Rev. Lett. 37, 1352 (1976).

[18]   D. S. Gemmell, E. P. Kanter, I. Plesser, and Z. Vager, private communication.

[19]   A. Breskin, A. Faibis, G. Goldring, M. Hass, R. Kaim, I. Plesser, Z. Vager, and N. Zwang, p. 1 of Ref. 1.

[20]   A. Breskin, A. Faibis, G. Goldring, M. Hass, R. Kaim, I. Plesser, Z. Vager, and N. Zwang, Nucl. Instrum. Methods 170, 93 (1980); 170, 99 (1980).

[21]   G. Goldring, Y. Eisen, P. Thieberger, and H. Wegner, p. 27 of Ref. 1.

[22]   M. Algranati, A. Faibis, R. Kaim, and Z. Vager, private communication.

[23]   S. T. Pratt and W. A. Chupka, Chem. Phys. 52, 443 (1980).

[24]   M. J. Gaillard, D. S. Gemmell, G. Goldring, I. Levine, W. J. Pietsch, J.-C. Poizat, A. J. Ratkowski, J. Remillieux, Z. Vager, and B. J. Zabransky, Phys. Rev. A 17, 1797 (1978).

[25]   G. D. Carney, Mol. Phys. 39, 923 (1980).

[26]   D. L. Smith and J. H. Futrell, J. Phys. B 8, 803 (1975).

[27]   G. D. Carney and R. N. Porter, J. Chem. Phys. 65, 3547 (1976).

[28]   D. S. Gemmell, E. P. Kanter, and W. J. Pietsch, Chem. Phys. Lett. 55, 331 (1978).

[29]   G. Herzberg, "Electronic Spectra of Polyatomic Molecules", Van Nostrand, Princeton, N.J., 1950.

[30]   D. S. Gemmell, E. P. Kanter, and W. J. Pietsch, J. Chem. Phys., 72, 1402 (1980).

[31]   R. N. Dixon, Mol. Phys. 20, 113 (1971).

[32]   N. Cue, A. K. Edwards, D. S. Gemmell, I. Plesser and J.-C. Poizat (to be published).

[33]   N. Cue, A. K. Edwards, D. S. Gemmell, E. P. Kanter, I. Plesser, and J.-C. Poizat (to be published).

[34]   H. Lew and I. Heiber, J. Chem. Phys. 58, 1246 (1973); H. Lew, Can. J. Phys., 54, 2028 (1976).

PHYSICS OF ELECTRONIC AND ATOMIC COLLISIONS
S. Datz (editor)
© North-Holland Publishing Company, 1982

RADIATION FROM CHANNELED LEPTONS

R.L. Swent and R.H. Pantell
Department of Electrical Engineering
Stanford University, CA 94305

S. Datz

Oak Ridge National Laboratory
Oak Ridge, TN 37830

M.J. Alguard

Measurex, Inc.
Cupertino, CA 95129

B.L. Berman, S.D. Bloom, R. Alvarez, and D.C. Hamilton

Lawrence Livermore Laboratory
Livermore, CA 94550

High-energy charged particles, when incident upon a single
crystal in a direction near a major crystallographic direction,
can be scattered coherently by the crystal lattice. Low-mass
particles emit radiation under these conditions, which is
called channeling radiation. Quantum-mechanically, this radiation is described as arising from transitions between bound
states in an averaged potential, which may be one- or
two-dimensional. By observing the radiation, one can obtain
information about the dynamics of the channeling process and
about the crystals in which the channeling occurs.

The subject of channeling has received considerable attention in the last fifteen
years.[1] Most of this work has been done with positive ions: protons, alpha particles, and heavier ions. In the last few years, more work has been done with
leptons. The term lepton in this context means positrons and electrons. Studies
with these light particles have been stimulated by the discovery that they emit
significant amounts of radiation when they are channeled.

The term channeling describes a process which occurs when a highly collimated beam
of charged particles is incident upon a single crystal in a direction that is
nearly parallel to a major crystallographic direction, that is, nearly parallel to
a row of atoms which constitute an axis, or nearly parallel to planes of atoms.
In this case, the particles will suffer only gentle, small-angle collisions with
the lattice atoms. If the collisions are sufficiently gentle, the particle trajectories will bend significantly only after traversing many thousands of lattice
sites. Because so many atoms are involved, the usual procedure is to perform an
average over the atomic potentials. For axial channeling, the potentials are
averaged along the row of atoms, and for planar channeling an average over the
planes is performed. This averaging results in a tremendous simplication, because
the longitudinal motion is separated from the transverse motion. The longitudinal
motion is essentially free (and relativistic), while the non-relativistic transverse motion is governed by the conservative transverse potential. The transverse
motion is one-dimensional for planar channeling and two-dimensional for axial
channeling.

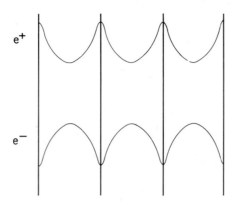

Figure 1
Illustration of planar-averaged potentials for positrons and
electrons. The vertical lines represent the atomic planes.

A typical planar-averaged potential is shown in Figure 1 for positrons and electrons. The minima for positrons lie in the open spaces between the planes of
atoms (indicated by vertical lines), while the electrons will be attracted to the
planes and will weave in and out of them. Classically, a particle will be trapped
in one of these wells if its transverse energy is less than the depth of the well.
Since the transverse energy is defined in terms of the transverse component of
momentum, $E_t = P_t^2/2m$, it is clear that only particles incident at sufficiently
small angles will be trapped.

Relativistic and quantum-mechanical effects complicate matters. Light particles
will have a small number of discrete bound states. Also, the effective mass in
the lab frame is $\gamma$ times the rest mass, where $\gamma$ is the Lorentz factor. Since the
number of bound states increases with the mass, this number is proportional to
the beam energy. The most interesting effects occur when the number of bound
states is small, somewhere between 2 and 10, and this defines the most interesting
energy range for a given potential. Because of the extra degrees of freedom,
there are more states in an axial well than a planar well at the same energy. For
channeling in silicon, the maximum energies for ten bound states are about 10 MeV
for axial channeling and about 100 MeV for planar channeling.

The capture probability for a given state is proportional to the square of the
overlap matrix element between the incident beam wavefunction and the channeling-
state wavefunction. Capture probabilities for the lowest six states for 56-MeV
electrons incident parallel to the {110} planes of Si are shown in Figure 2.

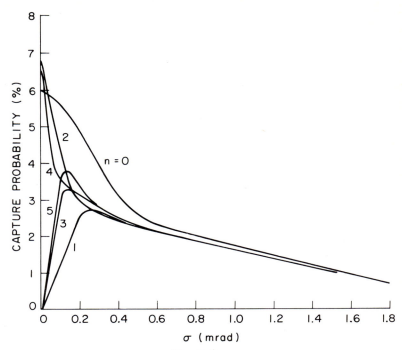

Figure 2
Capture probabilities for 56-MeV electrons for the {110} planes of Si
as a function of the rms angular spread of a Gaussian beam.

The probabilities are shown as a function of the angular divergence for a Gaussian
beam. The potential function has reflection symmetry about the atomic plane, and
the eigenfunctions therefore have definite parity with respect to reflection
through this plane. A Gaussian with zero divergence is a plane wave, and for this
case, the capture probabilities for all the odd-parity states are zero. As the
divergence is increased, the beam includes particles with non-zero transverse
momentum, and the odd states become populated. If the beam is broad enough, the
capture probabilities for all the excited states are equal.

We then have a situation where there exist discrete bound states, with the excited
states populated. Spontaneous radiative transitions can then occur, giving rise
to what is called channeling radiation. There are important relativistic factors
which affect the properties of this radiation. In the rest frame, which is an
inertial frame of reference moving at the longitudinal beam velocity, the particle
mass is the rest mass, but the potential is $\gamma$ times that in the lab frame. This
is due to the Lorentz contraction of the crystal. The averaged potential is proportional to the density of the atoms, and the contraction increases this by a
factor of $\gamma$. The energy level separation in the rest frame is then $\gamma\Delta E_L$, where

$\Delta E_L$ is the separation in the lab frame. In the rest frame, the photon energy is
equal to the difference in energy levels, but since the photon is observed in the
lab frame, a Doppler shift factor must be included. In the forward direction,
this factor is $2\gamma$. The net result is that the photon energy in the lab frame is
$2\gamma^2\Delta E_L$. This is quite important, because $\gamma$ can be very large. For $\gamma \simeq 100$, in a

planar potential with energy level separations of a few eV, the photons have
energies of many tens of keV.

The averaged potential for positrons can be approximated fairly well by a parabola
centered midway between the planes.[2] This gives rise to harmonic oscillator solutions for the tranverse motion, with equally spaced eigenvalues.  The actual
potential rises more steeply than a parabola, and this can be taken into account
by a perturbative treatment of terms proportional to $x^4$.  The result is that the
eigenvalue spacing increases as the quantum number increases, leading to a set of
closely spaced spectral lines rather than the single line of a harmonic oscillator.  Due to various line-broadening mechanisms, these closely-spaced lines blend
into one large line.

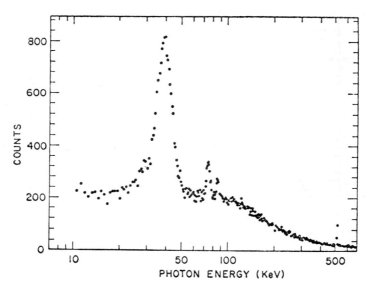

Figure 3
Photon counts versus photon energy for 54-MeV positrons channeled between the {110} planes of Si.

A photon spectrum[3] for 54-MeV positrons in {110} Si is shown in Figure 3.  The
large peak at 40 keV is the channeling radiation.  Also visible are a bremsstrahlung background, some Pb x-rays near 75 keV arising from lead shielding, and the
omnipresent annihilation radiation at 511 keV.  The full-width at half-maximum
linewidth of the channeling radiation is 26%.  Under ideal conditions, this could
be reduced to about 13%, with this residual linewidth arising mainly from the
anharmonic nature of the potential.  Positron channeling radiation has also been
observed from germanium, which gave a peak at 47 keV for the same beam energy.
Although the higher charge on the Ge nuclei makes the averaged potential well
much deeper than that of Si, the oscillator frequency is determined by the curvature of the potential midway between the planes, and this curvature doesn't
change much from Si to Ge.

A photon spectrum for 54-MeV electrons[4] channeled in {110} Si is shown in

Figure 4. The cusp-like shape of the electron potential gives rise to unequally spaced energy levels, with the lowest levels being the most widely spaced. The major peaks marked (1) through (5) are $\Delta n = 1$ peaks. Peak (5) is from the $n = 1$ to $n = 0$ transition, and peak (1) is from the $n = 5$ to $n = 4$ transition. The smaller peaks above 100 keV are the $5 \to 2$, $4 \to 1$ and $3 \to 0$ transitions, in order of increasing energy. The dipole matrix elements for $\Delta n = 3$ transitions are an order of magnitude smaller than those for $\Delta n = 1$. There is also bremsstrahlung in this spectrum. In addition to the bremsstrahlung and the $\Delta n = 1$ lines, there is a broad enhancement of photons in the energy range from about 20 keV to 120 keV. This spectral feature is officially known as "the bump." The exact nature of the bump is still uncertain, but it is believed to be due to free-to-bound transitions. That is, it is radiation from transitions whose initial states have transverse energies above the barrier, and whose final states are the bound states.

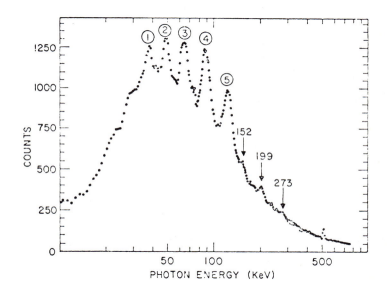

Figure 4
Photon spectrum for 54-MeV electrons channeled along the {110} planes of Si. The numbered peaks are discussed in the text.

Various scattering mechanisms limit the lifetimes of the channeling states, and if one multiplies the mean life by the velocity (the speed of light), one gets what is called the coherence length for the state. The coherence lengths of the initial and final state can be combined to give an effective coherence length for the transition, and this in turn can be related to the FWHM linewidth of the radiation:

$$\ell_{eff}^{-1} = \ell_i^{-1} + \ell_f^{-1}$$

$$FWHM = 2\gamma^2 \beta c \hbar / \ell_{eff}$$

Effective coherence lengths for the $\Delta n = 1$ transitions of Figure 4 are listed in
Table 1. It is seen that the coherence lengths increase with n. This is consistent with the supposition that scattering from thermally vibrating atoms is the
dominant lifetime-limiting mechanism. The lower-lying states have wavefunctions
which are concentrated closer to the atomic planes, leading to more scattering and
a shorter coherence length than the more loosely bound states.

Table I

| Transition | $\ell_{eff}(\mu m)$ |
|------------|---------------------|
| $1 \to 0$  | 0.24                |
| $2 \to 1$  | 0.33                |
| $3 \to 2$  | 0.45                |
| $4 \to 3$  | 0.61                |
| $5 \to 4$  | 0.67                |

Axial channeling of electrons has also been extensively studied in the energy
range from 2 to 4 MeV.[5] The row-averaged potential has cylindrical symmetry, and
looks somewhat like a 1/r potential. In fact, a 1/r potential was used in the
first analysis of this case.[6] The solutions are similar to those for the hydrogen
atom, but in two dimensions. There is a principal quantum number n, which is a
positive integer, and an azimuthal quantum number $\ell$, with $0 < \ell < n - 1$. This has
led to the use of spectroscopic notation for these states, e.g. 3p. All states
with $\ell > 0$ are doubly degenerate. The electron can have angular momentum $+\ell$ or
$-\ell$. In classical terms, it can spiral around the row in either a clockwise or
counterclockwise direction. The selection rule for dipole radiation is $\Delta \ell = \pm 1$.

Thermal vibrations smear the potential near the row, and cause it to be finite at
the origin, rather than singular, as the 1/r potential is. The effect of this
smearing is similar to the effect of a finite nuclear size on a hydrogen-like
atom, but the effect here is much larger. This and other deviations from a 1/r
potential cause higher angular momentum states to have lower energy. For example,
the 2p state lies considerably below the 2s state. A 1/r potential is not really
a very good starting point for calculations, but the notation has been retained.

Figure 5 shows the photon energies for several transitions in <111> Si as a function of beam energy.[5] As the beam energy increases, more states are bound, and
more transitions are observed. The photon energy for a given transition increases
rapidly with beam energy because of the $2\gamma^2$ factor.

In Figure 6 are two spectra[7] from 3.5-MeV electrons in <100> Si. The top one is
from a crystal 0.3 $\mu m$ thick, and the bottom one is from a 2.0 $\mu m$ crystal. The
spectra have been normalized to give equal yield per unit length. The plots show
photon courts on the vertical scale, photon energy on one horizontal scale, and
angle between the crystal axis and the beam axis on the other horizontal scale.
The most easily visible transitions, in order of increasing energy, are 3d - 2p,
2p - 1s and 3p - 1s. The notches at zero angle reflect the vanishing of the capture probabilities for the initial states under those conditions. As the tilt
angle is increased, the capture probability rises to a maximum, then falls off.
The difference in the relative magnitudes of the 3d - 2p and 2p - 1s transitions
for the two crystals is a consequence of the different lifetimes for the 3d and
2p states. The 2p state decays faster than the 3d state, and therefore produces
less radiation in the thicker crystal. The ridges seen in the top spectrum have
been identified as free-to-bound transitions, just as in the planar case.

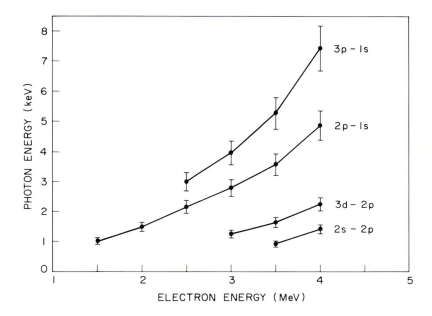

Figure 5
Photon energy as a function of electron energy for channeling radiation
along the <111> axis of Si (from Reference 5).

The atomic rows responsible for axial electron channeling are not isolated, but
have neighbors arranged in a symmetrical, periodic pattern. A row of the <100>
axis of Si has four nearest neighbors arranged in a square pattern. These neighboring strings affect the potential, and overlap between wavefunctions on adjacent
strings can cause energy shifts and the splitting of degenerate levels. In
Figure 7 are the results of calculations incorporating these neighbor effects. On
the left are the potential and energy levels for a single string. On the right
are the potential and energy levels for two multi-string approximations: the
tight-binding approximation (TB) and the many-beam approximation (MB). The tight-
binding approximation should be good as long as the overlaps are fairly small.
The many-beam approximation is an expansion of the wavefunctions in Fourier
series, and should be accurate even for large overlaps, if enough terms are used.
For the tightly bound states, the effect of neighbors is mostly a downward shift
in the potential and the energy levels, and the TB and MB calculations agree very
well. The 3d is the most loosely bound, and the two initially degenerate states
have been split by a considerable amount.

There are two other major axes in the silicon crystal, the <111> and the <110>.
The <111> axis has six equidistant and symmetrically arranged nearest neighbors.
The effects of neighboring strings will be qualitatively similar to those of the
<100> axis discussed above. The <110> axis, on the other hand, has strings which
come in pairs. The members of a pair are fairly close, and there are four
nearest-neighbor pairs at a larger distance. If a single string is like a hydrogen atom, a pair of strings is like a $H_2^+$ ion. Because of the large overlap for

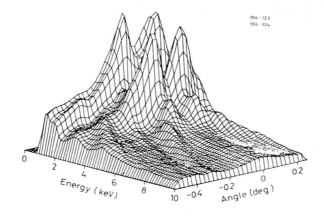

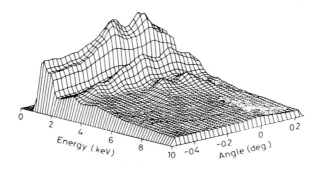

Figure 6
Photon counts versus photon energy and crystal tilt
angle for channeling of 3.5-MeV electrons along the
<100> axis of Si for two different crystals (see
text).

any but the most tightly bound states, "molecular orbitals" are the best starting
point for an analysis of the <110> states. This leads to classifying the states
as σ and π, gerade and ungerade, much as in three dimensions. This is by far the
most complicated of the channeling problems, and although experimental data are
available, the calculations are still in progress.

To summarize, channeling gives us a chance to study quantum mechanics in one and
two dimensions. It allows us to do spectroscopy of the energy levels of particles
of variable mass. The study of axial channeling is a peculiar combination of
high-energy physics, solid-state physics and atomic physics.

We are very grateful to E. Laegsgaard for providing us with so much information on
axial channeling. The planar positron and planar electron channeling work was
supported by the Division of Advanced Energy Projects, Office of Basic Energy
Sciences of the U.S. Department of Energy under contract No. DE-AT03-76ER70064;
by the U.S. Department of Energy under contracts No. W-7405-ENG-48 and

No. W-7405-ENG-26; and by the U.S. Air Force Office of Scientific Research under grant No. 81-0209.

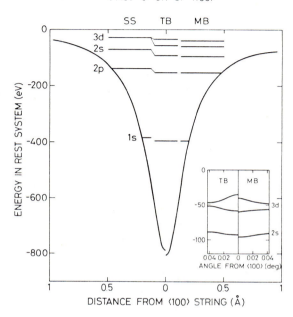

4 MeV e⁻ on Si ⟨100⟩

Figure 7
Radial potential and energy levels for the ⟨100⟩
axis of Si for the single-string case (left) and
two multi-string approaches (right), the tight-
binding approximation (TB) and the many-beam
approximation (MB).

References
(1) Gemmel, D.S., Rev. Mod. Phys. 46 (1974) 129.
(2) Pantell, R.H. and Alguard, M.J., J. Appl. Phys. 50 (1979) 798.
(3) Pantell, R.H. et al., IEEE Trans. Nuc. Sci. NS-28 (1981) 1152.
(4) Berman, B.L. et al., Phys. Lett. 82A (1981) 459.
(5) Anderson, J.U. and Laegsgaard, E., Phys. Rev. Lett. 44 (1980) 1079.
(6) Terhune, R.W. and Pantell, R.H., Appl. Phys. Lett. 30 (1977) 265.
(7) Laegsgaard, E., private communication.

AUTHOR INDEX